Physics of Optoelectronic Devices

WILEY SERIES IN PURE AND APPLIED OPTICS

Founded by Stanley S. Ballard, University of Florida

EDITOR: Joseph W. Goodman, Stanford University

Physics of Optoelectronic Devices

SHUN LIEN CHUANG
Professor of Electrical and Computer Engineering
University of Illinois at Urbana-Champaign

A Wiley-Interscience Publication

John Wiley & Sons, Inc.

New York / Chichester / Brisbane / Toronto / Singapore

Wiley Series in Pure and Applied Optics

The Wiley Series in Pure and Applied Optics publishes outstanding books in the field of optics. The nature of these books may be basic ("pure" optics) or practical ("applied" optics). The books are directed towards one or more of the following audiences: researchers in universities, government, or industrial laboratories; practitioners of optics in industry; or graduate-level courses in universities. The emphasis is on the quality of the book and its importance to the discipline of optics.

This text is printed on acid-free paper.

Copyright © 1995 by John Wiley & Sons, Inc.

All rights reserved. Published simultaneously in Canada.

Reproduction or translation of any part of this work beyond that permitted by Section 107 or 108 of the 1976 United States Copyright Act without the permission of the copyright owner is unlawful. Requests for permission or further information should be addressed to the Permissions Department, John Wiley & Sons, Inc., 605 Third Avenue, New York, NY 10158-0012.

Library of Congress Cataloging-in-Publication Data:

Chuang, S. L.
   Physics of optoelectronic devices / S.L. Chuang.
      p.   cm. — (Wiley series in pure and applied optics)
   "Wiley-Interscience publication."
   ISBN 0-471-10939-8 (alk. paper)
      1. Electrooptics.   2. Electrooptical devices.   3. Semiconductors.
   I. Title.   II. Series.
   QC673.C48   1995   94-24701
   621.381'045—dc20

Printed in the United States of America

10 9 8 7 6 5 4 3 2 1

To My Mother

Preface

This textbook is intended for graduate students and advanced undergraduate students in electrical engineering, physics, and materials science. It also provides an overview of the theoretical background for professional researchers in optoelectronic industries and research organizations. The book deals with the fundamental principles in semiconductor electronics, physics, and electromagnetics, and then systematically presents practical optoelectronic devices, including semiconductor lasers, optical waveguides, directional couplers, optical modulators, and photodetectors. Both bulk and quantum-well semiconductor devices are discussed. Rigorous derivations are presented and the author attempts to make the theories self-contained.

Research on optoelectronic devices has been advancing rapidly. To keep up with the progress in optoelectronic devices, it is important to grasp the fundamental physical principles. Only through a solid understanding of fundamental physics are we able to develop new concepts and design novel devices with superior performances. The physics of optoelectronic devices is a broad field with interesting applications based on electromagnetics, semiconductor physics, and quantum mechanics.

I have developed this book for a course on optoelectronic devices which I have taught at the University of Illinois at Urbana–Champaign for the past ten years. Many of our students are stimulated by the practical applications of quantum mechanics in semiconductor optoelectronic devices because many quantum phenomena can be observed directly using artifical materials such as quantum-well heterostructures with absorption or emission wavelengths determined by the quantized energy levels.

Scope

This book emphasizes the theory of semiconductor optoelectronic devices. Comparisons between theoretical and experimental results are also shown. The book starts with the fundamentals, including Maxwell's equations, the continuity equation, and the basic semiconductor equations of solid-state electronics. These equations are essential in learning semiconductor physics applied to optoelectronics. We then discuss the *propagation, generation, modulation, and detection of light*, which are the keys to understanding the physics behind the operation of optoelectronic devices. For example, knowledge of the generation and propagation of light is crucial for understanding how a semiconductor laser operates. The theory of gain coefficient of semiconductor lasers shows how light is amplified, and waveguide theory shows how light is confined to the waveguide in a laser cavity. An understanding of the modulation of light is useful in designing optical switches and modulators. The absorption coefficient of bulk and quantum-well semiconductors demonstrates how light is detected and leads to a discussion on the operating principles of photodetectors.

Features

- Important topics such as semiconductor heterojunctions and band structure calculations near the band edges for both bulk and quantum-well semiconductors are presented. Both Kane's model, assuming parabolic bands and Luttinger–Kohn's model, with valence-band mixing effects in quantum wells, are presented.
- Optical dielectric waveguide theory is discussed and applied to semiconductor lasers, directional couplers, and electrooptic modulators.
- Basic optical transitions, absorption, and gain are discussed with the time-dependent perturbation theory. The general theory for gain and absorption is then applied to studying interband and intersubband transitions in bulk and quantum-well semiconductors.
- Important semiconductor lasers such as double-heterostructure, stripe-geometry gain-guided semiconductor lasers, quantum-well lasers, distributed feedback lasers, coupled laser arrays, and surface-emitting lasers are discussed in great detail.
- High-speed modulation of semiconductor lasers using both linear and nonlinear gains is investigated systematically. The analytical theory for the laser spectral linewidth enhancement factor is derived.
- New subjects such as theories on the band structures of strained semiconductors and strained quantum-well lasers are investigated.
- The electroabsorptions, in bulk (Franz–Keldysh effects) and quantum-well semiconductors (quantum confined Stark effects), are discussed systematically including exciton effects. Both the bound and continuum states of excitons using the hydrogen atom model are discussed.
- Intersubband transitions in quantum wells, in addition to conventional interband absorptions for far-infrared photodetector applications, are presented.

Courses

A few possible courses for the use of this book are listed. Some background in undergraduate electromagnetics and modern physics is assumed. A background in quantum mechanics will be helpful but is not required, since all of the essentials are covered in the chapters on Fundamentals.

- **Overview of Optoelectronic Devices:** Chapter 1, Chapter 2, Chapter 3 (3.1, 3.2, 3.5, 3.7), Chapter 7 (7.1–7.3, 7.6), Chapter 8, Chapter 9 (9.1–9.6), Chapter 10 (10.1–10.3), Chapter 12, and Chapter 14.

- **Optoelectronic Device Physics:** Chapters 1–4, Chapter 7 (7.1, 7.5, 7.6), Chapters 9–14.
- **Electromagnetics and Optical Device Applications:** Chapter 2 (2.1–2.4), Chapters 5–8, Chapter 9 (9.1–9.3, 9.5, 9.6), Chapter 10 (10.1–10.3, 10.5–10.7), Chapter 12, Chapter 14.

The entire book (except for some advanced sections) can also be used for a two-semester course.

Acknowledgments

After receiving a rigorous training in my Ph.D. work on electromagnetics at Massachusetts Institute of Technology, I became interested in semiconductor optoelectronics because of recent developments in quantum-well devices with many applications in wave mechanics. I thank J. A. Kong, my Ph.D. thesis adviser, and many of my professors for their inspiration and insight.

Because of the significant number of research results appearing in the literature, it is difficult to list all of the important contributions in the field. For a textbook, only the fundamental principles are emphasized. I thank those colleagues who granted me permission to reproduce their figures. I apologize to all of my colleagues whose important contributions have not been cited. I am grateful to many colleagues and friends in the field, especially D. A. B. Miller, W. H. Knox, M. C. Nuss, A. F. J. Levi, J. O'Gorman, D. S. Chemla, and the late S. Schmitt-Rink, with whom I had many stimulating discussions on quantum-well physics during and after my sabbatical leave at AT&T Bell Laboratories. I would also like to thank many of my students who provided valuable comments, especially C. S. Chang and W. Fang, who proofread the manuscript. I thank many of my research assistants, especially D. Ahn, C. Y. P. Chao, and S. P. Wu, for their interaction on research subjects related to this book. The support of my research on quantum-well optoelectronic devices by the Office of Naval Research during the past years is greatly appreciated. I am grateful to L. Beck for reading the whole manuscript and K. C. Voyles for typing many revisions of the manuscript in the past years. The constant support and encouragement of my wife, Shu-Jung, are deeply appreciated. Teaching and conducting research have been the stimulus for writing this book; it was an enjoyable learning experience.

SHUN LIEN CHUANG

Illinois, March 1995

Contents

PART II   PROPAGATION OF LIGHT

PART V   DETECTION OF LIGHT

APPENDICES

Physics of Optoelectronic Devices

1

Introduction

Semiconductor optoelectronic devices, such as laser diodes, light-emitting diodes, optical waveguides, directional couplers, electrooptic modulators, and photodetectors, have important applications in optical communication systems. To understand the physics and the operational characteristics of these optoelectronic devices, we have to understand the fundamental principles. In this chapter, we discuss some of the basic concepts of optoelectronic devices, then present the overview of this book.

1.1 BASIC CONCEPTS

The basic idea is that for a semiconductor, such as GaAs or InP, many interesting optical properties occur near the band edges. For example, Table 1.1 shows part of the periodic table with many of the elements that are important for semiconductors [1, 2], including group IV, III–V, and II–VI compounds. For a III–V compound semiconductor such as GaAs, the gallium (Ga) and arsenic (As) atoms form a zinc-blende structure, which consists of two interpenetrating face-centered cubic lattices, one made of gallium atoms and the other made of arsenic atoms (Fig. 1.1). The Ga atom has an atomic number 31, which has an [Ar] $3d^{10}4s^24p^1$ configuration, i.e., three valence electrons on the outermost shell. (Here [Ar] denotes the configuration of Ar, which has an atomic number 18, and the 18 electrons are distributed as $1s^22s^22p^63s^23p^6$.) The As atom has an atomic number 33 with an [Ar] $3d^{10}4s^24p^3$ configuration or five valence electrons in the outermost shell. For a simplified view, we show a planar bonding diagram [3, 4] in Fig. 1.2a, where each bond between two nearby atoms is indicated with two dots representing two valence electrons. These valence electrons are contributed by either Ga or As atoms. The bonding diagram shows that each atom, such as Ga, is connected to four nearby As atoms by four valence bonds. If we assume that none of the bonds is broken, we have all of the electrons in the valence band and no free electrons in the conduction band. The energy band diagram as a function of position is shown in Fig. 1.2b, where E_c is the band edge of the conduction band and E_v is the band edge of the valence band.

When a light with an optical energy $h\nu$ above the bandgap E_g is incident on the semiconductor, optical absorption is significant. Here h is the Planck

Table 1.1 Part of the Periodic Table Containing Group II to VI Elements

Group II A	Group II B	Group III A	Group III B	Group IV A	Group IV B	Group V A	Group V B	Group VI A	Group VI B
4 Be $1s^22s^2$		5 B $1s^22s^22p^1$		6 C $1s^22s^22p^2$		7 N $1s^22s^22p^3$		8 O $1s^22s^22p^4$	
12 Mg [Ne]$3s^2$		13 Al [Ne]$3s^23p^1$		14 Si [Ne]$3s^23p^2$		15 P [Ne]$3s^23p^3$		16 S [Ne]$3s^23p^4$	
20 Ca [Ar]$4s^2$			21 Sc [Ar]$3d^14s^2$		22 Ti [AR]$3d^24s^2$		23 V [Ar]$3d^34s^2$		24 Cr [Ar]$3d^54s^1$
	30 Zn [Ar]$3d^{10}4s^2$	31 Ga [Ar]$3d^{10}4s^24p^1$		32 Ge [Ar]$3d^{10}4s^24p^2$		33 As [Ar]$3d^{10}4s^24p^3$		34 Se [Ar]$3d^{10}4s^24p^4$	
	38 Sr [Kr]$5s^2$		39 Y [Kr]$4d^15s^2$		40 Zr [Kr]$4d^25s^2$		41 Nb [Kr]$4d^45s^1$		42 Mo [Kr]$4d^55s^1$
	48 Cd [Kr]$4d^{10}5s^2$	49 In [Kr]$4d^{10}5s^25p^1$		50 Sn [Kr]$4d^{10}5s^25p^2$		51 Sb [Kr]$4d^{10}5s^25p^3$		52 Te [Kr]$4d^{10}5s^25p^4$	
56 Ba [Xe]$6s^2$	80 Hg [Xe]$4f^{14}5d^{10}6s^2$								

Note: [Ne] = $1s^22s^22p^6$
[Ar] = [Ne]$3s^23p^6$
[Kr] = [Ar]$3d^{10}4s^24p^6$
[Xe] = [Kr]$4d^{10}5s^25p^6$

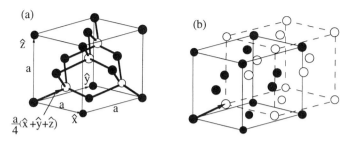

Figure 1.1. (a) A zinc-blende structure such as those of GaAs and InP semiconductors. (b) The zinc-blende structure in (a) consists of two interpenetrating face-centered cubic lattices separated by a constant vector $(a/4)(\hat{x} + \hat{y} + \hat{z})$, where a is the lattice constant of the semiconductor.

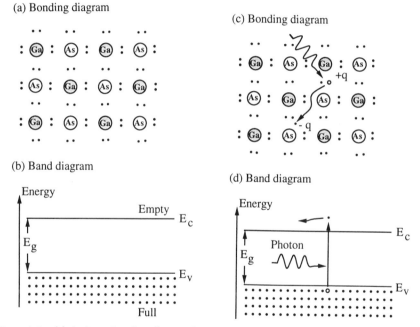

Figure 1.2. (a) A planar bonding diagram for a GaAs lattice. Each bond consists of two valence electrons shared by a gallium and an arsenic atom. (b) The energy-band diagram in real space shows the valence-band edge E_v below which all states are occupied, and the conduction band edge E_c above which all states are empty. The separation $E_c - E_v$ is the band gap E_g. (c) A bonding diagram showing a broken bond due to the absorption of a photon with an energy above the band gap. A free electron–hole pair is created. Note that the photogenerated electron is free to move around and the hole is also free to hop around at different bonds between the Ga and As atoms. (d) The energy-band diagram showing the energy levels of the electron and the hole.

constant and ν is the frequency of the photon,

$$h\nu = \frac{hc}{\lambda} = \frac{1.24}{\lambda} \quad (eV) \tag{1.1.1}$$

where c is the speed of light in free space and λ is wavelength in microns (μm). The absorption of a photon may break a valence bond and create an electron–hole pair, shown in Fig. 1.2c, where an empty position in the bond is represented by a hole. The same concept in the energy band diagram is illustrated in Fig. 1.2d, where the free electron propagating in the crystal is represented by a dot in the conduction band. It is equivalent to acquiring an energy larger than the band gap of the semiconductor, and the kinetic energy of the electron is that amount above the conduction-band edge. The reverse process can also occur if an electron in the conduction band recombines with a hole in the valence band; this excess energy may emerge as a photon, and the process is called spontaneous emission. In the presence of a photon propagating in the semiconductor with electrons in the conduction band and holes in the valence band, the photon may stimulate the downward transition of the electron from the conduction band to the valence band and emit another photon, which is called a stimulated emission process. Above the conduction-band edge or below the valence-band edge, we have to know the energy vs. momentum relation for the electrons or holes. These relations provide important information about the number of available states in the conduction band and in the valence band. We can imagine that by measuring the optical absorption spectrum as we tune the optical wavelength, we can somewhat map out the number of states per energy interval. This concept of joint density of states, which is discussed further in the following chapters, plays an important role in the optical absorption and gain processes in semiconductors.

The recent progress in modern crystal growth techniques [5] such as the molecular beam epitaxy (MBE) and the metal–organic chemical vapor deposition (MOCVD), has demonstrated that it is possible to grow semiconductors of different atomic compositions on top of another semiconductor substrate with monolayer precision. This opens up extremely exciting possibilities of the so-called "band-gap engineering." For example, aluminum arsenide (AlAs) has a similar lattice constant as gallium arsenide (GaAs). We can grow a few atomic layers of AlAs on top of a gallium arsenide substrate, then grow alternate layers of GaAs and AlAs. We can also grow a ternary compound such as $Al_xGa_{1-x}As$ (where the aluminum mole fraction x can be between 0 and 1) on a GaAs substrate and form a heterojunction (Fig. 1.3a). Interesting applications have been found using heterojunction structures. For example, when the wide-gap $Al_xGa_{1-x}As$ is doped by donors, the free electrons from the ionized donors tend to fall to the conduction band of the GaAs region because of the lower potential energy on that side; the band diagram is shown in Fig. 1.3b. (This band bending is investigated in Chapter 2.) An applied field in a direction parallel to the junction interface

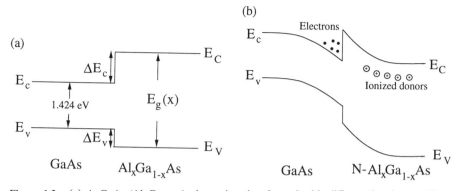

Figure 1.3.   (a) A GaAs/Al$_x$Ga$_{1-x}$As heterojunction formed with different band gaps. The band-edge discontinuities in the conduction band and the valence band are $\Delta E_c = 67\%\Delta E_g$ and $\Delta E_v = 33\%\Delta E_G$. (b) With n-type doping in the wide-gap Al$_x$Ga$_{1-x}$As region, the electrons ionized from the donors fall into the heterojunction surface layer on the GaAs side where the energy is smaller and an internal electric field pointing from the ionized (positive) donors in the Al$_x$Ga$_{1-x}$As region toward the electrons with negative charges creates a band bending, which looks like a triangular potential well to confine the electrons.

will create a conduction current. Since these electrons conduct in a channel on the GaAs region, which is undoped, the amount of impurity scatterings can be reduced. Therefore, the electron mobility can be enhanced. Based on this concept, the high-electron-mobility transistor (HEMT) has been realized.

For optoelectronic device applications, heterojunction structures [6] play important roles. For example, when semiconductor lasers were invented, they had to be cooled down to cryogenic temperature (77 K), and the lasers could lase only in a pulsed mode. These lasers had large threshold current densities, which means that a large amount of current has to be injected before the lasers could start lasing. With the introduction of the heterojunction semiconductor lasers, the concept of carrier and photon confinements makes room temperature cw operation possible, because the electrons and holes, once injected by the electrodes on both sides of the wide-band gap *P-N* regions (Fig. 1.4), will be confined in the central GaAs region where the band gap is smaller, resulting in a smaller potential energy for the electrons in the conduction band as well as a smaller potential energy for holes in the valence band. Note that the energy for the holes is measured downward, which is opposite to that of the electrons. For the photons, it turns out that the optical refractive index of the narrow band-gap material (GaAs) is larger than that of the wide-band gap material (Al$_x$Ga$_{1-x}$As). Therefore, the photons can be confined in the active region as well. This double confinement of both carriers and photons makes the stimulated emission process more efficient and leads to the room-temperature operation of laser diodes.

The control of the mole fractions of different atoms also makes the band-gap engineering extremely exciting. For optical communication systems,

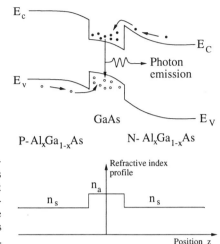

Figure 1.4. A double-heterojunction semiconductor laser structure, where the central GaAs region provides both the carrier confinement and optical confinement because of the conduction and valence-band profiles and the refractive index profile. This double confinement enhances stimulated emissions and the optical modal gain.

it has been found that minimum attenuation [7] in the fibers occurs at 1.33 and 1.55 μm. It is therefore natural to design sources such as light-emitting diodes and laser diodes, semiconductor modulators, and photodetectors operating at these desired wavelengths. For example, by controlling the mole fraction of gallium and indium in an $In_{1-x}Ga_xAs$ material, a wide tunable range of band gap is possible since InAs has a 0.354 eV band gap and GaAs has a 1.424 eV band gap at room temperature. At $x = 0.47$, the $In_xGa_{1-x}As$ alloy has a band gap of 0.75 eV and is lattice-matched to the InP substrate, because the lattice constant of the ternary alloy has a linear dependence on the mole fraction:

$$a(A_xB_{1-x}C) = xa(AC) + (1 - x)a(BC) \qquad (1.1.2)$$

where $a(AB)$ is the lattice constant of the binary compound AB and $a(BC)$ is that of the compound BC. This linear interpolation formula works very well for the lattice constant, but not for the band gap. For the band-gap dependence, a quadratic dependence on the mole fraction x is usually required (see Appendix K for some important material systems). For $Al_xGa_{1-x}As$ ternary compounds with $0 \leq x < 0.4$, the following linear formula is commonly used at room temperature:

$$E_g(Al_xGa_{1-x}As) = 1.424 + 1.247x \quad (eV) \qquad (1.1.3)$$

Most ternary compounds require a quadratic term. From the above formula, we can calculate the conduction and valence band-edge discontinuities between a GaAs and an $Al_xGa_{1-x}As$ heterojunction using $\Delta E_c = 67\%\Delta E_g$ and $\Delta E_v = 33\%\Delta E_g$, where $\Delta E_g = 1.247x$ (eV). When very thin layers of heterojunction structures are grown with a layer thickness thinner than the

(a) Field=0   **(b) Field>0**

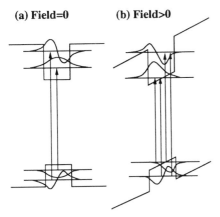

Figure 1.5. (a) A semiconductor quantum well without an applied electric field bias showing the quantized subbands and the corresponding wave functions. (b) With an applied electric field the tilted potential has the quantized energy levels shifted by the field and the wave functions are skewed from the previous even or odd symmetric wave functions.

coherent length of the conduction band electrons, quantum size effects occur. These include the quantization of the subband energies with corresponding wave functions (Fig. 1.5a).

The success in the growth of quantum-well structures makes a study of the introductory quantum physics realizable in these man-made semiconductor materials. Due to the quasi-two-dimensional confinement of electrons and holes in the quantum wells, many electronic and optical properties of these structures differ significantly from those of the bulk materials. Many interesting quantum mechanical phenomena using quantum-well structures and their applications have been predicted and confirmed experimentally [8]. For a simple quantum-well potential, we have the particle in a box (or well) model. These quantized energy levels appear in the optical absorption and gain spectra with exciting applications to electroabsorption modulators, quantum-well lasers, and photodetectors, because enhanced absorption occurs when the optical energy is close to the difference between the conduction and hole subband levels, as shown in Fig. 1.5a. The density of states in the quasi-two-dimensional structure is also different from that of bulk semiconductor. A significant discovery is that room temperature observation of these quantum mechanical phenomena can be observed. When an electric field bias is applied through the quantum-well region using a diode structure, the potential profiles are tilted and the positions of the quantized subbands are shifted (Fig. 1.5b). Therefore, the optical absorption spectrum can be changed by an electric field bias. This makes practical the applications of electroabsorption modulators using these quantum-well structures.

Experimental work on low threshold current quantum-well lasers [9] has been reported for different material systems, such as GaAs/AlGaAs, InGaAsP/InP, and InGaAs/InGaAsP. Advantages of the quantum-well lasers, such as a higher temperature stability, an improved linewidth enhancement factor, and wavelength tunability, have also been demonstrated. These devices are based on the band structure engineering concept using, for example,

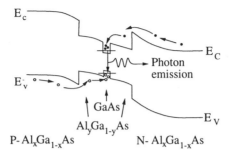

E_c

E_C

Photon
emission

E_v

GaAs

E_V

$Al_yGa_{1-y}As$

P- $Al_xGa_{1-x}As$ N- $Al_xGa_{1-x}As$

Figure 1.6. The energy-band diagram of a separated-confinement quantum-well laser structure. The active GaAs layer, which has a dimension around 100 Å, provides the carrier confinement and is sandwiched between two $Al_yGa_{1-y}As$ layers, where the aluminum mole fraction y is smaller than those (x) of the outermost $Al_xGa_{1-x}As$ cladding regions. The $Al_yGa_{1-y}As$ layers are of the order of submicrons (or an optical wavelength) and provide the optical confinement. The mole fraction y can also be graded such that it varies with the position along the crystal growth direction.

a separate-confinement heterostructure quantum-well structure to enhance the carrier and the optical confinements (Fig. 1.6).

The effect of the uniaxial stress perpendicular to the junction on the threshold current of GaAs double-heterostructure lasers was studied experimentally in the 1970s. The idea of using strained quantum wells [9, 10] by growing semiconductors with different lattices constants for tunable wavelength photodetectors and semiconductors was explored in the 1980s. Strained-layer quantum-well lasers have been investigated for low threshold current operation, polarization switching, and bistability applications. Important advantages using the strained-layer superlattice or quantum wells include the reduction of the threshold current density due to the raising of the heavy-hole band relative to the light-hole band, the elimination of the intervalence band absorption, and the reduction of the Auger recombination. Due to the selection rule for optical transitions, the polarization-dependent gains are also changed by the stress, since the optical gain is mainly TM polarized for the transition between the electron and the light-hole bands and TE polarized for the transition between the electron and the heavy-hole bands. Many of these details for valence subband electronic properties and polarization selection rules in quantum-well devices are explained in this book.

1.2 OVERVIEW

This book is divided into five parts: I, Fundamentals; II, Propagation; III, Generation; IV, Modulation; and V, Detection of Light. We start with the fundamentals on semiconductor electronics, quantum mechanics, solid state

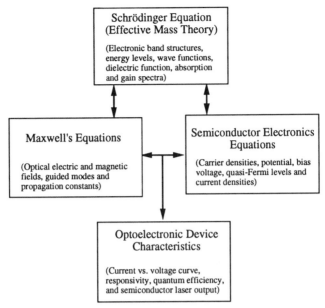

Figure 1.7.   Fundamental equations and their applications to optoelectronic device characteristics.

physics, and electromagnetics, with the emphasis on their applications to optoelectronic devices. In Fig. 1.7 we illustrate the important fundamental equations and their applications. In the presence of injection of electrons and holes using a voltage bias or an optical source, the semiconductor materials may change their absorptive properties to become gain media due to the carrier population effects. This implies that the optical dielectric function can be changed. This change can be modeled with the knowledge of the electronic band structures, which require the solutions of the Schrödinger equation or the so-called effective-mass equation for the given bulk or quantum-well semiconductors. Using Maxwell's equations, we obtain the optical electric and magnetic fields, assuming that we know the dielectric function of the semiconductors. The electronic band structure is also dependent on the static electric bias voltage, which determines the electron and hole current densities.

The semiconductor electronic equations governing the electron and hole concentrations and their corresponding current densities have to be solved. The device operation characteristics, such as the current–voltage relation in a *p-n* junction diode structure, the external quantum efficiency for the conversion of electric to optical power in a semiconductor laser, and the quantum efficiency for converting optical power to current in a photodetector, have to be investigated using these semiconductor electronic equations These fundamental equations actually are coupled to each other, and the most complete solution would require a self-consistent scheme, which would require heavy computations. Fortunately, with good understandings of most of the device physics, various approximation methods, such as the depletion approximation and perturbation theories for various device operation conditions, are possible. The validity of the models can be checked with full numerical solutions and confirmed with experimental observations.

After we explore the fundamentals, we investigate optoelectronic devices for propagation, generation, modulation, and detection of light. Below, we list some of the major issues of study in this book.

Part I Fundamentals

- Basic Semiconductor Electronics (Chapter 2)

    What are the fundamental semiconductor electronics equations governing the electron and hole concentrations and their corresponding current densities?

    How are the carrier densities affected by the presence of optical illumination or current injection?

    How are the energy-band diagrams drawn for heterojunctions such as P-n, N-p, p-N, or n-P junctions? Here a capital letter such as P refers to a wide-band-gap material doped P type, and a small p refers to a smaller band-gap semiconductor doped p type.

    Knowing the energy band bending, carrier injection, and current flow is a very important step to understanding the device operation characteristics.

- Basic Quantum Mechanics (Chapter 3)

    What are the eigenvalues and eigenfunctions of a rectangular quantum-well potential? These have important applications to semiconductor quantum-well devices.

    What are the bound- and continuum-state solutions of a hydrogen atom? The spherical harmonics solutions are useful when we talk about the band structures of the heavy-hole and light-hole bands of bulk and quantum-well semiconductors. The radial functions are useful for describing an exciton, formed by an electron–hole pair bounded by the Coulomb attractive potential, which is exactly the hydrogen model.

    What is Fermi's golden rule for the transition rate of a semiconductor in the presence of optical excitation? This rule is the basis for deriving the optical absorption and gain in semiconductor devices.

    What are the time-independent perturbation method and Löwdin's perturbation method? These have applications to Kane's model and Luttinger–Kohn's model for valence-band structures.

- Theory of Electronic Band Structures in Semiconductors (Chapter 4)

    What are the band structures near the band edges of a direct band-gap semiconductor?

    What is the effective-mass theory and how is it used to study the electronic properties of a semiconductor quantum-well structure?

    How do we calculate the electron and hole subband energies in a quantum well and their in-plane dispersion curves?

    What are the band structures of a strained bulk semiconductor and a strained quantum well?

- Electromagnetics (Chapter 5)

    What is the far-field radiation pattern if the modal field on the facet of a diode laser is known?

Part II Propagation of Light

- Light Propagation in Various Media (Chapter 6)

    What are the optical electric and magnetic fields of a laser light propagating in or reflected from a piece of semiconductor?

    What is a uniaxial medium? What are the basic concepts for a polaroid and quarter-wave plate?

- Optical Waveguide Theory (Chapter 7)

    How do we find the modes and their propagation constants in slab waveguides and rectangular dielectric waveguides?

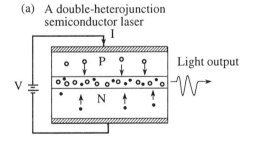

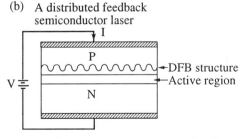

Figure 1.8. A cross section of (a) a double-heterojunction semiconductor laser structure in real space and (b) a distributed feedback semiconductor laser.

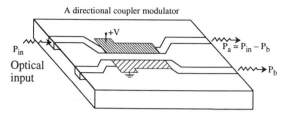

Figure 1.9. A directional coupler modulator whose output light power may be switched by an electric field bias.

How do the propagation constants change in the presence of absorption or gain in the waveguide? These waveguide modes are very important for applications in a semiconductor laser (Fig. 1.8a) and a directional coupler (Fig. 1.9).

- Waveguide Couplers and Coupled-Mode Theory (Chapter 8)

    What is the coupled-model theory and how do we design waveguide directional couplers?

    What are the solutions for the wave equation in a distributed feedback structure? The distributed feedback structure has applications in a semiconductor laser as well, Fig. 1.8b.

Part III Generation of Light

- Optical Processes in Semiconductors (Chapter 9)

    What are the basic formulas for optical absorption and gain in a semiconductor material?

    What is the difference in the optical absorption spectrum between a bulk and a quantum-well semiconductor?

    What is the difference between an interband and intersubband transition in a quantum-well structure?

- Semiconductor Lasers (Chapter 10)

    What are the operational principles of different types of semiconductor lasers?

    What determines the quantum efficiency and the threshold current of a diode laser?

Part IV Modulation of Light

- Direct Modulation of Semiconductor Lasers (Chapter 11)

    How do we directly modulate the light output from a semiconductor laser?

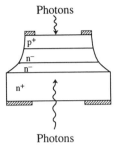

Figure 1.10. A *p-n* function photodiode with optical illumination from the top or from the bottom.

What determines the spectral linewidth of the semiconductor laser light?

- Electrooptic and Acoustooptic Modulators (Chapter 12)

    How do we modulate the intensity or phase of light?

- Electroabsorption Modulators (Chapter 13)

    How do we control the transmission of light passing through a bulk or a quantum-well semiconductor with an electric field bias?

Part V Detection of Light

- Photodetectors (Chapter 14)

    What are the different types of photodetectors and their operational characteristics? A simple example is a *p-n* junction photodiode as shown in Fig. 1.10. The absorption of photons and the conversion of optical energy to electric current will be investigated.

    What are quantum-well intersubband photodetectors?

Many of the above questions are still research issues that are under intensive investigation. The materials presented in this book emphasize the fundamental principles and analytical skills on the essentials of the physics of optoelectronic devices. The book was written with the hope that the readers of this book will acquire enough analytical power and knowledge for the physics of optoelectronic devices to analyze their research results, to understand more advanced materials in journal papers and research monographs, and to generate novel designs of optoelectronic devices. Many books that are highly recommended for further reading are listed in the bibliography.

PROBLEMS

1.1 (a) Calculate the band gap wavelength λ_g for Si, GaAs, InAs, InP, and GaP at 300 K.

Use the band gap energies in Appendix K.

(b) Find the optical energy corresponding to the wavelength 1.3μm and 1.55μm.

1.2 **(a)** Find the gallium mole fraction x for $In_{1-x}Ga_xAs$ compound semi-conductor such that its lattice constant equals that of InP. The lattice constants of a few binary compounds are listed in Appendix K.

(b) Find the alluminum mole fraction x for $Al_xIn_{1-x}As$ such that its lattice constant is the same as that of InP.

1.3 Calculate the band edge discontinuities ΔE_c and ΔE_v for GaAs/$Al_xGa_{1-x}As$ heterojunction if $x = 0.2$ and $x = 0.3$.

REFERENCES

1. C. Kittel, *Introduction to Solid State Physics*, Wiley, New York, 1976.
2. N. W. Ashcroft and N. D. Mermin, *Solid State Physics*, Holt, Rinehart & Winston, Saunders College, Philadelphia, 1976.
3. B. G. Streetman, *Solid State Electronic Devices*, Prentice Hall, Englewood Cliffs, NJ, 1980.
4. R. F. Pierret, *Semiconductor Fundamentals*, Vol. 1 in R. F. Pierret and G. W. Neudeck, Eds., *Modular Series on Solid State Devices*, Addison-Wesley, Reading, MA, 1983.
5. W. T. Tsang, Volume Ed., *Lightwave Communications Technology*, Vol. 22, Parts A–E, in R. K. Willardson and A. C. Beer, Eds., *Semiconductor and Semimetals*, Academic, New York, 1985.
6. H. C. Casey, Jr., and M. B. Panish, *Heterostructure Lasers*, Parts A and B, Academic, Orlando, FL, 1978.
7. T. Miya, Y. Terunuma, T. Hosaka, and T. Miyashita, "An ultimate low loss single mode fiber at 1.55 μm," *Electron. Lett.* **15**, 106–108 (1979).
8. D. S. Chemla and A. Pinczuk, Guest Ed., Special issues on Semiconductor Quantum Wells and Superlattices: Physics and Applications, *IEEE J. Quantum Electron.* **QE-22** (September 1986).
9. P. S. Zory, Jr., Ed., *Quantum Well Lasers*, Academic, San Diego, 1993.
10. T. P. Pearsall, Volume Ed., *Strained Layer Superlattices: Physics*, Vol. 32, 1990; and *Strained Layer Superlattices: Materials Science and Technology*, Vol. 33, 1991, in R. K. Willardson and A. C. Beer, Eds., *Semiconductor and Semimetals*, Academic, New York.

BIBLIOGRAPHY

General Semiconductor Optical and Electronic Physics

1. S. M. Sze, *Physics of Semiconductor Devices*, Wiley, New York, 1981.
2. S. Wang, *Fundamentals of Semiconductor Theory and Device Physics*, Prentice Hall, Englewood Cliffs, NJ, 1989.

3. B. R. Nag, *Theory of Electrical Transport in Semiconductors*, Pergamon, Oxford, UK, 1972.

4. B. R. Nag, *Electron Transport in Compound Semiconductors*, Springer, Berlin, 1980.

5. B. K. Ridley, *Quantum Processes in Semiconductors*, 2d ed., Clarendon, Oxford, UK, 1988.

6. N. W. Ashcroft and N. D. Mermin, *Solid State Physics*, Holt, Rinehard & Winston, New York, 1976.

7. H. Haken, *Light*, Vol. 1, *Waves, Photons, Atoms*, North-Holland, Amsterdam, 1986.

8. R. Loudon, *The Quantum Theory of Light*, 2d ed., Clarendon, Oxford, UK, 1986.

9. F. Bassani and G. P. Parravicini, *Electronic States and Optical Transitions in Solids*, Pergamon, Oxford, UK, 1975.

10. G. Bastard, *Wave Mechanics Applied to Semiconductor Heterostructures*, Halsted, New York, 1988.

11. H. Haug and S. W. Koch, *Quantum Theory of the Optical and Electronic Properties of Semiconductors*, World Scientific, Singapore, 1990.

12. K. Hess, *Advanced Theory of Semiconductor Devices*, Prentice Hall, Englewood Cliffs, NJ, 1988.

13. I. M. Tsidilkovski, *Band Structure of Semiconductors*, Pergamon, Oxford, UK, 1982.

14. K. Seeger, *Semiconductor Physics*, Springer, Berlin, 1982.

15. K. W. Böer, *Survey of Semiconductor Physics*, Van Nostrand Reinhold, New York, 1990.

16. C. Weisbuch and B. Vinter, *Quantum Semiconductor Structures*, Academic, New York, 1991.

17. R. K. Williardson and A. C. Beer, Eds., *Optical Properties of III-V Compounds*, Vol. 3 in *Semiconductors and Semimetals*, Academic, New York, 1967.

18. R. K. Williardson and A. C. Beer, Eds., *Modulation Techniques*, Vol. 9 in *Semiconductors and Semimetals*, Academic, New York, 1972.

19. M. Cardona, *Modulation Spectroscopy*, in *Solid State Phys.*, Suppl. 11, Academic, New York, 1969.

General Optical or Quantum Electronics

20. A. Yariv, *Quantum Electronics*, 3d ed., Wiley, New York, 1989.

21. A. Yariv, *Optical Electronics*, 3d ed., Holt, Rinehart & Winston. New York, 1985.

22. H. A. Haus, *Waves and Fields in Optoelectronics*, Prentice Hall, Englewood Cliffs, NJ, 1984.

23. J. T. Verdeyen, *Laser Electronics*, Prentice Hall, Englewood Cliffs, NJ, 1989.

24. B. E. A. Saleh and M. C. Teich, *Fundamentals of Photonics*, Wiley, New York, 1991.

25. A. K. Ghatak and K. Thyagarajan, *Optical Electronics*, Cambridge University Press, Cambridge, UK, 1989.

26. W. T. Tsang, Volume Ed., *Lightwave Communications Technology*, Vol. 22, Parts A–E in *Semiconductor and Semimetals*, R. K. Willardson and A. C. Beer, Eds., Academic, New York, 1985.

27. R. Dingle, *Applications of Multiquantum Wells, Selective Doping, and Superlattices*, Vol. 24 in *Semiconductor and Semimetals*, R. K. Willardson and A. C. Beer, Eds., Academic, New York, 1985.

28. J. Wilson and J. F. B. Hawkes, *Optoelectronics: An Introduction*, Prentice Hall, Englewood Cliffs, NJ, 1983.

29. R. G. Hunsperger, *Integrated Optics: Theory and Technology*, Springer, Berlin, 1984.

30. K. J. Ebeling, *Integrated Optoelectronics*, Springer, Berlin, 1993.

31. K. Chang, Ed., *Handbook of Microwave and Optical Components*, Vols. 3 and 4, Wiley, New York, 1991.

Semiconductor Lasers

32. G. H. B. Thompson, *Physics of Semiconductor Laser Devices*. Wiley, New York, 1980.

33. H. C. Casey, Jr., and M. B. Panish, *Heterostructure Lasers*, Parts A and B, Academic, Orlando, FL, 1978.

34. G. P. Agrawal and N. K. Dutta, *Long-Wavelength Semiconductor Lasers*, Van Nostrand Reinhold, New York, 1986.

35. P. S. Zory, Jr., Ed., *Quantum Well Lasers*, Academic, San Diego, 1993.

36. G. A. Evans and J. M. Hammer, Eds., *Surface Emitting Semiconductor Lasers and Arrays*, Academic, San Diego, 1993.

37. R. K. Williardson and A. C. Beer, Eds., *Lasers, Junctions, Transport*, Vol. 14 in *Semiconductors and Semimetals*, Academic, New York, 1979.

38. W. W. Chow, S. W. Koch, and M. Sargent III, *Semiconductor-Laser Physics*, Springer, Berlin, 1994.

39. J. K. Butler, Ed., *Semiconductor Injection Lasers*, IEEE Press, New York, 1980.

40. J. J. Coleman, Ed., *Selected Papers on Semiconductor Diode Lasers*, SPIE Milestone Series, Vol. MS50, SPIE Optical Engineering Press, Bellingham, WA, 1992.

Optical Waveguides and Modulators

41. H. Nishihara, M. Haruna, and T. Suhara, *Optical Integrated Circuits*, McGraw-Hill, New York, 1989.

42. T. Tamir, *Integrated Optics*, Springer, Berlin, 1979.

43. T. Tamir, *Guided Wave Optoelectronics*, 2d ed., Springer, Berlin, 1990.

44. D. Marcuse, *Theory of Dielectric Optical Waveguides*, Academic, New York, 1974.

45. A. W. Snyder and J. D. Love, *Optical Waveguide Theory*, Chapman & Hall, London, 1983.

46. A. B. Buckman, *Guided-Wave Photonics*, Saunders College, New York, 1992.

47. A. Yariv and P. Yeh, *Optical Waves in Crystals*, Wiley, New York, 1984.

Photodetectors

48. J. D. Vincent, *Fundamentals of Infrared Detector Operation and Testing*, Wiley, New York, 1990.

49. R. K. Willardson and A. C. Beer, Eds., *Infrared Detectors*, Vol. 5, in *Semiconductors and Semimetals*, Academic, New York, 1970.

50. R. K. Willardson and A. C. Beer, Eds., *Infrared Detectors II*, Vol. 12 in *Semiconductors and Semimetals*, Academic, New York, 1977.

51. R. K. Willardson and A. C. Beer, Eds., *Mercury Cadmium Telluride*, Vol. 18 in *Semiconductors and Semimetals*, Academic, New York, 1981.

52. A. Rogalsi, Ed., *Selected Papers on Semiconductor Infrared Detectors*, SPIE Milestone Series, Vol. MS66, SPIE Optical Engineering Press, Bellingham, WA, 1992.

53. M. O. Manasreh, Ed., *Semiconductor Quantum Wells and Superlattices for Long-Wavelength Infrared Detectors*, Artech House, Boston, MA 1993.

Nonlinear Optics

54. N. Bloembergen, *Nonlinear Optics*, Addison-Wesley, Redwood City, CA 1992. (Originally published by W.A. Benjamin, Inc., 1965).

55. R. W. Boyd, *Nonlinear Optics*, Academic, San Diego, 1992.

56. A. C. Newell and J. V. Moloney, *Nonlinear Optics*, Addison-Wesley, Redwood City, CA 1992.

57. Y. R. Shen, *The Principles of Nonlinear Optics*, Wiley, New York, 1984.

58. H. M. Gibbs, *Optical Bistability: Controlling Light with Light*, Academic, San Diego, 1985.

59. H. Haug, Ed., *Optical Nonlinearities and Instabilities in Semiconductors*, Academic, San Diego, 1988.

PART I

Fundamentals

2

Basic Semiconductor Electronics

In the study of semiconductor devices such as diodes and transistors, the characteristics of the devices are described by the voltage–current relations. The injection of electrons and holes by a voltage bias and their transport properties are studied. When optical injection or emission is involved, such as in laser diodes and photodetectors, we are interested in the optical field in the device as well as the light–matter interaction. In this case, we look for the light output versus the device bias current for a laser diode, or the change in the voltage–current relation due to the illumination of light in a photodetector. In general, it is useful to know the voltage, current, or quasi-static potentials and electric field in the electronic devices, and the optical electric and magnetic fields in the optoelectronic devices. Thus, a full understanding of the basic equations for the modeling of these devices is very important. In this chapter, we review the basic Maxwell's equations, semiconductor electronics equations, and boundary conditions. We also study the generation and recombination of carriers in semiconductors. The general theory for semiconductor heterojunctions and semiconductor/metal junctions is also investigated.

2.1 MAXWELL'S EQUATIONS AND BOUNDARY CONDITIONS [1, 2]

Maxwell's equations are the fundamental equations in electromagnetics. They were first established by James Clerk Maxwell in 1873 and were verified experimentally [3] by Heinrich Hertz in 1888. Maxwell unified all knowledge of electricity and magnetism, added a displacement current density term $\partial \mathbf{D}/\partial t$ in Ampère's law, and predicted electromagnetic wave motion. He explained light propagation as an electromagnetic wave phenomenon. Heinrich Hertz demonstrated experimentally the electromagnetic wave phenomenon using a spark-gap generator as a transmitter and a loop of wire with a very small gap as a receiver. He then set off a spark in the transmitter and showed that a spark at the receiver was produced. For a historical account of the classical and quantum theory of light, see Ref. 3, for example.

2.1.1 Maxwell's Equations in MKS Units

$$\nabla \times \mathbf{E} = -\frac{\partial}{\partial t}\mathbf{B} \qquad \text{Faraday's law} \qquad\qquad (2.1.1)$$

$$\nabla \times \mathbf{H} = \mathbf{J} + \frac{\partial \mathbf{D}}{\partial t} \qquad \text{Ampère's law} \qquad\qquad (2.1.2)$$

$$\nabla \cdot \mathbf{D} = \rho \qquad\qquad \text{Gauss's law} \qquad\qquad (2.1.3)$$

$$\nabla \cdot \mathbf{B} = 0 \qquad\qquad \text{Gauss's law} \qquad\qquad (2.1.4)$$

where $\mathbf{E}$ is the electric field (V/m), $\mathbf{H}$ is the magnetic field (A/m), $\mathbf{D}$ is the electric displacement flux density (C/m^2), and $\mathbf{B}$ is the magnetic flux density (V-s/m^2 or webers/m^2). The two source terms, the charge density ρ (C/m^3) and the current density $\mathbf{J}$(A/m^2), are related by the continuity equation

$$\nabla \cdot \mathbf{J} + \frac{\partial}{\partial t}\rho = 0 \qquad\qquad (2.1.5)$$

where no net generation or recombination of electrons is assumed. In the study of electromagnetics, one usually assumes that the source terms ρ and $\mathbf{J}$ are given quantities. It is noted that (2.1.4) is derivable from (2.1.1) by taking the divergence of (2.1.1) and noting that $\nabla \cdot (\nabla \times \mathbf{E}) = 0$ for any vector $\mathbf{E}$. Similarly, (2.1.3) is derivable from (2.1.2) using (2.1.5). Thus, we have only two independent vector equations, (2.1.1) and (2.1.2), or six scalar equations, since each vector has three components. However, there are $\mathbf{E}$, $\mathbf{H}$, $\mathbf{D}$, and $\mathbf{B}$, 12 scalar unknown components; thus we need six more scalar equations. These are the so-called constitutive relations which describe the properties of a medium. In isotropic media, they are given by

$$\mathbf{D} = \varepsilon \mathbf{E} \qquad \mathbf{B} = \mu \mathbf{H} \qquad\qquad (2.1.6)$$

In anisotropic media, they may be given by

$$\mathbf{D} = \bar{\bar{\varepsilon}} \cdot \mathbf{E} \qquad \mathbf{B} = \bar{\bar{\mu}} \cdot \mathbf{H} \qquad\qquad (2.1.7)$$

where $\bar{\bar{\varepsilon}}$ is the permittivity tensor and $\bar{\bar{\mu}}$ is the permeability tensor:

$$\bar{\bar{\varepsilon}} = \begin{bmatrix} \varepsilon_{xx} & \varepsilon_{xy} & \varepsilon_{xz} \\ \varepsilon_{yx} & \varepsilon_{yy} & \varepsilon_{yz} \\ \varepsilon_{zx} & \varepsilon_{zy} & \varepsilon_{zz} \end{bmatrix} \qquad \bar{\bar{\mu}} = \begin{bmatrix} \mu_{xx} & \mu_{xy} & \mu_{xz} \\ \mu_{yx} & \mu_{yy} & \mu_{yz} \\ \mu_{zx} & \mu_{zy} & \mu_{zz} \end{bmatrix} \qquad (2.1.8)$$

For electromagnetic fields at optical frequencies, $\rho = 0$ and $\mathbf{J} = 0$.

2.1.2 Boundary Conditions

By applying the first two Maxwell's equations over a small rectangular surface with a width δ (dashed line in Fig. 2.1a) across the interface of a boundary and using Stokes' theorem,

$$\oint_C \mathbf{E} \cdot d\ell = \int_S \nabla \times \mathbf{E} \cdot \hat{\mathbf{n}} \, dS = -\frac{d}{dt} \int_S \mathbf{B} \cdot \hat{\mathbf{n}} \, dS \tag{2.1.9a}$$

$$\oint_C \mathbf{H} \cdot d\ell = \int_S \nabla \times \mathbf{H} \cdot \hat{\mathbf{n}} \, dS = \int_S \mathbf{J} \cdot \hat{\mathbf{n}} \, dS + \frac{d}{dt} \int_S \mathbf{D} \cdot \hat{\mathbf{n}} \, dS \tag{2.1.9b}$$

the following boundary conditions can be derived by letting the width δ approach zero [1]:

$$\hat{\mathbf{n}} \times (\mathbf{E}_1 - \mathbf{E}_2) = 0 \tag{2.1.10}$$

$$\hat{\mathbf{n}} \times (\mathbf{H}_1 - \mathbf{H}_2) = \mathbf{J}_s \tag{2.1.11}$$

where $\mathbf{J}_s \ (= \lim_{\mathbf{J} \to \infty, \, \delta \to 0} \mathbf{J}\delta)$ is the surface current density (A/m). Note that the unit normal vector $\hat{\mathbf{n}}$ points from medium 2 to medium 1. Similarly, if we apply Gauss's laws (2.1.3) and (2.1.4) and integrate over a small volume (Fig. 2.1b) with a thickness δ and let δ approach zero, e.g.,

$$\oint_S \mathbf{D} \cdot \hat{\mathbf{n}} \, dS = \int_V \nabla \cdot \mathbf{D} \, dv = \int_V \rho \, dv = \rho \delta A$$

we obtain the following boundary conditions:

$$\hat{\mathbf{n}} \cdot (\mathbf{D}_1 - \mathbf{D}_2) = \rho_s \tag{2.1.12}$$

$$\hat{\mathbf{n}} \cdot (\mathbf{B}_1 - \mathbf{B}_2) = 0 \tag{2.1.13}$$

where $\rho_s \ (= \lim_{\rho \to \infty, \, \delta \to 0} \rho\delta)$ is the surface charge density (C/m^2). For an interface across two dielectric media, where no surface current or charge

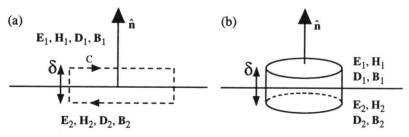

Figure 2.1. Geometry for deriving the boundary conditions across the interface of two media: (a) a rectangular surface is enclosed by the contour C (dashed line); and (b) a small volume with a thickness δ.

density can be supported, $\mathbf{J}_s = 0$ and $\rho_s = 0$, we have

$$\hat{\mathbf{n}} \times \mathbf{E}_1 = \hat{\mathbf{n}} \times \mathbf{E}_2 \tag{2.1.14a}$$

$$\hat{\mathbf{n}} \times \mathbf{H}_1 = \hat{\mathbf{n}} \times \mathbf{H}_2 \tag{2.1.14b}$$

$$\hat{\mathbf{n}} \cdot \mathbf{D}_1 = \hat{\mathbf{n}} \cdot \mathbf{D}_2 \tag{2.1.14c}$$

$$\hat{\mathbf{n}} \cdot \mathbf{B}_1 = \hat{\mathbf{n}} \cdot \mathbf{B}_2 \tag{2.1.14d}$$

For an interface between a dielectric medium and a perfect conductor,

$$\hat{\mathbf{n}} \times \mathbf{E}_1 = 0 \tag{2.1.15a}$$

$$\hat{\mathbf{n}} \times \mathbf{H}_1 = \mathbf{J}_s \tag{2.1.15b}$$

$$\hat{\mathbf{n}} \cdot \mathbf{D}_1 = \rho_s \tag{2.1.15c}$$

$$\hat{\mathbf{n}} \cdot \mathbf{B}_1 = 0 \tag{2.1.15d}$$

since the fields $\mathbf{E}_2$, $\mathbf{H}_2$, $\mathbf{D}_2$, and $\mathbf{B}_2$ inside the perfect conductor vanish. The surface charge density and the current density are supported by the perfect conductor surface.

2.1.3   Quasi-electrostatic Fields

For devices with a dc or low-frequency bias, since the time variation is very slow ($\partial/\partial t \simeq 0$), we usually have

$$\nabla \times \mathbf{E} = 0 \tag{2.1.16}$$

$$\nabla \cdot \mathbf{D} = \rho \tag{2.1.17}$$

and $\mathbf{H} \cong 0$, $\mathbf{B} \cong 0$ for the electronic devices for which no external magnetic fields are applied. In this case, the solution of the electric field can be put in the form of the gradient of an electrostatic potential ϕ:

$$\mathbf{E} = -\nabla\phi \tag{2.1.18}$$

and

$$\nabla \cdot (\varepsilon\nabla\phi) = -\rho \tag{2.1.19}$$

in an isotropic medium. Equation (2.1.19) is Poisson's equation. When the frequency becomes higher, for example, in a microwave transistor, one may include the displacement current density $\partial(\varepsilon\mathbf{E})/\partial t$ in the total current density in addition to the conduction current density $\mathbf{J}_{\text{con}}$:

$$\mathbf{J}_{\text{tot}} = \mathbf{J}_{\text{con}} + \frac{\partial}{\partial t}(\varepsilon\mathbf{E}) \tag{2.1.20}$$

2.2 SEMICONDUCTOR ELECTRONICS EQUATIONS

In this section, we present the basic semiconductor electronics equations, which are very useful in the modeling of semiconductor devices [4–7]. These equations are actually based on the Maxwell's equations and the charge continuity equations.

2.2.1 Poisson's Equation

As shown in the previous section, Poisson's equation in the semiconductor is given by (2.1.19):

$$\nabla \cdot (\varepsilon \nabla \phi) = -\rho \tag{2.2.1}$$

where ϕ is the electrostatic potential, and ρ is the charge density given by

$$\rho = q(p - n + C_0) \tag{2.2.2}$$

$$C_0 = N_D^+ - N_A^- \tag{2.2.3}$$

Here $q = 1.6 \times 10^{-19}$ C is the magnitude of a unit charge, p is the hole concentration, n is the electron concentration, N_D^+ is the ionized donor concentration, and N_A^- is the ionized acceptor concentration.

2.2.2 Continuity Equation

Since Ampère's law gives

$$\nabla \times \mathbf{H} = \mathbf{J}_{con} + \frac{\partial}{\partial t} \mathbf{D} \tag{2.2.4}$$

where the conduction current density is

$$\mathbf{J}_{con} = \mathbf{J}_p + \mathbf{J}_n \tag{2.2.5}$$

and $\mathbf{J}_p$ and $\mathbf{J}_n$ are the hole and electron current densities, respectively, we have

$$0 = \nabla \cdot (\nabla \times \mathbf{H}) = \nabla \cdot \mathbf{J}_{con} + \frac{\partial}{\partial t} \nabla \cdot \mathbf{D} \tag{2.2.6}$$

or

$$\nabla \cdot \mathbf{J}_{con} + \frac{\partial}{\partial t} \rho = 0 \tag{2.2.7}$$

Assuming that $C_0 = N_D^+ - N_A^-$ is independent of time, we have

$$\nabla \cdot (\mathbf{J}_p + \mathbf{J}_n) + q\frac{\partial}{\partial t}(p - n) = 0 \tag{2.2.8}$$

Thus, we may separate the above equation into two parts for electrons and holes:

$$\nabla \cdot \mathbf{J}_n - q\frac{\partial}{\partial t}n = +qR \tag{2.2.9}$$

$$\nabla \cdot \mathbf{J}_p + q\frac{\partial}{\partial t}p = -qR \tag{2.2.10}$$

where R is the net recombination rate ($\mathrm{cm}^{-3}\ \mathrm{s}^{-1}$) of electron–hole pairs. Sometimes it is convenient to write the generation rates (G_p and G_n) and recombination rates (R_p and R_n) explicitly:

$$R = R_n - G_n \tag{2.2.11}$$

for electrons and

$$R = R_p - G_p \tag{2.2.12}$$

for holes. Thus, we have the current continuity equations for the carriers:

$$\frac{\partial n}{\partial t} = G_n - R_n + \frac{1}{q}\nabla \cdot \mathbf{J}_n \tag{2.2.13}$$

$$\frac{\partial p}{\partial t} = G_p - R_p - \frac{1}{q}\nabla \cdot \mathbf{J}_p \tag{2.2.14}$$

2.2.3 Carrier Transport Equations

The carrier transport equations (assuming Boltzmann distributions for carriers) can be written as

$$\mathbf{J}_n = q\mu_n n\mathbf{E} + qD_n\nabla n \tag{2.2.15}$$

$$\mathbf{J}_p = q\mu_p p\mathbf{E} - qD_p\nabla p \tag{2.2.16}$$

where $\mathbf{E} = -\nabla\phi$ is the electric field, μ_n and μ_p are the electron and hole mobility, and D_n and D_p are the electron and hole diffusion coefficient, respectively. We may express the electric field in terms of the electrostatic potential in the carrier transport equation. We then have $\mathbf{J}_p, \mathbf{J}_n, \phi, p, n$, or nine scalar components as unknowns. We also have (2.2.1), (2.2.13), (2.2.14), (2.2.15), and (2.2.16), or nine scalar equations. We may also eliminate some of the unknown functions such as $\mathbf{J}_p$ and $\mathbf{J}_n$ and reduce the number of equations to three:

$$\frac{\partial n}{\partial t} = G_n - R_n + \frac{1}{q}\nabla \cdot [-q\mu_n n\nabla\phi + qD_n\nabla n] \qquad (2.2.17)$$

$$\frac{\partial p}{\partial t} = G_p - R_p - \frac{1}{q}\nabla \cdot [-q\mu_p p\nabla\phi - qD_p\nabla p] \qquad (2.2.18)$$

$$\nabla \cdot (\varepsilon\nabla\phi) = -q(p - n + C_0) \qquad (2.2.19)$$

with three unknowns n, p, and ϕ. In principle, these three unknowns can be solved using the above three equations once we specify the boundary conditions for a given device geometry.

2.2.4 Auxiliary Relations

Often it is convenient to introduce two auxiliary relations with two more functions, $F_n(\mathbf{r})$ and $F_p(\mathbf{r})$, the quasi-Fermi levels for the electrons and holes, respectively:

$$n(\mathbf{r}) = n_i \exp\left[\frac{F_n(\mathbf{r}) - E_i(\mathbf{r})}{k_B T}\right] \qquad (2.2.20)$$

$$p(\mathbf{r}) = n_i \exp\left[\frac{E_i(\mathbf{r}) - F_p(\mathbf{r})}{k_B T}\right] \qquad (2.2.21)$$

where the intrinsic carrier concentration n_i depends on the band edge concentration parameters N_c and N_v, the band gap, and the temperature

$$n_i^2 = N_c N_v\, e^{-E_g/k_B T} \qquad (2.2.22)$$

and the intrinsic energy level is

$$E_i(\mathbf{r}) = -q\phi(\mathbf{r}) + E_r \qquad (2.2.23)$$

Here E_r is a reference constant energy. We may substitute (2.2.20) and (2.2.21) into (2.2.17)–(2.2.19), and use only three unknowns, $\phi(\mathbf{r})$, $\phi_n(\mathbf{r})$, and $\phi_p(\mathbf{r})$, which have the same dimensions (V):

$$F_n(\mathbf{r}) = -q\phi_n(\mathbf{r}) + E_r \qquad (2.2.24)$$

$$F_p(\mathbf{r}) = -q\phi_p(\mathbf{r}) + E_r \qquad (2.2.25)$$

The two auxiliary relations are for nondegenerate semiconductors, for which the Maxwell–Boltzmann statistics are applicable. To take into account the effect of degeneracy, one may modify (2.2.20) and (2.2.21) simply by using the

Fermi–Dirac statistics together with the electron and hole density-of-state functions $\rho_e(E)$ and $\rho_h(E)$:

$$n = \int_{-\infty}^{\infty} f_n(E)\rho_e(E)\, dE \tag{2.2.26}$$

and

$$p = \int_{-\infty}^{\infty} f_p(E)\rho_h(E)\, dE \tag{2.2.27}$$

where

$$f_n(E) = \frac{1}{1 + e^{(E-F_n)/k_B T}} \tag{2.2.28}$$

is the Fermi–Dirac distribution for electrons and

$$f_p(E) = \frac{1}{1 + e^{(F_p-E)/k_B T}} \tag{2.2.29}$$

is the Fermi–Dirac distribution for the holes.

Density of States. The density of states for electrons, $\rho_e(E)$, is derived as follows. The number of electrons per unit volume is given by

$$n = \frac{2}{V} \sum_{k_x} \sum_{k_y} \sum_{k_z} f_n(E) \tag{2.2.30}$$

where the factor of 2 takes care of both spins of the electrons. For the electron states above the conduction band, we may assume that the electrons are in a box with a volume $L_x L_y L_z$ with wave numbers satisfying

$$k_x = m\frac{2\pi}{L_x}, \quad k_y = n\frac{2\pi}{L_y}, \quad k_z = \ell\frac{2\pi}{L_z}, \quad m, n, \ell = \text{integers}$$

Thus, the number of available states in a small cube $dk_x\, dk_y\, dk_z = d^3k$ in the $\mathbf{k}$ space is $d^3\mathbf{k}$ divided by the amount

$$\left(\frac{2\pi}{L_x}\right)\left(\frac{2\pi}{L_y}\right)\left(\frac{2\pi}{L_z}\right) = \frac{(2\pi)^3}{V} \tag{2.2.31}$$

for each state

$$\sum_{k_x} \sum_{k_y} \sum_{k_z} = \int \frac{d^3\mathbf{k}}{(2\pi)^3/V} \tag{2.2.32}$$

Thus

$$\frac{2}{V} \sum_{k_x} \sum_{k_y} \sum_{k_z} = \int \frac{d^3\mathbf{k}}{4\pi^3} = \int \frac{4\pi k^2 \, dk}{4\pi^3} = \int \frac{k^2 \, dk}{\pi^2} \tag{2.2.33}$$

If the parabolic band model is used,

$$E = E_c + \frac{\hbar^2 k^2}{2m_e^*} \tag{2.2.34}$$

where E_c is the conduction band edge, we obtain

$$\int \frac{k^2 \, dk}{\pi^2} = \int_{-\infty}^{\infty} dE \, \rho_e(E) \tag{2.2.35}$$

where $\rho_e(E)$ is called the density of states for the electrons in the conduction band

$$\rho_e(E) = \frac{1}{2\pi^2} \left(\frac{2m_e^*}{\hbar^2} \right)^{3/2} (E - E_c)^{1/2} \qquad \text{for } E > E_c \tag{2.2.36}$$

and $\rho_e(E)$ is zero if $E < E_c$. A similar expression holds for the density of states of the holes in the valence band,

$$\rho_h(E) = \frac{1}{2\pi^2} \left(\frac{2m_h^*}{\hbar^2} \right)^{3/2} (E_v - E)^{1/2} \qquad \text{for } E < E_v \tag{2.2.37}$$

and $\rho_h(E)$ is zero for $E > E_v$, where E_v is the valence-band edge. In the nondegenerate limit, it can be shown that (2.2.26) and (2.2.27) reduce to (2.2.20) and (2.2.21).

Using the Fermi–Dirac integral defined by

$$F_j(\eta) = \frac{1}{\Gamma(j+1)} \int_0^{\infty} \frac{x^j \, dx}{1 + e^{(x-\eta)}} \tag{2.2.38}$$

where Γ is the Gamma function, we may rewrite (2.2.26) and (2.2.27) as

$$n = N_c F_{1/2}\left(\frac{F_n - E_c}{k_B T}\right) \tag{2.2.39}$$

and

$$p = N_v F_{1/2}\left(\frac{E_v - F_p}{k_B T}\right) \tag{2.2.40}$$

where

$$N_c = 2\left(\frac{m_e^* k_B T}{2\pi \hbar^2}\right)^{3/2} = 2.51 \times 10^{19}\left(\frac{m_e^*}{m_0}\frac{T}{300}\right)^{3/2} \text{cm}^{-3} \tag{2.2.41}$$

$$N_v = 2\left(\frac{m_h^* k_B T}{2\pi \hbar^2}\right)^{3/2} = 2.51 \times 10^{19}\left(\frac{m_h^*}{m_0}\frac{T}{300}\right)^{3/2} \text{cm}^{-3} \tag{2.2.42}$$

and $\Gamma(3/2) = \sqrt{\pi}/2$ has been used.

Approximate Formula for the Fermi–Dirac Integral. An approximate analytic form for the Fermi–Dirac integral $F_j(\eta)$ valid for $-\infty < \eta < \infty$ is given by [5, 8–10]

$$F_j(\eta) = \frac{1}{e^{-\eta} + C_j(\eta)} \tag{2.2.43a}$$

where, for $j = \frac{1}{2}$, one uses either [8]

$$C_{1/2}(\eta) = \frac{3(\pi/2)^{1/2}}{\left[\eta + 2.13 + \left(|\eta - 2.13|^{12/5} + 9.6\right)^{5/12}\right]^{3/2}} \tag{2.2.43b}$$

or [9]

$$C_{1/2}(\eta) = \frac{3\pi^{1/2}/4}{\left(\eta^4 + 33.6\eta\left\{1 - 0.68\exp\left[-0.17(\eta + 1)^2\right]\right\} + 50\right)^{3/8}} \tag{2.2.43c}$$

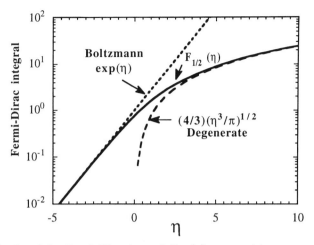

Figure 2.2. A plot of the Fermi–Dirac integral $F_{1/2}(\eta)$ vs. η and its asymptotic limits for $\eta \ll -1$ and $\eta \gg 1$.

The maximum errors of (2.2.43) are only 0.5 percent. We also see that

$$F_{1/2}(\eta) \sim \begin{cases} e^{\eta} & \eta \ll -1 \\ \dfrac{4}{3}\left(\dfrac{\eta^3}{\pi}\right)^{1/2} & \eta \gg 1 \end{cases} \tag{2.2.44}$$

These asymptotic limits are shown as dashed lines and compared with the exact numerical integration in Fig. 2.2.

Determination of the Fermi Level E_F. For a bulk semiconductor under thermal equilibrium, the Fermi level E_F $(= F_n = F_p)$ is determined by the charge neutrality condition:

$$n_0 + N_A^- = p_0 + N_D^+ \tag{2.2.45}$$

where N_A^- is the ionized acceptor concentration and N_D^+ is the ionized donor concentration. We have

$$N_A^- = \frac{N_A}{1 + g_A \exp[(E_A - E_F)/k_B T]} \tag{2.2.46}$$

where g_A is the ground-state degeneracy factor for acceptor levels. g_A equals 4 because in Si, Ge, and GaAs each acceptor level can accept one hole of either spin and the acceptor level is doubly degenerate as a result of the two degenerate valence bands at $k = 0$, and

$$N_D^+ = N_D \left\{ 1 - \cfrac{1}{1 + \cfrac{1}{g_D} \exp[(E_D - E_F)/k_B T]} \right\} \qquad (2.2.47)$$

where g_D is the ground state degeneracy of the donor impurity level and equals 2. For example, assuming that $N_A = 0$ and $N_D = 5 \times 10^{14}/\text{cm}^3$ for a GaAs sample at 300 K, Fig. 2.3a, we can plot n, p, and N_D^+ from (2.2.39), (2.2.40) and (2.2.47) vs. energy to determine the intersection point between the curve for $N_D^+ + p$ vs. energy and the curve of n vs. energy [4]. The horizontal reading of the intersection point gives the Fermi energy E_F.

In Fig. 2.3b, we repeat the same procedure for a larger donor concentration $N_D = 5 \times 10^{18}/\text{cm}^3$. We can see that the Fermi level E_F moves closer to the conduction band edge E_c. Notice that at thermal equilibrium, $n_0 p_0 = n_i^2$. For extrinsic semiconductors, we have two cases:

1. n-type, $N_D^+ - N_A^- \gg n_i$, therefore,

$$n_0 \simeq N_D^+ - N_A^- \text{ and } p_0 = \frac{n_i^2}{N_D^+ - N_A^-} \qquad (2.2.48a)$$

2. p-type, $N_A^- - N_D^+ \gg n_i$, therefore,

$$p_0 \simeq N_A^- - N_D^+ \text{ and } n_0 = \frac{n_i^2}{N_A^- - N_D^+} \qquad (2.2.48b)$$

So far, we assume that p is given by the heavy holes only, since the density of states of the heavy holes is usually much larger than that of the light holes. If we take into account the contribution due to light holes, the total hole concentration should be

$$p = p_{hh} + p_{lh} \qquad (2.2.49a)$$

where

$$p_{hh} = N_v^{hh} F_{1/2}\left(\frac{E_v - E_F}{k_B T}\right) \qquad N_v^{hh} = 2\left(\frac{m_{hh}^* k_B T}{2\pi\hbar^2}\right)^{3/2} \qquad (2.2.49b)$$

$$p_{lh} = N_v^{lh} F_{1/2}\left(\frac{E_v - E_F}{k_B T}\right) \qquad N_v^{lh} = 2\left(\frac{m_{lh}^* k_B T}{2\pi\hbar^2}\right)^{3/2} \qquad (2.2.49c)$$

Both terms have to be used in determining the Fermi level E_F using the charge neutrality condition.

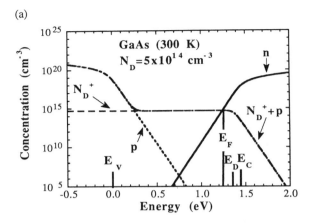

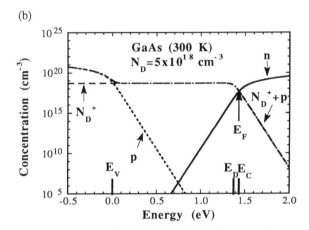

Figure 2.3. (a) A graphical approach to determine the Fermi level E_F from the charge neutrality condition $n = N_D^+ + p$. Both n and $N_D^+ + p$ vs. the horizontal variable E_F are plotted and the intersection of the two curves give the Fermi level of the system $N_D = 5 \times 10^{14}/\text{cm}^3$. (b) Same as (a) except that $N_D = 5 \times 10^{18}/\text{cm}^3$.

An inversion formula is sometimes convenient if we know the carrier concentration n. Define

$$u = \frac{n}{N_c} = F_{1/2}(\eta) \qquad \eta = \frac{E_F - E_c}{k_B T} \qquad (2.2.50)$$

Then, an approximate formula is [11]

$$\eta \approx \frac{\ln u}{1 - u^2} + \frac{v}{1 + (0.24 + 1.08v)^{-2}}$$

$$v = \left(\frac{3\sqrt{\pi} u}{4} \right)^{2/3} \qquad (2.2.51)$$

The maximum error [10, 11] of the above formula is 0.006. Further improvement by two orders of magnitude is possible [12].

An alternative approach for taking into account the degenerate semiconductor is to replace n_i in (2.2.20) and (2.2.21) by an effective intrinsic carrier concentration n_{ie} such that

$$n_{ie} = n_i^2 \, e^{\Delta E/k_B T} \tag{2.2.52}$$

where ΔE takes into consideration the effective band-gap narrowing, which has to be taken from electrical measurements.

2.2.5 Boundary Conditions [6]

On the surface of ideal ohmic contacts

$$\phi = \phi_0 + V_{\text{appl.}} \tag{2.2.53a}$$

$$n = n_0 \tag{2.2.53b}$$

$$p = p_0 \tag{2.2.53c}$$

The quasi-Fermi potentials satisfy

$$\phi_n = \phi_p = V_{\text{appl.}} \tag{2.2.54}$$

at the ohmic contacts. Here ϕ_0, n_0, and p_0 are the values of the corresponding variables for space-charge neutrality and at equilibrium. On an interface with surface recombination, such as that along a Si–SiO$_2$ interface in a MOSFET, one may have

$$\hat{\mathbf{n}} \cdot \mathbf{J}_n = -qR_s \tag{2.2.55a}$$

$$\hat{\mathbf{n}} \cdot \mathbf{J}_p = qR_s \tag{2.2.55b}$$

where R_s is the surface recombination rate. When interface charges ρ_s exist such as those of the effective oxide charges on the Si–SiO$_2$ interface, one has

$$-\hat{\mathbf{n}} \cdot (\varepsilon_1 \nabla \phi_1 - \varepsilon_2 \nabla \phi_2) = \rho_s \tag{2.2.56}$$

where $\hat{\mathbf{n}}$ points from region 2 to region 1. On a semiconductor–insulator interface where no surface recombination exists, one has

$$\hat{\mathbf{n}} \cdot \mathbf{J}_n = \hat{\mathbf{n}} \cdot \mathbf{J}_p = 0 \tag{2.2.57}$$

Example We assume the donor concentration N_D is 1×10^{17} cm^{-3} in a GaAs sample and these donors are ionized at room temperature. Since

$N_D \gg n_i = 2.1 \times 10^6$ the intrinsic carrier concentration, we have the electron concentration

$$n_0 = N_D = 1 \times 10^{17} \text{ cm}^{-3}$$

and

$$p_0 = \frac{n_i^2}{N_D} = 4.41 \times 10^{-5} \text{ cm}^{-3} \ll n_0$$

We can also calculate the band-edge concentration parameters using $m_e^* = 0.0665 \, m_0$ and $m_h^* = 0.50 \, m_0$ and $T = 300$ K:

$$N_c = 2 \left(\frac{m_e^* k_B T}{2\pi \hbar^2} \right)^{3/2}$$

$$= 2.51 \times 10^{19} \left(\frac{m_e^*}{m_0} \frac{T}{300} \right)^{3/2} \text{ cm}^{-3}$$

$$= 4.30 \times 10^{17} \text{ cm}^{-3}$$

$$N_v = 2.51 \times 10^{19} \left(\frac{m_h^*}{m_0} \frac{T}{300} \right)^{3/2} \text{ cm}^{-3}$$

$$= 8.87 \times 10^{18} \text{ cm}^{-3}$$

The Fermi level can be obtained from the inversion formula (2.2.51) using

$$u = \frac{n_0}{N_c} = \frac{1 \times 10^{17}}{4.30 \times 10^{17}} = 0.2326$$

and we find

$$\eta \simeq -1.381$$
$$E_F - E_c = \eta k_B T = -35.7 \text{ meV}$$

That is the Fermi level is 35.7 meV below the conduction band edge. If we use $\eta \simeq \ln u = -1.458$, we obtain $E_F - E_c = -37.7$ meV. The error is only 2 meV. ■

2.3 GENERATION AND RECOMBINATION IN SEMICONDUCTORS

In this section, we describe a phenomenological approach to the generation and recombination processes in semiconductors. A quantum mechanical approach can also be taken using the time-dependent perturbation theory,

Fermi's golden rule. The latter approach will be discussed when we study the optical absorption or gain in semiconductors in Chapter 9. In general, these generation–recombination processes can be classified as radiative or nonradiative. The radiative transitions involve the creation or annihilation of photons. The nonradiative transitions do not involve photons; they may involve the interaction with phonons or the exchange of energy and momentum with another electron or hole. The fundamental mechanisms can all be described using Fermi's golden rule with the energy and momentum conservation satisfied by these processes. The processes may also be discussed in terms of band-to-bound state transitions and band-to-band transitions.

2.3.1 Intrinsic Band-to-Band Generation–Recombination Processes

For band-to-band transitions [13–15] as shown in Fig. 2.4, the recombination rate of electrons and holes should be proportional to the product of the electron and hole concentrations

$$R_n = R_p = cnp \tag{2.3.1}$$

and the generation rate may be written as

$$G_n = G_p = e \tag{2.3.2}$$

where c is a capture coefficient and e is an emission rate. The net recombination rate is

$$R = R_n - G_n = R_p - G_p = cnp - e \tag{2.3.3}$$

If there is no external perturbation such as electric or optical injection of carriers, the net recombination rate should vanish at thermal equilibrium:

$$0 = cn_0 p_0 - e \tag{2.3.4}$$

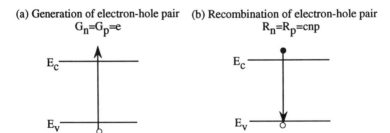

(a) Generation of electron-hole pair
$G_n = G_p = e$

(b) Recombination of electron-hole pair
$R_n = R_p = cnp$

Figure 2.4. The energy band diagrams for (a) generation and (b) recombination of an electron–hole pair.

where n_0 and p_0 are the electron and hole concentrations at thermal equilibrium $n_0 p_0 = n_i^2$. The net recombination rate can be written as

$$R = c(np - n_0 p_0) \tag{2.3.5}$$

If the carrier concentrations n and p deviate from their thermal equilibrium values by δn and δp, respectively,

$$n = n_0 + \delta n$$
$$p = p_0 + \delta p \tag{2.3.6}$$

we find that the recombination rate under the condition of low-level injection, $\delta n, \delta p \ll (n_0 + p_0)$, is given by

$$R = c(n_0 + p_0)\delta n = \frac{\delta n}{\tau} \tag{2.3.7}$$

where $\delta n = \delta p$, since electrons and holes are created in pairs for interband transitions, and the lifetime

$$\tau = \frac{1}{c(n_0 + p_0)} \tag{2.3.8}$$

has been used.

2.3.2 Extrinsic Shockley–Read–Hall (SRH) Generation–Recombination Processes [16–19]

There are basically four processes, [18, 19] as shown in Fig. 2.5. These processes may all be caused by the absorption or emission of phonons.

1. *Electron capture.* The recombination rate for the electrons is proportional to the density of electrons n, and the concentration of the traps

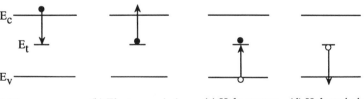

(a) Electron capture (b) Electron emission (c) Hole capture (d) Hole emission
$R_n = c_n n N_t (1-f_t)$ $G_n = e_n N_t f_t$ $R_p = c_p p N_t f_t$ $G_p = e_p N_t (1-f_t)$

Figure 2.5. The energy band diagrams for the Shockley–Read–Hall generation–recombination processes: (a) electron capture, (b) electron emission, (c) hole capture, and (d) hole emission.

N_t, multiplied by the probability that the trap is empty $(1 - f_t)$, where f_t is the occupation probability of the trap,

$$R_n = c_n n N_t (1 - f_t) \tag{2.3.9}$$

where c_n is the capture coefficient for the electrons.

2. *Electron emission.* The generation rate of the electrons due to this process is

$$G_n = e_n N_t f_t \tag{2.3.10}$$

where e_n is the emission coefficient, and $N_t f_t \equiv n_t$ is the density of the traps that are occupied by the electrons.

3. *Hole capture.* The recombination rate of the holes is given by the capture of holes by occupied traps; the number is given by $N_t f_t$:

$$R_p = c_p p N_t f_t \tag{2.3.11}$$

4. *Hole emission.* The generation rate for this process is proportional to the density of the traps that are empty (i.e., occupied by holes):

$$G_p = e_p N_t (1 - f_t) \tag{2.3.12}$$

where e_p is the emission coefficient for the holes.

Based on the principle of detail balancing, there should be zero net generation–recombination of electrons and holes, respectively, at thermal equilibrium. We have

$$R_n - G_n = c_n n_0 N_t (1 - f_{t_0}) - e_n N_t f_{t_0} = 0 \tag{2.3.13}$$

$$R_p - G_p = c_p p_0 N_t f_{t_0} - e_p N_t (1 - f_{t_0}) = 0 \tag{2.3.14}$$

where the subscript 0 denotes the thermal equilibrium values. Thus, we obtain a relation between c_n and e_n, c_p and e_p.

$$e_n = c_n n_0 \frac{1 - f_{t_0}}{f_{t_0}} \equiv c_n n_1 \tag{2.3.15}$$

$$e_p = c_p p_0 \frac{f_{t_0}}{1 - f_{t_0}} \equiv c_p p_1 \tag{2.3.16}$$

where for convenience we define n_1 and p_1 from the above equations, and $n_1 p_1 = n_0 p_0 = n_i^2$. The intrinsic concentration n_i may be replaced by an effective concentration n_{ie} for degenerate semiconductors.

The net recombination rates for electron and holes are

$$R_n - G_n = c_n N_t [n(1 - f_t) - n_1 f_t] \tag{2.3.17}$$

$$R_p - G_p = c_p N_t [pf_t - p_1(1 - f_t)] \tag{2.3.18}$$

We note that a combination of electron and hole captures (processes 1 and 3) destroys an electron–hole pair, while a combination of electron and hole emissions (processes 2 and 4) creates an electron hole pair. At equilibrium, these two rates are balanced; the fraction of occupied traps f_t is then given by

$$f_t = \frac{c_n n + c_p p_1}{c_n(n + n_1) + c_p(p + p_1)} \tag{2.3.19}$$

The net recombination rate is

$$R_n - G_n = R_p - G_p = \frac{c_n c_p np - e_n e_p}{c_n n + e_n + c_p p + e_p} N_t \tag{2.3.20}$$

$$= \frac{np - n_i^2}{\tau_p(n + n_1) + \tau_n(p + p_1)} \tag{2.3.21}$$

where

$$\tau_p = \frac{1}{c_p N_t} \tag{2.3.22}$$

$$\tau_n = \frac{1}{c_n N_t} \tag{2.3.23}$$

which are the hole and electron lifetimes, respectively. We note that the above capture and emission processes can be due to optical illumination at the proper photon energy in addition to thermal processes. Under the low-level injection condition, the net recombination rate (2.3.21) can be written in the same form as (2.3.7):

$$R = \frac{\delta n}{\tau_0}$$

where

$$\tau_0 = \frac{\tau_p(n + n_1) + \tau_n(p + p_1)}{n_0 + p_0} \tag{2.3.24}$$

2.3.3   Auger Generation–Recombination Processes

For band-to-bound state transitions, the following processes [18, 19] are possible (Fig. 2.6).

1. Electron captures with the released energy taken by an electron or a hole. The recombination rate is given by

$$R_n = (c_n^n n + c_n^p p) n N_t (1 - f_t) \qquad (2.3.25)$$

   where we note that the capture coefficient c_n in SRH recombinations has been replaced by $c_n^n n + c_n^p p$, since the two possible processes for electron captures are proportional to the concentration of the electron or hole that gains the energy released by the electron capture.

2. Electron emissions with the required energy supplied by an energetic electron or hole:

$$G_n = (e_n^n n + e_n^p p) N_t f_t \qquad (2.3.26)$$

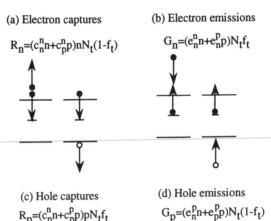

(a) Electron captures
$R_n = (c_n^n n + c_p^n p) n N_t (1-f_t)$

(b) Electron emissions
$G_n = (e_n^n n + e_n^p p) N_t f_t$

(c) Hole captures
$R_p = (c_p^n n + c_p^p p) p N_t f_t$

(d) Hole emissions
$G_p = (e_n^p n + e_p^p p) N_t (1-f_t)$

Figure 2.6. The energy band diagrams for the band-to-bound state Auger generation/ recombination processes: (a) electron capture, (b) electron emission, (c) hole capture, and (d) hole emission.

3. Hole captures with the released energy taken by an electron or a hole:

$$R_p = (c_p^n n + c_p^p p) p N_t f_t \tag{2.3.27}$$

4. Hole emissions with the required energy supplied by an energetic electron or hole:

$$G_p = (e_p^n n + e_p^p p) N_t (1 - f_t) \tag{2.3.28}$$

If all these processes of the impurity-assisted transitions due to thermal, optical, and Auger-impact ionization mechanisms exist, we use

$$c_n = c_n^t + c_n^o + c_n^n n + c_n^p p \tag{2.3.29}$$

$$e_n = e_n^t + e_n^o + e_n^n n + e_n^p p \tag{2.3.30}$$

$$c_p = c_p^t + c_p^o + c_p^n n + c_p^p p \tag{2.3.31}$$

$$e_p = e_p^t + e_p^o + e_p^n n + e_p^p p \tag{2.3.32}$$

The net recombination rate is given again by (2.3.20)

$$R_n - G_n = R_p - G_p = \frac{c_n c_p np - e_n e_p}{c_n n + e_n + c_p p + e_p} N_t \tag{2.3.33}$$

For band-to-band Auger-impact ionization processes, we have the four possible processes (Fig. 2.7):

1. *Electron capture.* An electron in the conduction band recombines with a hole in the valence band and releases its energy to a nearby electron.

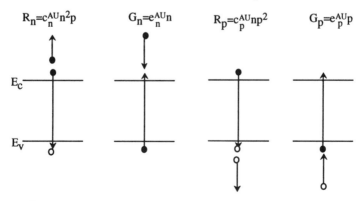

(a) Electron capture (b) Electron emission (c) Hole capture (d) Hole emission

$$R_n = c_n^{AU} n^2 p \qquad G_n = e_n^{AU} n \qquad R_p = c_p^{AU} np^2 \qquad G_p = e_p^{AU} p$$

Figure 2.7. The energy band diagrams for the band-to-band Auger generation/recombination processes: (a) electron capture, (b) electron emission, (c) hole capture, and (d) hole emission.

This process destroys an electron–hole pair. The recombination rate is

$$R_n = c_n^{AU} n^2 p \tag{2.3.34}$$

2. *Electron emission.* An electron in the valence band jumps to the conduction band because of the impact ionization of an incident energetic electron in the conduction band, which breaks a bond. This process creates an electron–hole pair. The generation rate is

$$G_n = e_n^{AU} n \tag{2.3.35}$$

3. *Hole capture.* An electron in the conduction band recombines with a hole in the valence band with the released energy taken up by a nearby hole. This process destroys an electron–hole pair. The recombination rate is

$$R_p = c_p^{AU} n p^2 \tag{2.3.36}$$

4. *Hole emission.* An electron in the valence band jumps to the conduction band (or the breaking of a bond to create an electron–hole pair) due to the impact of an energetic hole in the valence band. The generation rate is

$$G_p = e_p^{AU} p \tag{2.3.37}$$

At thermal equilibrium, no net generation/recombination exists. Thus process 1 and its reverse, process 2, balance each other, as do processes 3 and 4. Therefore,

$$R_n - G_n = 0 = c_n^{AU} n_0^2 p_0 - e_n^{AU} n_0 \tag{2.3.38}$$

or

$$e_n^{AU} = c_n^{AU} n_0 p_0 = c_n^{AU} n_i^2 \tag{2.3.39}$$

Similarly,

$$e_p^{AU} = c_p^{AU} n_i^2 \tag{2.3.40}$$

The total net Auger recombination rate is the sum of the net rates of electrons and holes, since each process creates or destroys one electron–hole pair. Therefore,

$$R = R_n - G_n + R_p - G_p = \left(c_n^{AU} n + c_p^{AU} p \right) \left(np - n_i^2 \right) \tag{2.3.41}$$

2.3.4 Impact Ionization Generation–Recombination Process [6]

This process is very much like the reverse Auger processes discussed above. However, the hot electron impact ionization processes usually depend on the incident current densities instead of the carrier concentrations. Microscopically, the processes are identical to the Auger-generation processes 2 and 4, which create an electron–hole pair due to an incident energetic electron or hole. These rates are usually given by

$$G_n = \alpha_n \frac{|J_n|}{q} \tag{2.3.42}$$

and

$$G_p = \beta_p \frac{|J_p|}{q} \tag{2.3.43}$$

where α_n and β_p are the ionization coefficients for electrons and holes, respectively. α_n is the number of electron–hole pairs generated per unit distance due to an incident electron. β_p is the number of electron–hole pairs created per unit distance due to an incident hole.

The total net recombination rate is

$$R = -G_n - G_p = -\alpha_n \frac{|J_n|}{q} - \beta_p \frac{|J_p|}{q} \tag{2.3.44}$$

Usually the ionization coefficients are related to the electric field E in the ionization region by the empirical formulas

$$\alpha_n(E) = \alpha_n^\infty \exp\left[-\left(\frac{E_{nc}}{E}\right)^\gamma\right] \tag{2.3.45}$$

$$\beta_p(E) = \beta_p^\infty \exp\left[-\left(\frac{E_{pc}}{E}\right)^\gamma\right] \tag{2.3.46}$$

where E_{nc} and E_{pc} are the critical fields, and usually $1 \leq \gamma \leq 2$. These impact ionization coefficients are used in the study of avalanche photodiodes in Chapter 14.

2.4 EXAMPLES AND APPLICATIONS TO OPTOELECTRONIC DEVICES

In this section, we consider a few simple examples to illustrate the optical generation–recombination processes and their effects on photodetectors. More details about photodetectors are discussed in Chapter 14.

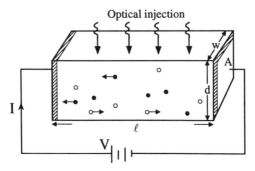

Figure 2.8. A homogeneous semiconductor under a uniform optical illumination from the side.

In the presence of optical injection, the continuity equations are

$$\frac{\partial n}{\partial t} = G_n - R_n + \frac{1}{q}\frac{\partial}{\partial x}J_n(x) \qquad (2.4.1)$$

$$\frac{\partial p}{\partial t} = G_p - R_p - \frac{1}{q}\frac{\partial}{\partial x}J_p(x) \qquad (2.4.2)$$

If a semiconductor is doped p-type with an acceptor concentration N_A, we know that

$$p_0 \simeq N_A \quad \text{and} \quad n_0 \simeq \frac{n_i^2}{N_A} \qquad (2.4.3)$$

2.4.1 Uniform Optical Injection

Suppose the optical generation rate is uniform across the semiconductor,

$$G_n = G_n(t) \qquad (2.4.4)$$

which is independent of the position (Fig. 2.8) and the recombination rate is $R_n = \delta n / \tau_n$. We then have

$$\frac{\partial n}{\partial t} = G_n(t) - \frac{\delta n}{\tau_n} \qquad (2.4.5)$$

The new carrier densities are

$$p = p_0 + \delta p \quad \text{and} \quad n = n_0 + \delta n \qquad (2.4.6)$$

where the excess carrier concentrations are equal $\delta p = \delta n$ for the interband optical generation process, since we assume that electrons and holes are created in pairs. Furthermore, for a constant light intensity,

$$G_n(t) = G_0 \qquad (2.4.7)$$

we find the steady-state solution

$$\delta n = G_0 \tau_n \qquad (2.4.8)$$

Therefore, the amount of the excess carrier concentration is simply the optical generation rate G_0 multiplied by the carrier lifetime. In the presence of an electric field E along the x direction, the conduction current density is

$$J_n = q\mu_n n E \qquad J_p = q\mu_p p E \qquad (2.4.9)$$

and the total current density is

$$J = J_n + J_p = \sigma E = \sigma \frac{V}{\ell} \qquad (2.4.10)$$

with the conductivity

$$\sigma = q(\mu_n n + \mu_p p) \qquad (2.4.11)$$

where V is the applied voltage, and ℓ is the length of the semiconductor. Therefore, at thermal equilibrium when there is no optical illumination,

$$\sigma_0 = q(\mu_n n_0 + \mu_p p_0) \qquad (2.4.12)$$

is the dark conductivity and

$$\Delta \sigma = q(\mu_n \delta n + \mu_p \delta p) \qquad (2.4.13)$$

is called the photoconductivity. We find

$$\Delta \sigma = q(\mu_n + \mu_p) G_0 \tau_n \qquad (2.4.14)$$

which is proportional to the generation rate and the carrier lifetime. The photocurrent is

$$\Delta I = \Delta \sigma \frac{A}{\ell} V = q(\mu_n + \mu_p) G_0 \tau_n \frac{A}{\ell} V \qquad (2.4.15a)$$

and A is the cross-sectional area of the semiconductor. For most semiconductors, the electron mobility is much larger than the hole mobility. Since $v_d = \mu_n V / \ell$ is the average drift velocity of the electrons, we can write the photocurrent as

$$\Delta I \simeq q(G_0 A \ell) \frac{\tau_n}{\tau_t} \qquad (2.4.15b)$$

where $\tau_t = \ell/v_d$ is the average transit time of the electrons across the photoconductor of a length ℓ and $(G_0 A \ell)$ is just the total number of generated electron–hole pairs in a volume $A\ell$. The ratio of the carrier recombination lifetime τ_n to the transit time τ_t is the photoconductive gain.

Example A homogeneous germanium photoconductor (Fig. 2.8) is illuminated with an optical beam with the wavelength $\lambda = 1.55~\mu$m and an optical power $P = 1$ mW. Assume that the optical beam is absorbed uniformly by the photoconductor and each photon creates one electron–hole pair, that is, the intrinsic quantum efficiency η_i is unity. The optical energy $h\nu$ is

$$h\nu = h\frac{c}{\lambda} = \frac{1.24}{\lambda} = 0.8 \quad (\text{eV})$$

where λ is in microns. The photon flux Φ is the number of photons injected per second:

$$\Phi = \frac{P}{h\nu} = \frac{1~\text{mW}}{0.8 \times (1.6 \times 10^{-19})\text{J}} = 7.81 \times 10^{15}~\text{s}^{-1}$$

Since we assume that all the photons are absorbed and the intrinsic quantum efficiency is unity, we have the generated rate per unit volume in $(wd\,\ell)$:

$$G_0 = \frac{\eta_i \Phi}{wd\,\ell} = \frac{7.81 \times 10^{15}}{wd\,\ell} \quad (\text{s}^{-1}~\text{cm}^{-3})$$

We use

$$\mu_n = 3900~\text{cm}^2~\text{V}^{-1}~\text{s}^{-1} \qquad \mu_p = 1800~\text{cm}^2~\text{V}^{-1}~\text{s}^{-1} \qquad \tau_n = 10^{-3}~\text{s}$$

$$w = 10~\mu\text{m} \qquad\qquad d = 1~\mu\text{m} \qquad\qquad \ell = 1~\text{mm}$$

The injected "primary" photocurrent is $q\eta_i\Phi = 1.6 \times 10^{-19} \times 7.81 \times 10^{15}$ A $= 1.25$ mA. The average electron transit time is

$$\tau_{tn} = \frac{\ell}{\mu_n \dfrac{V}{\ell}} = \frac{0.1~\text{cm}}{\left(3900\dfrac{\text{cm}^2}{\text{Vs}}\right)\left(\dfrac{1~\text{V}}{0.1~\text{cm}}\right)} = 2.56 \times 10^{-6}~\text{s}$$

and the average transit time for holes is

$$\tau_{tp} = \frac{\ell}{\mu_p \dfrac{V}{\ell}} = 5.56 \times 10^{-6}~\text{s}$$

The photocurrent in the photoconductor is

$$\Delta I = q(\mu_n + \mu_p)G_0\tau_n\frac{wd}{\ell}V = (q\eta_i\Phi)\left(\frac{\tau_n}{\tau_{tn}} + \frac{\tau_n}{\tau_{tp}}\right)$$

$$= 1.25 \text{ mA}\left(\frac{10^{-3}}{2.56 \times 10^{-6}} + \frac{10^{-3}}{5.56 \times 10^{-6}}\right)$$

$$= 0.71 \text{ A}$$

where a photoconductive gain of around 570 occurs in this example. ∎

2.4.2 Nonuniform Carrier Generation

If the carrier generation rate is not uniform, carrier diffusion will be important. For example, if the optical illumination on the semiconductor is limited
to only a width S, instead of the length ℓ (Fig. 2.9a), the excess carriers
generated in the illumination region will diffuse in both directions, assuming
there is no external field applied in the x direction [13]. Another example is
in a stripe-geometry semiconductor laser [20] with an intrinsic (i) region as
the active layer where the carriers are injected by a uniform current density
within a stripe width S (Fig. 2.9b). We ignore the current spreading outside
the region $|y| \leq S/2$.

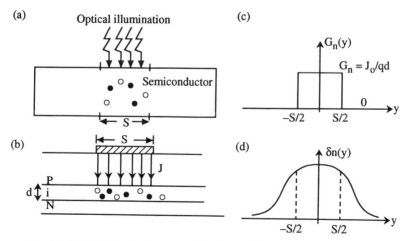

Figure 2.9. (a) Optical injection into a semiconductor. The illumination is over a width S.
(b) Current injection into the active (i) region with a thickness d of a stripe-geometry
semiconductor laser. (c) The carrier generation rate as a function of the position. (d) The excess
carrier distribution after diffusion.

The continuity equation for the electrons in the bulk semiconductor in Fig. 2.9a or in the intrinsic region in Fig. 2.9b is

$$\frac{\partial n}{\partial t} = G_n - \frac{\delta n}{\tau_n} + \frac{1}{q}\frac{\partial}{\partial y}J_n(y) \tag{2.4.16}$$

and the electron current density is given only by diffusion:

$$J_n(y) \simeq qD_n\frac{\partial n}{\partial y} \tag{2.4.17}$$

At steady state, $\partial/\partial t = 0$ and

$$D_n\frac{\partial^2}{\partial y^2}\delta n - \frac{\delta n}{\tau_n} = -G_n(y) \tag{2.4.18}$$

Here $G_n(y)$ is given by the optical intensity in Fig. 2.9a or, in the case of electric injection with a current density J_0 from the stripe S, Fig. 2.9c:

$$G_n(y) = \begin{cases} \dfrac{J_0}{qd} & |y| \le \dfrac{S}{2} \\ 0 & |y| > \dfrac{S}{2} \end{cases} \tag{2.4.19}$$

where $q = 1.6 \times 10^{-19}$ C, and d is the thickness of the active region. The solution to Eq. (2.4.16) is

$$\delta n(y) = \begin{cases} A\,e^{-y/L_n} + B\,e^{+y/L_n} + G_n\tau_n & |y| \le \dfrac{S}{2} \\ C\,e^{-(|y|-S/2)/L_n} & |y| \ge \dfrac{S}{2} \end{cases} \tag{2.4.20}$$

where $L_n = (D_n\tau_n)^{1/2}$ is the electron diffusion length. From the symmetry of the problem, we know that $\delta n(y)$ must be an even function of y. Therefore, the coefficients A and B must be equal. Another way to look at this is that the y component of the electron current density $J_n(y)$ must be zero at the symmetry plane $y = 0$. Therefore $(\partial/\partial y)\delta n(y) = 0$ at $y = 0$, which also gives $A = B$.

Matching the boundary conditions at $y = S/2$, in which $\delta n(y)$ and $J_n(y)$ are continuous, we find A and C. The final expressions for $\delta n(y)$ can be put in the form

$$
\delta n(y) =
\begin{cases}
G_0 \tau_n \left[1 - e^{-S/2L_n} \cosh\left(\dfrac{y}{L_n}\right) \right] & |y| \le \dfrac{S}{2} \\[2ex]
G_0 \tau_n \sinh\left(\dfrac{S}{2L_n}\right) e^{-|y|/L_n} & |y| \ge \dfrac{S}{2}
\end{cases}
\tag{2.4.21}
$$

The total carrier concentration $n(y) = n_0 + \delta n(y) \simeq \delta n(y)$; since the active region is undoped, n_0 is very small. This carrier distribution $n(y)$ (Fig. 2.9d), is proportional to the spontaneous emission profile from the stripe-geometry semiconductor laser measured experimentally. In Chapter 10, we discuss in more detail the stripe-geometry, gain-guided semiconductor laser, and take into account the current spreading effects.

2.5 SEMICONDUCTOR p-N AND n-P HETEROJUNCTIONS [20–25]

When two crystals of semiconductors with different energy gaps are combined, a heterojunction is formed. The conductivity type of the smaller energy gap crystal is denoted by a lower case n or p and that of the larger energy gap crystal is denoted by an upper case N or P. Here we discuss the basic Anderson model for the heterojunctions [21, 22]. It has been pointed out that a more fundamental approach using the bulk and interface properties of the semiconductors should be used for the heterojunction model [23–25]. In Appendix D, we also discuss a model-solid theory for the band lineups of semiconductor heterojunctions including the strain effects, which are convenient for the estimation of band-edge discontinuities.

2.5.1 Semiconductor p-N Heterojunction

Consider first a p-type narrow-gap semiconductor, such as GaAs, in contact with an N-type wide-band-gap semiconductor, such as $\mathrm{Al}_x\mathrm{Ga}_{1-x}\mathrm{As}$. Let χ be the electron affinity, which is the energy required to take an electron from the conduction band edge to the vacuum level, and let Φ be the work function, which is the energy difference between the vacuum level and the Fermi level. In each region, the Fermi level is determined by the charge neutrality condition

$$
n + N_A^- = p + N_D^+
\tag{2.5.1}
$$

where n and p are related to the quasi-Fermi levels F_n and F_p, respectively, through (2.2.39) or (2.2.40). For example, in the p region, $p \gg n$, we may denote $N_a \equiv N_A^- - N_D^+$ as the "net" acceptor concentration. If $N_a \gg n_i$, we then have

$$p \simeq N_a \tag{2.5.2}$$

which will determine the Fermi level F_p in the p region. Similarly, using $N_D \equiv N_D^+ - N_A^-$ as the "net" ionized donor concentration, we have

$$N \simeq N_D \tag{2.5.3}$$

which will determine the bulk Fermi level F_N in the N region before contact. When the two crystals are in contact, the Fermi level will line up to be a constant across the junction under thermal equilibrium conditions without any voltage bias. Thus there will be redistributions of electrons and holes such that a built-in electric field exists to prevent any current flow in the crystal. To find the band bending, we use the depletion approximation.

Depletion Approximation for an Unbiased p-N Junction. Since the charge density is

$$\rho(x) = q(p - n + N_D^+ - N_A^-) \tag{2.5.4}$$

and the free carriers p and n are depleted in the space charge region near the junction, we have

$$\rho(x) = \begin{cases} -qN_a & -x_p < x < 0 \\ +qN_D & 0 < x < x_N \end{cases} \tag{2.5.5}$$

where again N_a is the net acceptor concentration in the p side, and N_D is the net donor concentration on the N side.

From Gauss's law, we know that $\nabla \cdot (\varepsilon \mathbf{E}) = \rho$, which gives

$$\frac{d}{dx}E(x) = \begin{cases} -\dfrac{qN_a}{\varepsilon_p} & -x_p < x < 0 \\ +\dfrac{qN_D}{\varepsilon_N} & 0 < x < x_N \end{cases} \tag{2.5.6}$$

where ε_p and ε_N are the permittivity in the p and N regions, respectively. Gauss's law states that the slope of the $E(x)$ profile is given by the charge density divided by the permittivity. Thus the electric field is given by two linear functions

$$E(x) = \begin{cases} -\dfrac{qN_a(x + x_p)}{\varepsilon_p} & -x_p < x < 0 \\[2ex] +\dfrac{qN_D(x - x_N)}{\varepsilon_N} & 0 < x < x_N \end{cases} \quad (2.5.7)$$

in the depletion region and zero outside. The boundary condition states that the normal displacement vector $\mathbf{D} = \varepsilon \mathbf{E}$ is continuous at $x = 0$:

$$\varepsilon_p E(0_-) = \varepsilon_N E(0_+) \quad (2.5.8)$$

or

$$N_a x_p = N_D x_N \quad (2.5.9)$$

The electrostatic potential distribution $\phi(x)$ across the junction is related to the electric field by

$$E(x) = -\frac{d}{dx}\phi(x) \quad (2.5.10)$$

which means that the slope of the potential profile is given by the negative of the electric field profile. If we choose the reference potential to be zero for $x < -x_p$, we have

$$\phi(x) = \begin{cases} 0 & x \leq -x_p \\[2ex] \dfrac{qN_a}{2\varepsilon_p}(x + x_p)^2 & -x_p \leq x \leq 0 \\[2ex] \dfrac{qN_a x_p^2}{2\varepsilon_p} + \dfrac{qN_D}{2\varepsilon_N}(2xx_N - x^2) & 0 \leq x \leq x_N \\[2ex] V_0 & x_N \leq x \end{cases} \quad (2.5.11)$$

where

$$V_0 = V_{0p} + V_{0N}$$

$$V_{0p} = \phi(0) = \frac{qN_a x_p^2}{2\varepsilon_p}$$

$$V_{0N} = \frac{qN_D x_N^2}{2\varepsilon_N} \quad (2.5.12)$$

Here V_0 is the total potential drop across the junction, whereas V_{0p} is the portion of the voltage drop on the p side and V_{0N} is the portion of the voltage drop on the N side. The contact potential is evaluated using the bulk values of the Fermi levels F_p and F_N measured from the valence or conduction band edges E_{vp} and E_{CN}, respectively, before contact (see Fig. 2.10a):

$$V_0 = \frac{F_N - F_p}{q}$$

$$= \frac{E_{gp} + \Delta E_c - (F_p - E_{vp}) - (E_{CN} - F_N)}{q} \qquad (2.5.13)$$

and

$$\Delta E_c \simeq 0.67(E_{GN} - E_{gp}) \qquad \text{for the GaAs-Al}_x\text{Ga}_{1-x}\text{As system} \quad (2.5.14)$$

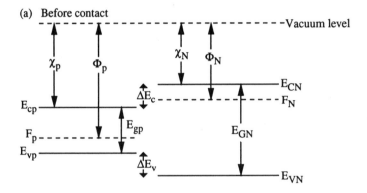

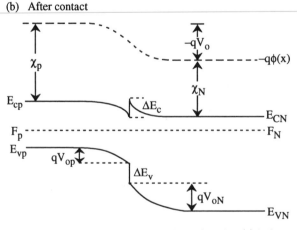

Figure 2.10. Energy band diagram for a $p\text{-}N$ heterojunction (a) before contact, and (b) after contact.

For nondegenerate semiconductors,

$$E_{CN} - F_N \simeq -k_B T \ln\left(\frac{N}{N_{CN}}\right) \qquad (2.5.15a)$$

$$F_p - E_{vp} \simeq -k_B T \ln\left(\frac{p}{N_{vp}}\right) \qquad (2.5.15b)$$

where N_{CN} and N_{vp} are evaluated from (2.2.41) and (2.2.42) for the N and p regions, respectively. Since the electron concentration on the N side, $N \simeq N_D$, and the hole concentration on the p side, $p \simeq N_a$, the contact potential V_0 can be evaluated from the above equations when the doping concentrations N_a and N_D are known. Using the two conditions (2.5.9) and (2.5.12) and the total width of the depletion region x_w from

$$x_w = x_p + x_N \qquad (2.5.16)$$

we find immediately that

$$x_p = \frac{N_D}{N_a + N_D} x_w \qquad x_N = \frac{N_a}{N_a + N_D} x_w \qquad (2.5.17)$$

Thus we may relate x_w to V_0 directly:

$$x_w = \left[\frac{2\varepsilon_p V_0}{qN_a N_D\left(N_D + \dfrac{\varepsilon_p}{\varepsilon_N} N_a\right)}\right]^{1/2} (N_a + N_D) \qquad (2.5.18)$$

The band edge $E_v(x)$ from the p side to the N side is given by (choosing $E_v(-\infty) = 0$ as the reference potential energy)

$$E_v(x) = \begin{cases} -q\phi(x) & p \text{ side} \\ -\Delta E_v - q\phi(x) & N \text{ side} \end{cases} \qquad (2.5.19)$$

or

$$E_v(x) = \begin{cases} 0 & x \leq -x_p \\ -\dfrac{q^2 N_a}{2\varepsilon_p}(x + x_p)^2 & -x_p \leq x < 0 \\ -\Delta E_v - \dfrac{q^2 N_a x_p^2}{2\varepsilon_p} - \dfrac{q^2 N_D}{2\varepsilon_N}(2xx_N - x^2) & 0 < x \leq x_N \\ -\Delta E_v - qV_0 & x_N \leq x \end{cases} \qquad (2.5.20)$$

(a) A p-N junction geometry

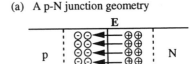

(b) Charge distribution

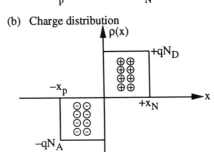

(c) Electric field

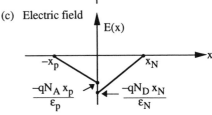

(d) Electrostatic potential

Figure 2.11. Illustrations of (a) a *p-N* junction geometry, (b) the charge distribution, (c) the electric field, and (d) the electrostatic potential based on the depletion approximation.

The conduction band edge $E_c(x)$ is above $E_v(x)$ by an amount E_{gp} on the p side and by an amount E_{GN} on the N side. $E_c(x)$ is always parallel to $E_v(x)$:

$$E_c(x) = \begin{cases} E_v(x) + E_{gp} & x < 0 \\ E_v(x) + E_{GN} & x > 0 \end{cases} \tag{2.5.21}$$

Illustrations for the charge density, the electric field, and the electrostatic potential are shown in Fig. 2.11. The final energy band diagram after contact is plotted in Fig. 2.10(b).

Example A p-GaAs/N-Al$_x$Ga$_{1-x}$As ($x = 0.3$) heterojunction is formed at thermal equilibrium without an external bias at room temperature. The doping concentration is $N_a = 1 \times 10^{18}$ cm^{-3} in the p side and $N_D = 2 \times 10^{17}$

cm^{-3} in the N side. Assume that the density-of-states hole effective mass for $Al_xGa_{1-x}As$ is

$$m_h^*(x) = (0.50 + 0.29x)m_0 \qquad 0 \le x \le 0.45$$

which accounts for both the heavy-hole and light-hole density of states. Other parameters are

$$m_e^*(x) = (0.0665 + 0.083x)m_0 \qquad E_g(x) = (1.424 + 1.247x) \quad (eV)$$

$$\varepsilon(x) = (13.1 - 3.0x)\varepsilon_0 \qquad (0 \le x \le 0.45)$$

where x is the mole fraction of aluminum.
(a)   We obtain for $x = 0.3$

$$p\text{-GaAs:} \quad m_e^* = 0.0665m_0, \; m_h^* = 0.50m_0,$$

$$\varepsilon_p = 13.1\varepsilon_0, \; E_{gp} = 1.424 \text{ eV}$$

$$N\text{-Al}_{0.3}Ga_{0.7}As: \quad m_e^* = 0.0914m_0, \; m_h^* = 0.587m_0,$$

$$\varepsilon_N = 12.2\varepsilon_0, \; E_{gN} = 1.798 \text{ eV}$$

The band-edge discontinuities are

$$\Delta E_g = 1.247 \times 0.3 = 0.374 \text{ eV} = 374 \text{ meV}$$

$$\Delta E_c = 0.67\Delta E_g = 250.6 \text{ meV} \qquad \Delta E_v = 0.33\Delta E_g = 123.4 \text{ meV}$$

(b)   We calculate the quasi-Fermi levels F_p and F_N for the bulk semiconductors for the given N_a and N_D separately.
       p-GaAs region

$$N_c = 2.51 \times 10^{19} \left(\frac{m_e^*}{m_0} \frac{T}{300} \right)^{3/2} cm^{-3} = 4.30 \times 10^{17} \; cm^{-3}$$

$$N_v = 2.51 \times 10^{19} \left(\frac{m_h^*}{m_0} \frac{T}{300} \right)^{3/2} cm^{-3} = 8.87 \times 10^{18} \; cm^{-3}$$

$$p = N_a = N_v F_{1/2} \left(\frac{E_{vp} - F_p}{k_B T} \right) \simeq N_v \exp \left(\frac{E_{vp} - F_p}{k_B T} \right)$$

$$F_p - E_{vp} = -k_B T \ln \frac{N_a}{N_v} = 56.4 \text{ meV}$$

$N\text{-}Al_{0.3}Ga_{0.7}As$ region

$$N_C = 2.51 \times 10^{19}(0.0914)^{3/2} = 6.94 \times 10^{17} \text{ cm}^{-3}$$

$$N_V = 2.51 \times 10^{19}(0.587)^{3/2} = 1.13 \times 10^{19} \text{ cm}^{-3}$$

$$N = N_D = N_C F_{1/2}\left(\frac{F_N - E_{CN}}{k_B T}\right) \simeq N_C \exp\left(\frac{F_N - E_{CN}}{k_B T}\right)$$

$$E_{CN} - F_N \simeq -k_B T \ln \frac{N_D}{N_C} = -25.85 \ln \frac{2 \times 10^{17}}{6.94 \times 10^{17}} = 32.2 \text{ meV}$$

(c) The contact potential is calculated using (2.5.13)

$$V_0 = \frac{E_{gp} + \Delta E_c - (F_p - E_{vp}) - (E_{CN} - F_N)}{q}$$

$$= (1424 + 250.6 - 56.4 - 32.2) \text{ mV} = 1586 \text{ mV}$$

(d) The depletion widths are

$$x_p = \left[\frac{2\varepsilon_p V_0}{qN_a N_D\left(N_D + \dfrac{\varepsilon_p}{\varepsilon_N}N_a\right)}\right]^{1/2} N_D$$

$$= \left[\frac{2 \times 13.1 \times 8.854 \times 10^{-14} \times 1.586}{1.6 \times 10^{-19} \times 10^{18} \times 2 \times 10^{17}\left(2 \times 10^{17} + \dfrac{13.1}{12.2} \times 10^{18}\right)}\right]^{1/2} \times 2 \times 10^{17}$$

$$= 0.019 \ \mu m$$

$$x_N = \frac{x_p}{N_D}N_a = 0.095 \ \mu m$$

$$x_w = x_p + x_N = 0.114 \ \mu m$$

The energy band diagram is plotted in Fig. 2.12.   ∎

Biased p-N Junction. With an applied voltage V_A across the diode (Fig. 2.13a), the potential barrier is reduced by V_A if V_A is positive. Note that our convention for the polarity of the bias voltage V_A is that the positive electrode is connected to the p side of the diode. Thus the depletion width x_w is reduced. Explicitly,

$$\phi(x) = \begin{cases} 0 & x \le -x_p \\[2mm] \dfrac{qN_a}{2\varepsilon_p}(x + x_p)^2 & -x_p \le x < 0 \\[3mm] \dfrac{qN_a x_p^2}{2\varepsilon_p} + \dfrac{qN_D}{2\varepsilon_N}(2xx_N - x^2) & 0 \le x \le x_N \\[3mm] V_0 - V_A & x_N \le x \end{cases} \qquad (2.5.22)$$

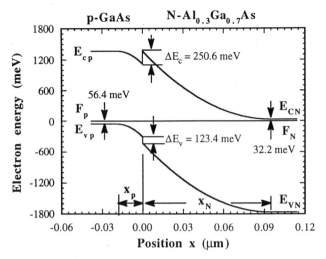

Figure 2.12. Band diagram of a p-GaAs$/N$-Al$_{0.3}$Ga$_{0.7}$As heterojunction with $N_a = 1 \times 10^{18}$ cm^{-3} in the p region and $N_D = 2 \times 10^{17}$ cm^{-3} in the N region.

The potential drop on the p side of the depletion region is $\phi(0)$ shown in Fig. 2.13b

$$V_{0p} - V_p = \frac{qN_a}{2\varepsilon_p}x_p^2 \qquad (2.5.23)$$

which is reduced from V_{0p} by an amount V_p. Similarly, the voltage drop across the N side of the depletion region is

$$V_{0N} - V_N = \frac{qN_D}{2\varepsilon_N}x_N^2 \qquad (2.5.24)$$

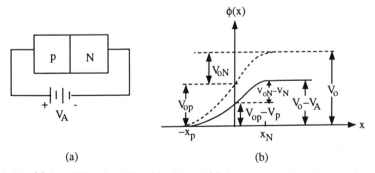

(a)   (b)

Figure 2.13. (a) A p-N junction with a bias V_A and (b) the corresponding electrostatic potential $\phi(x)$ (solid curve for $V_A > 0$ and dashed curve for $V_A = 0$).

which is reduced from V_{0N} by an amount V_N. The sum of V_p and V_N has to be equal to the bias voltage V_A:

$$V_p + V_N = V_A \tag{2.5.25}$$

Again the charge neutrality condition

$$N_a x_p = N_D x_N \tag{2.5.26}$$

and

$$x_w = x_p + x_N \tag{2.5.27}$$

together with

$$V_0 - V_A = \frac{qN_a}{2\varepsilon_p} x_p^2 + \frac{qN_D}{2\varepsilon_N} x_N^2 \tag{2.5.28}$$

lead to

$$x_w = \left[\frac{2\varepsilon_p(V_0 - V_A)}{qN_aN_D\left(N_D + \dfrac{\varepsilon_p}{\varepsilon_N}N_a\right)} \right]^{1/2} (N_a + N_D) \tag{2.5.29}$$

Thus the depletion region is reduced since x_w is proportional to $(V_0 - V_A)^{1/2}$. If the diode is reverse-biased, then V_A is negative and x_w becomes wider. The potential function $\phi(x)$ in (2.5.22) determines the band bendings $E_c(x)$ and $E_v(x)$ again through (2.5.19) and (2.5.21). The resultant band diagrams for a forward-biased and a reverse-biased diode are shown in Figs. 2.14a and b, respectively.

Quasi-Fermi Levels and Minority Carrier Injections. The quasi-Fermi levels $F_p(x)$ and $F_N(x)$ are separated by the bias voltage V_A multiplied by the magnitude of the electron charge $q = 1.6 \times 10^{-19}$ C.

$$F_N - F_p = qV_A \tag{2.5.30}$$

This fact can be seen clearly by comparing the band diagrams of the zero bias and forward bias cases in Figs. 2.10b and 2.14a, respectively. To find the variations of the quasi-Fermi levels $F_p(x)$ and $F_N(x)$ as functions of the position, we have to find the carrier concentrations everywhere in the diode. Assuming a low-level injection condition, which means that the amount of injected excess carriers due to the voltage bias is always much less than the majority carrier concentration in each region, the minority carrier concentration is much more strongly affected than the majority carrier concentration in each region. As a result of this, there are two different quasi-Fermi levels, $F_N(x)$ and $F_p(x)$, for the electron and the hole concentrations, respectively,

(a) Forward bias

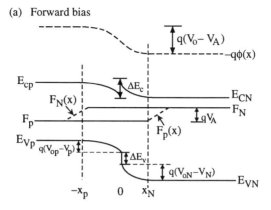

(b) Reverse bias

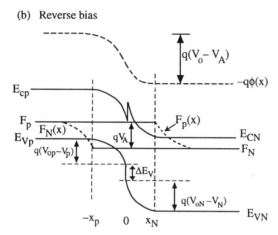

Figure 2.14. Energy band diagrams for a p-N heterojunction under (a) forward bias, and (b) reverse bias based on the depletion approximation.

near the junction region. Far away from the junction region, the quasi-Fermi level should approach its thermal equilibrium value in the bulk region, respectively, that is, $F_N(x \to +\infty) \to F_N$ (bulk value on the N side) and $F_p(x \to -\infty) \to F_p$ (bulk value on the p side), since the carrier concentrations are hardly affected away from the junction.

It is usually assumed that the quasi-Fermi levels stay as constants across the depletion region, that is, $F_N(x) = F_N$ for $-x_p \le x < +\infty$ and $F_p(x) = F_p$ for $-\infty < x < x_N$, as shown in Fig. 2.14. This is equivalent to the statement that the carrier distribution in energy for the same species of carrier across the depletion region stays the same in the depletion region. As a result of this assumption, we find that, for the Boltzmann distribution, on the p side

$$p_p(x) = N_{vp}\, e^{[E_v(x) - F_p(x)]/k_B T} \qquad (2.5.31a)$$

$$n_p(x) = N_{cp}\, e^{[F_N(x) - E_c(x)]/k_B T} \qquad (2.5.31b)$$

On the N side

$$P_N(x) = N_{VN} \, e^{[E_v(x) - F_p(x)]/k_B T} \tag{2.5.32a}$$

$$N_N(x) = N_{CN} \, e^{[F_N(x) - E_c(x)]/k_B T} \tag{2.5.32b}$$

Since at thermal equilibrium (i.e., no current injection, $V_A = 0$), $F_p(x) = F_N(x)$, we have

$$n_p p_p = N_{vp} N_{cp} \, e^{-(E_c - E_v)/k_B T} = n_{ip}^2 \tag{2.5.33}$$

where n_{ip} is the intrinsic carrier concentration on the p side. Thus, if $V_A \ne 0$, for $-x_p < x < 0$,

$$n_p p_p = n_{ip}^2 \, e^{(F_N - F_p)/k_B T} = n_{ip}^2 \, e^{qV_A/k_B T} \tag{2.5.34}$$

At the edge of the depletion region, $x = -x_p$, $p_p(-x_p) \simeq p_{p0} \simeq N_a$. Therefore,

$$n_p(-x_p) \cong \frac{n_{ip}^2}{N_a} \, e^{qV_A/k_B T} = n_{p0} \, e^{qV_A/k_B T} \tag{2.5.35}$$

Here the subscript 0 in p_{p0} and n_{p0} refers to their thermal equilibrium values. The minority carrier near the edge of the depletion region, $n_p(-x_p)$, differs by a factor $\exp(qV_A/k_B T)$ from its thermal equilibrium value n_{p0} due to the carrier injection. Similarly, at the edge of the depletion region on the N side, $x = x_N$, the minority carrier concentration is

$$P(x) = P_{N0} \, e^{qV_A/k_B T} \tag{2.5.36}$$

where $P_{N0} = n_{iN}^2/N_D$ and n_{iN} is the intrinsic carrier concentration on the N side.

If we compare the current density equations for the p side of the diode, assuming that n_{p0} and p_{p0} are independent of x ($\partial n/\partial x = \partial(\delta n)/\partial x$, $\partial p/\partial x = \partial(\delta p)/\partial x$),

$$j_n = q\mu_n nE + qD_n \frac{\partial}{\partial x} \delta n \tag{2.5.37a}$$

$$j_p = q\mu_p pE - qD_p \frac{\partial}{\partial x} \delta p \tag{2.5.37b}$$

we see that in general, $q\mu_n nE \ll q\mu_p pE$, since $n \ll p$ on the p side of the diode. If j_n and j_p are of the same order of magnitude and the charge neutrality condition, $\delta n(x) \simeq \delta p(x)$, applies in the quasi-neutral region (i.e., $x < -x_p$), where E is very small, we expect that the drift current is much less than the diffusion current for the electrons. Thus, the minority current density j_n is only dominated by the diffusion current density (this is not true for the majority current density as will be seen later):

$$j_n \simeq qD_n \frac{\partial}{\partial x} \delta n(x) \qquad (2.5.38)$$

Using the charge continuity equation for the electrons,

$$\frac{\partial}{\partial t} \delta n = -\frac{\delta n}{\tau_n} + \frac{1}{q} \frac{\partial}{\partial x} j_n \qquad (2.5.39)$$

We find, under steady-state conditions, $\partial(\delta n)/\partial t = 0$,

$$D_n \frac{\partial^2}{\partial x^2} \delta n(x) - \frac{\delta n(x)}{\tau_n} = 0 \qquad (2.5.40)$$

The solution with the boundary condition, $\delta n(x = -\infty) = 0$, has the form

$$\delta n(x) = \delta n(-x_p) e^{+(x+x_p)/L_n} \qquad (2.5.41)$$

where

$$\delta n(-x_p) = n_p(-x_p) - n_{p0}$$
$$= n_{p0}\left(e^{qV_A/k_BT} - 1\right) \qquad (2.5.42)$$

and $L_n = \sqrt{D_n \tau_n}$ is the diffusion length for electrons in the p region. The injection condition from (2.5.35) has been used. The total carrier concentrations on the p side for $x < -x_p$ are

$$n_p(x) = n_{p0} + \delta n(x)$$
$$= n_{p0} + n_{p0}\left(e^{qV_A/k_BT} - 1\right)e^{(x+x_p)/L_n} \qquad (2.5.43)$$

and

$$p_p(x) = p_{p0} + \delta p(x)$$
$$= p_{p0} + n_{p0}\left(e^{qV_A/k_BT} - 1\right)e^{(x+x_p)/L_n} \qquad (2.5.44)$$

where $\delta p(x) = \delta n(x)$, $p_{p0} = N_a$, and $n_{p0} = n_{ip}^2/N_a$.

Similarly, in the quasi-neutral region on the N side of the diode, $x > x_N$, the minority (hole) current density is approximately

$$j_P \simeq -qD_P \frac{\partial}{\partial x} \delta P(x) \qquad (2.5.45)$$

and

$$\frac{\partial}{\partial t}\,\delta P = -\frac{\delta P}{\tau_P} - \frac{1}{q}\frac{\partial}{\partial x}j_P \tag{2.5.46}$$

which lead to

$$\delta P(x) = \delta P(x_N)e^{-(x-x_N)/L_p} \tag{2.5.47}$$

where

$$\begin{aligned}\delta P(x_N) &= P(x_N) - P_{N0}\\ &= P_{N0}\bigl(e^{qV_A/k_BT} - 1\bigr)\end{aligned} \tag{2.5.48}$$

and $L_P = \sqrt{D_P\tau_P}$ is the diffusion length of the holes on the N side. The total carrier concentrations on the N side of the diode, $x > x_N$, are

$$\begin{aligned}P_N(x) &= P_{N0} + \delta P(x)\\ &= P_{N0} + P_{N0}\bigl(e^{qV_A/k_BT} - 1\bigr)e^{-(x-x_N)/L_p} \tag{2.5.49}\\ N_N(x) &= N_{N0} + \delta N(x)\\ &= N_{N0} + P_{N0}\bigl(e^{qV_A/k_BT} - 1\bigr)e^{-(x-x_N)/L_p} \tag{2.5.50}\end{aligned}$$

From the carrier concentration for $x < -x_p$ and $x > x_N$, the quasi-Fermi levels can be obtained from (2.5.31b) and (2.5.32a):

$$F_N(x) - E_{cp} = F_p - E_{cp} + k_BT\ln\bigl[1 + \bigl(e^{qV_A/k_BT} - 1\bigr)e^{(x+x_p)/L_n}\bigr] \qquad x \le -x_p \tag{2.5.51}$$

where $k_BT\ln(n_{p0}/N_{cp}) = F_p - E_{cp}$ on the p side has been used. Similarly,

$$E_{VN} - F_p(x) = E_{VN} - F_N + k_BT\ln\bigl[1 + \bigl(e^{qV_A/k_BT} - 1\bigr)e^{-(x-x_N)/L_p}\bigr] \qquad x > x_N \tag{2.5.52}$$

where $k_BT\ln(P_{N0}/N_{VN}) = E_{VN} - F_N$ has been used. From (2.5.51) and (2.5.52) we have

$$F_N(-x_p) = F_p + qV_A = F_N \tag{2.5.53a}$$
$$F_p(x_N) = F_N - qV_A = F_p \tag{2.5.53b}$$

and $F_N(x \to -\infty) \to F_p$, $F_p(x \to +\infty) \to F_N$.

For the forward bias case, $V_A > 0$, assuming that $\exp(qV_A/k_BT) \gg 1$, we have

$$F_N(x) = F_N + \frac{(x + x_p)k_BT}{L_n} \qquad \text{for } x \le -x_p \quad \text{and} \quad \left|\frac{x + x_p}{L_n}\right| \ll \frac{qV_A}{k_BT} \tag{2.5.54a}$$

and

$$F_p(x) = F_p + \frac{(x - x_N)k_BT}{L_p} \quad \text{for } x \geq x_N \quad \text{and} \quad \left| \frac{x - x_N}{L_p} \right| \ll \frac{qV_A}{k_BT}$$

$$(2.5.54b)$$

These results are plotted in Fig. 2.14a.

For the reverse bias, $V_A < 0$, assuming that $\exp(qV_A/k_BT) \ll 1$, we have from (2.5.51),

$$F_N(x) \simeq F_p + k_BT \ln\left[1 - e^{(x+x_p)/L_n}\right] \quad \text{for } x \leq -x_p \quad (2.5.55a)$$

$$\simeq F_p - k_BT \, e^{(x+x_p)/L_n} \quad \text{for } x + x_p \ll -L_n \quad (2.5.55b)$$

Similarly, for $F_p(x)$ on the N side,

$$F_p(x) \simeq F_N - k_BT \ln\left[1 - e^{-(x-x_N)/L_p}\right] \quad \text{for } x > x_N \quad (2.5.56a)$$

$$\simeq F_N + k_BT \, e^{(x-x_N)/L_p} \quad \text{for } x - x_N \gg L_p \quad (2.5.56b)$$

The quasi-Fermi levels are plotted in Fig. 2.14b.

Current Densities and I-V Characteristics. The current densities are obtained for the minority carriers first:

On the p side

$$j_n = qD_n\frac{\partial}{\partial x}\,\delta n(x)$$

$$= q\frac{D_n}{L_n}n_{p0}\left(e^{qV_A/k_BT} - 1\right)e^{(x+x_p)/L_n} \quad x \leq -x_p \quad (2.5.57)$$

On the N side

$$j_P = -qD_P\frac{\partial}{\partial x}\,\delta P(x)$$

$$= q\frac{D_P}{L_P}P_{N0}\left(e^{qV_A/k_BT} - 1\right)e^{-(x-x_N)/L_P} \quad x \geq x_N \quad (2.5.58)$$

Assuming that there is no generation or recombination current in the space charge region, that is, j_n and j_P are constant over the space charge region, the total current density is thus the sum of the two current densities:

$$j = j_n(-x_p) + j_P(x_N)$$

$$= q\left(\frac{D_n}{L_n}n_{p0} + \frac{D_P}{L_P}P_{N0}\right)\left(e^{qV_A/k_BT} - 1\right) \tag{2.5.59}$$

The total current is $I = jA$ with A the cross-sectional area of the diode:

$$I = I_0\left(e^{qV_A/k_BT} - 1\right) \tag{2.5.60a}$$

$$I_0 = q\left(\frac{D_n}{L_n}n_{p0} + \frac{D_P}{L_P}P_{N0}\right)A \tag{2.5.60b}$$

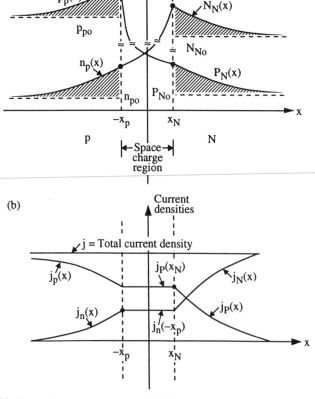

Figure 2.15. (a) The carrier concentrations and (b) the current densities as functions of position x in a forward biased p-N heterojunction diode using the depletion approximation.

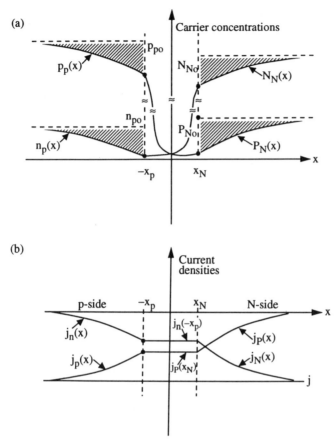

Figure 2.16. (a) The carrier concentrations and (b) the current densities as functions of x for a reverse biased p-N heterojunction diode using the depletion approximation.

From the total current density j, which is a constant across the diode, the majority carrier current densities $j_p(x)$ and $j_N(x)$ can be obtained from

$$j_p(x) = j - j_n(x) \quad \text{on the } p \text{ side} \qquad x \le -x_p \qquad (2.5.61)$$

$$j_N(x) = j - j_p(x) \quad \text{on the } N \text{ side} \qquad x \ge x_N \qquad (2.5.62)$$

The complete current distributions are plotted in Fig. 2.15 for the forward bias case and in Fig. 2.16 for the reverse bias case.

2.5.2 Semiconductor n-P Heterojunction

If the narrow gap semiconductor is doped n-type and the wide-gap semiconductor is doped P-type, we may have a band diagram such as that shown in Fig. 2.17a before contact. When the heterojunction is formed, a space charge region will exist due to the diffusion or redistribution of free carriers at

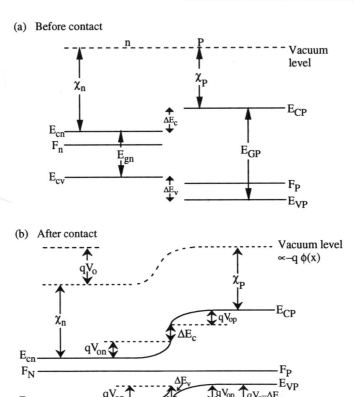

Figure 2.17. Energy band diagrams for an n-P heterojunction (a) before contact and (b) after contact.

thermal equilibrium. The excess electrons on the n side near the junction may spill over the P side and the holes on the P side near the junction may spill over the n side. Under the depletion approximation, we have a net charge distribution such as shown in Figs. 2.18a and b. The procedure to find the band diagram is similar to that for the p-N heterojunction in the previous section. We summarize the major results here.

The charge density $\rho(x)$ is

$$\rho(x) = \begin{cases} +qN_d & -x_n < x < 0 \\ -qN_A & 0 < x < x_P \end{cases} \tag{2.5.63}$$

The electric field $E(x)$ is (Fig. 2.18c)

$$E(x) = \begin{cases} \dfrac{qN_d(x + x_n)}{\varepsilon_n} & -x_n < x < 0 \\ -\dfrac{qN_A(x - x_P)}{\varepsilon_P} & 0 < x < x_P \end{cases} \tag{2.5.64}$$

(a) An n-P heterojunction

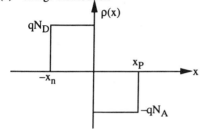

(b) Charge distribution

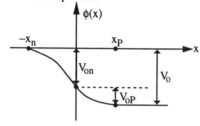

(c) Electric field

(d) Electrostatic potential

Figure 2.18. (a) An *n-P* heterojunction, (b) the charge distribution, (c) the electric field distribution, and (d) the electrostatic potential using the depletion approximation.

where the boundary condition at $x = 0$ gives

$$N_d x_n = N_A x_P \tag{2.5.65}$$

The electrostatic potential is

$$\phi(x) = \begin{cases} 0 & x \leq -x_n \\ -\dfrac{qN_d}{2\varepsilon_n}(x + x_n)^2 & -x_n \leq x \leq 0 \\ -\dfrac{qN_d}{2\varepsilon_n}x_n^2 - \dfrac{qN_A}{2\varepsilon_P}(2x_P x - x^2) & 0 \leq x \leq x_P \\ -V_0 & x_P \leq x \end{cases} \tag{2.5.66}$$

as shown in Fig. 2.18d, where

$$V_0 = V_{0n} + V_{0P}$$

$$V_{0n} = \frac{qN_d x_n^2}{2\varepsilon_n}$$

$$V_{0P} = \frac{qN_A x_P^2}{2\varepsilon_P} \qquad (2.5.67)$$

Again the total depletion width is

$$x_w = x_n + x_P \qquad (2.5.68)$$

$$x_n = \left[\frac{2\varepsilon_n V_0}{qN_d N_A \left(N_A + \dfrac{\varepsilon_n}{\varepsilon_P} N_d \right)} \right]^{1/2} N_A \qquad (2.5.69a)$$

$$x_P = \left[\frac{2\varepsilon_P V_0}{qN_d N_A \left(N_d + \dfrac{\varepsilon_P}{\varepsilon_n} N_A \right)} \right]^{1/2} N_d \qquad (2.5.69b)$$

and the contact potential V_0 is

$$V_0 = \frac{F_n - F_P}{q}$$

$$= \frac{E_{GP} - \Delta E_c - (E_{cn} - F_n) - (F_P - E_{VP})}{q} \qquad (2.5.70)$$

The band-edge function $E_c(x)$ or $E_v(x)$ can be obtained from $-q\phi(x)$:

$$E_v(x) = \begin{cases} -q\phi(x) & x < 0 \\ -q\phi(x) - \Delta E_v & x > 0 \end{cases} \qquad (2.5.71)$$

Thus

$$E_v(x) = \begin{cases} 0 & x < -x_n \\[2mm] \dfrac{q^2 N_d}{2\varepsilon_n}(x + x_n)^2 & -x_n \leq x < 0 \\[2mm] -\Delta E_v + \dfrac{q^2 N_d}{2\varepsilon_n}x_n^2 + \dfrac{q^2 N_A}{2\varepsilon_P}(2xx_P - x^2) & 0 < x \leq x_P \\[2mm] qV_0 - \Delta E_v & x_P \leq x \end{cases}$$

$$(2.5.72)$$

and

$$E_c(x) = \begin{cases} E_v(x) + E_{gn} & x < 0 \\ E_v(x) + E_{GP} & x > 0 \end{cases} \tag{2.5.73}$$

Example An n-GaAs/P-Al$_{0.3}$Ga$_{0.7}$As heterojunction is formed at room temperature and a zero bias. The doping concentrations are $N_d = 4 \times 10^{16}$ cm^{-3} in the n region and $N_A = 2 \times 10^{17}$ cm^{-3} in the P region. The parameters were given in the previous example of a p-N heterojunction.

1. n side (GaAs) $N_c = 2.5 \times 10^{19} \left(\dfrac{m_e^*}{m_0} \dfrac{T}{300} \right)^{3/2} = 4.30 \times 10^{17}$ cm^{-3}

$$E_{cn} - F_n = -k_B T \ln \frac{N_d}{N_c} = -25.85\{-2.3749\} = 61.4 \text{ meV}$$

P side (Al$_{0.3}$Ga$_{0.7}$As) $N_V = 2.5 \times 10^{19} \left(\dfrac{m_h^*}{m_0} \dfrac{T}{300} \right)^{3/2} = 1.13$ $\times 10^{19}$ cm^{-3}

$$P = N_A = N_V \exp \left(\frac{E_{VP} - F_P}{k_B T} \right)$$

$$F_P - E_{VP} = -k_B T \ln \frac{N_A}{N_V} = -25.85 \ln \frac{2 \times 10^{17}}{1.13 \times 10^{19}} = 104.3 \text{ meV}$$

2. The contact potential is calculated using (2.5.70):

$$V_0 = \frac{E_{GP} - \Delta E_c - (E_{cn} - F_n) - (F_P - E_{VP})}{q}$$

$$= 1798 - 250.6 - 61.4 - 104.3 = 1381.7 \text{ mV}$$

3. The depletion widths are

$$x_n = \left[\frac{2\varepsilon_n V_0}{q N_d N_A \left(N_A + \dfrac{\varepsilon_n}{\varepsilon_P} N_d \right)} N_A \right]^{1/2}$$

$$= \left[\frac{2 \times 13.1 \times 8.854 \times 10^{-14} \times 1.3817}{1.6 \times 10^{-19} \times 4 \times 10^{16} \times 2 \times 10^{17} \left(2 \times 10^{17} + \dfrac{13.1}{12.2} \times 4 \times 10^{16} \right)} \right]^{1/2}$$

$$\times 2 \times 10^{17} \text{ (cm)}$$
$$= 5.1 \text{ } \mu\text{m}$$

$$x_P = x_n \frac{N_d}{N_A} = 1.0 \text{ } \mu\text{m}$$

$$x_w = x_n + x_P = 6.1 \text{ } \mu\text{m}$$

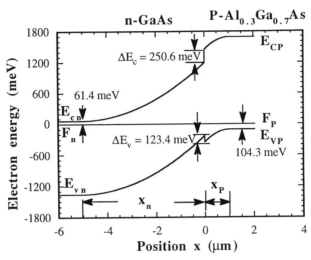

Figure 2.19. Band diagram of an unbiased n-GaAs$/P$-Al$_{0.3}$Ga$_{0.7}$As heterojunction with $N_d = 4 \times 10^{16}$ cm^{-3} in the n region and $N_A = 2 \times 10^{17}$ cm^{-3} in the P region.

4. The energy band diagram of the n-GaAs$/P$-Al$_{0.3}$Ga$_{0.7}$As junction is plotted in Fig. 2.19. ∎

Biased n-P Heterojunction. The band diagrams for the n-P heterojunction under forward $(V_A > 0)$ and reverse bias $(V_A < 0)$ conditions are shown in Figs. 2.20a and b. Note that the positive electrode of the voltage source is always connected to the P side of the diode for the definition of the polarity. The I-V_A curve of the n-P diode is similar to (2.5.60),

$$I = I_0\left(e^{qV_A/k_BT} - 1\right) \tag{2.5.74a}$$

$$I_0 = q\left(\frac{D_p}{L_p}p_{n0} + \frac{D_N}{L_N}N_{P0}\right)A \tag{2.5.74b}$$

where D_p, L_p, and p_{n0} refer to the quantities in the n region, while D_N, L_N, and N_{P0} are the quantities in the P region.

2.6 SEMICONDUCTOR n-N HETEROJUNCTIONS AND METAL–SEMICONDUCTOR JUNCTIONS

2.6.1 Semiconductor n-N Heterojunctions

For an n-N heterojunction, the theory is a bit more involved. With the development of the high electron mobility transistor (HEMT), also called the

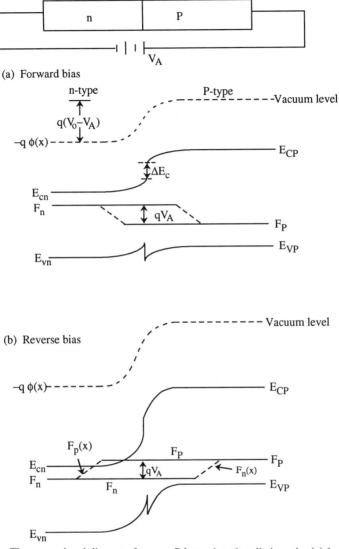

Figure 2.20.   The energy band diagrams for an *n-P* heterojunction diode under (a) forward bias and (b) reverse bias.

modulation doped field effect transistor (MODFET), the *n-N* heterojunction theory is very useful for modeling these devices [26]. Depending on the model, the results may be different. A qualitative approach is shown in Fig. 2.21. When the two crystals with *n*-type and *N*-type dopings are brought in contact, the charge redistribution produces a space charge region. Some electrons will spill over from the *N* region to the *n* region for the alignment

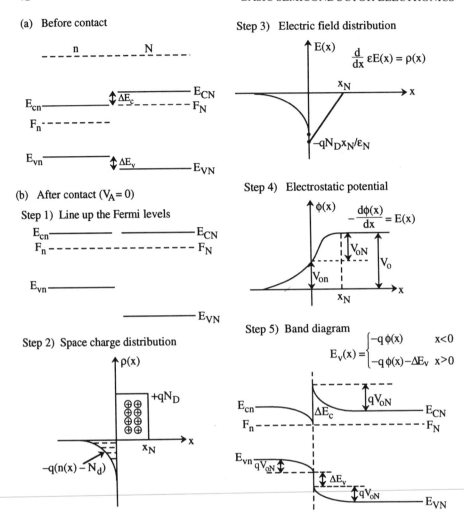

Figure 2.21. A step-by-step illustration for the energy band diagram of an *n-N* (isotope) heterojunction (a) before contact and (b) after contact showing five steps to obtain the steady-state energy band diagram.

of the Fermi levels. The charge distribution is

$$\rho(x) = \begin{cases} -q(n(x) - N_d) & x < 0 \\ qN_D & 0 < x < x_N \end{cases} \tag{2.6.1}$$

Note that far away from the junction $n(x \to -\infty) \to N_d$. The electron field satisfies

$$\frac{dE(x)}{dx} = \begin{cases} -\dfrac{q[n(x) - N_d]}{\varepsilon_n} & x < 0 \\[3mm] +\dfrac{qN_D}{\varepsilon_N} & 0 < x < x_N \end{cases} \tag{2.6.2}$$

Since $E(-\infty) = 0$, i.e., the electric field is zero far away from the junction, we integrate the above equation from $-\infty$ to 0_- and obtain

$$E(0_-) = -\frac{qn_s}{\varepsilon_n} \tag{2.6.3}$$

where

$$n_s = \int_{-\infty}^{0} (n(x) - N_d) \, dx \tag{2.6.4}$$

is the surface electron concentration. Similarly,

$$\varepsilon_N E(0_+) = -qN_D x_N = \varepsilon_n E(0_-) \tag{2.6.5}$$

or

$$n_s = N_D x_N \tag{2.6.6}$$

$$E(x) = \begin{cases} -\dfrac{q\int_{-\infty}^{x} [n(x') - N_d] \, dx'}{\varepsilon_n} & x < 0 \\[3mm] +\dfrac{qN_D(x - x_N)}{\varepsilon_N} & 0 < x < x_N \end{cases} \tag{2.6.7}$$

Judging from the charge distribution, the electric field and the potential profile, one sees that the band diagram in Fig. 2.21 (step 5) is very similar to that of the p-N heterojunction in Fig. 2.10b. Using the Boltzmann distribution

$$n(x) = N_{cn} \, e^{[F_n - E_c(x)]/k_B T} \tag{2.6.8}$$

$$N_d = N_{cn} \, e^{(F_n - E_{cn})/k_B T} \tag{2.6.9}$$

where $E_{cn} \equiv E_c(-\infty)$ is a constant, in (2.6.2) and

$$E(x) = -\frac{d\phi(x)}{dx} = \frac{1}{q}\frac{dE_c(x)}{dx} \tag{2.6.10}$$

we obtain a differential equation for $E_c(x)$:

$$\frac{d^2}{dx^2}E_c(x) = q\frac{dE(x)}{dx} = -q^2\frac{N_{cn}}{\varepsilon_n}\left(e^{-E_c(x)/k_BT} - e^{-E_{cn}/k_BT}\right)e^{F_n/k_BT}$$

$$\tag{2.6.11}$$

Using

$$\frac{1}{2}\frac{d}{dE_c}\left(\frac{dE_c}{dx}\right)^2 = \left(\frac{dE_c}{dx}\right)\frac{d}{dE_c}\left(\frac{dE_c}{dx}\right) = \frac{d^2}{dx^2}E_c \tag{2.6.12}$$

we can rewrite (2.6.11) as

$$d\left(\frac{dE_c}{dx}\right)^2 = -2q^2\frac{N_{cn}}{\varepsilon_n}\left(e^{-E_c(x)/k_BT} - e^{-E_{cn}/k_BT}\right)e^{F_n/k_BT}\,dE_c(x) \tag{2.6.13}$$

Integrating from $x = -\infty$ to 0_-, we obtain an expression for the electric field at $x = 0_-$:

$$[qE(0_-)]^2 = \left[\frac{dE_c(0_-)}{dx}\right]^2$$

$$= 2q^2\frac{N_{cn}}{\varepsilon_n}\left\{k_BT\left(e^{-E_c(0_-)/k_BT} - e^{-E_{cn}/k_BT}\right)\right.$$

$$\left. + [E_c(0_-) - E_{cn}]e^{-E_{cn}/k_BT}\right\}e^{F_n/k_BT} \tag{2.6.14}$$

Equation (2.6.14) can be further simplified to

$$E^2(0_-) = \frac{2k_BT}{\varepsilon_n}N_d\left[e^{[E_{cn}-E_c(0_-)]/k_BT} - 1 + \frac{E_c(0_-) - E_{cn}}{k_BT}\right]$$

$$= \frac{2k_BT}{\varepsilon_n}N_d\left(e^{qV_{0n}/k_BT} - 1 - \frac{qV_{0n}}{k_BT}\right) \tag{2.6.15}$$

where $qV_{0n} = E_{cn} - E_c(0_-)$.

Using the fact that the contact potential is given by

$$V_0 = \frac{F_N - F_n}{q} = \frac{(E_{cn} - F_n) + \Delta E_c - (E_{CN} - F_N)}{q} \tag{2.6.16}$$

and

$$\phi(x) = V_0 - \frac{q^2 N_D}{2\varepsilon_N}(x - x_N)^2 \qquad 0 \le x \le x_N \tag{2.6.17}$$

we obtain at $x = 0$

$$\phi(0) = V_0 - \frac{q^2 N_D}{2\varepsilon_N}x_N^2$$

$$= V_{0n} \tag{2.6.18}$$

$$V_{0N} = \frac{q^2 N_D}{2\varepsilon_N}x_N^2 = V_0 - V_{0n} \tag{2.6.19}$$

Therefore,

$$E(0_-) = \frac{\varepsilon_N}{\varepsilon_n}E(0_+) = \frac{-q}{\varepsilon_n}N_D x_N = \frac{-N_D}{\varepsilon_n}\sqrt{\frac{2\varepsilon_N V_{0N}}{N_D}}$$

$$= -\frac{1}{\varepsilon_n}\sqrt{2\varepsilon_N N_D(V_0 - V_{0n})} \tag{2.6.20}$$

Thus from (2.6.15) and (2.6.20)

$$k_B T N_d\left(e^{qV_{0n}/k_B T} - q\frac{V_{0n}}{k_B T} - 1\right) = \frac{\varepsilon_N N_D(V_0 - V_{0n})}{\varepsilon_n} \tag{2.6.21}$$

the only unknown V_{0n} is obtained by solving (2.6.21). After finding V_{0n}, we obtain V_{0N} and x_N from (2.6.19).

As an example, we show in Fig. 2.22 the energy band diagrams for a *P-n-N* heterojunction. The *n* region is assumed to be wide enough such that the two space charge regions near the two junctions do not merge. If they do, the flat portion in the center *n*-type semiconductor will not exist, and the band diagram is further distorted.

2.6.2   Metal–Semiconductor Junction

Consider a metal with a work function Φ_m brought in contact with an *n*-type semiconductor with an electron affinity χ. For an ideal contact, the barrier

(a) Before contact

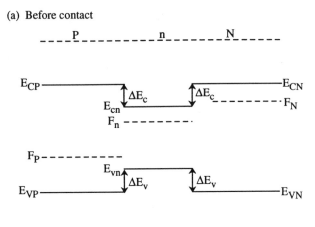

(b) After contact

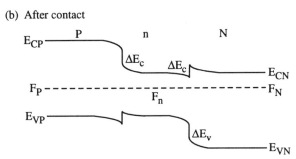

Figure 2.22. Energy band diagrams for a *P-n-N* heterojunction structure (a) before contact and (b) after contact. Assume thermal equilibrium.

height is given by (see Fig. 2.23)

$$\Phi_{ms} = \Phi_m - \chi \qquad (2.6.22)$$

and a negative charge distribution builds up at the metal surface with a space charge region formed in the semiconductor region.

The charge distribution in the semiconductor region is given by

$$\rho(x) = +q[N_D - n(x)] \qquad (2.6.23)$$

From Gauss's law, the electric field satisfies

$$\frac{dE(x)}{dx} = \frac{qN_D}{\varepsilon}\left(1 - e^{[E_{cn} - E_c(x)]/k_BT}\right) \qquad (2.6.24)$$

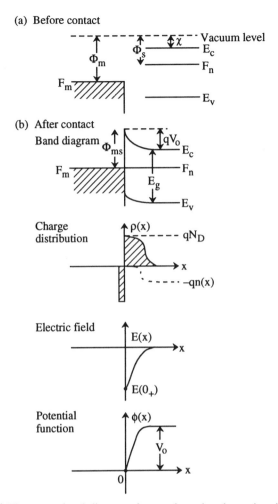

Figure 2.23. (a) The energy band diagram of a metal–semiconductor junction before contact. (b) The band diagram, charge distribution, electric field, and potential function after contact.

where $E_{cn} = E_c(\infty)$. We have

$$E_{cn} - E_c(x) = q[\phi(x) - V_0] \qquad (2.6.25)$$

where $V_0 = \phi(x \to \infty)$, with the reference potential $\phi(0) = 0$, and

$$E(x) = -\frac{d\phi(x)}{dx} \qquad (2.6.26)$$

Multiplying (2.6.24) by $E(x)$ on both sides

$$d\frac{E^2(x)}{2} = \frac{qN_D}{\varepsilon}\left(-1 + e^{[\phi(x) - V_0]q/k_BT}\right) d\phi(x) \qquad (2.6.27)$$

and integrating from 0_+ to $+\infty$, we obtain

$$E^2(0_+) = \frac{2qN_D}{\varepsilon}\left(V_0 - \frac{k_BT}{q} + \frac{k_BT}{q}e^{-qV_0/k_BT}\right) \qquad (2.6.28)$$

The contact potential is given by $V_0 = [\Phi_{ms} - (E_c - F_n)]/q$. The factor $\exp(-qV_0/k_BT)$ in the last expression in (2.6.28) is usually negligible for $V_0 > 0$.

When the metal semiconductor is forward biased with a voltage V and the barrier is reduced to $V_0 - V$, we have

$$|E(0_+)| = \sqrt{\frac{2qN_D}{\varepsilon}\left[V_0 - V - \frac{k_BT}{q}\left(1 - e^{-q(V_0 - V)/k_BT}\right)\right]}$$

$$\simeq \sqrt{\frac{2qN_D}{\varepsilon}\left(V_0 - V - \frac{k_BT}{q}\right)} \qquad (2.6.29)$$

If the electron concentration $n(x)$ is assumed to be depleted in a region of width x_n,

$$\frac{dE(x)}{dx} = \frac{qN_D}{\varepsilon} \qquad 0 < x < x_n \qquad (2.6.30)$$

Then

$$E(x) = \frac{qN_D(x - x_n)}{\varepsilon} \qquad (2.6.31)$$

and

$$E(0_+) = -\frac{qN_D x_n}{\varepsilon} \qquad (2.6.32)$$

An iterative approach to include the effect of the electron is to use $E(0_+)$ from (2.6.29), and obtain

$$x_n \simeq \sqrt{\frac{2\varepsilon}{qN_D}\left(V_0 - V - \frac{k_BT}{q}\right)} \qquad (2.6.33)$$

PROBLEMS

2.1 Derive the boundary conditions using the integral form of Maxwell's equations for a given charge density ρ_s and current density $\mathbf{J}_s$. Check the dimensions of all quantities appearing in the boundary conditions.

2.2 Consider an ideal case in which the electrons are distributed in a two-dimensional space with a surface carrier density n_s $(1/\text{cm}^2)$.

(a) Derive a two-dimensional density of states and plot it vs. the energy E.

(b) Find the relation between n_s and the Fermi level at a given temperature T.

(c) Repeat part (b) for $T = 0$.

2.3 Assume that the electrons are distributed in a "quantum-wire" geometry, i.e., an idealized one-dimensional space with a given "line" carrier density n_ℓ $(1/\text{cm})$.

(a) Derive a one-dimensional density of states and plot it vs. the energy E.

(b) Find the relation between n_ℓ and the Fermi level at a given temperature T.

(c) Repeat (b) for $T = 0$.

2.4 Use the inversion formula (2.2.51), and plot η vs. u. Compare this curve with the curve in Fig. 2.2. What is their relation?

2.5 Derive Eq. (2.3.24) for the average carrier lifetime τ_0.

2.6 Explain the band-to-band Auger recombination processes in k space by drawing the band structure including conduction (C), heavy-hole (H), light-hole (L), and spin-orbit split-off (S) bands [Ref. 27].

2.7 Suppose the carrier generation rate in Fig. 2.9 is

$$
G_n(y) = \begin{cases} G_0 \cos\left(\dfrac{\pi y}{S}\right) & |y| \le \dfrac{S}{2} \\[2mm] 0 & |y| > \dfrac{S}{2} \end{cases}
$$

(a) Find the excess carrier concentration $\delta n(y)$.

(b) Plot $G_n(y)$ and $\delta n(y)$ vs. y for $S = 10$ μm and $L_n = 4$ μm.

2.8 Plot the band diagram for a GaAs/Al$_{0.3}$Ga$_{0.7}$As p-N junction. The GaAs region is doped with an acceptor concentration $N_a = 1 \times 10^{18}/$ cm^3. The Al$_{0.3}$Ga$_{0.7}$As region is doped with a donor concentration $N_D = 1 \times 10^{18}/\text{cm}^3$. Assume zero bias. $\Delta E_c = 0.67\ \Delta E_g$ and $\Delta E_v = 0.33\ \Delta E_g$, where $\Delta E_g = 1.247x$ (eV) is the band-gap difference between

$Al_xGa_{1-x}As$ and GaAs. Evaluate the Fermi levels, the contact potential V_0, and the depletion width.

2.9 Repeat Problem 2.8 for a reverse bias of -2 V.

REFERENCES

1. J. A. Kong, *Electromagnetic Wave Theory*, 2d ed., Wiley, New York, 1990.

2. J. D. Jackson, *Classical Electrodynamics*, 2d ed., Wiley, New York, 1975.

3. M. Born and E. Wolf, *Principles of Optics*, Pergamon, Oxford, UK, 1975.

4. S. M. Sze, *Physics of Semiconductor Devices*, Wiley, New York, 1981.

5. S. Selberherr, *Analysis and Simulation of Semiconductor Devices*, Springer, New York, 1984.

6. W. L. Engl, H. K. Dirks, and B. Meinerzhagen, "Device modeling," *Proc. IEEE* **71**, 10–33 (1983).

7. A. H. Marshak and C. M. Van Vliet, "Electrical current and carrier density in degenerate materials with nonuniform band structure," *Proc. IEEE* **72**, 148–164 (1984).

8. X. Aymerich-Humett, F. Serra-Mestres, and J. Millan, "An analytical approximation for the Fermi–Dirac integral $F_{3/2}(x)$," *Solid-State Electron.* **24**, 981–982 (1981).

9. D. Bednarczyk and J. Bednarczyk, "The approximation of the Fermi–Dirac integral $F_{1/2}(\eta)$," *Phys. Lett.* **64A**, 409–410 (1978).

10. J. S. Blakemore, "Approximations for Fermi–Dirac integrals, especially the function $F_{1/2}(\eta)$ used to describe electron density in a semiconductor," *Solid-State Electron.* **25**, 1067–1076 (1982).

11. N. G. Nilsson, "An accurate approximation of the generalized Einstein relation for degenerate semiconductors," *Phys. Status Solidi: A* **19**, K75–K78 (1973).

12. T. Y. Chang and A. Izabelle, "Full range analytic approximations for Fermi energy and Fermi–Dirac integral $F_{-1/2}$ in terms of $F_{1/2}$," *J. Appl. Phys.* **65**, 2162–2164 (1989).

13. R. B. Adler, A. C. Smith, and R. L. Longini, *Introduction to Semiconductor Physics*, *Semiconductor Electronics Education Committee*, Vol. 1, Wiley, New York, 1964.

14. R. F. Pierret, *Semiconductor Fundamentals*, in R. F. Pierret and G. W. Neudeck, Eds., *Modular Series on Solid State Devices*, Vol. 1, Addison-Wesley, Reading, MA, 1983.

15. S. Wang, *Fundamentals of Semiconductor Theory and Device Physics*, Prentice Hall, Englewood Cliffs, NJ, 1989.

16. W. Shockley and W. T. Read, Jr., "Statistics of the recombinations of holes and electrons," *Phys. Rev.* **87**, 835–842 (1952).

17. R. N. Hall, "Electron–hole recombination in germanium," *Phys. Rev.* **87**, 387 (1952).

18. C. T. Sah, "The equivalent circuit model in solid-state electronics, Part I: The single energy level defect centers," *Proc. IEEE* **55**, 654–671 (1967); "Part II: The multiple energy level impurity centers," *Proc. IEEE* **55**, 672–684 (1967); "Part III: Conduction and displacement currents," *Solid-State Electron.* **13**, 1547–1575 (1970); "Equivalent circuit models in semiconductor transport for thermal, optical, Auger-impact, and tunnelling recombination-generation-trapping processes," *Phys. Status Solidi: A*, **7**, 541–559 (1971).

19. C. T. Sah, *Fundamentals of Solid-State Electronics*, World Scientific, Singapore, 1991.

20. H. C. Casey, Jr., and M. B. Panish, *Heterostructure Lasers, Part A: Fundamental Principles*; *Part B: Materials and Operating Characteristics*, Academic, Orlando, FL, 1978.

21. R. L. Anderson, "Germanium-gallium arsenide heterojunctions," *IBM J.* July, 283–287 (1960).

22. R. L. Anderson, "Experiments on Ge–GaAs heterojunctions," *Solid-State Electron.* **5**, 341–351 (1962).

23. W. R. Frensley and H. Kroemer, "Prediction of semiconductor heterojunction discontinuities from bulk band structures," *J. Vac. Sci. Technol.* **13**, 810–815 (1976).

24. W. A. Harrison, "Elementary theory of heterojunctions," *J. Vac. Sci. Technol.* **14**, 1016–1021 (1977).

25. H. Kroemer, "Heterostructure devices: A device physicist looks at interfaces," *Surf. Sci.* **132**, 543–576 (1983).

26. R. F. Pierret, "Extension of the approximate two-dimensional electron gas formulation," *IEEE Trans. Electron Devices* **ED-32**, 1279–1287 (1985).

27. A. Sugimara, "Band-to-band Auger effect in long wavelength multinary III-V alloy semiconductor lasers," *IEEE J. Quantum Electron.* **QE-18**, 352–363 (1982).

3
Basic Quantum Mechanics

In this chapter, we present some basic quantum mechanics, which in later chapters will help explain the physics of optoelectronic processes and devices. In quantum mechanics, there are three important potentials, which have exact analytical solutions and are presented in most undergraduate texts [1, 2] on modern physics or introductory quantum mechanics.

1. *A simple square well potential.* This potential is presented in Section 3.2. We take into account the fact that the effective masses inside and outside the quantum well are different for a semiconductor quantum well. More discussions on semiconductor quantum-well structures are presented in Chapter 4.

2. *A harmonic oscillator.* The solutions are Hermite–Gaussian functions [1–3] and are discussed in Section 3.3. A similar situation exists for the electric field of a waveguide with a parabolic permittivity or gain profile, since the solutions to the wave equation with a parabolic dependence on the position are also the Hermite–Gaussian functions.

3. *A hydrogen atom.* The solutions for both the bound states ($E < 0$) and the unbound (continuum) states ($E > 0$) are presented. Most textbooks discuss the bound states only, because they are important for hydrogen atoms. We use the hydrogen atom model to describe an exciton, which is formed by an electron–hole pair. The optical absorption process involving the Coulomb interaction between the electron and hole is called the excitonic absorption. The excitonic absorption spectrum is strongly dependent on the energies and wave functions of the exciton bound and continuum states. In bulk semiconductors, the electron–hole interaction is a Coulomb potential in a three-dimensional space. In quantum-well semiconductors, electrons and holes are confined to a quasi-two-dimensional structure; thus, their binding energy is different from the three-dimensional results. In Section 3.4 we summarize the analytical solutions for the hydrogen model in both three-dimensional and two-dimensional space for both the bound and unbound states [4–8]. The detailed derivations are presented in Appendix A. These results are used in Chapter 13 where we discuss excitonic absorption and quantum-confined Stark effects.

Since most potentials except the above three do not have exact, analytical solutions, we have to use perturbation methods to find their eigenfunctions and eigenenergies. We present a conventional time-independent perturbation theory in Section 3.5, then a slightly modified perturbation theory, called the Löwdin's method [9] in Section 3.6. The Löwdin's method is applied to derive the semiconductor band structures, such as the Luttinger–Kohn Hamiltonian in Chapter 4. In Section 3.7, we present the time-dependent perturbation theory and derive Fermi's golden rule, which is important to our understanding of the emission and absorption of photons.

3.1 SCHRÖDINGER EQUATION

The Schrödinger equation in a nonrelativistic quantum mechanical description of a single particle is

$$H\psi(\mathbf{r}, t) = i\hbar \frac{\partial}{\partial t} \psi(\mathbf{r}, t) \tag{3.1.1}$$

where the Hamiltonian H is

$$H = -\frac{\hbar^2}{2m} \nabla^2 + V(\mathbf{r}, t) \tag{3.1.2}$$

the potential energy function $V(\mathbf{r}, t)$ is real, $\hbar$ is the Planck constant h divided by 2π, and m is the mass of the particle. For a free particle in an unbounded space, $V(\mathbf{r}, t) = 0$, the solution is simply a plane wave:

$$\psi(\mathbf{r}, t) = \frac{1}{\sqrt{V}} e^{i\mathbf{k}\cdot\mathbf{r} - iEt/\hbar} \tag{3.1.3}$$

where the energy is

$$E = \frac{\hbar^2 k^2}{2m} \tag{3.1.4}$$

and V is the volume of the space. The probability density is defined as

$$\rho(\mathbf{r}, t) = \psi^*(\mathbf{r}, t)\psi(\mathbf{r}, t) \tag{3.1.5}$$

and the probability current density as

$$\mathbf{j}(\mathbf{r}, t) = \frac{\hbar}{2mi} [\psi^* \nabla \psi - \psi \nabla \psi^*] \tag{3.1.6}$$

Here $\psi^*\psi d^3\mathbf{r}$ is the probability of finding the particle in a volume $d^3\mathbf{r}$ near the position $\mathbf{r}$ at time t. The wave function is normalized such that

$\int_{\text{all space}} \psi^* \psi \, d^3 \mathbf{r} = 1$, that is, the probability of finding the particle in the whole space is unity. It is straightforward to show that

$$\nabla \cdot \mathbf{j} + \frac{\partial}{\partial t} \rho = 0 \tag{3.1.7}$$

which is the continuity equation or the conservation of probability density. It is analogous to the charge continuity equation in electromagnetics.

The expectation value of any physical quantity is given by

$$\langle O \rangle = \int_V \psi^*(\mathbf{r}, t) O \psi(\mathbf{r}, t) \, d^3 \mathbf{r} \tag{3.1.8}$$

where O is an operator for the physical quantity and the volume of integration is over the whole space. In the above real-space representation, that is, $\psi \equiv \psi(\mathbf{r}, t)$, the correspondence for the operators is

$$\text{Position} \qquad \mathbf{r}_{op} = \mathbf{r}$$

$$\text{Momentum} \quad \mathbf{p}_{op} = \frac{\hbar}{i} \nabla \tag{3.1.9}$$

If $V(\mathbf{r}, t)$ is independent of t, the solution $\psi(\mathbf{r}, t)$ can always be obtained using the separation of variables:

$$\psi(\mathbf{r}, t) = \psi(\mathbf{r}) \, e^{-iEt/\hbar} \tag{3.1.10}$$

and

$$\left[-\frac{\hbar^2}{2m} \nabla^2 + V(\mathbf{r}) \right] \psi(\mathbf{r}) = E \psi(\mathbf{r}) \tag{3.1.11}$$

which is the so-called time-independent Schrödinger equation. The solution may be in terms of quantized energy levels E_n with corresponding wave functions $\psi_n(\mathbf{r})$, or a continuous spectrum E with corresponding wave functions $\psi_E(\mathbf{r})$. In general, any solution of the Schrödinger equation may be constructed from the superposition of these stationary solutions:

$$\psi(\mathbf{r}, t) = \sum_n a_n \psi_n(\mathbf{r}) \, e^{-iE_n t/\hbar} + \int a_E \psi_E(\mathbf{r}) \, e^{-iEt/\hbar} \, dE \tag{3.1.12}$$

where $|a_n|^2$ gives the probability that the particle will be in the nth stationary state $\psi_n(\mathbf{r}, t)$ with an energy E_n. In studying a time-dependent potential problem, very often a perturbation approach is used if the time-dependent perturbing potential is small compared with the unperturbed Hamiltonian, and the above expansion in terms of the stationary states $\psi_n(\mathbf{r})$ or $\psi_E(\mathbf{r})$, which are solutions to the unperturbed problem, is very useful. In this case, a_n (and a_E) will be functions of time since the perturbation is time-dependent, and $|a_n(t)|^2$ will give the time-dependent probability that the particle is in state n of the unperturbed problem.

For a two-particle system, we use a wave function $\psi(\mathbf{r}_1, \mathbf{r}_2, t)$ to describe the particle 1 at $\mathbf{r}_1$, and particle 2 at $\mathbf{r}_2$, at time t. Here we assume that the two particles are not identical, such as those of a hydrogen atom. The Hamiltonian is

$$H = \frac{\mathbf{p}_1^2}{2m_1} + \frac{\mathbf{p}_2^2}{2m_2} + V(\mathbf{r}_1 - \mathbf{r}_2) \tag{3.1.13}$$

where

$$\mathbf{p}_1 = \frac{\hbar}{i} \nabla_1 \tag{3.1.14}$$

$$\mathbf{p}_2 = \frac{\hbar}{i} \nabla_2 \tag{3.1.15}$$

and ∇_i refers to the gradient operator with respect to $\mathbf{r}_i$ ($i = 1, 2$). This case will be solved in Section 3.4.

If we write the Fourier transforms of $\psi(\mathbf{r})$ and $V(\mathbf{r})$ in the single-particle Schrödinger equation (3.1.11),

$$\tilde{\psi}(\mathbf{k}) = \int \psi(\mathbf{r}) \, e^{-i\mathbf{k}\cdot\mathbf{r}} \, d^3\mathbf{r} \tag{3.1.16}$$

$$\tilde{V}(\mathbf{k}) = \int V(\mathbf{r}) \, e^{-i\mathbf{k}\cdot\mathbf{r}} \, d^3\mathbf{r} \tag{3.1.17}$$

we then obtain the momentum–space representation of the Schrödinger equation

$$\frac{\hbar^2 k^2}{2m} \tilde{\psi}(\mathbf{k}) + \int \frac{d^3\mathbf{k}'}{(2\pi)^3} \tilde{V}(\mathbf{k} - \mathbf{k}') \tilde{\psi}(\mathbf{k}') = E\tilde{\psi}(\mathbf{k}) \tag{3.1.18}$$

which becomes an integral equation. This integral equation is discussed in Chapter 13.

3.2   THE SQUARE WELL

We consider a square (or rectangular) quantum well with a barrier height V_0. In the one-dimensional case, the time-independent Schrödinger equation is

$$\left[-\frac{\hbar^2}{2m} \frac{d^2}{dz^2} + V(z) \right] \phi(z) = E\phi(z) \tag{3.2.1}$$

3.2.1   Infinite Barrier Model

First we assume that V_0 is infinitely high; thus, the wave function vanishes at the boundaries. The solution to the infinite barrier model satisfying the boundary conditions $\phi_n(0) = \phi_n(L) = 0$ is

$$\phi_n(z) = \sqrt{\frac{2}{L}} \sin\left(\frac{n\pi}{L} z\right) \qquad n = 1, 2, 3, \ldots \qquad (3.2.2)$$

with corresponding energy and wave number

$$E_n = \frac{\hbar^2}{2m}\left(\frac{n\pi}{L}\right)^2 \qquad k_z = \frac{n\pi}{L} \qquad (3.2.3)$$

i.e., the energy E_n and the wave number k_z in the z direction are quantized. The wave function has been normalized $\int_0^L |\phi_n(z)|^2\, dz = 1$.

If the origin of the coordinate $z = 0$ is chosen at the center of the well, $V(+z) = V(-z)$, the solution can always be put in terms of even or odd functions by parity consideration:

$$\phi_n(z) = \begin{cases} \sqrt{\dfrac{2}{L}} \sin\left(\dfrac{n\pi}{L} z\right) & n \text{ even} \\[2ex] \sqrt{\dfrac{2}{L}} \cos\left(\dfrac{n\pi}{L} z\right) & n \text{ odd} \end{cases} \qquad (3.2.4)$$

which can be obtained from (3.2.2) by replacing z by $z + L/2$ and discarding any minus signs, since we have the freedom to choose the phase of the wave function. The parity consideration is very useful since the symmetry properties of the system are employed; thus, the wave functions have associated symmetry properties.

In general, if $V(z) = V(-z)$, for any solution $\phi(z)$ of the Schrödinger equation (3.2.1), $\phi(-z)$ will also be a solution with the same eigenenergy, as can be seen by changing z to $-z$ in (3.2.1). If we form linear combinations of $\phi(z)$ and $\phi(-z)$, they will also be solutions. Specifically,

$$\phi_e(z) = \frac{1}{\sqrt{2}}\left[\phi(z) + \phi(-z)\right] \qquad (3.2.5)$$

$$\phi_o(z) = \frac{1}{\sqrt{2}}\left[\phi(z) - \phi(-z)\right] \qquad (3.2.6)$$

where $\phi_e(z)$ is an even function and $\phi_o(z)$ is an odd function.

The complete solution for the potential well $V(z)$ in a three-dimensional space is solved from

$$\left[-\frac{\hbar^2}{2m}\nabla^2 + V(z)\right]\psi(x, y, z) = E\psi(x, y, z) \qquad (3.2.7)$$

The normalized wave function is

$$\psi(x, y, z) = \frac{e^{ik_x x + ik_y y}}{\sqrt{A}}\phi_n(z) \qquad (3.2.8)$$

with a corresponding energy

$$E = \frac{\hbar^2}{2m}\left[k_x^2 + k_y^2 + \left(\frac{n\pi}{L}\right)^2\right] \qquad (3.2.9)$$

where $\phi_n(z)$ is the same as in (3.2.2), and A is the cross-sectional area in the x-y plane. This energy dispersion diagram E vs. k_x or k_y is plotted in Fig. 3.1b for $n = 1$ and 2. In semiconductors, m should be replaced by the effective mass m^* of the carriers.

Example For a GaAs/Al$_x$Ga$_{1-x}$As quantum-well structure, suppose the GaAs well width is 100 Å and the aluminum mole fraction x in the barrier region is large such that an infinite barrier model is applicable. We have the effective mass of the electrons $m_e^* = 0.0665m_0$ in the well region. The

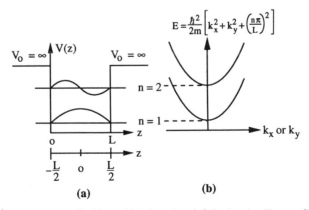

Figure 3.1. (a) A quantum well with a width L and an infinite barrier $V_0 = \infty$. (b) The energy dispersion in the k_x or k_y space using Eq. (3.2.9).

quantized electron subband energies are

$$E_n = \frac{\hbar^2}{2m_e^*}\left(\frac{n\pi}{L}\right)^2 = n^2 E_1 = E_1, 4E_1, 9E_1, \ldots$$

where

$$E_1 = \frac{\hbar^2}{2m_e^*}\left(\frac{\pi}{L}\right)^2 = \frac{1}{(m_e^*/m_0)L^2} \times 37.6 \text{ eV}$$

$$= \frac{1}{0.0665 \times 100^2} \times 37.6 \text{ eV} = 56.5 \text{ meV}$$

Therefore, we find the subband energies are

$$E_n = 56.5, 226, 508.5, \ldots \text{(meV)}. \qquad \blacksquare$$

Two-Dimensional Density of States. The electron concentration in a quantum well can be calculated using

$$n = \frac{2}{V}\sum_n \sum_{k_x}\sum_{k_y} f(E) \tag{3.2.10}$$

where the summation is over all the occupied subbands. Since the electron energy is quantized in the k_z quantum number, and the dependence on x and y is still plane-wave-like, we convert the summation over k_x and k_y into integrations:

$$\frac{2}{V}\sum_{k_x}\sum_{k_y} = \frac{2}{V}\int\frac{dk_x}{2\pi/L_x}\int\frac{dk_y}{2\pi/L_y} = \frac{2}{L_z}\int\frac{dk_x\,dk_y}{(2\pi)^2}$$

$$= \frac{2}{L_z}\int\frac{2\pi k_t\,dk_t}{(2\pi)^2} = \frac{m^*}{\pi\hbar^2 L_z}\int_{E_n}^{\infty}dE \tag{3.2.11}$$

where we have used $dk_x\,dk_y = d\phi_k k_t\,dk_t$ in the polar coordinate, $k_t^2 = k_x^2 + k_y^2$, $dE = \hbar^2 k_t\,dk_t/m^*$. We can then write the two-dimensional (2D) density of states as

$$\rho_{2D}(E) = \frac{m^*}{\pi\hbar^2 L_z}\sum_n H(E - E_n) \tag{3.2.12}$$

where $H(x)$ is a Heaviside step function $H(x) = 1$ for $x > 0$ and $H(x) = 0$

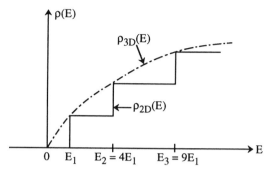

Figure 3.2. The electron density of states $\rho_{2D}(E)$ solid line for a two-dimensional quantum-well structure is compared with the three-dimensional density of states $\rho_{3D}(E)$ (dashed curve).

for $x < 0$. The electron density n is then given by

$$n = \int dE \rho_{2D}(E) f(E) \tag{3.2.13}$$

The density of states $\rho_{2D}(E)$ is plotted in Fig. 3.2. The step-like function with each step occurs wherever there is a new subband energy level E_1, $E_2 = 4E_1$, $E_3 = 9E_1$, etc. It is interesting to compare the two-dimensional density of states with the three-dimensional density of states (2.2.36):

$$\rho_{3D}(E) = \frac{1}{2\pi^2}\left(\frac{2m^*}{\hbar^2}\right)^{3/2}\sqrt{E} \tag{3.2.14}$$

At $E = E_n = n^2\hbar^2\pi^2/(2m^*L_z^2)$

$$\rho_{3D}(E_n) = \frac{1}{2\pi^2}\left(\frac{2m^*}{\hbar^2}\right)^{3/2}\left(\frac{\hbar^2 n^2\pi^2}{2m^*L_z^2}\right)^{1/2}$$

$$= n\left(\frac{m^*}{\pi\hbar^2 L_z}\right) \tag{3.2.15}$$

which is the same as $\rho_{2D}(E = E_n)$. The three-dimensional density of states $\rho_{3D}(E)$ is shown as the dashed curve in Fig. 3.2 with a smooth $\sqrt{E}$ behavior above the conduction band edge, while the two-dimensional $\rho_{2D}(E)$ shows a sharp step-like behavior above each subband edge.

Example As a numerical example, we calculate the electron density of states for a bulk GaAs semiconductor at an energy 0.1 eV above the conduction band edge:

$$\rho_{3D}(E - E_c = 0.1 \text{ eV}) = \frac{1}{2\pi^2}\left(\frac{2m_e^*}{\hbar^2}\right)^{3/2}(E - E_c)^{1/2}$$

$$= 3.69 \times 10^{19} \left(\text{cm}^{-3} \text{ eV}^{-1}\right)$$

For a quantum well with $L_z = 100$ Å, we estimate the density of states of the first step (with $E_1 = 56.5$ meV):

$$\rho_{2D}(E - E_1) = \frac{m_e^*}{\pi\hbar^2 L_z} = 2.78 \times 10^{19} \left(\text{cm}^{-3} \text{ eV}^{-1}\right)$$

We can also estimate the carrier concentration with an energy spread of $k_B T = 0.026$ eV at 300 K and obtain

$$n \sim 2.78 \times 10^{19} \times 0.026 = 7.2 \times 10^{17} \text{ cm}^{-3} \qquad \blacksquare$$

One-Dimensional Density of States. Similar to (3.2.10), the electron concentration in a one-dimensional (1D) quantum wire along the z direction can be obtained from

$$n = \frac{2}{V} \sum_{n_x, n_y} \sum_{k_z} f(E)$$

where the electron energy E is quantized in both the x and y directions,

$$E = E_{n_x} + E_{n_y} + \frac{\hbar^2}{2m^*} k_z^2$$

$$E_{n_x} = \frac{\hbar^2}{2m^*}\left(\frac{n_x \pi}{L_x}\right)^2 \qquad E_{n_y} = \frac{\hbar^2}{2m^*}\left(\frac{n_y \pi}{L_y}\right)^2$$

$n_x, n_y = 1, 2, 3, \ldots$, and k_z is a continuous variable. Using

$$\frac{2}{V} \sum_{n_x, n_y} \sum_{k_z} = \frac{2}{L_x L_y} \sum_{n_x, n_y} \int_{-\infty}^{\infty} \frac{dk_z}{2\pi}$$

$$= \frac{1}{\pi L_x L_y}\sqrt{\frac{2m^*}{\hbar^2}} \sum_{n_x, n_y} \int_{E_{n_x}+E_{n_y}}^{\infty} \frac{dE}{\sqrt{E - E_{n_x} - E_{n_y}}}$$

the 1D density of states $\rho_{1D}(E)$ is

$$\rho_{1D}(E) = \frac{1}{\pi L_x L_y}\sqrt{\frac{2m^*}{\hbar^2}} \sum_{n_x, n_y} \frac{1}{\sqrt{E - E_{n_x} - E_{n_y}}} \qquad E > E_{n_x} + E_{n_y}$$

3.2.2 Finite Barrier Model

For a finite barrier quantum well, as shown in Fig. 3.3, we have

$$V(z) = \begin{cases} V_0 & |z| \geq \dfrac{L}{2} \\[2mm] 0 & |z| < \dfrac{L}{2} \end{cases} \qquad (3.2.16)$$

We solve the Schrödinger equation

$$\left[-\frac{d}{dz}\frac{1}{m}\frac{d}{dz} + V(z) \right]\phi(z) = E\,\phi(z) \qquad (3.2.17)$$

where $m = m_w$ in the well and $m = m_b$ in the barrier region. We consider the bound-state solutions, which have energies in the range between 0 and V_0.

For the even wave functions, we have a solution of the form

$$\phi(z) = \begin{cases} C_1\,e^{-\alpha(|z|-L/2)} & |z| \geq \dfrac{L}{2} \\[2mm] C_2\cos kz & |z| < \dfrac{L}{2} \end{cases} \qquad (3.2.18)$$

where

$$k = \frac{\sqrt{2m_w E}}{\hbar} \qquad (3.2.19a)$$

$$\alpha = \frac{\sqrt{2m_b(V_0 - E)}}{\hbar} \qquad (3.2.19b)$$

Using boundary conditions in which the wave function ϕ and its first derivative divided by the effective mass $(1/m)(d\phi/dz)$ are continuous at the

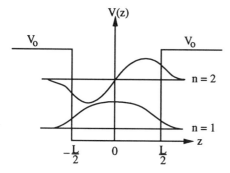

Figure 3.3. A quantum well with a width L and a finite barrier height V_0. The energy levels for $n = 1$ and $n = 2$ and their corresponding wave functions are shown.

interface between the barrier and the well, that is,

$$\phi\left(\frac{L^+}{2}\right) = \phi\left(\frac{L^-}{2}\right)$$

and

$$\frac{1}{m_b}\frac{d}{dz}\phi\left(\frac{L^+}{2}\right) = \frac{1}{m_w}\frac{d}{dz}\phi\left(\frac{L^-}{2}\right) \tag{3.2.20}$$

we obtain

$$C_1 = C_2 \cos k\frac{L}{2}$$

$$\frac{\alpha}{m_b}C_1 = \frac{k}{m_w}C_2 \sin k\frac{L}{2} \tag{3.2.21}$$

Eliminating C_1 and C_2, we obtain the eigenequation or the quantization condition

$$\alpha = \frac{m_b k}{m_w}\tan k\frac{L}{2} \tag{3.2.22}$$

The eigenenergy E can be found from the above equation by substituting k and α from (3.2.19a) and (3.2.19b) into (3.2.22). Similarly for odd wave functions, we have solutions of the form

$$\phi(z) = \begin{cases} C_1 e^{-\alpha(z-L/2)} & z > \dfrac{L}{2} \\[2ex] C_2 \sin kz & |z| \le \dfrac{L}{2} \\[2ex] -C_1 e^{+\alpha(z+L/2)} & z < -\dfrac{L}{2} \end{cases} \tag{3.2.23}$$

The boundary conditions (3.2.20) give

$$C_1 = C_2 \sin k\frac{L}{2}$$

$$-\frac{\alpha C_1}{m_b} = \frac{k}{m_w}C_2 \cos k\frac{L}{2} \tag{3.2.24}$$

Thus, the eigenequation is given by

$$\alpha = -\frac{m_b k}{m_w}\cot k\frac{L}{2} \tag{3.2.25}$$

which determines the eigenenergy E for the odd wave function, again using k and α in (3.2.19a) and (3.2.19b). In general, the solutions for the quantized eigenenergies can be obtained by finding $(\alpha L/2)$ and $(kL/2)$ directly from a graphical approach since from (3.2.19a) and (3.2.19b)

$$\left(k\frac{L}{2}\right)^2 + \frac{m_w}{m_b}\left(\alpha\frac{L}{2}\right)^2 = \frac{2m_w V_0}{\hbar^2}\left(\frac{L}{2}\right)^2 \tag{3.2.26}$$

and

$$\alpha\frac{L}{2} = \frac{m_b}{m_w}k\frac{L}{2}\tan k\frac{L}{2} \qquad \text{even solution} \tag{3.2.27a}$$

$$\alpha\frac{L}{2} = -\frac{m_b}{m_w}k\frac{L}{2}\cot k\frac{L}{2} \qquad \text{odd solution} \tag{3.2.27b}$$

The above equations can be solved by plotting them on the $\alpha L/2$ vs. $kL/2$ plane, as shown in Fig. 3.4. If $m_w = m_b$, Eq. (3.2.26) is a circle with a

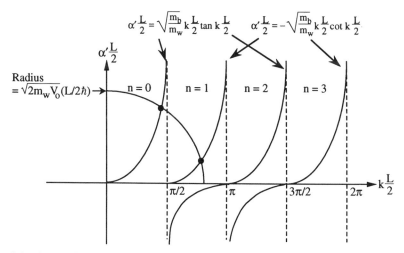

Figure 3.4. A graphical solution for the decaying constant α and wave number k of a finite quantum well. Here we define $\alpha' = \alpha\sqrt{m_w/m_b}$ to account for the different effective masses in the well and in the barrier.

radius $\sqrt{2m_w V_0}\,(L/2\hbar)$. If $m_w \neq m_b$, we define $\alpha' \equiv \alpha\sqrt{m_w/m_b}$, and rewrite

$$\alpha'\frac{L}{2} = \sqrt{\frac{m_b}{m_w}}\, k\frac{L}{2}\tan k\frac{L}{2} \tag{3.2.28a}$$

$$\alpha'\frac{L}{2} = -\sqrt{\frac{m_b}{m_w}}\, k\frac{L}{2}\cot k\frac{L}{2} \tag{3.2.28b}$$

On the $\alpha'L/2$ vs. $kL/2$ plane, Eq. (3.2.26) is still a circle with the same radius $\sqrt{2m_w V_0}\,(L/2\hbar)$. The solutions to α' and k are obtained from the intersection points between the circle (3.2.26) and either the tangent curve (3.2.28a) or the cotangent curve (3.2.28b), as shown in Fig. 3.4. It is obvious that if the radius of the circle is in the range

$$(N-1)\frac{\pi}{2} \leq \sqrt{2m_w V_0}\left(\frac{L}{2\hbar}\right) < N\frac{\pi}{2} \tag{3.2.29}$$

there are N solutions. After solving the eigenequations for α and k, we obtain the quantized energy E.

The bound-state solution ($E < V_0$) after normalization $\int_{-\infty}^{\infty}|\phi(z)|^2\,dz = 1$ is

$$\phi(z) = C\begin{cases} \cos k\dfrac{L}{2}\, e^{-\alpha(|z|-L/2)} & |z| > \dfrac{L}{2} \\[2mm] \cos kz & |z| < \dfrac{L}{2} \end{cases} \tag{3.2.30}$$

where

$$C = \sqrt{\dfrac{2}{L + \dfrac{2}{\alpha}\left(\cos^2 k\dfrac{L}{2} + \dfrac{m_b}{m_w}\sin^2 k\dfrac{L}{2}\right)}} \tag{3.2.31}$$

If $m_w = m_b$, we have $C = \sqrt{2/(L + 2/\alpha)}$. The length $L + (2/\alpha)$ is the well width plus twice the penetration depth $1/\alpha$ on each side of the well. The normalization factor is very similar to that of the infinite well (3.2.2) except that we have an effective well width $L_e = L + (2/\alpha)$. For odd solutions

$$\phi(z) = C\begin{cases} \sin k\dfrac{L}{2}\, e^{-\alpha(z-L/2)} & z \geq \dfrac{L}{2} \\[2mm] \sin kz & |z| \leq \dfrac{L}{2} \\[2mm] -\sin k\dfrac{L}{2}\, e^{+\alpha(z+L/2)} & z < -\dfrac{L}{2} \end{cases} \tag{3.2.32}$$

where

$$C = \sqrt{\cfrac{2}{L + \cfrac{2}{\alpha}\left(\sin^2\cfrac{kL}{2} + \cfrac{m_b}{m_w}\cos^2\cfrac{kL}{2}\right)}} \qquad (3.2.33)$$

Again C reduces to the expression $\sqrt{2/(L + (2/\alpha))}$ when $m_b = m_w$.

Alternatively, an effective well width L_{eff} can be defined by using the ground-state eigenenergy E obtained from the solution of the eigenequation (3.2.28a) and setting it equal to the ground-state energy ($n = 1$) of an infinite well:

$$E = \frac{\hbar^2}{2m}\left(\frac{\pi}{L_{\text{eff}}}\right)^2 \qquad (3.2.34)$$

The appropriate wave functions are then

$$\phi_n(z) = \sqrt{\frac{2}{L_{\text{eff}}}}\sin\left(\frac{n\pi}{L_{\text{eff}}}z\right) \qquad (3.2.35)$$

where the origin is set at the left boundary of the infinite well.

Example Consider a GaAs/Al$_x$Ga$_{1-x}$As quantum well shown in Fig. 3.5. Assume the following parameters

$$m_e^* = (0.0665 + 0.0835x)m_0 \qquad m_{hh}^* = (0.34 + 0.42x)m_0$$

$$m_{lh}^* = (0.094 + 0.043x)m_0 \qquad E_g(x) = 1.424 + 1.247x \ (\text{eV})$$

$$(0 \le x < 0.45, \text{ room temperature})$$

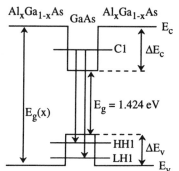

Figure 3.5. A GaAs/Al$_x$Ga$_{1-x}$As quantum-well structure showing the band gaps and the interband transition energies.

$\Delta E_g(x) = 1.247x$ (eV), $\Delta E_c = 0.67\Delta E_g$, and $\Delta E_v = 0.33\Delta E_g$, where m_0 is the free electron mass.

a. Consider the aluminum mole fraction $x = 0.3$ in the barrier regions ($x = 0$ in the well region) and the well width $L_w = 100$ Å. How many bound states are there in the conduction band? How many bound heavy-hole and light-hole subbands are there?

b. Find the lowest bound-state energies for the conduction subband (C1) and the heavy-hole (HH1) and the light-hole (LH1) subbands for a 100-Å GaAs quantum well in part (a). What are the C1–HH1 and C1–LH1 transition energies?

c. Assume that we define an effective well width L_{eff} using an infinite barrier model such that its ground-state energy is the same as the energy of the first conduction subband E_{c1} in (b). What is L_{eff}? If we repeat the same procedure for the HH1 and the LH1 subbands, what are L_{eff}?

SOLUTIONS. We tabulate the physical parameters as follows:

	m_e^*	m_{hh}^*	m_{lh}^*	E_g
Well	$0.0665m_0$	$0.34m_0$	$0.094m_0$	1.424 eV
Barrier ($x = 0.3$)	$0.0916m_0$	$0.466m_0$	$0.107m_0$	1.798 eV

$L_w = 100$ Å, $\Delta E_c = 0.67\Delta E_g = 0.2506$ eV, $\Delta E_v = 0.1235$ eV.

a. The number of bound states N is determined by $(N - 1)\pi/2 \leq \sqrt{2m_w^* V_0}(L/2\hbar) \leq N\pi/2$, $V_0 = \Delta E_c$ for electrons and ΔE_v for holes, respectively.

For electrons: $\sqrt{2m_e^* \Delta E_c}\left(\dfrac{L}{2\hbar}\right) = 3.30 < N\dfrac{\pi}{2}$ $N = 3$ bound states

For heavy holes: $\sqrt{2m_{hh}^* \Delta E_v}\left(\dfrac{L}{2\hbar}\right) = 5.25 < N\dfrac{\pi}{2}$ $N = 4$ bound states

For light holes: $\sqrt{2m_{lh}^* \Delta E_v}\left(\dfrac{L}{2\hbar}\right) = 2.76 < N\dfrac{\pi}{2}$ $N = 2$ bound states

b. The eigenenergy E is found by searching for the root in

$$\alpha = \frac{m_b^*}{m_w}k\tan\left(k\frac{L_w}{2}\right) \qquad \alpha = \sqrt{\frac{2m_b^*(V_0 - E)}{\hbar^2}} \qquad k = \sqrt{\frac{2m_w^* E}{\hbar^2}}$$

For electrons, $m_w^* = 0.0665m_0$, $m_b^* = 0.0916m_0$, and $V_0 = 250.6$ meV, the first subband energy is

$$E_{C1} = 30.7 \text{ meV}$$

Similarly, $E_{HH1} = 7.4$ meV and $E_{LH1} = 20.6$ meV. The transition energies are

$$E_{C1-HH1} = E_g + E_{C1} + E_{HH1} = 1462 \text{ meV}$$
$$E_{C1-LH1} = E_g + E_{C1} + E_{LH1} = 1475 \text{ meV}$$

c. The energy for an infinite barrier model is $E_1 = \pi^2 \hbar^2 / (2m^* L_{eff}^2)$

$$\text{For C1, } L_{eff} = \sqrt{\frac{\pi^2 \hbar^2}{2m_e^* E_{C1}}} = 136 \text{ Å}$$

$$\text{For HH1, } L_{eff} = \sqrt{\frac{\pi^2 \hbar^2}{2m_{hh}^* E_{HH1}}} = 122 \text{ Å}$$

$$\text{For LH1, } L_{eff} = \sqrt{\frac{\pi^2 \hbar^2}{2m_{lh}^* E_{LH1}}} = 139.6 \text{ Å} \qquad \blacksquare$$

3.3 THE HARMONIC OSCILLATOR

If an electron is in a parabolic potential of the form $V(z) = (K/2)z^2$, as shown in Fig. 3.6, the time-independent Schrödinger equation in one dimension is

$$\left(-\frac{\hbar^2}{2m} \frac{d^2}{dz^2} + \frac{K}{2} z^2 \right) \phi(z) = E \phi(z) \qquad (3.3.1)$$

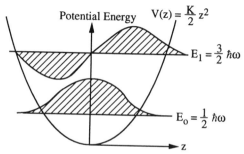

Figure 3.6. A parabolic quantum well with its quantized energy levels and wave functions.

If we define

$$\omega = \sqrt{\frac{K}{m}} \qquad (3.3.2)$$

and change the variable from z to ξ

$$\xi = \alpha z \qquad (3.3.3)$$

$$\alpha = \sqrt{\frac{m\omega}{\hbar}} \qquad (3.3.4)$$

Equation (3.3.1) becomes

$$\left[\frac{d^2}{d\xi^2} + \left(\frac{2E}{\hbar\omega} - \xi^2 \right) \right] \phi(\xi) = 0 \qquad (3.3.5)$$

The solutions are the Hermite–Gaussian functions

$$\phi_n(\xi) = \left(\frac{\alpha}{\sqrt{\pi}\, 2^n n!} \right)^{1/2} H_n(\xi)\, e^{-\xi^2/2} \qquad (3.3.6)$$

where $H_n(\xi)$ are the Hermite polynomials satisfying the differential equation [10]

$$\left(\frac{d^2}{d\xi^2} - 2\xi \frac{d}{d\xi} + 2n \right) H_n(\xi) = 0 \qquad (3.3.7)$$

and n is related to the energy E by

$$E = \left(n + \frac{1}{2} \right) \hbar\omega \qquad n = 0, 1, 2, 3, \ldots \qquad (3.3.8)$$

The Hermite polynomials can also be obtained from the generating function

$$e^{-t^2 + 2t\xi} = \sum_{n=0}^{\infty} \frac{H_n(\xi)}{n!} t^n \qquad (3.3.9)$$

Or

$$H_n(\xi) = \left(\frac{d^n}{dt^n} e^{-t^2 + 2t\xi} \right)_{t=0} = (-1)^n e^{\xi^2} \frac{d^n}{d\xi^n} e^{-\xi^2} \qquad (3.3.10)$$

The first few Hermite polynomials are

$$H_0(\xi) = 1$$
$$H_1(\xi) = 2\xi$$
$$H_2(\xi) = -2 + 4\xi^2$$
$$H_3(\xi) = -12\xi + 8\xi^3$$
$$H_4(\xi) = 12 - 48\xi^2 + 16\xi^4 \tag{3.3.11}$$

Another elegant way [3] to find the solutions for the harmonic oscillator is to use the matrix approach by defining the annihilation operator

$$a = \sqrt{\frac{m\omega}{2\hbar}}\, z + \frac{ip}{\sqrt{2m\hbar\omega}} \tag{3.3.12}$$

and the creation operator

$$a^+ = \sqrt{\frac{m\omega}{2\hbar}}\, z - \frac{ip}{\sqrt{2m\hbar\omega}} \tag{3.3.13}$$

Note the relation

$$zp - pz = i\hbar \tag{3.3.14}$$

which can be proved using $p = (\hbar/i)(\partial/\partial z)$ and

$$(zp - pz)\psi = z\frac{\hbar}{i}\frac{\partial}{\partial z}\psi - \frac{\hbar}{i}\frac{\partial}{\partial z}z\psi = i\hbar\psi$$

for any function $\psi(z)$. Therefore, we find

$$a^+a = \frac{m\omega}{2\hbar}z^2 + \frac{p^2}{2m\hbar\omega} + \frac{i}{2\hbar}(zp - pz)$$
$$= \frac{1}{\hbar\omega}\left(\frac{p^2}{2m} + \frac{m\omega^2}{2}z^2\right) - \frac{1}{2} \tag{3.3.15}$$

The Hamiltonian (3.3.1) can be rewritten in terms of a^+a:

$$H = \frac{p^2}{2m} + \frac{1}{2}m\omega^2 z^2$$
$$= \hbar\omega\left(a^+a + \frac{1}{2}\right) \tag{3.3.16}$$

Since the last term $\hbar\omega/2$ is a constant, the problem of solving the Schrödinger equation

$$H\phi_n = E_n\phi_n \qquad (3.3.17)$$

is the same as finding the eigenvector and the eigenvalue of the operator a^+a. Defining

$$N \equiv a^+a \qquad (3.3.18)$$

and noting that the Poisson bracket

$$[a, a^+] \equiv aa^+ - a^+a = \frac{i}{2\hbar}\{(pz - zp) - (zp - pz)\} = 1 \quad (3.3.19)$$

from (3.3.12) and (3.3.13), we find

$$Na = a^+aa = (aa^+ - 1)a = aN - a \qquad (3.3.20)$$

$$Na^+ = a^+aa^+ = a^+(a^+a + 1) = a^+(N + 1) \qquad (3.3.21)$$

For any eigenstate of H or N, say $\phi_n \equiv |n >$, we have

$$\langle n|N|n\rangle = \langle n|a^+a|n\rangle = \langle\psi|\psi\rangle \geq 0 \qquad (3.3.22)$$

since the wave function $|\psi\rangle \equiv a|n\rangle$ has a norm $\langle\psi|\psi\rangle$ which is always nonnegative. Let n be the eigenvalue of the operator N with the corresponding eigenvector $|n\rangle$, $N|n\rangle = n|n\rangle$. We have

$$Na|n\rangle = (aN - a)|n\rangle = (n - 1)a|n\rangle \qquad (3.3.23)$$

Thus $a|n\rangle$ is an eigenstate of the operator N with eigenvalue $n - 1$. Assume that the eigenfunctions are normalized for all n:

$$\langle n|n\rangle = 1 \qquad (3.3.24)$$

Then

$$n = \langle n|a^+a|n\rangle = (\langle n|a^+)(a|n\rangle) \qquad (3.3.25)$$

Since $(a|n\rangle)$ is an eigenstate of the operator N with eigenvalue $n - 1$ from (3.3.23) and its magnitude squared is n from (3.3.25), we have

$$a|n\rangle = \sqrt{n}|n - 1\rangle \qquad (3.3.26)$$

where the normalization condition $\langle n - 1 | n - 1 \rangle = 1$ has been used. Repeating the above procedure, we have

$$a^2 |n\rangle = \sqrt{n}\, a |n - 1\rangle = \sqrt{n}\, \sqrt{n - 1}\, |n - 2\rangle \qquad (3.3.27)$$

We find that the lowest state will be $|0\rangle$ with its eigenvalue $n = 0$ because of the nonnegative condition (3.3.22). Similarly, using (3.3.21), we find

$$a^+ |n\rangle = \sqrt{n + 1}\, |n + 1\rangle \qquad (3.3.28)$$

In general, the nth eigenstate can be obtained from the ground state $|0\rangle$ by the creation operator

$$|n\rangle = \frac{\left(a^+ \right)^n}{\sqrt{n!}} |0\rangle \qquad (3.3.29)$$

with

$$E_n = \left(n + \frac{1}{2} \right) \hbar\omega, \qquad n = 0, 1, 2, 3, \ldots \qquad (3.3.30)$$

The ground-state wave function can also be obtained using the fact that

$$a|0\rangle = 0$$

Therefore,

$$0 = \langle z | a | 0 \rangle = \langle z | \sqrt{\frac{m\omega}{2\hbar}}\, z + \sqrt{\frac{\hbar}{2m\omega}}\, \frac{\mathrm{d}}{\mathrm{d}z} | 0 \rangle$$

$$= \sqrt{\frac{\hbar}{2m\omega}} \left(\frac{\mathrm{d}}{\mathrm{d}z} + \frac{m\omega}{\hbar}\, z \right) \phi_0(z) \qquad (3.3.31)$$

The solution to the above first-order differential equation is

$$\phi_0(z) = \left(\frac{m\omega}{\pi\hbar} \right)^{1/4} e^{-m\omega z^2 /(2\hbar)} \qquad (3.3.32)$$

which has been normalized. From $\phi_0(z) = \langle z | 0 \rangle$ in (3.3.32), all the other eigenfunctions can be created sequentially using the creation operator in (3.3.28).

3.4   HYDROGEN ATOM (3D AND 2D EXCITON BOUND AND CONTINUUM STATES) [4–8]

In this section, we summarize the major results of the energies and wave functions for the hydrogen atom model with both bound $(E < 0)$ and continuum $(E > 0)$ state solutions. The hydrogen atom is a two-particle system for the positive nucleus (with a mass m_1) at a position $\mathbf{r}_1$ and an electron (with mass m_2) at a position $\mathbf{r}_2$. A general solution is to transform from $\mathbf{r}_1$ and $\mathbf{r}_2$ coordinates to the center-of-mass coordinates $\mathbf{R}$ and the difference coordinates $\mathbf{r}$:

$$\mathbf{R} = \frac{m_1\mathbf{r}_1 + m_2\mathbf{r}_2}{m_1 + m_2} \tag{3.4.1}$$

and

$$\mathbf{r} = \mathbf{r}_1 - \mathbf{r}_2 \tag{3.4.2}$$

The complete solution is of the form

$$\psi(\mathbf{r}_1,\mathbf{r}_2) = \frac{e^{i\mathbf{K}\cdot\mathbf{R}}}{\sqrt{V}}\psi(\mathbf{r}) \tag{3.4.3}$$

where V is the volume of the space and $\psi(\mathbf{r})$ satisfies

$$\left(-\frac{\hbar^2}{2m_r}\nabla_r^2 - \frac{e^2}{4\pi\varepsilon r}\right)\psi(\mathbf{r}) = E\psi(\mathbf{r}) \tag{3.4.4}$$

where m_r is the reduced mass, $1/m_r = 1/m_1 + 1/m_2$.

3.4.1   3D Solutions

In three-dimensional space the eigenfunctions can be expressed as

$$\psi(\mathbf{r}) = \begin{cases} R_{nl}(r)Y_{lm}(\theta,\varphi) & \text{bound states } (E < 0) \quad (3.4.5a) \\ R_{El}(r)Y_{lm}(\theta,\varphi) & \text{continuum states } (E > 0) \quad (3.4.5b) \end{cases}$$

where the radial functions $R_{nl}(r)$ and $R_{El}(r)$ are shown in Table 3.1. The spherical harmonics $Y_{lm}(\theta,\varphi)$ are tabulated in Appendix A. For bound-state solutions, the energy levels are quantized as

$$E_n = -\frac{1}{n^2}R_y \qquad (n = 1,2,3,\dots) \tag{3.4.6}$$

Table 3.1. The Discrete and Continuum States for 2D and 3D Hydrogen Atom (Exciton)

Three-Dimensional (3D) Case

Bound states: $\psi_{nlm}(\mathbf{r}) = R_{nl}(r)Y_{lm}(\theta,\varphi)$ $\quad \int_0^\infty R_{nl}(r)R_{nl}(r)r^2\,dr = 1$ $\quad a_0 = \dfrac{4\pi\varepsilon\hbar^2}{e^2 m_r}$ $\quad R_y = \dfrac{m_r e^4}{2(4\pi\varepsilon)^2\hbar^2}$

$$R_{nl}(r) = \left(\frac{2}{na_0}\right)^{3/2}\left[\frac{(n+l)!}{2n(n-l-1)!}\right]^{1/2}\frac{e^{-r/na_0}}{(2l+1)!}\left(\frac{2r}{na_0}\right)^l F\left[l+1-n,\,2l+2;\,\frac{2r}{na_0}\right]$$

$$(n = 1, 2, 3, \ldots, \text{ and } l \le n-1, |m| \le l)$$

$$E_n = -\frac{R_y}{n^2} \qquad 2|\psi_{n00}(\mathbf{r}=0)|^2 = \frac{2}{\pi a_0^3 n^3}$$

Continuum states: $\psi_{Elm}(\mathbf{r}) = R_{El}(r)Y_{lm}(\theta,\varphi)$ $\qquad \int_0^\infty R_{El}^*(r)R_{El}(r)r^2\,dr = \delta(E-E')$

$$R_{El}(r) = \frac{1}{(R_y a_0^3)^{1/2}}\left\{\left[\prod_{s=1}^l s^2+\frac{1}{(ka_0)^2}\right]\frac{e^{\pi/(ka_0)}}{\sinh(\pi/ka_0)}\right\}^{1/2}\frac{e^{-ikr}}{(2l+1)!}(2kr)^l F\left[l+1+\frac{i}{ka_0},\,2l+2,\,2ikr\right]$$

$$E = \frac{\hbar^2 k^2}{2m_r} \qquad 2|\psi_{E00}(\mathbf{r}=0)|^2 = \frac{1}{R_y a_0^3 2\pi}\left[\frac{e^{\pi/(ka_0)}}{\sinh(\pi/ka_0)}\right]$$

Table 3.1. (*Continued*)

Two-Dimensional (2D) Case

Bound States: $\psi_{nm}(\mathbf{r}) = R_{nm}(r)\dfrac{e^{im\varphi}}{\sqrt{2\pi}}$ $\displaystyle\int_0^\infty R_{nm}(r)R_{nm}(r)r\,dr = 1$ $a_0 = \dfrac{4\pi\varepsilon\hbar^2}{e^2 m_r}$ $R_y = \dfrac{m_r e^4}{2(4\pi\varepsilon)^2\hbar^2}$

$$R_{nm}(r) = \frac{4}{a_0}\left[\frac{(n+|m|-1)!}{(2n-1)^3(n-|m|-1)!}\right]^{1/2}\frac{e^{-r/[(n-1/2)a_0]}}{(2|m|)!}\left[\frac{2r}{\left(n-\frac{1}{2}\right)a_0}\right]^{|m|}F\left[|m|+1-n,\,2|m|+1;\,\frac{2r}{\left(n-\frac{1}{2}\right)a_0}\right]$$

$$(E_n) = -\frac{R_y}{\left(n-\frac{1}{2}\right)^2}\qquad 2|\psi_{n0}(\mathbf{r}=0)|^2 = \frac{2}{\pi a_0^2\left(n-\frac{1}{2}\right)^3}\qquad (n = 1,2,3,\ldots,\text{ and } |m| \le n-1)$$

Continuum States: $\psi_{Em}(\mathbf{r}) = R_{Em}(r)\dfrac{e^{im\varphi}}{\sqrt{2\pi}}$ $\displaystyle\int_0^\infty R^*_{Em}(r)R_{Em}(r)r\,dr = \delta(E-E')$

$$R_{Em}(r) = \frac{1}{a_0\sqrt{2R_y}}\left\{\prod_{s=1}^{|m|}\left[\left(s-\frac{1}{2}\right)^2 + \frac{1}{(ka_0)^2}\right]\frac{e^{\pi/(ka_0)}}{\cosh(\pi/ka_0)}\right\}^{1/2}\frac{e^{-ikr}}{(2|m|)!}(2kr)^{|m|}F\left[|m|+\frac{1}{2}+\frac{i}{(ka_0)},\,2|m|+1,\,2ikr\right]$$

$$E = \frac{\hbar^2 k^2}{2m_r}\qquad 2|\Psi_{E0}(\mathbf{r}=0)|^2 = \frac{1}{R_y a_0^2 2\pi}\left[\frac{e^{\pi/(ka_0)}}{\cosh(\pi/ka_0)}\right]$$

and the Rydberg energy for the hydrogen atom is

$$R_y = \frac{m_r e^4}{2(4\pi\varepsilon)^2\hbar^2} = \frac{\hbar^2}{2m_r} \frac{1}{a_0^2} \tag{3.4.7}$$

and the Bohr radius a_0 is

$$a_0 = \frac{4\pi\varepsilon\hbar^2}{e^2 m_r} \tag{3.4.8}$$

The wave function at the origin is

$$|\psi_{n00}(\mathbf{r} = 0)|^2 = \frac{1}{\pi a_0^3 n^3} \tag{3.4.9}$$

For continuum state solutions, the energy E is a continuous variable and the wave function at the origin is given by

$$|\psi_{E00}(\mathbf{r} = 0)|^2 = \frac{1}{R_y a_0^3 4\pi} \left[\frac{e^{\pi/(ka_0)}}{\sinh(\pi/ka_0)} \right] \tag{3.4.10}$$

where $E = \hbar^2 k^2/2m_r$. The expression in the square bracket of (3.4.10) is called the Sommerfeld enhancement factor for a 3D hydrogen atom.

3.4.2 2D Solutions

The solutions for the two-dimensional hydrogen atom problem are given by

$$\psi(\mathbf{r}) = \begin{cases} R_{nm}(r)\dfrac{e^{im\varphi}}{\sqrt{2\pi}} & \text{bound states } (E < 0) \tag{3.4.11a} \\[2em] R_{Em}(r)\dfrac{e^{im\varphi}}{\sqrt{2\pi}} & \text{continuum states } (E > 0) \tag{3.4.11b} \end{cases}$$

The eigenenergies for the bound states are quantized

$$E_n = -\frac{1}{(n - 1/2)^2} R_y \quad (n = 1, 2, 3, \dots) \tag{3.4.12}$$

We have

$$E_1 = -4R_y, \quad E_2 = -\frac{4}{9}R_y, \quad E_3 = -\frac{4}{25}R_y, \dots \tag{3.4.13}$$

It is noted that the binding energy for the 1s states $|E_1|$ is four times of that in the three-dimensional case. This enhancement of the binding energy is very useful in understanding the excitonic effects in semiconductor quantum wells and the observation of the excitonic optical absorption spectra, which we investigate in Chapter 13. The wave function at the origin is

$$|\psi_{n0}(\mathbf{r} = 0)|^2 = \frac{1}{\pi a_0^2 (n - 1/2)^3} \qquad (n = 1, 2, 3, \dots) \quad (3.4.14)$$

For continuum states, the energy E is a continuous variable and the wave function at the origin is

$$|\psi_{E0}(\mathbf{r} = 0)|^2 = \frac{1}{R_y a_0^3 4\pi} \left[\frac{e^{\pi/(ka_0)}}{\cosh(\pi/ka_0)} \right] \qquad (3.4.15)$$

where $E = \hbar^2 k^2 / 2m_r$. The expression inside the square bracket is the 2D Sommerfeld enhancement factor. The 2D and 3D solutions for excitons are summarized in Table 3.1.

3.5 TIME-INDEPENDENT PERTURBATION THEORY [10]

3.5.1 Perturbation Method

In most practical physical systems, the Schrödinger equations do not have exact or analytical solutions. It is always convenient to find the solutions using the perturbation method when the problem of interest can be separated into two parts: One part consists of an "unperturbed" Hamiltonian with known solutions, $H^{(0)}$, and the other part consists of a small perturbing potential, H':

$$H = H^{(0)} + H' \qquad (3.5.1)$$

The unperturbed wave functions $\phi_n^{(0)}$ and eigenvalues $E_n^{(0)}$ are assumed to be known:

$$H^{(0)}\phi_n^{(0)} = E_n^{(0)}\phi_n^{(0)} \qquad (3.5.2)$$

To find the solutions for the total Hamiltonian

$$H\psi = E\psi \qquad (3.5.3)$$

it is convenient to introduce a perturbation parameter λ (set $\lambda = 1$ later):

$$H = H^{(0)} + \lambda H' \qquad (3.5.4)$$

We look for the solutions of the form

$$E = E^{(0)} + \lambda E^{(1)} + \lambda^2 E^{(2)} + \cdots \tag{3.5.5a}$$

$$\psi = \psi^{(0)} + \lambda \psi^{(1)} + \lambda^2 \psi^{(2)} + \cdots \tag{3.5.5b}$$

Substituting the above expressions for H, E, and ψ into the Schrödinger equation, we find, to each order in λ,

Zeroth order $\quad H^{(0)}\psi^{(0)} = E^{(0)}\psi^{(0)}$ $\qquad$ (3.5.6)

First order $\quad H^{(0)}\psi^{(1)} + H'\psi^{(0)} = E^{(0)}\psi^{(1)} + E^{(1)}\psi^{(0)}$ $\qquad$ (3.5.7)

Second order $\quad H^{(0)}\psi^{(2)} + H'\psi^{(1)} = E^{(0)}\psi^{(2)} + E^{(1)}\psi^{(1)} + E^{(2)}\psi^{(0)}$ $\quad$ (3.5.8)

Zeroth-Order Solutions. It is clearly seen that the zeroth-order solutions are the unperturbed solutions:

$$\psi_n^{(0)} = \phi_n^{(0)} \tag{3.5.9a}$$

$$E_n^{(0)} = E_n^{(0)} \tag{3.5.9b}$$

First-Order Solutions. The first-order wave function $\psi^{(1)}$ may be expanded in terms of a linear combination of the unperturbed solutions:

$$\psi_n^{(1)} = \sum_m a_{mn}^{(1)}\phi_m^{(0)} \tag{3.5.10}$$

Thus

$$\left(H^{(0)} - E_n^{(0)}\right)\psi_n^{(1)} = E^{(1)}\phi_n^{(0)} - H'\phi_n^{(0)} \tag{3.5.11}$$

Multiplying the above equation by $\phi_m^{(0)*}$ and integrating over space, using

$$\langle \phi_m^{(0)} | \phi_n^{(0)} \rangle = \delta_{mn} \tag{3.5.12}$$

we obtain

$$E_n^{(1)} = H'_{nn} \tag{3.5.13a}$$

and

$$a_{mn}^{(1)} = \frac{H'_{mn}}{E_n^{(0)} - E_m^{(0)}} \qquad \text{if } m \neq n \tag{3.5.13b}$$

where

$$H'_{mn} = \int \phi_m^{(0)*} H' \phi_n^{(0)}\, d^3\mathbf{r} \tag{3.5.14}$$

To the first order in perturbation, we need to normalize the wave function

$$\int \left(\phi_n^{(0)} + \sum_m a_{mn}^{(1)} \phi_m^{(0)} \right)^* \left(\phi_n^{(0)} + \sum_m a_{mn}^{(1)} \phi_m^{(0)} \right) d^3\mathbf{r} = 1$$

or

$$1 + a_{nn}^{(1)*} + a_{nn}^{(1)} + \sum_m a_{mn}^{(1)} a_{mn}^{*(1)} = 1 \qquad (3.5.15)$$

Thus we have $a_{nn}^{(1)} = 0$ to the first order in perturbation. The result is therefore

$$\psi_n = \phi_n^{(0)} + \sum_{m \neq n} \frac{H'_{mn}}{E_n^{(0)} - E_m^{(0)}} \phi_m^{(0)} \qquad (3.5.16a)$$

$$E_n = E_n^{(0)} + H'_{nn} \qquad (3.5.16b)$$

Second-Order Solutions. Similarly, the second-order wave function $\psi^{(2)}$ can be expanded in terms of the zero-order solutions:

$$\psi_n^{(2)} = \sum_m a_{mn}^{(2)} \phi_m^{(0)} \qquad (3.5.17)$$

We find

$$E_n^{(2)} = \sum_{m \neq n} a_{mn}^{(1)} H'_{nm}$$

$$= \sum_{m \neq n} \frac{H'_{mn} H'_{nm}}{E_n^{(0)} - E_m^{(0)}} \qquad (3.5.18a)$$

and

$$a_{mn}^{(2)} = \sum_{k \neq n} \frac{H'_{mk} H'_{kn}}{\left(E_n^{(0)} - E_m^{(0)} \right)\left(E_n^{(0)} - E_k^{(0)} \right)} - \frac{H'_{mn} H'_{nn}}{\left(E_n^{(0)} - E_m^{(0)} \right)^2}, \qquad m \neq n$$

$$(3.5.18b)$$

To the second-order correction, the wave function may be normalized:

$$\int \left(\phi_n^{(0)} + \sum_{m \neq n} a_{mn}^{(1)} \phi_m^{(0)} + \sum_m a_{mn}^{(2)} \phi_m^{(0)} \right)^*$$

$$\times \left(\phi_n^{(0)} + \sum_{m \neq n} a_{mn}^{(1)} \phi_m^{(0)} + \sum_m a_{mn}^{(2)} \phi_m^{(0)} \right) d^3\mathbf{r} = 1 \qquad (3.5.19)$$

Neglecting terms of third and higher orders, we obtain

$$a_{nn}^{(2)} = -\frac{1}{2} \sum_{m \neq n} |a_{mn}^{(1)}|^2 \tag{3.5.20}$$

The normalized wave function ψ_n and its eigenenergy E_n, to the second order in perturbation, are given by

$$\psi_n = \phi_n^{(0)} + \sum_{m \neq n} \frac{H'_{mn}}{E_n^{(0)} - E_m^{(0)}} \phi_m^{(0)}$$

$$+ \sum_{m \neq n} \left\{ \left[\sum_{k \neq n} \frac{H'_{mk} H'_{kn}}{\left(E_n^{(0)} - E_m^{(0)}\right)\left(E_n^{(0)} - E_k^{(0)}\right)} - \frac{H'_{mn} H'_{nn}}{\left(E_n^{(0)} - E_m^{(0)}\right)^2} \right] \phi_m^{(0)} \right.$$

$$\left. - \frac{|H'_{mn}|^2}{2\left(E_n^{(0)} - E_m^{(0)}\right)^2} \phi_n^{(0)} \right\} \tag{3.5.21a}$$

$$E_n = E_n^{(0)} + H'_{nn} + \sum_{m \neq n} \frac{|H'_{nm}|^2}{E_n^{(0)} - E_m^{(0)}} \tag{3.5.21b}$$

Example

a. When an infinite quantum well has an applied electric field F in the z direction (see Fig. 3.7), the Hamiltonian can be written as

$$H = H_0 + eFz \tag{3.5.22}$$

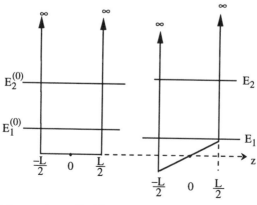

Figure 3.7. An infinite quantum well without an applied field (left figure) and with an applied field (right figure) where the potential due to the applied field eFz is treated as a perturbation.

where H_0 describes an electron in the infinite quantum well *without* the electric field. Treating the term eFz as a perturbation, show that the energy shift due to the applied electric field, $E - E_n^{(0)}$, can be written as

$$\Delta E = E - E_n^{(0)} = C_n \frac{\pi^2 (eFL)^2}{2E_1^{(0)}} \left(= C_n \frac{m^* e^2 F^2 L^4}{\hbar^2} \right) \quad (3.5.23)$$

where $E_1^{(0)}$ and $E_n^{(0)}$ are the energies of the ground state and the nth state of the infinite quantum well without an applied electric field. Find a general expression for C_n. All energies are measured from the center of the well.

b. Evaluate C_n for $n = 1$, 2, and 3 numerically by keeping the first two nonzero terms in the summation over the index m. Show that C_1 is negative while C_2 and C_3 are positive.

c. Compare numerically C_n, $n = 1, 2, 3$, with those in part (b), where C_n is given by

$$C_n = \frac{n^2 \pi^2 - 15}{24 n^4 \pi^4} \quad (3.5.24)$$

SOLUTION

a. We treat $H' = eFz$ as a perturbation from H_0. We write down the unperturbed wave functions and eigenvalues from the infinite barrier model.

Zeroth-order solutions

$$H_0 u_n^{(0)}(z) = \frac{-\hbar^2}{2m^*} \frac{d^2}{dz^2} u_n^{(0)}(z) = E_n^{(0)} u_n^{(0)}(z)$$

$$u_n^{(0)}(z) = \sqrt{\frac{2}{L}} \sin\left[\frac{n\pi}{L} \left(z + \frac{L}{2} \right) \right] \qquad E_n^{(0)} = n^2 \frac{\hbar^2 \pi^2}{2m^* L^2} \quad (3.5.25)$$

First-order perturbation

$$H'_{nn} = \langle u_n^{(0)} | eFz | u_n^{(0)} \rangle = eF \int_{-L/2}^{L/2} u_n^{(0)}(z) z u_n^{(0)}(z) \, dz = 0 \quad (3.5.26)$$

Therefore, the first-order correction in the energy vanishes because of the symmetry of the original quantum-well potential with well-defined even and odd parities of the wave functions.

Second-order perturbation
The energy to second-order perturbation is given by (3.5.21b). We need to evaluate H'_{nm}, $n \neq m$

$$H'_{nm} = \langle u_n^{(0)}|eFz|u_m^{(0)}\rangle = \langle u_n^{(0)}|eF\left(z + \frac{L}{2}\right)|u_m^{(0)}\rangle - eF\frac{L}{2}\langle u_n^{(0)}|u_m^{(0)}\rangle$$

$$= 2eFL\int_0^1 t\,\sin(n\pi t)\sin(m\pi t)\,dt$$

where we have changed the variable from z to t, $t = (z + L/2)/L$. Write

$$2\sin(n\pi t)\sin(m\pi t) = \cos[(n - m)\pi t] - \cos[(n + m)\pi t]$$

and use

$$\int_0^1 t\cos(M\pi t)\,dt = \begin{cases} \dfrac{(-1)^M - 1}{(M\pi)^2} & \text{for } M \neq 0 \\[2mm] \dfrac{1}{2} & \text{for } M = 0 \end{cases}$$

We obtain

$$H'_{nm} = \frac{eFL}{\pi^2}\left[\frac{(-1)^{n-m} - 1}{(n - m)^2} - \frac{(-1)^{n+m} - 1}{(n + m)^2}\right]$$

$$= \frac{eFL}{\pi^2}\left[(-1)^{n-m} - 1\right]\frac{4nm}{(n^2 - m^2)^2} \qquad (n \neq m) \quad (3.5.27)$$

The perturbation to second order gives

$$E = E_n^{(0)} + H'_{nn} + \sum_{m \neq n}\frac{|H'_{nm}|^2}{E_n^{(0)} - E_m^{(0)}}$$

$$= E_n^{(0)} + C_n\frac{\pi^2}{2}\frac{(eFL)^2}{E_1^{(0)}} \tag{3.5.28}$$

where we use $E_n^{(0)} - E_m^{(0)} = (n^2 - m^2)E_1^{(0)}$ and

$$C_n = \frac{32n^2}{\pi^6}\sum_{m \neq n}\frac{\left[(-1)^{n-m} - 1\right]^2 m^2}{(n^2 - m^2)^5} \tag{3.5.29}$$

The above results show that the energy shifts depend quadratically on the applied electric field, which is an important phenomenon when we study the quantum-confined Stark effects in Chapter 13.

b. If we only keep the first two nonvanishing terms in the above summation for C_n, we obtain

$$C_1 = \frac{32}{\pi^6} \times \left[\frac{4 \times 2^2}{\left(1 - 2^2\right)^5} + \frac{4 \times 4^2}{\left(1^2 - 4^2\right)^5} \right] = -2.194 \times 10^{-3}$$

$$C_2 = \frac{32 \times 2^2}{\pi^6} \times \left[\frac{4 \times 1^2}{\left(4 - 1\right)^5} + \frac{4 \times 3^2}{\left(4 - 9\right)^5} \right] = 6.578 \times 10^{-4}$$

$$C_3 = \frac{32 \times 3^2}{\pi^6} \times \left[\frac{4 \times 2^2}{\left(9 - 4\right)^5} + \frac{4 \times 4^2}{\left(9 - 16\right)^5} \right] = 3.931 \times 10^{-4}$$

We see that E_1 decreases with increasing field F since $C_1 < 0$ and both E_2 and E_3 increase slightly with increasing field F.

c. The above C_n in part (a) can be summed up to an analytical expression [11–13]:

$$C_n = \frac{n^2\pi^2 - 15}{24n^4\pi^5} \quad \text{or} \quad C_1 = -2.194 \times 10^{-3}$$

$$C_2 = 6.544 \times 10^{-4}$$

$$C_3 = 3.899 \times 10^{-4}$$

which are close to those in (b). These results using the second-order perturbation theory agree very well with those obtained from variational methods [13–16]. ∎

3.5.2 Matrix Formulation

Alternatively, the eigenvalue problem

$$H\psi = \left(H^{(0)} + H'\right)\psi = E\psi \tag{3.5.30}$$

can be solved directly by letting

$$\psi = \sum_m a_m \phi_m^{(0)} \tag{3.5.31}$$

where $\{\phi_m^{(0)}\}$ are the eigenfunctions of the unperturbed Hamiltonian,

$$H_0 \phi_m^{(0)} = E_m^{(0)} \phi_m^{(0)} \tag{3.5.32}$$

The second subscript n in a_{mn} is dropped for convenience. Here $\phi_m^{(0)}$, $m = 1, 2, \ldots, N$, may also be degenerate wave functions. Substituting (3.5.31) directly into (3.5.30) and taking the inner product with respect to $\phi_k^{(0)}$, $k = 1, \ldots, N$, gives

$$\sum_m (H_{km} - E\delta_{km}) a_m = 0 \qquad (3.5.33)$$

if $\{\phi_m^{(0)}\}$ form an orthonormal set. If $\{\phi_m^{(0)}, m = 1, 2, \ldots, N\}$ are not orthonormal to each other,

$$\langle \phi_k^{(0)} | \phi_m^{(0)} \rangle = N_{km} \qquad (3.5.34)$$

then, the above eigenequation becomes

$$\sum_m (H_{km} - EN_{km}) a_m = 0 \qquad (3.5.35)$$

The eigenequations (3.5.33) or (3.5.35) can be solved by letting

$$\det|H_{km} - E\delta_{km}| = 0 \qquad \text{for (3.5.33)} \qquad (3.5.36)$$

or

$$\det|H_{km} - EN_{km}| = 0 \qquad \text{for (3.5.35)} \qquad (3.5.37)$$

The above procedure is equivalent to diagonalizing the matrix H_{km} in (3.5.33) or simultaneously diagonalizing H_{km} and N_{km} in (3.5.35) (noting that N is also Hermitian). Finally, one finds the eigenvalues for E and the eigenvector a_m. The wave function is obtained from (3.5.31), which should be normalized.

3.5.3 Variational Principle

The above equations can also be derived from the variational principle by minimizing the functional for the energy

$$E\{\psi\} = \frac{\int \psi^* H \psi \, d^3\mathbf{r}}{\int \psi^* \psi \, d^3\mathbf{r}} = \frac{\sum_{m,n} a_m^* H_{mn} a_n}{\sum_{m,n} a_m^* N_{mn} a_n} \qquad (3.5.38)$$

assuming

$$\psi = \sum_{n=1}^N a_n \phi_n^{(0)} \qquad (3.5.39)$$

for the eigenfunction. Using

$$\delta E = \frac{\delta A - E \delta B}{B} \tag{3.5.40}$$

if $E = A/B$, we obtain

$$\sum_{n=1}^{N} (H_{mn} - E N_{mn}) a_n = 0 \tag{3.5.41}$$

When the basis functions $\{\phi_m^{(0)}\}$ are orthonormal to each other, the above equation reduces to

$$\sum_{n=1}^{N} (H_{mn} - E \delta_{mn}) a_n = 0 \tag{3.5.42}$$

where

$$H_{mn} = \int \phi_m^{(0)*} H \phi_n^{(0)} \, \mathrm{d}^3 \mathbf{r} = E_n^{(0)} \delta_{mn} + H'_{mn} \tag{3.5.43}$$

Equation (3.5.42) is identical to (3.5.33).

3.6 LÖWDIN'S RENORMALIZATION METHOD [9]

An interesting technique in perturbation theory is Löwdin's perturbation method. It may be useful to divide the eigenfunctions and energies into two classes, A and B. Suppose we are mainly interested in the states in class A, and we look for a formula for class A with the influence from the states in B as a perturbation. We start with eigenequation (3.5.42), assuming that the unperturbed states are orthonormalized. Equation (3.5.42) may be rewritten as

$$(E - H_{mm}) a_m = \sum_{n \neq m}^{A} H_{mn} a_n + \sum_{\alpha \neq m}^{B} H_{m\alpha} a_\alpha \tag{3.6.1}$$

or

$$a_m = \sum_{n \neq m}^{A} \frac{H_{mn}}{E - H_{mm}} a_n + \sum_{\alpha \neq m}^{B} \frac{H_{m\alpha}}{E - H_{mm}} a_\alpha \tag{3.6.2}$$

where the first sum on the right-hand side is over the states in class A only, while the second sum is over the states in class B. Since we are interested in the coefficients a_m for m in class A, we may eliminate those in class B by an iteration procedure and obtain

$$a_m = \sum_n^A \frac{U_{mn}^A - H_{mn}\delta_{mn}}{E - H_{mm}} a_n \tag{3.6.3}$$

and

$$U_{mn}^A = H_{mn} + \sum_{\alpha \neq m}^B \frac{H_{m\alpha}H_{\alpha n}}{E - H_{\alpha\alpha}} + \sum_{\substack{\alpha,\beta \neq m,n \\ \alpha \neq \beta}}^B \frac{H_{m\alpha}H_{\alpha\beta}H_{\beta n}}{(E - H_{\alpha\alpha})(E - H_{\beta\beta})} + \cdots \tag{3.6.4}$$

Or, equivalently, we solve the eigenvalue problems for a_n, $(n \in A)$:

$$\sum_n^A (U_{mn}^A - E\delta_{mn})a_n = 0 \qquad m \in A \tag{3.6.5}$$

and

$$a_\gamma = \sum_n^A \frac{U_{\gamma n}^A - H_{\gamma n}\delta_{\gamma n}}{E - H_{\gamma\gamma}} a_n \qquad \gamma \in B \tag{3.6.6}$$

When the coefficients a_n belonging to class A are determined from the eigenequation (3.6.5), the coefficients a_γ in class B can be found from (3.6.6). A necessary condition for the expansion of (3.6.4) to be convergent is

$$|H_{m\alpha}| \ll |E - H_{\alpha\alpha}| \qquad m \in A, \alpha \in B \tag{3.6.7}$$

In practice, to the second order in perturbation, we may replace E by E_A in (3.6.4), where E_A is an average energy of states in class A, and truncate the series (3.6.4) at the second-order term. For example, if class A consists of only a single nondegenerate state n, then class B consists of the rest. Equation (3.6.5) gives only one equation:

$$E = U_{nn}^A$$

$$= H_{nn} + \sum_{\alpha \neq n} \frac{H_{n\alpha}H_{\alpha n}}{E - H_{\alpha\alpha}} + \sum_{\substack{\alpha,\beta \neq n \\ \alpha \neq \beta}} \frac{H_{n\alpha}H_{\alpha\beta}H_{\beta n}}{(E - H_{\alpha\alpha})(E - H_{\beta\beta})} + \cdots \tag{3.6.8}$$

If we separate H into $H^{(0)}$ and a perturbation H'

$$H = H^{(0)} + H' \tag{3.6.9}$$

we obtain, to the second order in H',

$$E = E_n^{(0)} + H'_{nn} + \sum_{\alpha \neq n} \frac{H'_{n\alpha} H'_{\alpha n}}{E_n^{(0)} - E_\alpha^{(0)}} \tag{3.6.10}$$

where $E_n^{(0)} = H_{nn}^{(0)}$ has been used.

If, however, the states in class A are degenerate, the diagonal elements are exactly or almost the same:

$$H_m \simeq E_A \tag{3.6.11}$$

with differences of the first and higher orders. We have

$$U_{mn}^A = H_{mn} + \sum_\alpha^B \frac{H'_{m\alpha} H'_{\alpha n}}{E_A - H_{\alpha\alpha}} \tag{3.6.12}$$

Solving the eigenequation using U_{mn}^A above by letting

$$\det|U_{mn}^A - E\delta_{mn}| = 0 \tag{3.6.13}$$

the eigenvalues for E and eigenvectors for a_n are thus obtained, and the wave functions are given by

$$\psi = \sum_n a_n \phi_n^{(0)} \tag{3.6.14}$$

where a_n for $n \in$ class B are obtained from (3.6.6). The above wave function can be normalized directly by dividing the above expression by $\langle \sum_n a_n \phi_n^{(0)} | \sum_n a_n \phi_n^{(0)} \rangle$, or

$$\psi = \frac{\sum_n a_n \phi_n^{(0)}}{\sum_n |a_n|^2} \tag{3.6.15}$$

3.7 TIME-DEPENDENT PERTURBATION THEORY [10]

Consider the Schrödinger equation

$$H\psi(\mathbf{r}, t) = -\frac{\hbar}{i} \frac{\partial}{\partial t} \psi(\mathbf{r}, t) \tag{3.7.1}$$

where the Hamiltonian H consists of an unperturbed part H_0, which is time-independent, and a small perturbation $H'(\mathbf{r}, t)$, which depends on time:

$$H = H_0 + H'(\mathbf{r}, t) \tag{3.7.2}$$

The solution to the unperturbed part is assumed known:

$$H_0\phi_n(\mathbf{r}, t) = -\frac{\hbar}{i}\frac{\partial}{\partial t}\phi_n(\mathbf{r}, t) \tag{3.7.3}$$

$$\phi_n(\mathbf{r}, t) = \phi_n(\mathbf{r})e^{-iE_n t/\hbar} \tag{3.7.4}$$

The time-dependent perturbation is assumed to have the form

$$H'(\mathbf{r}, t) = \begin{cases} H'(\mathbf{r})e^{-i\omega t} + H'^{+}(\mathbf{r})e^{+i\omega t} & t \geq 0 \\ 0 & t < 0 \end{cases} \tag{3.7.5}$$

To find the solution $\psi(\mathbf{r}, t)$ to the time-dependent Schrödinger equation, we expand the wave function in terms of the unperturbed solutions:

$$\psi(\mathbf{r}, t) = \sum_n a_n(t)\phi_n(\mathbf{r})\, e^{-iE_n t/\hbar} \tag{3.7.6}$$

where $|a_n(t)|^2$ gives the probability that the electron is in the state n at time t. Substituting the expansion for ψ into the Schrödinger equation and using (3.7.3), we obtain

$$\sum_n \frac{da_n(t)}{dt}\phi_n(\mathbf{r})e^{-iE_n t/\hbar} = -\frac{i}{\hbar}\sum_n H'(\mathbf{r}, t)a_n(t)\phi_n(\mathbf{r})\, e^{-iE_n t/\hbar} \tag{3.7.7}$$

Taking the inner product with the wave function $\phi_m^*(\mathbf{r})$, and using the orthonormal property $\int d^3r\,\phi_m^*(\mathbf{r})\phi_n(\mathbf{r}) = \delta_{mn}$, we find

$$\frac{da_m(t)}{dt} = -\frac{i}{\hbar}\sum_n a_n(t)H'_{mn}(t)\, e^{i\omega_{mn}t} \tag{3.7.8}$$

where

$$H'_{mn}(t) = \langle m|H'(\mathbf{r}, t)|n\rangle$$

$$= \int \phi_m^*(\mathbf{r})H'(\mathbf{r}, t)\phi_n(\mathbf{r})\, d^3r$$

$$= H'_{mn}\, e^{-i\omega t} + H'^{+}_{mn}\, e^{+i\omega t} \tag{3.7.9}$$

$$\omega_{mn} = \frac{E_m - E_n}{\hbar} \tag{3.7.10}$$

and the matrix elements are defined as

$$H'_{mn} = \int \phi_m^*(\mathbf{r})H'(\mathbf{r})\phi_n(\mathbf{r})\, d^3r \tag{3.7.11}$$

Introducing the perturbation parameter λ (set $\lambda = 1$ later)

$$H = H_0 + \lambda H'(\mathbf{r}, t) \tag{3.7.12}$$

and letting

$$a_n(t) = a_n^{(0)}(t) + \lambda a_n^{(1)}(t) + \lambda^2 a_n^{(2)}(t) + \cdots \tag{3.7.13}$$

we obtain

$$\frac{d a_m^{(0)}}{dt} = 0$$

$$\frac{d}{dt} a_m^{(1)}(t) = -\frac{i}{\hbar} \sum_n a_n^{(0)}(t) H'_{mn}(t) \, e^{i\omega_{mn}t}$$

$$\frac{d}{dt} a_m^{(2)}(t) = -\frac{i}{\hbar} \sum_n a_n^{(1)}(t) H'_{mn}(t) \, e^{i\omega_{mn}t} \tag{3.7.14}$$

Thus, the zeroth-order solutions are constant. Let the electron be at state i, initially:

$$a_i^{(0)}(t = 0) = 1$$

$$a_m^{(0)} = 0 \qquad m \neq i \tag{3.7.15}$$

We have the zeroth-order solution:

$$a_i^{(0)}(t) = 1$$

$$a_m^{(0)}(t) = 0 \qquad m \neq i \tag{3.7.16}$$

Therefore, the electron stays at state i in the absence of any perturbation. The first-order solution is obtained from

$$\frac{d}{dt} a_m^{(1)}(t) = \frac{-i}{\hbar} H'_{mi}(t) \, e^{i\omega_{mi}t}$$

$$= \frac{-i}{\hbar} \left(H'_{mi} \, e^{i(\omega_{mi} - \omega)t} + H'^{+}_{mi} \, e^{i(\omega_{mi} + \omega)t} \right) \tag{3.7.17}$$

Suppose we are interested in a final state $m = f$; the above equation can be solved directly by integration:

$$a_f^{(1)}(t) = \frac{-1}{\hbar} \left(H'_{fi} \frac{e^{i(\omega_{fi} - \omega)t} - 1}{\omega_{fi} - \omega} + H'^{+}_{fi} \frac{e^{i(\omega_{fi} + \omega)t} - 1}{\omega_{fi} + \omega} \right) \tag{3.7.18}$$

If we consider the photon energy to be near resonance, either $\omega \sim \omega_{fi}$ or $\omega \sim -\omega_{fi}$, we find

$$|a_f^{(1)}(t)|^2 \simeq \frac{4|H_{fi}'|^2}{\hbar^2} \frac{\sin^2[(\omega_{fi} - \omega)t/2]}{(\omega_{fi} - \omega)^2} + \frac{4|H_{fi}'|^2}{\hbar^2} \frac{\sin^2[(\omega_{fi} + \omega)t/2]}{(\omega_{fi} + \omega)^2}$$

(3.7.19)

where the cross term has been dropped since it is small compared with either of the above two terms. When the interaction time is long enough, using

$$\frac{\sin^2(tx/2)}{x^2} \rightarrow \frac{\pi t}{2} \delta(x)$$

(3.7.20)

we find

$$|a_f^{(1)}(t)|^2 \simeq \frac{2\pi t}{\hbar^2}|H_{fi}'|^2\delta(\omega_{fi} - \omega) + \frac{2\pi t}{\hbar^2}|H_{fI}'|^2 \delta(\omega_{fi} + \omega)$$

(3.7.21)

The transition rate is given by

$$W_{i \to f} = \frac{d}{dt}|a_f^{(1)}(t)|^2$$

$$= \frac{2\pi}{\hbar}|H_{fi}'|^2\delta(E_f - E_i - \hbar\omega) + \frac{2\pi}{\hbar}|H_{fi}'|^2 \delta(E_f - E_i + \hbar\omega)$$

(3.7.22)

where the property $\delta(\hbar\omega) = \delta(\omega)/\hbar$ has been used. The first term corresponds to the absorption of a photon by an electron, since $E_f = E_i + \hbar\omega$, while the second corresponds to the emission of a photon, since $E_f = E_i - \hbar\omega$. These processes are illustrated in Fig. 3.8 for a two-level system.

In summary, we have Fermi's golden rule: For a time-harmonic perturbation, turned on at $t = 0$,

$$H'(\mathbf{r}, t) = H'(\mathbf{r})e^{-i\omega t} + H'^+(\mathbf{r})e^{+i\omega t} \qquad t \geq 0$$

(3.7.23)

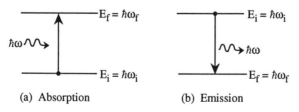

Figure 3.8. Diagrams showing the two processes in (3.7.22): (a) absorption and, (b) emission of a photon in a two-level system.

The transition rate of an electron from an initial state i with energy E_i to a final state with energy E_f is given by

$$W_{i \to f} = \frac{2\pi}{\hbar}|H'_{fi}|^2 \delta(E_f - E_i - \hbar\omega) + \frac{2\pi}{\hbar}|H'_{fi}|^2 \delta(E_f - E_i + \hbar\omega) \quad (3.7.24)$$

PROBLEMS

3.1 Consider an $In_{0.53}Ga_{0.47}As/InP$ quantum well. Assume the following parameters at 300 K:

$In_{0.53}Ga_{0.47}As$ region	InP region
$m_e^* = 0.041m_0$	$m_e^* = 0.077m_0$
$m_{hh}^* = 0.50m_0$	$m_{hh}^* = 0.60m_0$
$m_{lh}^* = 0.0503m_0$	$m_{lh}^* = 0.12m_0$
$E_g = 0.75$ eV	$E_g = 1.344$ eV

where m_0 is the free electron mass.

The band-edge discontinuity is $\Delta E_c = 0.40\Delta E_g$, and $\Delta E_v = 0.60\Delta E_g$.

(a) Consider a well width $L_w = 100$ Å. How many bound states are there in the conduction band? How many bound heavy-hole and light-hole subbands are there?

(b) Find the lowest bound-state energies for the conduction subband (C1) and the heavy-hole (HH1) and the light-hole (LH1) subbands in part (a). What are the C1-HH1 and C1-LH1 transition energies?

(c) Assume that we define an effective well width L_{eff} using an infinite barrier model such that its ground-state energy is the same as the energy of the first conduction subband E_{c1} in (b). What is L_{eff}? If we repeat the same procedure for the HH1 subband, what is L_{eff}?

3.2 In the infinite barrier model, the dispersion relation for the electron subband is given by

$$E = \frac{\hbar^2}{2m_e^*}\left[k_x^2 + k_y^2 + \left(\frac{n\pi}{L}\right)^2\right]$$

(a) Plot the dispersion relations E vs. $k_t = \sqrt{k_x^2 + k_y^2}$ for $n = 1, 2,$ and 3 on the same chart. Plot the corresponding wave functions $\phi_1(z)$, $\phi_2(z)$, and $\phi_3(z)$ with $k_t = 0$.

(b) Derive the electron density of states $\rho_e(E)$ vs. the energy E and plot $\rho_e(E)$ vs. E.

(c) Assume that there are n_s electrons per unit area in the quantum wells at $T = 0$. Find an expression for the Fermi-level position in terms of n_s.

3.3 Calculate the Fermi level for the electrons in a $GaAs/Al_{0.3}Ga_{0.7}As$ quantum well described in the example in Section 3.2 with a surface electron concentration $n_s = 1 \times 10^{12}/cm^2$ at $T = 0$ K. How many subbands are occupied by electrons

3.4 (a) Find an analytical expression relating the surface electron concentration $n_s(1/cm^2)$ to the Fermi level at a finite temperature T in a quantum well based on the infinite barrier model.

(b) What is the carrier concentration at the ith subband?

3.5 If a finite barrier model is used, what is the electron density of states function $\rho_e(E)$ in Problem 3.2. Plot $\rho_e(E)$ vs. the energy E.

3.6 Consider a graded quantum-well structure with a parabolic band-edge profile along the growth (z) axis in real space, as shown in Fig. 3.9.

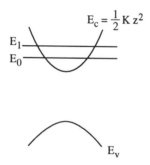

$$E_c = \frac{1}{2}K z^2$$

E_1

E_0

E_v

Figure 3.9.

(a) What are the general dispersion relations of the conduction subbands $E = E_n(k_x, k_y)$?

(b) What is the density of states of the electrons in the conduction band $\rho_e(E)$? Plot $\rho_e(E)$ vs. the energy E.

3.7 Show that the wave function in (3.3.32) is the solution of the differential equation (3.3.31) for the ground state of the harmonic oscillator problem.

3.8 An exciton consisting of an electron with an effective mass m_e^* and a hole with an effective mass m_h^* can be described using a hydrogen model.

(a) Calculate the Rydberg energy R_y for a GaAs semiconductor. Assume that $\varepsilon = 12.5\varepsilon_0$; the other parameters are given in the numerical example in Section 3.2.

(b) Calculate the 1s bound state energy for GaAs if the electron–hole pair is restricted in a pure two-dimensional space.

3.9 (a) Plot the continuum-state wave function at the origin $|\psi_{E00}(r = 0)|^2$ vs. the energy E for the three-dimensional hydrogen model.

(b) Plot the continuum-state wave function at the origin $|\psi_{E0}(r = 0)|^2$ vs. the energy E for the two-dimensional hydrogen model.

3.10 From the perturbation results in the example of Section 3.5, calculate the energy shifts at an applied field of 100 kV/cm using the effective well widths $L_{\text{eff}} =$ (a) 136 Å for electrons in C1 subband, (b) 122 Å for holes in HH1 subband, and (c) 139.6 Å for holes in LH1 subband in a $GaAs/Al_{0.3}Ga_{0.7}As$ quantum well.

3.11 Discuss the differences between the Löwdin's perturbation method in Section 3.6 and the conventional perturbation method in Section 3.5.

3.12 Derive Fermi's golden rule and discuss the regime of validity [10] for applying this rule.

REFERENCES

1. A. Goswami, *Quantum Mechanics*, Wm. C. Brown, Dubuque, IA, 1992.

2. R. P. Feynman, R. B. Leighton, and M. Sands, *The Feynman Lectures on Physics*, *Quantum Mechanics*, vol. 3, Addison-Wesley, Reading, MA, 1965.

3. R. P. Feynman, *Statistical Mechanics*: *A Set of Lectures*, Addison-Wesley, Redwood City, CA, 1972, Chapter 6.

4. For both bound and continuum state solutions in the three-dimensional case see L. D. Landau and E. M. Lifshitz, *Quantum Mechanics*, 3rd ed., Pergamon, Oxford, UK 1977, p. 117, and H. A. Bethe and E. E. Salpeter, *Quantum Mechanics of One- and Two-Electron Atoms*, Springer, Berlin, 1957.

5. For an *n*-dimensional space, $n \geq 2$, see M. Bander and C. Itzykson, "Group theory and the hydrogen atom (I) and (II)," *Rev. Mod. Phys.* **38**, 330–345 (1966), and **38**, 346–358 (1966).

6. C. Y. P. Chao and S. L. Chuang, "Analytical and numerical solutions for a two-dimensional exciton in momentum space," *Phys. Rev. B* **43**, 6530–6543 (1991).

7. E. Menzbacher, *Quantum Mechanics*, 2d ed., Wiley, New York, 1970.

8. H. Haug and S. W. Koch, *Quantum Theory of the Optical and Electronic Properties of Semiconductors*, World Scientific, Singapore, 1990.

9. P. Lödin, "A note on the quantum-mechanical perturbation theory," *J. Chem. Phys.* **19**, 1396–1401 (1951).

10. A. Yariv, *Quantum Electronics*, 3d ed., Wiley, New York, 1989.

11. A. Lukes, G. A. Ringwood, and B. Suprapto, "A particle in a box in the presence of an electric field and applications to disordered systems," *Physica* **84A**, 421–434 (1976).

12. F. M. Fernandez and E. A. Castro, "Hypervirial-perturbational treatment of a particle in a box in the presence of an electric field," *Physica* **111A**, 334–342 (1982).

13. M. Matsuura and T. Kamizato, "Subbands and excitons in a quantum well in an electric field," *Phys. Rev. B* **33**, 8385–8389 (1986).

14. G. Bastard, E. E. Mendez, L. L. Chang, and L. Esaki, "Variational calculations on a quantum well in an electric field," *Phys. Rev. B* **28**, 3241–3245 (1983).

15. D. Ahn and S. L. Chuang, "Variational calculations of subbands in a quantum well with uniform electric field: Gram–Schmidt orthogonalization approach," *Appl. Phys. Lett.* **49**, 1450–1452 (1986).

16. S. Nojima, "Electric field dependence of the exciton binding energy in GaAs/ $Al_xGa_{1-x}As$ quantum wells," *Phys. Rev. B* **37**, 9087–9088 (1988).

4

Theory of Electronic Band Structures in Semiconductors

To understand the optical properties of semiconductors, such as absorption or gain due to electronic transitions in the presence of an incident optical wave, we have to know the electronic band structure, including the energy band and the corresponding wave function. Knowing the initial and final states of the electrons, we may calculate the optical absorption using Fermi's golden rule (Section 3.7). In this chapter, we discuss the calculation of the band structures. For optical devices, most semiconductors have direct band gaps, and many physical phenomena near the band edges are of great interest. We focus on the conduction and valence band structures near the band edges, where the $\mathbf{k} \cdot \mathbf{p}$ method is extremely useful.

4.1 THE BLOCH THEOREM AND THE $\mathbf{k} \cdot \mathbf{p}$ METHOD FOR SIMPLE BANDS

For a periodic potential, the electronic band structure and the wave function can be derived from the Hamiltonian, which satisfies the symmetry of the semiconductor crystals. The general theory follows the Bloch theorem [1], which is discussed in this section. Numerical methods to find the band structures and the wave functions include the tight binding, the pseudopotential, the orthogonalized plane wave, the augmented plane wave, Green's function, and the cellular methods. Many texts on solid-state physics have detailed discussions on these methods [1, 2]. Our interest here is near the band edges of direct band-gap semiconductors, where the wave vector $\mathbf{k}$ deviates by a small amount from a vector $\mathbf{k}_0$ where a local minimum or maximum occurs. The $\mathbf{k} \cdot \mathbf{p}$ method was introduced by Bardeen [3] and Seitz [4]. The method has also been applied by many researchers to semiconductors [5–10]. Here we discuss Kane's model [7, 8], which takes into account the spin–orbit interaction, and Luttinger–Kohn's models [9] for degenerate bands. These models are very popular in studying bulk and quantum-well semiconductors and have been used during the past three decades. They are much easier to apply than most other numerical methods. For a general discussion, see Refs. 11–14.

For an electron in a periodic potential

$$V(\mathbf{r}) = V(\mathbf{r} + \mathbf{R}) \qquad (4.1.1)$$

where $\mathbf{R} = n_1\mathbf{a}_1 + n_2\mathbf{a}_2 + n_3\mathbf{a}_3$, and $\mathbf{a}_1, \mathbf{a}_2, \mathbf{a}_3$ are the lattice vectors, and n_1, n_2, and n_3 are integers, the electron wave function satisfies the Schrödinger equation

$$H\psi(\mathbf{r}) = \left[\frac{-\hbar^2}{2m_0} \nabla^2 + V(\mathbf{r}) \right] \psi(\mathbf{r}) = E(\mathbf{k})\psi(\mathbf{r}) \qquad (4.1.2)$$

The Hamiltonian is invariant under translation by the lattice vectors, $\mathbf{r} \rightarrow \mathbf{r} + \mathbf{R}$. If $\psi(\mathbf{r})$ describes an electron moving in the crystal, $\psi(\mathbf{r} + \mathbf{R})$ will also be a solution to (4.1.2). Thus, $\psi(\mathbf{r} + \mathbf{R})$ will differ from $\psi(\mathbf{r})$ at most by a constant. This constant must have a unity magnitude; otherwise, the wave function may grow to infinity if we repeat the translation $\mathbf{R}$ indefinitely. The general solution of the above equation is given by

$$\psi_{n\mathbf{k}}(\mathbf{r}) = e^{i\mathbf{k} \cdot \mathbf{r}} u_{n\mathbf{k}}(\mathbf{r}) \qquad (4.1.3a)$$

where

$$u_{n\mathbf{k}}(\mathbf{r} + \mathbf{R}) = u_{n\mathbf{k}}(\mathbf{r}) \qquad (4.1.3b)$$

is a periodic function. This result is the Bloch theorem. The wave function $\psi_{n\mathbf{k}}(\mathbf{r})$ is usually called the Bloch function. The energy is given by

$$E = E_n(\mathbf{k})$$

Here n refers to the band and $\mathbf{k}$ the wave vector of the electron. A full description of the band structure requires numerical methods [1, 2]. An example of the GaAs band structure calculated [15] by the pseudopotential method is shown in Fig. 4.1a, which represents the general energy bands along different $\mathbf{k}$ directions. Our interest will focus near the direct band gap (Γ valley) between the valence band edges and the conduction band edge, as shown in Fig. 4.1b, where the energy dispersions for a small $\mathbf{k}$ vector are considered.

The **k** · **p** method is a useful technique for analyzing the band structure near a particular point $\mathbf{k}_0$, especially when it is near an extremum of the band structure. Here we consider that the extremum occurs at the zone center where $\mathbf{k}_0 = 0$. This is a very useful case for III–V direct bandgap semiconductors.

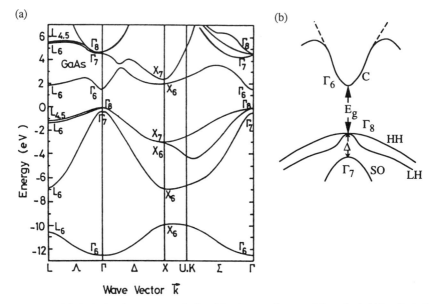

Figure 4.1. (a) GaAs band structure calculated by the pseudopotential method. (After Ref. 15.) (b) The band structure near the band edges of the direct band gap showing the conduction (C) heavy-hole (HH), light-hole (LH), and spin–orbit (SO) split-off bands.

Consider the general Schrödinger equation for an electron wave function $\psi_{nk}(\mathbf{r})$ in the nth band with a wave vector $\mathbf{k}$,

$$\left[\frac{p^2}{2m_0} + V(\mathbf{r})\right]\psi_{nk}(\mathbf{r}) = E_n(\mathbf{k})\psi_{nk}(\mathbf{r}) \qquad (4.1.4)$$

When written in terms of $u_{nk}(\mathbf{r})$, it becomes

$$\left[\frac{p^2}{2m_0} + \frac{\hbar}{m_0}\mathbf{k}\cdot\mathbf{p} + V(\mathbf{r})\right]u_{nk}(\mathbf{r}) = \left[E_n(\mathbf{k}) - \frac{\hbar^2 k^2}{2m_0}\right]u_{nk}(\mathbf{r}) \qquad (4.1.5)$$

The above equation can be expanded near a particular point $\mathbf{k}_0$ of interest in the band structure. When $\mathbf{k}_0 = 0$, the above equation is expanded near $E_n(0)$,

$$\left[H_0 + \frac{\hbar^2}{m_0}\mathbf{k}\cdot\mathbf{p}\right]u_{nk}(\mathbf{r}) = \left[E_n(\mathbf{k}) - \frac{\hbar^2 k^2}{2m_0}\right]u_{nk}(\mathbf{r}) \qquad (4.1.6)$$

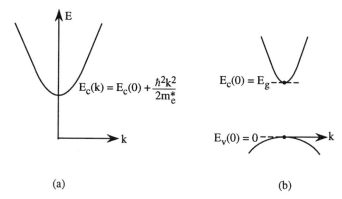

Figure 4.2. (a) A single-band model in the **k · p** theory. (b) The two-band model in the **k · p** theory.

where

$$H_0 = \frac{p^2}{2m_0} + V(\mathbf{r}) \tag{4.1.7a}$$

$$H_0 u_{n0}(\mathbf{r}) = E_n(0) u_{n0}(\mathbf{r}) \tag{4.1.7b}$$

4.1.1 The k · p Theory for a Single Band

If the band structure of interest is near a single band, such as the band edge of a conduction band (Fig. 4.2a) and the coupling to other bands is negligible, then the perturbation theory and Löwdin's method (as discussed in Section 3.6) give the same results. Here the particular band of interest, labeled n, is called class A, and class B consists of the rest of the bands, $n' \neq n$. The time-independent perturbation theory, Eq. (3.5.21b) in Section 3.5, gives the energy to second order in perturbation:

$$E_n(\mathbf{k}) = E_n(0) + \frac{\hbar^2 k^2}{2m_0} + \frac{\hbar}{m_0} \mathbf{k} \cdot \mathbf{p}_{nn} + \frac{\hbar^2}{m_0^2} \sum_{n' \neq n} \frac{|\mathbf{k} \cdot \mathbf{p}_{nn'}|^2}{E_n(0) - E_{n'}(0)} \tag{4.1.8}$$

and the wave function to the first order in perturbation (3.5.16a):

$$u_{n\mathbf{k}}(\mathbf{r}) = u_{n0}(\mathbf{r}) + \sum_{n' \neq n} \left[\frac{\hbar}{m_0} \frac{\mathbf{k} \cdot \mathbf{p}_{n'n}}{E_n(0) - E_{n'}(0)} \right] u_{n'0}(\mathbf{r})$$

$$\equiv \sum_{n'} a_{n'} u_{n'0}(\mathbf{r}) \tag{4.1.9a}$$

$$\psi_{n\mathbf{k}}(\mathbf{r}) = e^{i\mathbf{k} \cdot \mathbf{r}} u_{n\mathbf{k}}(\mathbf{r}) \tag{4.1.9b}$$

where the momentum matrix elements are defined as

$$\mathbf{p}_{nn'} = \int_{\substack{\text{unit} \\ \text{cell}}} u_{n0}^*(\mathbf{r})\mathbf{p}u_{n'0}(\mathbf{r})\, d^3\mathbf{r} \qquad (4.1.10)$$

and $u_{n\mathbf{k}}(\mathbf{r})$'s are normalized as

$$\int_{\substack{\text{unit} \\ \text{cell}}} u_{n0}^*(\mathbf{r})u_{n'0}(\mathbf{r})\, d^3\mathbf{r} = \delta_{nn'} \qquad (4.1.11)$$

If $\mathbf{k}_0$ is at an extremum of $E_n(\mathbf{k})$, then $E_n(\mathbf{k})$ must depend quadratically on $\mathbf{k}$ near $\mathbf{k}_0$ and $\mathbf{p}_{nn} = 0$. That is why we need to go to second-order perturbation theory for the energy correction and only the first-order correction is needed for the wave function. Since we set $\mathbf{k}_0$ to 0, we have

$$E_n(\mathbf{k}) - E_n(0) = \sum_{\alpha,\beta} D^{\alpha\beta} k_\alpha k_\beta = \frac{\hbar^2}{2} \sum_{\alpha,\beta} \left(\frac{1}{m^*}\right)_{\alpha\beta} k_\alpha k_\beta \qquad (4.1.12)$$

and

$$D^{\alpha\beta} = \frac{\hbar^2}{2m_0}\delta_{\alpha\beta} + \frac{\hbar^2}{2m_0^2} \sum_{n' \neq n} \frac{p_{nn'}^\alpha p_{n'n}^\beta + p_{nn'}^\beta p_{n'n}^\alpha}{E_n(0) - E_{n'}(0)} = \frac{\hbar^2}{2}\left(\frac{1}{m^*}\right)_{\alpha\beta} \qquad (4.1.13)$$

where $\alpha, \beta = x$, y, and z. It should be noted that the $D^{\alpha\beta}$ matrix in the quadratic form has been defined to be symmetric. The matrix $D^{\alpha\beta}$ is the inverse effective mass in matrix form multiplied by $\hbar^2/2$.

4.1.2   The k · p Theory for Two-Band (or Nondegenerate Multibands) Model

If only two (or multi-) strongly interacting nondegenerate bands are considered, we call them class A, as shown in Fig. 4.2b. To solve (4.1.6), we assume

$$u_{n\mathbf{k}}(\mathbf{r}) = \sum_{n'} a_{n'}(\mathbf{k})u_{n'0}(\mathbf{r}) \qquad (4.1.14)$$

Substituting the above expression into (4.1.6) for $u_{n\mathbf{k}}(\mathbf{r})$, and multiplying by $u_{n0}^*(\mathbf{r})$ and integrating over a unit cell, we have

$$\sum_{n'}\left\{\left[E_n(0) + \frac{\hbar^2 k^2}{2m_0}\right]\delta_{nn'} + \frac{\hbar}{m_0}\mathbf{k}\cdot\mathbf{p}_{nn'}\right\}a_{n'} = E_n(\mathbf{k})a_n \qquad (4.1.15)$$

where the orthogonality relation $\int u_{n0}^* u_{n'0}\, d^3\mathbf{r} = \delta_{nn'}$ has been used. For two coupled bands, labeled by n and n', the above equation can be solved from the determinantal equation:

$$
\begin{vmatrix}
E_n(0) + \dfrac{\hbar^2 k^2}{2m_0} - E & \dfrac{\hbar}{m_0} \mathbf{k} \cdot \mathbf{p}_{nn'} \\[3mm]
\dfrac{\hbar}{m_0} \mathbf{k} \cdot \mathbf{p}_{n'n} & E_{n'}(0) + \dfrac{\hbar^2 k^2}{2m_0} - E
\end{vmatrix} = 0 \qquad (4.1.16)
$$

The standard procedure is to find the eigenvalue E with the corresponding eigenvector. The two eigenvalues for (4.1.16) can also be compared with those obtained from a direct perturbation theory (see Problem 4.3).

4.2   KANE'S MODEL FOR BAND STRUCTURE: THE k · p METHOD WITH THE SPIN–ORBIT INTERACTION [7, 8]

In Kane's model for direct band semiconductors, the spin–orbit interaction is taken into account. Four bands—the conduction, heavy-hole, light-hole, and the spin–orbit split-off bands—are considered, which have double degeneracy with their spin counterparts (Fig. 4.3a).

4.2.1   The Schrödinger Equation for the Function $u_{n\mathbf{k}}(\mathbf{r})$

Consider the Hamiltonian near the zone center $\mathbf{k}_0 = 0$.

$$
H = H_0 + \frac{\hbar}{4m_0^2 c^2} \boldsymbol{\sigma} \cdot \nabla V \times \mathbf{p} \qquad (4.2.1a)
$$

$$
H_0 = \frac{p^2}{2m_0} + V(\mathbf{r}) \qquad (4.2.1b)
$$

where the second term in (4.2.1a) accounts for the spin–orbit interaction, and $\boldsymbol{\sigma}$ is the Pauli spin matrix with components

$$
\overline{\overline{\sigma}}_x = \begin{bmatrix} 0 & 1 \\ 1 & 0 \end{bmatrix} \qquad
\overline{\overline{\sigma}}_y = \begin{bmatrix} 0 & -i \\ i & 0 \end{bmatrix} \qquad
\overline{\overline{\sigma}}_z = \begin{bmatrix} 1 & 0 \\ 0 & -1 \end{bmatrix} \qquad (4.2.2)
$$

which, when operating on the spins,

$$
\uparrow \equiv \begin{bmatrix} 1 \\ 0 \end{bmatrix} \qquad \downarrow \equiv \begin{bmatrix} 0 \\ 1 \end{bmatrix} \qquad (4.2.3)
$$

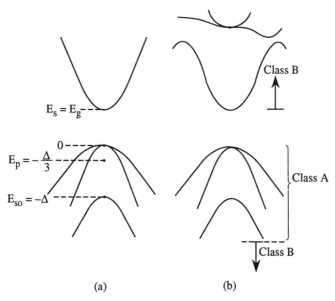

Figure 4.3. (a) The $\mathbf{k} \cdot \mathbf{p}$ method in Kane's model. Only a conduction band, a heavy-hole, a light-hole, and a spin–orbit split-off band with double degeneracy are considered. All other higher and lower bands are discarded. (b) Luttinger–Kohn's model. The heavy-hole, light-hole, and spin-split-off bands in double degeneracy are of interest and are called class A. All other bands are denoted as class B. The effects of bands in class B on those in class A are taken into account in this model.

give

$$\bar{\bar{\sigma}}_x \uparrow = \downarrow \qquad \bar{\bar{\sigma}}_y \uparrow = i\downarrow \qquad \bar{\bar{\sigma}}_z \uparrow = \uparrow$$

$$\bar{\bar{\sigma}}_x \downarrow = \uparrow \qquad \bar{\bar{\sigma}}_y \downarrow = -i\uparrow \qquad \bar{\bar{\sigma}}_z \downarrow = -\downarrow \qquad (4.2.4)$$

From the original Schrödinger equation for the Bloch function,

$$\left\{ \frac{p^2}{2m_0} + V(\mathbf{r}) + \frac{\hbar}{4m_0^2 c^2}[\nabla V \times \mathbf{p}] \cdot \boldsymbol{\sigma} \right\} \psi_{n\mathbf{k}}(\mathbf{r}) = E_n(\mathbf{k})\psi_{n\mathbf{k}}(\mathbf{r}) \quad (4.2.5)$$

The Schrödinger equation for the cell periodic function $u_{n\mathbf{k}}(\mathbf{r})$ is obtained:

$$\left\{ \frac{p^2}{2m_0} + V(\mathbf{r}) + \frac{\hbar}{m_0}\mathbf{k} \cdot \mathbf{p} + \frac{\hbar}{4m_0^2 c^2}[\nabla V \times \mathbf{p}] \cdot \boldsymbol{\sigma} + \frac{\hbar^2}{4m_0^2 c^2} \nabla V \times \mathbf{k} \cdot \boldsymbol{\sigma} \right\}$$

$$\times u_{n\mathbf{k}}(\mathbf{r}) = E' u_{n\mathbf{k}}(\mathbf{r}) \qquad (4.2.6)$$

where $E' = E_n(\mathbf{k}) - \hbar^2 k^2 / 2m_0$. The last term on the left-hand side is a $\mathbf{k}$-dependent spin–orbit interaction, which is small compared with the other terms because the crystal momentum $\hbar\mathbf{k}$ is very small compared with the atomic momentum $\mathbf{p}$ in the far interior of the atom where most of the spin–orbit interaction occurs. Thus, only the first four terms on the left-hand side are considered:

$$Hu_{n\mathbf{k}}(\mathbf{r}) \simeq \left(H_0 + \frac{\hbar}{m_0}\mathbf{k} \cdot \mathbf{p} + \frac{\hbar}{4m_0^2 c^2}\nabla V \times \mathbf{p} \cdot \boldsymbol{\sigma} \right) u_{n\mathbf{k}}(\mathbf{r}) = E' u_{n\mathbf{k}}(\mathbf{r})$$

$$(4.2.7)$$

4.2.2 Basis Functions and the Hamiltonian Matrix

We look for the eigenvalue E' with corresponding eigenfunction

$$u_{n\mathbf{k}}(\mathbf{r}) = \sum_{n'} a_{n'} u_{n'0}(\mathbf{r})$$

The band-edge functions $u_{n0}(\mathbf{r})$ are

> Conduction band: $|S\uparrow\rangle$, $|S\downarrow\rangle$ with corresponding eigenenergy E_s
> Valence band: $|X\uparrow\rangle$, $|Y\uparrow\rangle$, $|Z\uparrow\rangle$, $|X\downarrow\rangle$, $|Y\downarrow\rangle$, $|Z\downarrow\rangle$ with eigenenergy E_p

where the wave functions in each band are degenerate with respect to H_0. $H_0|S\uparrow\rangle = E_s|S\uparrow\rangle$, $H_0|S\downarrow\rangle = E_s|S\downarrow\rangle$, $H_0|X\uparrow\rangle = E_p|X\uparrow\rangle$, $H_0|Y\uparrow\rangle = E_p|Y\uparrow\rangle$, $H_0|Z\uparrow\rangle = E_p|Z\uparrow\rangle$, and so on. It is convenient to choose the basis functions

$$|iS\downarrow\rangle, \quad \left|\frac{X - iY}{\sqrt{2}}\uparrow\right\rangle, \quad |Z\downarrow\rangle, \quad \left|-\frac{X + iY}{\sqrt{2}}\uparrow\right\rangle$$

and

$$|iS\uparrow\rangle, \quad \left|-\frac{X + iY}{\sqrt{2}}\downarrow\right\rangle, \quad |Z\uparrow\rangle, \quad \left|\frac{X - iY}{\sqrt{2}}\downarrow\right\rangle$$

where the valence band basis functions are taken from the spherical harmonics $Y_{10} = |Z\rangle$, and $Y_{1\pm1} = \mp\frac{1}{\sqrt{2}}(X \pm iY)$, Eq. (A.32b) in Appendix A, for the p-state wave functions of a hydrogen atom. The reason for this choice is that the electron wave functions are p-like near the top of the valence band and s-like near the bottom of the conduction band. The first four basis functions are respectively degenerate with the last four basis functions. The

8×8 interaction matrix becomes

$$\begin{bmatrix} \overline{\overline{\mathbf{H}}} & 0 \\ 0 & \overline{\overline{\mathbf{H}}} \end{bmatrix}$$

where, assuming $\mathbf{k} = k\hat{z}$ (see Problem 4.4)

$$\overline{\overline{\mathbf{H}}} = \begin{bmatrix} E_s & 0 & kP & 0 \\ 0 & E_p - \dfrac{\Delta}{3} & \sqrt{2}\,\Delta/3 & 0 \\ kP & \sqrt{2}\,\Delta/3 & E_p & 0 \\ 0 & 0 & 0 & E_p + \dfrac{\Delta}{3} \end{bmatrix} \qquad (4.2.8)$$

and the Kane's parameter P and the spin–orbit split-off energy Δ are defined as

$$P \equiv -i\frac{\hbar}{m_0}\langle S|p_z|Z\rangle \qquad (4.2.9)$$

$$\Delta \equiv \frac{3\hbar i}{4m_0^2 c^2}\langle X|\frac{\partial V}{\partial x}p_y - \frac{\partial V}{\partial y}p_x|Y\rangle \qquad (4.2.10)$$

4.2.3 Solutions for the Eigenvalues and the Eigenfunctions of the Hamiltonian Matrix

Define the reference energy such that $E_p = -\Delta/3$, and $E_s = E_g$, which is the band gap energy. The Hamiltonian in (4.2.8) becomes

$$\overline{\overline{\mathbf{H}}} = \begin{bmatrix} E_g & 0 & kP & 0 \\ 0 & -\dfrac{2\Delta}{3} & \sqrt{2}\,\Delta/3 & 0 \\ kP & \sqrt{2}\,\Delta/3 & \dfrac{-\Delta}{3} & 0 \\ 0 & 0 & 0 & 0 \end{bmatrix} \qquad (4.2.11)$$

The determinantal equation $\det|\overline{\overline{\mathbf{H}}} - E'\overline{\overline{\mathbf{I}}}| = 0$ gives four eigenvalues for E'. We can see that the last band in (4.2.11) is decoupled from the first three bands:

(1) $E' = 0$ (i.e., band-edge energy is zero) $(4.2.12a)$

and

(2)
$$E'(E' - E_g)(E' + \Delta) - k^2 P^2 (E' + \tfrac{2}{3}\Delta) = 0 \qquad (4.2.12b)$$

The second equation gives three roots. Since k^2 is very small, we expect that the roots of Eq. (4.2.12b) will be very close to $E' = E_g$, $E' = 0$ and $E' = -\Delta$, the three band edges.

(i) Let $E' = E_g + \varepsilon(k^2)$ where $\varepsilon \ll \Delta$ and E_g. We find from (4.2.12b)

$$\varepsilon \simeq \frac{k^2 P^2 (E_g + 2\Delta/3)}{E_g(E_g + \Delta)} \qquad (4.2.13)$$

(ii) Let $E' = 0 + \varepsilon(k^2)$. Equation (4.2.12b) gives

$$\varepsilon \simeq -\frac{2k^2 P^2}{3E_g} \qquad (4.2.14)$$

(iii) Let $E' = -\Delta + \varepsilon(k^2)$. We find

$$\varepsilon = -\frac{k^2 P^2}{3(E_g + \Delta)} \qquad (4.2.15)$$

Since $E' = E_n(k) - \hbar^2 k^2 / 2m_0$, we obtain four eigenvalues from (4.2.12a) and (4.2.13)–(4.2.15). They are, starting from the highest energy level,

(4.2.13) $n = c$ $\qquad E_c(k) = E_g + \dfrac{\hbar^2 k^2}{2m_0} + \dfrac{k^2 P^2 (E_g + 2\Delta/3)}{E_g(E_g + \Delta)}$ $\qquad (4.2.16a)$

(4.2.12a) $n = hh$ $\qquad E_{hh}(k) = \dfrac{\hbar^2 k^2}{2m_0}$ $\qquad (4.2.16b)$

(4.2.14) $n = lh$ $\qquad E_{lh}(k) = \dfrac{\hbar^2 k^2}{2m_0} - \dfrac{2k^2 P^2}{3E_g}$ $\qquad (4.2.16c)$

(4.2.15) $n = so$ $\qquad E_{so}(k) = -\Delta + \dfrac{\hbar^2 k^2}{2m_0} - \dfrac{k^2 P^2}{3(E_g + \Delta)}$ $\qquad (4.2.16d)$

These results are not complete since the effects of higher bands have not been included; they will be considered when we discuss the Luttinger–Kohn model. Note that the above result gives an incorrect effective mass for the heavy-hole band.

The eigenfunctions can be obtained from (4.2.11) for the first 4×4 matrix,

$$\phi_{hh,\alpha} = \left| -\left(\frac{X + iY}{\sqrt{2}}\right)\uparrow \right\rangle \qquad\qquad \text{hh band} \qquad (4.2.17a)$$

$$\phi_{n,\alpha} = a_n |iS\downarrow\rangle + b_n \left|\frac{X - iY}{\sqrt{2}}\uparrow\right\rangle + c_n |Z\downarrow\rangle \qquad n = c, l\text{h, so} \quad (4.2.17b)$$

and the second 4×4 matrix,

$$\phi_{hh,\beta} = \left|\frac{X - iY}{\sqrt{2}}\downarrow\right\rangle \qquad\qquad \text{hh band} \qquad (4.2.18a)$$

$$\phi_{n,\beta} = a_n |iS\uparrow\rangle + b_n \left|-\frac{X + iY}{\sqrt{2}}\downarrow\right\rangle + c_n |Z\uparrow\rangle \qquad n = c, l\text{h, so} \quad (4.2.18b)$$

The eigenvectors are obtained by substituting each eigenvalue into the eigenequation

$$\begin{bmatrix} E_g - E_n' & 0 & kP \\ 0 & \dfrac{-2\Delta}{3} - E_n' & \sqrt{2}\dfrac{\Delta}{3} \\ kP & \sqrt{2}\dfrac{\Delta}{3} & -\dfrac{\Delta}{3} - E_n' \end{bmatrix} \begin{bmatrix} a_n \\ b_n \\ c_n \end{bmatrix} = 0 \qquad (4.2.19)$$

and then normalizing such that $(a_n^2 + b_n^2 + c_n^2)^{1/2} = 1$.

The results in the limit $k^2 \to 0$ give

$$n = c \quad a \cong 1, \quad b \cong 0, \quad c \simeq 0$$

$$n = l\text{h} \quad a \simeq 0, \quad b = \frac{1}{\sqrt{3}}, \quad c = \sqrt{\frac{2}{3}}$$

$$n = \text{so} \quad a \simeq 0, \quad b = \sqrt{\frac{2}{3}}, \quad c = -\sqrt{\frac{1}{3}} \qquad (4.2.20)$$

4.2.4 Summary of the Eigenenergies and Corresponding Band-Edge Basis Functions

We summarize the results below and in Fig. 4.4. The commonly used parabolic band models for the energy dispersions are also redefined in the parentheses.

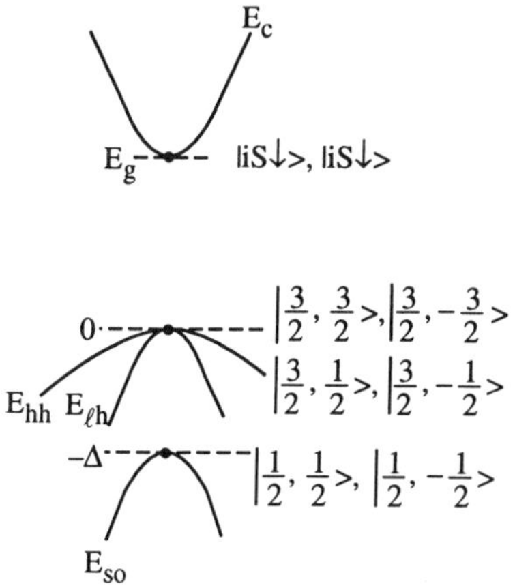

Figure 4.4. The band-edge energies E_g, 0, 0, and $-\Delta$ for the conduction, heavy-hole, light-hole, and spin split-off bands with their corresponding band-edge Bloch functions. Note that the dispersion relation $E - k$ for the heavy-hole band E_{hh} should curve down as shown and follow the result of the Luttinger–Kohn model.

Conduction band

$$E_c(k) = E_g + \frac{\hbar^2 k^2}{2m_0} + \frac{k^2 P^2}{3}\frac{(3E_g + 2\Delta)}{E_g(E_g + \Delta)} \left(\equiv E_g + \frac{\hbar^2 k^2}{2m_e^*} \right)$$

$$\phi_{c,\alpha} = |iS \downarrow\rangle$$
$$\phi_{c,\beta} = |iS \uparrow\rangle \tag{4.2.21}$$

Valence band

Heavy hole

$$E_{hh}(k) = \frac{\hbar^2 k^2}{2m_0} \left(\text{should be} - \frac{\hbar^2 k^2}{2m_{hh}^*} \right)$$

$$\phi_{hh,\alpha} = \frac{-1}{\sqrt{2}}|(X + iY)\uparrow\rangle \equiv \left|\frac{3}{2},\frac{3}{2}\right\rangle$$

$$\phi_{hh,\beta} = \frac{1}{\sqrt{2}}|(X - iY)\downarrow\rangle \equiv \left|\frac{3}{2},-\frac{3}{2}\right\rangle \tag{4.2.22}$$

Light hole

$$E_{lh}(k) = \frac{\hbar^2 k^2}{2m_0} - \frac{2k^2 P^2}{3E_g} \left(\equiv -\frac{\hbar^2 k^2}{2m_{lh}^*} \right)$$

$$\phi_{lh,\alpha} = \frac{1}{\sqrt{6}}|(X - iY)\uparrow\rangle + \sqrt{\frac{2}{3}}|Z\downarrow\rangle \equiv \left|\frac{3}{2},-\frac{1}{2}\right\rangle$$

$$\phi_{lh,\beta} = -\frac{1}{\sqrt{6}}|(X + iY)\downarrow\rangle + \sqrt{\frac{2}{3}}|Z\uparrow\rangle \equiv \left|\frac{3}{2},\frac{1}{2}\right\rangle \tag{4.2.23}$$

Spin-orbit split-off band

$$E_{so}(k) = -\Delta + \frac{\hbar^2 k^2}{2m_0} - \frac{k^2 P^2}{3(E_g + \Delta)} \left(\equiv -\Delta - \frac{\hbar^2 k^2}{2m_{so}^*} \right)$$

$$\phi_{so,\alpha} = \frac{1}{\sqrt{3}} |(X - iY)\uparrow\rangle - \frac{1}{\sqrt{3}} |Z\downarrow\rangle \equiv \left| \frac{1}{2}, \frac{-1}{2} \right\rangle$$

$$\phi_{so,\beta} = \frac{1}{\sqrt{3}} |(X + iY)\downarrow\rangle + \frac{1}{\sqrt{3}} |Z\uparrow\rangle \equiv \left| \frac{1}{2}, \frac{1}{2} \right\rangle \qquad (4.2.24)$$

The Kane's parameter P can also be related to the effective mass of the electron m_e^* using

$$E_c(k) - E_g = \frac{\hbar^2 k^2}{2m_0} + \frac{k^2 P^2 \left(E_g + \frac{2}{3}\Delta \right)}{E_g(E_g + \Delta)} = \frac{\hbar^2 k^2}{2m_e^*} \qquad (4.2.25)$$

or

$$P^2 = \left(1 - \frac{m_e^*}{m_0} \right) \frac{\hbar^2 E_g(E_g + \Delta)}{2m_e^*(E_g + 2\Delta/3)} \qquad (4.2.26)$$

Sometimes the $\hbar^2 k^2 / 2m_0$ term is ignored, since $m_e^* \simeq 0.067m \ll m_0$ for GaAs; therefore, the term m_e^*/m_0 in (4.2.26) is ignored.

We note that these wave functions in (4.2.21)–(4.2.24) are eigenvectors of the Hamiltonian

$$H = H_0 + \frac{\hbar^2 k^2}{2m_0} + \frac{\hbar}{m_0} \mathbf{k} \cdot \mathbf{p} + \frac{\hbar}{4m_0^2 c^2} \boldsymbol{\sigma} \cdot \nabla V \times \mathbf{p} \qquad (4.2.27)$$

with eigenenergies $E = E_g, 0, 0, -\Delta$ as $k \to 0$ for the conduction, heavy-hole, light-hole, and spin-orbit split-off bands, respectively.

4.2.5 General Coordinate Direction

If $\mathbf{k}$ is not along the z direction,

$$\mathbf{k} = k \sin\theta \cos\varphi\, \hat{x} + k \sin\theta \sin\varphi\, \hat{y} + k \cos\theta\, \hat{z} \qquad (4.2.28)$$

the following transformations can be used to find the basis functions in the general coordinate system:

$$
\begin{bmatrix} \uparrow' \\ \\ \downarrow' \end{bmatrix} = \begin{bmatrix} e^{-i\phi/2}\cos\dfrac{\theta}{2} & e^{i\phi/2}\sin\dfrac{\theta}{2} \\ \\ -e^{-i\phi/2}\sin\dfrac{\theta}{2} & e^{i\phi/2}\cos\dfrac{\theta}{2} \end{bmatrix} \begin{bmatrix} \uparrow \\ \\ \downarrow \end{bmatrix} \tag{4.2.29}
$$

$$
\begin{bmatrix} X' \\ Y' \\ Z' \end{bmatrix} = \begin{bmatrix} \cos\theta\cos\phi & \cos\theta\sin\phi & -\sin\theta \\ -\sin\phi & \cos\phi & 0 \\ \sin\theta\cos\phi & \sin\theta\sin\phi & \cos\theta \end{bmatrix} \begin{bmatrix} X \\ Y \\ Z \end{bmatrix} \tag{4.2.30}
$$

and the spherical symmetry function $S(r') = S(r)$, since $r' = r$, the length scale is preserved in a unitary transformation. The above transformation will be useful in Chapter 9 when we discuss optical matrix elements for quantum wells.

4.3 LUTTINGER–KOHN'S MODEL: THE k · p METHOD FOR DEGENERATE BANDS [9–13]

Suppose we are mainly interested in the six valence bands (the heavy-hole, light-hole, and spin-orbit split-off bands, all doubly degenerate) and ignore the coupling to the two degenerate conduction bands with both spins. It is convenient to use Löwdin's perturbation method and treat the six valence bands in class A and put the rest of the bands in class B (Fig. 4.3b). Note that Luttinger–Kohn's model can also be generalized to include both conduction bands in class A, especially for narrow band-gap semiconductors.

4.3.1 The Hamiltonian and the Basis Functions

Write the total Hamiltonian in (4.2.6) for $u_{\mathbf{k}}(\mathbf{r})$ (dropping the band index n for convenience):

$$
Hu_{\mathbf{k}}(\mathbf{r}) = E(\mathbf{k})u_{\mathbf{k}}(\mathbf{r}) \tag{4.3.1a}
$$

$$
H = H_0 + \frac{\hbar^2 k^2}{2m_0} + \frac{\hbar}{4m_0^2 c^2}\nabla V \times \mathbf{p} \cdot \boldsymbol{\sigma} + H' \tag{4.3.1b}
$$

where

$$
H_0 = \frac{p^2}{2m} + V(\mathbf{r}) \tag{4.3.1c}
$$

$$
H' = \frac{\hbar}{m_0}\mathbf{k} \cdot \boldsymbol{\Pi} \tag{4.3.1d}
$$

where

$$\Pi = \mathbf{p} + \frac{\hbar}{4m_0c^2}\sigma \times \nabla V \qquad (4.3.1e)$$

Note again that the last term in (4.2.6) is much smaller than the fourth term because $\hbar k \ll p = |\langle u_k|\mathbf{p}|u_k\rangle| \simeq \hbar/a$, since the electron velocity in the atomic orbit is much larger than the velocity of the wave packet with the wave vectors in the vicinity of $\mathbf{k}_0(= 0)$ [11].

We expand the function

$$u_\mathbf{k}(\mathbf{r}) = \sum_{j'}^A a_{j'}(\mathbf{k})u_{j'0}(\mathbf{r}) + \sum_\gamma^B a_\gamma(\mathbf{k})u_{\gamma 0}(\mathbf{r}) \qquad (4.3.2)$$

where j' is in class A and γ is in class B. Or specifically, we have in class A

$$u_{10}(\mathbf{r}) = \left|\frac{3}{2},\frac{3}{2}\right\rangle = \frac{-1}{\sqrt{2}}|(X+iY)\uparrow\rangle$$

$$u_{20}(\mathbf{r}) = \left|\frac{3}{2},\frac{1}{2}\right\rangle = \frac{-1}{\sqrt{6}}|(X+iY)\downarrow\rangle + \sqrt{\frac{2}{3}}|Z\uparrow\rangle$$

$$u_{30}(\mathbf{r}) = \left|\frac{3}{2},\frac{-1}{2}\right\rangle = \frac{1}{\sqrt{6}}|(X-iY)\uparrow\rangle + \sqrt{\frac{2}{3}}|Z\downarrow\rangle$$

$$u_{40}(\mathbf{r}) = \left|\frac{3}{2},\frac{-3}{2}\right\rangle = \frac{1}{\sqrt{2}}|(X-iY)\downarrow\rangle$$

$$u_{50}(\mathbf{r}) = \left|\frac{1}{2},\frac{1}{2}\right\rangle = \frac{1}{\sqrt{3}}|(X+iY)\downarrow\rangle + \frac{1}{\sqrt{3}}|Z\uparrow\rangle$$

$$u_{60}(\mathbf{r}) = \left|\frac{1}{2},\frac{-1}{2}\right\rangle = \frac{1}{\sqrt{3}}|(X-iY)\uparrow\rangle - \frac{1}{\sqrt{3}}|Z\downarrow\rangle \qquad (4.3.3)$$

from (4.2.22)–(4.2.24). At $\mathbf{k} = 0$, the band-edge functions (4.3.3) satisfy

$$H(\mathbf{k} = 0)u_{j0}(\mathbf{r}) = E_j(0)u_{j0}(\mathbf{r}) \qquad (4.3.4)$$

where

$$E_j(0) = E_p + \frac{\Delta}{3} = 0 \qquad \text{for } j = 1,2,3,4 \qquad (4.3.5a)$$

$$E_j(0) = E_p - \frac{2\Delta}{3} = -\Delta \qquad \text{for } j = 5,6 \qquad (4.3.5b)$$

since $E_p = -\Delta/3$. These band-edge energies and basis functions are also shown in Fig. 4.4.

4.3.2 Solution of the Hamiltonian Using Löwdin's Perturbation Method

With Löwdin's method we need to solve only the eigenequation

$$\sum_{j'}^{A} \left(U_{jj'}^{A} - E\delta_{jj'} \right) a_{j'}(\mathbf{k}) = 0 \qquad (4.3.6)$$

instead of

$$\sum_{j'}^{A} \left(H_{jj'} - E\delta_{jj'} \right) a_{j'}(\mathbf{k}) = 0 \qquad (4.3.7)$$

where

$$U_{jj'}^{A} = H_{jj'} + \sum_{\gamma \neq j, j'}^{B} \frac{H_{j\gamma} H_{\gamma j'}}{E_0 - E_\gamma} = H_{jj'} + \sum_{\gamma \neq j, j'}^{B} \frac{H'_{j\gamma} H'_{\gamma j'}}{E_0 - E_\gamma} \qquad (4.3.8a)$$

$$H_{jj'} = \langle u_{j0}|H|u_{j'0}\rangle = \left[E_j(0) + \frac{\hbar^2 k^2}{2m_0} \right]\delta_{jj'} \qquad (j, j' \in A) \qquad (4.3.8b)$$

$$H'_{j\gamma} = \langle u_{j0}|\frac{\hbar}{m_0}\mathbf{k}\cdot\mathbf{\Pi}|u_{\gamma 0}\rangle \cong \sum_{\alpha} \frac{\hbar k_\alpha}{m_0} p_{j\gamma}^{\alpha} \qquad (j \in A, \gamma \notin A) \quad (4.3.8c)$$

where we note that $\Pi_{jj'} = 0$, for $j, j' \in A$, and $\Pi_{j\gamma}^{\alpha} \simeq p_{j\gamma}^{\alpha}$ for $j \in A$ and $\gamma \notin A$. Since $\gamma \neq j$, adding the unperturbed part to the perturbed part in $H'_{j\gamma}$ does not affect the results, i.e., $H_{j\gamma} = H'_{j\gamma}$. We thus obtain

$$U_{jj'}^{A} = \left[E_j(0) + \frac{\hbar^2 k^2}{2m_0} \right]\delta_{jj'} + \frac{\hbar^2}{m_0^2}\sum_{\gamma \neq j, j'}^{B}\sum_{\alpha, \beta} \frac{k_\alpha k_\beta p_{j\gamma}^{\alpha} p_{\gamma j'}^{\beta}}{E_0 - E_\gamma} \qquad (4.3.9)$$

Let $U_{jj'}^{A} \equiv D_{jj'}$. We obtain the matrix of the form $D_{jj'}$

$$D_{jj'} = E_j(0)\,\delta_{jj'} + \sum_{\alpha, \beta} D_{jj'}^{\alpha\beta} k_\alpha k_\beta \qquad (4.3.10)$$

where $D_{jj'}^{\alpha\beta}$ is defined as

$$D_{jj'}^{\alpha\beta} = \frac{\hbar^2}{2m_0}\left[\delta_{jj'}\delta_{\alpha\beta} + \sum_{\gamma}^{B} \frac{p_{j\gamma}^{\alpha} p_{\gamma j'}^{\beta} + p_{j\gamma}^{\beta} p_{\gamma j'}^{\alpha}}{m_0(E_0 - E_\gamma)} \right] \qquad (4.3.11)$$

which is similar to (4.1.13), the single-band case (where $j = j' =$ the single-band index n). Here we have generalized (4.1.13) to include the degenerate bands.

4.3.3   Explicit Expression for the Luttinger–Kohn Hamiltonian Matrix $D_{jj'}$

To write out the matrix elements $D_{jj'}$ in (4.3.10) explicitly, we define

$$A_0 = \frac{\hbar^2}{2m_0} + \frac{\hbar^2}{m_0^2} \sum_\gamma^B \frac{p_{x\gamma}^x p_{\gamma x}^x}{E_0 - E_\gamma} \tag{4.3.12a}$$

$$B_0 = \frac{\hbar^2}{2m_0} + \frac{\hbar^2}{m_0^2} \sum_\gamma^B \frac{p_{x\gamma}^y p_{\gamma x}^y}{E_0 - E_\gamma} \tag{4.3.12b}$$

$$C_0 = \frac{\hbar^2}{m_0^2} \sum_\gamma^B \frac{p_{x\gamma}^x p_{\gamma y}^y + p_{x\gamma}^y p_{\gamma y}^x}{E_0 - E_\gamma} \tag{4.3.12c}$$

and define the band structure parameters γ_1, γ_2, and γ_3 as

$$-\frac{\hbar^2}{2m_0}\gamma_1 = \frac{1}{3}(A_0 + 2B_0)$$

$$-\frac{\hbar^2}{2m_0}\gamma_2 = \frac{1}{6}(A_0 - B_0)$$

$$-\frac{\hbar^2}{2m_0}\gamma_3 = \frac{C_0}{6} \tag{4.3.13}$$

We obtain the Luttinger–Kohn Hamiltonian $\overline{\overline{\mathbf{U}}}^A \equiv \overline{\overline{\mathbf{D}}}$, denoted as $\overline{\overline{\mathbf{H}}}^{LK}$, in the basis functions given by (4.3.3)

$$\overline{\overline{\mathbf{H}}}^{LK} = -\begin{bmatrix} P+Q & -S & R & 0 & -S/\sqrt{2} & \sqrt{2}R \\ -S^+ & P-Q & 0 & R & -\sqrt{2}Q & \sqrt{3/2}S \\ R^+ & 0 & P-Q & S & \sqrt{3/2}S^+ & \sqrt{2}Q \\ 0 & R^+ & S^+ & P+Q & -\sqrt{2}R^+ & -S^+/\sqrt{2} \\ -S^+/\sqrt{2} & -\sqrt{2}Q^+ & \sqrt{3/2}S & -\sqrt{2}R & P+\Delta & 0 \\ \sqrt{2}R^+ & \sqrt{3/2}S^+ & \sqrt{2}Q^+ & -S/\sqrt{2} & 0 & P+\Delta \end{bmatrix} \tag{4.3.14}$$

$$P = \frac{\hbar^2\gamma_1}{2m_0}\left(k_x^2 + k_y^2 + k_z^2\right)$$

$$Q = \frac{\hbar^2 \gamma_2}{2m_0} \left(k_x^2 + k_y^2 - 2k_z^2 \right)$$

$$R = \frac{\hbar^2}{2m_0} \left[-\sqrt{3}\,\gamma_2 \left(k_x^2 - k_y^2 \right) + i2\sqrt{3}\,\gamma_3 k_x k_y \right]$$

$$S = \frac{\hbar^2 \gamma_3}{m_0} \sqrt{3}\, (k_x - ik_y)k_z \qquad\qquad (4.3.15)$$

where the superscript $+$ means Hermitian conjugate.

4.3.4 Summary

In summary, for the valence hole subbands, we have to solve only for the eigenvalue equation

$$\sum_{j'=1}^{6} H_{jj'}^{\mathrm{LK}} a_{j'}(\mathbf{k}) = E a_j(\mathbf{k}) \qquad\qquad (4.3.16)$$

or for the eigenvalue E and the corresponding eigenvector *column* $[a_1, a_2, \ldots, a_6]$, where the matrix elements $H_{jj'}^{\mathrm{LK}} = E_j(0)\delta_{jj'} + \sum_{\alpha,\beta} D_{jj'}^{\alpha\beta} k_\alpha k_\beta$ are given in (4.3.14). The wave function $\psi_{n\mathbf{k}}(\mathbf{r})$ satisfying

$$\left[\frac{p^2}{2m_0} + V(\mathbf{r}) + \frac{\hbar}{4m_0^2 c^2} \nabla V \times \mathbf{p} \cdot \boldsymbol{\sigma} \right] \psi_{n\mathbf{k}}(\mathbf{r}) = E_n(\mathbf{k}) \psi_{n\mathbf{k}}(\mathbf{r}) \quad (4.3.17)$$

is then given by

$$\psi_{n\mathbf{k}}(\mathbf{r}) = e^{i\mathbf{k}\cdot\mathbf{r}} u_{n\mathbf{k}}(\mathbf{r}) \qquad u_{n\mathbf{k}}(\mathbf{r}) = \sum_{j=1}^{6} a_j(\mathbf{k}) u_{j0}(\mathbf{r}) \qquad (4.3.18)$$

and

$$E_n(\mathbf{k}) = E$$

4.4 THE EFFECTIVE MASS THEORY FOR A SINGLE BAND AND DEGENERATE BANDS

In this section, we summarize the effective mass theory for both a single band and degenerate bands in semiconductors. We outline a formal proof [9] of the effective mass equations in Appendix B.

4.4.1 The Effective Mass Theory for a Single Band

The most important conclusion of the effective mass theory (EMT) for a single band is as follows: If the energy dispersion relation for a single band n near $\mathbf{k}_0$ (assuming 0) is given by

$$E_n(\mathbf{k}) = E_n(0) + \sum_{\alpha, \beta} \frac{\hbar^2}{2} \left(\frac{1}{m^*} \right)_{\alpha\beta} k_\alpha k_\beta \qquad (4.4.1)$$

for the Hamiltonian H_0 with a periodic potential $V(\mathbf{r})$

$$H_0 = \frac{p^2}{2m_0} + V(\mathbf{r}) \qquad (4.4.2)$$

$$H_0 \psi_{n\mathbf{k}}(\mathbf{r}) = E_n(\mathbf{k})\psi_{n\mathbf{k}}(\mathbf{r}) \qquad (4.4.3)$$

then the solution for the Schrödinger equation with a perturbation $U(\mathbf{r})$ such as an impurity potential or a quantum-well potential

$$[H_0 + U(\mathbf{r})]\psi(\mathbf{r}) = E\psi(\mathbf{r}) \qquad (4.4.4)$$

is obtainable by solving

$$\left[\sum_{\alpha, \beta} \frac{\hbar^2}{2} \left(\frac{1}{m^*} \right)_{\alpha\beta} \left(-i \frac{\partial}{\partial x_\alpha} \right) \left(-i \frac{\partial}{\partial x_\beta} \right) + U(\mathbf{r}) \right] F(\mathbf{r}) = [E - E_n(0)] F(\mathbf{r})$$

$$(4.4.5)$$

for the envelope function $F(\mathbf{r})$ and the energy E. The wave function is approximated by

$$\psi(\mathbf{r}) = F(\mathbf{r}) u_{n\mathbf{k}_0}(\mathbf{r}) \qquad (4.4.6)$$

The most important result is that the periodic potential $V(\mathbf{r})$ determines the energy bands and the effective masses, $(1/m^*)_{\alpha\beta}$, and the effective mass equation (4.4.5) contains only the extra perturbation potential $U(\mathbf{r})$, since the effective masses already take into account the periodic potential (Fig. 4.5). The perturbation potential can also be a quantum-well potential, as in a semiconductor heterostructure such as GaAs/Al$_x$Ga$_{1-x}$As quantum wells.

4.4.2 The Effective Mass Theory for Degenerate Bands

In Section 4.3, we discuss the $\mathbf{k} \cdot \mathbf{p}$ method for degenerate bands such as the heavy-hole, light-hole, and the spin-orbit split-off bands of semiconductors. The effective mass theory for a perturbation potential $U(\mathbf{r})$ for degenerate

(a) A periodic crystal potential V(r)

(b) A periodic crystal potential V(r) with an impurity potential U(r)

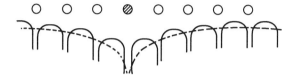

(c) An impurity potential U(r)

Figure 4.5. Illustrations of (a) the periodic potential $V(\mathbf{r})$ and (b) the sum of the periodic potential $V(\mathbf{r})$ and the impurity potential $U(\mathbf{r})$, and (c) only the impurity potential $U(\mathbf{r})$ for the effective mass theory.

bands is stated as follows [9, 11]. If the dispersion relation of a set of degenerate bands satisfying

$$H\psi_{n\mathbf{k}}(\mathbf{r}) = E(\mathbf{k})\psi_{n\mathbf{k}}(\mathbf{r}) \tag{4.4.7a}$$

$$H = \frac{p^2}{2m_0} + V(\mathbf{r}) + H_{so} \tag{4.4.7b}$$

$$H_{so} = \frac{\hbar}{4m_0^2 c^2} \nabla V \times \mathbf{p} \cdot \boldsymbol{\sigma} \tag{4.4.7c}$$

is given by

$$\sum_{j'=1}^{6} H_{jj'}^{LK} a_{j'}(\mathbf{k}) \equiv \sum_{j'=1}^{6} \left[E_j(0)\delta_{jj'} + \sum_{\alpha,\beta} D_{jj'}^{\alpha\beta} k_\alpha k_\beta \right] a_{j'}(\mathbf{k}) = E(\mathbf{k})a_j(\mathbf{k}) \tag{4.4.8}$$

then the solution $\psi(\mathbf{r})$ for the semiconductors in the presence of a perturbation $U(\mathbf{r})$,

$$[H + U(\mathbf{r})]\psi(\mathbf{r}) = E\psi(\mathbf{r}) \tag{4.4.9}$$

is given by

$$\psi(\mathbf{r}) = \sum_{j=1}^{6} F_j(\mathbf{r}) u_{jo}(\mathbf{r}) \qquad (4.4.10)$$

where $F_j(\mathbf{r})$ satisfies

$$\sum_{j'=1}^{6} \left[E_j(0)\delta_{jj'} + \sum_{\alpha,\beta} D_{jj'}^{\alpha\beta}\left(-i\frac{\partial}{\partial x_\alpha}\right)\left(-i\frac{\partial}{\partial x_\beta}\right) + U(\mathbf{r})\delta_{jj'} \right] F_{j'}(\mathbf{r}) = E F_j(\mathbf{r})$$

$$(4.4.11)$$

4.5 STRAIN EFFECTS ON BAND STRUCTURES

Strained-layer superlattices [16, 17] have been of great interest since the early 1980s. It has been demonstrated that it is possible to vary important material properties—lattice constant, band gap, and perpendicular transport effective mass—using ternary strained-layer superlattices. Applications of the strained-layer superlattices or quantum wells to long wavelength photodetectors [17] and semiconductor lasers [18] have been proposed and demonstrated. For example, strained quantum-well lasers have been shown to exhibit superior performance compared to that for conventional diode lasers [19, 20] in many aspects, which is discussed in Chapter 10. Detailed discussions on the semiconductor growth and the physics of strained-layer quantum wells can be found in Ref. 19.

When a crystal is under a uniform deformation, it may preserve the periodic property such that the Bloch theorem may still be applicable. The modulating part of the Bloch function remains periodic, with a period equal to that of the new elementary cell, since the elementary cell is also deformed. In this section, we derive the Hamiltonian for strained semiconductors and discuss their band structures.

4.5.1 The Pikus–Bir Hamiltonian [21, 22] for a Strained Semiconductor

Suppose that near the band extremum $\mathbf{k}_0 = 0$ of a semiconductor we have

$$[H_0 + V_0(\mathbf{r})]\psi_{n\mathbf{k}_0}(\mathbf{r}) = E_n(\mathbf{k}_0)\psi_{n\mathbf{k}_0}(\mathbf{r}) \qquad (4.5.1)$$

with the Bloch function $\psi_{n\mathbf{k}_0}(\mathbf{r}) = e^{i\mathbf{k}_0 \cdot \mathbf{r}} u_{n\mathbf{k}_0}(\mathbf{r})$, where $V_0(\mathbf{r})$ is a periodic potential in the undeformed crystal. Here we present a simple picture for the strain analysis [23–25]. As shown in Fig. 4.6, the unit vectors $\hat{x}$, $\hat{y}$, (and $\hat{z}$) (for simplicity, assuming they are basis vectors too) in the undeformed crystal

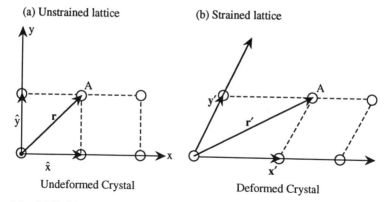

Figure 4.6. (a) Position vector **r** for atom A in an unstrained lattice. (b) Position vector **r'** for atom A in a strained lattice.

are related to **x'**, **y'**, (and **z'**) in the uniformly deformed crystal by

$$\mathbf{x'} = (1 + \varepsilon_{xx})\hat{x} + \varepsilon_{xy}\hat{y} + \varepsilon_{xz}\hat{z}$$
$$\mathbf{y'} = \varepsilon_{yx}\hat{x} + (1 + \varepsilon_{yy})\hat{y} + \varepsilon_{yz}\hat{z}$$
$$\mathbf{z'} = \varepsilon_{zx}\hat{x} + \varepsilon_{zy}\hat{y} + (1 + \varepsilon_{zz})\hat{z} \qquad (4.5.2)$$

Obviously **x'**, **y'**, and **z'** are not unit vectors anymore. We assume a homogeneous strain and $\varepsilon_{ij} = \varepsilon_{ji}$. We can define six strain components as

$$e_1 = \varepsilon_{xx} \qquad e_2 = \varepsilon_{yy} \qquad e_3 = \varepsilon_{zz}$$
$$e_4 = \mathbf{x'} \cdot \mathbf{y'} = \varepsilon_{xy} + \varepsilon_{yx}$$
$$e_5 = \mathbf{y'} \cdot \mathbf{z'} = \varepsilon_{yz} + \varepsilon_{zy}$$
$$e_6 = \mathbf{z'} \cdot \mathbf{x'} = \varepsilon_{zx} + \varepsilon_{xz} \qquad (4.5.3)$$

keeping only the linear terms in strain. To label a position A (or atom A) in the undeformed crystal, we have

$$\mathbf{r} = x\hat{x} + y\hat{y} + z\hat{z} \equiv (x, y, z) \qquad (4.5.4a)$$

The same atom in the deformed crystal can be labeled either as

$$\mathbf{r'} = x\mathbf{x'} + y\mathbf{y'} + z\mathbf{z'} \qquad (4.5.4b)$$

using the new basis vectors **x'**, **y'**, and **z'** or as

$$\mathbf{r'} = x'\hat{x} + y'\hat{y} + z'\hat{z} \equiv (x', y', z') \qquad (4.5.4c)$$

in the original bases in the undeformed crystal. An example is shown in Fig. 4.6. We have $\mathbf{r} = \hat{x} + \hat{y} \equiv (1, 1, 0)$ in the undeformed crystal and $\mathbf{r}' = \mathbf{x}' + \mathbf{y}'$ in the deformed crystal. We can also see that the change of the volume in the linear strain regime becomes

$$\frac{V + \delta V}{V} = \mathbf{x}' \cdot \mathbf{y}' \times \mathbf{z}' = 1 + (\varepsilon_{xx} + \varepsilon_{yy} + \varepsilon_{zz}) \qquad (4.5.5)$$

The quantity $\varepsilon_{xx} + \varepsilon_{yy} + \varepsilon_{zz}$ is the trace of the matrix $\bar{\bar{\varepsilon}}$, or $\mathrm{Tr}(\bar{\bar{\varepsilon}})$, which is exactly the fractional change of the volume $\delta V / V$ of the crystal under uniform deformation:

$$\frac{\delta V}{V} = \varepsilon_{xx} + \varepsilon_{yy} + \varepsilon_{zz} \qquad (4.5.6)$$

Major Results for the Pikus–Bir Hamiltonian for a Strained Semiconductor. Using the above relations (4.5.2)–(4.5.6) between the deformed coordinates and the undeformed coordinates, the Hamiltonian for a strained semiconductor can be derived; the details are shown in Appendix C. Here we summarize the major conclusions.

In the unstrained semiconductor, the full 6×6 Luttinger–Kohn Hamiltonian is given by (4.3.10)

$$D_{jj'} = H_{jj'}^{\mathrm{LK}} = E_j(0)\delta_{jj'} + \sum_{\alpha, \beta} D_{jj'}^{\alpha\beta} k_\alpha k_\beta \qquad (4.5.7)$$

and its full matrix form is shown in (4.3.14) and (4.3.15) explicitly in terms of the expressions P, Q, R, and S.

The strained Hamiltonian introduces extra terms denoted by

$$(H_\varepsilon)_{jj'} = \sum_{\alpha, \beta} \hat{D}_{jj'}^{\alpha\beta} \varepsilon_{\alpha\beta} \qquad (4.5.8)$$

due to the linear strain. We, therefore, use the correspondences between $D_{jj'}^{\alpha\beta}$ in $H_{jj'}^{\mathrm{LK}}$ and $\hat{D}_{jj'}^{\alpha\beta}$ in $(H_\varepsilon)_{jj'}$

$$k_\alpha k_\beta \leftrightarrow \varepsilon_{\alpha\beta} \qquad (4.5.9)$$

For Valence Band

$$\frac{\hbar^2 \gamma_1}{2m_0} \leftrightarrow D_v^d \equiv -a_v$$

$$\frac{\hbar^2 \gamma_2}{2m_0} \leftrightarrow \frac{D_u}{3} \equiv -\frac{b}{2}$$

$$\frac{\hbar^2 \gamma_3}{2m_0} \leftrightarrow \frac{D_u'}{3} \equiv -\frac{d}{2\sqrt{3}} \qquad (4.5.10)$$

Therefore, P_k, Q_k, R_k, and S_k can be added to their corresponding strain counterparts P_ε, Q_ε, R_ε, and S_ε shown explicitly in Appendix C. The energy parameters a_v, b, and d are called deformation potentials for the valence band and are tabulated in Appendix K for a few semiconductors.

For Conduction Band (isotropic case)

$$\frac{\hbar^2}{2m_e^*} \leftrightarrow a_c \qquad (4.5.11)$$

and the conduction band-edge dispersion is

$$E(\mathbf{k}) = E_c(0) + \frac{\hbar^2}{2} \sum_{\alpha, \beta} \left(\frac{1}{m^*} \right)_{\alpha\beta} k_\alpha k_\beta + a_c(\varepsilon_{xx} + \varepsilon_{yy} + \varepsilon_{zz}) \quad (4.5.12)$$

The inverse effective mass tensor is diagonal ($\alpha = \beta$) in the principal axis system. The strained Hamiltonian has been used extensively in the study of the strain effects [26–38] on the band structures of semiconductors.

4.5.2   Band Structures Without the Spin-Orbit Split-Off Band Coupling

Next we illustrate how the strain modifies the valence-band structures, including the band-edge energies and the effective masses, which are among the most important parameters characterizing any semiconductor materials. For most III–V semiconductors, the split-off bands are several hundred millielectron volts below the heavy-hole and light-hole bands. Since the energy range of interest is only several tens of millielectron volts, it is usual to assume that the split-off bands can be ignored. The band structures of the heavy-hole and light-hole bands are approximately described by the 4 × 4

Hamiltonian [37, 38]:

$$\overline{\overline{\mathbf{H}}} = - \begin{bmatrix} P+Q & -S & R & 0 \\ -S^+ & P-Q & 0 & R \\ R^+ & 0 & P-Q & S \\ 0 & R^+ & S^+ & P+Q \end{bmatrix} \begin{matrix} |\frac{3}{2},\frac{3}{2}\rangle \\ |\frac{3}{2},\frac{1}{2}\rangle \\ |\frac{3}{2},-\frac{1}{2}\rangle \\ |\frac{3}{2},-\frac{3}{2}\rangle \end{matrix} \qquad (4.5.13)$$

The Hamiltonian $\overline{\overline{\mathbf{H}}}$ in Eq. (C.24) in Appendix C is written for an arbitrary strain. For simplicity, we restrict ourselves to the special case of a biaxial strain, namely,

$$\varepsilon_{xx} = \varepsilon_{yy} \neq \varepsilon_{zz}$$

$$\varepsilon_{xy} = \varepsilon_{yz} = \varepsilon_{zx} = 0 \qquad (4.5.14)$$

Thus

$$R_\varepsilon = S_\varepsilon = 0$$

which essentially covers two of the most important strained systems: (1) a strained-layer semiconductor pseudomorphically grown on a (001)-oriented substrate and (2) a bulk semiconductor under an external uniaxial stress along the z direction. For the case of the lattice-mismatched strain, we obtain

$$\varepsilon_{xx} = \varepsilon_{yy} = \frac{a_0 - a}{a} \qquad (4.5.15a)$$

$$\varepsilon_{zz} = -\frac{2C_{12}}{C_{11}}\varepsilon_{xx} \qquad (4.5.15b)$$

where a_0 and a are the lattice constants of the substrate and the layer material (Fig. 4.7), and C_{11} and C_{12} are the elastic stiffness constants. Equations (4.5.15a) and (4.5.15b) can be derived by using the fact that in the plane of the heterojunction, the layered material is strained such that the

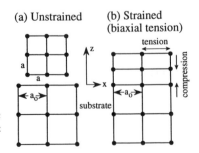

Figure 4.7. A layer material with a lattice constant a to be grown on a substrate with a lattice constant a_0: (a) unstrained; (b) strained.

lattice constant along the plane of the layer is equal to a_0. Therefore, $\varepsilon_{xx} = \varepsilon_{yy} = (a_0 - a)/a$. Since the stress tensor is related to strain by the elastic stiffness tensor with elements C_{ij},

$$\begin{bmatrix} \tau_{xx} \\ \tau_{yy} \\ \tau_{zz} \\ \tau_{xy} \\ \tau_{yz} \\ \tau_{zx} \end{bmatrix} = \begin{bmatrix} C_{11} & C_{12} & C_{12} & 0 & 0 & 0 \\ C_{12} & C_{11} & C_{12} & 0 & 0 & 0 \\ C_{12} & C_{12} & C_{11} & 0 & 0 & 0 \\ 0 & 0 & 0 & C_{44} & 0 & 0 \\ 0 & 0 & 0 & 0 & C_{44} & 0 \\ 0 & 0 & 0 & 0 & 0 & C_{44} \end{bmatrix} \begin{bmatrix} \varepsilon_{xx} \\ \varepsilon_{yy} \\ \varepsilon_{zz} \\ 0 \\ 0 \\ 0 \end{bmatrix} \qquad (4.5.16)$$

we find $\tau_{xy} = \tau_{yz} = \tau_{zx} = 0$. There should also be no stress in the z direction:

$$0 = \tau_{zz} = C_{12}(\varepsilon_{xx} + \varepsilon_{yy}) + C_{11}\varepsilon_{zz}$$

Therefore, $\varepsilon_{zz} = -(2C_{12}/C_{11})\varepsilon_{xx}$. For the case of an external uniaxial stress T along the z axis ($\tau_{zz} = T$ and $\tau_{xx} = \tau_{yy} = 0$), we have

$$\varepsilon_{xx} = \varepsilon_{yy} = \frac{-C_{12}}{C_{11}^2 + C_{11}C_{12} - 2C_{12}^2}T$$

$$\varepsilon_{zz} = \frac{C_{11} + C_{12}}{C_{11}^2 + C_{11}C_{12} - 2C_{12}^2}T \qquad (4.5.17)$$

The results and conclusions presented here can be generalized to other crystal orientations or stress directions. As an example, we discuss one of the most important systems: strained $In_{1-x}Ga_xAs$ on InP. High-quality and highly strained samples of this system have already been grown and widely studied for optoelectronics applications. All of the material parameters used are listed in Table 4.1, where m_e^* is the electron effective mass and m_0 is the free-electron mass (also see Appendix K). The parameters for $In_{1-x}Ga_xAs$ are taken as the linear interpolation of those of InAs and GaAs, except that for the energy gap, $E_g(In_{1-x}Ga_xAs) = 0.324 + 0.7x + 0.4x^2$. We have

$$a(x) = a(In_{1-x}Ga_xAs) = 5.6533x + 6.0584(1 - x) \qquad (4.5.18a)$$

$$\varepsilon_{xx} = \varepsilon_{yy} = \frac{a_0 - a(x)}{a(x)} \qquad (4.5.18b)$$

where $a_0 = 5.8688$ Å for the InP substrate.

At $x = 0.468 \simeq 0.47$, $a(0.468) = a_0$ and the strain is zero. In this case $In_{0.53}Ga_{0.47}As$ is lattice matched to InP. When $x > 0.468$, the gallium mole fraction is increased; therefore, the lattice constant is decreased and the

Table 4.1. Material Parameters

Parameters	GaAs	InAs	InP
a_0 (Å)	5.6533	6.0584	5.8688
E_g (eV)	1.424	0.36	1.344
γ_1	6.85	20.4	4.95
γ_2	2.1	8.3	1.65
γ_3	2.9	9.1	2.35
C_{11} (10^{11}dyn/cm^2)	11.879	8.329	10.11
C_{12}(10^{11}dyn/cm^2)	5.376	4.526	5.61
$a = a_c - a_v$ (eV)	-9.77	-6.0	-8.6
b (eV)	-1.7	-1.8	-2.0
m_e^*/m_0	0.067	0.025	0.077

In$_{1-x}$Ga$_x$As will be under biaxial tension (Fig. 4.7). Here biaxial tension means that the lattice in the parallel (xy) plane will experience a tensile strain with a simultaneous compressive strain along the growth (z) direction. On the other hand, if $x < 0.468$, we have $a(x) > a_0$, and we will have the case of biaxial compression.

At the zone center, $\mathbf{k} = 0$, we have only P_ε and Q_ε appearing in the diagonal terms of the matrix (4.5.13) nonvanishing. Therefore, we obtain the band-edge energies of the heavy-hole and light-hole bands:

$$
E_{HH}(\mathbf{k} = 0) = -P_\varepsilon - Q_\varepsilon = a_v(\varepsilon_{xx} + \varepsilon_{yy} + \varepsilon_{zz}) + \frac{b}{2}(\varepsilon_{xx} + \varepsilon_{yy} - 2\varepsilon_{zz})
$$
(4.5.19a)

$$
E_{LH}(\mathbf{k} = 0) = -P_\varepsilon + Q_\varepsilon = a_v(\varepsilon_{xx} + \varepsilon_{yy} + \varepsilon_{zz}) - \frac{b}{2}(\varepsilon_{xx} + \varepsilon_{yy} - 2\varepsilon_{zz})
$$
(4.5.19b)

On the other hand, the conduction band-edge energy of the electron is given by

$$
E_c(\mathbf{k} = 0) = E_g + a_c(\varepsilon_{xx} + \varepsilon_{yy} + \varepsilon_{zz})
$$
(4.5.20)

Note that both the conduction and valence band energies are defined to be positive for the upward direction of the energy. The net energy transitions will be

$$
E_{C-HH}(\mathbf{k} = 0) = E_g + a_c(\varepsilon_{xx} + \varepsilon_{yy} + \varepsilon_{zz}) + P_\varepsilon + Q_\varepsilon
$$

$$
= E_g + a(\varepsilon_{xx} + \varepsilon_{yy} + \varepsilon_{zz}) - \frac{b}{2}(\varepsilon_{xx} + \varepsilon_{yy} - 2\varepsilon_{zz})
$$
(4.5.21a)

for the conduction to heavy-hole band, and

$$E_{C-LH}(\mathbf{k} = 0) = E_g + a_c(\varepsilon_{xx} + \varepsilon_{yy} + \varepsilon_{zz}) + P_\varepsilon - Q_\varepsilon$$

$$= E_g + a(\varepsilon_{xx} + \varepsilon_{yy} + \varepsilon_{zz}) + \frac{b}{2}(\varepsilon_{xx} + \varepsilon_{yy} - 2\varepsilon_{zz})$$

$$(4.5.21b)$$

for the conduction to the light-hole band, where E_g is the band gap of the unstrained semiconductor, and

$$a = a_c - a_v \qquad (4.5.22)$$

is the hydrostatic deformation potential. Sometimes the hydrostatic and shear deformation energies, δE_{hy} and δE_{sh}, are defined, respectively, as

$$\delta E_{hy} = -a(\varepsilon_{xx} + \varepsilon_{yy} + \varepsilon_{zz}) \qquad (4.5.23)$$

and

$$\frac{1}{2}\delta E_{sh} = Q_\varepsilon = -\frac{b}{2}(\varepsilon_{xx} + \varepsilon_{yy} - 2\varepsilon_{zz}) \qquad (4.5.24)$$

The effective band gaps are given by

$$E_{C-HH} = E_g - \delta E_{hy} + \tfrac{1}{2}\delta E_{sh} \qquad (4.5.25a)$$

$$E_{C-LH} = E_g - \delta E_{hy} - \tfrac{1}{2}\delta E_{sh} \qquad (4.5.25b)$$

For the Hamiltonian in Eq. (4.5.13) or (C.24) in Appendix C, the valence-band structure of a bulk semiconductor is determined by the algebraic equation

$$\det\left[H_{ij}(\mathbf{k}) - \delta_{ij}E\right] = 0 \qquad (4.5.26)$$

where k is now interpreted as a real vector, and the envelope functions are taken as plane waves. For the 4×4 Hamiltonian, the solutions of Eq. (4.5.26) are simply

$$E_{HH}(\mathbf{k}) = -P_\varepsilon - P_k - \text{sgn}(Q_\varepsilon)\sqrt{(Q_\varepsilon + Q_k)^2 + |R_k|^2 + |S_k|^2} \quad (4.5.27a)$$

$$E_{LH}(\mathbf{k}) = -P_\varepsilon - P_k + \text{sgn}(Q_\varepsilon)\sqrt{(Q_\varepsilon + Q_k)^2 + |R_k|^2 + |S_k|^2} \quad (4.5.27b)$$

for the heavy holes and light holes, respectively. Each of the solutions is doubly degenerate. Note that it is important to include the sign factor $\text{sgn}(Q_\varepsilon)(= +1$ for $Q_\varepsilon > 0$ and $= -1$ for $Q_\varepsilon < 0)$ in front of the square root, because Q_ε can be either negative (compressive strain) or positive (tensile strain), while the square root of a positive quantity is conventionally taken as a positive. As $\mathbf{k}$ approaches zero, the band-edge energies of the heavy hole and light hole in (4.5.19) should be recovered. Note that for the unstrained case, the following expressions,

$$E(k) = -P_k \pm \sqrt{|Q_k|^2 + |R_k|^2 + |S_k|^2}$$
$$= Ak^2 \pm \sqrt{B^2k^2 + C^2(k_x^2k_y^2 + k_y^2k_z^2 + k_z^2k_x^2)} \quad (4.5.27c)$$

give the heavy-hole and light-hole dispersion relations.

The dispersion relations for the heavy-hole band $E_{HH}(\mathbf{k})$ and the light-hole band $E_{LH}(\mathbf{k})$ vs. the crystal growth direction k_z and the parallel direction k_x can be obtained analytically from (4.5.27a) and (4.5.27b), respectively. Along the parallel plane, e.g., the k_x direction ($k_y = k_z = 0$), we have for $Q_\varepsilon < 0$ (biaxial compression) and k_x is finite

$$E_{HH}(k_x) = -P_\varepsilon - \frac{\hbar^2\gamma_1}{2m_0}k_x^2 + \left[\left(Q_\varepsilon + \frac{\hbar^2\gamma_2}{2m_0}k_x^2\right)^2 + 3\left(\frac{\hbar^2\gamma_2}{2m_0}\right)^2 k_x^4\right]^{1/2}$$

$$(4.5.28a)$$

from (4.5.27a) and

$$E_{LH}(k_x) = -P_\varepsilon - \frac{\hbar^2\gamma_1}{2m_0}k_x^2 - \left[\left(Q_\varepsilon + \frac{\hbar^2\gamma_2}{2m_0}k_x^2\right)^2 + 3\left(\frac{\hbar^2\gamma_2}{2m_0}\right)^2 k_x^4\right]^{1/2}$$

$$(4.5.28b)$$

for the light hole from (4.5.27b).

In the case of biaxial tension ($Q_\varepsilon > 0$), we have

$$E_{HH}(k_x) = -P_\varepsilon - \frac{\hbar^2\gamma_1}{2m_0}k_x^2 - \left[\left(Q_\varepsilon + \frac{\hbar^2\gamma_2}{2m_0}k_x^2\right)^2 + 3\left(\frac{\hbar^2\gamma_2}{2m_0}\right)^2 k_x^4\right]^{1/2}$$

$$(4.5.29a)$$

$$E_{LH}(k_x) = -P_\varepsilon - \frac{\hbar^2\gamma_1}{2m_0}k_x^2 + \left[\left(Q_\varepsilon + \frac{\hbar^2\gamma_2}{2m_0}k_x^2\right)^2 + 3\left(\frac{\hbar^2\gamma_2}{2m_0}\right)^2 k_x^4\right]^{1/2}$$

$$(4.5.29b)$$

Again, at $k_x = 0$, $E_{HH}(0) + P_\varepsilon = -Q_\varepsilon < 0$ and $E_{LH}(0) + P_\varepsilon = +Q_\varepsilon > 0$, which agree with (4.5.19). Along the k_z direction, we obtain for both compression and tension,

$$E_{HH}(k_z) = -P_\varepsilon - \frac{\hbar^2}{2m_0}(\gamma_1 - 2\gamma_2)k_z^2 - Q_\varepsilon \quad (4.5.30a)$$

$$E_{LH}(k_z) = -P_\varepsilon - \frac{\hbar^2}{2m_0}(\gamma_1 + 2\gamma_2)k_z^2 + Q_\varepsilon \quad (4.5.30b)$$

The results of the valence-band structure for E_{HH} and E_{LH} versus k_x and k_z for both compression and tension are shown in Fig. 4.8 for $Ga_x In_{1-x} As$ grown on InP substrate. We can see clearly that the heavy-hole band has a lighter effective mass than the light-hole band in the k_x direction near $k_x = 0$ for the compression case ($Q_\varepsilon < 0$). The heavy hole still keeps its feature of a heavy effective mass along the k_z direction, and it is above the light-hole band at $k = 0$ for the compression case. On the other hand, the light-hole band is above the heavy hole band in the case of tension ($x > 0.468$).

For a finite and fixed strain, the small-k expansion of the above dispersion relation can be written as

$$E_{HH}(\mathbf{k}) \approx -P_\varepsilon - Q_\varepsilon$$

$$-\left(\frac{\hbar^2}{2m_0}\right)[(\gamma_1 + \gamma_2)k_t^2 + (\gamma_1 - 2\gamma_2)k_z^2] \qquad (k \to 0)$$

$$(4.5.31a)$$

$$E_{LH}(\mathbf{k}) \approx -P_\varepsilon + Q_\varepsilon$$

$$-\left(\frac{\hbar^2}{2m_0}\right)[(\gamma_1 - \gamma_2)k_t^2 + (\gamma_1 + 2\gamma_2)k_z^2] \qquad (k \to 0)$$

$$(4.5.31b)$$

where the transverse wave vector has a magnitude $k_t = \sqrt{k_x^2 + k_y^2}$. From (4.5.31) we immediately obtain the band-edge energies

$$E_{HH}(0) \approx -P_\varepsilon - Q_\varepsilon \qquad (4.5.32a)$$

$$E_{LH}(0) \approx -P_\varepsilon + Q_\varepsilon \qquad (4.5.32b)$$

and the effective masses parallel ($\parallel$ or t) or perpendicular ($\perp$ or z) to the xy plane

$$\frac{m_{hh}^z}{m_0} = \frac{1}{\gamma_1 - 2\gamma_2} \qquad \frac{m_{hh}^t}{m_0} = \frac{1}{\gamma_1 + \gamma_2}$$

$$\frac{m_{lh}^z}{m_0} = \frac{1}{\gamma_1 + 2\gamma_2} \qquad \frac{m_{lh}^t}{m_0} = \frac{1}{\gamma_1 - \gamma_2} \qquad (4.5.33)$$

These are the well-known results [26] when the coupling to the spin–orbit split-off band is neglected.

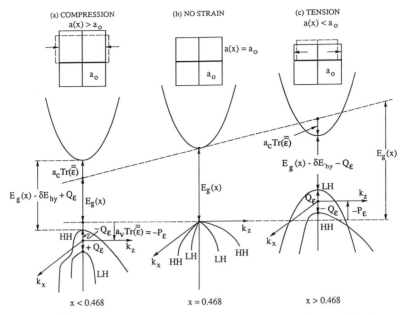

Figure 4.8. The energy-band structure in the momentum space for a bulk $Ga_x In_{1-x} As$ material under (a) biaxial compression, (b) lattice-matched condition, and (c) biaxial tension for different Ga mole fractions x. The heavy-hole band is above the light-hole band and its effective mass in the transverse plane (the k_x or k_y direction) is lighter than that of the light-hole band in the compressive strain case in (a). The light-hole band shifts above the heavy-hole band in the case of tension in (c). (After Ref. 37.)

4.5.3 Band Structures of Strained Semiconductors With Spin–Orbit Split-Off Bands Coupling [38]

If the coupling with the spin–orbit split-off bands is taken into account, the 6×6 Hamiltonian (C.24) in Appendix C has to be used. For the band edge at $\mathbf{k} = 0$, the Hamiltonian $\overline{\overline{H}}$ is simplified to

$$
\overline{\overline{H}}(\mathbf{k} = 0) = - \begin{bmatrix}
P_\varepsilon + Q_\varepsilon & 0 & 0 & 0 & 0 & 0 \\
0 & P_\varepsilon - Q_\varepsilon & 0 & 0 & -\sqrt{2}\,Q_\varepsilon & 0 \\
0 & 0 & P_\varepsilon - Q_\varepsilon & 0 & 0 & \sqrt{2}\,Q_\varepsilon \\
0 & 0 & 0 & P_\varepsilon + Q_\varepsilon & 0 & 0 \\
0 & -\sqrt{2}\,Q_\varepsilon & 0 & 0 & P_\varepsilon + \Delta & 0 \\
0 & 0 & \sqrt{2}\,Q_\varepsilon & 0 & 0 & P_\varepsilon + \Delta
\end{bmatrix}
$$

$$(4.5.34)$$

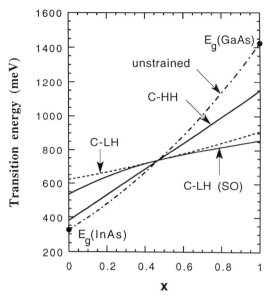

Figure 4.9. The energy band gap of a bulk $In_{1-x}Ga_xAs$ vs. the Ga mole fraction x: -··-, unstrained $In_{1-x}Ga_xAs$; —, transition energies from the conduction band (C) to the heavy-hole (HH) and light-hole (LH) bands for a bulk $In_{1-x}Ga_xAs$ pseudomorphically grown on InP; ---, the conduction to light-hole transition energy calculated without the spin–orbit (SO) split-off band coupling. (After Ref. 38.)

Clearly, the heavy-hole bands are decoupled from the rest of bands, while the light-hole bands ($|\frac{3}{2}, \pm \frac{1}{2}\rangle$) are coupled with the split-off bands ($|\frac{1}{2}, \pm \frac{1}{2}\rangle$) through the strain-dependent off-diagonal terms. This coupling would be totally unaccounted for in the 4×4 approximation. For $In_{1-x}Ga_xAs$ on InP, the transition energies from the heavy-hole and the light-hole bands to the conduction band with and without the SO coupling are shown [38] in Fig. 4.9. The comparison demonstrates how important it is to include the spin–orbit split-off bands, because the error in the light-hole energies could be as large as several tens of millielectron volts. The error is comparable to the heavy-hole and light-hole energy splits and is certainly too large to be ignored. As a consequence of the coupling, the eigenvectors corresponding to the energy $E(0)(= E_{LH}(0)$ or $E_{SO}(0))$, determined by

$$\begin{bmatrix} -P_\varepsilon + Q_\varepsilon & \pm\sqrt{2}Q_\varepsilon \\ \pm\sqrt{2}Q_\varepsilon & -P_\varepsilon - \Delta \end{bmatrix} \begin{bmatrix} F_{3/2,\,\pm 1/2} \\ F_{1/2,\,\pm 1/2} \end{bmatrix} = E(0) \begin{bmatrix} F_{3/2,\,\pm 1/2} \\ F_{1/2,\,\pm 1/2} \end{bmatrix} \qquad (4.5.35a)$$

$$|F_{3/2,\,\pm 1/2}|^2 + |F_{1/2,\,\pm 1/2}|^2 = 1 \qquad (4.5.35b)$$

are not a *pure* light-hole or split-off state, but an admixture of the light-hole and split-off states. The band-edge energies can be readily solved from (4.5.34) or (4.5.35a),

$$E_{HH}(0) = -P_\varepsilon - Q_\varepsilon \tag{4.5.36a}$$

$$E_{LH}(0) = -P_\varepsilon + \frac{1}{2}\left(Q_\varepsilon - \Delta + \sqrt{\Delta^2 + 2\Delta Q_\varepsilon + 9Q_\varepsilon^2}\right) \tag{4.5.36b}$$

$$E_{SO}(0) = -P_\varepsilon + \frac{1}{2}\left(Q_\varepsilon - \Delta - \sqrt{\Delta^2 + 2\Delta Q_\varepsilon + 9Q_\varepsilon^2}\right) \tag{4.5.36c}$$

If the split-off bands are included in the 6×6 Hamiltonian, the E-k relation determined by (4.5.26) becomes a sixth-order polynomial of E, which apparently can be decomposed into two identical cubic polynomials because of the symmetry property of the Hamiltonian. However, an attempt to expand and factor directly the determinantal equation is tedious. The details are given in Ref. 38.

The most important results are obtained from the series expansion of E up to the second order of k near the band edges:

$$E_{HH}(\mathbf{k}) \approx E_{HH}(0) - \left(\frac{\hbar^2}{2m_0}\right)\left[(\gamma_1 + \gamma_2)k_t^2 + (\gamma_1 - 2\gamma_2)k_z^2\right] \tag{4.5.37a}$$

$$E_{LH}(\mathbf{k}) \approx E_{LH}(0) - \left(\frac{\hbar^2}{2m_0}\right)\left[(\gamma_1 - f_+\gamma_2)k_t^2 + (\gamma_1 + 2f_+\gamma_2)k_z^2\right]$$
$$\tag{4.5.37b}$$

$$E_{SO}(\mathbf{k}) \approx E_{SO}(0) - \left(\frac{\hbar^2}{2m_0}\right)\left[(\gamma_1 - f_-\gamma_2)k_t^2 + (\gamma_1 + 2f_-\gamma_2)k_z^2\right]$$
$$\tag{4.5.37c}$$

where f_+ and f_- are dimensionless, strain-dependent factors

$$f_\pm(x) = \frac{2x\left[1 + \frac{3}{2}\left(x - 1 \pm \sqrt{1 + 2x + 9x^2}\right)\right] + 6x^2}{\frac{3}{4}\left(x - 1 \pm \sqrt{1 + 2x + 9x^2}\right)^2 + x - 1 \pm \sqrt{1 + 2x + 9x^2} - 3x^2} \tag{4.5.38}$$

where $x = Q_\varepsilon/\Delta$ and $k_t^2 = k_x^2 + k_y^2$. The band-edge energies in (4.5.37a)–(4.5.37c) are given in (4.5.36a)–(4.5.36c).

From (4.5.37a)–(4.5.37c) we obtain the effective masses perpendicular ($\perp = z$) and parallel ($\| = t$) to the x-y plane:

$$\frac{m_{hh}^z}{m_0} = \frac{1}{\gamma_1 - 2\gamma_2} \qquad \frac{m_{hh}^t}{m_0} = \frac{1}{\gamma_1 + \gamma_2}$$

$$\frac{m_{lh}^z}{m_0} = \frac{1}{\gamma_1 + 2f_+\gamma_2} \qquad \frac{m_{lh}^t}{m_0} = \frac{1}{\gamma_1 - f_+\gamma_2}$$

$$\frac{m_{so}^z}{m_0} = \frac{1}{\gamma_1 + 2f_-\gamma_2} \qquad \frac{m_{so}^t}{m_0} = \frac{1}{\gamma_1 - f_-\gamma_2} \qquad (4.5.39)$$

Note that $f_+ = 1$, $f_- = 0$ for the limiting case of zero strain ($Q_\varepsilon \to 0$, $x \to 0$), and the effective masses derived from the 6×6 Hamiltonian become identical to those derived from the 4×4 Hamiltonian.

4.6 ELECTRONIC STATES IN AN ARBITRARY ONE-DIMENSIONAL POTENTIAL

In this section we show how the Schrödinger equation for a one-dimensional potential profile with an arbitrary shape can be solved using a propagation-matrix approach, which is similar to that used in electromagnetic wave reflection or guidance in a multilayered medium [39, 40]. An arbitrary profile, $V(z)$, can always be approximated by a piecewise step profile, as shown in Fig. 4.10 as long as the original potential profile does not have singularities.

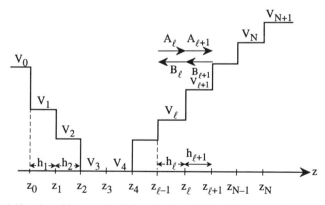

Figure 4.10. An arbitrary potential profile is subdivided into a stepwise potential.

4.6.1   Derivation of the Propagation Matrix Equation and Its Solution for the Eigenvalues

For a Schrödinger equation of the form

$$-\frac{\hbar^2}{2}\frac{d}{dz}\frac{1}{m(z)}\frac{d}{dz}\psi(z) + V(z)\psi(z) = E\psi(z) \qquad (4.6.1)$$

we find that in the region l, $z_{l-1} \leq z \leq z_l$

$$-\frac{\hbar^2}{2m_l}\frac{d^2}{dz^2}\psi_l(z) + V_l\psi_l(z) = E\psi_l(z) \qquad (4.6.2)$$

where $V(z) = V_l$ and $m(z) = m_l$ in region l. The solution can be written in the form

$$\psi_l(z) = A_l\,e^{ik_l(z-z_l)} + B_l\,e^{-ik_l(z-z_l)} \qquad \text{for } z_{l-1} \leq z \leq z_l \quad (4.6.3)$$

where the wave number in region l is

$$k_l = \sqrt{\frac{2m_l}{\hbar^2}(E - V_l)} \qquad (4.6.4)$$

Matching the boundary conditions in which $\psi(z)$ and $(1/m(z))d\psi(z)/dz$ are continuous at $z = z_l$, we find

$$A_l + B_l = A_{l+1}\,e^{ik_{l+1}(z_l-z_{l+1})} + B_{l+1}\,e^{-ik_{l+1}(z_l-z_{l+1})} \qquad (4.6.5a)$$

$$\frac{k_l}{m_l}(A_l - B_l) = \frac{k_{l+1}}{m_{l+1}}\left(A_{l+1}e^{ik_{l+1}(z_l-z_{l+1})} - B_{l+1}\,e^{-ik_{l+1}(z_l-z_{l+1})}\right) \quad (4.6.5b)$$

Define

$$P_{l(l+1)} = \frac{m_l k_{l+1}}{m_{l+1}k_l} \qquad (4.6.6)$$

and $z_{l+1} - z_l = h_{l+1}$ is the thickness of the region $l + 1$.

We can express A_{l+1} and B_{l+1} in terms of A_l and B_l in a matrix form:

$$\begin{bmatrix} A_{l+1} \\ B_{l+1} \end{bmatrix} = \overline{\overline{\mathbf{F}}}_{(l+1)l}\begin{bmatrix} A_l \\ B_l \end{bmatrix} \qquad (4.6.7a)$$

where we have defined a forward-propagation matrix

$$\overline{\overline{F}}_{(l+1)l} = \frac{1}{2}\begin{bmatrix} \left(1 + P_{(l+1)l}\right)e^{ik_{l+1}h_{l+1}} & \left(1 - P_{(l+1)l}\right)e^{ik_{l+1}h_{l+1}} \\ \left(1 - P_{(l+1)l}\right)e^{-ik_{l+1}h_{l+1}} & \left(1 + P_{(l+1)l}\right)e^{-ik_{l+1}h_{l+1}} \end{bmatrix} \quad (4.6.7b)$$

The relation can propagate from layer to layer:

$$\begin{bmatrix} A_{N+1} \\ B_{N+1} \end{bmatrix} = \overline{\overline{F}}_{(N+1)N}\overline{\overline{F}}_{N(N-1)} \cdots \overline{\overline{F}}_{10}\begin{bmatrix} A_0 \\ B_0 \end{bmatrix} \quad (4.6.8)$$

For bound-state solutions, we have $E < V_0$ and V_{N+1}. Therefore,

$$k_0 = i\alpha_0 \qquad \alpha_0 = \sqrt{\frac{2m_0}{\hbar^2}(V_0 - E)} \quad (4.6.9a)$$

$$k_{N+1} = i\alpha_{N+1} \qquad \alpha_{N+1} = \sqrt{\frac{2m_{N+1}}{\hbar^2}(V_{N+1} - E)} \quad (4.6.9b)$$

The solutions in region 0 and region $N + 1$ must be decaying solutions of the form

$$\psi_0(z) = B_0 e^{+\alpha_0(z-z_0)} \quad (4.6.10a)$$

and

$$\psi_{N+1}(z) = A_{N+1} e^{-\alpha_{N+1}(z-z_{N+1})} \quad (4.6.10b)$$

where $A_0 = 0$ and $B_{N+1} = 0$. Note that the last position z_{N+1} is introduced for the use of the propagation matrix and it is arbitrary (can be set equal to z_N).

We write the product of the matrices

$$\overline{\overline{F}}_{(N+1)N} \cdots \overline{\overline{F}}_{10} \equiv \begin{bmatrix} f_{11} & f_{12} \\ f_{21} & f_{22} \end{bmatrix} \quad (4.6.11)$$

and therefore

$$\begin{bmatrix} A_{N+1} \\ 0 \end{bmatrix} = \begin{bmatrix} f_{11} & f_{12} \\ f_{21} & f_{22} \end{bmatrix}\begin{bmatrix} 0 \\ B_0 \end{bmatrix} \quad (4.6.12)$$

For nontrivial solutions, we must have the eigenequation satisfied,

$$f_{22}(E) = 0 \quad (4.6.13)$$

since B_0 cannot be zero (otherwise, all field amplitudes are zero). Solving the eigenequation (4.6.13), we obtain the eigenvalues E. In practice, $f_{22}(E)$ is a complex function and the eigenvalues E are real. Therefore, a convenient method to find the eigenvalues is to search for the minima of $|f_{22}(E)|$ (or $\log|f_{22}(E)|$) in the range of energy given by the lowest and highest energies of the potential profile $V(z)$.

4.6.2   Self-consistent Solution for a Modulation-Doped Quantum Well

Consider a quantum-well structure with a built-in potential

$$
V_{\text{bi}}^e(z) = \begin{cases} \Delta E_c & |z| \geq \dfrac{L_w}{2} \\ 0 & |z| < \dfrac{L_w}{2} \end{cases}
\tag{4.6.14a}
$$

$$
V_{\text{bi}}^h(z) = \begin{cases} -\Delta E_v & |z| \geq \dfrac{L_w}{2} \\ 0 & |z| < \dfrac{L_w}{2} \end{cases}
\tag{4.6.14b}
$$

and a doping profile $N_D(z)$ and $N_A(z)$. The Schrödinger equations are

$$
\left[\frac{-\hbar^2}{2m_e^*} \frac{d^2}{dz^2} + V_e(z) \right] f(z) = E_e(0) f(z)
\tag{4.6.15a}
$$

$$
\left[\frac{+\hbar^2}{2m_h^*} \frac{d^2}{dz^2} + V_h(z) \right] g(z) = E_h(0) g(z)
\tag{4.6.15b}
$$

where the total potential profiles for the electrons and holes are, respectively,

$$
V_e(z) = V_{\text{bi}}^e(z) + |e|Fz + V_H(z)
\tag{4.6.16a}
$$

$$
V_h(z) = V_{\text{bi}}^h(z) + |e|Fz + V_H(z)
\tag{4.6.16b}
$$

and the Hartree potential is given by

$$
V_H(z) = -|e|\phi(z)
\tag{4.6.17}
$$

Note that we define $V_{\text{bi}}^h(z)$ as in (4.6.14b) such that all energies are measured upward. Therefore, the same expressions for the field-induced potential and the Hartree potential are used in (4.6.16a) and (4.6.16b) without any sign change. Here F is the externally applied field ($= 0$ here) and $\phi(z)$ is the electrostatic potential, which satisfies Gauss's law or Poisson's equation:

$$\nabla \cdot (\varepsilon \mathbf{E}) = \rho(z) \tag{4.6.18}$$

For a one-dimensional problem, we have the electric field

$$\mathbf{E} = \hat{z}E(z) \quad \text{and} \quad E(z) = -\frac{\partial \phi}{\partial z} \tag{4.6.19}$$

Therefore,

$$\frac{\partial}{\partial z}[\varepsilon E(z)] = \rho(z) \tag{4.6.20}$$

or

$$\frac{\partial}{\partial z}\left(\varepsilon \frac{\partial \phi}{\partial z}\right) = -\rho(z) \tag{4.6.21}$$

The charge distribution is given by

$$\rho(z) = |e|[p(z) - n(z) + N_D^+(z) - N_A^-(z)] \tag{4.6.22}$$

where $N_D^+(z)$ and $N_A^-(z)$ are the ionized donor and acceptor concentrations, respectively. The electron and hole concentrations, $n(z)$ and $p(z)$, are related to the wave functions of the nth conduction subband and the mth valence subband by

$$n(z) = \sum_n |f_n(z)|^2 N_{ns} \tag{4.6.23a}$$

$$p(z) = \sum_m |g_m(z)|^2 P_{ms} \tag{4.6.23b}$$

where the sums over n and m are only over the (lowest few) occupied subbands. Here the surface electron concentration in the nth conduction subband is

$$\begin{aligned}
N_{ns} &= \frac{2}{A} \sum_{k_t} \frac{1}{1 + e^{[E_{en}(k_t) - F_c]/k_B T}} \\
&= \frac{k_B T m_e^*}{\pi \hbar^2} \ln\left(1 + e^{[F_c - E_{en}(0)]/k_B T}\right) \tag{4.6.24a}
\end{aligned}$$

where

$$\frac{1}{A} \sum_{\mathbf{k}_t} = \int_0^\infty \frac{2\pi k_t \, dk_t}{(2\pi)^2}$$

and

$$\int \frac{dx}{1 + e^x} = -\ln(1 + e^{-x})$$

have been used. The surface hole concentration in the mth valence subband is

$$
\begin{aligned}
P_{ms} &= \frac{2}{A} \sum_{\mathbf{k}_t} \frac{1}{1 + e^{[F_v - E_{hm}(k_t)]/k_B T}} \\
&= \frac{k_B T m_h^*}{\pi \hbar^2} \ln\left(1 + e^{[E_{hm}(0) - F_v]/k_B T}\right)
\end{aligned}
\tag{4.6.24b}
$$

where $F_c = F_v = E_F$ in a modulation-doped sample without any external injection of carriers. The parabolic energy dispersion relations

$$E_{en}(k_t) = E_{en}(0) + \frac{\hbar^2 k_t^2}{2m_e} \tag{4.6.25a}$$

$$E_{hm}(k_t) = E_{hm}(0) - \frac{\hbar^2 k_t^2}{2m_h^*} \tag{4.6.25b}$$

have been used in the summation over the two-dimensional wave vector $\mathbf{k}$ in the x-y plane. The Fermi level is obtained from the charge neutrality condition:

$$\int_{-L/2}^{L/2} \rho(z) = 0 \tag{4.6.26a}$$

or

$$N_D^+ L_D + \sum_m P_{ms} = N_A^- L_A + \sum_n N_{ns} \tag{4.6.26b}$$

4.6.3   N-type Modulation-Doped Quantum Well

For an n-type modulation-doped quantum well (Fig. 4.11) we have $N_A = 0$ and $p(z) \simeq 0$. We only have to solve for the electron wave function $f(z)$ and the electrostatic potential $\phi(z)$ self-consistently. The electric field $E(z)$ is obtained by integration

$$E(z) = \frac{1}{\varepsilon} \int_{-L/2}^z \rho(z') \, dz' + E\left(-\frac{L}{2}\right) \tag{4.6.27}$$

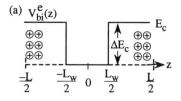

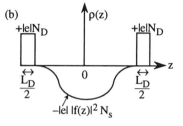

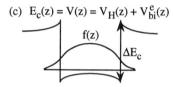

(c) $E_c(z) = V(z) = V_H(z) + V_{bi}^e(z)$

Figure 4.11. (a) The built-in potential, (b) the charge density, and (c) the screened electron potential energy profile of an n-type modulation doped quantum well.

For a symmetrically doped quantum well without any external bias, we have $E(-L/2) = 0$ by a symmetry consideration. Since $V_H(z) = -|e|\phi(z)$, we have

$$\frac{\partial V_H(z)}{\partial z} = +|e|E(z) \qquad (4.6.28a)$$

or

$$V_H(z) = |e|\int_{-L/2}^{z} E(z')\,dz' + V_H(0) \qquad (4.6.28b)$$

Here $V_H(0)$ can be chosen to be zero since it is only a reference potential energy.

Example We consider an n-type modulation-doped quantum well with doping widths $L_D/2 = 10$ Å at both ends (see Fig. 4.11), and $N_D = 4 \times 10^{18}$ cm^{-3} in a GaAs/Al$_{0.3}$Ga$_{0.7}$As quantum-well structure. The surface carrier concentration in a period is $N_D L_D = 8 \times 10^{11}$ cm^{-2}. The conduction band-edge discontinuity ΔE_c is 251 meV and the built-in potential profile $V_{bi}^e(z)$ is shown as the dashed line in Fig. 4.12a. The final conduction band potential profile $V_e(z) = V_{bi}^e(z) + V_H(z)$ after solving the Schrödinger equation and the Poisson's equation self-consistently is shown as the solid curves together with the corresponding eigenenergies of the lowest two states, E_1 and E_2,

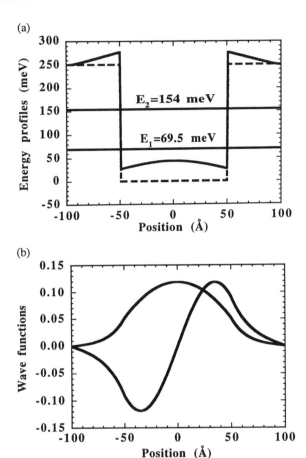

Figure 4.12. (a) Self-consistent potential profile (solid curve) $V_e(z)$ and the built-in potential profile (dashed) $V_{bi}^e(z)$ for a GaAs/Al$_{0.3}$Ga$_{0.7}$As quantum well with modulation dopings within 10 Å at both ends of the profile -100 Å $\leq z \leq -90$ Å and 90 Å $\leq z \leq 100$ Å. The two solid horizontal lines represent the energy levels E_1 and E_2 of the self-consistent potential. (b) The corresponding wave functions $f_1(z)$ and $f_2(z)$ of the self-consistent potential in (a).

plotted as two horizontal solid lines. The wave functions $f_1(z)$ and $f_2(z)$ for the self-consistent potential $V_e(z)$ are plotted in Fig. 4.12b. We can see that the effects of the charge distribution with positively ionized donors at the two ends and negative electrons in the wells create a net electric field pointing toward the center of the quantum well. Therefore, the band bending is curved upward since the slope of the potential energy profile gives the electric field and its direction. More examples of the modulation-doped potentials with an externally applied electric field F and their applications to intersubband photodetectors and resonant tunneling diodes can be found in Refs. 41–43. ■

(a)

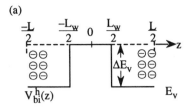

(b)

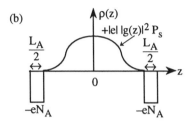

(c) $E_v(z) = V_{bi}^h(z) + V_H(z)$

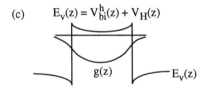

$E_v(z)$ **Figure 4.13.** Same as Fig. 4.11 except for p-type.

4.6.4 P-type Modulation-Doped Quantum Well

For the hole distribution in a P-type modulation-doped quantum well ($N_D^+ = 0$, $n(z) \simeq 0$), Fig. 4.13, it is also possible to take into consideration both the heavy-hole and the light-hole dispersion relations:

$$p(z) = \sum_{m=lh,\,hh} |g_m(z)|^2 P_{ms} \tag{4.6.29a}$$

$$P_{ms} = \frac{1}{A} \sum_{m=lh,\,hh} \sum_{\mathbf{k}_t} \frac{1}{1 + e^{(E_F - E_{hm}(k_t))/k_BT}}$$

$$= \frac{k_BT}{\pi\hbar^2} \left[m_{hh}^* \sum_{m=hh} \ln\left(1 + e^{(E_{hm}-E_F)/k_BT}\right) \right.$$

$$\left. + m_{lh}^* \sum_{m=lh} \ln\left(1 + e^{(E_{hm}-E_F)/k_BT}\right) \right] \tag{4.6.29b}$$

4.6.5 Populations by Both Electrons and Holes

In a laser structure, both electrons and holes are injected by an external electric or optical pumping. Two quasi-Fermi levels, F_c and F_v, can be used to describe $n(z)$ and $p(z)$, respectively. One usually uses the injection current density to find N_s, then

$$N_s = \sum_n N_{ns} \tag{4.6.30}$$

is used to determine F_c. The charge neutrality condition

$$P_s + \int N_D^+(z)\, dz = N_s + \int N_A^-(z)\, dz \tag{4.6.31}$$

is used to determine P_s. Then F_v is determined from

$$P_s = \sum_m P_{ms} \tag{4.6.32}$$

4.7 KRONIG–PENNEY MODEL FOR A SUPERLATTICE

In this section, we apply the propagation matrix approach discussed in Section 4.6 to study a one-dimensional periodic potential problem [44]. The results are the Kronig–Penney model for a superlattice structure [45–48].

The superlattice structure is shown in Fig. 4.14. Each period (or cell) consists of one barrier region with a width b and a well region with a width w. The period is $L = b + w$. The potential within a period $0 < z < L$ is given by

$$V(z) = \begin{cases} V_0 & 0 < z < b \\ 0 & b < z < b + w = L \end{cases} \tag{4.7.1}$$

and $V(z + nL) = V(z)$ for any integer n.

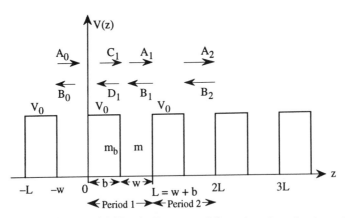

Figure 4.14. A periodic potential (Kronig–Penney model) consists of one barrier region with a thickness b and a quantum-well region with a width w. The period is $L = w + b$.

4.7.1 Derivation of the Propagation Matrix

The wave function in the period n is given by the plane wave solution of the wave equation:

$$\psi_n(z) = \begin{cases} A_n e^{ik(z-nL)} + B_n e^{-ik(z-nL)} & \text{for } nL - w \le z \le nL \\ C_n e^{ik_b[z-b-(n-1)L]} + D_n e^{-ik_b[z-b-(n-1)L]} \\ & \text{for } (n-1)L \le z \le (n-1)L + b \end{cases}$$

$$(4.7.2)$$

$$k = \sqrt{\frac{2m}{\hbar^2}E} \qquad k_b = \sqrt{\frac{2m_b}{\hbar^2}(E - V_0)} \qquad (4.7.3)$$

More specifically, we write

$$\psi(z) = \begin{cases} A_0 e^{ikz} + B_0 e^{-ikz} & -w \le z \le 0 \\ C_1 e^{ik_b(z-b)} + D_1 e^{-ik_b(z-b)} & 0 \le z \le b \\ A_1 e^{ik(z-L)} + B_1 e^{-ik(z-b)} & b \le z \le L \end{cases} \qquad (4.7.4)$$

Using the boundary conditions in which $\psi(z)$ and $(1/m)(\partial\psi/\partial z)$ are continuous at $z = 0$, we find the matrix equation for the coefficients of the wave functions in region 0 and the barrier b:

$$\begin{bmatrix} C_1 \\ D_1 \end{bmatrix} = \overline{\overline{\mathbf{F}}}_{b0} \begin{bmatrix} A_0 \\ B_0 \end{bmatrix} \qquad (4.7.5a)$$

where

$$\overline{\overline{\mathbf{F}}}_{b0} = \frac{1}{2} \begin{bmatrix} (1 + P_{b0}) e^{ik_b b} & (1 - P_{b0}) e^{ik_b b} \\ (1 - P_{b0}) e^{-ik_b b} & (1 + P_{b0}) e^{-ik_b b} \end{bmatrix} \qquad (4.7.5b)$$

$$P_{b0} = \frac{m_b k}{mk_b} \qquad (4.7.5c)$$

Similarly, the boundary conditions at $z = b$ give the matrix equation for region 1 and the barrier b

$$\begin{bmatrix} A_1 \\ B_1 \end{bmatrix} = \overline{\overline{\mathbf{F}}}_{1b} \begin{bmatrix} C_1 \\ D_1 \end{bmatrix} \qquad (4.7.6a)$$

$$\overline{\overline{\mathbf{F}}}_{1b} = \frac{1}{2} \begin{bmatrix} (1 + P_{1b}) e^{ikw} & (1 - P_{1b}) e^{ikw} \\ (1 - P_{1b}) e^{-ikw} & (1 + P_{1b}) e^{-ikw} \end{bmatrix} \qquad (4.7.6b)$$

where

$$P_{1b} = \frac{mk_b}{m_b k} \tag{4.7.6c}$$

Define

$$P = P_{1b} = \frac{mk_b}{m_b k} = \frac{1}{P_{b0}} \tag{4.7.7}$$

We have the transition matrix for one period consisting of one well and one barrier

$$\begin{bmatrix} A_1 \\ B_1 \end{bmatrix} = \overline{\overline{\mathbf{T}}} \begin{bmatrix} A_0 \\ B_0 \end{bmatrix} \tag{4.7.8a}$$

where

$$\overline{\overline{\mathbf{T}}} = \overline{\overline{\mathbf{F}}}_{1b} \overline{\overline{\mathbf{F}}}_{b0} = \begin{bmatrix} t_{11} & t_{12} \\ t_{21} & t_{22} \end{bmatrix} \tag{4.7.8b}$$

and the matrix elements are given by

$$t_{11} = e^{ikw} \left[\cos k_b b + \frac{i}{2} \left(P + \frac{1}{P} \right) \sin k_b b \right]$$

$$t_{12} = e^{ikw} \frac{i}{2} \left(P - \frac{1}{P} \right) \sin k_b b$$

$$t_{21} = e^{-ikw} \left(-\frac{i}{2} \right) \left(P - \frac{1}{P} \right) \sin k_b b$$

$$t_{22} = e^{-ikw} \left[\cos k_b b - \frac{i}{2} \left(P + \frac{1}{P} \right) \sin k_b b \right] \tag{4.7.8c}$$

Note that the determinant of the $\overline{\overline{\mathbf{T}}}$ matrix is unity:

$$\det |\overline{\overline{\mathbf{T}}}| = 1 \tag{4.7.9}$$

If we continue the relation (4.7.8a) to the nth period, we find

$$\begin{bmatrix} A_n \\ B_n \end{bmatrix} = \overline{\overline{\mathbf{T}}}^n \begin{bmatrix} A_0 \\ B_0 \end{bmatrix} \tag{4.7.10}$$

4.7.2 Solutions for the Eigenvalues and Eigenvectors

The eigenvalues and eigenvectors of the 2×2 matrix $\overline{\overline{\mathbf{T}}}$ are solutions of the determinantal equation:

$$\overline{\overline{\mathbf{T}}} \begin{bmatrix} A_0 \\ B_0 \end{bmatrix} = t \begin{bmatrix} A_0 \\ B_0 \end{bmatrix} \tag{4.7.11}$$

$$\det \begin{vmatrix} t_{11} - t & t_{12} \\ t_{21} & t_{22} - t \end{vmatrix} = 0 \tag{4.7.12}$$

We obtain a second-order polynomial equation for the eigenvalue t:

$$t^2 - (t_{11} + t_{22})t + \det \overline{\overline{\mathbf{T}}} = 0 \tag{4.7.13}$$

Since $\det |\overline{\overline{\mathbf{T}}}| = 1$, we obtain two roots:

$$t_{\pm} = \left(\frac{t_{11} + t_{22}}{2} \right) \pm \sqrt{\left(\frac{t_{11} + t_{22}}{2} \right)^2 - 1} \tag{4.7.14}$$

If $|(t_{11} + t_{22})/2| > 1$, t_+ and t_- are real and either $\lim_{n \to \infty} |t_{\pm}^n| \to \infty$ or $\lim_{n \to -\infty} |t_{\pm}^n| \to \infty$. In that case, the condition in which the wave function (i.e., $|A_n|$ and $|B_n|$) must remain finite as $n \to \pm\infty$ is violated. Hence, the eigenvalue E must satisfy the condition that

$$\left| \frac{t_{11} + t_{22}}{2} \right| = \left| \cos kw \cos k_b b - \frac{1}{2} \left(P + \frac{1}{P} \right) \sin kw \sin k_b b \right| \le 1 \tag{4.7.15}$$

The eigenvalues can be written in the forms

$$t_+ = e^{iqL} \quad \text{and} \quad t_- = e^{-iqL} \tag{4.7.16}$$

since $|t_{\pm}| = 1$ and $t_+ t_- = 1$. That is, the determinantal equation can be written as

$$(t - t_+)(t - t_-) = 0 \tag{4.7.17a}$$

or

$$t_+ + t_- = t_{11} + t_{22} \tag{4.7.17b}$$

and

$$t_+ t_- = \det |\overline{\overline{\mathbf{T}}}| = 1 \tag{4.7.17c}$$

We conclude that the eigenequation is

$$\cos qL = \cos kw \cos k_b b - \frac{1}{2}\left(P + \frac{1}{P}\right)\sin kw \sin k_b b \quad (4.7.18)$$

and the two eigenvectors

$$\begin{bmatrix} A_0^+ \\ B_0^+ \end{bmatrix} \quad \text{and} \quad \begin{bmatrix} A_0^- \\ B_0^- \end{bmatrix}$$

corresponding to eigenvalues, e^{iql} and e^{-iqL}, respectively, satisfy

$$\overline{\overline{T}}\begin{bmatrix} A_0^+ \\ B_0^+ \end{bmatrix} = e^{iqL}\begin{bmatrix} A_0^+ \\ B_0^+ \end{bmatrix} \quad (4.7.19a)$$

and

$$\overline{\overline{T}}\begin{bmatrix} A_0^- \\ B_0^- \end{bmatrix} = e^{-iqL}\begin{bmatrix} A_0^- \\ B_0^- \end{bmatrix} \quad (4.7.19b)$$

The eigenfunctions can be obtained from the following:

1. The ratio A_0^+/B_0^+ following (4.7.19a) and the normalization condition $\int_0^L |\psi(z)|^2\,dz = 1$,

$$\frac{A_0^+}{B_0^+} = \frac{t_{12}}{e^{iqL} - t_{11}}\left(= \frac{e^{iqL} - t_{22}}{t_{21}}\right)$$

$$= \frac{\frac{1}{2}(P - 1/P)\sin k_b b\, e^{ikw}}{\sin qL - \sin kw \cos k_b b - \frac{1}{2}(P + 1/P)\cos kw \sin k_b b} \quad (4.7.20)$$

where the eigenequation (4.7.18) has been used.

2. The ratio A_0^-/B_0^- from (4.7.19b) with $q \to -q$ in (4.7.20) and the normalization condition.

For bound state solutions, $E < V_0$, k_b is purely imaginary:

$$k_b = i\alpha_b \qquad \alpha_b = \sqrt{\frac{2m_b}{\hbar^2}(V_0 - E)} \quad (4.7.21)$$

The determinantal equation is given by

$$\cos qL = \cos kw \cosh \alpha_b b + \frac{1}{2}\left(\eta - \frac{1}{\eta}\right)\sin kw \sinh \alpha_b b \quad (4.7.22)$$

where η is defined in the following equation when P is purely imaginary:

$$P = \frac{mk_b}{m_b k} = i\frac{m\alpha_b}{m_b k} = i\eta \quad (4.7.23)$$

In summary, the Kronig–Penney model for a superlattice can be obtained from solving the determinantal equation

$$\cos qL = f(E) \quad (4.7.24a)$$

where the eigenequation is defined as

$$f(E) = \begin{cases} \cos kw \cosh \alpha_b b + \dfrac{1}{2}\left(\eta - \dfrac{1}{\eta}\right)\sin kw \sinh \alpha_b b & 0 < E < V_0 \\[2mm] \cos kw \cos k_b b - \dfrac{1}{2}\left(P + \dfrac{1}{P}\right)\sin kw \sin k_b b & V_0 < E \end{cases}$$

$$(4.7.24b)$$

A method to find the solution E is to plot $f(E)$ vs. $E \geq 0$. The region of E such that the condition $|f(E)| \leq 1$ is satisfied will be acceptable. The quantum number q is obtained from

$$qL = \cos^{-1} f(E) \quad (4.7.25)$$

where $f(E)$ is real. For a given E in the acceptable region (called the miniband), the solution q is used to find the eigenfunction from the ratio A_0^+/B_0^+ (or A_0^-/B_0^-) and the normalization condition of the wave function.

Example: A GaAs/Al$_{0.3}$Ga$_{0.7}$As Superlattice We consider a GaAs/ Al$_x$Ga$_{1-x}$As superlattice with $x = 0.3$, a well width $w = 100$ Å, and a barrier width $b = 20$ Å. The barrier height $V_0 = 0.2506$ eV is obtained from the conduction band discontinuity $V_0 = \Delta E_c = 0.67\Delta E_g(x)$, $\Delta E_g(x) = 1.247x$ (eV). We see in Fig. 4.15 that there are two minibands corresponding to the bound states ($E < V_0$) of the superlattice with $|f(E)| \leq 1$. The vertical axis is $f(E)$ and the horizontal axis is the electron energy E in millielectron volts.

Since $\cosh \alpha_b b \geq 1$ at $kw = N\pi$, where N is an integer, we obtain $f(E) = \cosh \alpha_b b \geq 1$ at $k = N\pi/w$. Therefore, $k = N\pi/w$ always occurs either in the forbidden miniband gap or at the edge of the miniband.

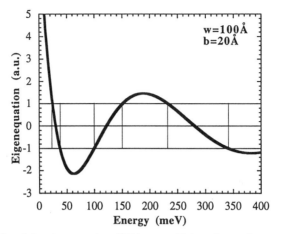

Figure 4.15. A plot of the eigenequation $f(E)(= \cos qL)$ from (4.7.24b) vs. the energy E. Only the miniband ranges such that $-1 \le f(E) \le 1$ are acceptable solutions. The plot is for the conduction minibands of a GaAs/Al$_{0.3}$Ga$_{0.7}$As superlattice with a well width $w = 100$ Å and a barrier width $b = 20$ Å. The barrier height is 250.6 meV.

In Fig. 4.16a we plot the energy spectrum for the electron minibands for a superlattice with a well width $w = 100$ Å and the barrier width b varying from 1 to 80 Å. We see that the width of the miniband energy decreases as the barrier width is increased because the coupling among wells becomes weaker for a thicker barrier. The heavy-hole and light-hole minibands for the same GaAs/Al$_{0.3}$Ga$_{0.7}$As superlattice are shown in Figs. 4.16b and c, respectively. We use $\Delta E_v = 0.33 \, \Delta E_g$, and $m_{hh}^z = m_0/(\gamma_1 - 2\gamma_2)$, $m_{lh}^z = m_0/(\gamma_1 + 2\gamma_2)$, where $\gamma_1 = 6.85$, and $\gamma_2 = 2.10$ for GaAs, $\gamma_1 = 3.45$, and $\gamma_2 = 0.68$ for AlAs. For Al$_x$Ga$_{1-x}$As, we use $\gamma_1(x) = 3.45x + 6.85(1 - x)$ and $\gamma_2(x) = 0.68x + 2.10(1 - x)$. More discussions on semiconductor superlattices using the Kronig–Penny model can be found in Refs. 44–50. ∎

4.7.3 Extension to an Arbitrary Periodic Profile

For a periodic profile with the potential function $V(z)$ within one period of arbitrary shape (Fig. 4.17), we proceed as follows. We divide a period L into N regions with a width $\Delta z = L/N$, and assign C_n and D_n for the forward and the backward propagating waves. Here

$$\begin{bmatrix} C_N \\ D_N \end{bmatrix} = \overline{\overline{\mathbf{F}}}_{N(N-1)} \overline{\overline{\mathbf{F}}}_{(N-1)(N-2)} \cdots \overline{\overline{\mathbf{F}}}_{21} \overline{\overline{\mathbf{F}}}_{10} \begin{bmatrix} C_0 \\ D_0 \end{bmatrix} \qquad (4.7.26)$$

where the matrices $\overline{\overline{\mathbf{F}}}_{(l+1)l}$ are obtained from (4.6.7b). Since the profile is

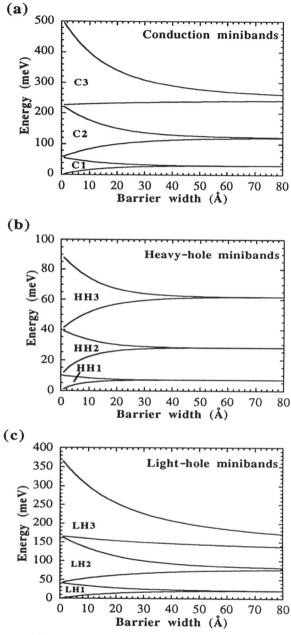

Figure 4.16. The minibands of a GaAs/Al$_{0.3}$Ga$_{0.7}$As superlattice with a well width $w = 100$ Å are plotted vs. the barrier width b. We show (a) the conduction minibands, (b) the heavy-hole minibands, and (c) the light-hole minibands.

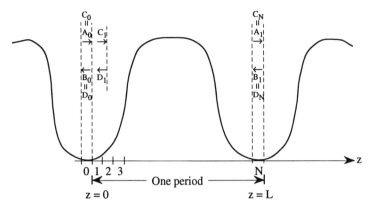

Figure 4.17. A periodic potential profile.

chosen such that the 0th region is identical to the Nth region, we have

$$\begin{bmatrix} A_1 \\ B_1 \end{bmatrix} \equiv \begin{bmatrix} C_N \\ D_N \end{bmatrix} = \bar{\bar{\mathbf{T}}} \begin{bmatrix} A_0 \\ B_0 \end{bmatrix} \qquad (4.7.27a)$$

where

$$\bar{\bar{\mathbf{T}}} = \bar{\bar{\mathbf{F}}}_{N(N-1)} \cdots \bar{\bar{\mathbf{F}}}_{10} \qquad (4.7.27b)$$

and $A_1 = C_N$, $B_1 = D_N$, $A_0 = C_0$, and $B_0 = D_0$ have been used (comparing Fig. 4.17 with Fig. 4.14).

The rest of the steps are the same as those for the Kronig–Penney model, that is, we solve the eigenvalue equation

$$\bar{\bar{\mathbf{T}}} \begin{bmatrix} A_0 \\ B_0 \end{bmatrix} = t \begin{bmatrix} A_0 \\ B_0 \end{bmatrix} \qquad (4.7.28)$$

and the eigenvalues have the form

$$t_+ = e^{iqL} \qquad t_- = e^{-iqL} \qquad (4.7.29)$$

and

$$\cos qL = t_+ + t_- = \frac{1}{2}\text{trace}\left(\bar{\bar{\mathbf{T}}}\right) = \frac{t_{11} + t_{22}}{2} \qquad (4.7.30)$$

For a periodic potential with $V_0 = V_N$, that is, the period L consists of N

subdivisions with $\Delta z = L/N$, we expect

$$\det|\bar{\bar{\mathbf{T}}}| = 1 \qquad (4.7.31)$$

which can be proved from a more general argument from the time-reversal condition of a real potential profile $V(z)$. It can be readily explained that if $\det|\bar{\bar{\mathbf{T}}}| \neq 1$, the infinite periodic potential will lead to $\lim_{n \to \infty}|t_{\pm}^{n}| \to \infty$, or $\lim_{n \to -\infty}|t_{\pm}^{n}| \to \infty$, which are not allowed solutions.

4.8 BAND STRUCTURES OF SEMICONDUCTOR QUANTUM WELLS

In Section 4.4, we discussed the effective mass theory for a single band and degenerate bands. In this section, we study the calculation of the band structures of semiconductor quantum wells. Many papers [51–67] on the recent development of the effective mass theory or the $\mathbf{k} \cdot \mathbf{p}$ theory for quantum-well structures have been published. Theoretical methods such as the tight-binding [68] and the bond–orbital models [69, 70] have also been introduced. Here we focus on the $\mathbf{k} \cdot \mathbf{p}$ method of the Luttinger–Kohn Hamiltonian together with the strain terms of Pikus and Bir, as discussed in Sections 4.4 and 4.5.

4.8.1 Conduction Band

The effective mass theory for the conduction band is obtained from the dispersion relation

$$E(k) = \frac{\hbar^2 k^2}{2m^*} \qquad (4.8.1)$$

where the effective mass of the electron in the conduction band is $m^* = m_b^*$ in the barrier region and $m^* = m_w^*$ in the quantum well. In the presence of the quantum-well potential,

$$V(z) = \begin{cases} V_0(= \Delta E_c) & |z| > \dfrac{L_w}{2} \\ 0 & |z| \leq \dfrac{L_w}{2} \end{cases} \qquad (4.8.2)$$

where the energies are all measured from the conduction band edge. The effective mass equation (4.4.5) for a single band is

$$\left[-\frac{\hbar^2}{2} \frac{\partial}{\partial z} \frac{1}{m(z)} \frac{\partial}{\partial z} + \frac{\hbar^2}{2m(z)} \nabla_t^2 + V(z) \right] \psi(\mathbf{r}) = E\psi(\mathbf{r}) \quad (4.8.3)$$

where $(1/m)(\partial/\partial z)$ appears inside $\partial/\partial z$ to ensure that the probability current density

$$j_z(z) \sim \frac{1}{m(z)} \left[\psi^* \frac{\partial}{\partial z} \psi - \psi \frac{\partial \psi^*}{\partial z} \right]$$

is continuous at the heterojunction.

In general, the wave function $\psi(\mathbf{r})$ can be written in the form

$$\psi(\mathbf{r}) = \frac{e^{i\mathbf{k}_t \cdot \mathbf{r}}}{\sqrt{A}} \psi(z) \quad (4.8.4)$$

and

$$-\frac{\hbar^2}{2} \frac{\partial}{\partial z} \frac{1}{m(z)} \frac{\partial}{\partial z} \psi(z) + V(z)\psi(z) = \left(E(k_t) - \frac{\hbar^2 k_t^2}{2m(z)} \right) \psi(z) \quad (4.8.5)$$

The eigenvalue and the eigenfunction are obtained from the above equation, following the procedures discussed in Chapter 3. Here we ignore the k_t dependence of $\psi(z)$. Equation (4.8.5) is usually solved at $k_t = 0$ for the nth subband energy $E_n(0)$ with a wave function $\psi(z) = f_n(z)$. Then we have $E_n(k_t) = E_n(0) + \hbar^2 k_t^2 / 2m_w$.

4.8.2 Valence Band

Band-Edge Energy. For a given quantum-well potential,

$$V_h(z) = \begin{cases} 0 & |z| \le \dfrac{L_w}{2} \\[2mm] -\Delta E_v & |z| > \dfrac{L_w}{2} \end{cases} \quad (4.8.6)$$

let us find the band-edge energy at $\mathbf{k}_t = 0$ first. The Luttinger–Kohn Hamiltonian (4.3.14) or (4.5.13) is diagonal for $k_x = k_y = 0$:

$$E_{HH}(k_z) = -\frac{\hbar^2}{2m_0}(\gamma_1 - 2\gamma_2)k_z^2 \quad (4.8.7a)$$

$$E_{LH}(k_z) = -\frac{\hbar^2}{2m_0}(\gamma_1 + 2\gamma_2)k_z^2 \quad (4.8.7b)$$

Define

$$m_{hh}^z \equiv \frac{m_0}{\gamma_1 - 2\gamma_2} \qquad (4.8.8a)$$

$$m_{lh}^z \equiv \frac{m_0}{\gamma_1 + 2\gamma_2} \qquad (4.8.8b)$$

Since the parameters γ_1 and γ_2 in the well are different from those in the barrier regions, we solve

$$\left[+ \frac{\hbar^2}{2} \frac{\partial}{\partial z} \frac{1}{m_{(m)}^z} \frac{\partial}{\partial z} + V_h(z) \right] g_{(m)}(z) = E g_{(m)}(z) \qquad (4.8.9)$$

where $(m) \equiv$ (hhm) or ($l$hm). Therefore, the band-edge energies for the hhm or lhm subbands can all be found from the equation identical to that for the parabolic band model. We only have to use the appropriate effective masses (4.8.8a) and (4.8.8b) in the corresponding regions.

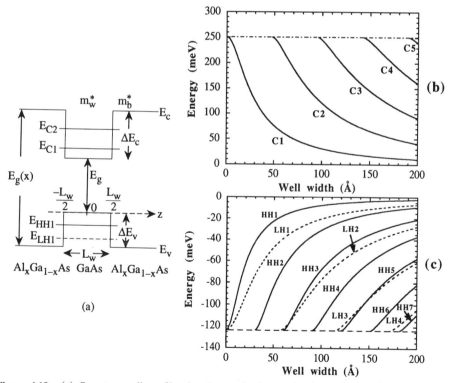

Figure 4.18. (a) Quantum-well profiles for the conduction and valence bands of a GaAs/ $Al_xGa_{1-x}As$ system. (b) Conduction subband energies, $E_{C1}, E_{C2}, \ldots,$ and (c) valence subband energies $E_{HH1}, E_{HH2}, \ldots,$ and $E_{LH1}, E_{LH2} \ldots$ vs. the well width L_w.

Example   As an example, we consider a $GaAs/Al_{0.3}Ga_{0.7}As$ quantum well as shown in Fig. 4.18a. The energies for the electron subbands (Fig. 4.18b), and for the heavy-hole and light-hole subbands (Fig. 4.18c) are plotted vs. the well width L_w. We can see that for the magnitudes all energy levels decrease with an increasing well width.   ∎

Valence Subbands Dispersion Relations. The effective mass equation for four degenerate valence bands (two heavy-hole and two light-hole bands) follows Eq. (4.4.11) for a quantum-well potential $V_h(z)$ given by (4.8.6):

$$\left[\overline{\overline{H}}^{LK}\left(k_x, k_y, k_z = -i\frac{\partial}{\partial z}\right) + V_h(z)\overline{\overline{I}}\right]\cdot\begin{bmatrix}F_1\\F_2\\F_3\\F_4\end{bmatrix} = E\begin{bmatrix}F_1\\F_2\\F_3\\F_4\end{bmatrix} \quad (4.8.10)$$

where $\overline{\overline{H}}^{LK}$ is from the first 4×4 portion of (4.3.14) or (4.5.13), and the envelope functions F_1, F_2, F_3, and F_4 can be written in the vector form

$$\mathbf{F}_k(\mathbf{r}) = \begin{bmatrix}F_1\\F_2\\F_3\\F_4\end{bmatrix} = \begin{bmatrix}g_{3/2}(k_x, k_y, z)\\g_{1/2}(k_x, k_y, z)\\g_{-1/2}(k_x, k_y, z)\\g_{-3/2}(k_x, k_y, z)\end{bmatrix}\frac{e^{ik_x x + ik_y y}}{\sqrt{A}} \quad (4.8.11)$$

The wave function in component form is expressed as

$$\psi_k(\mathbf{r}) = F_1\left|\tfrac{3}{2}, \tfrac{3}{2}\right\rangle + F_2\left|\tfrac{3}{2}, \tfrac{1}{2}\right\rangle + F_3\left|\tfrac{3}{2}, -\tfrac{1}{2}\right\rangle + F_4\left|\tfrac{3}{2}, -\tfrac{3}{2}\right\rangle$$
$$= \frac{e^{ik_x x + ik_y y}}{\sqrt{A}}\sum_\nu g_\nu(k_x, k_y, z)\left|\tfrac{3}{2}, \nu\right\rangle \quad (4.8.12)$$

where $\nu = \tfrac{3}{2}, \tfrac{1}{2}, -\tfrac{1}{2}$, and $-\tfrac{3}{2}$. Denote

$$\mathbf{k}_t = \hat{x}k_x + \hat{y}k_y \quad (4.8.13)$$

We write

$$\left[\overline{\overline{H}}^{LK}\left(\mathbf{k}_t, k_z = -i\frac{\partial}{\partial z}\right) + V_h(z)\overline{\overline{I}}\right]\cdot\begin{bmatrix}g_{3/2}(\mathbf{k}_t, z)\\g_{1/2}(\mathbf{k}_t, z)\\g_{-1/2}(\mathbf{k}_t, z)\\g_{-3/2}(\mathbf{k}_t, z)\end{bmatrix}$$
$$= E(\mathbf{k}_t)\begin{bmatrix}g_{3/2}(\mathbf{k}_t, z)\\g_{1/2}(\mathbf{k}_t, z)\\g_{-1/2}(\mathbf{k}_t, z)\\g_{-3/2}(\mathbf{k}_t, z)\end{bmatrix} \quad (4.8.14)$$

Theoretically, a model for a quantum well with infinite barriers has been used and many of the results can be expressed in analytical forms [60, 63, 64, 71].

4.8.3 Direct Experimental Measurements of the Subband Dispersions

Experimentally, low-temperature magnetoluminescent measurement [72], resonant magnetotunneling spectroscopy [73] and photoluminescent measurements of hot electrons recombining at neutral acceptors [74, 75] have been used to map out the hole subband dispersion curves. In Fig. 4.19, we

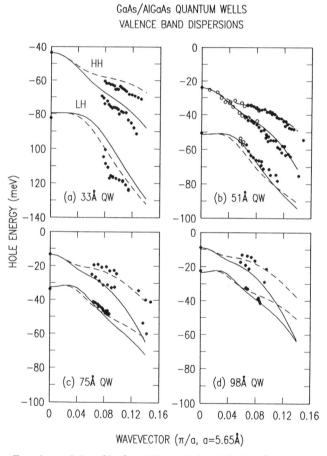

Figure 4.19. Experimental data (dots) and theoretical calculations (——, wavevector along the [100] direction; ————, wavevector along the [110] direction) for the hole subband energies of GaAs/Al$_x$Ga$_{1-x}$As quantum wells for (a) 33 Å quantum wells with $x = 0.325$ in the barriers, (b) 51 Å quantum wells with $x = 0.315$ in the barriers, (c) 75 Å quantum wells with $x = 0.32$, and (d) 98 Å quantum wells with $x = 0.38$. The horizontal axis is the wavevector normalized by π/a, where $a = 5.65$ Å is the lattice constant of GaAs. (After Ref. 75.)

show the experimental results (data points) of Ref. 75 compared with the theoretical calculations (solid curves [100] direction and dashed curves [110] direction). The well width L_w, the aluminum mole fraction x in the barriers, and the p-type doping concentration N_A in the wells are (a) $L_w = 33$ Å, $x = 0.325$, $N_A = 10^{18}$ cm^{-3}; (b) $L_w = 51$ Å, $x = 0.315$, $N_A = 10^{18}$ cm^{-3}, (c) $L_w = 75$ Å, $x = 0.32$, $N_A = 2 \times 10^{18}$ cm^{-3}, and (d) $L_w = 98$ Å, $x = 0.38$, $N_A = 3 \times 10^{17}$ cm^{-3}. Each well is Be-doped in the center 10 Å except for the 98 Å wells (doped in the center 35 Å) at the above indicated concentration N_A. It is noted that these nonparabolic subband dispersion curves exhibit the valence band mixing behavior in quantum wells.

We can see that the coupling between the heavy-hole HH1 and the light LH1 subbands leads to warping (or nonparabolicity) of the band structure. These valence band mixing effects cause the wave function to exhibit both the heavy-hole and the light-hole characteristics. Here the labeling of HH and LH is according to their wave functions at $k_x = k_y = 0$, since $|\frac{3}{2}, \pm \frac{3}{2}\rangle$ are assigned to that of the heavy hole, and $|\frac{3}{2}, \pm \frac{1}{2}\rangle$ are assigned to that of the light hole.

4.8.4   Block Diagonalization of the Luttinger–Kohn Hamiltonian

In this section, we consider a transformation that block diagonalizes [37, 54, 55] the 4×4 Hamiltonian (4.5.13), which includes the strain effects. Define the phases θ_R and θ_S of R and S by

$$R = |R|\,e^{i\theta_R} \qquad S = |S|\,e^{i\theta_S} \qquad (4.8.15)$$

The 4×4 Hamiltonian in (4.5.13) can be transformed to two 2×2 Hamiltonians $\overline{\overline{\mathbf{H}}}^U$ and $\overline{\overline{\mathbf{H}}}^L$

$$\overline{\overline{\mathbf{H}}} = \overline{\overline{\mathbf{U}}}\,\overline{\overline{\mathbf{H}}}^{\mathrm{LK}}\,\overline{\overline{\mathbf{U}}}^+ = -\begin{bmatrix} P+Q & \tilde{R} & 0 & 0 \\ \tilde{R}^+ & P-Q & 0 & 0 \\ 0 & 0 & P-Q & \tilde{R} \\ 0 & 0 & \tilde{R}^+ & P+Q \end{bmatrix} \begin{array}{l} |1\rangle \\ |2\rangle \\ |3\rangle \\ |4\rangle \end{array} \qquad (4.8.16)$$

where

$$\tilde{R} = |R| - i|S| \qquad \tilde{R}^+ = |R| + i|S| \qquad (4.8.17)$$

and the transformation between the old bases $|\frac{3}{2}, \nu\rangle (\nu = \pm \frac{3}{2}, \pm \frac{1}{2})$ and the new bases is

$$|1\rangle = \alpha|\tfrac{3}{2}, \tfrac{3}{2}\rangle - \alpha^*|\tfrac{3}{2}, -\tfrac{3}{2}\rangle$$
$$|2\rangle = -\beta^*|\tfrac{3}{2}, \tfrac{1}{2}\rangle + \beta|\tfrac{3}{2}, -\tfrac{1}{2}\rangle$$
$$|3\rangle = \beta^*|\tfrac{3}{2}, \tfrac{1}{2}\rangle + \beta|\tfrac{3}{2}, -\tfrac{1}{2}\rangle$$
$$|4\rangle = \alpha|\tfrac{3}{2}, \tfrac{3}{2}\rangle + \alpha^*|\tfrac{3}{2}, -\tfrac{3}{2}\rangle \qquad (4.8.18)$$

where

$$\alpha = \frac{1}{\sqrt{2}} \, e^{i[(\theta_S + \theta_R)/2 + \pi/4]} \tag{4.8.19a}$$

$$\beta = \frac{1}{\sqrt{2}} \, e^{i[(\theta_S - \theta_R)/2 + \pi/4]} \tag{4.8.19b}$$

and the transformation matrix

$$\overline{\overline{U}} = \begin{bmatrix} \alpha^* & 0 & 0 & -\alpha \\ 0 & -\beta & \beta^* & 0 \\ 0 & \beta & \beta^* & 0 \\ \alpha^* & 0 & 0 & \alpha \end{bmatrix} \tag{4.8.20}$$

In general, $k_z \rightarrow -i \, \partial/\partial z$ becomes an operator in (4.8.20), which is difficult to treat. For special cases such as an external stress $T \| [100]$, $[001]$, $[110]$, $\varepsilon_{xz} = \varepsilon_{yz} = 0$, θ_S is independent of k_z. Therefore, the Hamiltonian can be applied to the quantum-well problem directly.

For the case in which an external stress T is applied along the [110] direction or the case in which strain is caused by the lattice-mismatch accommodated elastic strain

$$\varepsilon_{xx} = \varepsilon_{yy} \neq \varepsilon_{zz} \qquad \varepsilon_{xy} = \varepsilon_{xz} = \varepsilon_{yz} = 0 \tag{4.8.21}$$

we have

$$\theta_S = -\phi \tag{4.8.22}$$

where $\phi = \tan^{-1}(k_y/k_x)$.

4.8.5 Axial Approximation for the Luttinger–Kohn Hamiltonian

If we approximate in the R_k term

$$R_k = -\frac{\hbar^2 \sqrt{3}}{2m_0} \left[\frac{\gamma_2 + \gamma_3}{2} (k_x - ik_y)^2 + \frac{\gamma_2 - \gamma_3}{2} (k_x + ik_y)^2 \right]$$

$$\simeq -\frac{\hbar^2 \sqrt{3}}{2m_0} \overline{\gamma} (k_x - ik_y)^2 \tag{4.8.23}$$

where $\overline{\gamma} = (\gamma_2 + \gamma_3)/2$ and using

$$k_x - ik_y = k_t \, e^{-i\phi} \tag{4.8.24}$$

we find that the energy subband dispersion relation is independent of the angle ϕ because

$$\tilde{R} = |R| - \mathrm{i}|S| = \frac{\hbar^2 \sqrt{3}}{2m_0} \bar{\gamma} k_t^2 - \mathrm{i} \frac{\hbar^2 \gamma_3}{2m_0} 2\sqrt{3}\, k_t k_z \qquad (4.8.25)$$

and the Hamiltonian depends only on the magnitude of the vector $\mathbf{k}_t$. This is called the axial approximation. Note that in this approximation, we assume that $\gamma_2 \simeq \gamma_3$ in the R_k term only, while we still use γ_2 and γ_3 in the other terms. We obtain

$$\theta_R = \pi - 2\phi \qquad (4.8.26a)$$

and

$$\alpha = \frac{1}{\sqrt{2}}\, e^{\mathrm{i}(3\pi/4 - 3\phi/2)} \qquad (4.8.26b)$$

$$\beta = \frac{1}{\sqrt{2}}\, e^{\mathrm{i}(\phi/2 - \pi/4)} \qquad (4.8.26c)$$

4.8.6   Numerical Approach for the Solutions of the Upper 2 × 2 Hamiltonian under Axial Approximation

Let us look at the upper 2 × 2 Hamiltonian in (4.8.16). The wave function for the hole subbands can be written generally as

$$\psi^U(\mathbf{k}_t, \mathbf{r}) = \frac{e^{\mathrm{i}\mathbf{k}_t \cdot \mathbf{r}_t}}{\sqrt{A}} \left[g^{(1)}(k_t, z)|1\rangle + g^{(2)}(k_t, z)|2\rangle \right]$$

$$= \begin{bmatrix} g^{(1)}(k_t, z) \\ g^{(2)}(k_t, z) \end{bmatrix} \frac{e^{\mathrm{i}\mathbf{k}_t \cdot \mathbf{r}_t}}{\sqrt{A}} \equiv \begin{bmatrix} F_1 \\ F_2 \end{bmatrix} \qquad (4.8.27)$$

This wave function satisfies the Hamiltonian equation

$$-\begin{bmatrix} P + Q - V_h(z) & \tilde{R} \\ \tilde{R}^+ & P - Q - V_h(z) \end{bmatrix} \begin{bmatrix} g^{(1)}(k_t, z) \\ g^{(2)}(k_t, z) \end{bmatrix} = E(k_t) \begin{bmatrix} g^{(1)}(k_t, z) \\ g^{(2)}(k_t, z) \end{bmatrix}$$

$$(4.8.28)$$

where P, Q, and $\tilde{R}$ are all differential operators and can be obtained from (C.25) and (4.8.25) with k_z replaced by $-\mathrm{i}(\partial/\partial z)$. These wave functions $g^{(1)}$ and $g^{(2)}$ depend on the magnitude of the wave vector k_t and position z, and are independent of the direction of the wave vector (or the angle ϕ):

$$P = P_\varepsilon + \frac{\hbar^2 \gamma_1}{2m_0}\left(k_t^2 - \frac{\partial^2}{\partial z^2}\right)$$

$$Q = Q_\varepsilon + \frac{\hbar^2 \gamma_2}{2m_0}\left(k_t^2 + 2\frac{\partial^2}{\partial z^2}\right)$$

$$\tilde{R} = \frac{\hbar^2\sqrt{3}}{2m_0}\bar{\gamma}k_t^2 - \frac{\hbar^2\gamma_3}{m_0}\sqrt{3}\,k_t\frac{\partial}{\partial z}$$

$$\tilde{R}^+ = \frac{\hbar^2\sqrt{3}}{2m_0}\bar{\gamma}k_t^2 + \frac{\hbar^2\gamma_3}{m_0}\sqrt{3}\,k_t\frac{\partial}{\partial z} \qquad (4.8.29)$$

The built-in quantum-well potential for the holes $V_h(z)$ has been incorporated into the diagonal terms of the Hamiltonian. Since $V_h(z)$ is a stepwise potential, boundary conditions between the well and the barrier interface have to be used properly. The basic idea is that the envelope function and the probability current density across the heterojunction should be continuous. Therefore, to ensure the Hermitian property of H, we have to write all operators of the form

$$A(z)\frac{\partial^2}{\partial z^2} \quad \text{as} \quad \frac{\partial}{\partial z}A(z)\frac{\partial}{\partial z} \qquad (4.8.30a)$$

and

$$B(z)\frac{\partial}{\partial z} \quad \text{as} \quad \frac{1}{2}\left[B(z)\frac{\partial}{\partial z} + \frac{\partial}{\partial z}B(z)\right] \qquad (4.8.30b)$$

We then have the following boundary conditions:

(1) $$\begin{bmatrix} F_1 \\ F_2 \end{bmatrix} = \text{continuous} \qquad (4.8.31a)$$

(2) $$\begin{bmatrix} \dfrac{\hbar^2}{2m_0}(\gamma_1 - 2\gamma_2)\dfrac{\partial}{\partial z} & \left(\dfrac{\hbar^2\gamma_3}{2m_0}\right)\sqrt{3}\,k_t \\[2ex] -\left(\dfrac{\hbar^2\gamma_3}{2m_0}\right)\sqrt{3}\,k_t & \dfrac{\hbar^2}{2m_0}(\gamma_1 + 2\gamma_2)\dfrac{\partial}{\partial z} \end{bmatrix}\begin{bmatrix} F_1 \\ F_2 \end{bmatrix} = \text{continuous}$$

$$(4.8.31b)$$

Note that the Luttinger parameters γ_1, γ_2, and γ_3 in the corresponding region have to be used.

Eq. (4.8.28) can be solved using a propagation-matrix method [37], which is very efficient, or using a finite-difference method [76]. The solution to the Schrödinger equation in matrix form (4.8.28) will be a set of subband energies:

$$E(\mathbf{k}_t) = E_m^U(\mathbf{k}_t)$$

where the subband index m refers to HH1, HH2,... and LH1, LH2,... subbands. The superscript U refers to the upper Hamiltonian.

4.8.7   Solutions for the Lower 2 × 2 Hamiltonian under Axial Approximation

Similar procedures hold for the lower 2 × 2 Hamiltonian of (4.8.16). The wave function is

$$\psi^L(k_t, \mathbf{r}) = \frac{e^{i\mathbf{k}_t \cdot \mathbf{r}_t}}{\sqrt{A}} \left[g^{(3)}(k_t, z)|3\rangle + g^{(4)}(k_t, z)|4\rangle \right]$$

$$= \begin{bmatrix} g^{(3)}(k_t, z) \\ g^{(4)}(k_t, z) \end{bmatrix} \frac{e^{i\mathbf{k}_t \cdot \mathbf{r}_t}}{\sqrt{A}} = \begin{bmatrix} F_3 \\ F_4 \end{bmatrix} \qquad (4.8.32)$$

which satisfies the Hamiltonian equation

$$-\begin{bmatrix} P - Q - V_h(z) & \tilde{R} \\ \tilde{R}^+ & P + Q - V_h(z) \end{bmatrix} \begin{bmatrix} g^{(3)}(k_t, z) \\ g^{(4)}(k_t, z) \end{bmatrix} = E^L(k_t) \begin{bmatrix} g^{(3)}(k_t, z) \\ g^{(4)}(k_t, z) \end{bmatrix}$$

$$(4.8.33)$$

The solution will be a set of dispersion curves labeled as m in $E_m^L(k_t)$. For a symmetrical potential, we find that $E_m^L(k_t)$ of the lower Hamiltonian is degenerate with that of the upper Hamiltonian $E_m^U(k_t)$ for each subband m. The wave functions $g^{(3)}(k_t, z)$ and $g^{(4)}(k_t, z)$ can also be related to $g^{(2)}(k_t, z)$ and $g^{(1)}(k_t, z)$, respectively, if we change $z \rightarrow -z$.

Example   The valence subband structures for a GaAs/Al$_{0.3}$Ga$_{0.7}$As quantum well with $L_w = 100$ Å and $L_w = 50$ Å are calculated using the propagation matrix method in Ref. [37] and are plotted in Figs. 4.20a and b, respectively. The horizontal wave vector k_t is normalized by $2\pi/a$, where $a = 5.6533$ Å is the lattice constant of GaAs. These subband dispersions can also be compared with the experimental data for similar GaAs quantum well structures with p-type dopings in Figs. 4.19d and b, respectively. The axial approximation gives very good results for a small k_t, and the results are exactly the same as those using the original 4 × 4 Hamiltonian at $k_t = 0$.

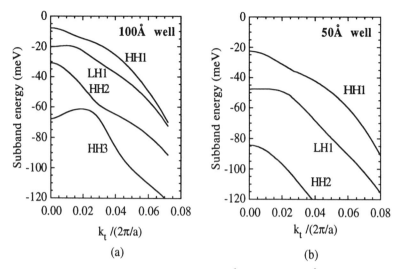

Figure 4.20. The valence subbands for (a) a 100-Å and (b) a 50-Å GaAs/Al$_{0.3}$Ga$_{0.7}$ As quantum well for k_t in the plane of the quantum well under the axial approximation. The vertical axis is the hole subband energy and the horizontal axis is the in-plane wave vector k_t normalized by $(2\pi/a)$, where a = 5.6533 Å is the lattice constant of GaAs. These results in (a) and (b) can be compared with the experimental data for similar GaAs quantum-well structures in Figs. 4.19d and b, respectively.

This approximation is attractive, since the ϕ dependence in the $k_x - k_y$ plane can be taken into account analytically in the basis functions, and the valence subband energies are independent of ϕ. These wave functions can be applied to study optical absorption and gain in quantum-well structures.

■

4.9 BAND STRUCTURES OF STRAINED SEMICONDUCTOR QUANTUM WELLS

For a strained quantum well, the conduction band edge is given by

$$E_c(k = 0) = a_c \text{Tr}(\overline{\overline{\varepsilon}}) = a_c(\varepsilon_{xx} + \varepsilon_{yy} + \varepsilon_{zz}) \qquad (4.9.1)$$

where $\varepsilon_{xx} = \varepsilon_{yy} = (a_0 - a)/a$ and $\varepsilon_{zz} = -2(C_{12}/C_{11})\varepsilon_{xx}$. Here a_0 is the lattice constant of the substrate, and a is the lattice constant of the quantum well. The energy is measured from the conduction band edge of an un-strained quantum well, as shown in Fig. 4.21. We assume that the barriers are lattice matched to the substrate; therefore, their strains are zero. For an unstrained quantum well, we have the potential energy for the electron given

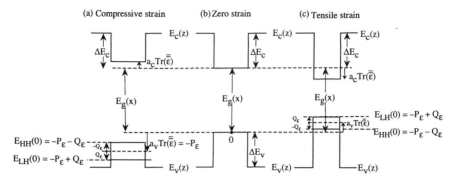

Figure 4.21. Band-edge profile for (a) a compressive strained, (b) an unstrained, and (c) a tensile strained quantum well.

by

$$
E_c^0(z) = \begin{cases} 0 & |z| \le \dfrac{L_w}{2} \\[2mm] \Delta E_c & |z| > \dfrac{L_w}{2} \end{cases}
\tag{4.9.2}
$$

For example, $E_g(\mathrm{In}_{1-x}\mathrm{Ga}_x\mathrm{As}) = 0.324 + 0.7x + 0.4x^2$ for an unstrained $\mathrm{In}_{1-x}\mathrm{Ga}_x\mathrm{As}$ compound, and $E_g(\mathrm{InP}) = 1.344$ eV at room temperature:

$$
\Delta E_g = 1.344 - E_g(\mathrm{In}_{1-x}\mathrm{Ga}_x\mathrm{As})
\tag{4.9.3}
$$

$$
a(x) = x\,a(\mathrm{GaAs}) + (1 - x)\,a(\mathrm{InAs})
$$

From previous experimental data, we take

$$
\Delta E_c = 0.4\,\Delta E_g(x)
$$

$$
\Delta E_v = 0.6\,\Delta E_g(x)
\tag{4.9.4}
$$

At $x = 0.468$, $\mathrm{In}_{1-x}\mathrm{Ga}_x\mathrm{As}$ is lattice matched to the InP substrate. Strain exists when $x \ne 0.468$.

4.9.1 Subband Energies in a Strained Quantum Well

The conduction band-edge energy for a strained quantum well is given by

$$
E_c(z) = \begin{cases} a_c(\varepsilon_{xx} + \varepsilon_{yy} + \varepsilon_{zz}) & |z| \le \dfrac{L_w}{2} \\[2mm] \Delta E_c & |z| > \dfrac{L_w}{2} \end{cases}
\tag{4.9.5}
$$

Similarly, the valence band-edge energy for the unstrained quantum well is defined as

$$V_h(z) = E_v^0(z) = \begin{cases} 0 & |z| \le \dfrac{L_w}{2} \\ -\Delta E_v & |z| > \dfrac{L_w}{2} \end{cases} \tag{4.9.6}$$

The valence band-edge energies for the strained quantum well are obtained from $E_{HH}(0)$ and $E_{LH}(0)$ in (4.5.36) ignoring the coupling with the spin–orbit split-off band.

$$E_{HH}(z) = \begin{cases} -P_\varepsilon - Q_\varepsilon & |z| \le \dfrac{L_w}{2} \\ -\Delta E_v & |z| > \dfrac{L_w}{2} \end{cases} \tag{4.9.7a}$$

$$E_{LH}(z) = \begin{cases} -P_\varepsilon + Q_\varepsilon & |z| \le \dfrac{L_w}{2} \\ -\Delta E_v & |z| > \dfrac{L_w}{2} \end{cases} \tag{4.9.7b}$$

where

$$P_\varepsilon = -a_v(\varepsilon_{xx} + \varepsilon_{yy} + \varepsilon_{zz}) \tag{4.9.8a}$$

$$Q_\varepsilon = \frac{-b}{2}(\varepsilon_{xx} + \varepsilon_{yy} - 2\varepsilon_{zz}) \tag{4.9.8b}$$

The subband edge energies can all be obtained from a simple quantum-well model using the band-edge effective masses:

$$m_e^* = \begin{cases} m_w^* & |z| \le \dfrac{L_w}{2} \\ m_b^* & |z| > \dfrac{L_w}{2} \end{cases} \tag{4.9.9a}$$

$$m_{hh}^z = \begin{cases} \dfrac{m_0}{\gamma_{1w} - 2\gamma_{2w}} & |z| \le \dfrac{L_w}{2} \\[2mm] \dfrac{m_0}{\gamma_{1b} - 2\gamma_{2b}} & |z| > \dfrac{L_w}{2} \end{cases} \tag{4.9.9b}$$

$$m_{lh}^z = \begin{cases} \dfrac{m_0}{\gamma_{1w} + 2\gamma_{2w}} & |z| \le \dfrac{L_w}{2} \\[2mm] \dfrac{m_0}{\gamma_{1b} + 2\gamma_{2b}} & |z| > \dfrac{L_w}{2} \end{cases} \tag{4.9.9c}$$

where the subscripts w and b in the effective mass parameters designate the well and the barrier region, respectively.

If the spin–orbit split-off band is included, we find that the band-edge energies for heavy hole $E_{HH}(z)$ and m_{hh}^z are not affected, but

$$
E_{LH}(z) = \begin{cases}
-P_\varepsilon + \dfrac{1}{2}\left(Q_\varepsilon - \Delta + \sqrt{\Delta^2 + 2\Delta Q_\varepsilon + 9Q_\varepsilon^2}\,\right) & |z| \le \dfrac{L_w}{2} \\[4mm]
-\Delta E_v & |z| > \dfrac{L_w}{2}
\end{cases}
$$

(4.9.10)

and

$$
m_{lh}^z = \begin{cases}
\dfrac{m_0}{\gamma_{1w} + 2f_+\gamma_{2w}} & |z| \le \dfrac{L_w}{2} \\[4mm]
\dfrac{m_0}{\gamma_{1b} + 2\gamma_{2b}} & |z| > \dfrac{L_w}{2}
\end{cases}
$$

(4.9.11)

where f_+ is defined in (4.5.38).

4.9.2   Valence Subband Energy Dispersions in a Strained Quantum Well

Using H from (4.5.13), we can solve the valence subband energies and eigenfunctions using (4.8.14):

$$
\left[\overline{\overline{\mathbf{H}}}\left(k_z = -i\frac{\partial}{\partial z}\right) + V_h(z)\overline{\overline{\mathbf{I}}}\right] \cdot
\begin{bmatrix}
g_{3/2}(\mathbf{k}_t, z) \\
g_{1/2}(\mathbf{k}_t, z) \\
g_{-1/2}(\mathbf{k}_t, z) \\
g_{-3/2}(\mathbf{k}_t, z)
\end{bmatrix}
= E(\mathbf{k}_t)
\begin{bmatrix}
g_{3/2}(\mathbf{k}_t, z) \\
g_{1/2}(\mathbf{k}_t, z) \\
g_{-1/2}(\mathbf{k}_t, z) \\
g_{-3/2}(\mathbf{k}_t, z)
\end{bmatrix}
$$

(4.9.12)

where $V_h(z)$ is given by the unstrained (built-in) potential $E_v^0(z)$ from (4.9.6), and the band-edge shifts due to $-P_\varepsilon$ and $\pm Q_\varepsilon$ have all been included in $\overline{\overline{\mathbf{H}}}$ for $|z| \le L_w/2$, and both P_ε and Q_ε are zero for $|z| > L_w/2$.

Example: An $In_{1-x}Ga_xAs/In_{1-x}Ga_xAs_yP_{1-y}$ Quantum Well   Again, a convenient way to find the valence subband energies and eigenfunctions is to use the block-diagonalized Hamiltonians (4.8.28) and (4.8.33), which simplify the numerical calculations by a significant amount. As a numerical example, we plot in Fig. 4.22 the energy dispersion curves calculated using an efficient

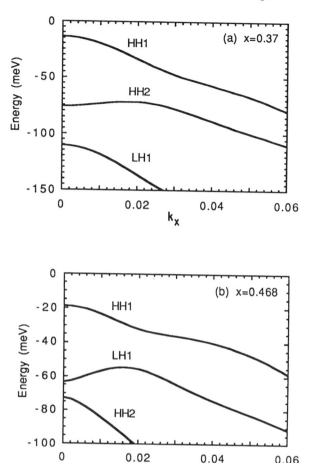

Figure 4.22. The valence subband dispersions of an $In_{1-x}Ga_xAs$ grown on $In_{1-x}Ga_xAs_yP_{1-y}$ quaternary barriers (with a band gap 1.3 μm) lattice matched to InP substrate. (a) $x = 0.37$ (biaxial compression), (b) $x = 0.468$ (unstrained), (c) $x = 0.55$ (small tension), and (d) $x = 0.60$ (large tension). Here the wave vector along the [100] direction k_x is normalized by $2\pi/a$, where $a = 5.6533$ Å is the lattice constant of GaAs. (After Ref. 37.)

propagation-matrix approach in Ref. 37 for the valence subbands of an $In_{1-x}Ga_xAs$ quantum well grown on an $In_{1-x}Ga_xAs_yP_{1-y}$ substrate with a band-gap wavelength 1.3-μm lattice matched to an InP substrate. The gallium model fraction x is varied: (a) $x = 0.37$ (biaxial compression), (b) $x = 0.468$ (unstrained), (c) $x = 0.55$ (small tension), and (d) $x = 0.60$ (large tension). These band structures will be useful to investigate the strain effects on the effective mass, the density of states, optical gain, and absorption in quantum wells. We discuss these effects and their applications further in Chapters 9 and 10. ■

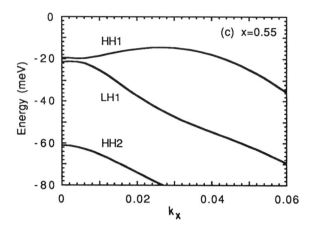

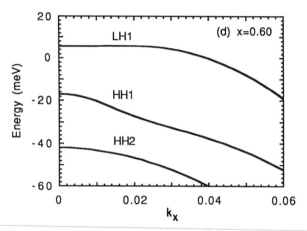

Figure 4.22. (*Continued*)

PROBLEMS

4.1 Derive Equations (4.1.6) and (4.1.7) for the wave function $u_{n\mathbf{k}}(\mathbf{r})$ in the $\mathbf{k} \cdot \mathbf{p}$ theory using (4.1.4) for $\psi_{n\mathbf{k}}(\mathbf{r})$.

4.2 Derive (4.1.8) using the perturbation theory in Section 3.5.

4.3 **(a)** Show that the eigenvalues of the determinantal equation (4.1.16) in a two-band model are

$$E = \frac{1}{2}\left[E_n + E_n' + \frac{\hbar^2}{m_0}k^2 \right] \pm \frac{1}{2}\left[(E_n - E_{n'})^2 + \frac{4\hbar^2}{m_0^2}|\mathbf{k} \cdot \mathbf{p}|^2 \right]^{1/2}$$

(b) Consider the two-band model, where one is the conduction band $(n = c)$ and the other is the heavy hole in the valence band $(n' = v)$. We have $\mathbf{k}_0 = 0$ for a direct band-gap material, $E_n = E_c$, and $E'_n = E_v$. Show that the equation

$$
\begin{bmatrix}
E_c + \dfrac{\hbar^2 k^2}{2m_0} - E & \dfrac{\hbar}{m_0}\mathbf{k} \cdot \mathbf{p}_{cv} \\[2ex]
\dfrac{\hbar}{m_0}\mathbf{k} \cdot \mathbf{p}_{vc} & E_v + \dfrac{\hbar^2 k^2}{2m_0} - E
\end{bmatrix}
\begin{bmatrix}
a_1 \\[2ex]
a_2
\end{bmatrix}
= 0
$$

leads to

$$
E = \frac{1}{2}\left(E_g + \frac{\hbar^2 k^2}{m_0} \right) \pm \frac{1}{2}\left[E_g^2 + 4\frac{\hbar^2}{m_0^2}|\mathbf{k} \cdot \mathbf{p}_{cv}|^2 \right]^{1/2}
$$

where $E_v = 0$, $E_c = E_g = $ the energy gap.

(c) Show that if we use the perturbation theory directly near each band for (4.1.15),

$$
E_n = E_n(0) + H'_{nn} + \sum_{n' \neq n} \frac{|H'_{nn}|^2}{E_n(0) - E_{n'}(0)}
$$

we obtain

$$
E = \begin{cases}
E_g + \dfrac{\hbar^2 k^2}{2m_0} + \dfrac{\hbar^2}{E_g m_0^2}|\mathbf{k} \cdot \mathbf{p}_{cv}|^2 & \text{for the conduction band} \\[3ex]
\dfrac{\hbar^2 k^2}{2m_0} - \dfrac{\hbar^2}{E_g m_0^2}|\mathbf{k} \cdot \mathbf{p}_{cv}|^2 & \text{for the valence band}
\end{cases}
$$

and

$$
u_{ck}(r) = u_{c0}(\mathbf{r}) + \frac{\hbar \mathbf{k} \cdot \mathbf{p}_{vc}}{m_0 E_g} u_{v0}(\mathbf{r})
$$

$$
u_{vk}(r) = u_{v0}(\mathbf{r}) - \frac{\hbar \mathbf{k} \cdot \mathbf{p}_{cv}}{m_0 E_g} u_{c0}(\mathbf{r})
$$

Check whether the energies using the perturbation theory are the same as those obtained by the Taylor-series expansion of the expressions in part (b), assuming that the $(\mathbf{k} \cdot \mathbf{p}_{cv})$ term is small.

4.4 The matrix $\overline{\overline{\mathbf{H}}}$ in (4.2.8) can be equivalently obtained from the symmetry properties of a cubic crystal using the wave functions $|S\rangle = S(r)$, $|X\rangle = xf(r)$, $|Y\rangle = yf(r)$ and $|Z\rangle = zf(r)$, and the properties of the operators $p = (\hbar/i)\nabla$, $\nabla = \hat{x}(\partial/\partial x) + \hat{y}(\partial/\partial y) + \hat{z}(\partial/\partial z)$. For example,

$$
\begin{aligned}
H_{11} &= \langle iS\downarrow|H_0 + \frac{\hbar}{m_0}\mathbf{k}\cdot\mathbf{p} + \frac{\hbar}{4m_0^2c^2}\sigma\cdot\nabla V\times\mathbf{p}|iS\downarrow\rangle \\
&= \langle S\downarrow|H_0|S\downarrow\rangle + \frac{\hbar}{m_0}\langle S\downarrow|\mathbf{k}\cdot\mathbf{p}|S\downarrow\rangle \\
&\quad + \langle S\downarrow|\frac{\hbar}{4m_0^2c^2}\sigma_z\left(\frac{\partial V}{\partial x}p_y - \frac{\partial V}{\partial y}p_x\right)|S\downarrow\rangle \\
&= E_S
\end{aligned}
\tag{1}
$$

where only the first term contributes (E_S), since the other two terms vanish.

$$
\langle S\downarrow|\mathbf{p}|S\downarrow\rangle = \langle S|\mathbf{p}|S\rangle = \int S(r)\frac{\hbar}{i}\nabla S(r)\,d^3r = 0
\tag{2}
$$

and

$$
\sigma\cdot\nabla V\times\mathbf{p} = \sigma_x(\nabla V\times\mathbf{p})_x + \sigma_y(\nabla V\times\mathbf{p})_y + \sigma_z(\nabla V\times\mathbf{p})_z
\tag{3}
$$

Noting that σ_x and σ_y change the spins from (4.2.4), we find

$$
\langle\downarrow|\sigma_x|\downarrow\rangle = \langle\downarrow|\sigma_y|\downarrow\rangle = 0
\tag{4}
$$

The last term in H_{11} is zero since it changes a sign if we exchange x with y. However, the symmetry of the crystal requires that the Hamiltonian matrix element should be the same with respect to the exchange of x with y. Thus, that term should vanish.

(a) Following the above procedure, show that

$$
\begin{aligned}
H_{13} &= \langle iS\downarrow|H_0 + \frac{\hbar}{m_0}\mathbf{k}\cdot\mathbf{p} + \frac{\hbar}{4m_0^2c^2}\sigma\cdot\nabla V\times\mathbf{p}|Z\downarrow\rangle \\
&= 0 - i\frac{\hbar}{m_0}k\langle S|\mathbf{p}_z|Z\rangle + 0 \\
&= kP
\end{aligned}
\tag{5}
$$

where P is defined in (4.2.9). The last term vanishes by using the symmetry properties of the wave functions.

(b) Show that

$$
H_{22} = \left\langle \frac{X - iY}{\sqrt{2}} \uparrow \left| H_0 + \frac{\hbar}{m_0} \mathbf{k} \cdot \mathbf{p} \right. \right.
$$

$$
\left. + \frac{\hbar}{4m_0^2 c^2} \boldsymbol{\sigma} \cdot \nabla V \times \mathbf{p} \left| \frac{X - iY}{\sqrt{2}} \uparrow \right\rangle \right.
$$

$$
= E_p + \frac{1}{2} \langle X - iY | \frac{\hbar}{m_0} \mathbf{k} \cdot \mathbf{p} | X - iY \rangle
$$

$$
+ \frac{\hbar}{8m_0^2 c^2} \langle X - iY | (\nabla V \times \mathbf{p})_z | X - iY \rangle
$$

$$
= E_p - \frac{\Delta}{3} \tag{6}
$$

where the second term in (6) vanishes, and the last term can be expanded into four terms ($\langle X | \cdots | X \rangle$, $\langle Y | \cdots | Y \rangle$, $\langle X | \cdots | -iY \rangle$, and $\langle -iY | \cdots | X \rangle$, of which the first two vanish.

(c) Derive the full matrix $\overline{\overline{\mathbf{H}}}$ in (4.2.8) for Kane's model.

4.5 **(a)** Show that the eigenvector for (4.2.19) can be written as

$$
a_n = \frac{kP(E_n' + 2\Delta/3)}{N}
$$

$$
b_n = \frac{(\sqrt{2}\,\Delta/3)(E_n' - E_g)}{N}
$$

$$
c_n = \frac{(E_n' - E_g)(E_n' + \frac{2}{3}\Delta)}{N}
$$

where N is obtained from the normalization condition

$$
\sqrt{a_n^2 + b_n^2 + c_n^2} = 1
$$

(b) Show that in the limit $k = 0$, the above eigenvectors lead to the wave functions in Eqs. (4.2.21)–(4.2.24).

4.6 Derive Eq. (4.2.26) for Kane's parameter P.

4.7 Tabulate all the band-edge energies and their corresponding wave functions in Kane's model for the conduction, heavy-hole, light-hole, and spin–orbit split-off bands.

4.8 Derive the Hamiltonian Eq. (4.3.1a)–(4.3.1e) for $u_\mathbf{k}(\mathbf{r})$ using (4.2.5).

4.9 Using the definition for $D_{jj'}^{\alpha\beta}$ in (4.3.11)–(4.3.13) show that the matrix representation of the Luttinger–Kohn Hamiltonian H^{LK} is given by (4.3.14) and (4.3.15).

4.10 Summarize the effects of the strain on the conduction band edge and the valence band edges from the results in Section 4.5.

4.11 Calculate the strains $(\varepsilon_{xx}, \varepsilon_{yy}, \varepsilon_{zz})$, P_ε, Q_ε, and the band-edge shifts for $In_{1-x}Ga_xAs$ quantum-well layers grown in InP substrate with (a) $x = 0.37$ and (b) $x = 0.57$.

4.12 Calculate the band-edge effective masses using (4.5.33) and (4.5.39) for Problem 4.11. Compare the values of the two structures.

4.13 Show that the eigenvalues for the matrix $H(\mathbf{k} = 0)$ in (4.5.34) are given by (4.5.36a)–(4.5.36c).

4.14 Calculate the band-edge energies with spin–orbit coupling effects using (4.5.36a)–(4.5.36c) and compare with those using (4.5.19) without the spin–orbit coupling effects for the $In_{1-x}Ga_xAs/InP$ system with (a) $x = 0.37$, (b) $x = 0.47$, and (c) $x = 0.57$.

4.15 Show that the eigenvalues of the matrix (4.5.13) are given by (4.5.27a) and (4.5.27b).

4.16 Derive the propagation matrix (4.6.7) from the boundary conditions.

4.17 Discuss in the infinite barrier model (i.e., $V_0 = \infty$ and $V_{N+1} = \infty$) (a) how to modify the boundary conditions in the propagation-matrix method in Section 4.6.1 and (b) how to find the eigenenergy E and its corresponding wave function.

4.18 Summarize the procedures for solving (a) an n-type and (b) a p-type modulation-doped quantum well self-consistently in Sections 4.6.2–4.6.4.

4.19 Discuss qualitatively how Fig. 4.12 should be modified if the donor profile is within 20 Å at the center of the quantum well $N_D(z) = N_D$ for $|z| \leq 20$ Å and $N_D(z) = 0$ otherwise.

4.20 Derive (4.7.24a) and (4.7.24b) in the Kronig–Penney model.

4.21 (a) Discuss the procedure for determining the number of bands that can be considered as "bound states" of the superlattice from Fig. 4.15.

(b) What happens if the barrier height V_0 falls within a miniband of Fig. 4.15?

4.22 Compare the miniband energies of the $GaAs/Al_{0.3}Ga_{0.7}As$ superlattice in Fig. 4.16 with those calculated for a single isolated quantum well with the same well width $w = 100$ Å from Chapter 3, Section 3.2.

4.23 (a) Using the band-edge discontinuity rules discussed in appendix D, design an $In_{1-x}Ga_xAs/InP$ heterojunction such that $\Delta E_c = 0.3$ eV. Assume that strain occurs only in the $In_{1-x}Ga_xAs$ layer, since the substrate InP is thick.

(b) Calculate ΔE_v, $E_g(In_{1-x}Ga_xAs)$, P_ε, Q_ε, and so on for part (a).

4.24 Comment on the results from the subband energies in Figs. 4.18b and c.

4.25 Check the subband energies at the band edges ($k_t = 0$) in Fig. 4.19 with your calculated values. Comment on the discrepancies if there are any.

4.26 Derive the block-diagonalized Hamiltonian (4.8.16) using the axial approximation and the basis transformation (4.8.18).

4.27 (a) Find the eigenvalues and the corresponding eigenvectors for the 2×2 Hamiltonian $\overline{\overline{H}}^U$ in (4.8.16) in the plane-wave representation:

$$\overline{\overline{H}}^U \begin{bmatrix} g^{(1)} \\ g^{(2)} \end{bmatrix} = E \begin{bmatrix} g^{(1)} \\ g^{(2)} \end{bmatrix} \qquad \overline{\overline{H}}^U = -\begin{bmatrix} P+Q & \tilde{R} \\ \tilde{R}^+ & P-Q \end{bmatrix}$$

(b) Repeat part (a) for

$$\overline{\overline{H}}^L = -\begin{bmatrix} P-Q & \tilde{R} \\ \tilde{R}^+ & P+Q \end{bmatrix}$$

4.28 (a) Express the wave function $\Psi^U(k_t, r)$ in (4.8.27), which is in bases $|1\rangle$ and $|2\rangle$ in the original basis $|\frac{3}{2}, \frac{3}{2}\rangle, |\frac{3}{2}, \frac{1}{2}\rangle, |\frac{3}{2}, -\frac{1}{2}\rangle$ and $|\frac{3}{2}, -\frac{3}{2}\rangle$.

(b) Express the bases $|\frac{3}{2}, \nu\rangle$ ($\nu = \frac{3}{2}, \frac{1}{2}, -\frac{1}{2}, -\frac{3}{2}$) in terms of bases $|1\rangle$, $|2\rangle$, $|3\rangle$, and $|4\rangle$.

4.29 Label all the energies in Figs. 4.21a–c for $In_{1-x}Ga_xAs/InP$ quantum wells with (a) $x = 0.37$, (b) $x = 0.47$, and (c) $x = 0.57$.

4.30 Calculate the first quantized subband energies for the electron, heavy-hole, and light-hole subbands in Problem 4.29 using the method discussed in Section 3.2.

REFERENCES

1. N. W. Ashcroft and N. D. Mermin, *Solid State Physics*, Holt, Rinehart & Winston, New York, 1976.

2. F. Bassani and G. Pastori Parravicini, *Electronic States and Optical Transitions in Solids*, Pergamon, Oxford, UK, 1975.

3. J. Bardeen, "An improved calculation of the energies of metallic Li and Na," *J. Chem. Phys.* **6**, 367–371 (1938).

4. F. Seitz, *The Modern Theory of Solids*, McGraw Hill, New York, 1940, p. 352.

5. W. Shockley, "Energy band structures in semiconductors," *Phys. Rev.* **78**, 173–174 (1950).

6. G. Dresselhaus, A. F. Kip, and C. Kittel, "Cyclotron resonance of electrons and holes in silicon and germanium crystals," *Phys. Rev.* **98**, 368–384 (1955).

7. E. O. Kane, "Band structure of indium antimonide," *J. Phys. Chem. Solids*, **1**, 249–261 (1957).

8. E. O. Kane, "The **k · p** method," Chapter 3, in R. K. Willardson and A. C. Beer, Eds., *Semiconductors and Semimetals*, Vol. 1, Academic, New York, 1966.

9. J. M. Luttinger and W. Kohn, "Motion of electrons and holes in perturbed periodic fields," *Phys. Rev.* **97**, 869–883 (1955).

10. J. M. Luttinger, "Quantum theory of cyclotron resonance in semiconductors: General theory," *Phys. Rev.* **102**, 1030–1041 (1956).

11. I. M. Tsidilkovski, *Band Structure of Semiconductors*, Pergamon, Oxford, UK, 1982.

12. B. R. Nag, *Electron Transport in Compound Semiconductors*, Springer, Berlin, 1980.

13. C. R. Pidgeon, "Free carrier optical properties of semiconductors," Chapter 5 in M. Balkanski, Ed., *Handbook on Semiconductors*, Vol. 2, North-Holland, Amsterdam, 1980.

14. For some recent discussion on the effective-mass theory, see M. G. Burt, "The justification for applying the effective-mass approximation to microstructures," *J. Phys.: Condens. Matter* **4**, 6651–6690 (1992); M. G. Burt, "An exact formulation of the envelope function method for the determination of electronic states in semiconductor microstructures," *Semicond. Sci. Technol.* **3**, 739–753 (1988); M. G. Burt, "The evaluation of the matrix element for interband optical transitions in quantum wells using envelope functions," *J. Phys.: Condens. Matter* **5**, 4091–4098 (1993).

15. J. R. Chelikowsky and M. L. Cohen, "Nonlocal pseudopotential calculations for the electronic structure of eleven diamond and zinc-blende semiconductors," *Phys. Rev. B* **14**, 556–582 (1976).

16. G. C. Osbourn, "$In_xGa_{1-x}As$-$In_yGa_{1-y}As$ strained-layer superlattices: A proposal for useful, new electronic materials," *Phys. Rev. B* **27**, 5126–5128 (1983).

17. G. C. Osbourn, "InAsSb strained-layer superlattices for long wavelength detector applications," *J. Vac. Sci. Technol.* **B2**, 176–178 (1984).

18. E. Yablonovitch, "Band structure engineering of semiconductor lasers for optical communications," *J. Lightwave Technol.* **6**, 1292–1299 (1988).

19. T. P. Pearsall, Volume Editor, "Strained-layer superlattices: Physics," in R. K. Willardson and A. C. Beer, Eds., *Semiconductors and Semimetals*, Vol. 32, Academic, New York, 1990.

20. P. S. Zory, *Quantum Well Lasers*, Academic, New York, 1993.

21. G. E. Pikus and G. L. Bir, "Effects of deformation on the hole energy spectrum of germanium and silicon," *Sov. Phys.-Solid State* **1**, 1502–1517 (1960).

22. G. L. Bir and G. E. Pikus, *Symmetry and Strain-Induced Effects in Semiconductors*, Wiley, New York, 1974.

23. R. P. Feynman, R. B. Leighton, and M. Sands, *The Feynman Lectures on Physics*, Chapter 39, Addison-Wesley, Reading, MA, 1964.

24. J. F. Nye, *Physical Properties of Crystals*, Oxford University Press, Oxford, UK, 1985.

25. K. Seeger, *Semiconductor Physics*, Springer, Berlin, 1982.

26. J. C. Hensel and G. Feher, "Cyclotron resonance experiments in uniaxially stressed silicon: Valence band inverse mass parameters and deformation potentials," *Phys. Rev.* **129**, 1041–1062 (1963).

27. H. Hasegawa, "Theory of cyclotron resonance in strained silicon crystals," *Phys. Rev.* **129**, 1029–1040 (1963).

28. A. Gavini and M. Cardona, "Modulated piezoreflectance in semiconductors," *Phys. Rev. B* **1** 672–682 (1970).

29. L. D. Laude, F. H. Pollak, and M. Cardona, "Effects of uniaxial stress on the indirect exciton spectrum of silicon," *Phys. Rev. B* **3**, 2623–2636 (1971).

30. K. Suzuki and J. C. Hensel, "Quantum resonances in the valence bands of germanium, I: Theoretical considerations," *Phys. Rev. B* **9**, 4184–4218 (1974).

31. J. C. Hensel and K. Suzuki, "Quantum resonances in the valence bands of germanium, II: Cyclotron resonances in uniaxially stressed crystals," *Phys. Rev. B* **9**, 4219–4257 (1974).

32. M. Chandrasekhar and F. H. Pollak, "Effects of uniaxial stress on the electroreflectance spectrum of Ge and GaAs," *Phys. Rev. B* **15**, 2127–2144 (1977).

33. D. E. Aspnes and M. Cardona, "Strain dependence of effective masses in tetrahedral semiconductors," *Phys. Rev. B* **17**, 726–740 (1978).

34. H. Mathieu, P. Merle, E. I. Ameziane, B. Archilla, J. Camassel, and G. Poiblaud, "Deformation potentials of the direct and indirect absorption edges of GaP," *Phys. Rev. B* **19**, 2209–2223 (1979).

35. D. L. Smith and C. Mailhiot, "Theory of semiconductor superlattice electronic structure," *Rev. Mod. Phys.* **62**, 173–234 (1990).

36. D. L. Smith and C. Mailhiot, "Strained-layer semiconductor superlattices," *Crit. Rev. Solid State Mater. Sci.* **16**, 131–160 (1990).

37. S. L. Chuang, "Efficient band-structure calculations of strained quantum wells using a two-by-two Hamiltonian," *Phys. Rev. B* **43**, 9649–9661 (1991). (Note that a_v in this paper should be taken as $-a_v$ to compare with data in the literature.)

38. C. Y. P. Chao and S. L. Chuang, "Spin–orbit-coupling effects on the valence-band structure of strained semiconductor quantum wells," *Phys. Rev. B* **46**, 4110–4122 (1992).

39. J. A. Kong, *Electromagnetic Wave Theory*, 2d ed., Wiley, New York, 1990.

40. P. Yeh, *Optical Waves in Layered Media*, Wiley, New York, 1988.

41. E. J. Roan and S. L. Chuang, "Linear and nonlinear intersubband electroabsorptions in a modulation-doped quantum well," *J. Appl. Phys.* **69**, 3249–3260 (1991).

42. K. T. Kim, S. S. Lee, and S. L. Chuang, "Inter-miniband optical absorption in a modulation-doped $Al_xGa_{1-x}As/GaAs$ superlattice," *J. Appl. Phys.* **69**, 6617–6624 (1991).

43. F. Stern and S. Das Sarma, "Electron energy levels in GaAs-Ga$_{1-x}$Al$_x$As heterojunctions," *Phys. Rev. B* **30**, 840–848 (1984).

44. E. Merzbacher, *Quantum Mechanics*, 2d ed., Wiley, New York, 1970.

45. C. Kittel, *Introduction to Solid State Physics*, Wiley, New York, 1976.

46. G. Bastard, "Superlattice band structure in the envelope-function approximation," *Phys. Rev. B* **24**, 5693–5697 (1981).

47. G. Bastard and J. A. Brum, "Electronic states in semiconductor heterostructures," *IEEE J. Quantum Electron.* **QE-22**, 1625–1644 (1986).

48. G. Bastard, *Wave Mechanics Applied to Semiconductor Heterostructures*, Halsted, New York, 1988.

49. K. H. Yoo, L. R. Ram-Mohan, and D. F. Nelson, "Effect of nonparabolicity in GaAs/Ga$_{1-x}$Al$_x$As semiconductor quantum wells," *Phys. Rev. B* **39**, 12808–12813 (1989).

50. D. Mukherji and B. R. Nag, "Band structure of semiconductor superlattices," *Phys. Rev. B* **12**, 4338–4345 (1975).

51. C. Wetzel, B. K. Meyer, and P. Omling, "Electron effective mass in direct-band-gap GaAs$_{1-x}$P$_x$ alloys," *Phys. Rev. B* **47**, 15588–15592 (1993).

52. C. Hermann and C. Weisbuch, "$\mathbf{k} \cdot \mathbf{p}$ perturbation theory in III–V compounds and alloys: a reexamination," *Phys. Rev. B* **15**, 823–833 (1977).

53. W. Zawadzki, P. Pfeffer, and H. Sigg, "Five-level $\mathbf{k} \cdot \mathbf{p}$ model for conduction electrons in GaAs. Description of cyclotron resonance experiments," *Solid State Commun.* **53**, 777–782 (1985).

54. D. A. Broido and L. J. Sham, "Effective masses of holes at GaAs-AlGaAs heterojunctions," *Phys. Rev. B* **31**, 888–892 (1985); and **31**, 6831 (1985).

55. A. Twardowski and C. Hermann, "Variational calculation of polarization of quantum-well photoluminescence," *Phys. Rev. B* **35**, 8144–8153 (1987).

56. R. Eppenga, M. F. H. Schuurmans, and S. Colak, "New $\mathbf{k} \cdot \mathbf{p}$ theory for GaAs/Ga$_{1-x}$Al$_x$As-type quantum wells," *Phys. Rev. B* **36**, 1554–1564 (1987).

57. M. Altarelli, "Band structure, impurities and excitons in superlattices," pp. 12–37 in G. Allan, G. Bastard, N. Boccara, M. Lannoo, and M. Voos, Eds., *Proceedings of Les Houches Winterschool Semiconductor Superlattices and Heterojunctions*, Springer, Berlin, 1986.

58. D. L. Smith and C. Mailhiot, "$\mathbf{k} \cdot \mathbf{p}$ theory of semiconductor superlattice electronic structure, I: Formal results," *Phys. Rev. B* **33**, 8345–8359 (1986).

59. C. Mailhiot and D. L. Smith, "$\mathbf{k} \cdot \mathbf{p}$ theory of semiconductor superlattice electronic structure, II: Application to Ga$_{1-x}$As-Al$_{1-y}$In$_y$As[100] superlattices," *Phys. Rev. B* **33**, 8360–8372 (1986).

60. L. C. Andreani, A. Pasquarello, and F. Bassani, "Hole subbands in strained GaAs-Ga$_{1-x}$Al$_x$As quantum wells: Exact solution of the effective-mass equation," *Phys. Rev. B* **36**, 5887–5894 (1987).

61. L. R. Ram-Mohan, K. H. Yoo, and R. L. Aggarwal, "Transfer-matrix algorithm for the calculation of the band structure of semiconductor superlattices," *Phys. Rev. B* **38**, 6151–6159 (1988).

62. G. Y. Wu, T. C. McGill, C. Mailhiot, and D. L. Smith, "$\mathbf{k} \cdot \mathbf{p}$ theory of semiconductor superlattice electronic structure in an applied magnetic field," *Phys. Rev. B* **39**, 6060–6070 (1989).

63. B. K. Ridley, "The in-plane effective mass in strained-layer quantum wells," *J. Appl. Phys.* **68**, 4667–4673 (1990).

64. I. Suemune, "Band-edge hole mass in strained quantum-well structures," *Phys. Rev. B* **43**, 14099–14106 (1991).

65. M. Silver, W. Batty, A. Ghiti, and E. P. O'Reilly, "Strain-induced valence-subband splitting in III–V semiconductors," *Phys. Rev. B* **46**, 6781–6788 (1992).

66. T. Manku and A. Nathan, "Valence energy-band structure for strained group-IV semiconductors," *J. Appl. Phys.* **73**, 1205–1213 (1993).

67. D. Munzar, "Heavy hole–light splitting in GaAs/AlAs superlattices," *Phys. Status Solidi B* **175**, 395–401 (1993).

68. J. N. Schulman and Y. C. Chang, "Band mixing in semiconductor superlattices," *Phys. Rev. B* **31**, 2056–2068 (1985).

69. Y. C. Chang, "Bond–orbital models for superlattices," *Phys. Rev. B* **37**, 8215–8222 (1988).

70. M. P. Houng, Y. C. Chang, and W. I. Wang, "Orientation dependence of valence-subband structures in GaAs-Ga$_{1-x}$Al$_x$As quantum-well structures," *J. Appl. Phys.* **64**, 4609–4613 (1988).

71. S. S. Nedorezov, "Space quantization in semiconductor films," *Sov. Phys.–Solid State* **12**, 1814–1819 (1971).

72. E. D. Jones, S. K. Lyo, I. J. Fritz, J. F. Klem, J. E. Schirber, C. P. Tigges, and T. J. Drummond, "Determination of energy-band dispersion curves in strained-layer structures," *Appl. Phys. Lett.* **54**, 2227–2229 (1989).

73. R. K. Hayden, L. Eaves, M. Henini, T. Takamasu, N. Miura, and U. Ekenberg, "Investigating the cubic anisotropy of the confined hole subbands of an AlAs/GaAs/AlAs quantum well using resonant magnetotunneling spectroscopy," *Appl. Phys. Lett.* **61**, 84–86 (1992).

74. J. A. Kash, M. Zachau, and M. A. Tischer, "Anisotropic valence bands in quantum wells: Quantitative comparison of theory and experiment," *Phys. Rev. Lett.* **69**, 2260–2263 (1992).

75. J. A. Kash, M. Zachau, M. A. Tischlet, and U. Ekenberg, "Optical measurements of warped valence bands in quantum wells," *Surf. Sci.* **305**, 251–255 (1994).

76. D. Ahn, S. L. Chuang, and Y. C. Chang, "Valence-band mixing effects on the gain and the refractive index change of quantum-well lasers," *J. Appl. Phys.* **64**, 4056–4064 (1988).

5

Electromagnetics

In this chapter, we discuss a few important results using electromagnetic theory. In Section 5.1 we present the general solutions to Maxwell's equations for a given current density **J** and a charge density ρ which satisfy the continuity equation. We discuss the gauge transformation including the Lorentz gauge and the Coulomb gauge; the latter will be used in the Hamiltonian to account for the interaction between the electrons and the photon field in semiconductors in Chapter 9. The time-harmonic fields and the duality principle, which will be useful in studying the propagation of light in waveguides and laser cavities, are presented in Section 5.2. The duality principle allows us to obtain the guided wave solutions for the transverse magnetic (TM) polarization of light once we obtain the solutions for the transverse electric (TE) polarization of light, as we show in Chapter 7.

The plane wave reflection from a layered medium is investigated in Section 5.3 for both TE and TM polarizations. These results are useful for understanding the ray-optics approach to the optical waveguide theory. The radiation of the electromagnetic field is presented in Section 5.4 with the aim that the far-field pattern from a diode laser and laser arrays will be derived once we know the laser mode on the facet of the laser cavity. In Appendix E, we discuss the Kramers–Kronig relation, which relates the real part of a permittivity function to its imaginary part. This is useful in optical materials because often the absorption coefficients of the semiconductors are measured, then the real parts of the refractive indices are calculated based on the Kramers–Kronig relation. The Poynting's theorem for power conservation and the reciprocity theorem are included in Appendix F.

5.1 GENERAL SOLUTIONS TO MAXWELL'S EQUATIONS AND GAUGE TRANSFORMATIONS [1]

In this section, we study the general solutions to Maxwell's equations:

$$\nabla \times \mathbf{E} = -\frac{\partial}{\partial t}\mathbf{B} \tag{5.1.1}$$

$$\nabla \times \mathbf{H} = \mathbf{J} + \frac{\partial}{\partial t}\mathbf{D} \tag{5.1.2}$$

$$\nabla \cdot \mathbf{D} = \rho \tag{5.1.3}$$

$$\nabla \cdot \mathbf{B} = 0 \tag{5.1.4}$$

where the source terms ρ and $\mathbf{J}$ satisfy the continuity equation

$$\nabla \cdot \mathbf{J} + \frac{\partial}{\partial t} \rho = 0 \tag{5.1.5}$$

Since

$$\nabla \cdot (\nabla \times \mathbf{A}) \equiv 0 \tag{5.1.6}$$

for any vector $\mathbf{A}$, we can write $\mathbf{B}$ in the form

$$\mathbf{B} = \nabla \times \mathbf{A} \tag{5.1.7}$$

However, this equation does not uniquely specify $\mathbf{A}$. If we add $\nabla \xi$ to $\mathbf{A}$, where ξ is an arbitrary function, we find that

$$\mathbf{A}' = \mathbf{A} + \nabla \xi \tag{5.1.8}$$

still satisfies $\nabla \times \mathbf{A}' = \mathbf{B}$. From the fundamental theorem of vector analysis, to uniquely define a vector $\mathbf{A}$, we have to specify both its curl and its divergence.

Substituting (5.1.7) into (5.1.1), we obtain

$$\nabla \times \left(\mathbf{E} + \frac{\partial}{\partial t} \mathbf{A} \right) = 0 \tag{5.1.9}$$

Since $\nabla \times (-\nabla \phi) \equiv 0$ for any function ϕ, we may write

$$\mathbf{E} = -\frac{\partial}{\partial t} \mathbf{A} - \nabla \phi \tag{5.1.10}$$

Assuming an isotropic homogeneous medium

$$\mathbf{D} = \varepsilon \mathbf{E} \tag{5.1.11}$$

$$\mathbf{B} = \mu \mathbf{H} \tag{5.1.12}$$

and substituting the expressions for $\mathbf{B}$ and $\mathbf{E}$ from (5.1.7) and (5.1.10) into (5.1.2), we find

$$\left(\nabla^2 \mathbf{A} - \mu\varepsilon \frac{\partial^2}{\partial t^2} \mathbf{A} \right) - \nabla \left(\nabla \cdot \mathbf{A} + \mu\varepsilon \frac{\partial \phi}{\partial t} \right) = -\mu \mathbf{J} \tag{5.1.13}$$

where the vector identity

$$\nabla \times (\nabla \times \mathbf{A}) = \nabla(\nabla \cdot \mathbf{A}) - \nabla^2 \mathbf{A} \tag{5.1.14}$$

has been used. Gauss's law for the electric field (5.1.3) gives

$$\nabla^2 \phi + \frac{\partial}{\partial t}(\nabla \cdot \mathbf{A}) = -\frac{\rho}{\varepsilon} \tag{5.1.15}$$

Thus, generally speaking, we should find the solutions for the scalar potential ϕ and the vector potential $\mathbf{A}$ from (5.1.13) and (5.1.15) in terms of the sources ρ and $\mathbf{J}$. We still have to choose a gauge to uniquely specify the potentials ϕ and $\mathbf{A}$. Since for any ϕ and $\mathbf{A}$ satisfying (5.1.7) and (5.1.10), the following gauge transformations

$$\mathbf{A}' = \mathbf{A} + \nabla \xi \tag{5.1.16}$$

$$\phi' = \phi - \frac{\partial}{\partial t}\xi \tag{5.1.17}$$

also satisfy (5.1.7) and (5.1.10), where ξ is an arbitrary function. Thus, we have to specify $\nabla \cdot \mathbf{A}$. There are two common ways to specify $\nabla \cdot \mathbf{A}$: the Lorentz gauge and the Coulomb gauge.

Lorentz Gauge. In the Lorentz gauge, we choose

$$\nabla \cdot \mathbf{A} = -\mu\varepsilon\frac{\partial \phi}{\partial t} \tag{5.1.18}$$

We then have

$$\nabla^2 \mathbf{A} - \mu\varepsilon\frac{\partial^2}{\partial t^2}\mathbf{A} = -\mu\mathbf{J} \tag{5.1.19}$$

$$\nabla^2 \phi - \mu\varepsilon\frac{\partial^2}{\partial t^2}\phi = -\frac{\rho}{\varepsilon} \tag{5.1.20}$$

That is, the vector and scalar potentials satisfy the wave equations with sources $\mathbf{J}$ and ρ, respectively. The solutions can be written generally as

$$\mathbf{A}(\mathbf{r}, t) = \frac{\mu}{4\pi}\int \frac{d^3\mathbf{r}'\mathbf{J}(\mathbf{r}', t - |\mathbf{r} - \mathbf{r}'|/c)}{|\mathbf{r} - \mathbf{r}'|} \tag{5.1.21}$$

$$\phi(\mathbf{r}, t) = \frac{1}{4\pi\varepsilon}\int \frac{d^3\mathbf{r}'\rho(\mathbf{r}', t - |\mathbf{r} - \mathbf{r}'|/c)}{|\mathbf{r} - \mathbf{r}'|} \tag{5.1.22}$$

where the time delay due to propagation, $|\mathbf{r} - \mathbf{r}'|/c$, appears explicitly in the arguments of the source terms.

Coulomb Gauge [2]. In the Coulomb gauge, we choose

$$\nabla \cdot \mathbf{A} = 0 \tag{5.1.23}$$

Therefore,

$$\nabla^2 \phi = -\frac{\rho}{\varepsilon} \tag{5.1.24}$$

which is the Poisson's equation. The solution is

$$\phi(\mathbf{r}, t) = \frac{1}{4\pi\varepsilon} \int \frac{d^3\mathbf{r}'\rho(\mathbf{r}', t)}{|\mathbf{r} - \mathbf{r}'|} \tag{5.1.25}$$

The solution for $\mathbf{A}$ has to be found from (5.1.13) and (5.1.23). For an optical field, $\rho = 0$, we also find $\phi = 0$ and $E = -\partial\mathbf{A}/\partial t$.

5.2 TIME-HARMONIC FIELDS AND DUALITY PRINCIPLE [1]

Often, the excitation sources $\mathbf{J}$ and ρ depend on time sinusoidally. We may write the source terms as

$$\mathbf{J}(\mathbf{r}, t) = \mathrm{Re}\left[\mathbf{J}(\mathbf{r}, \omega)e^{-i\omega t}\right] \tag{5.2.1}$$

$$\rho(\mathbf{r}, t) = \mathrm{Re}\left[\rho(\mathbf{r}, \omega)e^{-i\omega t}\right] \tag{5.2.2}$$

and similar expressions for the fields $\mathbf{E}$, $\mathbf{H}$, $\mathbf{B}$, and $\mathbf{D}$, where Re [ ] means the real part of the function in the bracket. For example, an electric field in the time domain of the form

$$\mathbf{E}(\mathbf{r}, t) = \hat{x}E_x \cos(\omega t - \phi_x) + \hat{y}E_y \cos(\omega t - \phi_y) + \hat{z}E_z \cos(\omega t - \phi_z) \tag{5.2.3a}$$

can be expressed in terms of a phasor $\mathbf{E}(\mathbf{r}, \omega)$ or simply $\mathbf{E}(\mathbf{r})$:

$$\mathbf{E}(\mathbf{r}, t) = \mathrm{Re}\left[\mathbf{E}(\mathbf{r}, \omega)\, e^{-i\omega t}\right] \tag{5.2.3b}$$

$$\mathbf{E}(\mathbf{r}, \omega) = \hat{x}E_x\, e^{i\phi_x} + \hat{y}E_y\, e^{i\phi_y} + \hat{z}E_z\, e^{i\phi_z} \tag{5.2.3c}$$

We then obtain Maxwell's equations in the frequency domain:

$$\nabla \times \mathbf{E}(\mathbf{r}) = i\omega\mathbf{B}(\mathbf{r}) \tag{5.2.4}$$

$$\nabla \times \mathbf{H}(\mathbf{r}) = \mathbf{J}(\mathbf{r}) - i\omega\mathbf{D}(\mathbf{r}) \tag{5.2.5}$$

$$\nabla \cdot \mathbf{D}(\mathbf{r}) = \rho(\mathbf{r}) \tag{5.2.6}$$

$$\nabla \cdot \mathbf{B}(\mathbf{r}) = 0 \tag{5.2.7}$$

where all quantities $\mathbf{E}$, $\mathbf{H}$, $\mathbf{B}$, $\mathbf{D}$, ρ, and $\mathbf{J}$ are in phasor representations and their dependencies on the angular frequency ω are implicitly assumed.

The relation of $\mathbf{D}(\mathbf{r}, t)$ to $\mathbf{E}(\mathbf{r}, t)$ is

$$\mathbf{D}(\mathbf{r}, t) = \int_{-\infty}^{t} \varepsilon(\mathbf{r}, t - t')\mathbf{E}(\mathbf{r}, t')\, dt' \tag{5.2.8}$$

or

$$\mathbf{D}(\mathbf{r}, t) = \varepsilon_0 \mathbf{E}(\mathbf{r}, t) + \mathbf{P}(\mathbf{r}, t) = \varepsilon_0 \mathbf{E}(\mathbf{r}, t) + \varepsilon_0 \int_{-\infty}^{t} \chi(\mathbf{r}, t - t')\mathbf{E}(\mathbf{r}, t')\, dt'$$

$$\tag{5.2.9}$$

In the frequency domain, we have

$$\mathbf{D}(\mathbf{r}, \omega) = \varepsilon(\mathbf{r}, \omega)\mathbf{E}(\mathbf{r}, \omega) \tag{5.2.10}$$

$$\varepsilon(\mathbf{r}, \omega) = \varepsilon_0\big[1 + \chi(\mathbf{r}, \omega)\big] \tag{5.2.11}$$

In a source free region, we find

$$\nabla \times \mathbf{E} = i\omega\mu\mathbf{H}$$
$$\nabla \times \mathbf{H} = -i\omega\varepsilon\mathbf{E}$$
$$\nabla \cdot \varepsilon\mathbf{E} = 0$$
$$\nabla \cdot \mu\mathbf{H} = 0 \tag{5.2.12}$$

If we make the following changes

$$\mathbf{E} \rightarrow \mathbf{H} \qquad \mathbf{H} \rightarrow -\mathbf{E} \qquad \mu \rightarrow \varepsilon \qquad \varepsilon \rightarrow \mu \tag{5.2.13}$$

then Maxwell's equations stay unchanged. Thus, for a given solution $(\mathbf{E}, \mathbf{H})$ in a medium (μ, ε), we have a corresponding solution $(\mathbf{H}, -\mathbf{E})$ in a medium (ε, μ). This is the duality principle. Since the boundary conditions in which the tangential components of the electric and magnetic fields are continuous across the dielectric interfaces and stay the same in both media, we have a dual situation such that the simultaneous replacements in (5.2.13) give the solutions to Maxwell's equations as well. If including both electric and magnetic sources, we write

$$\nabla \times \mathbf{E} = i\omega\mu\mathbf{H} - \mathbf{M}$$
$$\nabla \times \mathbf{H} = -i\omega\varepsilon\mathbf{E} + \mathbf{J}$$
$$\nabla \cdot \varepsilon\mathbf{E} = \rho_e$$
$$\nabla \cdot \mu\mathbf{H} = \rho_M \tag{5.2.14}$$

Once we make the following identification

$$\mathbf{E} \to \mathbf{H} \qquad \mathbf{H} \to -\mathbf{E} \qquad \mathbf{J} \to \mathbf{M} \qquad \mathbf{M} \to -\mathbf{J}$$
$$\mu \to \varepsilon \qquad \varepsilon \to \mu \qquad \rho_e \to \rho_M \qquad \rho_M \to -\rho_e \qquad (5.2.15)$$

we find that Maxwell's equations are invariant. As an application, if we know the solutions $(\mathbf{E}, \mathbf{H})$ to the original Maxwell equations in a medium (μ, ε) due to the electric sources ρ_e and $\mathbf{J}$, we can obtain the "dual" set of solutions due to the magnetic sources ρ_M and $\mathbf{M}$ in a medium (ε, μ) with "dual" boundary conditions. For example, tangential magnetic fields are continuous vs. tangential electric fields continuous in the original problem with electric sources. Applications of (5.2.13) are discussed in Chapter 7, where we explore the optical waveguide theory for both the TE and TM polarizations. The idea is that once we obtain the field solutions for the TE polarization in a dielectric waveguide from Maxwell equations, we can use the duality principle to write down the TM solutions simply by inspection.

5.3 PLANE WAVE REFLECTION FROM A LAYERED MEDIUM [1, 3]

In this section, we first discuss plane wave reflection from a single planar interface between two dielectric media. We then present a propagation-matrix approach to plane wave reflection from a multiple layered medium.

5.3.1 TE Polarization

Consider a plane wave incident on a planar boundary, as shown in Fig. 5.1a, with the electric field given by

$$\mathbf{E}_i = \hat{y} E_y = \hat{y} E_0 \, e^{-ik_{1x}x + ik_{1z}z} \qquad (5.3.1)$$

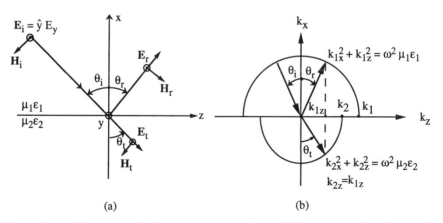

(a)   (b)

Figure 5.1. (a) Plane wave reflection from a planar surface. (b) The phase-matching diagram for plane wave reflection and transmission.

and the magnetic field given by

$$\mathbf{H}_i = \frac{1}{i\omega\mu_1} \nabla \times \mathbf{E}_i$$

$$= \frac{-1}{\omega\mu_1}(k_{1z}\hat{x} + k_{1x}\hat{z})E_0 e^{-ik_{1x}x + ik_{1z}z} \tag{5.3.2}$$

where the components of the wave vector satisfy the dispersion relation:

$$k_{1x}^2 + k_{1z}^2 = \omega^2\mu_1\varepsilon_1 = k_0^2 n_1^2 \tag{5.3.3}$$

Here $k_0 = \omega\sqrt{\mu_0\varepsilon_0} = \omega/c$ is the wave number in free space, and $n_1 = \sqrt{\mu_1\varepsilon_1/(\mu_0\varepsilon_0)}$ is the refractive index of the material. Usually $\mu_1 = \mu_0$ for dielectric materials.

The reflected fields are given by

$$\mathbf{E}_r = \hat{y}rE_0 e^{ik_{1x}^r x + ik_{1z}^r z} \tag{5.3.4}$$

$$\mathbf{H}_r = \frac{1}{\omega\mu_1}(-k_{1z}^r\hat{x} + k_{1x}^r\hat{z})rE_0 e^{ik_{1x}^r x + ik_{1z}^r z} \tag{5.3.5}$$

The transmitted fields are

$$\mathbf{E}_t = \hat{y}tE_0 e^{-ik_{2x}x + ik_{2z}z} \tag{5.3.6}$$

$$\mathbf{H}_t = \frac{-1}{\omega\mu_2}(k_{2z}\hat{x} + k_{2x}\hat{z})tE_0 e^{-ik_{2x}x + ik_{2z}z} \tag{5.3.7}$$

where

$$k_{2x}^2 + k_{2z}^2 = \omega^2\mu_2\varepsilon_2 = k_0^2 n_2^2 \tag{5.3.8}$$

and $n_2 = \sqrt{\mu_2\varepsilon_2/\mu_0\varepsilon_0}$. Matching the boundary conditions in which the tangential electric field (E_y) is continuous at $x = 0$,

$$E_0 e^{ik_{1z}z} + rE_0 e^{ik_{1z}^r z} = tE_0 e^{ik_{2z}z} \tag{5.3.9}$$

for all z, we obtain

$$k_{1z} = k_{1z}^r = k_{2z} \tag{5.3.10}$$

$$1 + r = t \tag{5.3.11}$$

Equation (5.3.10) is Snell's law or the phase-matching condition (Fig. 5.1b):

$$k_1 \sin\theta_i = k_1 \sin\theta_r = k_2 \sin\theta_t \tag{5.3.12}$$

or

$$\theta_i = \theta_r \quad \text{and} \quad n_1 \sin \theta_i = n_2 \sin \theta_t$$

where $k_i = \omega\sqrt{\mu_i \varepsilon_i} = k_0 n_i$, $i = 1, 2$. The other boundary condition in which the tangential magnetic field (H_z) is continuous leads to

$$1 - r = \frac{\mu_1 k_{2x}}{\mu_2 k_{1x}} t \tag{5.3.13}$$

Solving (5.3.11) and (5.3.13) for r and t, we obtain the reflection coefficient

$$r = \frac{1 - P_{12}}{1 + P_{12}} \qquad P_{12} = \frac{\mu_1 k_{2x}}{\mu_2 k_{1x}} \tag{5.3.14}$$

and the transmission coefficient

$$t = \frac{2}{1 + P_{12}} \tag{5.3.15}$$

for the electric field. We note that when $k_1 > k_2$, the total internal reflection occurs at an angle of incidence larger than the critical angle θ_c, given by the condition when $\theta_t = 90°$:

$$k_1 \sin \theta_c = k_2 \tag{5.3.16}$$

or $\theta_c = \sin^{-1}(n_2/n_1)$. When $\theta_i > \theta_c$, $k_{1z} = k_1 \sin \theta_i > k_2$, we have

$$k_{2x}^2 = k_2^2 - k_{2z}^2 = k_2^2 - k_{1z}^2 < 0 \tag{5.3.17}$$

Thus, k_{2x} is purely imaginary, $k_{2x} = +i\alpha_2$, $\alpha_2 > 0$, and the field is decaying along the $-x$ direction:

$$E_y = tE_0 \, e^{\alpha_2 x + ik_{1z} z} \tag{5.3.18}$$

The reflection coefficient becomes

$$r = \frac{1 - i(\mu_1 \alpha_2 / \mu_2 k_{1x})}{1 + i(\mu_1 \alpha_2 / \mu_2 k_{1x})} = e^{-i2\phi_{12}} \tag{5.3.19a}$$

where

$$\phi_{12} = \tan^{-1}\left(\frac{\mu_1 \alpha_2}{\mu_2 k_{1x}}\right) \tag{5.3.19b}$$

The fraction of the power reflected from its boundary is given by the reflectivity R:

$$R = \frac{(1/2)\mathrm{Re}(\mathbf{E}_r \times \mathbf{H}_r^*)_x}{(-1/2)\mathrm{Re}(\mathbf{E}_i \times \mathbf{H}_i^*)_x}\bigg|_{x=0} = |r|^2 \tag{5.3.20}$$

The transmissivity is given by

$$T = \frac{(-1/2)\mathrm{Re}(\mathbf{E}_t \times \mathbf{H}_t^*)_x}{(-1/2)\mathrm{Re}(\mathbf{E}_i \times \mathbf{H}_i^*)_x}\bigg|_{x=0} = \frac{\mu_1}{k_{1x}}\mathrm{Re}\left(\frac{k_{2x}^*}{\mu_2}\right)|t|^2 \tag{5.3.21}$$

Power conservation requires $R + T = 1$, which can also be checked by substituting (5.3.14) and (5.3.15) into (5.3.20) and (5.3.21). It is clear that if $\theta_i > \theta_c$, total internal reflection occurs; the transmissivity is zero since k_{2x} is purely imaginary, and the reflectivity is unity, $|r|^2 = |e^{-i2\phi_{12}}| = 1$. However, the reflected field experiences an optical phase shift of an amount $-2\phi_{12}$. This angle is called the Goos–Hänchen phase shift.

Special Case for Normal Incidence. At normal incidence, $\theta_i = 0°$, $k_{2x} = k_2 = k_0 n_2$ and $k_{1x} = k_1 = k_0 n_1$. We find the reflection and transmission coefficients for the field are, respectively,

$$r = \frac{n_1 - n_2}{n_1 + n_2} \qquad t = \frac{2n_1}{n_1 + n_2} \tag{5.3.22}$$

5.3.2   TM Polarization

The result of TM polarization can be obtained by the duality principle using the exchange of the physical quantities in Eq. (5.2.13). The results are

$$\mathbf{H}_i = \hat{y}H_y$$

$$= \hat{y}H_0\,e^{-ik_{1x}x + ik_{1z}z} \tag{5.3.23}$$

$$\mathbf{H}_r = \hat{y}r_{\mathrm{TM}}H_0\,e^{+ik_{1x}x + ik_{1z}z} \tag{5.3.24}$$

$$\mathbf{H}_t = \hat{y}t_{\mathrm{TM}}H_0\,e^{-ik_{2x}x + ik_{2z}z} \tag{5.3.25}$$

where $k_{2z} = k_{1z}$ and

$$r_{\mathrm{TM}} = \frac{1 - P_{12}}{1 + P_{12}} \qquad P_{12} = \frac{\varepsilon_1 k_{2x}}{\varepsilon_2 k_{1x}} \tag{5.3.26}$$

$$t_{\mathrm{TM}} = \frac{2}{1 + P_{12}} \tag{5.3.27}$$

For total internal reflection, $k_{2x} = i\alpha_2$, and $r_{TM} = e^{-i2\phi_{12}^{TM}}$, where

$$\phi_{12}^{TM} = \tan^{-1}\left(\frac{\varepsilon_1\alpha_2}{\varepsilon_2 k_{1x}}\right) \tag{5.3.28}$$

The magnetic field experiences a Goos–Hänchen phase shift of an amount $-2\phi_{12}^{TM}$ when the angle of incidence is larger than the critical angle θ_c.

5.3.3 Propagation Matrix Approach for Plane Wave Reflection From a Multilayered Medium

If the medium is inhomogeneous along the x direction $\varepsilon = \varepsilon(x)$ and $\mu = \mu(x)$, we can discretize the permittivity and permeability and use the propagation matrix method, as shown in Section 4.6 for the Schrödinger equation. We use

$$\mu_\ell = \mu(x_\ell)$$
$$\varepsilon_\ell = \varepsilon(x_\ell) \qquad \text{for } -d_{\ell-1} \geq x \geq -d_\ell$$

as shown in Fig. 5.2. For a TE polarized incident wave

$$\mathbf{E}_i = \hat{y}E_0\, e^{-ik_{0x}x + ik_{0z}z} \tag{5.3.29}$$

we have the reflected wave

$$\mathbf{E}_r = \hat{y}rE_0\, e^{ik_{0x}x + ik_{0z}z} \tag{5.3.30}$$

In the ℓth layer, $x_{\ell-1} \geq x \geq x_\ell$, the electric field is given by $\mathbf{E}_\ell = \hat{y}E_y^\ell$, where

$$E_y^\ell = \left(A_\ell e^{-ik_{\ell x}(x+d_\ell)} + B_\ell e^{+ik_{\ell x}(x+d_\ell)}\right) e^{ik_{\ell z}z}$$

$$H_z^\ell = \frac{1}{i\omega\mu_\ell}\frac{\partial}{\partial x}E_y^\ell$$

$$= \frac{-k_{\ell x}}{\omega\mu_\ell}\left(A_\ell e^{-ik_{\ell x}(x+d_\ell)} - B_\ell e^{ik_{\ell x}(x+d_\ell)}\right) e^{ik_{\ell z}z} \tag{5.3.31}$$

and $k_{\ell x} = \sqrt{\omega^2\mu_\ell\varepsilon_\ell - k_{0z}^2}$, $k_{\ell z} = k_{0z}$ for all ℓ. The boundary conditions in which E_y and H_z are continuous at $x = -d_\ell$ lead to

$$A_\ell + B_\ell = A_{\ell+1}e^{-ik_{(\ell+1)x}(-d_\ell+d_{\ell+1})} + B_{\ell+1}e^{ik_{(\ell+1)x}(-d_\ell+d_{\ell+1})}$$

$$A_\ell - B_\ell = \frac{\mu_\ell k_{(\ell+1)x}}{\mu_{\ell+1}k_{\ell x}}\left(A_{\ell+1}e^{-ik_{(\ell+1)x}(-d_\ell+d_{\ell+1})} - B_{\ell+1}e^{ik_{(\ell+1)x}(-d_\ell+d_{\ell+1})}\right)$$

$$\tag{5.3.32}$$

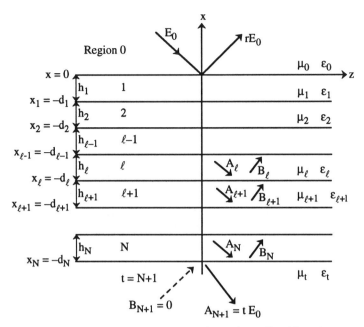

Figure 5.2. A plane wave with TE polarization $\mathbf{E} = \hat{y}E_y = \hat{y}E_0 e^{-ik_{0x}x + ik_{0z}z}$ is incident on a multilayered medium.

Again, similar to the propagation method in a one-dimensional potential in the Schrödinger equation, we define

$$P_{\ell(\ell+1)} = \frac{\mu_\ell k_{(\ell+1)x}}{\mu_{\ell+1} k_{\ell x}} \tag{5.3.33}$$

and the thickness of the region $\ell + 1$

$$h_{\ell+1} = d_{\ell+1} - d_\ell \tag{5.3.34}$$

We have

$$\begin{bmatrix} A_\ell \\ B_\ell \end{bmatrix} = \overline{\overline{\mathbf{B}}}_{\ell(\ell+1)} \begin{bmatrix} A_{\ell+1} \\ B_{\ell+1} \end{bmatrix} \tag{5.3.35a}$$

where the backward-propagation matrix $\overline{\overline{\mathbf{B}}}_{\ell(\ell+1)}$ is defined as

$$\overline{\overline{\mathbf{B}}}_{\ell(\ell+1)} = \frac{1}{2} \begin{bmatrix} (1 + P_{\ell(\ell+1)}) e^{-ik_{(\ell+1)x}h_{\ell+1}} & (1 - P_{\ell(\ell+1)}) e^{ik_{(\ell+1)x}h_{\ell+1}} \\ (1 - P_{\ell(\ell+1)}) e^{-ik_{(\ell+1)x}h_{\ell+1}} & (1 + P_{\ell(\ell+1)}) e^{ik_{(\ell+1)x}h_{\ell+1}} \end{bmatrix}$$

$$\tag{5.3.35b}$$

Alternatively, we can define a forward-propagation matrix $\bar{\bar{F}}_{(\ell+1)\ell}$ similar to that for the quantum potential problem (4.6.7):

$$\begin{bmatrix} A_{\ell+1} \\ B_{\ell+1} \end{bmatrix} = \bar{\bar{F}}_{(\ell+1)\ell} \begin{bmatrix} A_\ell \\ B_\ell \end{bmatrix} \tag{5.3.36a}$$

$$\bar{\bar{F}}_{(\ell+1)\ell} = \frac{1}{2} \begin{bmatrix} (1 + P_{(\ell+1)\ell})e^{ik_{(\ell+1)x}h_{\ell+1}} & (1 - P_{(\ell+1)\ell})e^{ik_{(\ell+1)x}h_{\ell+1}} \\ (1 - P_{(\ell+1)\ell})e^{-ik_{(\ell+1)x}h_{\ell+1}} & (1 + P_{(\ell+1)\ell})e^{-ik_{(\ell+1)x}h_{\ell+1}} \end{bmatrix} \tag{5.3.36b}$$

Note that $P_{(\ell+1)\ell} = 1/P_{\ell(\ell+1)}$ and the product $\bar{\bar{B}}_{\ell(\ell+1)}\bar{\bar{F}}_{(\ell+1)\ell}$ is a unity matrix.

The amplitudes of the incident and reflected waves are related to those amplitudes in the transmitted region $N + 1$ by

$$\begin{bmatrix} E_0 \\ rE_0 \end{bmatrix} = \bar{\bar{B}}_{01}\bar{\bar{B}}_{12}\bar{\bar{B}}_{23} \cdots \bar{\bar{B}}_{N(N+1)} \begin{bmatrix} A_{N+1} \\ B_{N+1} \end{bmatrix}$$

$$= \begin{bmatrix} b_{11} & b_{12} \\ b_{21} & b_{22} \end{bmatrix}\begin{bmatrix} tE_0 \\ 0 \end{bmatrix} \tag{5.3.37}$$

where $B_{N+1} = 0$, since there is no incident wave from the bottom region in Fig. 5.2. We obtain the transmission coefficient of the multilayered medium

$$t = \frac{1}{b_{11}} \tag{5.3.38}$$

and the reflection coefficient of the multilayered medium

$$r = \frac{b_{21}}{b_{11}} \tag{5.3.39}$$

The electric field in the transmitted region is given by

$$E_y = tE_0\, e^{-ik_{(N+1)x}(x-d_{N+1})}\, e^{ik_{(N+1)z}z} \tag{5.3.40}$$

Since we use d_N as the bottom boundary, the last region $N + 1$ can be chosen such that $h_{N+1} = d_{N+1} - d_N = 0$ for convenience. It means that the field will be measured with $d_{N+1} = d_N$.

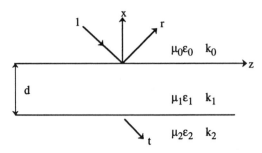

Figure 5.3. A plane wave reflection from a dielectric slab with a thickness d.

As an example, we consider a dielectric slab with a thickness d as shown in Fig. 5.3. The backward-propagation matrices $\overline{\overline{\mathbf{B}}}_{01}$ and $\overline{\overline{\mathbf{B}}}_{12}$ are

$$\overline{\overline{\mathbf{B}}}_{01} = \frac{1}{2}\begin{bmatrix} (1 + P_{01})\,e^{-ik_{1x}d} & (1 - P_{01})\,e^{ik_{1x}d} \\ (1 - P_{01})\,e^{-ik_{1x}d} & (1 + P_{01})\,e^{ik_{1x}d} \end{bmatrix} \qquad (5.3.41)$$

$$\overline{\overline{\mathbf{B}}}_{12} = \frac{1}{2}\begin{bmatrix} (1 + P_{12}) & (1 - P_{12}) \\ (1 - P_{12}) & (1 + P_{12}) \end{bmatrix} \qquad (5.3.42)$$

We calculate the matrix product $\overline{\overline{\mathbf{B}}}_{01}\overline{\overline{\mathbf{B}}}_{12}$, then the transmission coefficient

$$t = \frac{1}{b_{11}} = \frac{4e^{ik_{1x}d}}{(1 + P_{01})(1 + P_{12})\left(1 + r_{01}r_{12}\,e^{i2k_{1x}d}\right)} \qquad (5.3.43)$$

and the reflection coefficient

$$r = \frac{b_{21}}{b_{11}} = \frac{r_{01} + r_{12}\,e^{i2k_{1x}d}}{1 + r_{01}r_{12}\,e^{i2k_{1x}d}} \qquad (5.3.44)$$

where

$$r_{\ell(\ell+1)} = \frac{1 - P_{\ell(\ell+1)}}{1 + P_{\ell(\ell+1)}} \qquad (5.3.45)$$

is the reflection coefficient of a planar surface between medium ℓ and medium $\ell + 1$. Note that $r_{\ell(\ell+1)} = -r_{(\ell+1)\ell}$ since $P_{\ell(\ell+1)} = 1/P_{(\ell+1)\ell}$.

For TM polarization, we use the duality principle again: exchange $\mathbf{E} \to \mathbf{H}$, $\mathbf{H} \to -\mathbf{E}$, $\mu_\ell \to \varepsilon_\ell$ and $\varepsilon_\ell \to \mu_\ell$.

Example A common formula used in determining the quantum efficiency of a photodetector is the absorbence A defined as $A = 1 - R - T$, where R is the total reflectivity of the power from the slab of an absorption depth d, and T is the transmissivity of the power. Here $R + T + A = 1$ by power conservation. Consider normal incidence $k_{1x} = k_1 = k_{1r} + i(\alpha/2)$, where α is the absorption coefficient of the optical intensity and a factor of $\frac{1}{2}$ accounts for the absorption coefficient for the optical electric field.

Assuming that the transmission region 2 is the same as region 0 (Fig. 5.3), we have

$$r_{12} = r_{10} = -r_{01} \tag{5.3.46}$$

Equation (5.3.20) and (5.3.21) also lead to (neglecting the small imaginary part of P_{01})

$$1 = R_{01} + T_{01} = |r_{01}|^2 + P_{01}|t_{01}|^2 \tag{5.3.47}$$

The power reflectivity of the slab is found by using (5.3.44):

$$R = |r|^2 = \frac{R_{01}\left(1 - e^{-\alpha d}\right)^2}{\left(1 - R_{01}e^{-\alpha d}\right)^2} \tag{5.3.48}$$

if we ignore the phase coherence $ik_{1r}d$ in the exponent, assuming that the absorption is the dominant effect. The transmitted power is found by using (5.3.43):

$$T = |t|^2 = \frac{16P_{01}^2\, e^{-\alpha d}}{\left(1 + P_{01}\right)^4\left(1 - R_{01}\, e^{-\alpha d}\right)^2} = \frac{\left(1 - R_{01}\right)^2 e^{-\alpha d}}{\left(1 - R_{01}\, e^{-\alpha d}\right)^2} \tag{5.3.49}$$

Therefore, the absorbance of the slab absorption region with a thickness d is

$$A = 1 - R - T = \frac{\left(1 - R_{01}\right)\left(1 - e^{-\alpha d}\right)}{\left(1 - R_{01}\, e^{-\alpha d}\right)^2}\left(1 + R_{01}\, e^{-\alpha d}\right) \tag{5.3.50}$$

Sometimes, the multiple reflection is negligible if $\alpha d \gg 1$ and the above absorbance is approximated by

$$A \simeq \left(1 - R_{01}\right)\left(1 - e^{-\alpha d}\right) \tag{5.3.51}$$

which is also used commonly in photodetector applications, as we discuss in Chapter 14. ∎

5.4 RADIATION AND FAR-FIELD PATTERN [1]

5.4.1 General Expressions for Radiation Fields

We write (5.1.19) and (5.1.20) in the Lorentz gauge ($\nabla \cdot \mathbf{A} = -\mu\varepsilon\partial\phi/\partial t$) in the frequency domain:

$$\nabla^2\mathbf{A}(\mathbf{r}) + \omega^2\mu\varepsilon\mathbf{A}(\mathbf{r}) = -\mu\mathbf{J}(\mathbf{r}) \tag{5.4.1}$$

$$\nabla^2\phi(\mathbf{r}) + \omega^2\mu\varepsilon\phi(\mathbf{r}) = -\frac{\rho(r)}{\varepsilon} \tag{5.4.2}$$

We find

$$\mathbf{A}(\mathbf{r}) = \frac{\mu}{4\pi} \int \frac{e^{ik|\mathbf{r}-\mathbf{r}'|}\mathbf{J}(\mathbf{r}')}{|\mathbf{r}-\mathbf{r}'|} d^3\mathbf{r}' \tag{5.4.3}$$

and

$$\phi(\mathbf{r}) = \frac{1}{4\pi\varepsilon} \int \frac{e^{ik|\mathbf{r}-\mathbf{r}'|}\rho(\mathbf{r}')}{|\mathbf{r}-\mathbf{r}'|} d^3\mathbf{r}' \tag{5.4.4}$$

where $k = \omega\sqrt{\mu\varepsilon}$. From the charge continuity equation, we have

$$\rho(\mathbf{r}) = \frac{1}{i\omega}\nabla \cdot \mathbf{J}(\mathbf{r}) \tag{5.4.5}$$

The electric field in the time domain is given by

$$\mathbf{E} = -\frac{\partial}{\partial t}\mathbf{A} - \nabla\phi \tag{5.4.6}$$

and in the frequency domain is

$$\begin{aligned}
\mathbf{E}(\mathbf{r}) &= i\omega\mathbf{A}(\mathbf{r}) - \nabla\phi(\mathbf{r}) \\
&= i\omega\frac{\mu}{4\pi}\int\frac{e^{ik|\mathbf{r}-\mathbf{r}'|}}{|\mathbf{r}-\mathbf{r}'|}\mathbf{J}(\mathbf{r}')\,d^3\mathbf{r}' - \nabla\frac{1}{4\pi\varepsilon}\int\frac{e^{ik|\mathbf{r}-\mathbf{r}'|}\nabla'\cdot\mathbf{J}(\mathbf{r}')\,d^3\mathbf{r}'}{i\omega|\mathbf{r}-\mathbf{r}'|} \\
&= i\omega\mu\left[\bar{\mathbf{I}} + \frac{1}{k^2}\nabla\nabla\right]\cdot\int\frac{e^{ik|\mathbf{r}-\mathbf{r}'|}}{4\pi|\mathbf{r}-\mathbf{r}'|}\mathbf{J}(\mathbf{r}')\,d^3\mathbf{r}'
\end{aligned} \tag{5.4.7}$$

where

$$\begin{aligned}
\int g(\mathbf{r}-\mathbf{r}')\nabla'\cdot\mathbf{J}(\mathbf{r}')\,d^3\mathbf{r}' &= \int\nabla'\cdot(g\mathbf{J})\,d^3\mathbf{r}' - \int\nabla'g(\mathbf{r}-\mathbf{r}')\cdot\mathbf{J}(\mathbf{r}')\,d^3\mathbf{r}' \\
&= +\nabla\cdot\int g(\mathbf{r}-\mathbf{r}')\mathbf{J}(\mathbf{r}')\,d^3\mathbf{r}'
\end{aligned}$$

has been used. The term $\int \nabla' \cdot (g\mathbf{J})\mathrm{d}^3r' = \oint g\mathbf{J} \cdot \mathrm{d}S' = 0$ since the surface integral at infinity should vanish, assuming there is no source at infinity and $g(\mathbf{r} - \mathbf{r}') = \exp(ik|\mathbf{r} - \mathbf{r}'|)/4\pi|\mathbf{r} - \mathbf{r}'| \to 0$ as $\mathbf{r}'$ approaches infinity. Here $\bar{\mathbf{I}}$ is a unity matrix:

$$\bar{\mathbf{I}} = \begin{bmatrix} 1 & 0 & 0 \\ 0 & 1 & 0 \\ 0 & 0 & 1 \end{bmatrix} = \hat{x}\hat{x} + \hat{y}\hat{y} + \hat{z}\hat{z} = \hat{r}\hat{r} + \hat{\theta}\hat{\theta} + \hat{\phi}\hat{\phi}$$

Defining a dyadic Green's function as

$$\bar{\bar{\mathbf{G}}}(\mathbf{r},\mathbf{r}') \equiv \left[\bar{I} + \frac{1}{k^2}\nabla\nabla\right]\frac{e^{ik|\mathbf{r}-\mathbf{r}'|}}{4\pi|\mathbf{r} - \mathbf{r}'|} \tag{5.4.8}$$

we have

$$\mathbf{E}(\mathbf{r}) = i\omega\mu\int\bar{\bar{\mathbf{G}}}(\mathbf{r},\mathbf{r}') \cdot \mathbf{J}(\mathbf{r}')\,\mathrm{d}^3r' \tag{5.4.9}$$

and the magnetic field is given by

$$\mathbf{H} = \frac{1}{i\omega\mu}\nabla \times \mathbf{E}(\mathbf{r})$$

$$= \nabla \times \int\bar{\bar{\mathbf{G}}}(\mathbf{r},\mathbf{r}') \cdot \mathbf{J}(\mathbf{r}')\,\mathrm{d}^3r' \tag{5.4.10}$$

Suppose we consider an equivalent magnetic current source $\mathbf{M}$ and an equivalent magnetic charge density ρ_M:

$$\nabla \times \mathbf{E} = i\omega\mu\mathbf{H} - \mathbf{M}$$
$$\nabla \times \mathbf{H} = -i\omega\varepsilon\mathbf{E}$$
$$\nabla \cdot \varepsilon\mathbf{E} = 0$$
$$\nabla \cdot \mu\mathbf{H} = \rho_M \tag{5.4.11}$$

where

$$\nabla \cdot \mathbf{M} - i\omega\rho_M = 0 \tag{5.4.12}$$

Equivalently, from the duality principle, we can find

$$\mathbf{H}(\mathbf{r}) = i\omega\varepsilon\int\bar{\bar{\mathbf{G}}}(\mathbf{r},\mathbf{r}') \cdot \mathbf{M}(\mathbf{r}')\,\mathrm{d}^3r' \tag{5.4.13}$$

$$\mathbf{E}(\mathbf{r}) = -\nabla \times \int\bar{\bar{\mathbf{G}}}(\mathbf{r},\mathbf{r}') \cdot \mathbf{M}(\mathbf{r}')\,\mathrm{d}^3r' \tag{5.4.14}$$

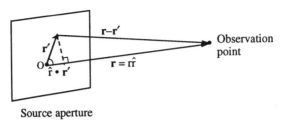

Figure 5.4. Far-field approximation $|\mathbf{r} - \mathbf{r}'| \simeq r - \hat{r} \cdot \mathbf{r}'$, where $\hat{r}$ is a unit vector along the direction of $\mathbf{r}$. The length $|\mathbf{r} - \mathbf{r}'|$ is approximately r subtracted by the distance of the projection of $\mathbf{r}'$ on $\hat{r}$.

Thus, in the presence of both electric and magnetic sources, $\mathbf{J}$ and $\mathbf{M}$, we find

$$\mathbf{E}(\mathbf{r}) = \int \left[i\omega\mu \overline{\overline{\mathbf{G}}}(\mathbf{r}, \mathbf{r}') \cdot \mathbf{J}(\mathbf{r}') - \nabla \times \overline{\overline{\mathbf{G}}}(\mathbf{r}, \mathbf{r}') \cdot \mathbf{M}(\mathbf{r}') \right] d^3r' \quad (5.4.15)$$

$$\mathbf{H}(\mathbf{r}) = \int \left[\nabla \times \overline{\overline{\mathbf{G}}}(\mathbf{r}, \mathbf{r}') \cdot \mathbf{J}(\mathbf{r}') + i\omega\varepsilon \overline{\overline{\mathbf{G}}}(\mathbf{r}, \mathbf{r}') \cdot \mathbf{M}(\mathbf{r}') \right] d^3r' \quad (5.4.16)$$

5.4.2 Far-Field Approximation

In many cases of radiation of the fields, for example, if the source distribution is over a finite region and the observation point $\mathbf{r}$ is far away from the source, we may approximate

$$|\mathbf{r} - \mathbf{r}'| \simeq r - \hat{r} \cdot \mathbf{r}' \quad (5.4.17)$$

where $\hat{r}$ is a unit vector along the $\mathbf{r}$ direction, as shown in Fig. 5.4. The distance $\hat{r} \cdot \mathbf{r}'$ is the projection of $\mathbf{r}'$ on $\hat{r}$ and the distance $|\mathbf{r} - \mathbf{r}'|$ is approximately the difference between the length r and $\hat{r} \cdot \mathbf{r}'$. With this approximation, we further approximate

$$\frac{e^{ik|\mathbf{r} - \mathbf{r}'|}}{4\pi|\mathbf{r} - \mathbf{r}'|} \simeq \frac{e^{ikr}}{4\pi r} e^{-i\mathbf{k} \cdot \mathbf{r}'} \quad (5.4.18)$$

where $\mathbf{k} = k\hat{r}$ is along the direction of observation. We keep the term $\hat{r} \cdot \mathbf{r}'$ only in the phase factor, since any small variation in $\mathbf{r}'$ comparable with wavelength causes a big change in the phase of $e^{-i\mathbf{k} \cdot \mathbf{r}'}$, but a negligible change in the denominator of (5.4.18), since $\mathbf{r}$ is very large in the far-field region. We consider radiation into the free space, $k = k_0 = \omega\sqrt{\mu_0\varepsilon_0}$. Therefore,

$$\overline{\overline{\mathbf{G}}}(\mathbf{r}, \mathbf{r}') \simeq \left[\overline{\overline{\mathbf{I}}} - \hat{r}\hat{r} \right] \frac{e^{ikr}}{4\pi r} e^{-i\mathbf{k} \cdot \mathbf{r}'} = \left(\hat{\theta}\hat{\theta} + \hat{\phi}\hat{\phi} \right) \frac{e^{ikr}}{4\pi r} e^{-i\mathbf{k} \cdot \mathbf{r}'} \quad (5.4.19)$$

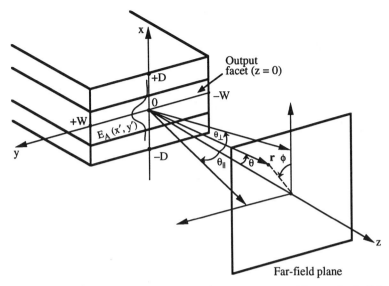

Figure 5.5. The coordinates for calculating the far-field pattern at position $\mathbf{r} = (r, \theta, \phi)$ from an aperture field $E_A(x', y')$ at the laser output facet plane $z = 0$. The angle $\theta_\perp$ is measured from the z axis along the x direction, and $\theta_\parallel$ is measured from the z axis toward the y direction.

and

$$\nabla \times \overline{\overline{\mathbf{G}}}(\mathbf{r}, \mathbf{r}') \simeq \mathrm{i}k\hat{r} \times \overline{\overline{\mathbf{G}}}(\mathbf{r}, \mathbf{r}') \tag{5.4.20}$$

Thus, using $\hat{r} \times (\overline{\overline{\mathbf{I}}} - \hat{r}\hat{r}) \cdot \mathbf{M} = \hat{r} \times (\hat{\theta}M_\theta + \hat{\phi}M_\phi) = \hat{r} \times \mathbf{M}$ since $\hat{r} \times \hat{r} = 0$, we find

$$\mathbf{E}(\mathbf{r}) = \frac{e^{\mathrm{i}kr}}{4\pi r} \left\{ \mathrm{i}\omega\mu\left(\hat{\theta}\hat{\theta} + \hat{\phi}\hat{\phi}\right) \cdot \int \mathrm{d}^3\mathbf{r}'\, e^{-\mathrm{i}\mathbf{k}\cdot\mathbf{r}'}\mathbf{J}(\mathbf{r}') \right.$$

$$\left. - \mathrm{i}k\hat{r} \times \int \mathrm{d}^3\mathbf{r}'\, e^{-\mathrm{i}\mathbf{k}\cdot\mathbf{r}'}\mathbf{M}(\mathbf{r}') \right\} \tag{5.4.21}$$

As an example, for a semiconductor laser, shown in Fig. 5.5, if the aperture field at the output facet of the laser is given by $\mathbf{E}_A(\mathbf{r}')$, we may use an "equivalent source" by simply taking [1]

$$\mathbf{J}(\mathbf{r}') = 0 \tag{5.4.22}$$

and the magnetic current source is a current sheet from the tangential component of $\mathbf{E}_A$ and its image with

$$\mathbf{M}_S(\mathbf{r}') = -2\hat{n} \times \mathbf{E}_A(\mathbf{r}') = -2\hat{z} \times \mathbf{E}_A(\mathbf{r}') \tag{5.4.23}$$

where $\hat{n}$ is a unit normal to the aperture surface and a factor of two accounts for the image source. We then integrate over the surface of the aperture to find the radiation field using Eq. (5.4.21):

$$\mathbf{E}(\mathbf{r}) = \frac{-ik\,e^{ikr}}{4\pi r}\hat{r} \times \iint_{\text{Aperture}} dx'\,dy'\,e^{-ik\hat{r}\cdot\mathbf{r}'}\mathbf{M}_S(\mathbf{r}') \quad (5.4.24)$$

The radiated power density is

$$P = \frac{1}{2\eta_0}|\mathbf{E}(\mathbf{r})|^2 \quad (5.4.25)$$

where $\eta_0 = \sqrt{\mu_0/\varepsilon_0} = 377\ \Omega$ in free space, and

$$\hat{r} = \sin\theta\cos\phi\,\hat{x} + \sin\theta\sin\phi\,\hat{y} + \cos\theta\,\hat{z} \quad (5.4.26)$$

is a unit vector along the direction of observation $\mathbf{r}$, which is the same as the direction of wave propagation in the far field $\mathbf{k}$. If we consider the aperture field to be due to a TE mode in the active region of the semiconductor laser

$$\mathbf{E}_A = \hat{y}E_A(x, y) \quad (5.4.27)$$

then the equivalent magnetic current sheet is

$$\begin{aligned}\mathbf{M}_S &= -2\hat{n} \times \mathbf{E}_A \\ &= 2\hat{x}E_A(x, y)\end{aligned} \quad (5.4.28)$$

Consider two cases.

1. $\theta_\perp$ *measurement.* If the far-field pattern is measured along the x direction in the far-field plane, $\phi = 0$ (or $\phi = \pi$), and $\theta = \theta_\perp$, we obtain

$$\hat{r} = \sin\theta_\perp\,\hat{x} + \cos\theta_\perp\,\hat{z} \quad (5.4.29)$$

$$\mathbf{r}' = x'\hat{x} + y'\hat{y} \qquad \text{since } z' = 0 \text{ on the aperture} \quad (5.4.30)$$

$$\begin{aligned}\mathbf{E}(\mathbf{r}) &= \frac{-ik\,e^{ikr}}{4\pi r}\hat{r} \times 2\hat{x}\int_{-D}^{D} dx'\int_{-W}^{W} dy'\,e^{-ik\sin\theta_\perp x'}E_A(x', y') \\ &= \hat{y}\left(\frac{-ik\,e^{ikr}}{2\pi r}\right)\cos\theta_\perp\int_{-D}^{D} dx'\int_{-W}^{W} dy'\,e^{-ik\sin\theta_\perp x'}E_A(x', y')\end{aligned} \quad (5.4.31)$$

The optical power intensity pattern is

$$P(\theta_\perp) \propto \cos^2\theta_\perp\left|\int_{-D}^{D} dx'\int_{-W}^{W} dy'\,e^{-ik\sin\theta_\perp x'}E_A(x', y')\right|^2 \quad (5.4.32)$$

where the extra factor $\cos^2\theta_\perp$ is necessary due to the polarization of the aperture field.

2. $\theta_\parallel$ *measurement.* If the far-field pattern is measured along the y direction in the far-field plane, $\phi = \pi/2$, and $\theta = \theta_\parallel$, we obtain

$$\hat{r} = \sin\theta_\parallel \hat{y} + \cos\theta_\parallel \hat{z} \tag{5.4.33}$$

$$\mathbf{E}(\mathbf{r}) = \frac{-ike^{ikr}}{2\pi r}\left(\cos\theta_\parallel \hat{y} - \sin\theta_\parallel \hat{z}\right)\int_{-D}^{D}dx'\int_{-W}^{W}dy'\, e^{-ik\sin\theta_\parallel y'}E_A(x',y') \tag{5.4.34}$$

and the radiation power intensity pattern is

$$P(\theta_\parallel) \propto \left|\int_{-D}^{D}dx'\int_{-W}^{W}dy'\, e^{-ik\sin\theta_\parallel y'}E_A(x',y')\right|^2 \tag{5.4.35}$$

The above results for the far-field patterns are useful for studying the beam divergence for a single-element semiconductor laser and coupled laser arrays. A simple scalar formulation can be found in Ref. 4, where numerical results for various diode lasers are shown.

PROBLEMS

5.1 (a) Take the Fourier transforms of $\mathbf{A}(\mathbf{r}, t)$ and $\phi(\mathbf{r}, t)$ with respect to time in (5.1.21) and (5.1.22) and express $\mathbf{A}(\mathbf{r}, \omega)$ and $\phi(\mathbf{r}, \omega)$ in terms of $\mathbf{J}(\mathbf{r}', \omega)$ and $\rho(\mathbf{r}', \omega)$.

(b) Take the Fourier transforms of $\mathbf{A}(\mathbf{r}, t)$ and $\phi(\mathbf{r}, t)$ with respect to both the spatial and time variables $\mathbf{r}$ and t and express $\mathbf{A}(\mathbf{k}, \omega)$ and $\phi(\mathbf{k}, \omega)$ in terms of $\mathbf{J}(\mathbf{k}, \omega)$ and $\rho(\mathbf{k}, \omega)$.

5.2 Check the duality principle using (5.2.15).

5.3 Derive the Kramers–Kronig relations between the real and imaginary parts of the complex refractive index, $\bar{n}(\omega) = n(\omega) + i\kappa(\omega)$, similar to Eqs. (E.14) and (E.15) in Appendix E.

5.4 Derive the reciprocity relation (F.17) in Appendix F.

5.5 (a) A plane wave is reflected between the free space and a bulk GaAs semiconductor with a refractive index assumed to be $n = 3.5$. Calculate the reflection and the transmission coefficients of the field, r and t, at normal incidence.

(b) Repeat part (a) if the wave is incident from the GaAs region onto the GaAs/air interface at normal incidence.

(c) For oblique incidence in part (b), find the critical angle and the Brewster angle.

5.6 For a plane wave incident from an InP region ($n = 3.16$ and $\lambda_0 = 1.55$ μm) to the InP/air surface and reflected with an angle of incidence $\theta_i = 15°$, calculate (a) the reflection and the transmission coefficients, r and t, of the optical field for both TE and TM polarizations, and (b) the reflectivity R and the transmissivity T for both polarizations.

5.7 Calculate the Goos–Hänchen phase shifts for an angle of incidence $\theta_i = 45°$ between an InP/air interface for both TE and TM polarizations ($n = 3.16$ and $\lambda_0 = 1.55$ μm).

5.8 (a) Calculate the reflection and transmission coefficients, r and t, for a plane wave normally incident on a slab of GaAs sample with a thickness $d = 10$ μm, refractive index $n = 3.5$ at a wavelength $\lambda_0 = 1$ μm. (The background is free space.)

(b) Find the power reflectivity and transmissivity in part (a).

5.9 Find the far-field radiation patterns $P(\theta_\perp)$ and $P(\theta_\parallel)$ in Fig. 5.5, assuming a uniform aperture field $E_A(x, y) = E_0$ for $|x| \leq D$ and $|y| \leq W$, and $E_A(x, y) = 0$ otherwise.

5.10 If the aperture field is a Gaussian function in both the x and y directions, $E_A(x, y) \approx E_D \exp[-(x/x_0)^2 - (y/y_0)^2]$ in Fig. 5.5, find the far-field radiation patterns $P(\theta_\perp)$ and $P(\theta_\parallel)$.

REFERENCES

1. J. A. Kong, *Electromagnetic Wave Theory*, 2d ed., Wiley, New York, 1990.

2. J. D. Jackson, *Classical Electrodynamics*, 2d ed., Wiley, New York, 1975.

3. L. C. Shen and J. A. Kong, *Applied Electromagnetism*, Brooks/Cole Engineering Division, Monterey, CA, 1983.

4. H. C. Casey, Jr., and M. B. Panish, *Heterostructure Lasers, Part A: Fundamental Principles*, Academic, Orlando, FL, 1978.

PART II

Propagation of Light

6

Light Propagation in Various Media

The propagation of light is an interesting topic because many common phenomena such as refraction of light, polarization properties of light, and scattering of light can be observed every day. The Rayleigh scattering of light by water molecules has been used to explain why the sky is blue in the daytime and red in the evening. In this chapter, we discuss some basic properties of the propagation of electromagnetic waves in isotropic and uniaxial media [1–2]. For light propagation in gyrotropic media and the magnetooptic effects, see Appendix G.

6.1 PLANE WAVE SOLUTIONS FOR MAXWELL'S EQUATIONS IN HOMOGENEOUS MEDIA

We shall investigate solutions to Maxwell's equations in regions where $\mathbf{J} = \rho = 0$. Maxwell's equations for a harmonic field become

$$\nabla \times \mathbf{E} = i\omega\mathbf{B} \qquad (6.1.1a)$$

$$\nabla \times \mathbf{H} = -i\omega\mathbf{D} \qquad (6.1.1b)$$

$$\nabla \cdot \mathbf{B} = 0 \qquad (6.1.1c)$$

$$\nabla \cdot \mathbf{D} = 0 \qquad (6.1.1d)$$

We also assume that the media are homogeneous; therefore, we have plane wave solutions of the form $\exp(i\mathbf{k} \cdot \mathbf{r})$. Since all complex field vectors $\mathbf{E}$, $\mathbf{H}$, $\mathbf{B}$, and $\mathbf{D}$ have the same spatial dependence $\exp(i\mathbf{k} \cdot \mathbf{r})$, we obtain the basic relations for plane waves propagation in source-free homogeneous media:

$$\mathbf{k} \times \mathbf{E} = \omega\mathbf{B} \qquad (6.1.2a)$$

$$\mathbf{k} \times \mathbf{H} = -\omega\mathbf{D} \qquad (6.1.2b)$$

$$\mathbf{k} \cdot \mathbf{B} = 0 \qquad (6.1.2c)$$

$$\mathbf{k} \cdot \mathbf{D} = 0 \qquad (6.1.2d)$$

We see that the wave vector $\mathbf{k}$ is always perpendicular to $\mathbf{B}$ and $\mathbf{D}$ from (6.1.2c) and (6.1.2d). But we cannot say that $\mathbf{k}$ is always perpendicular to $\mathbf{H}$ or

E unless the media are isotropic. In isotropic media, $\mathbf{D} = \varepsilon\mathbf{E}$ and $\mathbf{B} = \mu\mathbf{H}$. Therefore, **D** and **B** are parallel to **E** and **H**, respectively. Otherwise, the media are called anisotropic. Thus, the time-averaged Poynting vector, which is given by $\frac{1}{2}\mathrm{Re}(\mathbf{E} \times \mathbf{H}^*)$, does not necessarily point in the direction of **k** in an anisotropic medium. In other words, the direction of power flow of a plane wave may not always be in the direction of the wave vector **k**.

In the following sections, we discuss how characteristic polarizations and propagation constants in a given medium are obtained. A characteristic mode or normal mode is defined as a wave with a polarization, which does not change while propagating in the homogeneous medium. This polarization is called the characteristic polarization of the medium.

6.2 LIGHT PROPAGATION IN ISOTROPIC MEDIA

In isotropic media, the constitutive relations are simply

$$\mathbf{D} = \varepsilon\mathbf{E} \qquad (6.2.1)$$

$$\mathbf{B} = \mu\mathbf{H} \qquad (6.2.2)$$

Thus, **D** is always parallel to **E**, and **B** is parallel to **H**. Any polarization direction of **E** can be a characteristic polarization of the isotropic medium. Equations (6.1.2a)–(6.1.2d) become

$$\mathbf{k} \times \mathbf{E} = \omega\mu\mathbf{H} \qquad (6.2.3a)$$

$$\mathbf{k} \times \mathbf{H} = -\omega\varepsilon\mathbf{E} \qquad (6.2.3b)$$

$$\mathbf{k} \cdot \mathbf{H} = 0 \qquad (6.2.3c)$$

$$\mathbf{k} \cdot \mathbf{E} = 0 \qquad (6.2.3d)$$

These are the equations satisfied by plane waves in isotropic media. One sees clearly that

1. **k** is perpendicular to both **E** and **H**.
2. $\mathbf{k} \times \mathbf{E}$ points in the direction of **H**, and $\mathbf{k} \times \mathbf{H}$ points in the direction of $-\mathbf{E}$.

Thus, the three vectors **E**, **H**, and **k** are perpendicular to each other.

Let us take the cross product of (6.2.3a) by **k** from the left-hand side:

$$\mathbf{k} \times (\mathbf{k} \times \mathbf{E}) = \omega\mu\mathbf{k} \times \mathbf{H} \qquad (6.2.4)$$

The left-hand side can be simplified using the vector identity

$$\mathbf{k} \times (\mathbf{k} \times \mathbf{E}) = (\mathbf{k} \cdot \mathbf{E})\mathbf{k} - k^2\mathbf{E}$$
$$= -k^2\mathbf{E} \qquad (6.2.5)$$

where we have made use of (6.2.3d). We obtain

$$-k^2\mathbf{E} = \omega\mu\mathbf{k} \times \mathbf{H}$$
$$= \omega\mu(-\omega\varepsilon\mathbf{E}) \qquad (6.2.6)$$

using (6.2.3b). The result is simply

$$(k^2 - \omega^2\mu\varepsilon)\mathbf{E} = 0 \qquad (6.2.7)$$

The solutions to the above equation are either a trivial solution, $\mathbf{E} = 0$ (no field, which is not of interest), or

$$k^2 = \omega^2\mu\varepsilon \qquad (6.2.8)$$

which is the dispersion relation in an isotropic medium. We obtain the wave number $k = \omega\sqrt{\mu\varepsilon}$. If we define the refractive index $n = \sqrt{\mu\varepsilon/(\mu_0\varepsilon_0)}$, where $\mu_0 = 4\pi \times 10^{-7}\text{H/m}$, and $\varepsilon_0 = 8.854 \times 10^{-12}\text{F/m} \simeq (1/36\pi) \times 10^{-9}\text{F/m}$, we have

$$k = \frac{\omega}{c}n$$

and $c = 1/\sqrt{\mu_0\varepsilon_0} = 3 \times 10^8$ m/s is the speed of light in free space. Let us define the unit vectors $\hat{e}$, $\hat{h}$, and $\hat{k}$ such that

$$\mathbf{E} = \hat{e}E \qquad \mathbf{H} = \hat{h}H \qquad \mathbf{k} = \hat{k}k \qquad (6.2.9)$$

Substituting (6.2.9) into (6.2.3a), we obtain

$$\hat{k}k \times \hat{e}E = \omega\mu\hat{h}H \qquad (6.2.10)$$

Or simply $\hat{k} \times \hat{e}$ is in the direction of $\hat{h}$. Since $\hat{k}$, $\hat{e}$, and $\hat{h}$ are all unit vectors and $\hat{k}$ is perpendicular to $\hat{e}$ (6.2.10) leads to

$$\hat{k} \times \hat{e} = \hat{h} \qquad kE = \omega\mu H \qquad (6.2.11)$$

Using $k = \omega\sqrt{\mu\varepsilon}$, we have

$$E = \eta H \qquad \eta = \sqrt{\mu/\varepsilon} \qquad (6.2.12)$$

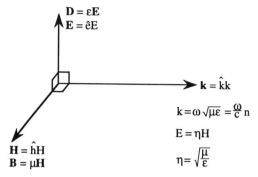

Figure 6.1. In isotropic media, $\mathbf{D} = \varepsilon\mathbf{E}$, $\mathbf{B} = \mu\mathbf{H}$. The three vectors $\mathbf{E}$, $\mathbf{H}$, and $\mathbf{k}$, or $\mathbf{D}$, $\mathbf{B}$, and $\mathbf{k}$ form a right-handed rectangular coordinate system.

where η is called the characteristic impedance of the medium. We conclude, as shown in Fig. 6.1, that

1. The three unit vectors $\hat{e}$, $\hat{h}$, and $\hat{k}$ construct a right-hand rectangular coordinate system.
2. The magnitude of the electric field is equal to the magnitude of the magnetic field multiplied by the characteristic impedance of the medium η.

Propagation Constant and Optical Refractive Index in Semiconductors. In most semiconductors, $\mu = \mu_0$, and the permittivity function is complex:

$$\varepsilon(\omega) = \varepsilon'(\omega) + i\varepsilon''(\omega) \tag{6.2.13}$$

where the real and imaginary parts of the permittivity function satisfy the Kramers–Kronig relations (Appendix E). The complex propagation constant $k = k' + ik''$ can be written in terms of the complex refractive index $\bar{n}(\omega) = n(\omega) + i\kappa(\omega)$,

$$k = \frac{\omega}{c}\bar{n}(\omega) = \frac{\omega}{c}[n(\omega) + i\kappa(\omega)] \tag{6.2.14}$$

where the imaginary part of the complex propagation constant is half of the absorption coefficient α:

$$\alpha(\omega) = 2\,\mathrm{Im}\,k = 2\frac{\omega}{c}\kappa(\omega) = \frac{4\pi}{\lambda}\kappa(\omega) \tag{6.2.15}$$

where λ is the wavelength in free space. For a plane wave propagating in the $+z$ direction with a polarization along $\hat{x}$, the electric field behaves as follows:

$$\mathbf{E} = \hat{x}E_0\,e^{ikz} = \hat{x}E_0\exp\left(i\frac{2\pi}{\lambda}nz - \frac{\alpha}{2}z\right) \qquad (6.2.16a)$$

$$\mathbf{H} = \hat{y}\frac{E_0}{\eta}\,e^{ikz} = \hat{y}\frac{\bar{n}E_0}{\eta_0}\exp\left(i\frac{2\pi}{\lambda}nz - \frac{\alpha}{2}z\right) \qquad (6.2.16b)$$

where $\eta_0 = \sqrt{\mu_0/\varepsilon_0} = 120\pi$ is the characteristic impedance of the free space. Its time-averaged Poynting vector for the optical power density is

$$\mathbf{S} = \frac{1}{2}\,\mathrm{Re}[\mathbf{E}\times\mathbf{H}^*] = \hat{z}\frac{n}{2\eta_0}|E_0|^2\,e^{-\alpha z} \qquad (6.2.17)$$

which decays exponentially as the wave propagates farther along the z direction with a decay constant determined by the absorption coefficient.

At an optical energy below the band gap of most semiconductors, the absorption is usually small or negligible. Above the band gap, the optical absorption is important. When a plane wave is normally incident from the air to a semiconductor surface, the reflectivity of the power is

$$R = \left|\frac{n_0 - \bar{n}}{n_0 + \bar{n}}\right|^2 = \frac{(n - 1)^2 + \kappa^2}{(n + 1)^2 + \kappa^2} \qquad (6.2.18)$$

which takes into account the absorption effect when the refractive index $\bar{n}$ is complex and n_0 of the air is 1.

Numerical Example   For InP material, the dispersive effects of the real and the imaginary parts $n(\omega)$ and $\kappa(\omega)$ are given in Appendix J as a function of the photon energy $\hbar\omega$ or the free-space wavelength λ. At an energy $\hbar\omega = 2.0$ eV ($\lambda = 0.62\ \mu$m), which is close to that of a HeNe laser wavelength, we have

$$n = 3.549 \qquad \kappa = 0.317$$

The absorption coefficient is

$$\alpha = \frac{4\pi}{\lambda}\kappa = \frac{4\pi}{0.62\ \mu m}\times 0.317 = 6.43\times 10^4\ \mathrm{cm}^{-1}$$

The reflectivity for the reflected power from the semiconductor is

$$R = \frac{(3.549 - 1)^2 + 0.317^2}{(3.549 + 1)^2 + 0.317^2} = 0.317 \qquad\blacksquare$$

6.3   LIGHT PROPAGATION IN UNIAXIAL MEDIA [1, 2]

Consider the case of a uniaxial medium described by

$$\mathbf{D} = \bar{\bar{\varepsilon}} \cdot \mathbf{E} \tag{6.3.1a}$$

$$\bar{\bar{\varepsilon}} = \begin{bmatrix} \varepsilon & 0 & 0 \\ 0 & \varepsilon & 0 \\ 0 & 0 & \varepsilon_z \end{bmatrix} \tag{6.3.1b}$$

$$\mathbf{B} = \mu\mathbf{H} \tag{6.3.2}$$

in the principal coordinate system, which means that the coordinate system has been chosen such that the permittivity matrix $\bar{\bar{\varepsilon}}$ has been diagonalized. Two of the three diagonal values are equal in a uniaxial medium. The medium is positive uniaxial if $\varepsilon_z > \varepsilon$, and negative uniaxial if $\varepsilon_z < \varepsilon$. Here, the z axis is called the optical axis.

6.3.1   Field Solutions

Since the medium described by (6.3.1) and (6.3.2) is invariant under any rotation around the z axis, we can always choose the plane containing the wave vector $\mathbf{k}$ and the z axis to be the x-z plane, i.e.,

$$\mathbf{k} = \hat{x}k_x + \hat{z}k_z \qquad k_x = k \sin\theta, \quad k_z = k\cos\theta$$

without loss of generality, as shown in Fig. 6.2. In Table 6.1, we summarize the forms of vectors and dyadics and their corresponding matrix representations. Using (6.1.2a) and (6.1.2b),

$$\mathbf{k} \times (\mathbf{k} \times \mathbf{E}) = \mathbf{k} \times \omega\mathbf{B} \tag{6.3.3a}$$

$$= \omega\mu\mathbf{k} \times \mathbf{H} \tag{6.3.3b}$$

$$= -\omega^2\mu\bar{\bar{\varepsilon}} \cdot \mathbf{E} \tag{6.3.3c}$$

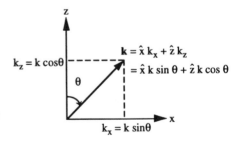

Figure 6.2.   The wave vector $\mathbf{k}$ in the x-z plane.

Table 6.1 Representations of Vectors, Dyadics, and Their Corresponding Matrix Forms

Vectors and Dyadics	Matrices
$\mathbf{a} = a_x \hat{x} + a_y \hat{y} + a_z \hat{z}$	$\mathbf{a} = \begin{bmatrix} a_x \\ a_y \\ a_z \end{bmatrix}$
$\mathbf{a} \cdot \mathbf{b} = a_x b_x + a_y b_y + a_z b_z$ $(\mathbf{a} \cdot \mathbf{b})(\mathbf{c} \cdot \mathbf{d}) = \mathbf{a} \cdot (\mathbf{bc}) \cdot \mathbf{d}$	$\mathbf{a}^t \mathbf{b}$ $(\mathbf{a}^t \mathbf{b})(\mathbf{c}^t \mathbf{d}) = \mathbf{a}^t(\mathbf{bc}^t)\mathbf{d}$
where	
$\mathbf{bc} = b_x c_x \hat{x}\hat{x} + b_x c_y \hat{x}\hat{y} + b_x c_z \hat{x}\hat{z}$	$\mathbf{bc}^t = \begin{bmatrix} b_x \\ b_y \\ b_z \end{bmatrix} [c_x \quad c_y \quad c_z]$
$+ b_y c_x \hat{y}\hat{x} + b_y c_y \hat{y}\hat{y} + b_y c_z \hat{y}\hat{z}$	
$+ b_z c_x \hat{z}\hat{x} + b_z c_y \hat{z}\hat{y} + b_z c_z \hat{z}\hat{z}$	$= \begin{bmatrix} b_x c_x & b_x c_y & b_x c_z \\ b_y c_x & b_y c_y & b_y c_z \\ b_z c_x & b_z c_y & b_z c_z \end{bmatrix}$
$\mathbf{k} \times \mathbf{E} = \begin{vmatrix} \hat{x} & \hat{y} & \hat{z} \\ k_x & k_y & k_z \\ E_x & E_y & E_z \end{vmatrix}$	$\mathbf{k} \times \mathbf{E} = \begin{bmatrix} 0 & -k_z & k_y \\ k_z & 0 & -k_x \\ -k_y & k_x & 0 \end{bmatrix}\begin{bmatrix} E_x \\ E_y \\ E_z \end{bmatrix}$
$= \hat{x}(k_y E_z - k_z E_y)$ $+ \hat{y}(k_z E_x - k_x E_z)$ $+ \hat{z}(k_x E_y - k_y E_x)$	$= \begin{bmatrix} (k_y E_z - k_z E_y) \\ (k_z E_x - k_x E_z) \\ (k_x E_y - k_y E_x) \end{bmatrix}$
$\hat{z} \times$	$\hat{z} \times \equiv \begin{bmatrix} 0 & -1 & 0 \\ 1 & 0 & 0 \\ 0 & 0 & 0 \end{bmatrix}$
$(k_x \hat{x} + k_z \hat{z}) \times$	$(k_x \hat{x} + k_z \hat{z}) \times \equiv \begin{bmatrix} 0 & -k_z & 0 \\ k_z & 0 & -k_x \\ 0 & k_x & 0 \end{bmatrix}$

and the matrix representation (see Table 6.1) for the cross product $\mathbf{k} \times$, we have

$$\begin{bmatrix} 0 & -k_z & 0 \\ k_z & 0 & -k_x \\ 0 & k_x & 0 \end{bmatrix}^2 \begin{bmatrix} E_x \\ E_y \\ E_z \end{bmatrix} = -\omega^2 \mu \begin{bmatrix} \varepsilon & 0 & 0 \\ 0 & \varepsilon & 0 \\ 0 & 0 & \varepsilon_z \end{bmatrix} \begin{bmatrix} E_x \\ E_y \\ E_z \end{bmatrix} \tag{6.3.4}$$

The left-hand side of (6.3.4) is

$$\begin{bmatrix} -k_z^2 & 0 & k_x k_z \\ 0 & -k_x^2 - k_z^2 & 0 \\ k_x k_z & 0 & -k_x^2 \end{bmatrix} \begin{bmatrix} E_x \\ E_y \\ E_z \end{bmatrix}$$

The matrix equation (6.3.4) can be written as

$$\begin{bmatrix} k_z^2 - \omega^2 \mu \varepsilon & 0 & -k_x k_z \\ 0 & k_x^2 + k_z^2 - \omega^2 \mu \varepsilon & 0 \\ -k_x k_z & 0 & k_x^2 - \omega^2 \mu \varepsilon_z \end{bmatrix} \begin{bmatrix} E_x \\ E_y \\ E_z \end{bmatrix} = 0 \quad (6.3.5)$$

To have nontrivial solutions for the electric field, we require that the determinant of the matrix in the above equation be zero. That is,

$$\left(k_x^2 + k_z^2 - \omega^2 \mu \varepsilon \right) \left[\left(k_z^2 - \omega^2 \mu \varepsilon \right) \left(k_x^2 - \omega^2 \mu \varepsilon_z \right) - k_x^2 k_z^2 \right] = 0 \quad (6.3.6)$$

There are two possible solutions.

Solution 1.

$$k_x^2 + k_z^2 - \omega^2 \mu \varepsilon = 0 \tag{6.3.7}$$

If (6.3.7) is true, we find from (6.3.5), which contains three algebraic equations, that

$$E_y \text{ can be arbitrary except zero (for nontrivial solution)} \quad (6.3.8a)$$

and

$$E_x = E_z = 0 \tag{6.3.8b}$$

That is, if the electric field is polarized only in the y direction, the wave vector must satisfy the dispersion relation given by Equation (6.3.7). The solutions of the fields are, therefore,

$$\mathbf{E} = \hat{y} E_0 e^{i \mathbf{k} \cdot \mathbf{r}} \tag{6.3.9a}$$

$$\mathbf{H} = \frac{1}{\omega \mu} \mathbf{k} \times \mathbf{E} = \frac{1}{\omega \mu} \left(-\hat{x} k_z + \hat{z} k_x \right) E_0 e^{i \mathbf{k} \cdot \mathbf{r}} \tag{6.3.9b}$$

$$\mathbf{B} = \mu \mathbf{H} = \frac{1}{\omega} \left(-\hat{x} k_z + \hat{z} k_x \right) E_0 e^{i \mathbf{k} \cdot \mathbf{r}} \tag{6.3.9c}$$

$$\mathbf{D} = \bar{\bar{\varepsilon}} \cdot \mathbf{E} = \hat{y} \varepsilon E_0 e^{i \mathbf{k} \cdot \mathbf{r}} \tag{6.3.9d}$$

where

$$\mathbf{k} \cdot \mathbf{r} = k_x x + k_z z = k(x \sin \theta + z \cos \theta) \tag{6.3.9e}$$

We can see that for an electric field polarized in the y direction, it will propagate with a wave number $k = \omega\sqrt{\mu\varepsilon}$ and its polarization remains y-polarized. Therefore, this polarization (6.3.9a) is a characteristic polarization of the uniaxial medium.

Solution 2.

$$\frac{k_x^2}{\omega^2 \mu \varepsilon_z} + \frac{k_z^2}{\omega^2 \mu \varepsilon} = 1 \tag{6.3.10}$$

which is obtained by setting the square bracket in (6.3.6) to zero. If (6.3.10) is true, we find immediately from (6.3.5) that

$$E_y = 0 \quad \left(\text{since } k_x^2 + k_z^2 - \omega^2 \mu \varepsilon \neq 0 \text{ if (6.3.10) is true}\right) \tag{6.3.11}$$

and

$$\left(k_z^2 - \omega^2 \mu \varepsilon\right) E_x - k_x k_z E_z = 0 \tag{6.3.12a}$$

Using (6.3.10) and (6.3.12a), we obtain

$$\left(-\frac{\varepsilon}{\varepsilon_z} k_x^2\right) E_x - k_x k_z E_z = 0 \tag{6.3.12b}$$

or simply

$$k_x \varepsilon E_x + k_z \varepsilon_z E_z = 0 \tag{6.3.12c}$$

which is

$$\mathbf{k} \cdot \mathbf{D} = 0 \tag{6.3.12d}$$

The complete solutions of the fields obeying (6.3.11) and (6.3.12c) are given by

$$\mathbf{E} = \hat{x} E_x + \hat{z} E_z = \left(\hat{x} - \hat{z}\frac{k_x \varepsilon}{k_z \varepsilon_z}\right) E_{x0}\, e^{i\mathbf{k}\cdot\mathbf{r}} \tag{6.3.13a}$$

$$\mathbf{H} = \frac{1}{\omega\mu} \mathbf{k} \times \mathbf{E} = \frac{1}{\omega\mu} \begin{vmatrix} \hat{x} & \hat{y} & \hat{z} \\ k_x & 0 & k_z \\ E_x & 0 & E_z \end{vmatrix} = \hat{y}\frac{\omega\varepsilon}{k_z} E_{x0}\, e^{i\mathbf{k}\cdot\mathbf{r}} \tag{6.3.13b}$$

$$\mathbf{B} = \mu\mathbf{H} = \hat{y}\frac{\omega\mu\varepsilon}{k_z} E_{x0}\, e^{i\mathbf{k}\cdot\mathbf{r}} \tag{6.3.13c}$$

$$\mathbf{D} = \bar{\bar{\varepsilon}} \cdot \mathbf{E} = \hat{x}\varepsilon E_x + \hat{z}\varepsilon_z E_z$$

$$= \left(\hat{x} - \hat{z}\frac{k_x}{k_z}\right) \varepsilon E_{x0}\, e^{i\mathbf{k}\cdot\mathbf{r}} \tag{6.3.13d}$$

where

$$\mathbf{k} \cdot \mathbf{r} = k_x x + k_z z = k(x \sin \theta + z \cos \theta) \tag{6.3.13e}$$

The polarization of the electric field given by Eq. (6.3.13a) is another characteristic polarization of the uniaxial medium since it is a characteristic mode of the medium from the fact that the fields (6.3.13a)–(6.3.13d) satisfy all of Maxwell's equations. This wave propagates with a wave vector determined by Eq. (6.3.10). We note again that

$$\mathbf{k} \cdot \mathbf{D} = 0 \tag{6.3.14a}$$

$$\mathbf{k} \cdot \mathbf{B} = 0 \tag{6.3.14b}$$

$$\mathbf{k} \cdot \mathbf{H} = 0 \quad \text{since } (\mathbf{B} \| \mathbf{H}) \tag{6.3.14c}$$

$$\mathbf{k} \cdot \mathbf{E} \neq 0! \tag{6.3.14d}$$

Thus the wave vector $\mathbf{k}$ is not perpendicular to the electric field $\mathbf{E}$. The complex Poynting vector, $\mathbf{E} \times \mathbf{H}^*$, is therefore not in the direction of $\mathbf{k}$. The solutions of the fields (6.3.13a)–(6.3.13d) are called extraordinary waves. The dispersion relation of the extraordinary waves is given by (6.3.10). On the other hand, the solutions of the fields (6.3.9a)–(6.3.9d), for which the wave vector $\mathbf{k}$ is perpendicular to $\mathbf{E}$, $\mathbf{H}$, $\mathbf{D}$, and $\mathbf{B}$, are called ordinary waves. The complex Poynting vector points in the direction of $\mathbf{k}$. The ordinary waves behave the same as those in isotropic media with a permittivity ε. A summary of results of the ordinary and extraordinary waves are shown in Fig. 6.3.

6.3.2 k Surfaces

We may plot the dispersion relations such as (6.3.7) and (6.3.10) in the $\mathbf{k}$ space. For the ordinary waves in case (1), Eq. (6.3.7) is a circle with a radius $\omega \sqrt{\mu \varepsilon}$ in the k_x-k_z plane. For the extraordinary waves in case 2, Eq. (6.3.10) is an ellipse in the k_x-k_z plane. We summarize the results as follows.

1. *Ordinary wave surface*

$$k_x^2 + k_z^2 = \omega^2 \mu \varepsilon \quad (\mathbf{E} \text{ must be polarized in the } y \text{ direction}) \tag{6.3.15}$$

In other words, $\mathbf{E}$ must be polarized in the direction perpendicular to the plane containing both the optical (z) axis and the wave vector $\mathbf{k}$ for ordinary waves. The wave number $k = \omega \sqrt{\mu \varepsilon} = \omega n_o / c$ and the refractive index $n_0 = \sqrt{\mu \varepsilon / \mu_0 \varepsilon_0}$.

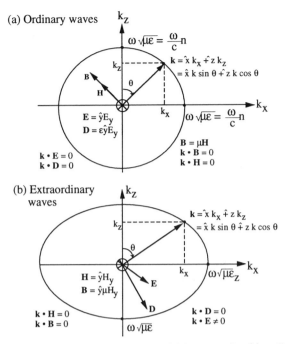

Figure 6.3. A summary of the **k** vector and the field vectors for (a) ordinary waves and (b) extraordinary waves.

2. *Extraordinary wave surface*

$$\frac{k_x^2}{\omega^2 \mu \varepsilon_z} + \frac{k_z^2}{\omega^2 \mu \varepsilon} = 1 \quad (\mathbf{E} \text{ must lie on the } x\text{-}z \text{ plane}) \quad (6.3.16)$$

Here $\omega\sqrt{\mu\varepsilon_z} = \omega n_e/c$. The extraordinary wave surface has two principal axes given by $\omega n_o/c$ and $\omega n_e/c$. The electric field **E** must be polarized in the plane containing both the optical axis and the wave vector **k**. The results are shown in Figs. 6.3 and 6.4.

Remember we have made use of the symmetry of the media with respect to the z axis (the optical axis). The plane containing the optical axis and the wave vector **k** is chosen to be the x-z plane. If we allow the wave vector **k** to rotate around the z axis, we would expect the following:

1. The ordinary waves will have the electric field perpendicular to the wave vector **k** and the optical axis, and the dispersion relation is a spherical surface in the **k** space:

$$k_x^2 + k_y^2 + k_z^2 = \omega^2 \mu \varepsilon \quad (6.3.17)$$

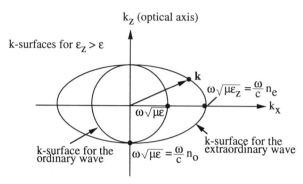

Figure 6.4. A plot of the k surfaces for the ordinary wave and the extraordinary wave of a uniaxial medium.

2. The extraordinary waves will have the electric field polarized in the plane containing the optical axis and the wave vector **k**. The dispersion relation is given by an ellipsoid in the **k** space:

$$\frac{k_x^2}{\omega^2 \mu \varepsilon_z} + \frac{k_y^2}{\omega^2 \mu \varepsilon_z} + \frac{k_z^2}{\omega^2 \mu \varepsilon} = 1 \qquad (6.3.18)$$

which is a result of rotating the ellipse of (6.3.16) around the optical axis.

Special Cases

1. If the wave propagates perpendicularly to the optical axis, for example, $\mathbf{k} = \hat{x}k$, $k_y = k_z = 0$, we have from (6.3.9a) and (6.3.13a) either

$$\mathbf{E} = \hat{y}E_0\, e^{i\mathbf{k}\cdot\mathbf{r}} = \hat{y}E_0\, e^{ik_o x} \qquad k_o = \omega\sqrt{\mu\varepsilon} \qquad \text{(ordinary waves)} \quad (6.3.19)$$

or

$$\mathbf{E} = \hat{z}E_0\, e^{i\mathbf{k}\cdot\mathbf{r}} = \hat{z}E_0\, e^{ik_e x} \qquad k_e = \omega\sqrt{\mu\varepsilon_z} \qquad \text{(extraordinary waves)}$$
$$(6.3.20)$$

Note that when $k_z = 0$, one finds $E_x = 0$ from (6.3.12c), and (6.3.13a) should reduce to (6.3.20).

2. If the wave propagates parallel to the optical axis, for example, $\mathbf{k} = \hat{z}k$, $k_x = k_y = 0$, we have from (6.3.9a) and (6.3.13a) again

$$\mathbf{E} = \hat{y}E_0\, e^{ik_o z} \qquad \text{(for } \mathbf{E} \text{ polarized in the } y \text{ direction)} \qquad (6.3.21)$$
$$\mathbf{E} = \hat{x}E_0\, e^{ik_o z} \qquad \text{(for } \mathbf{E} \text{ polarized in the } x\text{-}z \text{ plane)} \qquad (6.3.22)$$

Both electric fields propagate with the same wave number $k_o = \omega\sqrt{\mu\varepsilon}$ and both are perpendicular to the direction of propagation! The ordinary wave and the extraordinary wave become degenerate (both are ordinary waves now).

3. In general, the wave propagates in a direction with an angle θ to the optical axis. We have the dispersion relation

$$k^2 = k^2 \sin^2\theta + k^2 \cos^2\theta = \omega^2\mu\varepsilon \qquad (6.3.23)$$

for the ordinary waves. The phase velocity

$$v_p = \frac{\omega}{k} = \frac{1}{\sqrt{\mu\varepsilon}} \qquad (6.3.24)$$

is independent of the angle θ for ordinary waves (remember that the polarization of the electric field is perpendicular to **k** and the optical axis). For the extraordinary waves, for which **E** is on the x-z plane, we have

$$\frac{k^2}{\omega^2}\left(\frac{\sin^2\theta}{\mu\varepsilon_z} + \frac{\cos^2\theta}{\mu\varepsilon}\right) = 1 \qquad (6.3.25)$$

The phase velocity is

$$v_p = \frac{\omega}{k} = \left(\frac{\sin^2\theta}{\mu\varepsilon_z} + \frac{\cos^2\theta}{\mu\varepsilon}\right)^{1/2} = \frac{1}{\sqrt{\mu\varepsilon}}\left(\frac{\varepsilon}{\varepsilon_z}\sin^2\theta + \cos^2\theta\right)^{1/2} \qquad (6.3.26)$$

which depends on the angle of propagation.

6.3.3 Index Ellipsoid

For a given permittivity matrix $\bar{\bar{\varepsilon}}$, we define an impermeability matrix $\bar{\bar{\mathbf{K}}}$ as its inverse

$$\bar{\bar{\mathbf{K}}} = \bar{\bar{\varepsilon}}^{-1} \qquad (6.3.27)$$

An index ellipsoid is defined as

$$\varepsilon_0 \sum_{i,j=1}^{3} K_{ij}x_i x_j = 1 \qquad (6.3.28)$$

where K_{ij} are the elements of the matrix $\bar{\bar{\mathbf{K}}}$ and $x_1 = x$, $x_2 = y$, $x_3 = z$. In the principal axis system such that $\bar{\bar{\varepsilon}}$ is diagonalized

$$\bar{\bar{\varepsilon}} = \varepsilon_0 \begin{bmatrix} n_x^2 & 0 & 0 \\ 0 & n_y^2 & 0 \\ 0 & 0 & n_z^2 \end{bmatrix} \qquad (6.3.29)$$

the index ellipsoid equation (6.3.28) becomes

$$\frac{x^2}{n_x^2} + \frac{y^2}{n_y^2} + \frac{z^2}{n_z^2} = 1 \qquad (6.3.30)$$

which is plotted in Fig. 6.5. If $n_x = n_y = n_o$ and $n_z = n_e$ in a uniaxial medium as described by (6.3.1), we have

$$\frac{x^2}{n_o^2} + \frac{y^2}{n_o^2} + \frac{z^2}{n_e^2} = 1 \qquad (6.3.31)$$

It should be noted that the index ellipsoid is directly defined from the inverse of the permittivity function, and it contains information about the refractive indices of the characteristic polarizations. On the other hand, the k surfaces depend on the polarization and the direction of wave propagation. For a uniaxial medium, there are two k surfaces, one for the ordinary wave and the other for the extraordinary wave, with only one index ellipsoid to define the medium. The concept of index ellipsoid will be used again when we discuss electroptical effects in Chapter 12.

6.3.4 Applications

Quarter-Wave Plate. Based on (6.3.19) and (6.3.20), we assume a plane wave incident from the free space to a slab of uniaxial medium with

$$\bar{\bar{\varepsilon}} = \begin{bmatrix} \varepsilon & 0 & 0 \\ 0 & \varepsilon & 0 \\ 0 & 0 & \varepsilon_z \end{bmatrix} \qquad (6.3.32)$$

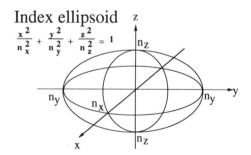

Figure 6.5. An index ellipsoid in the principal axis system for which the permittivity matrix is diagonalized.

Consider an incident electric field from free space given by

$$\mathbf{E} = \left(\hat{y} \frac{1}{\sqrt{2}} + \hat{z} \frac{1}{\sqrt{2}} \right) E_0 \, e^{ik_0 x} \tag{6.3.33}$$

which is linearly polarized and $k_0 = \omega \sqrt{\mu_0 \varepsilon_0}$ is the propagation constant in free space with a subscript "zero" (Fig. 6.6). We also assume that the permeability of the uniaxial medium μ is the same as that of the free space μ_0. The electric field makes an angle of 45 degrees with the y and z axes. After passing through the uniaxial medium, the electric field will be (ignoring the reflections at the boundaries)

$$\mathbf{E} = \hat{y} \frac{E_0}{\sqrt{2}} \, e^{ik_0 d} + \hat{z} \frac{E_0}{\sqrt{2}} \, e^{ik_e d} \tag{6.3.34}$$

since the y component of the electric field will be an ordinary wave that propagates with a wave number $k_o = \omega \sqrt{\mu \varepsilon}$ (the subscript "oh" stands for ordinary wave), and the z component of the electric field will be an extraordinary wave with a wave number $k_e = \omega \sqrt{\mu \varepsilon_z}$. If we choose the thickness d such that the phase difference between the two components of the electric field to be 90 degrees or multiplied by an odd integer (assuming $\varepsilon_z > \varepsilon$),

$$(k_e - k_o)d = \frac{\pi}{2}, \frac{3\pi}{2}, \dots, \frac{(2n+1)\pi}{2}, \text{etc.} \quad (n \text{ an integer}) \tag{6.3.35}$$

we obtain

$$\mathbf{E} = \left[\hat{y} + \hat{z} \, e^{i(k_e - k_o)d} \right] \frac{E_0}{\sqrt{2}} \, e^{ik_0 d}$$

$$= (\hat{y} \pm i z) \frac{E_0}{\sqrt{2}} \, e^{ik_0 d} \quad \text{for } (k_e - k_o)d = \begin{cases} \dfrac{\pi}{2}, & \dfrac{5\pi}{2}, & \dfrac{9\pi}{2}, & \cdots \\[2mm] \dfrac{3\pi}{2}, & \dfrac{7\pi}{2}, & \dfrac{11\pi}{2}, & \cdots \end{cases}$$

$$\tag{6.3.36}$$

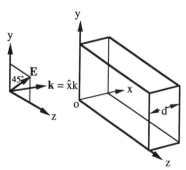

Figure 6.6. A plane wave $\mathbf{E}$ incident on a uniaxial medium with a thickness d. The electric field vector makes an angle of 45° with the two principal (y and z) axes.

which becomes circularly polarized at the output end $x = d$. Using

$$k_e = \frac{2\pi}{\lambda_e} = \frac{\omega}{c} n_e \tag{6.3.37a}$$

$$k_o = \frac{2\pi}{\lambda_o} = \frac{\omega}{c} n_o \tag{6.3.37b}$$

and defining

$$k_e - k_o \equiv \frac{2\pi}{\lambda_d} \quad \text{or} \quad \lambda_d = \frac{\lambda_0}{|n_e - n_o|} \tag{6.3.37c}$$

where λ_0 is the wavelength in free space ($\omega/c = 2\pi/\lambda_0$), we have

$$d = \frac{\lambda_d}{4}, \frac{3\lambda_d}{4}, \frac{5\lambda_d}{4}, \ldots \tag{6.3.38}$$

Thus the plate, which can transform an incident linearly polarized wave into a circularly polarized wave, is called a quarter-wave plate. Note that the value λ_d is neither the wavelength in the free space nor the wavelength in the uniaxial medium; it corresponds to λ_0 divided by the difference in the two refractive indices n_e and n_o. In many crystals such as lithium niobate (LiNbO$_3$) or KDP (KH$_2$PO$_4$), the refractive indices have the property $n_e > n_o$. In some crystals such as quartz (SiO$_2$), $n_e < n_o$. The velocities of the two characteristic polarizations c/n_e and c/n_o are not equal. The axis along which the polarization propagates faster is called the fast axis and the other axis is called the slow axis.

Polaroid. If we have a uniaxial medium given by

$$\bar{\bar{\varepsilon}} = \begin{bmatrix} \varepsilon & 0 & 0 \\ 0 & \varepsilon & 0 \\ 0 & 0 & \varepsilon_z - i\dfrac{\sigma_z}{\omega} \end{bmatrix} \tag{6.3.39}$$

we see that for an incident wave propagating in the x direction, if the electric field is polarized in the y direction, it will propagate through with a propagation constant $k_o = \omega\sqrt{\mu\varepsilon}$, which is real. However, if $\mathbf{E}$ is polarized in the z direction, it will propagate with a complex propagation constant

$$k_e = \omega\sqrt{\mu\left(\varepsilon_z - i\frac{\sigma_z}{\omega}\right)} \simeq \omega\sqrt{-i\mu\frac{\sigma_z}{\omega}} \quad \text{for} \quad \frac{\sigma_z}{\omega} \gg \varepsilon_z \tag{6.3.40}$$

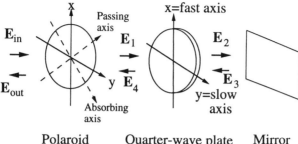

Figure 6.7. An experimental setup with a polaroid, a quarter-wave plate and a mirror for complete absorption of the light incident from the left side.

and the wave will be attenuated significantly in the medium. Thus, if the thickness of the plate is large enough, an incident field with an arbitrary polarization will have its z component attenuated when passing through the plate. The transmitted field will be essentially polarized in the y direction only, which is linearly polarized.

Example An experimental setup for an absorber of laser light uses a polaroid, a quarter-wave plate, and a mirror, as shown in Fig. 6.7. For an incident light with an optical field $\mathbf{E}_{in}$ randomly polarized and propagating along the z direction, the transmitted field $\mathbf{E}_1$ after passing through the polaroid is linearly polarized along the direction of the passing axis, which makes 45° with the x and y axes:

$$\mathbf{E}_1 = \left(\frac{\hat{x} + \hat{y}}{\sqrt{2}}\right) E_0 \, e^{ik_0 z}$$

where k_0 is the propagation constant in free space. The wave $\mathbf{E}_1$ also makes a 45° angle with the fast (x) and slow (y) axes of the quarter-wave plate. The transmitted field at $z = d$ of the quarter-wave plate is

$$\mathbf{E}_2 = \hat{x} \frac{E_0}{\sqrt{2}} \exp\left(i\frac{\omega}{c} n_f d\right) + \hat{y} \frac{E_0}{\sqrt{2}} \exp\left(i\frac{\omega}{c} n_s d\right)$$

$$= (\hat{x} + \hat{y} \, e^{i\pi/2}) \frac{E_0}{\sqrt{2}} \exp\left(i\frac{\omega}{c} n_f d\right)$$

where $(\omega/c)(n_s - n_f)d = \pi/2$. The reflected wave from the mirror is

$$\mathbf{E}_3 = -(\hat{x} + \hat{y} \, e^{i\pi/2}) \frac{E_0}{\sqrt{2}} \exp\left(i\frac{\omega}{c} n_f d\right) \exp(-ik_0 z)$$

propagating in the $-z$ direction. Upon impinging on the quarter-wave plate at $z = d$, the x component of the electric field propagates with a propagation constant $\omega n_f/c$, and the y component propagates with a propagation constant $\omega n_s/c$ again with an additional phase difference of $\pi/2$:

$$\mathbf{E}_4 = -(\hat{x} + \hat{y}\,e^{i\pi})\frac{E_0}{\sqrt{2}}\exp\left(i\frac{\omega}{c}n_f 2d\right)\exp(-ik_0 z)$$

Therefore, $\mathbf{E}_4$ is linearly polarized in the direction $(-\hat{x} + \hat{y})$ which is along the absorbing axis of the polaroid. The final output light $\mathbf{E}_{out} = 0$ since $\mathbf{E}_4$ is absorbed by the polaroid. This setup actually works as a light absorber, which can be used to absorb laser light to avoid stray light reflections in the laboratory. ■

PROBLEMS

6.1 For a laser light with a power 1 mW propagating in a GaAs semiconductor ($n = 3.5$) waveguide with a cross section 10μm $\times$ 1 μm, find (a) the power density if we assume that the intensity is uniform within the waveguide cross section, and (b) the electric field strength and the magnetic field strength, assuming that the plane wave is uniform.

6.2 Consider a LiNbO$_3$ crystal with the ordinary and extraordinary refractive indices, $n_o = 2.297$, $n_e = 2.208$. Write down the dispersion relations for the ordinary waves and the extraordinary waves.

6.3 For a LiNbO$_3$ crystal with the permittivity matrix given by

$$\bar{\bar{\varepsilon}} = \varepsilon_0 \begin{pmatrix} n_0^2 & 0 & 0 \\ 0 & n_0^2 & 0 \\ 0 & 0 & n_e^2 \end{pmatrix} \quad \text{where } n_o = 2.297 \text{ and } n_e = 2.208$$

(a) Find the two characteristic polarizations with corresponding propagation constants for a wave propagating in the y direction.

(b) If a plane wave propagates in $+z$ direction, what are the possible characteristic polarizations?

6.4 In Problem 6.3, if a plane wave propagates in a direction $\mathbf{k} = k(\hat{x}\sin\theta + \hat{z}\cos\theta)$ on the x-z plane, find the explicit expressions of the fields $\mathbf{E}$, $\mathbf{H}$, $\mathbf{D}$, and $\mathbf{B}$ for (a) the ordinary wave and (b) the extraordinary wave.

6.5 **(a)** Calculate the thickness $\lambda_d/4$ for a quarter-wave plate for a LiNbO$_3$ assuming $n_o = 2.297$, $n_e = 2.208$ at a wavelength $\lambda = 0.633$ μm.
(b) Repeat part (a) for a KDP crystal with $n_o = 1.5074$, $n_e = 1.4669$ at the same wavelength.

6.6 If we replace the mirror by a polaroid in Fig. 6.7 with the passing axis perpendicular to that of the entrance polaroid, find the optical transmitted field passing through this new polaroid.

6.7 Two polaroid analyzers are arranged with their passing axes perpendicular to each other and a quarter-wave plate is inserted between them. Initially, the fast axis is assumed to be aligned with the passing axis of the entrance polaroid. Find the optical transmission power intensity for an incident randomly polarized light as a function of the rotation angle θ between the fast axis and the passing axis of the entrance polaroid. Plot the transmission power vs. θ.

REFERENCES

1. J. A. Kong, *Electromagnetic Wave Theory*, 2d. ed., Wiley, New York, 1990.
2. M. Born and E. Wolf, *Principles of Optics*, Pergamon, Oxford, UK, 1975.

7

Optical Waveguide Theory

This chapter covers topics on optical dielectric waveguide theory, which will be useful for studying heterojunction semiconductor lasers. We present the fundamental results in Section 7.1 for transverse electric (TE) and transverse magnetic (TM) modes in a symmetric dielectric waveguide including the propagation constant, the optical field pattern, the optical confinement factor, and the cutoff conditions. We then discuss an asymmetric dielectric waveguide in Section 7.2. A ray optics approach is shown in Section 7.3, which explains the waveguide behavior. More practical structures, such as rectangular dielectric guides used in index-guidance diode lasers, are then discussed using an approximate Marcatili's method in Section 7.4 and the effective index method in Section 7.5. The results for a laser cavity with gain and loss in the media are derived to obtain the lasing conditions of a Fabry–Perot semiconductor laser cavity. Reference books on general waveguide theory can be found in Refs. 1–4 if the readers are interested in more extensive treatment of the subject.

7.1 SYMMETRIC DIELECTRIC SLAB WAVEGUIDES

The dielectric slab waveguide theory is very important since it provides almost all of the basic principles for general dielectric waveguides and the guidelines for more complicated cross sections such as the optical fibers. Let us consider a slab waveguide as shown in Fig. 7.1, where the width w is much larger than the thickness d, and the field dependence on y is negligible $(\partial/\partial y \equiv 0)$. From the wave equation

$$(\nabla^2 + \omega^2\mu\varepsilon)\mathbf{E} = 0 \tag{7.1.1}$$

we shall find the solutions for the fields everywhere.

We assume that the waveguide is symmetric, that is, the permittivity and the permeability are ε and μ, respectively, for $|x| \geq d/2$, and ε_1 and μ_1 for $|x| < d/2$. The origin has been chosen to be at the center of the guide, since the waveguide is symmetric; therefore, we have even-mode and odd-mode solutions. We separate the solutions of the fields into two classes: TE polarization and TM polarization. Here, a waveguide mode or a normal

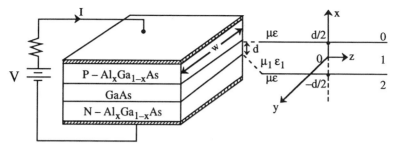

Figure 7.1. A simplified heterojunction diode laser structure for waveguide analysis.

mode is defined as the wave solution to Maxwell's equations with all of the boundary conditions satisfied; the transverse spatial profile and its polarization remain unchanged while propagating down the waveguide.

7.1.1 Derivations of Electric Fields and Guidance Conditions for TE Polarization

For TE polarization, the electric field only has an E_y component:

$$\mathbf{E} = \hat{y}E_y \qquad (E_x = E_z = 0)$$

$$\left(\frac{\partial^2}{\partial x^2} + \frac{\partial^2}{\partial z^2} + \omega^2 \mu\varepsilon\right)E_y = 0 \tag{7.1.2}$$

TE Even Modes. We obtain the guided mode solutions by examining wave equation (7.1.2), which has solutions of the form $\{\exp(ik_z z), \exp(-ik_z z)\} \times \{\exp(ik_x x), \exp(-ik_x x), \text{ or } \cos k_x x, \sin k_x x\}$. Since the guide is translationally invariant along the z direction, we choose $\exp(ik_z z)$ for a guided mode propagating in the $+z$ direction. We then choose the standing-wave solution $\cos k_x x$ or $\sin k_x x$ inside the guide and $\exp(-\alpha x)$ or $\exp(+\alpha x)$ outside the waveguide since the wave is a guided mode solution.

With the above observations, the electric field for the even modes can be written in the form

$$E_y = e^{ik_z z}\begin{cases} C_0 \, e^{-\alpha(x-d/2)} & x \geq \dfrac{d}{2} \\[2mm] C_1 \cos k_x x & |x| \leq \dfrac{d}{2} \\[2mm] C_0 \, e^{+\alpha(x+d/2)} & x \leq -\dfrac{d}{2} \end{cases} \tag{7.1.3}$$

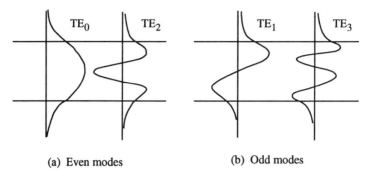

Figure 7.2. (a) The electric field profiles of the TE even modes. (b) The electric field profiles of the TE odd modes.

as shown in Fig. 7.2a, where k_x, k_z, and α satisfy

$$k_x^2 + k_z^2 = \omega^2 \mu_1 \varepsilon_1 \qquad (7.1.4a)$$

$$-\alpha^2 + k_z^2 = \omega^2 \mu \varepsilon \qquad (7.1.4b)$$

which are obtained by substituting (7.1.3) into the wave equation (7.1.2). Matching the boundary conditions in which E_y and H_z are continuous at $x = d/2$ and $x = -d/2$, we obtain

$$C_0 = C_1 \cos\left(k_x \frac{d}{2}\right) \qquad (7.1.5a)$$

$$\frac{\alpha}{\mu} C_0 = C_1 \frac{k_x}{\mu_1} \sin\left(k_x \frac{d}{2}\right) \qquad (7.1.5b)$$

where $H_z = (1/i\omega\mu)(\partial E_y/\partial x)$ has been used. Note that the permeability is μ_1 for $|x| \le d/2$. Eliminating C_0 and C_1, from (7.1.5a) and (7.1.5b), we find the eigenequation or the guidance condition:

$$\alpha = \frac{\mu}{\mu_1} k_x \tan\left(k_x \frac{d}{2}\right) \qquad (7.1.6)$$

The above guidance condition also shows that in order for E_y of the form (7.1.3) to be a guided mode solution, the standing-wave pattern along the transverse x direction or the amount of oscillations (determined by $k_x d$) has to match the decay constant α outside the guide. An interesting limit is to consider that the decay rate α is infinitely large, E_y will be zero outside the guide, and (7.1.6) gives $k_x d/2 = \pi/2, 3\pi/2, \ldots$ or $k_x d = m\pi$, which is the same as the guidance condition for a metallic waveguide, of which the tangential electric field vanishes at the surfaces of the perfectly conducting waveguide.

TE Odd Modes. The electric field can be written in the form

$$
E_y = e^{ik_z z}\begin{cases} C_0 e^{-\alpha(x-d/2)} & x \geq \dfrac{d}{2} \\[2mm] C_1 \sin k_x x & |x| \leq \dfrac{d}{2} \\[2mm] -C_0 e^{\alpha(x+d/2)} & x \leq -\dfrac{d}{2} \end{cases} \tag{7.1.7}
$$

as shown in Fig. 7.2b. Matching the tangential field components E_y and H_z at $x = d/2$ or $-d/2$, we obtain

$$
C_0 = C_1 \sin\left(k_x \frac{d}{2}\right) \tag{7.1.8a}
$$

$$
-\frac{\alpha}{\mu} C_0 = C_1 \frac{k_x}{\mu_1} \cos\left(k_x \frac{d}{2}\right) \tag{7.1.8b}
$$

Thus

$$
\alpha = -\frac{\mu}{\mu_1} k_x \cot\left(k_x \frac{d}{2}\right) \tag{7.1.9}
$$

7.1.2 Graphical Solution for the Guidance Conditions

To find the propagation constant k_z of the guided mode, we have to solve for k_x, k_z, and α from (7.1.4a), (7.1.4b), (7.1.6), and (7.1.9). We rewrite the eigenequations in the following forms:

$$
\alpha \frac{d}{2} = \frac{\mu}{\mu_1}\left(k_x \frac{d}{2}\right)\tan\left(k_x \frac{d}{2}\right) \qquad \text{TE even modes} \tag{7.1.10a}
$$

$$
\alpha \frac{d}{2} = -\frac{\mu}{\mu_1}\left(k_x \frac{d}{2}\right)\cot\left(k_x \frac{d}{2}\right) \qquad \text{TE odd modes} \tag{7.1.10b}
$$

Subtract (7.1.4b) from (7.1.4a) to eliminate k_z and multiply the result by $(d/2)^2$,

$$
\left(\alpha \frac{d}{2}\right)^2 + \left(k_x \frac{d}{2}\right)^2 = \omega^2(\mu_1 \varepsilon_1 - \mu\varepsilon)\left(\frac{d}{2}\right)^2 \tag{7.1.11}
$$

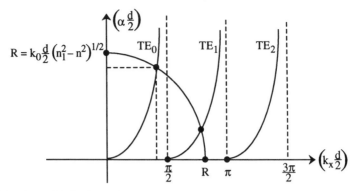

Figure 7.3. A graphical solution for the eigenequations to determine $\alpha d/2$ and $k_x d/2$ for the even modes, TE_0, TE_2, etc., and the odd modes, TE_1, TE_3, and so on.

We have only two unknowns, k_x and α. The solutions for α and k_x can be obtained from a graphical approach by looking at the $(\alpha d/2)$ vs. $(k_x d/2)$ plane [4, 5].

Equation (7.1.11) appears as a circle with a radius

$$R = \omega\sqrt{\mu_1\varepsilon_1 - \mu\varepsilon}\left(\frac{d}{2}\right) = \left(k_0\frac{d}{2}\right)\sqrt{n_1^2 - n^2} \qquad (7.1.12)$$

in Fig. 7.3 and the two eigenequations (7.1.10a) and (7.1.10b) are also plotted on the same plane. Their interceptions give the solutions for $(\alpha d/2)$ and $(k_x d/2)$. Here the refractive index inside the guide $n_1 = \sqrt{\mu_1\varepsilon_1/(\mu_0\varepsilon_0)}$ and outside $n = \sqrt{\mu\varepsilon/(\mu_0\varepsilon_0)}$ have been used and $k_0 = \omega\sqrt{\mu_0\varepsilon_0} = \omega/c$ is the free-space wave number. Given the waveguide dimension d, we find the propagation constant k_z from either (7.1.4a) or (7.1.4b) after k_x or α is determined.

7.1.3 Cutoff Condition

It is clear from the graph that the cutoff condition occurs at $R = m\pi/2$ for the TE_m mode

$$k_0\frac{d}{2}\sqrt{n_1^2 - n^2} = m\frac{\pi}{2} \qquad m = 0, 1, 2, \ldots \qquad (7.1.13)$$

Thus the TE_0 mode has no cutoff frequency. (*Note:* Here we assume that the refractive indices are independent of the frequency. In semiconductors, the frequency-dependent refractive indices should be used.) For

$$m\frac{\pi}{2} < k_0\frac{d}{2}\sqrt{n_1^2 - n^2} < (m + 1)\frac{\pi}{2} \tag{7.1.14}$$

there are $(m + 1)$ guided TE modes in the dielectric slab.

For a single mode operation, the condition $(k_0 d/2)\sqrt{n_1^2 - n^2} < \pi/2$ is required. This puts a limit on ω, d, or $n_1^2 - n^2$. For example, given the wavelength λ_0 in free space and the waveguide dimension d,

$$n_1^2 - n^2 < \frac{(\pi/2)^2}{(k_0 d/2)^2} = \left(\frac{\lambda_0}{2d}\right)^2 \tag{7.1.15}$$

If n_1 is almost equal to n, as is the case in an $Al_xGa_{1-x}As/GaAs/Al_xGa_{1-x}As$ waveguide, we have

$$\Delta n < \frac{1}{8n_1}\left(\frac{\lambda_0}{d}\right)^2 \tag{7.1.16}$$

using $n_1^2 - n^2 \simeq 2n_1\Delta n$. For example, at $n_1 \simeq 3.6$, we have $\Delta n < 0.035$ at $\lambda_0 = d$. Single-mode operation is achieved with a small difference in the refractive index $\Delta n < 0.035$.

7.1.4 Low- and High-Frequency Limits

1. In the low-frequency limit, the waveguide modes can be cut off. At cutoff, $(k_0 d/2)\sqrt{n_1^2 - n^2} = m\pi/2$, and $\alpha = 0$, as can be seen from the graphical solution. (The cutoff condition can be achieved by reducing the frequency, or d, or $n_1^2 - n^2$.) We find $k_x d/2 = m\pi/2$ and from (7.1.4b)

$$k_z = \omega\sqrt{\mu\varepsilon} = \frac{\omega n}{c} \tag{7.1.17}$$

since $\alpha = 0$. Therefore, the propagation constant of the guided mode k_z approaches the propagation constant $\omega n/c$ outside the waveguide. This is expected since the decaying constant $\alpha = 0$ at cutoff, which means that the waveguide mode does not decay outside the guide. Almost all of the mode power propagates outside the guide; therefore, the velocity of the mode will equal the speed of light outside the guide (Fig. 7.4b).

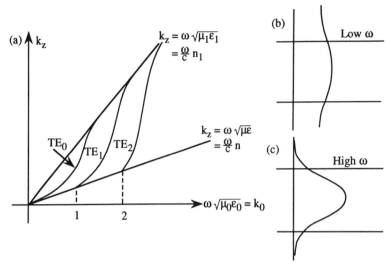

Figure 7.4. The dispersion curves of the TE_m modes. (b) The field profile in the low-frequency limit. (c) The field profile in the high-frequency limit.

2. In the high-frequency limit, $R \to \infty$. The graphical solution in Fig. 7.3 shows that $(k_x d/2) \to (m + 1)\pi/2$ and $(\alpha d/2) \to \infty$ for the TE_m mode. Since m is a fixed mode number, we find that $(\alpha d/2) \to R \to \infty$, as can also be seen from the graphical solution. Therefore,

$$k_z^2 = \omega^2 \mu_1 \varepsilon_1 - k_x^2 \to \omega^2 \mu_1 \varepsilon_1 \qquad (7.1.18)$$

or $k_z = \omega\sqrt{\mu_1 \varepsilon_1} = \omega n_1/c$. The propagation constant k_z approaches that in the waveguide since the mode decays rapidly outside the guide ($\alpha \to \infty$). Most of the power is guided inside the waveguide, Fig. 7.4c.

7.1.5 Propagation Constant k_z and the Effective Index n_{eff}

From Eqs. (7.1.17) and (7.1.18), we see that the propagation constant k_z starts from $\omega\sqrt{\mu\varepsilon}$ at cutoff and increases to $\omega\sqrt{\mu_1\varepsilon_1}$ as the frequency goes to infinity. The dispersion curves k_z vs. $\omega\sqrt{\mu_0\varepsilon_0}$ are plotted in Fig. 7.4a. An effective index for the guided mode is defined as

$$n_{eff} = \frac{k_z}{k_0} \qquad (7.1.19)$$

Again we have the lower bound $\omega\sqrt{\mu\varepsilon}/k_0 = n$ and the upper bound $\omega\sqrt{\mu_1\varepsilon_1}/k_0 = n_1$ at the low- and high-frequency limits, and the dispersion curves in Fig. 7.4a map to a set of effective index curves as shown in Fig. 7.5a.

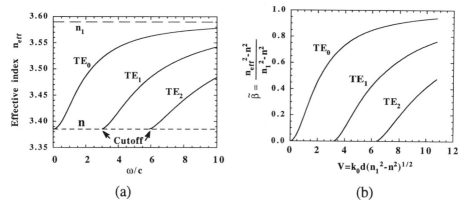

Figure 7.5. (a) The effective index $n_{\text{eff}} = k_z/k_0$ of the TE_m modes vs. the free-space wave number ω/c. (b) The normalized propagation parameter $\tilde{\beta}$ vs. the normalized parameter V number for the TE_m modes.

Very often a normalized propagation parameter

$$\tilde{\beta} = \frac{k_z^2 - \omega^2\mu\varepsilon}{\omega^2\mu_1\varepsilon_1 - \omega^2\mu\varepsilon} = \frac{n_{\text{eff}}^2 - n^2}{n_1^2 - n^2} \tag{7.1.20}$$

is also plotted vs. the V number or normalized frequency

$$V = k_0 d \left(n_1^2 - n^2\right)^{1/2} \tag{7.1.21}$$

in Fig. 7.5b, showing a set of universal curves for the lowest three TE modes.

Notice that the above discussions assume that n and n_1 are independent of the frequency. In semiconductors, the dispersion effect of the refractive index near the band edge can be significant. It is important to know the refractive index at the operation wavelength $n(\lambda)$. The effective index $n_{\text{eff}}(\lambda)$ still falls between $n(\lambda) < n_{\text{eff}}(\lambda) < n_1(\lambda)$ when the material dispersion is taken into account.

7.1.6 Refractive Index of $Al_x Ga_{1-x} As$ System

The refractive index of this system and many III–V materials [6–18] below the direct band gap can be approximately given by the form

$$n(\omega) \simeq \sqrt{\frac{\varepsilon'(\omega)}{\varepsilon_0}} \tag{7.1.22}$$

where $\varepsilon'(\omega)$ is the real part of the permittivity function and the imaginary part $\varepsilon''(\omega)$ is negligible for optical energy below the band gap [12, 13, 15]:

$$\frac{\varepsilon'(\omega)}{\varepsilon_0} \simeq A(x)\left\{f(y) + \frac{1}{2}\left[\frac{E_g(x)}{E_g(x) + \Delta(x)}\right]^{3/2} f(y_{SO})\right\} + B(x)$$

(7.1.23a)

$$f(y) = \frac{1}{y^2}\left[2 - (1 + y)^{1/2} - (1 - y)^{1/2}\right]$$

$$y = \frac{\hbar\omega}{E_g(x)}$$

$$y_{SO} = \frac{\hbar\omega}{E_g(x) + \Delta(x)}$$

(7.1.23b)

Here $E_g(x)$ is the band-gap energy and $\Delta(x)$ is the spin–orbit splitting energy. For $Al_xGa_{1-x}As$ ternaries below the band gap, the parameters as functions of the aluminum mole fraction x are

$$E_g(x) = 1.424 + 1.266x + 0.26x^2 \quad (eV)$$

$$\Delta(x) = 0.34 - 0.5x \quad (eV)$$

$$A(x) = 6.64 + 16.92x$$

$$B(x) = 9.20 - 9.22x$$

(7.1.24)

These results have been compared with experimental data with a very good agreement, as shown in Fig. 7.6a [11] for the real part $\varepsilon_1(\omega) = \varepsilon'(\omega)/\varepsilon_0$ below the band gap. The refractive index $n(\omega)$ below the band gap is plotted in Fig. 7.6b. In Fig. 7.6c, we show the refractive index above and below the band gap for $Al_xGa_{1-x}As$ with various aluminum mole fractions x. For comparison, the refractive index of a high-purity GaAs is shown as the top dashed curve and that of a p-type doped GaAs is shown as the top solid curve. We can see the features near the band gap of the $Al_xGa_{1-x}As$ alloys. More data and theoretical fits on GaAs, $Al_xGa_{1-x}As$, $In_{1-x}Ga_xA_yP_{1-y}$, InP, and other material systems are documented in Refs. 9–19. It is noted that the refractive index is generally dispersive near or above the band gap with a fast variation near various interband transition energies. It also depends slightly on the doping concentration of the semiconductors, the temperature, and the strain. Above the band gap, both the real and the imaginary parts of the permittivity function are important, and Eqs. (E.22a) and (E.22b) in Appendix E should be used to relate $\varepsilon'(\omega)$ and $\varepsilon''(\omega)$ to the complex refractive index $n + i\kappa$.

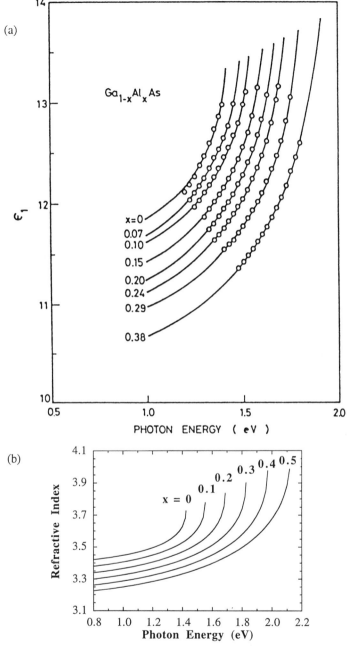

Figure 7.6. (a) The real part of the relative permittivity function $\varepsilon_1(\omega) = \varepsilon'(\omega)/\varepsilon_0$ vs. the photon energy $\hbar\omega$ (eV) for $Ga_{1-x}Al_xAs$ alloy. Solid lines are calculated using (7.1.23) and (7.1.24). The circles are experimental data. (After Ref. 11.) (b) The theoretical refractive index $n(\omega) \simeq \sqrt{\varepsilon'(\omega)/\varepsilon_0}$ below the band gap calculated using (7.1.22)–(7.1.24). (c) Experimental data for the refractive index of $Al_xGa_{1-x}As$ as a function of photon energy $h\nu$. The top dashed curve is the refractive index of a high-purity GaAs and the top solid curve is that of a silicon-doped p-type GaAs with $p = 2 \times 10^{18}$ cm^{-3}. (After Ref. 14.)

(c)

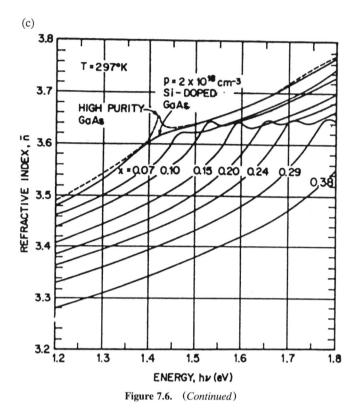

Figure 7.6. (*Continued*)

7.1.7 Normalization Constant for the Optical Mode

The undetermined coefficient C_1 is determined by the normalization condition in which the total guided power is assumed to be 1. That is,

$$P = \frac{1}{2}\text{Re}\int_{-\infty}^{\infty} (\mathbf{E} \times \mathbf{H}^*) \cdot \hat{z}\, dx = 1 \qquad (7.1.25)$$

Since for TE modes,

$$H_x = -\frac{k_z}{\omega\mu}E_y$$

$$P = \frac{-1}{2}\int E_y H_x^*\, dx = \frac{1}{2}\int_{-\infty}^{\infty} \frac{k_z}{\omega\mu}|E_y|^2\, dx \qquad (7.1.26)$$

where the permeability is μ_1 inside the guide and μ outside the guide. We find the normalization coefficient C_1 to be

$$C_1 = \left[\frac{4\omega\mu}{k_z(d + 2/\alpha)} \right]^{1/2} \quad \text{(for } \mu = \mu_1) \quad (7.1.27)$$

for both TE even and odd modes. The above coefficient can be derived using the expressions for E_y (7.1.3) and (7.1.7), and the eigenequations (7.1.10a) and (7.1.10b) for the even and odd modes, respectively.

7.1.8 Optical Confinement Factor Γ

An important quantity, called the optical confinement factor, is defined as the fraction of power guided in the waveguide:

$$\Gamma = \frac{\frac{1}{2}\int_{\text{inside}} \text{Re}(\mathbf{E} \times \mathbf{H}^*) \cdot \hat{z}\, dx}{\frac{1}{2}\int_{\text{total}} \text{Re}(\mathbf{E} \times \mathbf{H}^*) \cdot \hat{z}\, dx} \quad (7.1.28)$$

Using the expression in (7.1.3) for TE even modes, we find

$$\Gamma = \frac{(k_z/2\omega\mu_1)\int_{|x|<d/2}|E_y|^2\, dx}{(k_z/2\omega\mu_1)\int_{|x|<d/2}|E_y|^2\, dx + (k_z/2\omega\mu)\int_{|x|>d/2}|E_y|^2\, dx}$$

$$= \left[1 + \frac{\mu_1 \cos^2(k_x d/2)}{(\mu\alpha d/2)(1 + (\sin k_x d/k_x d))} \right]^{-1} \quad (7.1.29)$$

It is obvious that $\Gamma < 1$ always. As the waveguide dimension becomes smaller, $d \to 0$ or $k_x d \to 0$, we find that for TE_0 mode

$$\alpha \frac{d}{2} = \frac{\mu}{\mu_1}\left(k_x \frac{d}{2}\right)\tan\left(k_x \frac{d}{2}\right) \to \frac{\mu}{\mu_1}\left(k_x \frac{d}{2}\right)^2 \ll 1 \quad (7.1.30)$$

and (7.1.29) gives

$$\Gamma \to \frac{\mu}{\mu_1}\alpha d \to 2\left(\frac{\mu}{\mu_1}\right)^2\left(k_x \frac{d}{2}\right)^2$$

Since $(k_x d/2)^2 + (\alpha d/2)^2 = R^2$, we have $\alpha d/2 \to (\mu/\mu_1)(k_x d/2)^2 \ll k_x d/2$. Thus $k_x d/2 \to R$. Therefore,

$$\Gamma \to 2\left(\frac{\mu}{\mu_1}\right)^2 \left(k_0 \frac{d}{2}\right)^2 (n_1^2 - n^2) \qquad (7.1.31)$$

For $\mu = \mu_1$ and $k_0 = 2\pi/\lambda_0$, the following formula is useful in estimating the optical confinement factor of the fundamental TE_0 mode in the weak guidance limit:

$$\Gamma \to 2\left(\frac{\pi d}{\lambda_0}\right)^2 (n_1^2 - n^2) \qquad (7.1.32)$$

7.1.9 Summary of TE and TM Modes

TE Polarization

$$\mathbf{E} = \hat{y} E_y \qquad \mathbf{H} = \frac{1}{i\omega\mu_i}\left(-\hat{x}\, ik_z E_y + \hat{z}\frac{\partial}{\partial x} E_y\right)$$

$(\mu_i = \mu_1$ inside, and $\mu_i = \mu$ outside the guide) $\qquad (7.1.33)$

1. *TE even modes*

$$E_y = C_1 e^{ik_z z}\begin{cases} \cos\left(k_x \frac{d}{2}\right) e^{-\alpha(x-d/2)} & x \ge \frac{d}{2} \\[2mm] \cos k_x x & |x| \le \frac{d}{2} \\[2mm] \cos\left(k_x \frac{d}{2}\right) e^{\alpha(x+d/2)} & x < -\frac{d}{2} \end{cases} \qquad (7.1.34a)$$

$$C_1 = \left(\frac{4\omega\mu_1}{k_z d}\right)^{1/2} \Bigg/ \left[1 + \frac{\sin k_x d}{k_x d} + \left(\frac{\mu_1}{\mu}\right)\left(\frac{2}{\alpha d}\right)\cos^2(k_x d/2)\right]^{1/2} \qquad (7.1.34b)$$

The eigenequation is

$$\alpha\frac{d}{2} = \frac{\mu}{\mu_1}\left(k_x\frac{d}{2}\right)\tan\left(k_x\frac{d}{2}\right) \qquad (7.1.34c)$$

The optical confinement factor is

$$\Gamma = \left[1 + \left(\frac{\mu_1}{\mu}\right)\left(\frac{2}{\alpha d}\right)\frac{\cos^2(k_x d/2)}{(1 + \sin k_x d/k_x d)}\right]^{-1} \qquad (7.1.34d)$$

2. TE odd modes

$$
E_y = C_1 e^{ik_z z}
\begin{cases}
\sin\left(k_x \dfrac{d}{2}\right) e^{-\alpha(x-d/2)} & x \geq \dfrac{d}{2} \\[2mm]
\sin k_x x & |x| \leq \dfrac{d}{2} \\[2mm]
-\sin\left(k_x \dfrac{d}{2}\right) e^{\alpha(x+d/2)} & x < -\dfrac{d}{2}
\end{cases}
\tag{7.1.35a}
$$

$$
C_1 = \left(\frac{4\omega\mu_1}{k_z d}\right)^{1/2} \bigg/ \left[1 - \frac{\sin k_x d}{k_x d} + \left(\frac{\mu_1}{\mu}\right)\left(\frac{2}{\alpha d}\right)\sin^2(k_x d/2)\right]^{1/2}
\tag{7.1.35b}
$$

$$
\alpha\frac{d}{2} = -\frac{\mu}{\mu_1}\left(k_x\frac{d}{2}\right)\cot\left(k_x\frac{d}{2}\right)
\tag{7.1.35c}
$$

$$
\Gamma = \left[1 + \left(\frac{\mu_1}{\mu}\right)\left(\frac{2}{\alpha d}\right)\frac{\sin^2(k_x d/2)}{(1 - \sin k_x d/k_x d)}\right]^{-1}
\tag{7.1.35d}
$$

TM Polarization. For TM polarization, we may obtain the results by the duality principle: replacing the field solutions **E** and **H** of the TE modes by **H** and $-\mathbf{E}$, respectively, μ by ε and ε by μ as discussed in Section 5.2. We obtain

$$
\mathbf{H} = \hat{y}H_y \qquad \mathbf{E} = \frac{1}{i\omega\varepsilon_i}\left(\hat{x}ik_z H_y - \hat{z}\frac{\partial}{\partial x}H_y\right)
$$

$$
(\varepsilon_i = \varepsilon_1 \text{ inside, } \varepsilon_i = \varepsilon \text{ outside the guide})
\tag{7.1.36}
$$

1. TM even modes

$$
H_y = C_1 e^{ik_z z}
\begin{cases}
\cos\left(k_x \dfrac{d}{2}\right) e^{-\alpha(x-d/2)} & x \geq \dfrac{d}{2} \\[2mm]
\cos k_x x & |x| \leq \dfrac{d}{2} \\[2mm]
\cos\left(k_x \dfrac{d}{2}\right) e^{\alpha(x+d/2)} & x < -\dfrac{d}{2}
\end{cases}
\tag{7.1.37a}
$$

$$
C_1 = \left(\frac{4\omega\varepsilon_1}{k_z d}\right)^{1/2} \bigg/ \left[1 + \frac{\sin k_x d}{k_x d} + \left(\frac{\varepsilon_1}{\varepsilon}\right)\left(\frac{2}{\alpha d}\right)\cos^2(k_x d/2)\right]^{1/2}
\tag{7.1.37b}
$$

The eigenequation is

$$\alpha \frac{d}{2} = \frac{\varepsilon}{\varepsilon_1}\left(k_x\frac{d}{2}\right)\tan\left(k_x\frac{d}{2}\right) \qquad (7.1.37c)$$

The optical confinement factor is

$$\Gamma = \left[1 + \left(\frac{\varepsilon_1}{\varepsilon}\right)\left(\frac{2}{\alpha d}\right)\frac{\cos^2 k_x d/2}{(1 + \sin k_x d/k_x d)}\right]^{-1} \qquad (7.1.37d)$$

2. *TM odd modes*

$$H_y = C_1 e^{ik_z z}\begin{cases} \sin\left(k_x\frac{d}{2}\right)e^{-\alpha(x-d/2)} & x \geq \frac{d}{2} \\[2mm] \sin k_x x & |x| \leq \frac{d}{2} \\[2mm] -\sin\left(k_x\frac{d}{2}\right)e^{\alpha(x+d/2)} & x \leq -\frac{d}{2} \end{cases} \qquad (7.1.38a)$$

$$C_1 = \left(\frac{4\omega\varepsilon_1}{k_z d}\right)^{1/2}\bigg/\left[1 - \frac{\sin k_x d}{k_x d} + \left(\frac{\varepsilon_1}{\varepsilon}\right)\left(\frac{2}{\alpha d}\right)\sin^2(k_x d/2)\right]^{1/2}$$

$$\qquad (7.1.38b)$$

The eigenequation is

$$\alpha \frac{d}{2} = -\frac{\varepsilon}{\varepsilon_1}\left(k_x\frac{d}{2}\right)\cot\left(k_x\frac{d}{2}\right) \qquad (7.1.38c)$$

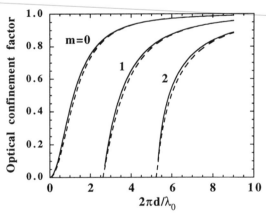

Figure 7.7. Optical confinement factors for the TE_m (solid curves) and the TM_m (dashed curves) modes as a function of $2\pi d/\lambda_0$. The refractive indices are $n_1 = 3.590$ and $n = 3.385$.

The optical confinement factor is

$$\Gamma = \left[1 + \left(\frac{\varepsilon_1}{\varepsilon}\right)\left(\frac{2}{\alpha d}\right)\frac{\sin^2(k_x d/2)}{(1 - \sin k_x d/k_x d)}\right]^{-1} \qquad (7.1.38d)$$

For dielectric materials, $\mu = \mu_1 = \mu_0$, we can replace all factors containing μ/μ_1 by 1 and $\varepsilon/\varepsilon_1$ by n^2/n_1^2 in the expressions for the eigenequations and the optical confinement factors for both TE and TM polarizations. In Fig. 7.7 we plot the optical confinement factor Γ for the TE (solid curves) and TM (dashed curves) modes. All optical confinement factors start from zero at cutoff and approach unity at the high-frequency limit.

7.2 ASYMMETRIC DIELECTRIC SLAB WAVEGUIDES

7.2.1 TE Polarization, $E = \hat{y}E_y$

If the dielectric constant in the substrate ε_2 is different from that of ε in the top medium, or $\mu_2 \neq \mu$, the structure becomes asymmetric (Fig. 7.8). In this case, the field solutions cannot be either even or odd modes. The general solution can be put in the form

$$E_y = e^{ik_z z}\begin{cases} C_0 e^{-\alpha x} & x \geq 0 \\ C_1\cos(k_{1x}x + \phi) & -d \leq x \leq 0 \\ C_2 e^{\alpha_2(x+d)} & x \leq -d \end{cases} \qquad (7.2.1)$$

Again, the transverse mode profile is decaying away from the guide with different decay constants α and α_2 in region 0 and region 2, respectively. The standing-wave solution inside the guide is written as $C_1\cos(k_{1x}x + \phi)$, which is equivalent to $A \cos k_{1x}x + B \sin k_{1x}x$. We choose this form because it is easier to match the boundary conditions later and the phase angle ϕ is also related to the Goos–Hänchen phase shift (discussed in Section 5.3) when

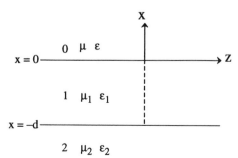

Figure 7.8. An asymmetric dielectric slab waveguide.

total internal reflection occurs. $H_z = (1/i\omega\mu)(\partial E_y/\partial x)$ is given by

$$H_z = e^{ik_z z} \begin{cases} \dfrac{-\alpha}{i\omega\mu}C_0 e^{-\alpha x} & x \geq 0 \\[2mm] \dfrac{-k_{1x}}{i\omega\mu_1}C_1\sin(k_{1x}x + \phi) & -d \leq x \leq 0 \\[2mm] \dfrac{\alpha_2}{i\omega\mu_2}C_2 e^{\alpha_2(x+d)} & x \leq -d \end{cases} \qquad (7.2.2)$$

where

$$-\alpha^2 + k_z^2 = \omega^2\mu\varepsilon \qquad (7.2.3\text{a})$$
$$k_{1x}^2 + k_z^2 = \omega^2\mu_1\varepsilon_1 \qquad (7.2.3\text{b})$$
$$-\alpha_2^2 + k_z^2 = \omega^2\mu_2\varepsilon_2 \qquad (7.2.3\text{c})$$

The next step is to match the boundary conditions.

1. E_y is continuous at $x = 0$ and $x = -d$:

$$C_0 = C_1\cos\phi \qquad (7.2.4\text{a})$$
$$C_1\cos(-k_{1x}d + \phi) = C_2 \qquad (7.2.4\text{b})$$

2. H_z is continuous at $x = 0$ and $x = -d$:

$$\frac{\alpha}{\mu}C_0 = \frac{1}{\mu_1}k_{1x}C_1\sin\phi \qquad (7.2.5\text{a})$$

$$-\frac{C_1}{\mu_1}k_{1x}\sin(-k_{1x}d + \phi) = \frac{1}{\mu_2}\alpha_2 C_2 \qquad (7.2.5\text{b})$$

Dividing (7.2.5a) by (7.2.4a) and (7.2.5b) by (7.2.4b), we obtain

$$\alpha = \frac{\mu}{\mu_1}k_{1x}\tan\phi \qquad (7.2.6\text{a})$$

$$\alpha_2 = -\frac{\mu_2}{\mu_1}k_{1x}\tan(-k_{1x}d + \phi) \qquad (7.2.6\text{b})$$

or equivalently

$$\phi = \tan^{-1}\frac{\mu_1\alpha}{\mu k_{1x}} \qquad (7.2.7\text{a})$$

$$k_{1x}d - \phi = \tan^{-1}\frac{\mu_1\alpha_2}{\mu_2 k_{1x}} + m\pi \qquad (7.2.7\text{b})$$

We add $m\pi$ in (7.2.7b) since there can be multiple solutions to (7.2.6b) especially when d is thick. We do not add any $m'\pi$ in (7.2.7a) simply because it will be convenient to choose $m' = 0$ for the eigenequation.

Eliminating ϕ, we obtain the eigenequation for the TE_m mode:

$$k_{1x}d = \tan^{-1}\frac{\mu_1\alpha}{\mu k_{1x}} + \tan^{-1}\frac{\mu_1\alpha_2}{\mu_2 k_{1x}} + m\pi \quad (m = 0,1,2,\ldots) \quad (7.2.8)$$

Alternatively,

$$\tan k_{1x}d = \frac{(\mu_1\alpha/\mu k_{1x}) + (\mu_1\alpha_2/\mu_2 k_{1x})}{1 - (\mu_1\alpha/\mu k_{1x})(\mu_1\alpha_2/\mu_2 k_{1x})} \quad (7.2.9)$$

If $\mu = \mu_1 = \mu_2$,

$$\tan k_{1x}d = \frac{(\alpha + \alpha_2)k_{1x}}{k_{1x}^2 - \alpha\alpha_2} \quad (7.2.10)$$

After solving for α, α_2, k_{1x}, and k_z from (7.2.3) and (7.2.9), the complete electric field can be obtained from the normalization condition

$$P = \frac{-1}{2}\,\mathrm{Re}\int_{-\infty}^{\infty}(E_y H_x^*)\,dx = 1 \quad (7.2.11)$$

We obtain

$$E_y = C_1\,e^{ik_z z}\begin{cases}\cos\phi e^{-\alpha x} & x \geq 0\\ \cos(k_1 x + \phi) & -d \leq x \leq 0\\ \cos(-k_{1x}d + \phi)\,e^{\alpha_2(x+d)} & x \leq -d\end{cases} \quad (7.2.12)$$

where $C_1 = [4\omega\mu/k_z(d + 1/\alpha + 1/\alpha_2)]^{1/2}$ if $\mu = \mu_1 = \mu_2$; otherwise, it can be cast in a more complicated analytical expression.

Cutoff Condition. For the wave to be guided, ε_1 has to be larger than both ε_2 and ε. Let us assume that $\varepsilon_2 > \varepsilon(\mu = \mu_1 = \mu_2)$. When reducing the frequency until $k_z = \omega\sqrt{\mu_2\varepsilon_2}, = k_0 n_2$, the decay constant in region 2, α_2, will vanish before α does. Thus, at cutoff

$$k_{1x}d = \tan^{-1}\frac{\mu_1\alpha}{\mu k_{1x}} + m\pi \quad (7.2.13)$$

for the TE_m mode at $\alpha_2 = 0$, and $k_z = k_0 n_2$. We also have

$$\alpha = \sqrt{k_z^2 - \omega^2\mu\varepsilon} = k_0\sqrt{n_2^2 - n^2} \quad (7.2.14a)$$

$$k_{1x} = k_0\sqrt{n_1^2 - n_2^2} \quad (7.2.14b)$$

at cutoff. The cutoff frequency is determined from

$$k_0 d\sqrt{n_1^2 - n_2^2} = \tan^{-1}\frac{\mu_1\sqrt{n_2^2 - n^2}}{\mu\sqrt{n_1^2 - n_2^2}} + m\pi \quad (7.2.15)$$

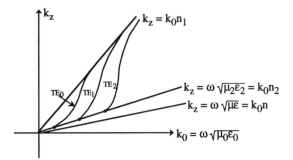

Figure 7.9.   the dispersion relations for an asymmetric dielectric waveguide.

where $k_0 = \omega\sqrt{\mu_0\varepsilon_0} = \omega/c$. The dispersion curves for the TE_m modes are plotted in Fig. 7.9. They start from the cutoff condition $k_z = k_0 n_2$ and approach the upper limit $k_z \to k_0 n_1$ when the frequency is increased. Note that if $n_1 > n_2 \gg n$, then

$$\frac{\alpha}{k_{1x}} = \frac{\sqrt{n_2^2 - n^2}}{\sqrt{n_1^2 - n_2^2}} \gg 1$$

We obtain

$$k_0 d\sqrt{n_1^2 - n_2^2} = \left(m + \frac{1}{2}\right)\pi \tag{7.2.16}$$

which is equivalent to the cutoff condition of the $TE_{(2m+1)}$ mode in a symmetric waveguide with a $2d$ thickness. This can be easily understood from a comparison of the two electric field profiles in Fig. 7.10. Since the decaying

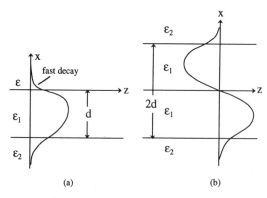

Figure 7.10.   (a) An asymmetric waveguide with $\varepsilon_1 > \varepsilon_2 \gg \varepsilon$ and a thickness d. (b) A symmetric waveguide with a thickness $2d$. The cutoff condition for the TE_m mode in (a) is equivalent to that of the $TE_{(2m+1)}$ mode in (b).

constant α in region ε decays very fast in Fig. 7.10a, its field profile looks very much like half of the field profile of Fig. 7.10b.

7.2.2 TM Polarization, $\mathbf{H} = \hat{y}H_y$

We again obtain the solutions for the TM polarization using the duality principle. We obtain from (7.2.12)

$$
H_y = C_1\, e^{ik_z z}\begin{cases} \cos\phi\, e^{-\alpha x} & x \geq 0 \\ \cos(k_{1x}x + \phi) & -d \leq x \leq 0 \\ \cos(-k_{1x}d + \phi)e^{\alpha_2(x+d)} & x \leq -d \end{cases} \quad (7.2.17)
$$

where the constant C_1 can be found from the normalization condition, noting that $\varepsilon \neq \varepsilon_1 \neq \varepsilon_2$. The guidance condition is obtained from (7.2.8) after replacing μ_i by ε_i. We find

$$
k_{1x}d = \tan^{-1}\frac{\varepsilon_1\alpha}{\varepsilon k_{1x}} + \tan^{-1}\frac{\varepsilon_1\alpha_2}{\varepsilon_2 k_{1x}} + m\pi \quad (7.2.18)
$$

for the TM_m mode.

We summarize the guidance conditions in terms of the refractive indices n, n_1, n_2 and use $\mu = \mu_1 = \mu_2$ for dielectric materials using (7.2.8) and (7.2.18):

$$
k_{1x}d = \tan^{-1}\frac{\alpha}{k_{1x}} + \tan^{-1}\frac{\alpha_2}{k_{2x}} + m\pi \qquad (\mathrm{TE}_m \text{ mode}) \quad (7.2.19a)
$$

$$
k_{1x}d = \tan^{-1}\frac{n_1^2\alpha}{n^2 k_{1x}} + \tan^{-1}\frac{n_1^2\alpha_2}{n_2^2 k_{1x}} + m\pi \qquad (\mathrm{TM}_m \text{ mode}) \quad (7.2.19b)
$$

7.3 RAY OPTICS APPROACH TO THE WAVEGUIDE PROBLEMS

An efficient method to find the eigenequation for the dielectric slab waveguide problem is to use the ray optics picture. We know from Section 5.3 that when a plane wave is incident on a planar dielectric boundary with an angle of incidence θ larger than the critical angle, the reflection coefficient r_{12} has a Goos–Hänchen phase shift $-2\phi_{12}$.

For a mode to be guided, the total field after reflecting back and forth (from point A to B) by the two boundaries should repeat itself (Fig. 7.11):

$$
\left(r_{12}e^{ik_{1x}d}\right)r_{10}e^{ik_{1x}d} = 1 \quad (7.3.1)
$$

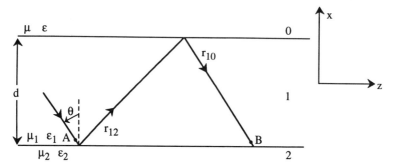

Figure 7.11. A ray optics picture for a guided mode in a slab waveguide.

where the phase $2k_{1x}$ accounts for the round-trip optical path delay along the x direction.

The above condition is also called the transverse resonance condition. In terms of the Goos–Hänchen phase shifts, we obtain

$$e^{-i2\phi_{12}-i2\phi_{10}+i2k_{1x}d} = 1 \qquad (7.3.2)$$

where for TE modes

$$\phi_{12} = \tan^{-1}\frac{\mu_1\alpha_2}{\mu_2 k_{1x}} \qquad (7.3.3a)$$

$$\phi_{10} = \tan^{-1}\frac{\mu_1\alpha}{\mu k_{1x}} \qquad (7.3.3b)$$

Therefore, we find

$$2k_{1x}d - 2\phi_{12} - 2\phi_{10} = 2m\pi \qquad (7.3.4)$$

The total phase delay (noting that the Goos–Hänchen phase shifts are negative, $-2\phi_{12} - 2\phi_{10}$, or advance phase shifts) should be an even multiple of π. Equation (7.3.4) gives the guidance condition for the TE$_m$ mode:

$$k_{1x}d = \tan^{-1}\frac{\mu_1\alpha_2}{\mu_2 k_{1x}} + \tan^{-1}\frac{\mu_1\alpha}{\mu k_{1x}} + m\pi \qquad (7.3.5)$$

If $\mu_2 = \mu$ and $\varepsilon_2 = \varepsilon$, i.e., the slab waveguide is symmetric; we have $\phi_{12} = \phi_{10}$ and

$$k_{1x}d = m\pi + 2\phi_{10} \qquad (7.3.6)$$

which leads to

$$\tan\left(k_{1x}\frac{d}{2}\right) = \begin{cases} \tan\phi_{10} = \dfrac{\mu_1\alpha}{\mu k_{1x}} & \text{if } m \text{ is even} \\[3mm] -\cot\phi_{10} = -\dfrac{\mu}{\mu_1}\dfrac{k_{1x}}{\alpha} & \text{if } m \text{ is odd} \end{cases} \qquad (7.3.7)$$

or

$$\alpha = \frac{\mu}{\mu_1}k_{1x}\tan\left(k_{1x}\frac{d}{2}\right) \qquad m = \text{even} \qquad (7.3.8a)$$

$$\alpha = -\frac{\mu}{\mu_1}k_{1x}\cot\left(k_{1x}\frac{d}{2}\right) \qquad m = \text{odd} \qquad (7.3.8b)$$

which are the same as (7.1.10a) and (7.1.10b), as expected.

7.4 RECTANGULAR DIELECTRIC WAVEGUIDES

The rectangular dielectric waveguide theory is very useful since most waveguides have finite dimensions in both the x and y directions. There are two possible classes of modes. The analyses are approximate here, assuming $w \geq d$, as shown in Fig. 7.12. The method we use here is sometimes called Marcatili's method for rectangular dielectric waveguides [20, 21]. These modes are

1. HE_{pq} modes or $E^y_{(p+1)(q+1)}$ modes: H_x and E_y are the dominant components.
2. EH_{pq} modes or $E^x_{(p+1)(q+1)}$ modes: E_x and H_y are the dominant components.

In general, we have $\partial/\partial z = ik_z$ and Maxwell's equations in the following forms will be used as the starting equations in the later analysis for both

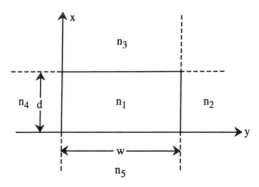

Figure 7.12. A rectangular dielectric waveguide.

HE_{pq} and EH_{pq} modes:

$$H = \frac{1}{i\omega\mu}\nabla \times E$$

$$= \frac{1}{i\omega\mu}\left[\hat{x}\left(\frac{\partial}{\partial y}E_z - ik_zE_y\right) + \hat{y}\left(ik_zE_x - \frac{\partial}{\partial x}E_z\right) + \hat{z}\left(\frac{\partial}{\partial x}E_y - \frac{\partial}{\partial y}E_x\right)\right]$$

$$(7.4.1)$$

$$E = \frac{1}{-i\omega\varepsilon}\nabla \times H$$

$$= \frac{-1}{i\omega\varepsilon}\left[\hat{x}\left(\frac{\partial}{\partial y}H_z - ik_zH_y\right) + \hat{y}\left(ik_zH_x - \frac{\partial}{\partial x}H_z\right) + \hat{z}\left(\frac{\partial}{\partial x}H_y - \frac{\partial}{\partial y}H_x\right)\right]$$

$$(7.4.2)$$

$$\frac{\partial}{\partial x}E_x + \frac{\partial}{\partial y}E_y + ik_zE_z = 0 \tag{7.4.3}$$

$$\frac{\partial}{\partial x}H_x + \frac{\partial}{\partial y}H_y + ik_zH_z = 0 \tag{7.4.4}$$

Here we have assumed that the permittivity is a constant in each region.

7.4.1 HE_{pq} Modes (or $E^y_{(p+1)(q+1)}$ modes)

Suppose we start with the HE_{pq} modes. The large components are H_x and E_y. The other components E_x, E_z, H_y, and H_z are assumed to be small. In principle, we can assume that $H_y \simeq 0$, and express E_x, E_y, E_z, and H_z in terms of H_x using Maxwell's equations. These are called the HE_{pq} modes or E^y_{pq} modes. Alternatively, we can also assume $E_x \simeq 0$, and express E_z, H_x, H_y, and H_z in terms of E_y. Since these are all approximate solutions, they have to be compared with the exact solutions [22, 23]. We find that both approximations give essentially the same eigenequations with minor differences. We call these modes HE_{pq} modes, where p labels the number of nodes for a wave function in the x direction and q for the y direction. The solution procedure is as follows.

1. Solve the wave equation for H_x everywhere:

$$\left(\frac{\partial^2}{\partial x^2} + \frac{\partial^2}{\partial y^2} - k_z^2 + \omega^2\mu\varepsilon\right)H_x = 0 \tag{7.4.5}$$

2. Use (7.4.4),

$$H_z = \frac{1}{-ik_z}\frac{\partial}{\partial x}H_x \tag{7.4.6}$$

3.
$$E_z = \frac{1}{i\omega\varepsilon}\frac{\partial}{\partial y}H_x \tag{7.4.7}$$

4. Since $H_y = 0$, we obtain

$$E_x = \frac{-1}{i\omega\varepsilon}\frac{\partial}{\partial y}H_z = \frac{-1}{\omega\varepsilon k_z}\frac{\partial}{\partial x}\frac{\partial}{\partial y}H_x \tag{7.4.8}$$

5.
$$E_y = \frac{1}{-i\omega\varepsilon}\left(ik_z H_x - \frac{\partial}{\partial x}H_z\right) = \frac{1}{\omega\varepsilon k_z}\left(-k_z^2 + \frac{\partial^2}{\partial x^2}\right)H_x$$

$$= \frac{1}{\omega\varepsilon k_z}\left(-\omega^2\mu\varepsilon - \frac{\partial^2}{\partial y^2}\right)H_x \tag{7.4.9}$$

Once H_x is found, all other components (except that $H_y = 0$) can be expressed in terms of H_x using Eqs. (7.4.6)–(7.4.9).

The expression for H_x can be written approximately as

$$H_x = e^{ik_z z}\begin{cases} C_1\cos(k_x x + \phi_x)\cos(k_y y + \phi_y) & \text{Region 1} \\ C_2\cos(k_x x + \phi_x)e^{-\alpha_2(y-w)} & \text{Region 2} \\ C_3 e^{-\alpha_3(x-d)}\cos(k_y y + \phi_y) & \text{Region 3} \\ C_4\cos(k_x x + \phi_x)e^{\alpha_4 y} & \text{Region 4} \\ C_5 e^{\alpha_5 x}\cos(k_y y + \phi_y) & \text{Region 5} \end{cases} \tag{7.4.10}$$

where $k_x^2 + k_y^2 + k_z^2 = \omega^2\mu_1\varepsilon_1$, $k_x^2 - \alpha_2^2 + k_z^2 = \omega^2\mu_2\varepsilon_2$, etc., which are obtained by substituting the field expressions (7.4.10) into the wave equation (7.4.5) using the material parameters μ_i and ε_i for region i. The choice of the standing-wave solutions $\cos(k_x x + \phi_x)$ or decaying solutions is made simply by inspecting Fig. 7.12 and keeping in mind that we are looking for the guided mode solution. The same x dependence is used for regions 1 and 2 because both regions share the same boundary at $y = w$, and the phase-matching condition or Snell's law requires the same dependence on the spatial variable x. Similar rules apply to the field expressions in other regions.

Boundary Conditions. To derive the guidance conditions and the relations between the field coefficients $C_1 \cdots C_5$, we match the boundary conditions.

1. At $x = 0$ and $x = d$, H_z and E_y are continuous. The other boundary condition for E_z continuous is ignored since E_z is small.
   At $x = 0$,

$$-k_x C_1 \sin \phi_x = \alpha_5 C_5 \tag{7.4.11a}$$

$$C_1 \cos \phi_x \left(\frac{\omega^2 \mu_1 \varepsilon_1 - k_y^2}{\varepsilon_1} \right) = C_5 \left(\frac{\omega^2 \mu_5 \varepsilon_5 - k_y^2}{\varepsilon_5} \right) \tag{7.4.11b}$$

At $x = d$,

$$k_x C_1 \sin(k_x d + \phi_x) = \alpha_3 C_3 \tag{7.4.12a}$$

$$C_1 \cos(k_x d + \phi_x) \left(\frac{\omega^2 \mu_1 \varepsilon_1 - k_y^2}{\varepsilon_1} \right) = C_3 \left(\frac{\omega^2 \mu_3 \varepsilon_3 - k_y^2}{\varepsilon_3} \right) \tag{7.4.12b}$$

Eliminating C_1, C_3, and C_5, we find

$$\alpha_5 = \frac{-\varepsilon_1}{\varepsilon_5} \left(\frac{\omega^2 \mu_5 \varepsilon_5 - k_y^2}{\omega^2 \mu_1 \varepsilon_1 - k_y^2} \right) k_x \tan \phi_x \tag{7.4.13a}$$

$$\alpha_3 = \frac{\varepsilon_1}{\varepsilon_3} \left(\frac{\omega^2 \mu_3 \varepsilon_3 - k_y^2}{\omega^2 \mu_1 \varepsilon_1 - k_y^2} \right) k_x \tan(k_x d + \phi_x) \tag{7.4.13b}$$

Eliminating ϕ_x again, we find

$$k_x d = \tan^{-1} \frac{\varepsilon_3 (\omega^2 \mu_1 \varepsilon_1 - k_y^2) \alpha_3}{\varepsilon_1 (\omega^2 \mu_3 \varepsilon_3 - k_y^2) k_x} + \tan^{-1} \frac{\varepsilon_5 (\omega^2 \mu_1 \varepsilon_1 - k_y^2) \alpha_5}{\varepsilon_1 (\omega^2 \mu_5 \varepsilon_5 - k_y^2) k_x} + p\pi \tag{7.4.14a}$$

The above formula can be further simplified for $k_y \ll k_1$, k_3, and k_5:

$$k_x d = p\pi + \tan^{-1} \frac{\mu_1 \alpha_3}{\mu_3 k_x} + \tan^{-1} \frac{\mu_1 \alpha_5}{\mu_5 k_x} \tag{7.4.14b}$$

2. At $y = 0$ and $y = w$, match H_x and E_z. The other conditions that H_z and E_x are continuous give identical results as H_x and E_z
   At $y = 0$

$$C_1 \cos \phi_y = C_4 \tag{7.4.15a}$$

$$-C_1 \frac{k_y}{\varepsilon_1} \sin \phi_y = C_4 \frac{\alpha_4}{\varepsilon_4} \tag{7.4.15b}$$

At $y = w$

$$C_1 \cos(k_y w + \phi_y) = C_2 \tag{7.4.16a}$$

$$C_1 \frac{k_y}{\varepsilon_1} \sin(k_y w + \phi_y) = C_2 \frac{\alpha_2}{\varepsilon_2} \tag{7.4.16b}$$

Thus,

$$\alpha_4 = \frac{-\varepsilon_4}{\varepsilon_1} k_y \tan \phi_y \tag{7.4.17a}$$

$$\alpha_2 = \frac{\varepsilon_2}{\varepsilon_1} k_y \tan(k_y w + \phi_y) \tag{7.4.17b}$$

Therefore,

$$k_y w = q\pi + \tan^{-1} \frac{\varepsilon_1 \alpha_2}{\varepsilon_2 k_y} + \tan^{-1} \frac{\varepsilon_1 \alpha_4}{\varepsilon_4 k_y} \tag{7.4.18}$$

In summary, the two important guidance conditions are

$$k_x d = p\pi + \tan^{-1} \frac{\mu_1 \alpha_3}{\mu_3 k_x} + \tan^{-1} \frac{\mu_1 \alpha_5}{\mu_5 k_x} \tag{7.4.19}$$

$$k_y w = q\pi + \tan^{-1} \frac{\varepsilon_1 \alpha_2}{\varepsilon_2 k_y} + \tan^{-1} \frac{\varepsilon_1 \alpha_4}{\varepsilon_4 k_y} \tag{7.4.20}$$

which can also be seen from the ray optics approach. An intuitive way to understand the approximate boundary conditions is that for (i) at $x = 0$ and $x = d$, the optical power density of interest is its x component bouncing back and forth between the two interfaces at $x = 0$ and d. We therefore need $E_y H_z^*$ and $H_y^* E_z$. However, $H_y \simeq 0$. We need only E_y and H_z and drop E_z since it is small. For boundary conditions (ii) at $y = 0$ and w, we use only H_z and E_x, or H_x and E_z for the y component of the optical power density. Both pairs give the same results since H_y is not of concern.

7.4.2 EH_{pq} Modes (or $E_{(p+1)(q+1)}^x$ modes)

For the EH_{pq} mode, the dominant field components are E_x and H_y. We can assume that $H_x \simeq 0$ and H_y have exactly the same form as in (7.4.10). Expressing all other components in terms of H_y only, we then match the boundary conditions. We use

$$H_z = \frac{1}{-ik_z} \frac{\partial}{\partial y} H_y \qquad E_z = \frac{1}{-i\omega\varepsilon} \frac{\partial}{\partial x} H_y \qquad E_y = \frac{1}{\omega\varepsilon k_z} \frac{\partial}{\partial y} \frac{\partial}{\partial x} H_y$$

$$E_x = \frac{1}{\omega\varepsilon k_z} \left(\omega^2 \mu\varepsilon + \frac{\partial^2}{\partial x^2} \right) H_y$$

which are derivable from Eqs. (7.4.1)–(7.4.4). The following guidance conditions are found after matching the boundary conditions,

$$k_x d = p\pi + \tan^{-1} \frac{\varepsilon_1 \alpha_3}{\varepsilon_3 k_x} + \tan^{-1} \frac{\varepsilon_1 \alpha_5}{\varepsilon_5 k_x} \qquad (7.4.21)$$

$$k_y w = q\pi + \tan^{-1} \left[\frac{\varepsilon_2 (\omega^2 \mu_1 \varepsilon_1 - k_x^2) \alpha_2}{\varepsilon_1 (\omega^2 \mu_2 \varepsilon_2 - k_x^2) k_y} \right] + \tan^{-1} \left[\frac{\varepsilon_4 (\omega^2 \mu_1 \varepsilon_1 - k_x^2) \alpha_4}{\varepsilon_1 (\omega^2 \mu_4 \varepsilon_4 - k_x^2) k_y} \right]$$
$$\qquad (7.4.22)$$

where (7.4.22) can be further simplified for $k_x \ll k_1$, k_2, and k_4:

$$k_y w = q\pi + \tan^{-1} \frac{\mu_1 \alpha_2}{\mu_2 k_y} + \tan^{-1} \frac{\mu_1 \alpha_4}{\mu_4 k_y} \qquad (7.4.23)$$

Alternatively, we may assume that $E_y \simeq 0$ and E_x is of the form (7.4.10), and then express all the other components in terms of the E_x component using Maxwell's equations. These results will give almost the same eigenequations and mode patterns.

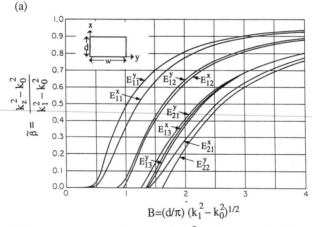

(a)

$$B = (d/\pi) (k_1^2 - k_0^2)^{1/2}$$

Figure 7.13. (a) Normalized propagation parameter $\tilde{\beta}$ vs. the normalized frequency $B = (d/\pi)(k_1^2 - k_0^2)^{1/2}$ of a rectangular waveguide with an aspect ratio $w/d = 2$. The refractive index in the guide is $n_1 = 1.5$ and outside is $n_0 = 1$. (Reprinted from Ref. 23, ©1989, published by VSP, The Netherlands.) Here the width w is along the y direction and the thickness d is along the x direction. (b) Plots of the transverse field components $\mathbf{E}_S^{TE}$ and $\mathbf{H}_S^{TM}$ for the lowest nine modes for a rectangular waveguide in (a). Note the duality relation between the transverse electric field of the HE_{pq} mode and the transverse magnetic field of the EH_{pq} mode. (Reprinted from Ref. 23, ©1989, published by VSP, The Netherlands.) (c) The schematic intensity patterns for the modes in part (b). Since the intensity patterns of the HE_{pq} mode are almost indistinguishable from that of the EH_{pq} mode, we only show one set of patterns here.

(b)

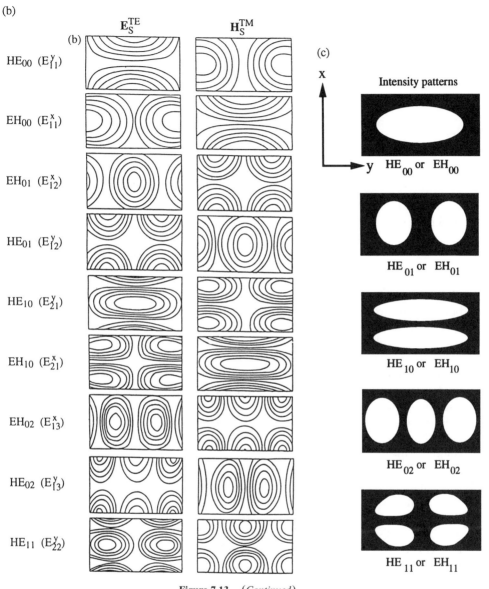

Figure 7.13. (*Continued*)

Example We consider a rectangular waveguide where the aspect ratio $w/d = 2$, the refractive index of the waveguide n_1 is 1.5, and the background medium is free space $n_0 = 1$. We plot in Fig. 7.13a, after Ref. 23, the normalized propagation constant $\tilde{\beta}$ vs. a normalized frequency $B = V/\pi$,

$$\tilde{\beta} = \frac{k_z^2 - k_0^2}{k_1^2 - k_0^2} = \frac{n_{eff}^2 - n_0^2}{n_1^2 - n_0^2} \qquad (7.4.24)$$

and

$$B = \frac{V}{\pi} = \left(\frac{d}{\pi}\right)\left(k_1^2 - k_0^2\right)^{1/2} \qquad (7.4.25)$$

where the V number is defined the same as that of a slab waveguide with a thickness d,

$$V = k_0 d \left(n_1^2 - n_0^2\right)^{1/2} \qquad (7.4.26)$$

The corresponding transverse electric and magnetic field patterns inside the dielectric waveguide are also plotted in Fig. 7.13b. Note the dual relations between the transverse electric field $\mathbf{E}_S^{TE}$ of the HE_{pq} mode and the transverse magnetic field $\mathbf{H}_S^{TM}$ of the EH_{pq} mode. We can see that the field patterns provide the information on the intensity variations of the E_x, E_y, H_x, and H_y components of the waveguide modes (Fig. 7.13c). The intensity pattern of the HE_{pq} mode is almost indistinguishable from that of the EH_{pq} mode. Therefore, only one set of intensity patterns is shown. These mode patterns should be interesting to compare with those of the metallic waveguides [24]. The results in Fig. 7.13a have been compared with those of Goell [22] with a good agreement. The original labels of the modes in Ref. 23 are slightly different. In Figs. 7.13a and b, we use our convention for the $HE_{pq}(= E_{(p+1)(q+1)}^y)$ modes and the $EH_{pq}(= E_{(p+1)(q+1)}^x)$ modes, and the coordinate system is shown in Fig. 7.12, where the vertical axis is the x axis and the horizontal axis is the y axis. This coordinate system is chosen such that when the width $w \to \infty$, the results here approach those of the slab waveguides discussed in Sections 7.1–7.3 using the same coordinate system. Our convention is the same as those for Marcatili [20] and Goell [22] using the $E_{(p+1)(q+1)}^x$ and $E_{(p+1)(q+1)}^y$ modes, except that our aspect ratio is $w/d > 1$, where w is the width along the y direction instead of the x direction. ∎

7.5 THE EFFECTIVE INDEX METHOD

The effective index method is a useful technique to find the propagation constant of a dielectric waveguide. As an example, assume that we are

(a) A rectangular dielectric waveguide

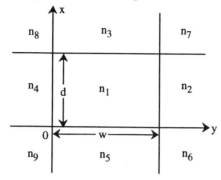

(b) Step one: solve for each y the effective index $n_{\text{eff}}(y)$.

i)__n_8__   ii)__n_3__   iii)__n_7__

___n_4___   ___n_1___   ___n_2___

___n_9___   ___n_5___   ___n_6___

(c) Step two: solve the slab waveguide problem using $n_{\text{eff}}(y)$

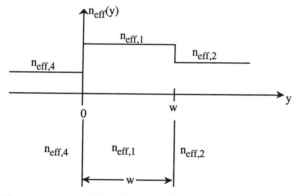

Figure 7.14. (a) A rectangular dielectric waveguide to be solved using the effective index method. (b) Solve the slab waveguide problem at each fixed y and obtain an effective index profile $n_{\text{eff}}(y)$. (c) Solve the slab waveguide problem with the effective index profile $n_{\text{eff}}(y)$.

interested in the solutions for a rectangular dielectric waveguide as in Section 7.4. Consider first the $E^y_{(p+1)(q+1)}$ modes for which E_y is the dominant component (see Fig. 7.14a).

Step 1. We first solve a slab waveguide problem with refractive index n_1 inside and n_3, n_5 outside the waveguide (Fig. 7.14b). The eigenequation for the E_y component will be that of the TE modes of a slab guide discussed in Section 7.1, since E_y is parallel to the boundaries:

$$k_{1x}d = \tan^{-1}\frac{\mu_1\alpha_3}{\mu_3 k_{1x}} + \tan^{-1}\frac{\mu_1\alpha_5}{\mu_5 k_{1x}} + p\pi \qquad (7.5.1)$$

where α_i is the decaying constant (outside the waveguide) and μ_i is the permeability in the ith region. From the solutions of the slab waveguide problem, we obtain the propagation constant, therefore, the effective index $n_{\text{eff},1}$. For $y > w$ or $y < 0$, similar equations hold if $n_2 > n_6$, n_7, and $n_4 > n_8$, n_9. Otherwise, approximations have to be made by assuming that $n_{\text{eff},2} \simeq n_2$ and $n_{\text{eff},4} \simeq n_4$ when the fields in regions 6, 7, 8, and 9 are negligible.

Step 2. We then solve the slab waveguide problem as shown in Fig. 7.14c, and obtain

$$k_{1y}w = \tan^{-1}\frac{\varepsilon_{\text{eff},1}\alpha_2}{\varepsilon_{\text{eff},2} k_{1y}} + \tan^{-1}\frac{\varepsilon_{\text{eff},1}\alpha_4}{\varepsilon_{\text{eff},4} k_{1y}} + q\pi \qquad (7.5.2)$$

where $\varepsilon_{\text{eff},i} = n_{\text{eff},i}^2\varepsilon_0$. Note that the E_y component is perpendicular to the slab boundaries; therefore, we should use the guidance condition for the TM modes of the slab waveguide problem. In step 1, for each y, we find a modal distribution as a function of x. This field distribution will also vary as y changes. Thus we call this function $F(x, y)$. In step 2, the solution for the modal distribution will be a function of only y, called $G(y)$. Thus the total electric field E_y can be written as

$$E_y(x, y) \simeq F(x, y)G(y) \qquad (7.5.3)$$

It is interesting to compare the two eigenequations (7.5.1) and (7.5.2) obtained here with Eqs. (7.4.19) and (7.4.20) derived in Section 7.4. The difference is that the "effective permittivities" appear in (7.5.2) in the effective index method, while the material permittivities are used in (7.4.20) in the previous approximate Marcatili method. When comparing the propagation constants with the numerical approach from the full-wave analysis, the results of both Marcatili's method in Section 7.4 and the effective index method agree very well with those of the numerical method, generally speaking. The effective index method usually has a better agreement with the numerical method, especially near cutoff.

The effective index method has also been applied [25, 26] to the analysis of diode lasers with lateral variations in thickness of the active region or in the complex permittivity. The resultant analytical formulas in many cases are helpful in understanding the semiconductor laser operation characteristics.

7.6 WAVE GUIDANCE IN A LOSSY OR GAIN MEDIUM

In an unbounded medium, the propagation constant k is given by

$$k = \omega\sqrt{\mu\varepsilon} \tag{7.6.1}$$

when the permittivity is complex,

$$\varepsilon = \varepsilon' + i\varepsilon'' \tag{7.6.2}$$

that is, the medium is lossy ($\varepsilon'' > 0$) or it has gain ($\varepsilon'' < 0$), and the propagation constant becomes

$$k = \omega\sqrt{\mu(\varepsilon' + i\varepsilon'')} = \frac{\omega}{c}(n + i\kappa)$$

$$= k' + ik'' \tag{7.6.3}$$

where the real and imaginary parts of the complex refractive index, n and κ, have been tabulated and plotted in Appendix J for GaAs and InP semiconductors. For a wave propagating in the $+z$ direction, the field solution is (consider $\mathbf{E} = \hat{y}E_y$)

$$E_y = E_0 e^{ikz}$$

$$= E_0 e^{ik'z} e^{-k''z} \tag{7.6.4}$$

The field decays as it propagates in the $+z$ direction for a lossy medium and grows if the medium has a gain mechanism such as in the active region of a laser.

For a waveguide that may contain gains in the active region and losses in the passive region, the propagation constant is generally complex. A typical

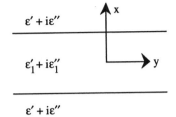

Figure 7.15. A lossy (gain) waveguide structure.

analysis is based on the perturbation theory or a variational method. Consider a dielectric slab waveguide that has complex permittivities everywhere, as shown in Fig. 7.15:

$$\varepsilon(x) = \begin{cases} \varepsilon_1' + i\varepsilon_1'' & \text{inside} \\ \varepsilon' + i\varepsilon'' & \text{outside} \end{cases}$$

Inside the guide,

$$k_1 = \omega\sqrt{\mu(\varepsilon_1' + i\varepsilon_1'')} \simeq \omega\sqrt{\mu\varepsilon'}\left(1 + i\frac{\varepsilon_1''}{2\varepsilon_1'}\right)$$

$$= k_1' - i\frac{g}{2} \tag{7.6.5}$$

where $k_1' = \omega\sqrt{\mu\varepsilon_1'} = k_0 n_1$, and $g = -k_1'\varepsilon_1''/\varepsilon_1'$ is the gain coefficient, if $\varepsilon_1'' < 0$. Outside the waveguide we have

$$k = \omega\sqrt{\mu(\varepsilon' + i\varepsilon'')} \cong \omega\sqrt{\mu\varepsilon'}\left(1 + i\frac{\varepsilon''}{2\varepsilon'}\right)$$

$$= k' + i\frac{\alpha}{2} \tag{7.6.6}$$

where $\alpha = k'\varepsilon''/\varepsilon'$ is the decaying constant of the absorption coefficient for the optical intensity. Suppose we try to solve the TE mode

$$\mathbf{E} = \hat{y}E_y = \hat{y}\phi(x)\,e^{ik_z z} \tag{7.6.7}$$

$$\left[\frac{d^2}{dx^2} - k_z^2 + \omega^2\mu\varepsilon(x)\right]\phi(x) = 0 \tag{7.6.8}$$

A simple method is to use the variational approach:

$$k_z^2 = \frac{\int_{-\infty}^{\infty} \phi^*(x) \left[d^2/dx^2 + \omega^2\mu\varepsilon(x) \right] \phi(x)\, dx}{\int_{-\infty}^{\infty} \phi^*(x)\phi(x)\, dx} \tag{7.6.9}$$

When we write the permittivity as the sum of a lossless part $\varepsilon^{(0)}(x)$ and a part due to gain or loss $\Delta\varepsilon(x)$, we obtain

$$\varepsilon^{(0)}(x) = \begin{cases} \varepsilon_1' & \text{inside} \\ \varepsilon' & \text{outside} \end{cases} \qquad \Delta\varepsilon(x) = \begin{cases} i\varepsilon_1'' & \text{inside} \\ i\varepsilon'' & \text{outside} \end{cases} \tag{7.6.10}$$

where $\Delta\varepsilon(x)$ is considered a perturbation. The unperturbed solution for $\varepsilon^{(0)}(x)$ has been discussed previously and the solution $\phi^{(0)}(x)$ is just that of a lossless slab waveguide. Note that

$$\left[\frac{d^2}{dx^2} - k_z^{(0)2} + \omega^2\mu\varepsilon^{(0)}(x) \right]\phi^{(0)}(x) = 0 \tag{7.6.11}$$

If we approximate the trial function $\phi(x)$ in the variational expression (7.6.9) by the unperturbed solution $\phi^{(0)}(x)$, we obtain

$$k_z^2 = k_z^{(0)2} + i\omega^2\mu\varepsilon_1''\Gamma + i\omega^2\mu\varepsilon''(1 - \Gamma) \tag{7.6.12}$$

where

$$\Gamma = \frac{\int_{\text{inside}} |\phi^{(0)}(x)|^2\, dx}{\int_{-\infty}^{\infty} |\phi^{(0)}(x)|^2\, dx} \tag{7.6.13}$$

is the optical confinement factor.

When ε_1' is close to ε', the propagation constant can be written in the form

$$k_z \simeq k_z^{(0)} - i\Gamma\frac{g}{2} + i(1 - \Gamma)\frac{\alpha}{2} \tag{7.6.14}$$

which is obtained by taking the square roots of both sides of (7.6.12) and

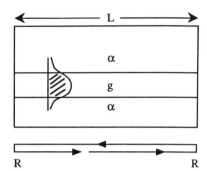

Figure 7.16. A semiconductor laser cavity with a net gain g in the active region and an absorption coefficient α outside the active region. The cavity length is L and the reflectivity is R at both ends.

using the binomial expansion assuming that the imaginary part is small compared with the real part, and $k_z^{(0)}$ is close to k' and k_1'. When applying (7.6.14) to a laser cavity, as shown in Fig. 7.16 we have the wave bouncing back and forth in a round trip of distance $2L$:

$$e^{ik_z 2L} r^2 = 1 \tag{7.6.15}$$

where r is the reflection coefficient of the electric field. The reflectivity R is related to r by $R = |r|^2$. Ignoring the small imaginary part of r, which is a good approximation for most cases, we have

$$e^{i2k_z^{(0)}L} e^{\Gamma g L - (1-\Gamma)\alpha L} R = 1 \tag{7.6.16}$$

Thus

$$2k_z^{(0)}L = 2m\pi \qquad m = \text{an integer} \tag{7.6.17}$$

which determines the longitudinal mode spectrum of the semiconductor laser, and

$$\Gamma g = (1 - \Gamma)\alpha + \frac{1}{L}\ln\frac{1}{R} \tag{7.6.18}$$

which determines the threshold gain condition. Here Γg is the modal gain of the dielectric waveguide. Strictly speaking, if the modal dispersion is important, $k_z^{(0)}$ of the guided mode solution should be used. For most semiconductor waveguides with small index differences, we can assume $k_z^{(0)} \simeq (\omega/c)n_1$, and (7.6.17) can also be approximated by

$$\frac{\omega}{c}n_1 L = m\pi \tag{7.6.19}$$

The frequency spectrum is

$$f_m = m \frac{c}{2n_1 L} \tag{7.6.20}$$

and the Fabry–Perot frequency spacing is

$$\Delta f = \frac{c}{2n_1 L} \left(1 + \frac{f}{n_1} \frac{\partial n_1}{\partial f} \right)^{-1} \tag{7.6.21}$$

Alternatively, we write in terms of the free-space wavelength λ

$$m\lambda = 2n_1 L \tag{7.6.22}$$

and

$$\Delta\lambda = \frac{-\lambda^2 \Delta m}{2n_1 L[1 - (\lambda/n_1)(dn_1/d\lambda)]} \tag{7.6.23}$$

when the dispersion effect $(dn_1/d\lambda \neq 0)$ is taken into account. For two nearby Fabry–Perot modes, $\Delta m = 1$ and $\Delta\lambda$ given in (7.6.23) gives the wavelength spacing of the Fabry–Perot modes at λ:

$$\left| \frac{\Delta\lambda}{\lambda} \right| = \frac{\lambda}{2n_1 L[1 - (\lambda/n_1)(dn_1/d\lambda)]} \tag{7.6.24}$$

In Fig. 7.17, we show the amplified spontaneous emission spectrum of a InGaAsP/InGaAsP strained quantum-well laser at 300 K. The cavity length

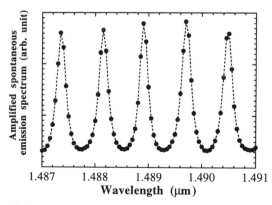

Figure 7.17. Amplified spontaneous emission spectrum of a 370-μm semiconductor laser cavity at room temperature below threshold. The injection current is $I = 8$ mA. (Above the laser threshold current $I_{th} = 13.5$ mA, lasing starts.)

L is 370 μm. The threshold current is I_{th} = 13.5 mA at this temperature. Using $n_1 \simeq 3.395$ and $dn_1/d\lambda \simeq -0.264$ $(\mu m)^{-1}$ near λ = 1.49 μm for the barrier $In_{1-x}Ga_xAs_yP_{1-y}$ (λ_g = 1.28 μm, y = 0.55 and x = 0.25, see Appendix K) [10, 16], we obtain the wavelength spacing $\Delta\lambda \simeq 7.9$ Å, which agrees with the spacing shown in the figure.

PROBLEMS

7.1 Consider a $Ga_xIn_{1-x}As$ waveguide with a thickness d confined between two InP regions, where the gallium mole fraction x is 0.47, as shown in Fig. 7.18.

(a) If the free-space wavelength λ_0 is 1.65 μm, what is the photon energy (in eV)?

(b) Find the maximum thickness d_0 (in μm) so that only the TE$_0$ mode is guided at the wavelength λ_0 given in part (a). Assume that the waveguide loss is negligible at this wavelength.

(c) If the thickness d of the waveguide is 0.3 μm, find the following parameters approximately (in 1/μm) for the TE$_0$ mode using the graphical approach in the text: (i) the decay constant α in the InP regions, (ii) k_x, and (iii) k_z in the waveguide.

(d) Calculate the optical confinement factor for the TE$_0$ mode in part (c).

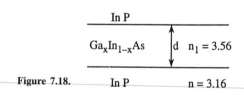

Figure 7.18.

7.2 Repeat Problems 7.1(b) and (c) for the TM polarization. Are the answers the same as those for the TE polarization?

7.3 Calculate the Goos–Hänchen phase shifts for the TE$_0$ and TM$_0$ guided modes in Problem 7.1(c).

7.4 Show that for a symmetric dielectric waveguide with the V number

$$V = k_0 d\sqrt{n_1^2 - n^2}$$

approaching zero, the propagation constant k_z of the TE$_0$ mode can be written as

$$k_z = k_0 n + C_0 V^4$$

Find the expression for C_0. (*Hint*: Start with the guidance condition and the dispersion relation for $V \to 0$.)

7.5 Consider a symmetrical slab waveguide with a thickness d (Fig. 7.19a) and a symmetric quantum well with a well width d (Fig. 7.19b). Consider the guided TE even modes for the optical waveguide and the general even bound states for the heterojunction quantum-well structure. Summarize in a table an exact one-to-one correspondence of the wave functions, material parameters, eigenequations (guidance conditions), propagation constants, and decay constants, for the two physical problems above.

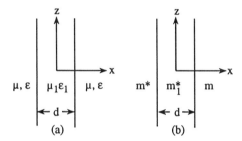

Figure 7.19.

7.6 Consider a symmetric optical slab waveguide with a refractive index inside the guide 3.5 and a refractive index outside the guide 3.2, the waveguide dimension is 2μm, and assume that the magnetic permeability is the same everywhere.

(a) For a light with a wavelength 1.5 μm in free space, find the number of guided TE modes in this waveguide.

(b) Using the ray optics approach, find the angle θ_m of the guided TE_m mode bouncing back and forth between the two boundaries of the waveguide (Fig. 7.20).

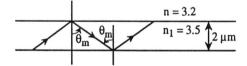

Figure 7.20.

(c) (Independent of parts (a) and (b).) Find the cutoff wavelength λ_m of the TE_m mode. Evaluate the decay constant α outside the waveguide and the propagation constant k_z at the cutoff condition.

7.7 (a) Plot the refractive index n vs. the optical energy $\hbar\omega$ using Eqs. (7.1.22) to (7.1.24) for $x = 0$, 0.1, 0.2, and 0.3.

(b) Design a GaAs/$Al_xGa_{1-x}As$ waveguide using the above results such that only the fundamental TE_0 mode can be guided at a wavelength λ_0 near the GaAs bandgap wavelength at room temperature. Ignore the absorption effects for simplicity. Specify the possible parameters such as the aluminum mole fraction x and the waveguide thickness d.

7.8 (a) Derive the normalization constant C_1 for the electric field of the TE mode in an asymmetric dielectric slab waveguide.

(b) Repeat (a) for the magnetic field of the TM mode.

7.9 (a) Find the analytical expressions for k_x, k_z, and α in a symmetric waveguide for the TE_m mode at the cutoff condition.

(b) Repeat (a) for the TM_m mode.

7.10 (a) Find the analytical expressions for k_x, k_z, α, and α_2 in an asymmetric waveguide for the TE_m mode at the cutoff condition.

(b) Repeat (a) for the TM_m mode.

7.11 For an $In_{0.53}Ga_{0.47}As$/InP dielectric waveguide with a cross section ($2\mu m \times 1\ \mu m$), find the propagation constant of the lowest HE_{00} (or E_{11}^y) mode. You may use the graph in Fig. 7.13, assuming $n_1 = 3.56$ and $n = 3.16$ at a wavelength $\lambda_0 = 1.65\ \mu m$. What is the effective index of this mode?

7.12 Repeat Problem 7.11 using the effective index method. Check the accuracy of the result compared with that of the numerical solution from Fig. 7.13.

7.13 List all of the assumptions made in order to derive the threshold condition

$$\Gamma g = (1 - \Gamma)\alpha + \frac{1}{L}\ln\frac{1}{R}$$

7.14 Show that if the reflectivities from the two end facets of a Fabry–Perot cavity are different, $R_1 \neq R_2$, the threshold condition is given by

$$\Gamma g = (1 - \Gamma)\alpha + \frac{1}{2L}\ln\left(\frac{1}{R_1 R_2}\right)$$

7.15 Consider the laser structure shown in Fig. 7.21 with losses described by the absorption coefficient $\alpha_s = 10\ cm^{-1}$ in the substrates and $\alpha = 5\ cm^{-1}$ in the active region. The reflectivity R is 0.3 at both ends and

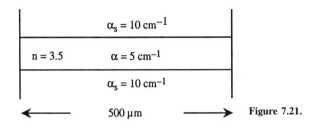

Figure 7.21.

the waveguide length is 500 μm. Assume, for simplicity, that the propagation constant of the guided mode is approximately $\omega n/c$, where $n = 3.5$.

(a) What is the threshold gain g_{th} of the structure if the optical confinement factor of the guided mode is 0.9?

(b) What is the longitudinal mode spectrum of this laser structure? Calculate the frequency spacing Δf and the wavelength spacing $\Delta \lambda$ of two nearby longitudinal modes at $\lambda_0 \simeq 0.8$ μm and ignore the dispersion effect for simplicity.

REFERENCES

1. D. Marcuse, *Theory of Dielectric Optic Waveguides*, Academic, New York, 1974.

2. A. W. Snyder and J. D. Love, *Optical Waveguide Theory*, Chapman & Hall, London, 1983.

3. A. Yariv, *Optical Electronics*, 3d ed., Holt-Rinehart & Winston, New York, 1985.

4. H. C. Casey, Jr., and M. B. Panish, *Heterostructure Lasers, Part A: Fundamental Principles*, Academic, Orlando, FL, 1978.

5. J. A. Kong, *Electromagnetic Wave Theory*, Wiley, New York, 1990.

6. M. A. Afromowitz, "Refractive index of $Ga_{1-x}Al_xAs$," *Solid State Commun.* **15**, 59–63 (1974).

7. J. P. van der Ziel, M. Ilegems, and R. M. Mikulyak, "Optical birefringence of thin GaAs–AlAs multilayer films," *Appl. Phys. Lett.* **28**, 735–737 (1976).

8. J. P. van der Ziel and A. C. Gossard, "Absorption, refractive index, and birefringence of AlAs-GaAs monolayers," *J. Appl. Phys.* **48**, 3018–3023 (1977).

9. J. S. Blakemore, "Semiconducting and other major properties of gallium arsenide," *J. Appl. Phys.* **53**, R123–R181 (1982).

10. S. Adachi, "Refractive indices of III-V compounds: Key properties of InGaAsP relevant to device design," *J. Appl. Phys.* **53**, 5863–5869 (1982).

11. S. Adachi and K. Oe, "Internal strain and photoelastic effects in $Ga_{1-x}Al_xAs/$ GaAs and $In_{1-x}Ga_xAs_yP_{1-y}/InP$ crystals," *J. Appl. Phys.* **54**, 6620–6627 (1983).

12. S. Adachi, "GaAs, AlAs, and $Ga_{1-x}Al_xAs$: Material parameters for use in research and device applications," *J. Appl. Phys.* **58**, R1–R29 (1985).

13. S. Adachi, "Optical properties of $Al_xGa_{1-x}As$ alloys," *Phys. Rev. B* **38**, 12345–12352 (1988).

14. H. C. Casey, Jr., D. D. Sell, and M. B. Panish, "Refractive index of $Al_xGa_{1-x}As$ between 1.2 and 1.8 eV," *Appl. Phys. Lett* **24**, 63–65 (1974).

15. M. Amiotti and G. Landgren, "Ellipsometric determination of thickness and refractive index at 1.3, 1.55 and 1.7 μm for $In_{(1-x)}Ga_xAs_yP_{(1-y)}$ films on InP," *J. Appl. Phys.* **73**, 2965–2971 (1993).

16. S. Adachi, *Physical Properties of III–V Semiconductor Compounds*, Wiley, New York, 1992.

17. S. Adachi, *Properties of Indium Phosphide*, INSPEC, The Institute of Electrical Engineers, London, 1991.

18. K. H. Hellwege, Ed., *Landolt-Börnstein Numerical Data and Functional Relationships in Science and Technology*, New Series, Group III **17a**, Springer, Berlin, 1982; Groups III–V **22a**, Springer, Berlin, 1986.

19. O. Madelung, Ed., *Semiconductors, Group IV Elements and III–V Compounds*, in P. Poerschke, Ed., *Data in Science and Technology*, Springer, Berlin, 1991.

20. E. A. J. Marcatili, "Dielectric rectangular waveguide and directional coupler for integrated optics," *Bell Syst. Tech. J.* **48**, 2071–2102 (1969).

21. E. A. J. Marcatili, "Slab-coupled waveguides," *Bell Syst. Tech. J.* **53**, 645–674 (1974).

22. J. E. Goell, "A circular-harmonic computer analysis of rectangular dielectric waveguide," *Bell Syst. Tech. J.* **48**, 2133–2160 (1969).

23. W. C. Chew, "Analysis of optical and millimeter wave dielectric waveguide," *J. Electromagnetic Waves Appl.* **3**, 359–377 (1989).

24. C. S. Lee, S. W. Lee, and S. L. Chuang, "Plot of modal field distribution in rectangular and circular waveguides," *IEEE Trans. Microwave Theory Tech.* **MTT-33**, 271–274 (1985).

25. W. Streifer, R. D. Burnham and D. R. Scifres, "Analysis of diode lasers with lateral spatial variations in thickness," *Appl. Phys. Lett.* **37**, 121–123 (1980).

26. J. Buus, "The effective index method and its application to semiconductor lasers," *IEEE J. Quantum Electron.* **QE-18**, 1083–1089 (1982).

8

Waveguide Couplers and Coupled-Mode Theory

In this chapter, we discuss how a light can be coupled into and out of a waveguide, and how it can be coupled between two parallel waveguides or coupled within the same waveguide with a distributed feedback coupling. We present the coupled-mode theory for wave propagation in parallel waveguides. An analogy of this coupling between two propagation modes in dielectric waveguides is a coupled pendulum system or a coupled LC resonator system. The coupling of two resonators leads to energy transfer between two resonators at a frequency determined by half of the beat frequency of the two system modes: the in-phase mode and the out-of-phase mode. For coupling of propagation modes between two parallel waveguides, a beat length can also be defined, which is twice the minimum distance required for the exchange of the guided power between the waveguides. Many interesting physical systems exhibit the phenomena of the coupling of modes, which can be understood from a simple model of two coupled pendulums.

8.1 WAVEGUIDE COUPLERS [1–3]

We discuss transverse couplers, prism couplers, and grating couplers and point out the significance of the phase-matching condition in this section.

8.1.1 Transverse Couplers

Direct Focusing. As shown in Fig. 8.1, a lens is used to focus and shape the beam profile so that it is matched to the modal distribution in the waveguide. The coupling efficiency depends on the overlap integral between the field profile of the incident wave and the modal profile of the guided mode in the thin film. This configuration can achieve nearly 100% efficiency. However, it may not be suitable for general integrated optics applications because of the nonplanar configuration.

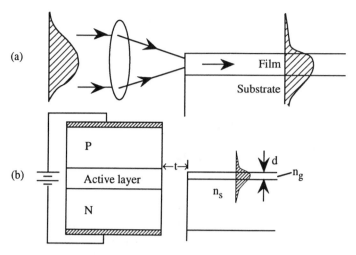

Figure 8.1. (a) Direct focusing. (b) End-butt coupling.

End-butt Coupling. As shown in Fig. 8.1b, the efficiency of an end-butt coupling can be controlled by the gap width t, the waveguide dimension d and refractive indices of the waveguide and the substrate, n_g and n_s, respectively.

8.1.2 Prism Couplers

For a waveguide mode, the propagation constant β along the z direction has to satisfy the conditions, $k_0 n_0 < \beta < k_0 n_1$ and $k_0 n_s < \beta < k_0 n_1$, where $k_0 = 2\pi/\lambda_0$ is the wave number and λ_0 is the wavelength in free space. For an incident plane wave from region 0, its horizontal wave number is $k_z = k_0 n_0 \sin \theta_i < k_0 n_0$ for any real incident angle θ_i. Thus, the wave cannot be phase matched to the guided mode since $k_0 n_0 < \beta$, Fig. 8.2b. The excitation of the guided mode is very weak. However, using a prism above the waveguide with a small gap (assuming $n_p > n_0$ and $n_p > n_s$), Fig. 8.2a, the phase-matching condition can be satisfied: $k_0 n_p \sin \theta_p = \beta$, Fig. 8.2c. This angle of incidence θ_p will be larger than the critical angle for the interface between the prism and region 0,

$$\theta_c = \sin^{-1} \frac{n_0}{n_p} \tag{8.1.1}$$

since $\beta > k_0 n_0$. Therefore, the transmitted wave in the gap region is an evanescent type and can be resonantly coupled to the guided mode. In general, for a particular TE_m mode with a propagation constant β_m, we can

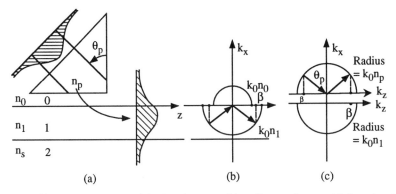

Figure 8.2. (a) A prism coupler. (b) The phase-matching diagram for a guided mode with a propagation constant $\beta > k_0 n_0$. (c) The phase-matching diagram for an incident wave from the prism region and coupling to the guided mode with $k_0 n_p \sin \theta_p = \beta$.

choose θ_p such that

$$k_0 n_p \sin \theta_p = \beta_m \tag{8.1.2}$$

Other important parameters are the incident beam profile, coupling length, and gap dimension.

8.1.3 Grating Couplers

Consider first a plane wave incident on a periodic grating which is assumed to be a perfect conductor with the surface profile described by $x = h(z)$ with a period Λ, as shown in Fig. 8.3.

$$\mathbf{E}_i = \hat{y} E_0 \, e^{-ik_{x0}x + ik_{z0}z}$$

$$k_{x0} = k \cos \theta_i \qquad k_{z0} = k \sin \theta_i \tag{8.1.3}$$

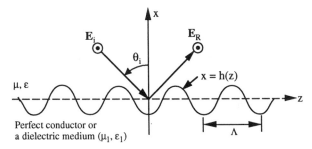

Figure 8.3. A plane wave $\mathbf{E}_i$ with the polarization along the y direction is incident on a grating surface with a profile $x = h(z)$ and a period Λ.

The reflected wave is $\mathbf{E}_R = \hat{y}E_R(x, z)$. Since the boundary condition requires that the tangential electric field vanish at the conductor surface, the reflected wave on the surface of the grating should satisfy

$$E_R(x = h(z), z) = -E_0 e^{-ik_{x0}h(z) + ik_{z0}z}$$

$$= E_0 e^{ik_{z0}z} \sum_{m=-\infty}^{\infty} A_m e^{im(2\pi/\Lambda)z} \qquad (8.1.4a)$$

$$A_m = \frac{1}{\Lambda} \int_0^{\Lambda} dz \left(-e^{-ik_{x0}h(z)} \right) e^{-im(2\pi/\Lambda)z} \qquad (8.1.4b)$$

where we have made use of the Fourier series expansion, since $e^{-ik_{x0}h(z)}$ is a periodic function with a period Λ. In the region above the grating, $E_R(x, z)$ satisfies the wave equation and should have solutions of the form $\exp(ik_x x + ik_z z)$ as long as $k_x^2 + k_z^2 = \omega^2 \mu \varepsilon$. Therefore, we see that a general solution for $E_R(x, z)$ will be

$$E_R(x, z) = E_0 \sum_{m=-\infty}^{\infty} R_m e^{ik_{xm}x + ik_{zm}z} \qquad (8.1.5)$$

where the horizontal wave numbers are

$$k_{zm} = k_{z0} + m\frac{2\pi}{\Lambda} \qquad (8.1.6)$$

and their corresponding vertical components can be obtained from the dispersion relation

$$k_{xm} = \sqrt{\omega^2 \mu \varepsilon - k_{zm}^2} \qquad (8.1.7)$$

The horizontal wave number k_{zm} for a particular m differs from its nearby value by $2\pi/\Lambda$. We can generalize the above discussions to the case of a penetrable dielectric grating. A phase-matching diagram is shown in Fig. 8.4. These plane waves for the reflected and transmitted fields are called the space harmonics. It can be seen that for a particular m such that $k_{zm} > \omega\sqrt{\mu\varepsilon}$, k_{xm} becomes purely imaginary and we choose the imaginary part of k_{xm} to be positive such that it decays away from the surface. In the transmission region with a refractive index $n_1 = \sqrt{\mu_1 \varepsilon_1 / (\mu_0 \varepsilon_0)}$, the transmitted wave vectors $\mathbf{k}_m^t$ will have the same horizontal wave numbers k_{zm} as

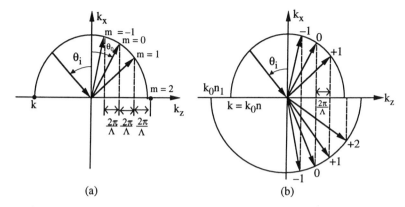

Figure 8.4. Phase-matching diagram for a plane wave (a) reflection and (b) both reflection and transmission from a grating structure with a period Λ.

those of the reflected waves (8.1.6):

$$\mathbf{k}'_m = \hat{x} k'_{xm} + \hat{z} k'_{zm} \tag{8.1.8a}$$

$$k'_{zm} = k_{zm} = k_{z0} + m \frac{2\pi}{\Lambda} \tag{8.1.8b}$$

$$k'_{xm} = \sqrt{\omega^2 \mu_1 \varepsilon_1 - k^2_{zm}} \tag{8.1.8c}$$

For a general theory to find the reflection and transmission coefficients of the space harmonics, see Refs. 4 and 5.

Phase Matching Condition for Grating Couplers. A dielectric grating can therefore be used as a waveguide input or output coupler, as shown in Fig. 8.5. Suppose we have a dielectric slab waveguide with a propagation constant β for the fundamental TE$_0$ mode. If the incident angle θ_i and the grating period Λ are chosen such that

$$k_0 n_0 \sin \theta_i + m \frac{2\pi}{\Lambda} = \beta \tag{8.1.9}$$

for a particular integer m, then the incident wave will be coupled to the guided TE$_0$ mode (Fig. 8.5b). The coupling strength depends on the incident beam profile and its distance D from the edge of the grating, as shown in Fig. 8.5a. If D is too long, the incident power coupled into the waveguide with the grating section can leak back into the incident region. If D is too short, half of the beam profile will be reflected from the planar interface in the section $z > 0$, and the coupling into the guided mode in the region $z > 0$

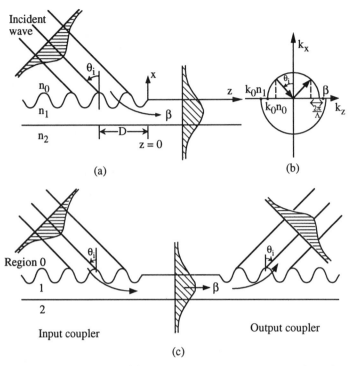

Figure 8.5. (a) A grating coupler. (b) Phase-matching diagram for $\beta = k_0 n_0 \sin \theta_i + 2\pi/\Lambda$. (c) Grating input and output couplers.

will be small. We can use the grating as an output coupler based on the reciprocity concept. The guided wave in the center waveguide section in Fig. 8.5c can radiate back into region 0 in the right grating section while propagating along the waveguide. The exit angle can be determined using (8.1.9).

8.2   COUPLING OF MODES IN THE TIME DOMAIN [6]

Consider first a resonator consisting of an inductor and a capacitor, as shown in Fig. 8.6a. The differential equations for the voltage $v(t)$ and the current $i(t)$ are

$$v(t) = L \frac{di(t)}{dt}$$

$$i(t) = -C \frac{dv(t)}{dt} \tag{8.2.1}$$

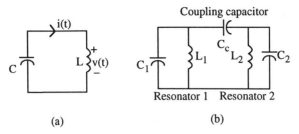

Figure 8.6.   (a) An LC resonator. (b) Two coupled resonators.

or in terms of only $v(t)$ (or $i(t)$)

$$\frac{d^2}{dt^2}v(t) + \frac{1}{LC}v(t) = 0 \qquad (8.2.2)$$

Their solutions are

$$v(t) = V_0\cos(\omega_0 t + \phi)$$

$$i(t) = \sqrt{\frac{C}{L}}\,V_0\sin(\omega_0 t + \phi) \qquad (8.2.3)$$

where the resonance frequency is $\omega_0 = 1/\sqrt{LC}$. If we define the two new variables $a_+(t)$ and $a_-(t)$

$$a_\pm(t) = \sqrt{\frac{C}{2}}\left(v(t) \pm j\sqrt{\frac{L}{C}}\,i(t)\right) = \sqrt{\frac{C}{2}}\,V_0\,e^{\pm j(\omega_0 t + \phi)} \qquad (8.2.4)$$

we see immediately that

$$\frac{d}{dt}a_+(t) = j\omega_0 a_+(t) \qquad \frac{d}{dt}a_-(t) = -j\omega_0 a_-(t) \qquad (8.2.5)$$

Therefore, we have two independent equations. The squares of the magnitudes are

$$|a_\pm(t)|^2 = \frac{C}{2}V_0^2$$

$$= \left\langle\frac{C}{2}v^2(t)\right\rangle + \left\langle\frac{L}{2}i^2(t)\right\rangle \qquad (8.2.6)$$

which is the total time-averaged energy stored in the circuit. Since the two variables $a_+(t)$ and $a_-(t)$ are decoupled, we may choose $a(t) = a_+(t)$ to describe the resonator.

8.2.1 Energy Conservation Condition

Let us consider two resonators, 1 and 2, coupled through a capacitor as shown in Fig. 8.6b. Without the coupling capacitor, each variable $a_1(t)$ or $a_2(t)$ satisfies the differential equation of the form (8.2.5) with a resonant frequency ω_1 or ω_2, respectively. With the coupling capacitor, they satisfy

$$\frac{da_1}{dt} = j\omega_1 a_1 + jK_{12}a_2 \qquad (8.2.7a)$$

$$\frac{da_2}{dt} = j\omega_2 a_2 + jK_{21}a_1 \qquad (8.2.7b)$$

This assumes that the coupling is weak; thus the coupling coefficients $|K_{12}| \ll \omega_1$ and $|K_{21}| \ll \omega_2$, and the energy stored in the coupling capacitor is negligible. A similar coupling system, a coupled pendulum system, is shown in Fig. 8.7.

Therefore, from energy conservation in which the total energy in two resonators should stay as a constant (neglecting the energy storage in the weak coupling capacitor),

$$\frac{d}{dt}\left(|a_1|^2 + |a_2|^2\right) = 0 \qquad (8.2.8)$$

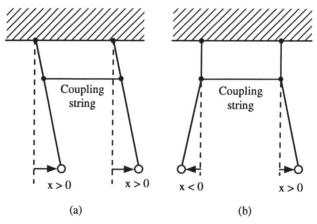

Figure 8.7. A coupled pendulum system: (a) in-phase mode with a resonant frequency ω_-; (b) out-of-phase mode with a resonant frequency ω_+.

we find that

$$a_1 a_2^* j(K_{21} - K_{12}^*) + a_1^* a_2 j(K_{12} - K_{21}^*) = 0 \qquad (8.2.9)$$

Since a_1 and a_2 are arbitrary (defined by the energy and phase in each resonator), we conclude that the two coupling coefficients must be related by

$$K_{12} = K_{21}^* \qquad (8.2.10)$$

8.2.2 Eigenstate Solutions

To find the eigenstate solution of the coupled resonator, we assume the solution to be

$$a_1(t) = A_1 e^{j\omega t} \qquad a_2(t) = A_2 e^{j\omega t} \qquad (8.2.11)$$

where ω is the new natural resonant frequency of the entire coupled system. Substituting (8.2.11) into (8.2.7), we obtain

$$(\omega - \omega_1)A_1 - K_{12}A_2 = 0$$
$$- K_{21}A_1 + (\omega - \omega_2)A_2 = 0 \qquad (8.2.12)$$

Since the determinant of the above linear equations should vanish for nontrivial solutions for A_1 and A_2, we find

$$(\omega - \omega_1)(\omega - \omega_2) - K_{12}K_{21} = 0 \qquad (8.2.13)$$

The second-order polynomial equation has two solutions:

$$\omega_\pm = \frac{\omega_1 + \omega_2}{2} \pm \sqrt{\left(\frac{\omega_1 - \omega_2}{2}\right)^2 + |K_{12}|^2} \qquad (8.2.14)$$

where $K_{21} = K_{12}^*$ has been used. Let

$$\phi = \frac{\omega_1 + \omega_2}{2} \qquad \Omega = \sqrt{\Delta^2 + |K_{12}|^2} \qquad \Delta = \frac{\omega_2 - \omega_1}{2} \qquad (8.2.15)$$

The energy splittings between ω_+ and ω_- for $\omega_1 \neq \omega_2$ and $\omega_1 = \omega_2$ are plotted in Figs. 8.8a and b. The ratio of the two amplitudes A_1 and A_2 satisfies

$$\frac{A_1}{A_2} = \frac{K_{12}}{\omega - \omega_1} = \frac{\omega - \omega_2}{K_{21}} \qquad (8.2.16)$$

where $\omega = \omega_+$ or ω_-. Thus, the solutions $a_1(t)$ and $a_2(t)$ can be represented

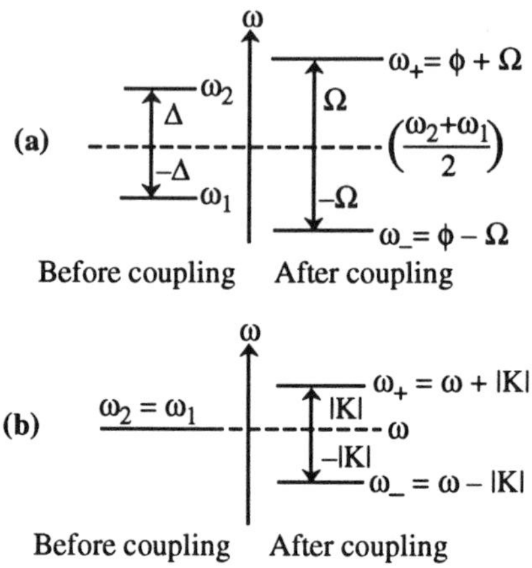

Figure 8.8. The frequency splitting in (a) asynchronous coupling and (b) synchronous coupling.

as a summation of the two eigensolutions. Each solution is called a system mode with a corresponding resonant frequency. We obtain the ratios of the two amplitudes for both system modes:

$$\text{For } \omega = \omega_+ \qquad \frac{A_1^+}{A_2^+} = \frac{K_{12}}{\Delta + \Omega} = \frac{-\Delta + \Omega}{K_{21}} \qquad (8.2.17a)$$

$$\text{For } \omega = \omega_- \qquad \frac{A_1^-}{A_2^-} = \frac{K_{12}}{\Delta - \Omega} = \frac{-\Delta - \Omega}{K_{21}} \qquad (8.2.17b)$$

which determine the eigenvectors $\begin{bmatrix} A_1^+ \\ A_2^+ \end{bmatrix}$ and $\begin{bmatrix} A_1^- \\ A_2^- \end{bmatrix}$. We can write the general solution as

$$\begin{bmatrix} a_1(t) \\ a_2(t) \end{bmatrix} = c_1 \begin{bmatrix} K_{12} \\ \Delta + \Omega \end{bmatrix} e^{j\omega_+ t} + c_2 \begin{bmatrix} K_{12} \\ \Delta - \Omega \end{bmatrix} e^{j\omega_- t} \qquad (8.2.18)$$

which also implies that the total solution consists of the beating of the two natural modes ω_+ and ω_-. Suppose initially that $a_1(t = 0) = a_1(0)$, and $a_2(t = 0) = a_2(0)$. We can find c_1 and c_2 in terms of $a_1(0)$ and $a_2(0)$ from (8.2.18), $a_1(0) = (c_1 + c_2) K_{12}$, $a_2(0) = c_1(\Delta + \Omega) + c_2(\Delta - \Omega)$, and obtain

$$\begin{bmatrix} a_1(t) \\ a_2(t) \end{bmatrix} = e^{j\phi t} \begin{bmatrix} \cos \Omega t - j\dfrac{\Delta}{\Omega} \sin \Omega t & j\dfrac{K_{12}}{\Omega} \sin \Omega t \\ j\dfrac{K_{21}}{\Omega} \sin \Omega t & \cos \Omega t + j\dfrac{\Delta}{\Omega} \sin \Omega t \end{bmatrix} \begin{bmatrix} a_1(0) \\ a_2(0) \end{bmatrix}$$

$$(8.2.19)$$

If initially $a_2(0) = 0$, we have

$$a_1(t) = a_1(0)\left[\cos \Omega t - j\frac{\Delta}{\Omega}\sin \Omega t\right]e^{j\phi t}$$

$$a_2(t) = a_1(0)\frac{jK_{21}}{\Omega}\sin \Omega t \, e^{j\phi t} \qquad (8.2.20)$$

Let us consider two cases, assuming that $|a_1(0)|^2 = 1$ for simplicity.

1. *Asynchronous coupling* $(\omega_1 \neq \omega_2)$. Equation (8.2.20) gives

$$|a_2(t)|^2 = \left|\frac{K_{21}}{\Omega}\right|^2 \sin^2 \Omega t \qquad (8.2.21)$$

and $|a_1(t)|^2 = 1 - |a_2(t)|^2$. Since $|K_{21}/\Omega| < 1$, we find the peak energy transferred to resonator 2 is $|K_{21}/\Omega|^2$, which means that the energy transfer from resonator 1 to 2 is never 100%. The energies $|a_1(t)|^2$ and $|a_2(t)|^2$ are plotted in Fig. 8.9a. We see that the energy transfers periodically between two resonators with a period $T = \pi/\Omega = 2\pi/(\omega_+ - \omega_-)$. The "beat" angular frequency $\omega_B = \omega_+ - \omega_- = 2\Omega$ determines the energy exchange rate. This frequency splitting, as shown in Fig. 8.8a, is larger than the original energy difference before coupling, $\omega_2 - \omega_1 = 2\Delta$. In other words, the two frequencies appear to "repel" each other in the coupled resonator system.

2. *Synchronous coupling* $(\omega_1 = \omega_2)$. We find $\Delta = 0$, $\Omega = |K_{12}|(\equiv K)$, and

$$|a_1(t)|^2 = \cos^2 Kt \qquad |a_2(t)|^2 = \sin^2 Kt \qquad (8.2.22)$$

Therefore, at $t = (n + \frac{1}{2})\pi/K$, where n is an integer, the energy is completely transferred from resonator 1 to 2. At $t = n\pi/K$, the energy is transferred back to resonator 1. The energies $|a_1(t)|^2$ and $|a_2(t)|^2$ are shown in Fig. 8.9b. The time constant $T = \pi/K$ is called one beat period, and the beat angular frequency is $\omega_B = \omega_+ - \omega_- = 2K$, as shown in Fig. 8.8b. The frequency splits from a degenerate system $(\omega_1 = \omega_2)$ into a nondegenerate system with two system resonant frequencies $(\omega_+ \neq \omega_-)$. The energy splitting is $2|K|$, twice the magnitude of the coupling coefficient.

One way to determine the phase of K_{12} is to consider the two system modes. Assume $K_{12} < 0$:

For $\omega = \omega_+$ $\qquad \dfrac{A_1^+}{A_2^+} = \dfrac{K_{12}}{|K_{12}|} = -1$ (out-of-phase mode)

For $\omega = \omega_-$ $\qquad \dfrac{A_1^-}{A_2^-} = \dfrac{K_{12}}{-|K_{12}|} = +1$ (in-phase mode)

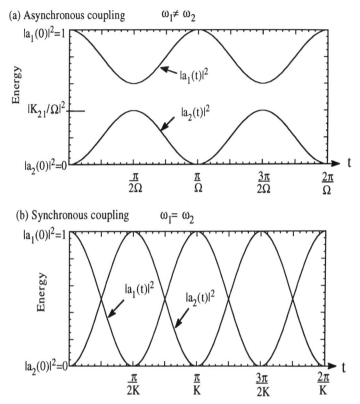

Figure 8.9. The energies in resonators 1 and 2 as functions of time, assuming that at $t = 0$ only resonator 1 is excited. (a) Asynchronous coupling ($\omega_1 \neq \omega_2$). The energy transfer from resonator 1 to 2 is incomplete. (b) Synchronous coupling ($\omega_1 = \omega_2$). The energy transfer from resonator 1 to 2 can be complete at $t = \pi/(2K)$, $3\pi/(2K)$, and so on.

We find experimentally that for the in-phase mode as shown in Fig. 8.7a the resonant frequency is smaller than that of the out-of-phase mode. Therefore, the assumption $K_{12} < 0$ is true. An analogous system to the coupled resonators is a coupled quantum-well system. The in-phase mode for two coupled quantum wells has a lower eigenenergy than that of the out-of-phase mode, as shown in Ref. 7.

8.3 COUPLED OPTICAL WAVEGUIDES [6, 8–32]

The coupled-mode theory is also useful in understanding the coupling mechanisms in parallel waveguides, as shown in Fig. 8.10. The guidance is along the z direction. Waveguides a and b can be two rectangular or slab dielectric guides. For a guided mode in the $+z$ direction in a single

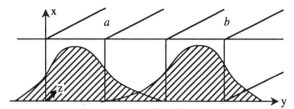

Figure 8.10. Two coupled optical waveguides a and b.

waveguide a, the electric and magnetic fields in the frequency domain (Chapter 5) can be expressed as

$$\mathbf{E}(x, y, z) = \mathbf{E}^{(a)}(x, y)a(z) \tag{8.3.1a}$$

$$\mathbf{H}(x, y, z) = \mathbf{H}^{(a)}(x, y)a(z) \tag{8.3.1b}$$

$$a(z) = a_0 e^{i\beta_a z} \tag{8.3.1c}$$

where $\mathbf{E}^{(a)}(x, y)$ and $\mathbf{H}^{(a)}(x, y)$ are the modal distributions in the xy plane, which satisfy the normalization condition:

$$\frac{1}{2}\mathrm{Re} \int\int \mathbf{E}^{(a)}(x, y) \times \mathbf{H}^{(a)*}(x, y) \cdot \hat{z}\, dx\, dy = 1 \tag{8.3.2}$$

Note that the time conversion $\exp(-i\omega t)$ in Maxwell's equations has been adopted. We also see that

$$\frac{d}{dz}a(z) = i\beta_a a(z) \tag{8.3.3}$$

The total guided power for the fields in (8.3.1) and (8.3.2) is

$$P = \frac{1}{2}\mathrm{Re} \int\int \mathbf{E}(x, y, z) \times \mathbf{H}^*(x, y, z) \cdot \hat{z}\, dx\, dy$$

$$= |a(z)|^2 \tag{8.3.4}$$

If the mode is guided in the $-z$ direction, then $a(z) = a_0 \exp(-i\beta_a z)$:

$$\frac{d}{dz}a(z) = -i\beta_a a(z) \tag{8.3.5}$$

and the Poynting vector $\mathbf{E} \times \mathbf{H}^*$ will be in the $-\hat{z}$ direction. The relations between the field vectors $(\mathbf{E}^+, \mathbf{H}^+)$ of the forward propagation mode and those of the backward propagation mode are shown in Appendix H. If we still define the power to be the $+\hat{z}$ component of the guided power, we then obtain

$$P = -|a(z)|^2 \tag{8.3.6}$$

for the guided power in the $-z$ direction for (8.3.5).

Consider two parallel waveguides, a and b, for which the total field solutions can be written as linear combinations of the individual waveguide modes:

$$\mathbf{E}(x, y, z) = a(z)\mathbf{E}^{(a)}(x, y) + b(z)\mathbf{E}^{(b)}(x, y)$$
$$\mathbf{H}(x, y, z) = a(z)\mathbf{H}^{(a)}(x, y) + b(z)\mathbf{H}^{(b)}(x, y) \tag{8.3.7}$$

The amplitudes $a(z)$ and $b(z)$ satisfy

$$\frac{d}{dz}a = i\beta_a a + iK_{ab}b$$

$$\frac{d}{dz}b = iK_{ba}a + i\beta_b b \tag{8.3.8}$$

where K_{ab} and K_{ba} are the coupling coefficients. The expressions for K_{ab} and K_{ba} will be derived in Appendix H. The total guided power is

$$P = \frac{1}{2}\text{Re}\int\int\mathbf{E}(x, y, z) \times \mathbf{H}^*(x, y, z) \cdot \hat{z}\,dx\,dy$$

$$= s_a|a(z)|^2 + s_b|b(z)|^2 + \text{Re}[a(z)b^*(z)C_{ba} + b(z)a^*(z)C_{ab}] \tag{8.3.9}$$

where the cross overlap integrals

$$C_{pq} = \frac{1}{2}\int\int_{-\infty}^{\infty}\mathbf{E}^{(q)}(x, y) \times \mathbf{H}^{(p)*}(x, y) \cdot \hat{z}\,dx\,dy \tag{8.3.10}$$

and $s_a, s_b = +1$ for propagation in the $+z$ direction, and -1 for propagation in the $-z$ direction.

If we assume that the coupling of the two waveguide modes is very weak so that the cross overlap integrals C_{ab} and C_{ba} are negligible, the total power is then

$$P = s_a|a(z)|^2 + s_b|b(z)|^2 \tag{8.3.11}$$

If the system is lossless, power conservation requires

$$\frac{dP}{dz} = 0 \tag{8.3.12}$$

Substituting (8.3.11) into (8.3.12) and using the coupled-mode equations (8.3.8), we find

$$K_{ab} = K_{ba}^* \qquad \text{if } s_a s_b > 0 \quad \text{(codirectional coupling)}$$

$$K_{ab} = -K_{ba}^* \qquad \text{if } s_a s_b < 0 \quad \text{(contradirectional coupling)} \quad (8.3.13)$$

The coupled-mode equations can be expressed in a matrix form

$$\frac{d}{dz}\begin{bmatrix} a \\ b \end{bmatrix} = i\overline{\overline{M}}\begin{bmatrix} a \\ b \end{bmatrix} \tag{8.3.14a}$$

where

$$\overline{\overline{M}} = \begin{bmatrix} \beta_a & K_{ab} \\ K_{ba} & \beta_b \end{bmatrix} \tag{8.3.14b}$$

8.3.1 Eigenstate Solutions

The solution can be obtained by substituting

$$\begin{bmatrix} a(z) \\ b(z) \end{bmatrix} = \begin{bmatrix} A \\ B \end{bmatrix} e^{i\beta z} \tag{8.3.15}$$

into (8.3.14). We find

$$\left[\overline{\overline{M}} - \beta \overline{\overline{I}}\right]\begin{bmatrix} A \\ B \end{bmatrix} = 0$$

where $\overline{\overline{I}}$ is an identity matrix, or equivalently,

$$\begin{bmatrix} \beta_a - \beta & K_{ab} \\ K_{ba} & \beta_b - \beta \end{bmatrix}\begin{bmatrix} A \\ B \end{bmatrix} = 0 \tag{8.3.16}$$

For nontrivial solutions to (8.3.16), the determinant must vanish:

$$\det|\overline{\overline{M}} - \beta\overline{\overline{I}}| = (\beta_a - \beta)(\beta_b - \beta) - K_{ab}K_{ba} = 0 \tag{8.3.17}$$

There are two eigenvalues for β:

$$\beta = \frac{\beta_a + \beta_b}{2} \pm \psi \tag{8.3.18}$$

where

$$\psi = \sqrt{\Delta^2 + K_{ab}K_{ba}} \tag{8.3.19}$$

$$\Delta = \frac{\beta_b - \beta_a}{2} \tag{8.3.20}$$

For codirectional coupling, $K_{ba} = K_{ab}^*$, ψ is real for a lossless system. As an example, we assume two arbitrary dispersion relations

$$\beta_a \equiv \beta_a(\omega) \qquad \beta_b \equiv \beta_b(\omega)$$

The $\beta_a(\omega)$ and $\beta_b(\omega)$ vs. ω curves are shown as dashed lines in Fig. 8.11a. Both curves have positive slopes and intercept each other at (ω_0, β_0). The eigensolutions β_+ and β_- are the solid curves. At $\omega = \omega_0$, $\beta_a = \beta_b$, $\psi = |K_{ab}|(\equiv K)$, and the splitting $(\beta_+ - \beta_-)$ is $2K$.

Since

$$\frac{A}{B} = \frac{K_{ab}}{\beta - \beta_a} = \frac{\beta - \beta_b}{K_{ba}} \tag{8.3.21}$$

we find if $K_{ab} > 0$ (the expressions for K_{ab} and K_{ba} will be derived in Appendix H):

$$\text{For } \beta = \beta_+ \qquad \frac{A^+}{B^+} = \frac{K_{ab}}{\Delta + \psi} = \frac{-\Delta + \psi}{K_{ba}} > 0 \tag{8.3.22}$$

$$\text{For } \beta = \beta_- \qquad \frac{A^-}{B^-} = \frac{K_{ab}}{\Delta - \psi} = \frac{-\Delta - \psi}{K_{ba}} < 0 \tag{8.3.23}$$

Furthermore, if the two guides have the same propagation constants, $\beta_a = \beta_b$, we find (for real K_{ab} and codirectional coupling)

$$\frac{A^+}{B^+} = 1 \qquad \text{for } \beta = \beta_+ \tag{8.3.24}$$

$$\frac{A^-}{B^-} = -1 \qquad \text{for } \beta = \beta_- \tag{8.3.25}$$

We thus have two possible system modes, the β_+ (in-phase) mode and the β_- (out-of-phase) mode, which are plotted in Fig. 8.12.

For contradirectional coupling (Fig. 8.11b), $\psi = \sqrt{\Delta^2 - |K_{ab}|^2}$ the propagation constants β_+ and β_- are complex in the region:

$$\left| \frac{\beta_a - \beta_b}{2} \right| < |K_{ab}|$$

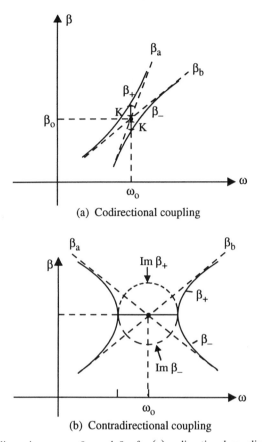

(a) Codirectional coupling

(b) Contradirectional coupling

Figure 8.11.   The dispersion curves β_+ and β_- for (a) codirectional coupling and (b) contradirectional coupling.

The imaginary parts Im β_+ and Im β_- are shown as the dashed semicircles. We also see a stop band with its bandwidth $\Delta\omega$ determined by $|(\beta_a(\omega) - \beta_b(\omega))/2| = |K_{ab}|$.

8.3.2   General Solution of the Coupled Waveguides

After finding the eigenvectors $\mathbf{v}_1$ and $\mathbf{v}_2$, where

$$\mathbf{v}_1 = \begin{bmatrix} K_{ab} \\ \Delta + \psi \end{bmatrix} \text{ or } \begin{bmatrix} -\Delta + \psi \\ K_{ba} \end{bmatrix} \quad \text{for } \beta_+ = \frac{\beta_a + \beta_b}{2} + \psi \quad (8.3.26a)$$

$$\mathbf{v}_2 = \begin{bmatrix} K_{ab} \\ \Delta - \psi \end{bmatrix} \text{ or } \begin{bmatrix} -\Delta - \psi \\ K_{ba} \end{bmatrix} \quad \text{for } \beta_- = \frac{\beta_a + \beta_b}{2} - \psi \quad (8.3.26b)$$

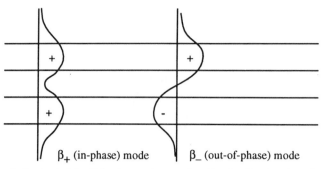

Figure 8.12. A simple sketch of the two system modes: the β_+ (in-phase) mode and the β_- (out-of-phase) mode.

the solution for $a(z)$ and $b(z)$, given the initial condition $a(0)$ and $b(0)$, can be obtained in the same way as (8.2.19). It can also be derived using the eigenmatrix $\overline{\overline{\mathbf{V}}}$, where its columns are the eigenvectors of the matrix $\overline{\overline{\mathbf{M}}}$:

$$\overline{\overline{\mathbf{V}}} = [\mathbf{v}_1 \vdots \mathbf{v}_2] \tag{8.3.27}$$

The general solution is

$$\begin{bmatrix} a(z) \\ b(z) \end{bmatrix} = \overline{\overline{\mathbf{V}}} \begin{bmatrix} e^{i\beta_+ z} & 0 \\ 0 & e^{i\beta_- z} \end{bmatrix} \overline{\overline{\mathbf{V}}}^{-1} \begin{bmatrix} a(0) \\ b(0) \end{bmatrix} \tag{8.3.28}$$

The derivation is straightforward. The general solution is a linear combination of the eigensolutions (see (8.2.18) also):

$$\begin{aligned} \begin{bmatrix} a(z) \\ b(z) \end{bmatrix} &= c_1 \begin{bmatrix} K_{ab} \\ \Delta + \Omega \end{bmatrix} e^{i\beta_+ z} + c_2 \begin{bmatrix} K_{ab} \\ \Delta - \Omega \end{bmatrix} e^{i\beta_- z} \\ &= [\mathbf{v}_1 \vdots \mathbf{v}_2] \begin{bmatrix} e^{i\beta_+ z} & 0 \\ 0 & e^{i\beta_- z} \end{bmatrix} \begin{bmatrix} c_1 \\ c_2 \end{bmatrix} \end{aligned} \tag{8.3.29}$$

Therefore, at $z = 0$, we find

$$\begin{bmatrix} c_1 \\ c_2 \end{bmatrix} = \overline{\overline{\mathbf{V}}}^{-1} \begin{bmatrix} a(0) \\ b(0) \end{bmatrix} \tag{8.3.30}$$

Substituting (8.3.30) into (8.3.29), we obtain the solution (8.3.28). The solutions can also be expressed in terms of a matrix $\overline{\overline{\mathbf{S}}}(z)$ and the initial conditions:

$$\begin{bmatrix} a(z) \\ b(z) \end{bmatrix} = \overline{\overline{S}}(z) \begin{bmatrix} a(0) \\ b(0) \end{bmatrix} \qquad (8.3.31)$$

where

$$\overline{\overline{S}}(z) = \overline{\overline{V}} \begin{bmatrix} e^{i\beta_{+}z} & 0 \\ 0 & e^{i\beta_{-}z} \end{bmatrix} \overline{\overline{V}}^{-1}$$

$$= \begin{bmatrix} \cos \psi z - i\dfrac{\Delta}{\psi}\sin \psi z & i\dfrac{K_{ab}}{\psi}\sin \psi z \\[3mm] i\dfrac{K_{ba}}{\psi}\sin \psi z & \cos \psi z + i\dfrac{\Delta}{\psi}\sin \psi z \end{bmatrix} e^{i(\beta_{a}+\beta_{b})z/2} \qquad (8.3.32)$$

If at $z = 0$, the optical power is incident only in waveguide 1, ($a(0) = 1$, $b(0) = 0$), we find

$$|b(z)|^2 = \left| \frac{K_{ba}}{\psi} \right|^2 \sin^2 \psi z \qquad (8.3.33)$$

Therefore, at $\psi z = \pi/2, 3\pi/2, \ldots, (2n + 1)\pi/2 \,(n = \text{an integer})$, the power transfer from guide a to guide b is maximum. Since

$$\left| \frac{K_{ba}}{\psi} \right|^2 = \frac{|K_{ba}|^2}{[(\beta_b - \beta_a)/2]^2 + |K_{ba}|^2} < 1 \qquad (8.3.34)$$

for $\beta_a \neq \beta_b$, the power transfer is never complete (Fig. 8.13a). The minimum distance, $L_B = \pi/\psi$, at which the output power reappears in the same guide as the input is called a beat length. If $\beta_a = \beta_b$, we have $\psi = |K_{ab}|$, and the solutions are

$$a(z) = \cos Kz \, e^{i\beta z} \qquad b(z) = i \sin Kz \, e^{i\beta z} \qquad (8.3.35)$$

where $K = K_{ab} = K_{ba}$, $\beta = \beta_a = \beta_b$ have been used. Complete power transfer occurs for synchronous coupling (Fig. 8.13b). For $K\ell = (2n + 1)\pi/2$, complete power transfer occurs; this is called a cross $\otimes$ state. For $K\ell = n\pi$, there is no power transfer from guide a to guide b, this is called a parallel $\ominus$ state.

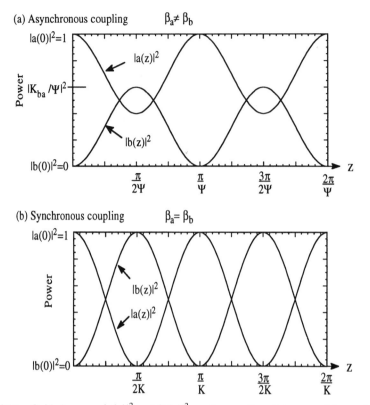

Figure 8.13. Guided powers $|a(z)|^2$ and $|b(z)|^2$ vs. the coupling distance z: (a) asynchronous coupling $\beta_a \neq \beta_b$, (b) synchronous coupling $\beta_a = \beta_b$.

8.4 IMPROVED COUPLED-MODE THEORY AND ITS APPLICATIONS [18–32]

A detailed formulation of the improved-coupled mode theory to account for either asymmetry ($K_{ab} \neq K_{ba}$) or the finite overlap integrals C_{ab} and C_{ba} is shown in Appendix H. The major results are that we have the same form in the coupled-mode equations:

$$\frac{d}{dz}a(z) = i\gamma_a a(z) + ik_{ab}b(z) \tag{8.4.1a}$$

$$\frac{d}{dz}b(z) = ik_{ba}a(z) + i\gamma_b b(z) \tag{8.4.1b}$$

where the coefficients γ_a, γ_b, k_{ab}, and k_{ba} are all defined in (H.30). The transverse fields have been assumed to be

$$\mathbf{E}_t = a(z)\mathbf{E}_t^{(a)}(x, y) + b(z)\mathbf{E}_t^{(b)}(x, y) \tag{8.4.2a}$$

$$\mathbf{H}_t = a(z)\mathbf{H}_t^{(a)}(x, y) + b(z)\mathbf{H}_t^{(b)}(x, y) \tag{8.4.2b}$$

and all the transverse field components are real for a lossless system, we find γ_a, γ_b, k_{ab}, and k_{ba} to be real. Power conservation leads to

$$
\begin{aligned}
0 &= \frac{d}{dz}P(z) = \frac{d}{dz}\frac{1}{2}\mathrm{Re}\int \mathbf{E}_t \times \mathbf{H}_t^* \cdot \hat{z}\,dx\,dy \\
&= \frac{d}{dz}\mathrm{Re}[aa^* + ab^*C_{ba} + ba^*C_{ab} + bb^*] \\
&= \frac{d}{dz}[aa^* + (ab^* + ba^*)C + bb^*] \\
&= ab^*\mathrm{i}(\gamma_a C - \gamma_b C + k_{ba} - k_{ab}) \\
&\quad + ba^*\mathrm{i}(\gamma_b C - \gamma_a C + k_{ab} - k_{ba})
\end{aligned}
\tag{8.4.3}
$$

Here the overlap integrals C_{ab}, C_{ba}, and $C = (C_{ab} + C_{ba})/2$ have all been defined in (H.11). Since both a and b are arbitrary, we conclude that the coefficients in front of ab^* and ba^* must vanish. Therefore,

$$k_{ab} - k_{ba} = (\gamma_a - \gamma_b)C \tag{8.4.4}$$

which is identical to the *exact* relation (H.31). What is important is that when the overlap coefficient C is not small, the total guided power should be given by

$$p(z) = |a|^2 + |b|^2 + (ab^* + ba^*)C \tag{8.4.5}$$

where the last two terms due to the cross powers between $\mathbf{E}^{(a)}$ and $\mathbf{H}^{(b)}$ or $\mathbf{E}^{(b)}$ and $\mathbf{H}^{(a)}$ are not negligible anymore.

8.4.1   Applications of the Improved Coupled-Mode Theory [23, 24, 29]

To illustrate the applications of the improved coupled-mode theory, we consider two parallel waveguides that are not necessarily identical. We define the asynchronism factor δ in terms of the parameters γ_a, γ_b, k_{ab}, and k_{ba}:

$$\delta = \frac{\gamma_b - \gamma_a}{2\sqrt{k_{ab}k_{ba}}} \tag{8.4.6}$$

For the initial excitation at $z = 0$ of a two-coupled waveguide, $a(0) = 1$, $b(0) = 0$, we obtain from (8.3.31) and (8.3.32)

$$a(\ell) = \left(\cos \psi\ell - i\frac{\Delta}{\psi}\sin \psi\ell\right)e^{i\phi\ell} \qquad (8.4.7a)$$

$$b(\ell) = \frac{ik_{ba}}{\psi}\sin \psi\ell \; e^{i\phi\ell} \qquad (8.4.7b)$$

where

$$\phi = \frac{\gamma_b + \gamma_a}{2} \qquad (8.4.7c)$$

$$\psi = \sqrt{\Delta^2 + k_{ab}k_{ba}} \qquad (8.4.7d)$$

$$\Delta = \frac{\gamma_b - \gamma_a}{2} \qquad (8.4.7e)$$

We can show from the exact relation (8.4.4)

$$\delta = \frac{1}{2C}\left(\sqrt{\frac{k_{ba}}{k_{ab}}} - \sqrt{\frac{k_{ab}}{k_{ba}}}\right) \qquad (8.4.8)$$

or

$$\sqrt{\frac{k_{ba}}{k_{ab}}} = C\delta + \sqrt{1 + C^2\delta^2} = e^{\sinh^{-1}C\delta} \qquad (8.4.9)$$

The solutions (8.4.7a) and (8.4.7b) can be written as

$$a(\ell) = \left\{\cos\left[\sqrt{k_{ab}k_{ba}}\,\ell\left(1 + \delta^2\right)^{1/2}\right]\right.$$

$$\left. - i\frac{\delta}{\left(1 + \delta^2\right)^{1/2}}\sin\left[\sqrt{k_{ab}k_{ba}}\,\ell\left(1 + \delta^2\right)^{1/2}\right]\right\}e^{i\phi\ell} \qquad (8.4.10)$$

$$b(\ell) = i\frac{e^{\sinh^{-1}C\delta}}{\sqrt{1 + \delta^2}}\sin\left[\sqrt{k_{ab}k_{ba}}\,\ell\left(1 + \delta^2\right)^{1/2}\right]e^{i\phi\ell} \qquad (8.4.11)$$

The output power P_a in waveguide a when waveguide b terminates at $z = \ell$

is obtained using

$$E_t(x, y, z = \ell) \simeq a(\ell)E_t^{(a)}(x, y) + b(\ell)E_t^{(b)}(x, y) \quad (8.4.12a)$$

$$= \sum_{n=1}^{\infty} u_n^{(a)}E_t^{(a)n}(x, y) \quad (8.4.12b)$$

$$H_t(x, y, z = \ell) \simeq a(\ell)H_t^{(a)}(x, y) + b(\ell)H_t^{(b)}(x, y) \quad (8.4.13a)$$

$$= \sum_{n=1}^{\infty} v_n^{(a)}H_t^{(a)n}(x, y) \quad (8.4.13b)$$

where the expansion in (8.4.12a) or (8.4.13a) is in terms of individual waveguide modes, and in (8.4.12b) or (8.4.13b) is in terms of all the guided and radiation modes of waveguide a alone since they form a complete set. Here $n = 1$ refers to the fundamental guided mode $E_t^{(a)}$ and $H_t^{(a)}$ used in (8.4.12a) and (8.4.13a). Cross-multiplying (8.4.12) by $H_t^{(a)}$ and integrating over the cross section, one obtains [23, 29]

$$u_1^{(a)} = a(\ell) + C_{ab}b(\ell) \quad (8.4.14a)$$

Similarly, one finds

$$v_1^{(a)} = a(\ell) + C_{ba}b(\ell) \quad (8.4.14b)$$

where the overlap integrals, C_{ab} and C_{ba}, are defined in (8.3.10). The guided power due to the first mode β_1 in waveguide a is

$$P_a = \frac{1}{2}\text{Re}\left(u_1^{(a)}v_1^{(a)*}\frac{1}{2}\int\int E_t^{(a)1} \times H_t^{(a)1} \cdot \hat{z}\,dx\,dy\right)$$

$$= 1 - \left(\frac{1 - C_{ab}C_{ba}}{1 + \delta^2}\right)e^{2\sinh^{-1}C\delta}\sin^2\left[\sqrt{k_{ab}k_{ba}}\ell\left(1 + \delta^2\right)^{1/2}\right] \quad (8.4.15)$$

A similar procedure for the output power in waveguide b when waveguide a is terminated at $z = \ell$ leads to

$$P_b = \text{Re}\left[(C_{ba}a + b)(C_{ab}^*a^* + b^*)\right]$$

$$= C_{ab}C_{ba} + \frac{1 - C_{ab}C_{ba}}{1 + \delta^2}\sin^2\left(\sqrt{k_{ab}k_{ba}}\ell\left(1 + \delta^2\right)^{1/2}\right) \quad (8.4.16)$$

These results are very similar to those in Ref. 24 except that the parameters are defined in terms of the more accurate parameters γ_a, γ_b, k_{ab}, and k_{ba}.

8.4.2 Numerical Results for Two Strongly Coupled Waveguides

In Fig. 8.14 we show numerical results for two coupled Ti-diffused LiNbO$_3$ channel waveguides modeled as two coupled slab waveguides (which is possible using the effective index method [28]) with the refractive index in

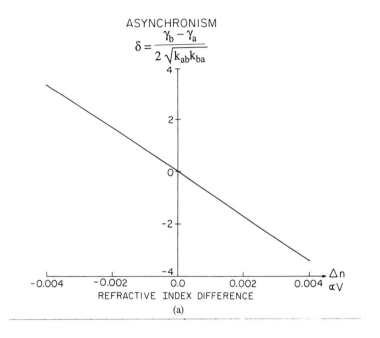

ASYNCHRONISM

$$\delta = \frac{\gamma_b - \gamma_a}{2\sqrt{k_{ab}k_{ba}}}$$

REFRACTIVE INDEX DIFFERENCE

(a)

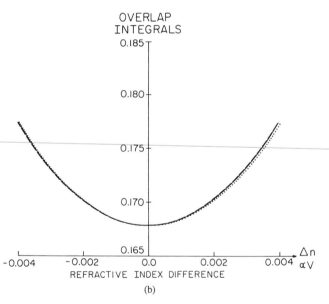

OVERLAP
INTEGRALS

REFRACTIVE INDEX DIFFERENCE

(b)

Figure 8.14. (a) The asynchronism δ is plotted versus the refractive index difference of two-coupled waveguides, $\Delta n = n_a - n_b$, which is proportional to the applied voltage V. (b) The overlap integrals C_{ab} (dashed line), C_{ba} (dotted line), and $C = (C_{ab} + C_{ba})/2$ (solid line) are shown and *are almost identical*.

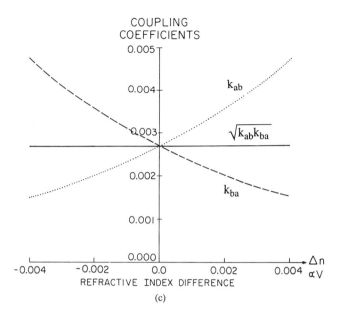

COUPLING
COEFFICIENTS

(c)

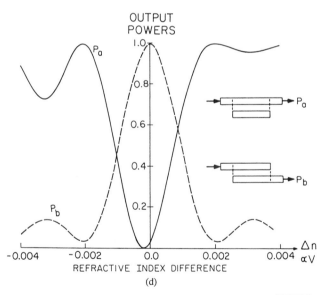

OUTPUT
POWERS

(d)

Figure 8.14. (*Continued*) (c) The coupling coefficients k_{ab}, k_{ba}, and $\sqrt{k_{ab}k_{ba}}$ are plotted versus Δn. (d) The output powers in guide a, P_a (solid curve) and in guide b, P_b (dashed curve) are plotted versus Δn. The parameters are $d_a = d_b = 2$ μm, waveguide edge-to-edge separation $t = 1.9$ μm, wavelength $\lambda = 1.06$ μm. $n_a = 2.2 + \Delta n/2$, $n_b = 2.2 - \Delta n/2$. The outside refractive index is $n_0 = 2.19$. The coupler length is 0.5811 mm. (After Ref. 29, © 1987 IEEE.)

waveguide a, $n_a = 2.2 + \Delta n/2$, the effective refractive index in waveguide
b, $n_b = 2.2 - \Delta n/2$, where the refractive index difference

$$\Delta n = n_a - n_b$$

is proportional to the externally applied voltage V across the two waveguides.
The refractive index outside the two waveguides is assumed to be constant,
$n_0 = 2.19$. The waveguide dimensions are $d_a = d_b = 2$ μm; the edge-to-edge
separation $t = 1.9$ μm. The wavelength λ is 1.06 μm. In Fig. 8.14a, the
asynchronism δ is plotted versus the refractive index difference. We see
clearly that $|\delta|$ is linearly proportional to $|\Delta n|$. The overlap integrals C_{ab}
(dashed line) and C_{ba} (dotted line) with their average C (solid line) are
shown in Fig. 8.14b, where they vary between 0.168 at $\Delta n = 0$ to around
0.178. The coupling coefficients k_{ab}, k_{ba}, and $\sqrt{k_{ab}k_{ba}}$ are shown in
Fig. 8.14c, and agree well with the qualitative results of Ref. 23. The output
powers P_a (solid curve) and P_b (dashed curve) are shown in Fig. 8.14d. Note
that the minimum of P_a does not occur right at $\Delta n = 0$ (where $P_b = 1.0$) due
to the crosstalk. The asymmetry of P_a and the symmetric properties of P_b
versus Δn or the applied voltage agree very well with the experimental data
of Marcatili et al. [24], shown in Fig. 8.15. Note that the polarity of the
voltage V in Fig. 8.15 is opposite to that used in Fig. 8.14d. More discussions
and experimental results on three coupled waveguides and the crosstalk can
be found in Refs. 33–38. Extensions of the coupled-mode theory for
anisotropic waveguides or multiple quantum-well waveguides are reported in
Refs. 39–41.

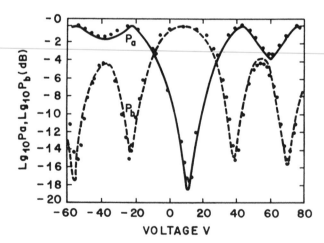

Figure 8.15. Solid and dashed curves are the power outputs of the modulators for P_a and P_b in
the inlet of Fig. 8.14d, respectively. They are measured as functions of the voltages V applied to
the electrodes. The dots are the results predicted by the improved coupled-mode theory in Ref.
23. (After Ref. 24.)

8.5 APPLICATIONS OF OPTICAL WAVEGUIDE COUPLERS

In this section, we discuss some applications of the coupled waveguide theory.

8.5.1 Tunable Filter [42–44]

Consider two parallel slab waveguides a and b with different refractive indices $n_a > n_b$ and thicknesses $d_a < d_b$, as shown in Fig. 8.16. The dispersion relations of the TE_0 waveguides of guides a and b will have their low-frequency asymptotes approaching $\beta = k_0 n_s$ and their high-frequency asymptotes approaching either $\beta = k_0 n_a$ and $\beta = k_0 n_b$, respectively. For $d_a < d_b$ the two dispersion curves intersect each other at (ω_0, β_0). Therefore, for an incident optical wave from waveguide b at $z = 0$, $b(0) = 1$, $a(0) = 0$, we find

$$a(\ell) = i\frac{K_{ab}}{\psi}\sin\psi\ell \exp\left(i\frac{\gamma_a + \gamma_b}{2}\ell\right) \tag{8.5.1}$$

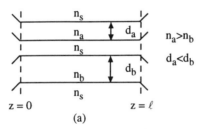

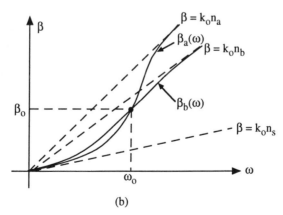

Figure 8.16. (a) Two coupled dielectric slab waveguides a and b with the refractive indices, $n_a > n_b$ and dimensions of $d_a < d_b$. (b) The dispersion curves $\beta_a(\omega)$ and $\beta_b(\omega)$ for the structures in (a) intersect at a point (ω_0, β_0).

where $\psi = \sqrt{\Delta^2 + K_{ab}K_{ba}}$ and $\Delta = (\gamma_b - \gamma_a)/2$. Note that $\gamma_a \simeq \beta_a$, $\gamma_b \simeq \beta_b$, $k_{ab} \simeq K_{ab}$, and $k_{ba} \simeq K_{ba}$ have been assumed here. In general, we can always use the more accurate parameters γ_a, γ_b, k_{ab}, and k_{ba} if necessary. The power transfer from waveguide b to waveguide a is determined by

$$|a(\ell)|^2 = \frac{|K_{ab}|^2}{\Delta^2 + K_{ab}K_{ba}} \sin^2\psi\ell \qquad (8.5.2)$$

Therefore, we have complete power transfer at $K\ell = \pi/2$, $3\pi/2$, $5\pi/2$, etc., only at $\omega = \omega_0$, $\beta_a = \beta_b$, $\Delta = 0$ and $\psi = K_{ab} = K_{ba}(\equiv K)$. For a fixed coupler length $\ell = \pi/2K$, for example, an incident light with a wide spectrum on waveguide b, the output optical power in waveguide a will contain signals of frequences close to ω_0. The directional coupler acts as an optical filter. Furthermore, if the materials are electrooptical so that we may apply a voltage to change the refractive indices n_a and n_b, the intersection frequency ω_0 will be tunable.

Optical wavelength filters using waveguide couplers have been reported for Ti:LiNbO$_3$ and InGaAsP/InP materials. The In$_{1-x}$Ga$_x$As$_y$P$_{1-y}$/InP material system is especially interesting because of its applications at 1.3 or 1.55 μm wavelengths and its potential for optoelectronic integrated circuits. The experiment reported in Ref. 44 has two-coupled waveguides: one has a narrower guide width $d_a = 0.42$ μm, but a larger refractive index n_a than n_b (obtained following Ref. 45) with $y_a = 0.127$; the other has a guide width $d_b = 0.91$ μm and $y_b = 0.078$. The gallium mole fraction x_a (or x_b) depends on the arsenic mole fraction y_a (or y_b) for lattice matching. The input power is assumed to be 1 in waveguide b. The output power from waveguide a, P_a, peaks at 1.12 μm. We have applied the improved coupled-mode theory and compared our theoretical results (solid and dashed lines) with the experimental results (open circles) in Fig. 8.17. The parameters reported in Ref. 44 used for the theoretical calculations are within the measurement accuracy. The possible reasons of discrepancy may be that (1) there is still some difference between the theoretical model in Ref. 45 and the experimental values for the refractive index, and (2) the losses in the waveguides are not taken into account. However, the comparison shown in Fig. 8.17 shows very good agreement in general.

8.5.2 Optical Waveguide Switch

Consider a directional coupler with an incident light into waveguide a, as shown in Fig. 8.18. Assume $a(0) = 1$. The output power from waveguide b is

$$P_b = |b(\ell)|^2 \simeq \frac{K^2}{\Delta^2 + K^2} \sin^2(\psi\ell) \qquad (8.5.3)$$

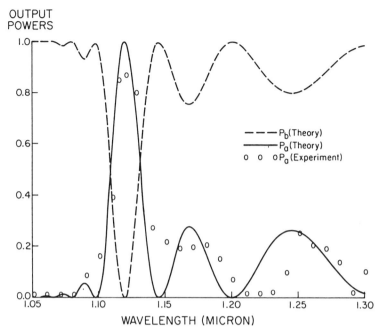

Figure 8.17. The output powers of two coupled $In_xGa_{1-x}As_yP_{1-y}/$ InP waveguides used as an optical wavelength filter. The input power is assumed to be 1 in waveguide b. The theoretical results for output powers at waveguide a, P_a (solid curve), at waveguide b, P_b (dashed curve), are compared with the experimental data of Ref. 44 (open circles) for P_a. (After Ref. 29, © 1987 IEEE.)

No power is transferred to waveguide b at the exit if

$$\psi \ell = n\pi \qquad n = 1, 2, 3, \ldots$$

We can plot $\Delta \ell$ and $K\ell$ in the plane to represent this state, called the parallel state $\ominus$ since all input power into guide a at $z = \ell$ exits from the same waveguide. In other words, the equations

$$(\Delta \ell)^2 + (K\ell)^2 = (n\pi)^2 \qquad n = 1, 2, 3, \ldots \qquad (8.5.4)$$

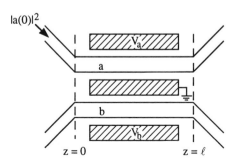

Figure 8.18. An optical waveguide switch.

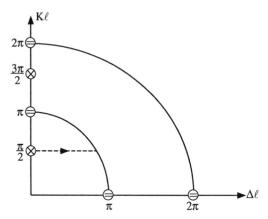

Figure 8.19. A switching diagram for an optical waveguide switch. The $\ominus$ states are represented by the solid curves and the $\otimes$ states are discrete points on the $K\ell$ axis. The horizontal dashed line shows a switching from a $\otimes$ state toward the $\ominus$ state.

represent a set of circles on the $K\ell$ vs. $\Delta\ell$ plane, as shown in the switching diagram, Fig. 8.19. On the other hand, complete power transfer can occur only if $\Delta^2 = 0$ and $K\ell = (2m + 1)\pi/2$, where $m = 0, 1, 2, \ldots$. We call this a cross state $\otimes$ when the total power is transferred from one waveguide to the other. These cross states are represented by only discrete points labeled as $\otimes$ on the vertical $(K\ell)$ axis.

Using the electrooptical effects, which are discussed further in Chapter 12, the refractive indices in the waveguides can be changed by an applied voltage bias. The refractive index difference between two waveguides, or the mismatch factor $\Delta = (\beta_a - \beta_b)/2$, can be tuned by the voltage. Therefore, we can switch the directional coupler from a cross state at $K\ell = \pi/2$ (and $\Delta = 0$) to a parallel state, shown as the horizontal dashed line in Fig. 8.19.

8.5.3   The $\Delta\beta$ Coupler

It is difficult to achieve switching in the design of a single-section waveguide coupler since the coupling coefficient is a function of the spacing and the material parameters; these parameters are harder to adjust in the fabrication processes. It is generally easier to control coupling by an applied voltage to change $\Delta\beta = \beta_b - \beta_a$.

A two-section waveguide coupler has been proposed [46] to achieve switching with more flexibility, as shown in Fig. 8.20. The amplitudes $a(z)$ and $b(z)$ for a coupled waveguide configuration as shown in the first half section of Fig. 8.20 are related to the initial conditions by

$$\begin{bmatrix} a(z) \\ b(z) \end{bmatrix} = \overline{\overline{S}}(z; \beta_a, \beta_b) \begin{bmatrix} a(0) \\ b(0) \end{bmatrix} \qquad (8.5.5a)$$

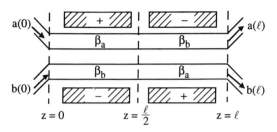

Figure 8.20. A $\Delta\beta$ coupler with two sections where the propagation constants β_a and β_b exchange positions because of the change in the applied voltages.

where we keep the order β_a and β_b in the arguments of the $\overline{\overline{S}}$ matrix:

$$
\overline{\overline{S}}(z; \beta_a, \beta_b) = e^{i\phi z}
\begin{bmatrix}
\cos \psi z - i\dfrac{\Delta}{\psi} \sin \psi z & i\dfrac{K_{ab}}{\psi} \sin \psi z \\[2ex]
i\dfrac{K_{ba}}{\psi} \sin \psi z & \cos \psi z + i\dfrac{\Delta}{\psi} \sin \psi z
\end{bmatrix}
\tag{8.5.5b}
$$

$$
\phi = \frac{\beta_b + \beta_a}{2} \qquad \psi = \sqrt{\Delta^2 + K_{ab}K_{ba}} \qquad \Delta = \frac{\beta_b - \beta_a}{2} \tag{8.5.6}
$$

The output mode amplitudes at the exit, $z = \ell$, are obtained by the product of two matrices because of the cascade connection:

$$
\begin{bmatrix} a(\ell) \\ b(\ell) \end{bmatrix} = \overline{\overline{S}}\left(\frac{\ell}{2}; \beta_b, \beta_a\right) \overline{\overline{S}}\left(\frac{\ell}{2}; \beta_a, \beta_b\right) \begin{bmatrix} a(0) \\ b(0) \end{bmatrix}
$$

$$
= e^{i\phi\ell}
\begin{bmatrix}
\cos^2 \dfrac{\psi\ell}{2} + \dfrac{\Delta^2 - K_{ab}^2}{\psi^2} \sin^2 \dfrac{\psi\ell}{2} & i\dfrac{K_{ab} + K_{ba}}{\psi} \sin \dfrac{\psi\ell}{2}\left(\cos \dfrac{\psi\ell}{2} + i\dfrac{\Delta}{\psi} \sin \dfrac{\psi\ell}{2}\right) \\[3ex]
i\dfrac{K_{ab} + K_{ba}}{\psi} \sin \dfrac{\psi\ell}{2}\left(\cos \dfrac{\psi\ell}{2} - i\dfrac{\Delta}{\psi} \sin \dfrac{\psi\ell}{2}\right) & \cos^2 \dfrac{\psi\ell}{2} + \dfrac{\Delta^2 - K_{ba}^2}{\psi^2} \sin^2 \dfrac{\psi\ell}{2}
\end{bmatrix}
$$

$$
\cdot \begin{bmatrix} a(0) \\ b(0) \end{bmatrix} \tag{8.5.7}
$$

If $K_{12} \simeq K_{21} \equiv K$, $a(0) = 1$, and $b(0) = 0$, we find

$$a(\ell) = \left(\cos^2 \frac{\psi \ell}{2} + \frac{\Delta^2 - K^2}{\psi^2} \sin^2 \psi \frac{\ell}{2} \right) e^{i\phi\ell} \qquad (8.5.8a)$$

$$b(\ell) = i \frac{2K}{\psi} \sin \frac{\psi \ell}{2} \left(\cos \frac{\psi \ell}{2} - i \frac{\Delta}{\psi} \sin \frac{\psi \ell}{2} \right) e^{i\phi\ell} \qquad (8.5.8b)$$

The full power transfer (cross state) occurs when $a(\ell) = 0$, or

$$\cot^2 \frac{\psi \ell}{2} = \frac{K^2 - \Delta^2}{K^2 + \Delta^2} \qquad (8.5.9)$$

For example, if $\Delta = 0$, we find $\psi = K$, and $\cot^2(K\ell/2) = 1$. Therefore, $K\ell = \pi/2, 3\pi/2, 5\pi/2$, etc. No power transfer (parallel state) occurs if $|b(\ell)|^2 = 0$. This can happen in either of the following cases:

$$\text{Case 1} \qquad \sin \frac{\psi \ell}{2} = 0 \qquad (8.5.10)$$

or

$$\text{Case 2} \qquad \cos^2 \frac{\psi \ell}{2} + \frac{\Delta^2}{\psi^2} \sin^2 \frac{\psi \ell}{2} = 0 \qquad (8.5.11)$$

For case 1, $\psi \ell = 2m\pi$, or

$$(\Delta \ell)^2 + (K\ell)^2 = (2m\pi)^2 \qquad (8.5.12)$$

which are circles in the $K\ell$ vs. $\Delta\ell$ plane for $m = $ a nonzero integer. For case 2, it is satisfied only when $\Delta = 0$, and $\cos(K\ell/2) = 0$, i.e., $K\ell = (2m + 1)\pi$, $m = $ an integer. A plot of the switching diagram is shown in Fig. 8.21. The switching diagram shows the degree of freedom to switch from a parallel $\ominus$ to a cross $\otimes$ state or vice versa.

The theory for coupled waveguides has also been extended to gain media [47, 48], coupled fiber Fabry–Perot resonators [49, 50], laser amplifiers [51, 52], and coupled laser arrays [53–55]. In Section 10.5, we discuss coupled semiconductor laser arrays, and the coupled-mode theory is used to investigate the supermode behaviors.

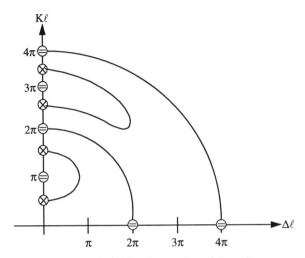

Figure 8.21. Switching diagram for a $\Delta\beta$ coupler.

8.6   DISTRIBUTED FEEDBACK STRUCTURES [6, 8, 56–64]

Consider a corrugated waveguide structure as shown in Fig. 8.22. The permittivity function $\varepsilon(\mathbf{r})$ can be written as an unperturbed part $\varepsilon^{(0)}(\mathbf{r})$ and a perturbed part $\Delta\varepsilon(\mathbf{r})$:

$$\varepsilon(\mathbf{r}) = \varepsilon^{(0)}(\mathbf{r}) + \Delta\varepsilon(\mathbf{r}) \qquad (8.6.1)$$

For the unperturbed part, assume that a particular guided mode with a propagation constant β_s or $-\beta_s$ is of interest. The amplitude for the forward

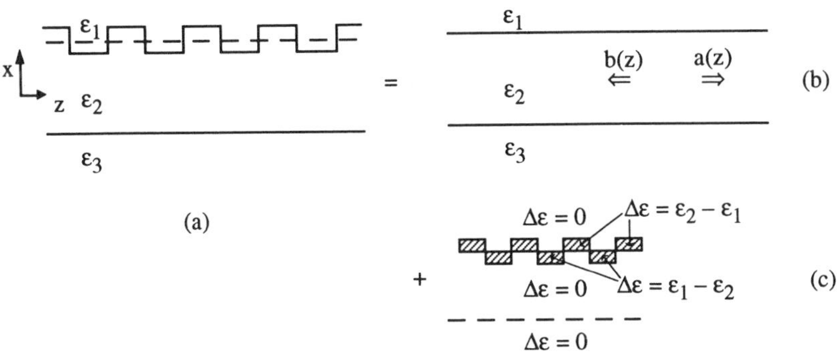

Figure 8.22.   (a) A corrugated waveguide with a permittivity function $\varepsilon(\mathbf{r})$ can be written as the sum of (b) an unperturbed part $\varepsilon^{(0)}(\mathbf{r})$, which describes a uniform slab waveguide, and (c) a perturbed part $\Delta\varepsilon = \varepsilon_2 - \varepsilon_1$ or $\Delta\varepsilon = \varepsilon_1 - \varepsilon_2$ in the corrugation regions.

propagating mode $a(z)$ is $a_0 e^{i\beta_s z}$ and backward propagating mode $b(z)$ is $b_0 e^{-i\beta_s z}$. Since $\varepsilon(\mathbf{r})$ is periodic in z, we may write, for Fig. 8.22, using the Fourier series expansion

$$\Delta\varepsilon(x, z) = \sum_{p=-\infty}^{\infty} d_p(x) e^{ip(2\pi/\Lambda)z} \qquad (8.6.2)$$

where Λ is the period. For a lossless structure, $\Delta\varepsilon(x, z)$ is real; therefore,

$$d_{-p}(x) = d_p^*(x) \qquad (8.6.3)$$

8.6.1 Derivations of the Coupled-Mode Equations for Distributed Feedback Structures

From Maxwell's equations, we find, for $\mathbf{E} = \hat{y}E_y$,

$$\nabla^2 E_y - \mu_0\varepsilon^{(0)}(x, z)\frac{\partial^2}{\partial t^2}E_y = \mu_0\Delta\varepsilon(x, z)\frac{\partial^2}{\partial t^2}E_y \qquad (8.6.4)$$

For a time harmonic field with an $e^{-i\omega t}$ dependence, we find

$$\left[\frac{\partial^2}{\partial x^2} + \frac{\partial^2}{\partial z^2} + \omega^2\mu\varepsilon^{(0)}(x, z)\right]E_y = -\omega^2\mu\Delta\varepsilon(x, z)E_y \qquad (8.6.5)$$

We expand the solution E_y in terms of the unperturbed waveguide modes

$$E_y = \sum_{m=-\infty}^{\infty} A_m(z)E_y^{(m)}(x) e^{i\beta_m z} \qquad (8.6.6)$$

where m refers to the mode numbers, TE_0, TE_1, TE_2, and so on, and $E_y^{(m)}(x)$ is the transverse modal distribution with a corresponding propagation constant β_m:

$$\left[\frac{\partial^2}{\partial x^2} + \frac{\partial^2}{\partial z^2} + \omega^2\mu\varepsilon^{(0)}(x, z)\right]E_y^{(m)}(x) e^{i\beta_m z} = 0 \qquad (8.6.7)$$

We substitute the general expression (8.6.6) in the wave equation and ignore the $\partial^2 A_m(z)/\partial z^2$ terms, assuming that $A_m(z)$ are slowly varying functions. We find

$$\sum_m 2\frac{\partial A_m(z)}{\partial z} i\beta_m E_y^{(m)}(x) e^{i\beta_m z}$$

$$= -\omega^2\mu\sum_m \sum_{p=-\infty}^{\infty} d_p(x)A_m(z)E_y^{(m)}(x) e^{i\beta_m z + ip(2\pi/\Lambda)z} \qquad (8.6.8)$$

If we multiply both sides of (8.6.8) by $(\beta_s/2\omega\mu)E_y^{(s)*}(x)$, which is related to the x component of the magnetic field of the guided sth mode by $H_x^{(s)}(x) = (-\beta_s/\omega\mu)E_y^{(s)}(x)$, and make use of the normalization condition,

$$\frac{1}{2}\int \mathbf{E}^{(m)} \times \mathbf{H}^{(s)*} \cdot \hat{z}\, dx = \frac{\beta_s}{2\omega\mu}\int \left|E_y^{(s)}(x)\right|^2 dx\, \delta_{ms} = \delta_{ms} \quad (8.6.9)$$

we find

$$2\frac{\partial A_s(z)}{\partial z}i\beta_s\, e^{i\beta_s z} = -\omega^2\mu\sum_m A_m(z)\sum_{p=-\infty}^{\infty}\frac{\beta_s}{2\omega\mu}\int_{-\infty}^{\infty}d_p(x)E_y^{(s)*}(x)E_y^{(m)}(x)\, dx$$

$$\times e^{i(\beta_m+p2\pi/\Lambda)z} \quad (8.6.10)$$

where we have assumed that $\varepsilon^{(0)}(x, z)$ is real; hence, β_s is also real. If we focus on the coupling of the forward propagating β_s mode to the backward propagating $(m = -s)$ mode so that $\beta_{-s} + p(2\pi/\Lambda)$ is close to β_s, for a particular $p = \ell$ where $\beta_{-s} = -\beta_s$, assuming the couplings to the other modes $m \neq -s$ are negligible, we then have

$$\frac{dA_s(z)}{dz} = iK_{ab}A_{-s}(z)\, e^{-i2(\beta_s-\ell\pi/\Lambda)z} \quad (8.6.11)$$

where the coupling coefficient K_{ab} is

$$K_{ab} = \frac{\omega}{4}\int_{-\infty}^{\infty}d_\ell(x)\left|E_y^{(s)}(x)\right|^2 dx \quad (8.6.12)$$

A phase-matching diagram is shown in Fig. 8.23, where we consider the first-order grating $\ell = \pm 1$ such that $\beta_s = -\beta_s + 2\pi/\Lambda$ and $A_{-s}(z)$ is coupled to $A_s(z)$ by (8.6.11). Equation (8.6.12) defines the coupling coefficient for TE modes in a corrugated waveguide. Extensive research work has been done on both TE and TM mode coupling coefficients in distributed feedback structures and applications to DFB lasers [58–64]. If we multiply (8.6.8) by the wave function $H_x^{(-s)}(x)$ for the $-s$ mode with a propagation constant $\beta_{-s} = -\beta_s$, we find the second coupled-mode equation

$$\frac{dA_{-s}(z)}{dz} = iK_{ba}A_s(z)\, e^{i2(\beta_s-\ell\pi/\Lambda)z} \quad (8.6.13)$$

where the coupling coefficient K_{ba} is

$$K_{ba} = -\frac{\omega}{4}\int_{-\infty}^{\infty}d_{-\ell}(x)\left|E_y^{(s)}(x)\right|^2 dx \quad (8.6.14)$$

We also see that the condition for contradirectional coupling in a lossless

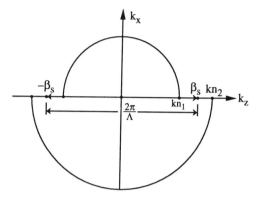

Figure 8.23. Phase-matching diagram for a forward propagating mode with a propagation constant β_s coupling to a backward propagating mode with a propagation constant $\beta_{-s}\,(= -\beta_s)$ in the negative z direction. $\beta_s \simeq -\beta_s + 2\pi/\Lambda$ is the phase-matching condition for the z components of the propagation wave vectors. The difference $\Delta\beta = (\beta_s - \pi/\Lambda)$ is the detuning parameter.

coupler

$$K_{ba} = -K_{ab}^* \qquad (8.6.15)$$

is satisfied since $d_{-\ell} = d_\ell^*$. As shown in Fig. 8.23, the forward propagation amplitude $A_s(z)$ is coupled to the backward propagation amplitude $A_{-s}(z)$ by (8.6.13). Therefore, the phase-matching condition requires $\beta_{-s} \simeq \beta_s - 2\pi/\Lambda$. Define

$$\Delta\beta = \beta_s - \frac{\ell\,\pi}{\Lambda} \qquad (8.6.16)$$

$$A(z) = A_s(z)\,e^{i\Delta\beta z}$$
$$B(z) = A_{-s}(z)\,e^{-i\Delta\beta z} \qquad (8.6.17)$$

We obtain

$$\frac{d}{dz}\begin{bmatrix} A(z) \\ B(z) \end{bmatrix} = i\begin{bmatrix} \Delta\beta & K_{ab} \\ K_{ba} & -\Delta\beta \end{bmatrix}\begin{bmatrix} A(z) \\ B(z) \end{bmatrix} \qquad (8.6.18)$$

8.6.2 Eigenstate Solutions of the Coupled-Mode Equations

We follow a similar procedure as before for coupling in two parallel waveguides by assuming the eigensolution of the form

$$\begin{bmatrix} A(z) \\ B(z) \end{bmatrix} = \begin{bmatrix} A_0 \\ B_0 \end{bmatrix} e^{iqz} \qquad (8.6.19)$$

The eigenequations become (using $K \equiv K_{ab}$)

$$\begin{bmatrix} \Delta\beta - q & K \\ -K^* & -\Delta\beta - q \end{bmatrix} \begin{bmatrix} A_0 \\ B_0 \end{bmatrix} = 0 \qquad (8.6.20)$$

The eigenvalues are

$$q_\pm = \pm\sqrt{(\Delta\beta)^2 - |K|^2} = \pm iS \qquad (8.6.21)$$

where

$$S = \sqrt{|K|^2 - (\Delta\beta)^2} \qquad (8.6.22)$$

In the original expression for the electric field (8.6.6), we keep only two counterpropagating modes, $m = s, -s$ (for the forward and backward propagating TE_0 modes):

$$E_y(x) = A_s(z)\, e^{i\beta_s z} E_y^{(s)}(x) + A_{-s}(z)\, e^{-i\beta_s z} E_y^{(-s)}(x)$$
$$= a(z) E_y^{(s)}(x) + b(z) E_y^{(-s)}(x) \qquad (8.6.23)$$

The magnitude of the electric field for the forward propagating mode is

$$a(z) = A_s(z)\, e^{i\beta_s z} = A(z)\, e^{-i\Delta\beta z}\, e^{i\beta_s z} = A(z)\, e^{i(\ell\pi/\Lambda)z} \quad (8.6.24a)$$

The magnitude of the backward propagating mode is

$$b(z) = A_{-s}(z)\, e^{-i\beta_s z} = B(z)\, e^{+i\Delta\beta z}\, e^{-i\beta_s z} = B(z)\, e^{-i(\ell\pi/\Lambda)z} \quad (8.6.24b)$$

The total electric field (8.6.23) can be written as

$$E_y(x) = A(z)\, e^{i(\ell\pi/\Lambda)z} E_y^{(s)}(x) + B(z)\, e^{-i(\ell\pi/\Lambda)z} E_y^{(-s)}(x) \quad (8.6.25)$$

The propagation constants of the system modes are

$$\beta_\pm = \frac{\ell\pi}{\Lambda} \pm \sqrt{[\beta_s(\omega) - (\ell\pi/\Lambda)]^2 - |K|^2}$$

$$= \frac{\ell\pi}{\Lambda} \pm i\sqrt{|K|^2 - |\beta_s(\omega) - (\ell\pi/\Lambda)|^2} \qquad (8.6.26)$$

Far away from the point $\beta_s(\omega_0) = \ell\pi/\Lambda$, when $|\beta_s(\omega) - (\ell\pi/\Lambda)| > |K|$, $\beta\Lambda$

is real, and the propagation constant approaches the following limits:

$$\beta_\pm \to \frac{\ell \pi}{\Lambda} \pm \left(\beta_s - \frac{\ell \pi}{\Lambda} \right) \to \begin{cases} \beta_s(\omega) \\ -\beta_s(\omega) + \dfrac{2\ell \pi}{\Lambda} \end{cases} \qquad (8.6.27)$$

At $\omega = \omega_0$ defined by $\beta_s(\omega_0) = \ell \pi/\Lambda$, the real and imaginary parts of the propagation constants $\beta_\pm$ satisfy

$$\mathrm{Re}(\beta_\pm \Lambda) = \ell \pi \qquad \mathrm{Im}(\beta_\pm \Lambda) = \pm |K| \Lambda \qquad (8.6.28)$$

and $S = |K|$. For ω in the range $|\beta_s(\omega) - \ell \pi/\Lambda| < |K|$, we have

$$\mathrm{Re}(\beta_\pm \Lambda) = \ell \pi$$

$$\mathrm{Im}(\beta_\pm \Lambda) = \pm S\Lambda = \pm \sqrt{|K|^2 \Lambda^2 - [\beta_s(\omega)\Lambda - \ell \pi]^2} \qquad (8.6.29)$$

In Fig. 8.24, we plot the dispersion relations for $\ell = 1$. We see that the stop band occurs at the vicinity of the frequency ω_0.

For the two eigenvalues $\pm iS$, their corresponding eigenvectors are determined by

$$q = +iS \qquad \frac{A_0}{B_0} = \frac{-K}{\Delta\beta - iS} = \frac{\Delta\beta + iS}{-K^*} \qquad (8.6.30)$$

$$q = -iS \qquad \frac{A_0}{B_0} = \frac{-K}{\Delta\beta + iS} = \frac{\Delta\beta - iS}{-K^*} \qquad (8.6.31)$$

The general solution can be written as

$$\begin{bmatrix} A(z) \\ B(z) \end{bmatrix} = c_1 \begin{bmatrix} -K \\ \Delta\beta - iS \end{bmatrix} e^{-Sz} + c_2 \begin{bmatrix} -K \\ \Delta\beta + iS \end{bmatrix} e^{+Sz} \qquad (8.6.32)$$

8.6.3 Reflection and Transmission of a Distributed Feedback Structure

For a wave incident from the left side of the distributed feedback structure at $z = 0$, as shown in Fig. 8.25, we may find the general solution using the fact that at $z = L$, $B(L) = 0$, since there is no incident wave from the right-hand side. Therefore,

$$c_1(\Delta\beta - iS) e^{-SL} + c_2(\Delta\beta + iS)e^{SL} = B(L) = 0 \qquad (8.6.33)$$

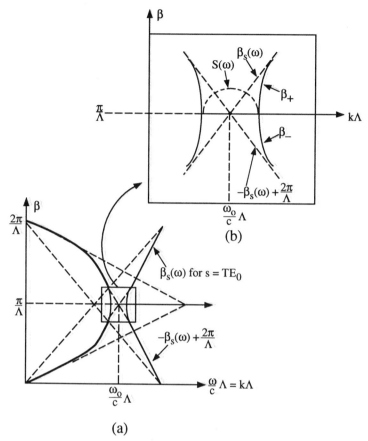

Figure 8.24.   (a) The dispersion curves for the contradirectional coupling of modes in a periodic waveguide structure; (b) the enlarged portion of the intersection region.

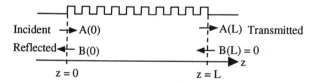

Figure 8.25.   A periodic corrugated waveguide structure with an incident wave at $z = 0$. $A(0)$ is the field amplitude of the incident wave and $B(0)$ is the field amplitude of the reflected wave. $A(L)$ is the field amplitude of the transmitted wave and $B(L) = 0$ since we assume no wave is incident from the right-hand side.

For a given $A(0)$, we have

$$-K(c_1 + c_2) = A(0) \tag{8.6.34}$$

Solving for c_1 and c_2 in terms of $A(0)$ and substituting into (8.6.32), we obtain

$$A(z) = \frac{-\Delta\beta \sinh S(z - L) + iS \cosh S(z - L)}{\Delta\beta \sinh SL + iS \cosh SL} A(0) \tag{8.6.35}$$

$$B(z) = \frac{K^* \sinh S(z - L)}{\Delta\beta \sinh SL + iS \cosh SL} A(0) \tag{8.6.36}$$

The reflection coefficient at $z = 0$ is

$$r(0) = \frac{B(0)}{A(0)}$$

$$= \frac{-K^* \sinh SL}{\Delta\beta \sinh SL + iS \cosh SL} \tag{8.6.37}$$

At $\Delta\beta = 0$, $S = |K|$

$$|r(0)| = \tanh|KL| \tag{8.6.38}$$

which approaches 1 if $|KL|$ is large enough. Also, when $|\Delta\beta| > |K|$, S becomes purely imaginary and $|r(0)|^2$ vanishes when

$$\sin^2\sqrt{(\Delta\beta L)^2 - |KL|^2} = 0, \quad \text{or} \quad |\Delta\beta/K| = \sqrt{1 + (n\pi/|KL|)^2}.$$

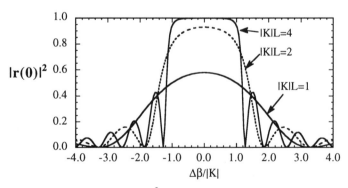

Figure 8.26. A plot of the reflectivity $|r(0)|^2$ vs. the ratio $\Delta\beta/|K|$ for three different values of $|KL|$. For a large value of $|KL|$, the bandwidth for $\Delta\beta$ is about $2|K|$.

A plot of the reflectivity $|r(0)|^2$ vs. $\Delta\beta/|K|$ for three different values of $|KL|$ (i.e., three lengths of DFB reflectors) is shown in Fig. 8.26. We see that if $|KL|$ increases, the reflectivity for small $\Delta\beta\ell$ increases. We should note that when $|S| \to 0$, $|\Delta\beta| \to |K|$, and $|r(0)|^2 \to |KL|^2/[|KL|^2 + 1]$. The transmission coefficient at $z = L$ is

$$t(L) = \frac{A(L)}{A(0)} = \frac{iS}{\Delta\beta \sinh SL + iS \cosh SL} \qquad (8.6.39)$$

The theory developed in this section will be further applied to the study of distributed feedback lasers in Section 10.6.

PROBLEMS

8.1 A TE light ($\lambda = 0.8$ μm) is normally incident on a prism from air as shown in Fig. 8.27. The waveguide width d is chosen such that the light can be resonantly coupled into the waveguide, thereby, the phase-matching condition is satisfied.

(a) Find the propagation constant k_z.

(b) Find the Goos–Hänchen phase shifts for the guided mode for both the air–waveguide and substrate–waveguide interfaces.

(c) From the ray optics picture, determine the width d from the propagation constant k_z obtained in (a).

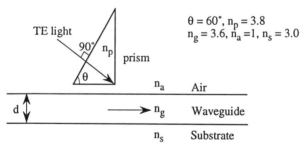

Figure 8.27.

8.2 Consider an InP/ $In_{0.53}Ga_{0.47}As$ / InP dielectric waveguide with a dimension $d = 0.3$ μm. The wavelength in free space λ_0 is 1.65 μm. The refractive indices are $n_1(In_{0.53}Ga_{0.47}As) = 3.56$ and $n = 3.16$. The propagation constant for the TE_0 mode is found to be $\beta = 12.7$ $(\mu m)^{-1}$.

(a) If a grating coupler is designed using the same materials, Fig. 8.28, so that the first-order space harmonic mode ($m = 1$) is coupled to the TE_0 mode at $\theta_i = 45°$, find the grating period Λ.

(b) Repeat part (a) for $m = 2$. Draw a phase matching diagram to show all of the propagating (nonevanescent) space harmonics (called Floquet modes) in the InP region for this case and show their reflection angles measured from the x axis.

(c) Consider a grating output coupler shown in Fig. 8.28. If the output coupler is designed so that there is one and only one Floquet mode propagating in the x direction (this configuration also has applications for surface emitting diode lasers), what should the period Λ_2 be?

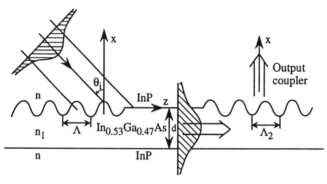

Figure 8.28.

8.3 Consider a three-parallel-waveguide coupler shown in Fig. 8.29. Using the coupled-mode theory, the electric field of the three-guide coupler

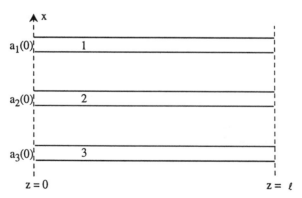

Figure 8.29.

can be written as

$$E(x, z) = \hat{y}\left[a_1(z)E_y^{(1)}(x) + a_2(z)E_y^{(2)}(x) + a_3(z)E_y^{(3)}(x)\right]$$

where $E_y^{(j)}(x)$ describes the x dependence of the TE_0 mode of waveguide j ($j = 1, 2, 3$) in the absence of the other two guides. The coupled-mode equation can be written as

$$\frac{d}{dz}\begin{bmatrix} a_1 \\ a_2 \\ a_3 \end{bmatrix} = i\begin{bmatrix} \beta & \kappa & 0 \\ \kappa & \beta & \kappa \\ 0 & \kappa & \beta \end{bmatrix}\begin{bmatrix} a_1 \\ a_2 \\ a_3 \end{bmatrix}$$

where the coupling between guides 1 and 3 has been neglected since they are far apart.

(a) Find the three eigenvalues and the corresponding eigenvectors of the three-waveguide coupler.

(b) Sketch approximately the mode profiles of the three eigenmodes as a function of x.

(c) Given the initial values, $a_1(0)$, $a_2(0)$, and $a_3(0)$, find $a_1(z)$, $a_2(z)$, and $a_3(z)$ in terms of their initial values. Put your results in a matrix form.

8.4 The coupled-mode equations for N identical equally spaced waveguides, assuming only adjacent waveguide coupling, can be written as

$$\frac{d}{dz}A(z) = i\overline{\overline{M}}A(z)$$

where

$$A(z) = \begin{bmatrix} a_1 \\ a_2 \\ \vdots \\ a_N \end{bmatrix} \qquad \overline{\overline{M}} = \begin{bmatrix} \beta & K & & & 0 \\ K & \beta & K & & \\ & \ddots & \ddots & \ddots & \\ & & K & \beta & K \\ 0 & & & K & \beta \end{bmatrix}$$

(a) Show that the general solutions for the normalized eigenvectors and their eigenvalues for an arbitrary N are

$$A^{(\ell)}(z) = \begin{bmatrix} a_1^\ell \\ a_2^\ell \\ \vdots \\ a_N^\ell \end{bmatrix} e^{i\beta_\ell z} \qquad \ell = 1, 2, \ldots, N$$

where

$$\beta_\ell = \beta + 2K \cos\left(\frac{\ell\pi}{N+1}\right)$$

$$a_k^\ell = \sqrt{\frac{2}{N+1}} \sin\left(\frac{k\ell\pi}{N+1}\right) \qquad k = 1, 2, \ldots, N$$

The order of β_ℓ is chosen such that $\beta_1 > \beta_2 > \beta_3 > \cdots > \beta_N$.
Hint:

$$\det\begin{bmatrix} 2\cos\theta & 1 & & & 0 \\ 1 & 2\cos\theta & 1 & & \\ & \ddots & \ddots & \ddots & \\ & & 1 & 2\cos\theta & 1 \\ 0 & & & 1 & 2\cos\theta \end{bmatrix} = \frac{\sin[(N+1)\theta]}{\sin\theta}$$

where N is the dimension of the matrix.

(b) Find the eigenvalues and their corresponding eigenvectors for $N = 4$, and plot the transverse field profile $E(x) = \sum_{i=1}^{4} a_i E_i(x)$ for all four eigenvalues.

8.5 **(a)** Sketch qualitatively the dispersion relation β_a of the TE_0 mode vs. ω for a single slab waveguide in Fig. 8.30a with a fixed refractive index n_a at two different thicknesses $(d_1 < d_2)$. Explain your sketch of β vs. ω curves using the graphical solution for $\alpha d/2$ vs. $k_x d/2$.

(b) Consider the two coupled slab waveguides (Fig. 8.30b). With $n_a > n_b$, derive a condition between d_a and d_b such that the dispersion curves $\beta_a(\omega)$ intercept at a frequency ω_0.

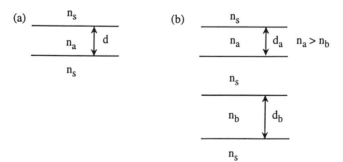

Figure 8.30.

8.6 Consider a distributed feedback structure as shown in Fig. 8.31a, where

$$\Delta\varepsilon(x,z) = \begin{cases} \varepsilon - \varepsilon_1 & \text{if } -h < x < 0 \quad \text{and} \quad 0 < z < \dfrac{\Lambda}{2} \\[2ex] \varepsilon_1 - \varepsilon & \text{if } 0 < x < h \quad \text{and} \quad \dfrac{\Lambda}{2} < z < \Lambda \end{cases}$$

and $\Delta\varepsilon(x,z)$ is periodic in z, Fig. 8.31b.

(a) Find the Fourier series expansion of $\Delta\varepsilon(x,z)$, i.e., $d_p(x)$, where

$$\Delta\varepsilon(x,z) = \sum_{p=-\infty}^{\infty} d_p(x)\exp\!\left(i\frac{2\pi}{\Lambda}pz\right)$$

(b) Evaluate the coupling coefficients K_{ab} and K_{ba} for the TE_0 mode. Assume that the first orders $p = \pm 1$ provide the coupling.

(c) Let $K \equiv K_{ab} = -K_{ba}^*$. Consider an incident wave from $z = 0$ with an amplitude $A(0)$. Derive the general expressions for $B(0)$ and $A(\ell)$.

(d) Plot the reflectivity $|B(0)/A(0)|^2$ vs. $|\Delta\beta/K|$ for $K\ell = 4$.

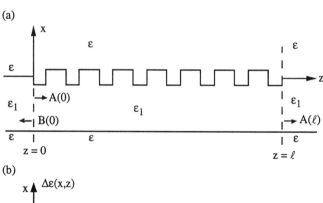

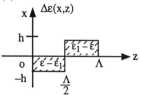

Figure 8.31.

8.7 Consider a distributed feedback structure as shown in Fig. 8.32 where the height function is

$$
h(z) = \begin{cases}
-h_0 + h_0(z - n\Lambda)\dfrac{4}{\Lambda} & n\Lambda \le z \le n\Lambda + \dfrac{\Lambda}{2} \\[2mm]
h_0 - h_0\left(z - n\Lambda - \dfrac{\Lambda}{2}\right)\dfrac{4}{\Lambda} & n\Lambda + \dfrac{\Lambda}{2} \le z \le (n+1)\Lambda \\[2mm]
& n = 0,1,2,\ldots,N-1 \\[1mm]
0 & z > \ell,\, z < 0
\end{cases}
$$

Let the unperturbed $\varepsilon^{(0)}(x, z)$ be $\varepsilon^{(0)} = \begin{cases} \varepsilon_1 & -d < x < 0 \\ \varepsilon & \text{otherwise} \end{cases}$

Repeat Problem 8.6(a)–(d) for this sawtooth grating structure.

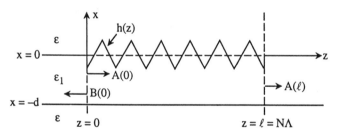

Figure 8.32.

8.8 (a) Describe how a $\Delta\beta$ coupler works and its advantages compared with a single-section directional coupler.

(b) Show the switching diagram ($K\ell$ vs. $\Delta\ell$) with as much detail as possible.

(c) Consider an input with $a(0) = 1$, $b(0) = 0$, $\Delta\ell = \pi/2$, and $K\ell = \pi$, where ℓ is the total length of the $\Delta\beta$ coupler. Find the output powers $|a(\ell)|^2$ and $|b(\ell)|^2$.

8.9 (a) Write down the expression for the coupling coefficient K for two identical parallel waveguides. Consider only the TE_0 mode. Discuss qualitatively all possible ways and trade-offs to increase the coupling coefficient.

(b) Repeat (a) for the coupling coefficient K for a distributed feedback structure.

(c) Compare the coupling coefficients in (a) and (b). What are their similarities and differences?

8.10 (a) Consider the slab waveguide in Fig. 8.33a with a TE_0 mode profile approximated by $E_y(x) = E_0 e^{-x^2/d^2}$ at the facet $z = 0$, and E_0 is assumed such that the guided power is normalized. Find the far-field radiation pattern using an approximate formula,

$$P(\theta) = \left| \int_{-\infty}^{\infty} dx\, e^{-ik\sin\theta x} E_y(x) \right|^2$$

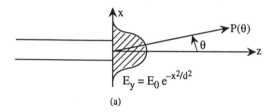

$$E_y = E_0 e^{-x^2/d^2}$$

(a)

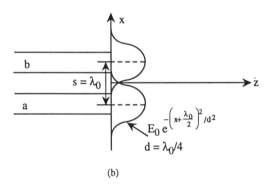

$$s = \lambda_0$$

$$E_0 e^{-\left(x+\frac{\lambda_0}{2}\right)^2/d^2}$$

$$d = \lambda_0/4$$

(b)

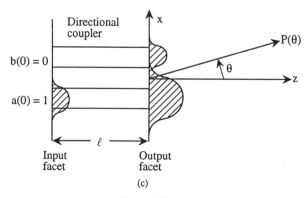

(c)

Figure 8.33.

where we have ignored all prefactors. Find the angle $\theta_{1/2}$ where the radiated power is half of its peak value at $\theta = 0$, that is, $P(\theta = \theta_{1/2}) = \frac{1}{2}P(\theta = 0)$. Assume that $d = \lambda_0/4$, and λ_0 is the wavelength in free space. (*Note*: $\int_{-\infty}^{\infty} e^{-ax^2}\,dx = \sqrt{\pi/a}$.)

(b) Suppose we use the approximation in part (a) for the TE_0 mode in two coupled identical waveguides with the center-to-center separation $s = \lambda_0$, where λ_0 is the laser wavelength in free space (Fig. 8.33b). Calculate the radiation pattern for the in-phase eigenmode of the coupled waveguides. Roughly plot the far-field radiation pattern of the in-phase mode by specifying the angles for the zeroes and maxima.

(c) Suppose we shine a light to excite the TE_0 mode in guide a only on the input facet of the coupler and look at the electric field profile at the output facet at a distance ℓ away from the input facet (Fig. 8.33c). For $\ell = \pi/4K$, where K is the coupling coefficient of the two identical guides, find the expressions for the electric field profile at the output facet and its far-field radiation pattern. Use the approximates for the TE_0 mode profile in (a) and (b) for simplicity. Explain how one may control the radiation pattern using the directional coupler.

8.11 **(a)** Find the far-field pattern due to the TE_0 mode using the numerical values from Problem 7.1 of Chapter 7 on the output facet of a diode in the $\theta_\perp$ direction. For simplicity, assume that the TE_0 mode is uniform in the y direction from $-W$ to $+W$, and D is very large (see Fig. 5.5). Plot the power pattern vs. $\theta_\perp$ roughly.

(b) Consider a directional coupler with three identical guides as in part (a). Using the eigenvalues and eigenvectors from Problem 8.3, plot (i) the three electric field profiles of the coupled waveguide system and (ii) the far-field radiation pattern $P(\theta_\perp)$ of the three eigenmodes, assuming that the waveguide center-to-center spacing is $\lambda_0/2$, as shown in Fig. 8.34.

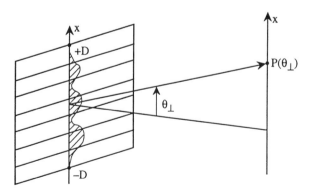

Figure 8.34.

REFERENCES

1. R. G. Hunsperger, *Integrated Optics: Theory and Technology*, Springer, Berlin, 1984.

2. T. Tamir, *Integrated Optics*, Springer, Berlin, 1979.

3. T. Tamir, *Guided Wave Optoelectronics*, Springer, Berlin, 1990.

4. J. A. Kong, *Electromagnetic Wave Theory*, 2d ed., Wiley, New York, 1990.

5. S. L. Chuang and J. A. Kong, "Wave scattering and guidance by dielectric waveguides with periodic surfaces," *J. Opt. Soc. Am.* **73**, 669–679 (1983); *Addendum* **73**, 1823–1824 (1983).

6. H. A. Haus, *Waves and Fields in Optoelectronics*, Prentice Hall, Englewood Cliffs, NJ, 1984.

7. S. L. Chuang and B. Do, "Electron states in two coupled quantum wells: A strong coupling of modes approach," *J. Appl. Phys.* **62**, 1290–1297 (1987).

8. A. Yariv, *Optical Electronics*, Holt, Rinehart & Winston, New York, 1985.

9. A. Yariv, "Coupled-mode theory for guided-wave optics," *IEEE J. Quantum Electron.* **QE-9**, 919–933 (1973).

10. H. F. Taylor and A. Yariv, "Guided wave optics," *Proc. IEEE* **62**, 1044–1060 (1974).

11. H. Nishihara, M. Haruna, and T. Suhara, *Optical Integrated Circuits*, McGraw-Hill, New York, 1989.

12. E. Marom, O. G. Ramer, and S. Ruschin, "Relation between normal-mode and coupled-mode analyses of parallel waveguides," *IEEE J. Quantum Electron.* **QE-20**, 1311–1319 (1984).

13. J. E. Watson, M. A. Milbrodt, and T. C. Rice, "A polarization-independent 1×16 guided-wave optical switch integrated on lithium niobate," *J. Lightwave Technol.* **LT-4**, 1717–1721 (1986).

14. L. A. Molter-Orr and H. A. Haus, "$N \times N$ coupled waveguide switch," *Opt. Lett.* **9**, 466–467 (1984).

15. N. Kishi, E. Yamashita, and K. Atsuki, "Modal and coupling field analysis of optical fibers with circularly distributed multiple cores and a central core," *J. Lightwave Technol.* **LT-4**, 991–996 (1986).

16. D. B. Mortimore and J. W. Arkwright, "Theory and fabrication of wavelength flattened $1 \times N$ single-mode couplers," *Appl. Opt.* **29**, 1814–1818 (1990).

17. Y. Suematsu and K. Kishino, "Coupling coefficient in strongly coupled dielectric waveguides," *Radio Science* **12**, 587–592 (1977).

18. A. Hardy and W. Streifer, "Coupled mode theory of parallel waveguides," *J. Lightwave Technol.* **LT-3**, 1135–1146 (1985).

19. A. Hardy and W. Streifer, "Coupled mode solutions of multiwaveguide systems," *IEEE J. Quantum Electron.* **QE-22**, 528–534 (1986).

20. J. R. Qian and W. P. Huang, "Coupled-mode theory for LP modes," *J. Lightwave Technol.* **LT-4**, 619–625 (1986).

21. J. R. Qian and W. P. Huang, "LP modes and ideal modes on optical fibers," *J. Lightwave Technol.* **LT-4**, 626–630 (1986).

22. D. Marcuse, "Directional couplers made of nonidentical asymmetric slabs. Part I: synchronous couplers," *J. Lightwave Technol.* **LT-5**, 113–118 (1987).

23. E. A. J. Marcatili, "Improved coupled-mode equations for dielectric guides," *IEEE J. Quantum Electron.* **QE-22**, 988–993 (1986).

24. E. A. J. Marcatili, L. L. Buhl, and R. C. Alferness, "Experimental verification of the improved coupled-mode equations," *Appl. Phys. Lett.* **49**, 1692–1693 (1986).

25. W. Streifer, M. Osiński, and A. Hardy, "Reformation of the coupled-mode theory of multiwaveguide systems," *J. Lightwave Technol.* **LT-5**, 1–4 (1987).

26. H. A. Haus, W. P. Huang, S. Kawakami, and N. A. Whitaker, "Coupled-mode theory of optical waveguides," *J. Lightwave Technol.* **LT-5**, 16–23 (1987).

27. S. L. Chang, "A coupled-mode formulation by reciprocity and a variational principle," *J. Lightwave Technol.* **LT-5**, 5–15 (1987).

28. S. L. Chuang, "A coupled-mode theory for multiwaveguide systems satisfying the reciprocity theorem and power conservation," *J. Lightwave Technol.* **LT-5**, 174–183 (1987).

29. S. L. Chuang, "Application of the strongly coupled-mode theory to integrated optical devices," *IEEE J. Quantum Electron.* **QE-23**, 499–509 (1987).

30. A. Ankiewicz, A. W. Snyder, and X. H. Zheng, "Coupling between parallel optical fiber cores—critical examination," *J. Lightwave Technol.* **LT-4**, 1317–1323 (1986).

31. A. W. Snyder, A. Ankiewicz, and A. Altintas, "Fundamental error of recent coupled mode formulations," *Electron. Lett.* **23**, 1097–1098 (1987).

32. H. C. Chang, "Coupled-mode equations for dielectric waveguides based on projection and partition modal amplitudes," *IEEE J. Quantum Electron.* **QE-23**, 1929–1937 (1987).

33. J. P. Donnelly, N. L. DeMeo, Jr., and G. A. Ferrante, "Three-guide optical couplers in GaAs," *J. Lightwave Technol.* **LT-1**, 417–424 (1983).

34. K. L. Chen and S. Wang, "Cross-talk problems in optical directional couplers," *Appl. Phys. Lett.* **44**, 166–168 (1984).

35. J. P. Donnelly, N. L. DeMeo, Jr., G. A. Ferrante, K. B. Nichols, and F. J. O'Donnell, "Optical guided-wave gallium arsenide monolithic interferometer," *Appl. Phys. Lett.* **45**, 360–362 (1984).

36. K. L. Chen and S. Wang, "The cross talk in three-waveguide optical directional couplers," *IEEE J. Quantum Electron.* **QE-22**, 1039–1041 (1986).

37. L. McCaughan and S. K. Korotky, "Three-electrode Ti:LiNbO$_3$ optical switch," *J. Lightwave Technol.* **LT-4**, 1324–1327 (1986).

38. J. P. Donnelly, "Limitations on power-transfer efficiency in three-guide optical couplers," *IEEE J. Quantum Electron.* **QE-22**, 610–616 (1986).

39. D. Marcuse, "Coupled-mode theory for anisotropic optical waveguides," *Bell Syst. Tech. J.* **54**, 985–995 (1975).

40. A. Hardy, W. Streifer, and M. Osiński, "Coupled-mode equations for multimode waveguide systems in isotropic or anisotropic media," *Opt. Lett.* **11**, 742–744 (1986).

41. L. Tsang and S. L. Chuang, "Coupled GaAs multiple-quantum-well channel waveguides including quadratic electrooptic effect," *J. Lightwave Technol.* and *IEEE J. Quantum Electron.* **6**, 832–836 (1988).

42. R. C. Alferness and R. V. Schmidt, "Tunable optical waveguide directional coupler filter," *Appl. Phys. Lett.* **33**, 161–163 (1978).

43. R. C. Alferness and J. J. Veselka, "Tunable Ti:LiNbO$_3$ waveguide filter for long-wavelength ($\lambda = 1.3 - 1.6$ μm) multiplexing/demultiplexing," *Tech. Dig. Conf. Lasers Electrooptics* (Anaheim, CA), 230–231 (1984).

44. B. Broberg, B. S. Lindgren, M. G. Öberg, and H. Jiang, "A novel integrated optics wavelength filter in InGaAsP-InP," *J. Lightwave Technol.* **LT-4**, 196–203 (1986).

45. B. Broberg and S. Lindgren, "Refractive index of In$_{1-x}$Ga$_x$As$_y$P$_{1-y}$ layers and InP in the transparent wavelength region," *J. Appl. Phys.* **55**, 3376–3381 (1984).

46. H. Kogelnik and R. V. Schmidt, "Switched directional couplers with alternating $\Delta\beta$," *IEEE J. Quantum Electron.* **QE-12**, 396–401 (1976).

47. C. J. Setterlind and L. Thylen, "Directional coupler switches with optical gain," *IEEE J. Quantum Electron.* **QE-22**, 595–602 (1986).

48. G. H. B. Thompson, "Analysis of optical directional couplers that include gain or loss and their application to semiconductor slab dielectric guides," *J. Lightwave Technol.* **LT-4**, 1678–1693 (1986).

49. P. Urquhart, "Transversely coupled fiber Fabry–Perot resonator: Theory," *Appl. Opt.* **26**, 456–463 (1987).

50. M. Brierley and P. Urquhart, "Transversely coupled fiber Fabry–Perot resonators: performance characteristics," *Appl. Opt.* **26**, 4841–4845 (1987).

51. M. J. Adams, "Theory of twin-guide Fabry–Perot laser amplifiers," *IEE Proc.* **136**, 287–292 (1989).

52. M. J. Adams, D. A. H. Mace, J. Singh, and M. A. Fisher, "Optical switching in the twin-guide Fabry–Perot laser amplifier," *IEEE J. Quantum Electron.* **26**, 1764–1771 (1990).

53. W. J. Fader and G. E. Palma, "Normal modes of N coupled lasers," *Opt. Lett.* **10**, 381–383 (1985).

54. E. Kapon, J. Katz, and A. Yariv, "Supermode analysis of phase-locked arrays of semiconductor lasers," *Opt. Lett.* **10**, 125–127 (1984); "Erratum," *Opt. Lett.* **9**, 318 (1984).

55. A. Hardy and W. Streifer, "Analysis of phased-array diode lasers," *Opt. Lett.* **10**, 335–337 (1985).

56. H. Kogelnik, "Coupled wave theory for thick hologram gratings," *Bell Syst. Tech. J.* **48**, 2909–2947 (1969).

57. W. S. C. Chang, "Periodic structures and their application in integrated optics," *IEEE Trans. Microwave Theory Tech.* **MTT-21**, 775–785 (1973).

58. W. Streifer, D. R. Scifres, and R. D. Burnham, "Coupling coefficients for distributed feedback single- and double-heterostructure diode lasers," *IEEE J. Quantum Electron.* **QE-11**, 867–873 (1975).

59. W. Streifer, D. R. Scifres, and R. D. Burnham, "TM-mode coupling coefficients in guided-wave distributed feedback lasers," *IEEE J. Quantum Electron.* **QE-12**, 74–78 (1976).

60. P. H. H. Hsieh and C. G. Fonstad, "Coupling coefficient calculations for lead-tin telluride distributed feedback lasers," *IEEE J. Quantum Electron.* **QE-13**, 17–23 (1977).

61. Y. Yamamoto, T. Kamiya, and H. Yanai, "Improved coupled mode analysis of corrugated waveguides and lasers," *IEEE J. Quantum Electron.* **QE-14**, 245–258 (1978).

62. Y. Yamamoto, T. Kamiya, and H. Yanai, "Improved coupled mode analysis of corrugated waveguides and lasers-II: TM mode," *IEEE J. Quantum Electron.* **QE-14**, 620–624 (1978).

63. S. H. Kim and C. G. Fonstad, "Tunable narrow-band thin-film waveguide grating filters," *IEEE J. Quantum Electron.* **QE-15**, 1405–1408 (1979).

64. E. Kapon, A. Hardy, and A. Katzir, "TE and TM coupling coefficients in distributed feedback double-heterostructure PbSnTe lasers," *IEEE J. Quantum Electron.* **QE-17**, 391–397 (1981).

PART III

Generation of Light

9

\

Optical Processes in Semiconductors

The quantum theory of light has been an intriguing subject since 1900 when Planck [1] proposed the idea of quantization of the electromagnetic energy. In this theory, the light or the electromagnetic wave with an oscillating frequency ω is quantized into indivisible packets of energy $\hbar\omega$. In 1905, Einstein [2] proposed that the photoelectric effect could be explained using a corpuscle concept of the electromagnetic radiation. With the invention of lasers, quantum theory of light also plays an important role in our understanding of many optical and electronic processes in materials such as semiconductors. Rigorous in-depth treatments of the quantum theory of light can be found in Refs. 3 and 4.

In this chapter, we study the optical transitions, absorptions, and gains in semiconductors. We start with the Fermi's golden rule, using the time-dependent perturbation theory derived in Section 3.7. We discuss the Einstein A and B coefficients for spontaneous and stimulated emissions and show how they can be calculated quantum mechanically. A formal density-matrix approach can also be used to derive the optical absorption and gain coefficients in semiconductors and is shown in Appendix I.

We then discuss the optical processes in bulk and quantum-well semiconductors. The optical matrix elements are derived based on Kane's model for the semiconductor conduction and valence-band structures. For semiconductor quantum wells, valence-band mixing effects play an important role in the optical gain of a quantum-well laser. We show how these heavy-hole and light-hole band mixing effects can be taken into account in the modeling of gain spectrum in quantum-well structures.

9.1 OPTICAL TRANSITIONS USING FERMI'S GOLDEN RULE

Consider a semiconductor illuminated by light. The interaction between the photons and the electrons in the semiconductor can be described by the Hamiltonian [5]

$$H = \frac{1}{2m_0}(\mathbf{p} - e\mathbf{A})^2 + V(\mathbf{r}) \qquad (9.1.1)$$

where m_0 is the free electron mass, $e = -|e|$ for electrons, $\mathbf{A}$ is the vector potential accounting for the presence of the electromagnetic field, and $V(\mathbf{r})$ is the periodic crystal potential. The above Hamiltonian can be understood because taking the partial derivatives of H with respect to the momentum variable $\mathbf{p}$ and the position vector $\mathbf{r}$ will lead to the classical equation of motion described by the Lorentz force equation [5] in the presence of an electric and a magnetic field (see Problem 9.1).

9.1.1 The Electron–Photon Interaction Hamiltonian

The Hamiltonian can be expanded into

$$H = \frac{\mathbf{p}^2}{2m_0} + V(\mathbf{r}) - \frac{e}{2m_0}(\mathbf{p}\cdot\mathbf{A} + \mathbf{A}\cdot\mathbf{p}) + \frac{e^2\mathbf{A}^2}{2m_0}$$

$$\simeq H_0 + H' \qquad\qquad (9.1.2)$$

where H_0 is the unperturbed Hamiltonian and H' is considered as a perturbation due to light:

$$H_0 = \frac{\mathbf{p}^2}{2m_0} + V(\mathbf{r}) \qquad\qquad (9.1.3)$$

$$H' \simeq -\frac{e}{m_0}\mathbf{A}\cdot\mathbf{p} \qquad\qquad (9.1.4)$$

The Coulomb gauge (discussed in Section 5.1)

$$\nabla \cdot \mathbf{A} = 0 \qquad\qquad (9.1.5)$$

has been used such that $\mathbf{p}\cdot\mathbf{A} = \mathbf{A}\cdot\mathbf{p}$, noting that $\mathbf{p} = (\hbar/i)\nabla$. The last term $e^2\mathbf{A}^2/2m_0$ is much smaller than the terms linear in $\mathbf{A}$, since $|e\mathbf{A}| \ll |\mathbf{p}|$ for most practical optical field intensities. This can be checked using $\mathbf{p} \simeq \hbar k \simeq \hbar\pi/a_0$, where $a_0 \sim 5.5$ Å is the lattice constant. Assuming the vector potential for the optical electric field of the form

$$\mathbf{A} = \hat{e}A_0\cos(\mathbf{k}_{op}\cdot\mathbf{r} - \omega t)$$

$$= \hat{e}\frac{A_0}{2}e^{i\mathbf{k}_{op}\cdot\mathbf{r}}e^{-i\omega t} + \hat{e}\frac{A_0}{2}e^{-i\mathbf{k}_{op}\cdot\mathbf{r}}e^{+i\omega t} \qquad (9.1.6)$$

where $\mathbf{k}_{op}$ is the wave vector, ω is the optical angular frequency, and $\hat{e}$ is a unit vector in the direction of the optical electric field, we have

$$E(\mathbf{r}, t) = -\frac{\partial \mathbf{A}}{\partial t} \tag{9.1.7}$$

$$= -\hat{e}\omega A_0 \sin(\mathbf{k}_{op} \cdot \mathbf{r} - \omega t) \tag{9.1.8}$$

$$H(\mathbf{r}, t) = \frac{1}{\mu} \nabla \times \mathbf{A}$$

$$= -\frac{1}{\mu} \mathbf{k}_{op} \times \hat{e} A_0 \sin(\mathbf{k}_{op} \cdot \mathbf{r} - \omega t) \tag{9.1.9}$$

since the scalar potential φ vanishes ($\rho = 0$) for the optical field. Thus the Poynting vector for the power intensity (W/cm^2) is given by

$$S(\mathbf{r}, t) = E(\mathbf{r}, t) \times H(\mathbf{r}, t)$$

$$= \mathbf{k}_{op} \frac{\omega A_0^2}{\mu} \sin^2(\mathbf{k}_{op} \cdot \mathbf{r} - \omega t) \tag{9.1.10}$$

which is pointing along the direction of wave propagation $\mathbf{k}_{op}$. The time average of the Poynting flux is simply

$$\langle S(\mathbf{r}, t) \rangle = \frac{\omega A_0^2}{2\mu} \mathbf{k}_{op} \tag{9.1.11}$$

noting that the time average of the $\sin^2(\cdot)$ function is $1/2$. Thus, the magnitude of the optical intensity is

$$S = |\langle S(\mathbf{r}, t) \rangle| = \frac{\omega A_0^2}{2\mu} k_{op} = \frac{n_r \omega^2 A_0^2}{2\mu c} = \frac{n_r c \varepsilon_0 \omega^2 A_0^2}{2} \tag{9.1.12}$$

where the permeability, $\mu = \mu_0$, c is the speed of light in free space, and n_r is the refractive index of the material. The interaction Hamiltonian $H'(\mathbf{r}, t)$ can be written as

$$H'(\mathbf{r}, t) = -\frac{e}{m_0} \mathbf{A}(\mathbf{r}, t) \cdot \mathbf{p}$$

$$= H'(\mathbf{r})e^{-i\omega t} + H'^{+}(\mathbf{r})e^{+i\omega t} \tag{9.1.13}$$

where $H'(\mathbf{r})$ is

$$H'(\mathbf{r}) = -\frac{eA_0 e^{i\mathbf{k}_{op} \cdot \mathbf{r}}}{2m_0} \hat{e} \cdot \mathbf{p} \tag{9.1.14}$$

and the superscript $+$ means the Hermitian adjoint operator of $H'(\mathbf{r})$.

9.1.2 Transition Rate due to Electron–Photon Interaction

The transition rate for the absorption of a photon (Fig. 9.1a), assuming an electron is initially at state E_a, is given by Fermi's golden rule and has been derived in Section 3.7 using the time-dependent perturbation theory:

$$W_{\text{abs}} = \frac{2\pi}{\hbar}|\langle b|H'(\mathbf{r})|a\rangle|^2\,\delta(E_b - E_a - \hbar\omega) \qquad (9.1.15)$$

where $E_b > E_a$ has been assumed. Similarly the transition rate for the emission of a photon (Fig. 9.1b) if an electron is initially at state E_b is

$$W_{\text{ems}} = \frac{2\pi}{\hbar}|\langle a|H'(\mathbf{r})|b\rangle|^2\,\delta(E_a - E_b + \hbar\omega) \qquad (9.1.16)$$

The total upward transition rate per unit volume ($\text{s}^{-1}\,\text{cm}^{-3}$) in the crystal, taking into account the probability that state a is occupied and state b is empty, is

$$R_{a\to b} = \frac{2}{V}\sum_{\mathbf{k}_a}\sum_{\mathbf{k}_b}\frac{2\pi}{\hbar}|H'_{ba}|^2\,\delta(E_b - E_a - \hbar\omega)f_a(1 - f_b) \quad (9.1.17)$$

where we sum over the initial and final states and assume that the Fermi–Dirac distribution

$$f_a = \frac{1}{1 + e^{(E_a - E_F)/k_B T}} \qquad (9.1.18)$$

is the probability that the state a is occupied. A similar expression holds for f_b, where E_a is replaced by E_b, and $(1 - f_b)$ is the probability that the state b

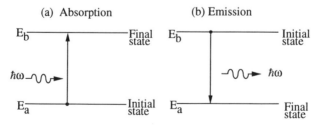

Figure 9.1. (a) The absorption and (b) the emission of a photon with the corresponding electron transitions.

is empty. The prefactor 2 takes into account the sum over spins, and the matrix element H'_{ba} is given by

$$H'_{ba} \equiv \langle b|H'(\mathbf{r})|a\rangle$$

$$= \int \Psi_b^*(\mathbf{r}) H'(\mathbf{r}) \Psi_a(\mathbf{r}) \, d^3\mathbf{r} \qquad (9.1.19)$$

with $|H'_{ba}| = |H'_{ab}|$. The downward transition rate per unit volume (s^{-1} cm^{-3}) in the crystal can be written:

$$R_{b \to a} = \frac{2}{V} \sum_{\mathbf{k}_a} \sum_{\mathbf{k}_b} \frac{2\pi}{\hbar} |H'_{ba}|^2 \, \delta(E_a - E_b + \hbar\omega) f_b (1 - f_a) \quad (9.1.20)$$

Noting the even property of the delta function, $\delta(-x) = \delta(x)$, the net upward transition rate per unit volume can be written as

$$R = R_{a \to b} - R_{b \to a}$$

$$= \frac{2}{V} \sum_{\mathbf{k}_a} \sum_{\mathbf{k}_b} \frac{2\pi}{\hbar} |H'_{ba}|^2 \, \delta(E_b - E_a - \hbar\omega)(f_a - f_b) \qquad (9.1.21)$$

9.1.3   Optical Absorption Coefficient

The absorption coefficient α (1/cm) in the crystal is the fraction of photons absorbed per unit distance:

$$\alpha = \frac{\text{No. of photons absorbed per unit volume per second}}{\text{No. of photons injected per unit area per second}} \qquad (9.1.22)$$

The injected number of photons per unit area per second is the optical intensity S (W/cm^2) divided by the energy of a photon $\hbar\omega$; therefore,

$$\alpha(\hbar\omega) = \frac{R}{(S/\hbar\omega)}$$

$$= \frac{\hbar\omega}{\left(n_r \omega^2 A_0^2 / 2\mu c\right)} \frac{2}{V} \sum_{\mathbf{k}_a} \sum_{\mathbf{k}_b} \frac{2\pi}{\hbar} |H'_{ba}|^2 \, \delta(E_b - E_a - \hbar\omega)(f_a - f_b)$$

$$(9.1.23)$$

Using the dipole (long wavelength) approximation that $\mathbf{A}(\mathbf{r}) = \mathbf{A}\, e^{i\mathbf{k}_{op}\cdot\mathbf{r}} \simeq \mathbf{A}$, we find that the matrix elements can be written in terms of the momentum matrix element:

$$H'_{ba} = -\frac{e}{m_0}\mathbf{A} \cdot \langle b|\mathbf{p}|a \rangle = -\frac{eA_0}{2m_0}\hat{e} \cdot \mathbf{p}_{ba} \qquad (9.1.24)$$

The absorption coefficient (9.1.23) becomes

$$\alpha(\hbar\omega) = \frac{\pi e^2}{n_r c \varepsilon_0 m_0^2 \omega} \frac{2}{V} \sum_{\mathbf{k}_a} \sum_{\mathbf{k}_b} |\hat{e} \cdot \mathbf{p}_{ba}|^2 \delta(E_b - E_a - \hbar\omega)(f_a - f_b)$$

$$(9.1.25)$$

The Hamiltonian can also be written in terms of the electric dipole moment:

$$\boldsymbol{\mu}_{ba} = e\langle b|\mathbf{r}|a \rangle = e\mathbf{r}_{ba} \qquad (9.1.26)$$

$$H'_{ba} = -\langle b|e\mathbf{r} \cdot \mathbf{E}|a \rangle = -\boldsymbol{\mu}_{ba} \cdot \mathbf{E} \qquad (9.1.27)$$

because

$$\mathbf{p} = m_0 \frac{\mathrm{d}}{\mathrm{d}t}\mathbf{r} = \frac{m_0}{i\hbar}(\mathbf{r}H_0 - H_0\mathbf{r}) \qquad (9.1.28)$$

and

$$H'_{ba} = \left\langle b \left| \frac{-e}{m_0}\mathbf{A}(\mathbf{r}) \cdot \mathbf{p} \right| a \right\rangle$$

$$\simeq \frac{-e}{i\hbar}\mathbf{A} \cdot \langle b|\mathbf{r}H_0 - H_0\mathbf{r}|a \rangle$$

$$= -\frac{e(E_a - E_b)}{i\hbar}\mathbf{A} \cdot \langle b|\mathbf{r}|a \rangle$$

$$\simeq -i\omega\mathbf{A} \cdot \langle b|e\mathbf{r}|a \rangle = -\boldsymbol{\mu}_{ba} \cdot \mathbf{E} \qquad (9.1.29)$$

where $E_b - E_a \simeq \hbar\omega$ has been used, and $\mathbf{E} = +i\omega\mathbf{A}$ for the first term in $\mathbf{A}(\mathbf{r},t)$ with $\exp(-i\omega t)$ dependence from (9.1.6) and (9.1.7).

In terms of the dipole moment, we write the absorption coefficient as

$$\alpha(\hbar\omega) = \frac{\pi\omega}{n_r c \varepsilon_0} \frac{2}{V} \sum_{\mathbf{k}_a} \sum_{\mathbf{k}_b} |\hat{e} \cdot \boldsymbol{\mu}_{ba}|^2 \delta(E_b - E_a - \hbar\omega)(f_a - f_b) \quad (9.1.30)$$

We can see that the factors containing A_0^2 are canceled since the linear optical absorption coefficient is independent of the optical intensity, which is proportional to A_0^2. When the scattering relaxation is included, the delta function may be replaced by a Lorentzian function with a linewidth Γ:

$$\delta(E_b - E_a - \hbar\omega) \rightarrow \frac{\Gamma/(2\pi)}{(E_b - E_a - \hbar\omega)^2 + (\Gamma/2)^2} \quad (9.1.31)$$

where a factor π has been included such that the area under the function is properly normalized:

$$\int \delta(E_b - E_a - \hbar\omega) \, \mathrm{d}(\hbar\omega) = 1 \quad (9.1.32)$$

9.1.4 Real and Imaginary Parts of the Permittivity Function

The absorption coefficient α is related to the imaginary part of the permittivity function by

$$\alpha = 2 \operatorname{Im} \omega\sqrt{\mu(\varepsilon_1 + i\varepsilon_2)}$$

$$\simeq 2 \operatorname{Im} \frac{\omega}{c} n_r \left(1 + i\frac{\varepsilon_2}{2\varepsilon_1}\right)$$

$$= \frac{\omega}{n_r c} \frac{\varepsilon_2}{\varepsilon_0} \quad (9.1.33)$$

assuming that $|\varepsilon_2| \ll \varepsilon_1$. Here a factor of 2 accounts for the fact that α refers to the absorption coefficient of the optical intensity, not the electric field. We obtain the imaginary part of the permittivity function:

$$\varepsilon_2(\omega) = \frac{n_r c \varepsilon_0}{\omega} \alpha(\hbar\omega)$$

$$= \frac{\pi e^2}{m_0^2 \omega^2} \frac{2}{V} \sum_{\mathbf{k}_a} \sum_{\mathbf{k}_b} |\hat{e} \cdot \mathbf{p}_{ba}|^2 \delta(E_b - E_a - \hbar\omega)(f_a - f_b) \quad (9.1.34)$$

In semiconductors, the conduction and valence-band structures determine the energy-momentum relations $E_b = E(\mathbf{k}_b)$ and $E_a = E(\mathbf{k}_a)$, respectively. An important part of calculating the absorption coefficient is to find the band structures and the wave functions. The optical matrix element $\mathbf{p}_{ba}$ is calculated from the wave functions based on the parabolic band model. We usually start from bulk semiconductors for which the band structures are known. Then we use the $\mathbf{k} \cdot \mathbf{p}$ perturbation method and the effective mass theory near the band edges, as discussed in Chapter 4, to study the optical processes near the band edges of bulk and quantum-well semiconductor structures. Using the Kramers–Kronig relation derived in Appendix E,

$$\varepsilon_1(\omega) = \varepsilon_0 + \frac{2}{\pi} \mathbf{P} \int_0^\infty \frac{\omega' \varepsilon_2(\omega')}{\omega'^2 - \omega^2} d\omega' \qquad (9.1.35)$$

where $\mathbf{P}$ denotes the principal value of the integral, we obtain

$$\varepsilon_1(\omega) = \varepsilon_0 + \left(\frac{2e^2\hbar^2}{m_0^2} \right) \frac{2}{V} \sum_{\mathbf{k}_a} \sum_{\mathbf{k}_b} |\hat{e} \cdot \mathbf{p}_{ba}|^2 \frac{f_a - f_b}{(E_b - E_a)\left[(E_b - E_a)^2 - (\hbar\omega)^2\right]}$$

$$(9.1.36)$$

If, instead, Eq. (9.1.30) with the dipole moment matrix is used, we have

$$\varepsilon_1(\omega) = \varepsilon_0 + \frac{2}{V} \sum_{\mathbf{k}_a} \sum_{\mathbf{k}_b} |\hat{e} \cdot \boldsymbol{\mu}_{ba}|^2 \frac{2(E_b - E_a)}{(E_b - E_a)^2 - (\hbar\omega)^2} (f_a - f_b) \quad (9.1.37)$$

and

$$\varepsilon_2(\omega) = \pi \frac{2}{V} \sum_{\mathbf{k}_a} \sum_{\mathbf{k}_b} |\hat{e} \cdot \boldsymbol{\mu}_{ba}|^2 \, \delta(E_b - E_a - \hbar\omega)(f_a - f_b) \quad (9.1.38)$$

In practice, linewidth broadening caused by scatterings has to be considered to compare the theoretical absorption spectrum with experimental data. We replace the delta function by a Lorentzian function in (9.1.34) using (9.1.31),

$$\varepsilon_2(\omega) = \frac{e^2}{m_0^2\omega^2} \frac{2}{V} \sum_{\mathbf{k}_a} \sum_{\mathbf{k}_b} |\hat{e} \cdot \mathbf{p}_{ba}|^2 \frac{\Gamma/2}{(E_b - E_a - \hbar\omega)^2 + (\Gamma/2)^2} (f_a - f_b)$$

$$(9.1.39)$$

The real part of the permittivity function $\varepsilon_1(\omega)$ can be obtained using (9.1.36) for a zero linewidth with the factor $1/(E_b - E_a - \hbar\omega)$ replaced by

$$\mathrm{Re}\left(\frac{1}{E_b - E_a - \hbar\omega - i\Gamma/2} \right) = \frac{E_b - E_a - \hbar\omega}{(E_b - E_a - \hbar\omega)^2 + (\Gamma/2)^2}$$

and we obtain

$$\varepsilon_1(\omega) = \varepsilon_0 + \left(\frac{2e^2\hbar^2}{m_0^2}\right)\frac{2}{V}\sum_{\mathbf{k}_a}\sum_{\mathbf{k}_b}|\hat{e}\cdot\mathbf{p}_{ba}|^2$$

$$\times\frac{f_a - f_b}{(E_b - E_a)(E_b - E_a + \hbar\omega)}\frac{E_b - E_a - \hbar\omega}{(E_b - E_a - \hbar\omega)^2 + (\Gamma/2)^2}$$

$$(9.1.40)$$

Similar procedures apply to (9.1.37) and (9.1.38).

9.2 SPONTANEOUS AND STIMULATED EMISSIONS

In Section 9.1, we presented a quantum mechanical deviation for optical absorption of semiconductors in the presence of a monochromatic electromagnetic field using Fermi's golden rule. Here we discuss the spontaneous and stimulated emissions in semiconductors and derive the relations between the spontaneous emission, stimulated emission, and absorption spectra [4, 6, 7].

First we consider a *discrete* two-level system in the presence of an electromagnetic field with a broad spectrum (Fig. 9.2a). The total transition rate per unit volume $(\text{s}^{-1}\text{cm}^{-3})$ is given by

$$R_{12} = \frac{1}{V}\sum_k\frac{2\pi}{\hbar}|H'_{12}|^2\,\delta(E_2 - E_1 - \hbar\omega_k)2n_{\text{ph}}\qquad(9.2.1)$$

where H'_{12} is the matrix element of the interaction Hamiltonian due to the electromagnetic field and $\hbar\omega_k$ is the energy of a photon with a wave vector $\mathbf{k}$.

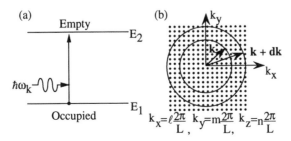

Figure 9.2. (a) A photon incident on a discrete two-level system where level 1 is occupied and level 2 is empty. (b) The k-space diagram for the density of photon states. A dot represents one state with two possible polarizations.

The expression

$$n_{ph} = \frac{1}{e^{\hbar\omega_k/k_BT} - 1} \tag{9.2.2}$$

is the number of photons per state following the Bose–Einstein statistics for identical particles (photons), and a factor of 2 accounts for two polarization states for each **k** vector. The photon field can be described by a plane wave for simplicity as

$$e^{i\mathbf{k}\cdot\mathbf{r}} = e^{ik_x x + ik_y y + ik_z z} \tag{9.2.3}$$

9.2.1 Density of States for the Photons

To define the density of states for the photon field, we use the periodic boundary conditions that the wave function should be periodic in the x, y, and z directions with a period L. Therefore,

$$k_x = \ell\frac{2\pi}{L} \qquad k_y = m\frac{2\pi}{L} \qquad k_z = n\frac{2\pi}{L} \tag{9.2.4}$$

The volume of a state in the k space is therefore $(2\pi/L)^3$ (Fig. 9.2b). Using the dispersion relation for the photon,

$$\omega_k = \frac{kc}{n_r} \tag{9.2.5}$$

where c/n_r is the speed of light in the medium with a refractive index n_r, we can change the sum over the **k** vector to an integral.

Let us look at the integral using the number of states with a differential volume in the k space $d^3k/(2\pi/L)^3 = k^2\,dk\,d\Omega/(2\pi/L)^3$, where $d\Omega$ is the differential solid angle,

$$N(E_{21}) = \frac{2}{V}\sum_{\mathbf{k}}\delta(E_2 - E_1 - E_k) = 2\int\frac{k^2\,dk\,d\Omega}{(2\pi)^3}\delta(E_2 - E_1 - E_k) \tag{9.2.6}$$

V is the volume of the space,

$$E_k = \hbar\omega_k = \frac{\hbar kc}{n_r} \tag{9.2.7}$$

is the photon energy, and the integration over the solid angle is 4π. We find

$$N(E_{21}) = \frac{8\pi n_r^3 E_{21}^2}{h^3 c^3} \tag{9.2.8}$$

which is the number of states with photon energy E_{21} per unit volume per energy interval, $cm^{-3}(eV)^{-1}$, and $E_{21} = E_2 - E_1$ is the energy spacing between the two levels. Note that the Planck constant h instead of $\hbar$ is used in (9.2.8).

9.2.2 Spontaneous and Stimulated Emissions: Einstein's A and B Coefficients

Define

$$B_{12} = \frac{2\pi}{\hbar}|H'_{12}|^2 \tag{9.2.9}$$

as the transition rate per incident photon within an energy interval, (eV/s). We obtain the upward transition rate per unit volume (s^{-1} cm^{-3}) for a broad spectrum or incoherent light (blackbody radiation)

$$R_{12} = B_{12}P(E_{21}) \tag{9.2.10}$$

where

$$P(E_{21}) = N(E_{21})n_{ph} \tag{9.2.11}$$

is the number of photons per unit volume per energy interval with a dimension of $cm^{-3}(eV)^{-1}$, and

$$n_{ph} = \frac{1}{e^{E_{21}/k_B T} - 1} \tag{9.2.12}$$

is the average number of photons per state at an optical energy E_{21}.

If we take into account the occupation probabilities of level 1 and level 2, f_1 and f_2, respectively, the expression for R_{12} is slightly modified (Fig 9.3):

$$R_{12}(= r_{12}(E)\,dE) = \frac{1}{V}\sum_k B_{12}\,\delta(E_2 - E_1 - \hbar\omega_k)2n_{ph}f_1(1 - f_2)$$

$$= B_{12}f_1(1 - f_2)P(E_{21}) \tag{9.2.13}$$

where $r_{12}(E)\,dE$ means that the upward transition rate per unit volume has been integrated for a light with a spectral width dE near $E = E_{21}$. $r_{12}(E)$ is the number of $1 \rightarrow 2$ transitions per second per unit volume per energy interval (s^{-1} cm^{-3} eV^{-1}).

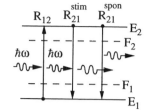

Figure 9.3. Schematic diagram for the stimulated absorption rate R_{12}, stimulated emission rate R_{21}^{stim}, and the spontaneous emission rate R_{21}^{spon} in the presence of two levels with quasi-Fermi levels, F_1 and F_2, respectively.

Similarly, a stimulated emission rate per unit volume can be given:

$$R_{21}^{\text{stim}} = r_{21}^{\text{stim}}(E)\,dE = B_{21}f_2(1 - f_1)P(E_{21}) \qquad (9.2.14)$$

The spontaneous emission rate per unit volume is independent of the photon density, and is given by

$$R_{21}^{\text{spon}} = r_{21}^{\text{spon}}(E)\,dE = A_{21}f_2(1 - f_1) \qquad (9.2.15)$$

At thermal equilibrium, there is only one Fermi level, therefore $F_1 = F_2$. We balance the total downward transition rate with the upward transition rate:

$$R_{12} = R_{21}^{\text{stim}} + R_{21}^{\text{spon}} \qquad (9.2.16)$$

or

$$B_{12}f_1(1 - f_2)P(E_{21}) = B_{21}f_2(1 - f_1)P(E_{21}) + A_{21}f_2(1 - f_1) \quad (9.2.17)$$

We obtain

$$\frac{A_{21}}{B_{12}\,e^{E_{21}/k_B T} - B_{21}} = N(E_{21})\frac{1}{e^{E_{21}/k_B T} - 1} \qquad (9.2.18)$$

Therefore, we find

$$B_{12} = B_{21} \qquad (9.2.19)$$

and

$$\frac{A_{21}}{B_{21}} = N(E_{21}) = \frac{8\pi n_r^3 E_{21}^2}{h^3 c^3} \qquad (9.2.20)$$

The ratio of the stimulated and spontaneous emission rates is

$$\frac{R_{21}^{\text{stim}}}{R_{21}^{\text{spon}}} = \frac{B_{21}P(E_{21})f_2(1 - f_1)}{A_{21}f_2(1 - f_1)} = n_{\text{ph}} \qquad (9.2.21)$$

which is the number of photons per state (9.2.12).

9.2.3 Derivation of the Optical Gain and Spontaneous Emission Spectrum

The net absorption rate per unit volume within a spectral width dE is

$$
\begin{aligned}
R_{net}^{abs} &= r_{net}^{abs}(E_{21}) \, dE \\
&= B_{12} f_1 (1 - f_2) P(E_{21}) - B_{21} f_2 (1 - f_1) P(E_{21}) \\
&= B_{12}(f_1 - f_2) P(E_{21})
\end{aligned}
\tag{9.2.22}
$$

Therefore the absorption spectrum within a spectral width dE can be written as

$$
\begin{aligned}
\alpha(E_{21}) \, dE &= \frac{r_{net}^{abs}(E_{21}) \, dE}{P(E_{21})(c/n_r)} \\
&= \frac{n_r}{c} B_{12}(f_1 - f_2)
\end{aligned}
\tag{9.2.23}
$$

The ratio of the spontaneous emission spectrum and the absorption spectrum is

$$
\begin{aligned}
\frac{r_{21}^{spon}(E_{21})}{\alpha(E_{21})} &= \frac{A_{21} f_2 (1 - f_1)}{(n_r/c) B_{12}(f_1 - f_2)} \\
&= \frac{c}{n_r} N(E_{21}) \frac{1}{e^{(E_{21} - \Delta F)/k_B T} - 1}
\end{aligned}
\tag{9.2.24}
$$

or

$$
r_{21}^{spon}(E_{21}) = \left(\frac{8 \pi n_r^2 E_{21}^2}{h^3 c^2} \right) \frac{1}{e^{(E_{21} - \Delta F)/k_B T} - 1} \alpha(E_{21})
\tag{9.2.25}
$$

and $\Delta F = F_2 - F_1$ is the quasi-Fermi level separation.

If we consider the net upward transition for a monochromatic light (a single photon with $E = \hbar\omega$) instead of a light spectrum, we use $r_{net}^{abs}(E)$, that is, the net upward transition rate per unit volume per energy interval from (9.2.22) with $n_{ph} = 1$,

$$
r_{net}^{abs}(E) \, dE = B_{12}(f_1 - f_2) N(E)
\tag{9.2.26a}
$$

We change $1/dE$ to a delta function for a pair of discrete states E_2 and E_1,

$$
r_{net}^{abs}(E) = N(E) B_{12} \, \delta(E_2 - E_1 - E)(f_1 - f_2)
\tag{9.2.26b}
$$

This has to be summed over all initial states 1 (using $\mathbf{k}_a$ as the quantum number) and final states 2 (using $\mathbf{k}_b$) for all the electron wave vectors,

$$r_{\text{net}}^{\text{abs}}(E = \hbar\omega) = 2N(E) \sum_{\mathbf{k}_a} \sum_{\mathbf{k}_b} B_{ab} \delta(E_b - E_a - \hbar\omega)(f_a - f_b) \quad (9.2.27)$$

Similarly from (9.2.23),

$$\alpha(\hbar\omega) = \frac{n_r}{c} 2 \sum_{\mathbf{k}_a} \sum_{\mathbf{k}_b} B_{ab} \delta(E_b - E_a - \hbar\omega)(f_a - f_b) \quad (9.2.28)$$

where

$$B_{ab} = B_{ba} = \frac{2\pi}{\hbar} \left| \left\langle b \left| -\frac{eA_0}{2m_0} \hat{e} \cdot \mathbf{p} \right| a \right\rangle \right|^2 \quad (9.2.29)$$

for each incident photon with an energy $\hbar\omega$. Therefore, we have the Poynting vector or the photon intensity for a photon incident into a volume V with a speed c/n_r:

$$S = \hbar\omega \frac{c}{n_r} \frac{1}{V}$$

$$= \frac{n_r c \varepsilon_0 \omega^2 A_0^2}{2} \quad (9.2.30)$$

Thus, we obtain

$$A_0^2 = \frac{2\hbar}{n_r^2 \varepsilon_0 \omega} \frac{1}{V} \quad (9.2.31)$$

and

$$B_{ba} = \frac{2\pi}{\hbar} \frac{e^2 A_0^2}{4m_0^2} |\hat{e} \cdot \mathbf{p}_{ba}|^2 = \frac{\pi e^2}{n_r^2 \varepsilon_0 m_0^2 \omega} |\hat{e} \cdot \mathbf{p}_{ba}|^2 \frac{1}{V} \quad (9.2.32)$$

which is proportional to the optical matrix element, and the absorption coefficient is

$$\alpha(\hbar\omega) = C_0 \frac{2}{V} \sum_{\mathbf{k}_a} \sum_{\mathbf{k}_b} |\hat{e} \cdot \mathbf{p}_{ba}|^2 \delta(E_b - E_a - \hbar\omega)(f_a - f_b) \quad (9.2.33a)$$

$$C_0 = \frac{\pi e^2}{n_r c \varepsilon_0 m_0^2 \omega} \quad (9.2.33b)$$

The above result is identical to (9.1.25).

Similarly, the spontaneous emission rate per unit volume per unit energy interval (s^{-1} cm^{-3} eV^{-1}) is given by

$$r^{spon}(E = \hbar\omega) = \left(\frac{8\pi n_r^2 E^2}{h^3 c^2}\right) C_0 \frac{2}{V} \sum_{\mathbf{k}_a} \sum_{\mathbf{k}_b} |\hat{e} \cdot \mathbf{p}_{ba}|^2 \delta(E_b - E_a - E) f_b(1 - f_a)$$

(9.2.34)

and the net stimulated emission rate per unit volume per energy interval is given by

$$r_{net}^{stim}(E = \hbar\omega) = \left(\frac{8\pi n_r^2 E^2}{h^3 c^2}\right) C_0 \frac{2}{V} \sum_{\mathbf{k}_a} \sum_{\mathbf{k}_b} |\hat{e} \cdot \mathbf{p}_{ba}|^2 \delta(E_b - E_a - E)(f_b - f_a)$$

(9.2.35)

When $\alpha(\hbar\omega)$ becomes negative, we have gain in the medium:

$$g(E) = -\alpha(E) \tag{9.2.36}$$

We can also write

$$r^{spon}(E) = \left(\frac{8\pi n_r^2 E^2}{h^3 c^2}\right) \frac{1}{1 - e^{[E - (F_2 - F_1)]/k_B T}} g(E) \tag{9.2.37}$$

$$r_{net}^{stim}(E) = \left(\frac{8\pi n_r^2 E^2}{h^3 c^2}\right) g(E) \tag{9.2.38}$$

$$\frac{r_{net}^{stim}(E)}{r^{spon}(E)} = 1 - e^{[E - (F_2 - F_1)]/k_B T} \tag{9.2.39}$$

The spontaneous emission spectrum $r^{spon}(E)$ and the gain spectrum $g(E)$ are plotted in Fig. 9.4. The emission spectrum is always a positive quantity, while the gain changes sign to become absorption [8] when the optical energy is larger than the quasi-Fermi level separation $F_2 - F_1$. Expressions similar to Eqs. (9.2.34) and (9.2.37) have been compared with experimental data with general good agreement [9]. Note that the spontaneous emission intensity per unit volume per energy interval should be $\hbar\omega r^{spon}(\hbar\omega)$ which will be dis-

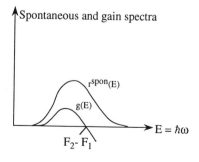

Spontaneous and gain spectra

$r^{spon}(E)$

$g(E)$

$E = \hbar\omega$

$F_2 - F_1$

Figure 9.4. The spontaneous emission spectrum $r^{spon}(E)$ and the gain spectrum $g(E)$ are plotted. The gain changes to absorption when the optical energy E is larger than the difference between the quasi-Fermi level difference $F_2 - F_1$.

cussed further in Section 10.1. Some effects of the optical matrix elements [10] due to doping effects or conduction-band nonparabolicity [9] have been discussed. The crossing point $F_2 - F_1$ in the gain spectrum serves as a good way to estimate the carrier density in a laser diode or a light-emitting diode in addition to matching both the gain and the spontaneous emission spectra. More details on the gain spectrum and the population inversion condition, $\hbar\omega < F_2 - F_1$, are given in Sections 9.3 and 9.4.

9.3 INTERBAND ABSORPTION AND GAIN

For interband transitions between the valence band and conduction band of a semiconductor, we have to evaluate the optical matrix element, using either

$$H'_{ba} = \left\langle b \left| \frac{-e\mathbf{A}(\mathbf{r})}{m_0} \cdot \mathbf{p} \right| a \right\rangle \tag{9.3.1}$$

or

$$H'_{ba} = \langle b| - e\mathbf{r} \cdot \mathbf{E}|a\rangle \tag{9.3.2}$$

9.3.1 Evaluation of the Interband Optical Matrix Element and the k-Selection Rule

The vector potential for the optical field is $\mathbf{A}(\mathbf{r}) = \mathbf{A}\, e^{i\mathbf{k}_{op}\cdot\mathbf{r}} = (\hat{e}A_0/2)\, e^{i\mathbf{k}_{op}\cdot\mathbf{r}}$. The Bloch functions for electrons in the valence band E_a and the conduction band E_b are, respectively,

$$\psi_a(\mathbf{r}) = u_v(\mathbf{r})\frac{e^{i\mathbf{k}_v\cdot\mathbf{r}}}{\sqrt{V}} \tag{9.3.3a}$$

$$\psi_b(\mathbf{r}) = u_c(\mathbf{r})\frac{e^{i\mathbf{k}_c\cdot\mathbf{r}}}{\sqrt{V}} \tag{9.3.3b}$$

where $u_v(\mathbf{r})$ and $u_c(\mathbf{r})$ are the periodic parts of the Bloch functions, and the remainders are the envelope functions (plane waves) for a free electron. The momentum matrix element is

$$
\begin{aligned}
H'_{ba} &= -\frac{eA_0}{2m_0}\hat{e}\cdot\int\psi_b^*\, e^{i\mathbf{k}_{op}\cdot\mathbf{r}}\mathbf{p}\psi_a\, d^3\mathbf{r}\\
&= -\frac{eA_0}{2m_0}\hat{e}\cdot\int u_c^*(\mathbf{r})\, e^{-i\mathbf{k}_c\cdot\mathbf{r}}\, e^{i\mathbf{k}_{op}\cdot\mathbf{r}}\left[\left(\frac{\hbar}{i}\nabla u_v(\mathbf{r})\right)e^{i\mathbf{k}_v\cdot\mathbf{r}} + \hbar\mathbf{k}_v u_v(\mathbf{r})e^{i\mathbf{k}_v\cdot\mathbf{r}}\right]\frac{d^3\mathbf{r}}{V}\\
&\simeq -\frac{eA_0}{2m_0}\hat{e}\cdot\int_\Omega u_c^*(\mathbf{r})\frac{\hbar}{i}\nabla u_v(\mathbf{r})\frac{d^3\mathbf{r}}{\Omega}\int_V e^{i(-\mathbf{k}_c+\mathbf{k}_{op}+\mathbf{k}_v)\cdot\mathbf{r}}\frac{d^3\mathbf{r}}{V}\\
&= -\frac{eA_0}{2m_0}\hat{e}\cdot\mathbf{p}_{cv}\,\delta_{\mathbf{k}_c,\mathbf{k}_v+\mathbf{k}_{op}}
\end{aligned}\tag{9.3.4}
$$

$$\mathbf{p}_{cv} = \int_\Omega u_c^*(\mathbf{r})\frac{\hbar}{i}\nabla u_v(\mathbf{r})\frac{d^3\mathbf{r}}{\Omega} \tag{9.3.5}$$

where we noted that $[u_c^*(\mathbf{r})(\hbar/i)\nabla u_v(\mathbf{r})]$ and $[u_c^*(\mathbf{r})u_v(\mathbf{r})]$ are fast varying functions over a unit cell and are periodic, while the envelope functions are slowly varying functions over a unit cell. Therefore, the integral over $d^3\mathbf{r}$ can be separated into the product of two integrals, one over the unit cell Ω for the fast varying part, and the other over the slowly varying part. Alternatively, for an integral

$$I = \int_V \left[u_c^*(\mathbf{r}) \frac{\hbar}{i} \nabla u_v(\mathbf{r}) \right] F(\mathbf{r})\, d^3\mathbf{r} \tag{9.3.6}$$

where $F(\mathbf{r})$ is slowly varying over a unit cell, we may use the periodic property of the Bloch periodic functions

$$u_c^*(\mathbf{r}) \frac{\hbar}{i} \nabla u_v(\mathbf{r}) = \sum_G C_G\, e^{i\mathbf{G}\cdot\mathbf{r}} \tag{9.3.7}$$

where the $\mathbf{G}$'s are the reciprocal lattice vectors:

$$I = \int_V \sum_G C_G\, e^{i\mathbf{G}\cdot\mathbf{r}} F(\mathbf{r})\, d^3\mathbf{r}$$

$$= \sum_R \sum_G C_G \int_\Omega e^{i\mathbf{G}\cdot(\mathbf{r}+\mathbf{R})} F(\mathbf{r}+\mathbf{R})\, d^3\mathbf{r} \tag{9.3.8}$$

the $\mathbf{R}$'s are the lattice vectors, and $\exp[i\mathbf{G}\cdot\mathbf{R}] = 1$. Since $F(\mathbf{r})$ is slowly varying over a unit cell, we may approximate $F(\mathbf{r}+\mathbf{R}) \simeq F(\mathbf{R})$ and put it outside of the integral over a unit cell. Therefore,

$$I \simeq \sum_R F(\mathbf{R})\Omega \int_\Omega \sum_G C_G\, e^{i\mathbf{G}\cdot\mathbf{r}} \frac{d^3\mathbf{r}}{\Omega}$$

$$= \int_V F(\mathbf{r})\, d^3\mathbf{r} \int_\Omega u_c^*(\mathbf{r}) \frac{\hbar}{i} \nabla u_v(\mathbf{r}) \frac{d^3\mathbf{r}}{\Omega} \tag{9.3.9}$$

where Ω is the differential volume $d^3\mathbf{r}$ when we consider the sum of the envelope function over the whole crystal. Note that the orthogonal property $\int_\Omega u_c^* u_v\, d^3\mathbf{r} = 0$ has been used in (9.3.4). Also note that from $\langle \psi_a | \psi_a \rangle = \langle \psi_b | \psi_b \rangle = 1$, the normalization conditions for u_c and u_v are $(1/\Omega)\int_\Omega u_c^* u_c\, d^3\mathbf{r} = (1/\Omega)\int u_v^* u_v\, d^3\mathbf{r} = 1$. From the matrix element (9.3.4), we see that the momentum conservation

$$\hbar \mathbf{k}_c = \hbar \mathbf{k}_v + \hbar \mathbf{k}_{op} \tag{9.3.10}$$

is obeyed. The electron at the final state has its crystal momentum $\hbar \mathbf{k}_c$ equal to its initial momentum $\hbar \mathbf{k}_v$ plus the photon momentum $\hbar \mathbf{k}_{op}$. Since

$k_{op} \sim 2\pi/\lambda_0$, and the magnitudes k_c, k_v are of the order $2\pi/a_0$, where a_0 is the lattice constant of the semiconductors, which is typically of the order of 5.5 Å and is much smaller than λ_0, we may ignore $\mathbf{k}_{op}$ and obtain

$$H'_{ba} \simeq -\frac{eA_0}{2m_0}\hat{e} \cdot \mathbf{p}_{cv}\,\delta_{\mathbf{k}_c,\mathbf{k}_v} \qquad (9.3.11)$$

which is the **k**-selection rule in the optical transitions. Notice that the interband momentum matrix element, $\mathbf{p}_{cv}$, depends only on the periodic parts (u_c and u_v) of the Bloch functions and is derived from the original optical momentum matrix element $\mathbf{p}_{ba}$, which, on the other hand, depends on the full Bloch functions (i.e., including the envelope functions).

9.3.2   Absorption

Using the **k**-selection rule, we find that the absorption coefficient (9.1.25) for a bulk semiconductor is

$$\alpha(\hbar\omega) = C_0\frac{2}{V}\sum_{\mathbf{k}}|\hat{e} \cdot \mathbf{p}_{cv}|^2\,\delta(E_c - E_v - \hbar\omega)(f_v - f_c) \quad (9.3.12a)$$

$$C_0 = \frac{\pi e^2}{n_r c \varepsilon_0 m_0^2 \omega} \qquad (9.3.12b)$$

where the Fermi–Dirac distributions for the electrons in the valence band and in the conduction band are, respectively

$$f_v(\mathbf{k}) = \frac{1}{1 + e^{(E_v(\mathbf{k}) - F_v)/k_B T}}$$

$$f_c(\mathbf{k}) = \frac{1}{1 + e^{(E_c(\mathbf{k}) - F_c)/k_B T}} \qquad (9.3.13)$$

and F_v, F_c are the quasi-Fermi levels. We use **k** to represent both $\mathbf{k}_c$ and $\mathbf{k}_v$. In the case of thermal equilibrium, $F_v = F_c = E_F$. We assume that the semiconductor is undoped, the valence band is completely occupied, and the conduction band is empty, $f_v = 1$ and $f_c = 0$. We further assume that the matrix element $(\hat{e} \cdot \mathbf{p}_{cv})^2$ is independent of **k** and denote the absorption at thermal equilibrium:

$$\alpha_0(\hbar\omega) = C_0|\hat{e} \cdot \mathbf{p}_{cv}|^2 \int \frac{2d^3\mathbf{k}}{(2\pi)^3}\delta\left(E_g + \frac{\hbar^2 k^2}{2m_r} - \hbar\omega\right) \qquad (9.3.14)$$

where

$$E_c = E_g + \frac{\hbar^2 k^2}{2m_e^*} \qquad E_v = -\frac{\hbar^2 k^2}{2m_h^*}$$

$$\frac{1}{m_r} = \frac{1}{m_e^*} + \frac{1}{m_h^*} \tag{9.3.15}$$

Notice that all energies are measured from *the top of the valence band*. Therefore, both E_c and F_c contain the band-gap energy E_g. The integral can be carried out analytically:

$$\alpha_0(\hbar\omega) = C_0 |\hat{e} \cdot \mathbf{p}_{cv}|^2 \rho_r(\hbar\omega - E_g) \tag{9.3.16a}$$

$$C_0 = \frac{\pi e^2}{n_r c \varepsilon_0 m_0^2 \omega} \tag{9.3.16b}$$

Therefore, the absorption coefficient depends on the momentum-matrix element and the reduced density of states:

$$\rho_r(\hbar\omega - E_g) = \frac{1}{2\pi^2} \left(\frac{2m_r}{\hbar^2} \right)^{3/2} (\hbar\omega - E_g)^{1/2} \tag{9.3.16c}$$

If we use the dipole moment (9.3.2) and define

$$\boldsymbol{\mu}_{cv} = \langle u_c | e\mathbf{r} | u_v \rangle \tag{9.3.17}$$

we can relate $\boldsymbol{\mu}_{cv}$ to $\mathbf{p}_{cv}$ by

$$\mathbf{p}_{cv} = im_0 \omega_{cv} \mathbf{r}_{cv}$$

$$= \frac{im_0 \omega_{cv} \boldsymbol{\mu}_{cv}}{e} \tag{9.3.18}$$

where $\omega_{cv} = (E_c - E_v)/\hbar \simeq \omega$. The absorption coefficient $\alpha_0(\hbar\omega)$ is then given by

$$\alpha_0(\hbar\omega) = \left(\frac{\pi\omega}{n_r c \varepsilon_0} \right) |\hat{e} \cdot \boldsymbol{\mu}_{cv}|^2 \frac{1}{2\pi^2} \left(\frac{2m_r}{\hbar^2} \right)^{3/2} (\hbar\omega - E_g)^{1/2} \tag{9.3.19}$$

The absorption spectrum is plotted in Fig. 9.5. We can see the dominant square-root behavior of the optical energy above the band gap, $(\hbar\omega - E_g)^{1/2}$, representing the joint density of states. Below the band-gap energy E_g, the absorption does not occur since the photons see a forbidden band gap.

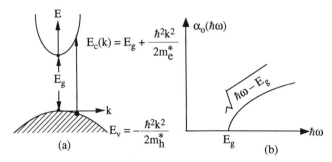

Figure 9.5. (a) Optical absorption in a direct-band-gap semiconductor. (b) The absorption spectrum due to the interband transitions.

9.3.3 Gain

Under current injection or optical pumping, there will be both electrons and holes in the semiconductors. Let us assume that a quasi-equilibrium state has been reached such that we have two quasi-Fermi levels, F_c and F_v, for the electrons and holes, respectively. Carrying out the integration as before, we get

$$\alpha(\hbar\omega) = C_0(\hat{e} \cdot \mathbf{p}_{cv})^2 \int \frac{2d^3\mathbf{k}}{(2\pi)^3} \delta\left(E_g + \frac{\hbar^2 k^2}{2m_r} - \hbar\omega\right)[f_v(k) - f_c(k)] \tag{9.3.20}$$

Let

$$X = E_g + \frac{\hbar^2 k^2}{2m_r} - \hbar\omega \qquad E = \frac{\hbar^2 k^2}{2m_r}$$

We find, by a change of variables, the integration is the same as in (9.3.14) with the contribution at $X = 0$, and $E = \hbar\omega - E_g$:

$$\alpha(\hbar\omega) = \alpha_0(\hbar\omega)[f_v(k_0) - f_c(k_0)] \tag{9.3.21}$$

where

$$k_0 = \sqrt{\frac{2m_r}{\hbar^2}(\hbar\omega - E_g)}$$

We note that $\alpha(\hbar\omega)$ becomes negative if $f_v(k) - f_c(k) < 0$. Since

$$f_v(k) - f_c(k) = \frac{e^{(E_c - F_c)/k_B T} - e^{(E_v - F_v)/k_B T}}{\left(1 + e^{(E_v - F_v)/k_B T}\right)\left(1 + e^{(E_c - F_c)/k_B T}\right)} < 0 \tag{9.3.22}$$

leads to

$$e^{(E_c - F_c)/k_B T} < e^{(E_v - F_v)/k_B T}$$

or

$$F_c - F_v > E_c - E_v = \hbar\omega \qquad (9.3.23)$$

The population inversion condition is satisfied if (9.3.23) holds and the absorption becomes negative or, equivalently, there is gain in the medium. The above condition (9.3.23) is called the Bernard–Duraffourg inversion condition [11]. Plots of $\alpha(\hbar\omega)$ and $(f_v - f_c)$ are illustrated [12] in Fig. 9.6. We can see that gain exists in the optical spectral region $E_g < \hbar\omega < F_c - F_v$. The gain spectrum is determined by the absorption spectrum $\alpha_0(\hbar\omega)$ of the semiconductors without optical excitations multiplied by the Fermi–Dirac inversion factor $(f_v - f_c)$, which accounts for the population inversion probability. This factor, $f_v - f_c$, has a lower limit -1 at small optical energies and

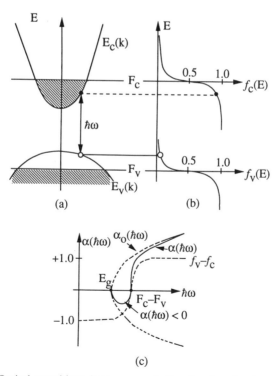

Figure 9.6. (a) Optical transitions between conduction band and valence band. (b) The Fermi–Dirac distributions $f_v(E)$ and $f_c(E)$. (c) The absorption spectrum $\alpha(\hbar\omega)$. Note that it is negative for $E_g < \hbar\omega < F_c - F_v$, where gain exists.

has a zero crossing at the quasi-Fermi level separation, $F_c - F_v$, then approaches $+1$ at large energies, as shown in Fig. 9.6. The increasing absorption coefficient with increasing energy $\hbar\omega$ and the decreasing magnitude of $|f_v - f_c|$ (dashed line) give a peak gain, which depends on the temperature and the effective masses of the electrons and holes near the band edges.

9.3.4 Summary

We summarize the final explicit expressions for the gain spectrum $g(\hbar\omega) = -\alpha(\hbar\omega)$.

1. Zero linewidth (assume no scatterings)

$$g(\hbar\omega) = C_0(\hat{e} \cdot \mathbf{p}_{cv})^2 \rho_r(E = \hbar\omega - E_g)$$
$$\times \left[f_c(E = \hbar\omega - E_g) - f_v(E = \hbar\omega - E_g) \right] \quad (9.3.24)$$

2. A finite linewidth Γ

$$g(\hbar\omega) = C_0(\hat{e} \cdot \mathbf{p}_{cv})^2 \int_0^\infty dE \rho_r(E) \frac{\Gamma/(2\pi)}{\left(E_g + E - \hbar\omega \right)^2 + (\Gamma/2)^2}$$
$$\times \left[f_c(E) - f_v(E) \right] \quad (9.3.25)$$

where

$$C_0 = \frac{\pi e^2}{n_r C \varepsilon_0 m_0^2 \omega} \quad (9.3.26a)$$

$$\rho_r(E) = \frac{1}{2\pi^2} \left(\frac{2m_r}{\hbar^2} \right)^{3/2} E^{1/2} \quad (9.3.26b)$$

$$f_c(E) = \frac{1}{1 + \exp\{\left[E_g + (m_r/m_e^*)E - F_c \right]/k_B T\}} \quad (9.3.26c)$$

$$f_v(E) = \frac{1}{1 + \exp\{\left[-(m_r/m_h^*)E - F_v \right]/k_B T\}} \quad (9.3.26d)$$

9.4 INTERBAND ABSORPTION AND GAIN IN A QUANTUM-WELL STRUCTURE

In this section we consider the interband absorption and gain in a quantum well, ignoring the exciton effects due to the Coulomb interaction between electrons and holes. The exciton effects are considered in Section 13.3.

The interband absorption or gain can be calculated from (9.1.25) generally if scattering relaxation is not considered:

$$\alpha(\hbar\omega) = C_0\frac{2}{V}\sum_{\mathbf{k}_a}\sum_{\mathbf{k}_b}|\hat{e}\cdot\mathbf{p}_{ba}|^2\,\delta(E_b - E_a - \hbar\omega)(f_a - f_b) \qquad (9.4.1a)$$

$$C_0 = \frac{\pi e^2}{n_r c\varepsilon_0 m_0^2\omega} \qquad (9.4.1b)$$

9.4.1 Interband Optical Matrix Element of a Quantum Well

Within a two-band model, the Bloch wave functions can be described by

$$\psi_a(\mathbf{r}) = u_v(\mathbf{r})\frac{e^{i\mathbf{k}_t\cdot\boldsymbol{\rho}}}{\sqrt{A}}g_m(z) \qquad (9.4.2)$$

for a hole wave function in the heavy-hole or a light-hole subband m, and

$$\psi_b(\mathbf{r}) = u_c(\mathbf{r})\frac{e^{i\mathbf{k}_t'\cdot\boldsymbol{\rho}}}{\sqrt{A}}\phi_n(z) \qquad (9.4.3)$$

for an electron in the conduction subband n. The momentum matrix $\mathbf{p}_{ba}$ is given by

$$\mathbf{p}_{ba} = \langle\psi_b|\mathbf{p}|\psi_a\rangle$$
$$\simeq \langle u_c|\mathbf{p}|u_v\rangle\,\delta_{\mathbf{k}_t,\mathbf{k}_t'}I_{hm}^{en} \qquad (9.4.4a)$$

where

$$I_{hm}^{en} = \int_{-\infty}^{\infty} dz\,\phi_n^*(z)g_m(z) \qquad (9.4.4b)$$

is the overlap integral of the conduction- and valence-band envelope functions. The $\mathbf{k}$-selection rule in the plane of the quantum well is still satisfied. The polarization dependent matrix element $\mathbf{p}_{cv} = \langle u_c|\mathbf{p}|u_v\rangle$ is discussed in Section 9.5. Here we have to take into account the quantizations of the electron and hole energies E_a and E_b:

$$E_a = E_{hm} - \frac{\hbar^2 k_t^2}{2m_h^*} \qquad (9.4.5a)$$

$$E_b = E_g + E_{en} + \frac{\hbar^2 k_t^2}{2m_e^*} \qquad (9.4.5b)$$

Note that $E_{hm} < 0$ and

$$E_b - E_a = E_{hm}^{en} + \frac{\hbar^2 k_t^2}{2m_r} \equiv E_{hm}^{en}(k_t) \qquad (9.4.6a)$$

where

$$E_{hm}^{en} = E_g + E_{en} - E_{hm} \qquad (9.4.6b)$$

is the band-edge transition energy ($k_t = 0$). The summations over the quantum numbers $\mathbf{k}_a$ and $\mathbf{k}_b$ become summations over $(\mathbf{k}_t', m)$ and $(\mathbf{k}_t, n)$. Since $\mathbf{k}_t = \mathbf{k}_t'$ in the matrix element (9.4.4a), we find

$$\alpha(\hbar\omega) = C_0 \sum_{n,m} |I_{hm}^{en}|^2 \frac{2}{V} \sum_{\mathbf{k}_t} |\hat{e} \cdot \mathbf{p}_{cv}|^2 \, \delta(E_{hm}^{en}(k_t) - \hbar\omega)(f_v^m - f_c^n) \quad (9.4.7)$$

9.4.2 Joint Density of States and Optical Absorption Spectrum

Change the variable from $\mathbf{k}_t$ to E_t in (9.4.7)

$$E_t = \frac{\hbar^2 k_t^2}{2m_r}$$

and use the two-dimensional reduced density of states ρ_r^{2D}

$$\frac{2}{V} \sum_{\mathbf{k}_t} = \frac{2A}{V} \int \frac{d^2 \mathbf{k}_t}{(2\pi)^2} = \frac{1}{\pi L_z} \int_0^\infty k_t \, dk_t = \int_0^\infty dE_t \rho_r^{2D} \qquad (9.4.8a)$$

$$\rho_r^{2D} = \frac{m_r}{\pi \hbar^2 L_z} \qquad (9.4.8b)$$

where A is the area of the cross section, $AL_z = V$, L_z is an effective period of the quantum wells, and V is a volume of a period. We obtain

$$\alpha(\hbar\omega) = C_0 \frac{2}{L_z} \sum_{n,m} |I_{hm}^{en}|^2 \int_0^\infty \frac{k_t \, dk_t}{2\pi} |\hat{e} \cdot \mathbf{p}_{cv}|^2 \, \delta(E_{hm}^{en}(k_t) - \hbar\omega)(f_v^m - f_c^n)$$

$$(9.4.9a)$$

$$= C_0 \sum_{n,m} |I_{hm}^{en}|^2 \int_0^\infty dE_t \rho_r^{2D} |\hat{e} \cdot \mathbf{p}_{cv}|^2 \, \delta(E_{hm}^{en} + E_t - \hbar\omega)(f_v^m - f_c^n)$$

$$(9.4.9b)$$

Therefore, the contribution is at $E_{hm}^{en} + E_t = \hbar\omega$ and the absorption edges occur at $\hbar\omega = E_{hm}^{en}$. For an unpumped semiconductor, $f_v^m = 1$ and $f_c^n = 0$, we have the absorption spectrum at thermal equilibrium $\alpha_0(\hbar\omega)$:

$$\alpha_0(\hbar\omega) = C_0 \sum_{n,m} |I_{hm}^{en}|^2 |\hat{e} \cdot \mathbf{p}_{cv}|^2 \rho_r^{2D} H(\hbar\omega - E_{hm}^{en}) \qquad (9.4.10)$$

where H is the Heaviside step function, which has the property that $H(x) = 1$ for $x > 0$, and 0 for $x < 0$. For a symmetric quantum well, we find $I_{hm}^{en} = \delta_{nm}$ using an infinite well model and $\alpha_0(\hbar\omega)/(C_0|\hat{e} \cdot \mathbf{p}_{cv}|^2)$ is given by

$$\frac{\alpha_0(\hbar\omega)}{C_0|\hat{e} \cdot \mathbf{p}_{cv}|^2} = \begin{cases} \dfrac{m_r}{\pi\hbar^2 L_z} & \text{for } E_{h1}^{e1} < \hbar\omega < E_{h2}^{e2} \\[2mm] 2\dfrac{m_r}{\pi\hbar^2 L_z} & \text{for } E_{h2}^{e2} < \hbar\omega < E_{h3}^{e3} \\[2mm] 3\dfrac{m_r}{\pi\hbar^2 L_z} & \text{for } E_{h3}^{e3} < \hbar\omega < E_{h4}^{e4} \\[2mm] \text{etc.} \end{cases} \qquad (9.4.11)$$

which is the joint density of states of a two-dimensional structure. A plot of the above function is shown in Fig. 9.7.

With carrier injection, the Fermi–Dirac population inversion factor, $f_v^m - f_c^n$, has to be included, and we obtain the absorption coefficient

$$\alpha(\hbar\omega) = \alpha_0(\hbar\omega)\left[f_v^m(E_t = \hbar\omega - E_{hm}^{en}) - f_c^n(E_t = \hbar\omega - E_{hm}^{en}) \right] \qquad (9.4.12)$$

The above gain process can also be understood from Fig. 9.8 and the following analysis for carrier populations in quantum wells.

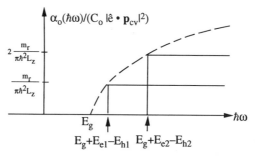

Figure 9.7.   The stepwise absorption spectrum for a quantum-well structure.

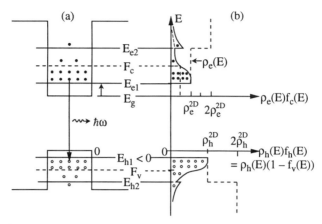

Figure 9.8. (a) Population inversion in a quantum well such that $F_c - F_v > \hbar\omega > E_g + E_{e1} - E_{h1}$. Here F_c is measured from the valence band edge where the energy level is chosen to be zero. (b) The products of the density of states and the occupation probability for electrons in the conduction band $\rho_e(E)f_c(E)$ and holes in the valence band $\rho_h(E)f_h(E) = \rho_h(E)[1 - f_v(E)]$ are plotted vs. the energy E in the vertical scale.

9.4.3 Determination of the Quasi-Fermi Levels

For a given injected electron density N, which is usually determined by the injection current and the background doping, the quasi-Fermi level F_c for the electrons can be determined using

$$N = \sum_{\substack{n = \text{occupied} \\ \text{subbands}}} N_n = \sum_n \int dE \rho_e^{2D}(E) f_c^n(E) \qquad (9.4.13)$$

where N_n is the electron density in the nth conduction subband. The last expression in (9.4.13) shows that the carrier concentration is just the area below the function $\rho_e^{2D}(E)f_c(E)$ in the top figure of Fig. 9.8b. Here $E = E_{en} + (\hbar^2 k_t^2/2m_e^*)$ is the total electron energy in the nth subband:

$$N_n = \frac{2}{V}\sum_{\mathbf{k}_t} f_c^n \qquad f_c^n = \frac{1}{e^{(E_{en} + \hbar^2 k_t^2/2m_e^* - F_c)/k_B T} + 1}$$

$$\frac{2}{V}\sum_{\mathbf{k}_t} = \frac{2}{AL_z}\frac{A}{(2\pi)^2}\int d^2\mathbf{k}_t = \int_0^\infty dE_t \rho_e^{2D}$$

$$E_t = \frac{\hbar^2 k_t^2}{2m_e^*} \qquad \rho_e^{2D} = \frac{m_e^*}{\pi\hbar^2 L_z} \qquad (9.4.14)$$

Using

$$\int dx \frac{1}{1 + e^x} = -\ln(1 + e^{-x}) \tag{9.4.15}$$

we obtain

$$N_n = \frac{m_e^* k_B T}{\pi \hbar^2 L_z} \ln\left(1 + e^{(F_c - E_{en})/k_B T}\right) \tag{9.4.16}$$

The quasi-Fermi level for the holes, F_v, can be determined from

$$P = N + N_A^- - N_D^+ = \sum_m P_m$$

$$= \sum_m \frac{2}{V} \sum_{k_t} [1 - f_v^m(k_t)] = \sum_m \int_0^\infty dE_t \rho_h(E_t) f_h(E_t)$$

$$= \sum_m \frac{m_h^* k_B T}{\pi \hbar^2 L_z} \ln\left(1 + e^{(E_{hm} - F_v)/k_B T}\right) \tag{9.4.17}$$

Again, the hole concentration is just the area below the function $\rho_h(E) f_h(E)$ in the bottom figure of Fig. 9.8b.

9.4.4   Summary of the Gain Spectrum

We can write the gain spectrum from (9.4.10) and (9.4.12). The following is a summary of the expressions for the gain spectrum:

1. *Zero linewidth*

$$g(\hbar\omega) = C_0 \sum_{n,m} |I_{hm}^{en}|^2 |\hat{e} \cdot \mathbf{p}_{cv}|^2 [f_c^n(E_t = \hbar\omega - E_{hm}^{en})$$

$$- f_v^m(E_t = \hbar\omega - E_{hm}^{en})] \rho_r^{2D} H(\hbar\omega - E_{hm}^{en}) \tag{9.4.18}$$

2. *Finite linewidth* Γ

$$g(\hbar\omega) = C_0 \sum_{n,m} |I_{hm}^{en}|^2 \int_0^\infty dE_t \rho_r^{2D} |\hat{e} \cdot \mathbf{p}_{cv}|^2$$

$$\times \frac{\Gamma/(2\pi)}{[E_{hm}^{en}(0) + E_t - \hbar\omega]^2 + (\Gamma/2)^2} [f_c^n(E_t) - f_v^m(E_t)]$$

$$\tag{9.4.19}$$

where

$$C_0 = \frac{\pi e^2}{n_r c \varepsilon_0 m_0^2 \omega} \tag{9.4.20a}$$

$$\rho_r^{2D} = \frac{m_r}{\pi \hbar^2 L_z} \tag{9.4.20b}$$

$$I_{hm}^{en} = \int_{-\infty}^{\infty} dz \phi_n(z) g_m(z) \tag{9.4.20c}$$

$$f_c^n(E_t) = \frac{1}{1 + \exp\{[E_g + E_{en} + (m_r/m_e^*)E_t - F_c]/k_B T\}} \tag{9.4.20d}$$

$$f_v^m(E_t) = \frac{1}{1 + \exp\{[E_{hm} - (m_r/m_h)E_t - F_v]/k_B T\}} \tag{9.4.20e}$$

The proper momentum matrix elements $|\hat{e} \cdot \mathbf{p}_{cv}|^2$ for bulk and quantum wells are discussed in Section 9.5. It is important to note that the momentum matrix element is isotropic for bulk cubic semiconductors, and it is polarization dependent for quantum-well semiconductors. Theory and experiments on polarization-dependent gain in quantum-well structures based on Kane's model [13] have been investigated [14–20]. The momentum matrix $|\hat{e} \cdot \mathbf{p}_{cv}|^2$ should be used with care [16] since various missing factors exist in the literature.

9.4.5 Theoretical Gain Spectrum and Comparison with Experiments

The gain spectra using (9.4.18) for a zero linewidth and using (9.4.19) for a finite linewidth are plotted in Fig. 9.9 for transitions (a) between a pair of

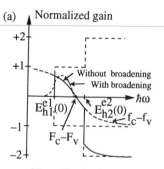

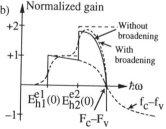

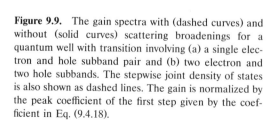

Figure 9.9. The gain spectra with (dashed curves) and without (solid curves) scattering broadenings for a quantum well with transition involving (a) a single electron and hole subband pair and (b) two electron and two hole subbands. The stepwise joint density of states is also shown as dashed lines. The gain is normalized by the peak coefficient of the first step given by the coefficient in Eq. (9.4.18).

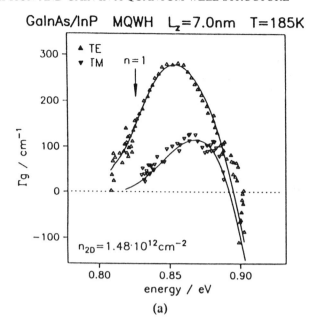

(a)

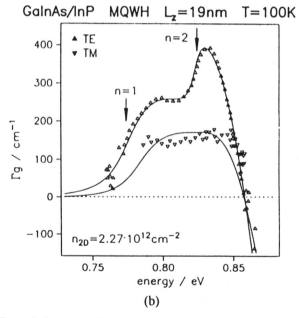

(b)

Figure 9.10. Theoretical and experimental modal gain spectra Γg for both the TE and TM polarizations of an $In_{0.53}Ga_{0.47}As/InP$ multiple-quantum-well laser (a) with a well width 70 Å at $T = 185$ K. The sheet carrier density is estimated to be $1.48 \times 10^{12}/cm^2$; (b) with a well width 190 Å at $T = 100$ K and the sheet carrier density $2.27 \times 10^{12}/cm^2$. The second electron and hole subbands are occupied and they contribute to the second peak at the high-energy end (After Ref. 19 © 1989 IEEE.)

electron and hole subbands and (b) between two pairs of electron and hole subbands. Experimental data for quantum-well lasers have been shown to compare very well with the above theory in general [19–20], as can be seen from Fig. 9.10. The materials are $In_{0.53}Ga_{0.47}As$ quantum wells lattice-matched to InP barriers and substrate. We can see at a small carrier concentration, Fig. 9.10a, that only $n = 1$ electron and hole subbands contribute to the gain. The gain for TE polarization is larger than that of the TM polarization since most of the holes occupy the heavy hole subband. At a higher carrier concentration in a quantum well with a larger well width in Fig. 9.10b, the second electron and hole subbands are occupied and features like double steps appear in the TE gain spectrum.

9.5 MOMENTUM MATRIX ELEMENTS OF BULK AND QUANTUM-WELL SEMICONDUCTORS

In a bulk semiconductor, the optical matrix element is usually isotropic. However, in a quantum-well or superlattice structure, the optical matrix element will depend on the polarization of the optical electromagnetic field [14]. In Kane's model for the semiconductor band structures near the band edges, as discussed in Section 4.2, the electron wave vector is originally assumed to be in the z direction. The wave functions at the band edges have also been obtained as $|\frac{3}{2}, \pm \frac{3}{2}\rangle$ for heavy holes, $|\frac{3}{2}, \pm \frac{1}{2}\rangle$ for light holes, and $|\frac{1}{2}, \pm \frac{1}{2}\rangle$ for spin-orbit split-off holes.

9.5.1 Coordinate Transformation

If the electron wave vector $\mathbf{k}$ has a general direction specified by (k, θ, ϕ) in spherical coordinates,

$$\mathbf{k} = k \sin\theta \cos\phi \, \hat{x} + k \sin\theta \sin\phi \, \hat{y} + k \cos\theta \hat{z}$$

we find the band-edge wave functions are as follows, using the coordinate transformations in (4.2.32) and (4.2.33).

Conduction Band

$$|iS \downarrow'\rangle \quad \text{and} \quad |iS \uparrow'\rangle \qquad (9.5.1)$$

Heavy-Hole Band

$$\left|\frac{3}{2},\frac{3}{2}\right\rangle' = \frac{-1}{\sqrt{2}}\left|(X'+iY')\uparrow'\right\rangle$$

$$= \frac{-1}{\sqrt{2}}\left|(\cos\theta\cos\phi - i\sin\phi)X\right.$$

$$+ (\cos\theta\sin\phi + i\cos\phi)Y - \sin\theta Z\rangle\left|\uparrow'\right\rangle$$

$$\left|\frac{3}{2},-\frac{3}{2}\right\rangle' = \frac{1}{\sqrt{2}}\left|(X'-iY')\downarrow'\right\rangle$$

$$= \frac{1}{\sqrt{2}}\left|(\cos\theta\cos\phi + i\sin\phi)X\right.$$

$$+ (\cos\theta\sin\phi - i\cos\phi)Y - \sin\theta Z\rangle\left|\downarrow'\right\rangle \quad (9.5.2)$$

Light-Hole Band

$$\left|\frac{3}{2},\frac{1}{2}\right\rangle' = \frac{-1}{\sqrt{6}}\left|(X'+iY')\downarrow'\right\rangle + \sqrt{\frac{2}{3}}\left|Z'\uparrow'\right\rangle$$

$$= \frac{-1}{\sqrt{6}}\left|(\cos\theta\cos\phi - i\sin\phi)X\right.$$

$$+ (\cos\theta\sin\phi + i\cos\phi)Y - \sin\theta Z\rangle\left|\downarrow'\right\rangle$$

$$+ \sqrt{\frac{2}{3}}\left|\sin\theta\cos\phi X + \sin\theta\sin\phi Y + \cos\theta Z\rangle\left|\uparrow'\right\rangle$$

$$\left|\frac{3}{2},-\frac{1}{2}\right\rangle' = \frac{1}{\sqrt{6}}\left|(X'-iY')\uparrow'\right\rangle + \sqrt{\frac{2}{3}}\left|Z'\downarrow'\right\rangle$$

$$= \frac{1}{\sqrt{6}}\left|(\cos\theta\cos\phi + i\sin\phi)X\right.$$

$$+ (\cos\theta\sin\phi - i\cos\phi)Y - \sin\theta Z\rangle\left|\uparrow'\right\rangle$$

$$+ \sqrt{\frac{2}{3}}\left|\sin\theta\cos\phi X + \sin\theta\sin\phi Y + \cos\theta Z\rangle\left|\downarrow'\right\rangle \quad (9.5.3)$$

Spin–Orbit Split-off Band

$$\left|\frac{1}{2},\frac{1}{2}\right\rangle' = \frac{1}{\sqrt{3}}\left|(X' + iY')\downarrow'\right\rangle + \frac{1}{\sqrt{3}}\left|Z'\uparrow'\right\rangle$$

$$= \frac{1}{\sqrt{3}}\left|(\cos\theta\cos\phi - i\sin\phi)X\right.$$

$$+ (\cos\theta\sin\phi + i\cos\phi)Y - \sin\theta Z\rangle\left|\downarrow'\right\rangle$$

$$+ \frac{1}{\sqrt{3}}\left|\sin\theta\cos\phi X + \sin\theta\sin\phi Y + \cos\theta Z\right\rangle\left|\uparrow'\right\rangle$$

$$\left|\frac{1}{2},-\frac{1}{2}\right\rangle' = \frac{1}{\sqrt{3}}\left|(X' - iY')\uparrow'\right\rangle - \frac{1}{\sqrt{3}}\left|Z'\downarrow'\right\rangle$$

$$= \frac{1}{\sqrt{3}}\left|(\cos\theta\cos\phi + i\sin\phi)X\right.$$

$$+ (\cos\theta\sin\phi - i\cos\phi)Y - \sin\theta Z\rangle\left|\uparrow'\right\rangle$$

$$- \frac{1}{\sqrt{3}}\left|\sin\theta\cos\phi X + \sin\theta\sin\phi Y + \cos\theta Z\right\rangle\left|\downarrow'\right\rangle \quad (9.5.4)$$

We keep the primes in $|\downarrow'\rangle$ and $|\uparrow'\rangle$ for spins in the new coordinate system for ease of calculation of the dipole matrix elements, as will be seen later.

Consider a quantum-well structure with the growth axis along the z direction. In general [14], the z axis for the p-state functions does not have to coincide with the z axis for the quantum-well structures. However, when averaged over the angle in the plane of the quantum well, it is found that the same matrix elements for both polarizations give the same results as those obtained assuming that both z axes coincide with each other.

To calculate the optical momentum matrix element

$$\mathbf{M} = \langle u_c|\mathbf{p}|u_v\rangle = \mathbf{p}_{cv}$$

$$= \hat{x}\left\langle u_c\left|\frac{\hbar}{i}\frac{\partial}{\partial x}\right|u_v\right\rangle + \hat{y}\left\langle u_c\left|\frac{\hbar}{i}\frac{\partial}{\partial y}\right|u_v\right\rangle + \hat{z}\left\langle u_c\left|\frac{\hbar}{i}\frac{\partial}{\partial z}\right|u_v\right\rangle \quad (9.5.5)$$

with polarization dependence, we evaluate $\hat{e}\cdot\mathbf{M}$ for $\hat{e} = \hat{x}$ or $\hat{y}$ for TE polarization, and $\hat{e} = \hat{z}$ for TM polarization. Here u_c is $|iS\uparrow'\rangle$ or $|iS\downarrow'\rangle$ and u_v can be u_{hh} or u_{lh}.

Define Kane's parameter P and a momentum-matrix parameter P_x as

$$P = \frac{\hbar}{m_0}\langle iS|P_z|Z\rangle = \frac{\hbar}{m_0}P_x$$

$$P_x = \langle iS|p_x|X\rangle = \langle iS|p_y|Y\rangle = \langle iS|p_z|Z\rangle \tag{9.5.6}$$

Then for conduction to heavy-hole transitions, we obtain $\mathbf{M}_{c-hh}$ as

$$\left\langle iS\uparrow'|\mathbf{p}|\frac{3}{2},\frac{3}{2}\right\rangle = -\left[(\cos\theta\cos\phi - i\sin\phi)\hat{x}\right.$$

$$\left. + (\cos\theta\sin\phi + i\cos\phi)\hat{y} - \sin\theta\hat{z}\right]\frac{P_x}{\sqrt{2}}$$

$$\left\langle iS\downarrow'|\mathbf{p}|\frac{3}{2},-\frac{3}{2}\right\rangle = \left[(\cos\theta\cos\phi + i\sin\phi)\hat{x}\right.$$

$$\left. + (\cos\theta\sin\phi - i\cos\phi)\hat{y} - \sin\theta\hat{z}\right]\frac{P_x}{\sqrt{2}}$$

$$\left\langle iS\uparrow'|\mathbf{p}|\frac{3}{2},-\frac{3}{2}\right\rangle = 0$$

$$\left\langle iS\downarrow'|\mathbf{p}|\frac{3}{2},\frac{3}{2}\right\rangle = 0 \tag{9.5.7}$$

For conduction to light-hole transitions, $\mathbf{M}_{c-lh}$ are

$$\left\langle iS\uparrow'|\mathbf{p}|\frac{3}{2},\frac{1}{2}\right\rangle = \sqrt{\frac{2}{3}}\left(\sin\theta\cos\phi\hat{x} + \sin\theta\sin\phi\hat{y} + \cos\theta\hat{z}\right)P_x$$

$$\left\langle iS\downarrow'|\mathbf{p}|\frac{3}{2},-\frac{1}{2}\right\rangle = \sqrt{\frac{2}{3}}\left(\sin\theta\cos\phi\hat{x} + \sin\theta\sin\phi\hat{y} + \cos\theta\hat{z}\right)P_x$$

$$\left\langle iS\uparrow'|\mathbf{p}|\frac{3}{2},-\frac{1}{2}\right\rangle = \frac{1}{\sqrt{6}}\left[(\cos\theta\cos\phi + i\sin\phi)\hat{x}\right.$$

$$\left. + (\cos\theta\sin\phi - i\cos\phi)\hat{y} - \sin\theta\hat{z}\right]P_x$$

$$\left\langle iS\downarrow'|\mathbf{p}|\frac{3}{2},\frac{1}{2}\right\rangle = \frac{-1}{\sqrt{6}}\left[(\cos\theta\cos\phi - i\sin\phi)\hat{x}\right.$$

$$\left. + (\cos\theta\sin\phi + i\cos\phi)\hat{y} - \sin\theta\hat{z}\right]P_x \tag{9.5.8}$$

9.5.2   Momentum Matrix Element of a Bulk Semiconductor

For a bulk semiconductor, we take the average of the momentum matrix element with respect to the solid angle $d\Omega$. For example, the momentum matrix element for TE polarization ($\hat{e} = \hat{x}$) due to the transitions from the conduction band with one spin (say $< iS \uparrow '|$) to both heavy hole bands, $|\frac{3}{2}, \frac{3}{2}\rangle'$ and $|\frac{3}{2}, -\frac{3}{2}\rangle'$ (one of the two transitions is zero) is

$$|\hat{e} \cdot \mathbf{p}_{cv}|^2 \equiv \langle |\hat{e} \cdot \mathbf{M}_{c-hh}|^2 \rangle = \frac{1}{4\pi} \int |\hat{x} \cdot \mathbf{M}_{c-hh}|^2 \sin\theta \, d\theta \, d\phi$$

$$= \frac{1}{4\pi} \int_0^\pi \sin\theta \, d\theta \int_0^{2\pi} d\phi (\cos^2\theta \cos^2\phi + \sin^2\phi) \frac{P_x^2}{2}$$

$$= \frac{1}{3} P_x^2 \equiv M_b^2 \tag{9.5.9}$$

where

$$M_b^2 = \frac{1}{3} P_x^2 = \frac{m_0^2}{3\hbar^2} P^2$$

$$= \left(\frac{m_0}{m_e^*} - 1\right) \frac{m_0 E_g (E_g + \Delta)}{6(E_g + \frac{2}{3}\Delta)} \tag{9.5.10}$$

is the bulk momentum matrix element (squared). Notice that the above result is the same for the other spin in the conduction band. If we take $\hat{e} = \hat{y}$ or $\hat{z}$, and average over the solid angle, we still obtain M_b^2 since the bulk crystal is isotropic. We could also take $|\langle iS \downarrow '|ex|\frac{3}{2}, \frac{1}{2}\rangle|^2 + |\langle iS \downarrow '|ex|\frac{3}{2}, -\frac{1}{2}\rangle|^2$ and average over the solid angle and obtain M_b^2 as expected. In practice, an energy parameter E_p for the matrix element is defined: $E_p = (2m_0/\hbar^2)P^2$ or $M_b^2 = (m_0/6)E_p$; E_p is tabulated in Table K.2 in Appendix K.

9.5.3   Momentum Matrix Elements of Quantum Wells

Next, we consider quantum-well structures. The optical matrix elements will become polarization dependent. The theory and experimental data on the polarization-dependent gain are discussed in Refs. 14–18 based on the parabolic band model. Improvement of the optical matrix element using the valence band-mixing model in quantum wells is discussed in Section 9.7.

TE Polarization. Let $\hat{e} = \hat{x}$, that is, the optical field is polarized along the x (or y) direction. The optical dipole matrix element is averaged over the azimuthal angle ϕ in the plane of quantum wells. We obtain for $\langle iS\uparrow'|$

$$|\hat{e}\cdot\mathbf{p}_{cv}|^2 \equiv \langle|\hat{e}\cdot\mathbf{M}_{c-hh}|^2\rangle = \frac{1}{2\pi}\int_0^{2\pi}d\phi|\hat{x}\cdot\mathbf{M}_{c-hh}|^2$$

$$= \frac{1}{2\pi}\int_0^{2\pi}d\phi(\cos^2\theta\cos^2\phi + \sin^2\phi)\frac{P_x^2}{2}$$

$$= \frac{3}{4}(1 + \cos^2\theta)M_b^2 \tag{9.5.11}$$

and the same matrix element is obtained for the other spin $\langle iS\downarrow'|$ in the conduction band. Similarly, for the sum of transitions of an electron with spin $\downarrow'$ from the conduction band to both light-hole bands, we find

$$\langle|\hat{e}\cdot\mathbf{M}_{c-lh}|^2\rangle = \frac{1}{2\pi}\int_0^{2\pi}d\phi\left(\left|\left\langle iS\downarrow'|p_x|\frac{3}{2},\frac{1}{2}\right\rangle\right|^2 + \left|\left\langle iS\downarrow'|p_x|\frac{3}{2},-\frac{1}{2}\right\rangle\right|^2\right)$$

$$= \left(\frac{2}{3}\sin^2\theta\langle\cos^2\phi\rangle + \frac{1}{6}\cos^2\theta\langle\cos^2\phi\rangle + \frac{1}{6}\langle\sin^2\phi\rangle\right)P_x^2$$

$$= \left[\sin^2\theta + \frac{1}{4}(\cos^2\theta + 1)\right]M_b^2$$

$$= \left(\frac{5}{4} - \frac{3}{4}\cos^2\theta\right)M_b^2 \tag{9.5.12}$$

The above result is the same for both spins $\langle iS\downarrow'|$ and $\langle iS\uparrow'|$. Note that a sum rule exists:

$$\langle|\hat{e}\cdot\mathbf{M}_{c-hh}|^2\rangle + \langle|\hat{e}\cdot\mathbf{M}_{c-lh}|^2\rangle = 2M_b^2 \tag{9.5.13}$$

which is independent of the angle θ.

TM Polarization. $\hat{e} = \hat{z}$

$$\langle|\hat{e}\cdot\mathbf{M}_{c-hh}|^2\rangle = \frac{1}{2\pi}\int_0^{2\pi}d\phi|\hat{z}\cdot\mathbf{M}_{c-hh}|^2 = \frac{3}{2}\sin^2\theta M_b^2 \tag{9.5.14}$$

$$\langle|\hat{e}\cdot\mathbf{M}_{c-lh}|^2\rangle = \frac{1}{2\pi}\int_0^{2\pi}d\phi\left(\left|\left\langle iS\downarrow'|ez|\frac{3}{2},\frac{1}{2}\right\rangle\right|^2 + \left|\left\langle iS\downarrow'|ez|\frac{3}{2},-\frac{1}{2}\right\rangle\right|^2\right)$$

$$= \left(\frac{1}{6}\sin^2\theta + \frac{2}{3}\cos^2\theta\right)P_x^2$$

$$= \frac{1 + 3\cos^2\theta}{2}M_b^2 \tag{9.5.15}$$

The sum of (9.5.14) and (9.5.15) for $\hat{e} = \hat{z}$ is still $2M_b^2$. The above results, (9.5.14) and (9.5.15), are the same for both spins in the conduction band. The angular factor $\cos^2 \theta$ can be related to the electron or hole wave vectors by

$$\cos^2 \theta = \frac{k_z^2}{k_x^2 + k_y^2 + k_z^2} = \frac{E_{en}}{E_{en} + \left(\hbar^2 k_t^2 / 2m_e^* \right)} \qquad (9.5.16)$$

for the electron wave vector $\mathbf{k} = \hat{\rho} k_t + \hat{z} k_{zn}$, with $E_{en} = \hbar^2 k_{zn}^2 / 2m_e^*$ or

$$\cos^2 \theta = \frac{|E_{hm}|}{|E_{hm}| + \left(\hbar^2 k_t^2 / 2m_h^* \right)} \qquad (9.5.17)$$

for the hole wave vector $\mathbf{k} = \hat{\rho} k_t + \hat{z} k_{zm}$ with $E_{hm} = -\hbar k_{zm}^2 / 2m_h^*$. Note that the hole energy E_{hm} is defined to be negative so all of the energies are measured upward. An alternative approximation is

$$\cos^2 \theta_{nm} = \frac{E_{eh} + |E_{hm}|}{E_{en} + |E_{hm}| + \left(\hbar k_t^2 / 2m_r \right)} \qquad (9.5.18)$$

where the reduced effective mass m_r, defined by

$$\frac{1}{m_r} = \frac{1}{m_e^*} + \frac{1}{m_h^*} \qquad (9.5.19)$$

has been used. The results of the momentum matrix elements are summarized in Table 9.1. It is noted at the subband edges, where $k_t = 0$, that for

Table 9.1   Summary of the Momentum Matrix Elements in Parabolic Band Model ($|\hat{e} \cdot \mathbf{p}_{cv}|^2 = |\hat{e} \cdot \mathbf{M}|^2$)

Bulk $|\hat{x} \cdot \mathbf{p}_{cv}|^2 = |\hat{y} \cdot \mathbf{p}_{cv}|^2 = |\hat{z} \cdot \mathbf{p}_{cv}|^2 = M_b^2 = \dfrac{m_0}{6} E_p$

Quantum Well

TE Polarization ($\hat{e} = \hat{x}$ or $\hat{y}$)

$\langle |\hat{e} \cdot \mathbf{M}_{c-hh}|^2 \rangle = \frac{3}{4}(1 + \cos^2 \theta) M_b^2$

$\langle |\hat{e} \cdot \mathbf{M}_{c-lh}|^2 \rangle = (\frac{5}{4} - \frac{3}{4}\cos^2 \theta) M_b^2$

TM Polarization ($\hat{e} = \hat{z}$)

$\langle |\hat{e} \cdot \mathbf{M}_{c-hh}|^2 \rangle = \frac{3}{2}\sin^2 \theta M_b^2$

$\langle |\hat{e} \cdot \mathbf{M}_{c-lh}|^2 \rangle = \frac{1}{2}(1 + 3\cos^2 \theta) M_b^2$

Conservation Rule

$\langle |\hat{x} \cdot \mathbf{M}_{c-h}|^2 \rangle + \langle |\hat{y} \cdot \mathbf{M}_{c-h}|^2 \rangle + \langle |\hat{z} \cdot \mathbf{M}_{c-h}|^2 \rangle = 3M_b^2, (h = hh \text{ or } lh)$

$\langle |\hat{e} \cdot \mathbf{M}_{c-hh}|^2 \rangle + \langle |\hat{e} \cdot \mathbf{M}_{c-lh}|^2 \rangle = 2M_b^2$

TE polarization, the matrix elements are

$$\left\langle |\hat{x} \cdot \mathbf{M}_{c-hh}|^2 \right\rangle_{k_t=0} = \frac{3}{2} M_b^2 \qquad (9.5.20)$$

$$\left\langle |\hat{x} \cdot \mathbf{M}_{c-lh}|^2 \right\rangle_{k_t=0} = \frac{1}{2} M_b^2 \qquad (9.5.21)$$

and for TM polarization, the matrix elements are

$$\left\langle |\hat{z} \cdot \mathbf{M}_{c-hh}|^2 \right\rangle_{k_t=0} = 0 \qquad (9.5.22)$$

$$\left\langle |\hat{z} \cdot \mathbf{M}_{c-lh}|^2 \right\rangle_{k_t=0} = 2 M_b^2 \qquad (9.5.23)$$

These matrix elements are useful in studying the excitonic absorptions in quantum wells and they are used in Chapter 13 where we discuss electroabsorption modulators using quantum-confined Stark effects.

9.6 INTERSUBBAND ABSORPTION

In an n-type doped quantum-well structure, intersubband absorption is of interest because of its applications to far-infrared photodetectors [21–23], for example. In this section, we study the infrared absorption [24] due to intersubband transitions in the conduction band of a quantum-well structure with modulation doping. We assume that the doping is not large enough so that screening or many-body effects due to the electron–electron Coulomb interaction [25] can be ignored. The far-infrared photodetector applications using intersubband transitions are discussed in Section 14.5.

9.6.1 Intersubband Dipole Moment

Consider the intersubband transition between the ground state

$$\psi_a(\mathbf{r}) = u_c(\mathbf{r}) \frac{e^{i\mathbf{k}_t \cdot \boldsymbol{\rho}}}{\sqrt{A}} \phi_1(z) \qquad (9.6.1)$$

and the first excited state

$$\psi_b(\mathbf{r}) = u_{c'}(\mathbf{r}) \frac{e^{i\mathbf{k}'_t \cdot \boldsymbol{\rho}}}{\sqrt{A}} \phi_2(z) \qquad (9.6.2)$$

where the transverse wave vectors $\mathbf{k}_t = k_x \hat{x} + k_y \hat{y}, \mathbf{k}'_t = k'_x \hat{x} + k'_y \hat{y}$ and the position vector $\boldsymbol{\rho} = x\hat{x} + y\hat{y}$ in the quantum-well plane have been used. The optical dipole moment is given by

$$\mu_{ba} = \langle \psi_b | e\mathbf{r} | \psi_a \rangle$$

$$\simeq \langle u_c | u_{c'} \rangle \left\langle \frac{e^{i\mathbf{k}_t' \cdot \boldsymbol{\rho}}}{\sqrt{A}} \phi_2 \middle| e\mathbf{r} \middle| \frac{e^{i\mathbf{k}_t \cdot \boldsymbol{\rho}}}{\sqrt{A}} \phi_1 \right\rangle$$

$$\simeq \delta_{\mathbf{k}_t, \mathbf{k}_t'} \langle \phi_2 | ez | \phi_1 \rangle \hat{z} \qquad (9.6.3)$$

which has only a z component, where we have used the orthonormal conditions

$$\langle \phi_2 | \phi_1 \rangle = 0 \quad \text{and} \quad \langle u_c | u_{c'} \rangle \simeq 1$$

9.6.2 Intersubband Absorption Spectrum

The energies of the initial state and the final states are, respectively,

$$E_a = E_1 + \frac{\hbar^2 k_t^2}{2m_e^*} \qquad E_b = E_2 + \frac{\hbar^2 k_t^2}{2m_e^*} \qquad (9.6.4)$$

as shown in Fig. 9.11c. Using (9.1.30) and (9.1.31), we obtain nonzero absorption coefficient $\alpha(\hbar\omega)$ only for $\hat{e} = \hat{z}$ (TM polarization) since the x

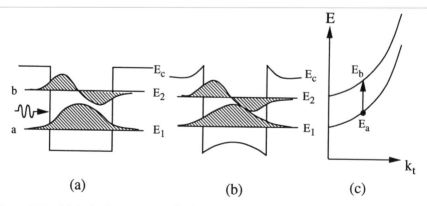

Figure 9.11. (a) A simple quantum well with a small doping concentration. (b) A modulation-doped quantum well with a significant amount of screening due to a large doping concentration. (c) The subband energy diagram in the k_t space. A direct vertical transition occurs because of the k-selection rule in the plane of quantum wells.

and y components of the intersubband dipole moment in (9.6.3) are zero:

$$\alpha(\hbar\omega) = \left(\frac{\omega}{n_r c\varepsilon_0}\right)\frac{2}{V}\sum_{\mathbf{k}_t}\sum_{\mathbf{k}_t'}\frac{|\hat{e}\cdot\mu_{ba}|^2(\Gamma/2)}{(E_b - E_a - \hbar\omega)^2 + (\Gamma/2)^2}(f_a - f_b)$$

$$= \left(\frac{\omega}{n_r c\varepsilon_0}\right)\frac{2}{V}\sum_{\mathbf{k}_t}\frac{|\mu_{21}|^2(\Gamma/2)}{(E_2 - E_1 - \hbar\omega)^2 + (\Gamma/2)^2}(f_a - f_b)$$

$$= \left(\frac{\omega}{n_r c\varepsilon_0}\right)\frac{|\mu_{21}|^2(\Gamma/2)}{(E_2 - E_1 - \hbar\omega)^2 + (\Gamma/2)^2}(N_1 - N_2) \qquad (9.6.5)$$

where

$$\mu_{21} = \langle \phi_2 | ez | \phi_1 \rangle = \int \phi_2^*(z) ez \phi_1(z) \, dz \qquad (9.6.6)$$

is the intersubband dipole moment and N_i is the number of electrons per unit volume in the ith subband, which has been derived in (9.4.16):

$$N_i = \frac{m_e^* k_B T}{\pi \hbar^2 L_z}\ln\left(1 + e^{(E_F - E_i)/k_B T}\right) \qquad (9.6.7a)$$

Or, in terms of the sheet carrier density $(1/\text{cm}^2)$,

$$N_{is} = N_i L_z = \frac{m_e^* k_B T}{\pi \hbar^2}\ln\left(1 + e^{(E_F - E_i)/k_B T}\right) \qquad (9.6.7b)$$

The absorption coefficient is, therefore,

$$\alpha(\hbar\omega) = \left(\frac{\omega}{n_r c\varepsilon_0}\right)\frac{|\mu_{21}|^2(\Gamma/2)}{(E_2 - E_1 - \hbar\omega)^2 + (\Gamma/2)^2}\left(\frac{m_e^* k_B T}{\pi \hbar^2 L}\right)\ln\left(\frac{1 + e^{(E_F - E_1)/k_B T}}{1 + e^{(E_F - E_2)/k_B T}}\right)$$

$$(9.6.8)$$

which is a Lorentzian shape with a linewidth Γ. In the low-temperature limit such that $(E_F - E_i) \gg k_B T$, we obtain

$$N_i = \frac{m_e^*}{\pi \hbar^2 L_z}(E_F - E_i) \qquad (9.6.9)$$

and for occupations of the first two levels,

$$\alpha(\hbar\omega) = \left(\frac{\omega}{n_r c \varepsilon_0}\right) \frac{|\mu_{21}|^2 (\Gamma/2)}{(E_2 - E_1 - \hbar\omega)^2 + (\Gamma/2)^2} \left(\frac{m_e^*}{\pi \hbar^2 L_z}\right)(E_2 - E_1)$$

$$(9.6.10)$$

An integrated absorbence is given by

$$A = \int_0^\infty \alpha(\hbar\omega)\, d(\hbar\omega) \simeq \left(\frac{\omega}{n_r c \varepsilon_0}\right)|\mu_{21}|^2 \pi (N_1 - N_2) \qquad (9.6.11)$$

where we have replaced the integration limits $(0, \infty)$ by $(-\infty, \infty)$ for the Lorentzian function in (9.6.5) with an integrated area equal to π.

Example Let us estimate the peak absorption coefficient in an n-type doped GaAs quantum well using an infinite barrier model with a well width $L_z = 100$ Å. The first two subband energies and wave functions are ($m_e^* = 0.0665 m_0$)

$$E_1 = \frac{\hbar^2}{2m_e^*}\left(\frac{\pi}{L_z}\right)^2 = 56.5 \text{ meV} \qquad \phi_1(z) = \sqrt{\frac{2}{L_z}}\, \sin\left(\frac{\pi}{L_z} z\right)$$

$$E_2 = 4E_1 = 226 \text{ meV} \qquad\qquad \phi_2(z) = \sqrt{\frac{2}{L_z}}\, \sin\left(\frac{2\pi}{L_z} z\right)$$

The intersubband dipole moment is

$$\mu_{21} = e \int_0^{L_z} \phi_2(z)\, z \phi_1(z)\, dz = -\frac{16}{9\pi^2} eL_z$$

$$= -18 \text{ eÅ} = -2.88 \times 10^{-28} \text{ C} \cdot \text{m}$$

If the carrier concentration N is 1×10^{18} cm^{-3}, we can assume that only the first subband is occupied and check whether the second subband population N_2 is indeed small. We calculate

$$E_F - E_1 = k_B T \left[\exp\left(\frac{NL_z}{N_s}\right) - 1\right] = 78 \text{ meV}$$

where

$$N_s = \frac{m_e^* k_B T}{\pi \hbar^2} = 7.19 \times 10^{11} \text{ cm}^{-2}$$

We can check that $N_2 = (N_s/L_z)\ln\{1 + \exp[(E_F - E_2)/k_BT]\} = 2.4 \times 10^{15}$ cm$^{-3} \ll N$. If N_2 is not negligible, we have to use $N_1 + N_2 = N$ to determine the Fermi level E_F assuming that the first two subbands are occupied.

The peak absorption coefficient occurs at $\hbar\omega \approx E_2 - E_1 = 170$ meV. The peak wavelength is $\lambda \approx 1.24/0.170 = 7.3$ μm. Assume that the linewidth is $\Gamma = 30$ meV and the refractive index $n_r = 3.3$. We find

$$\alpha_{\text{peak}} = \frac{\omega}{n_r c \varepsilon_0} \frac{|\mu_{21}|^2}{(\Gamma/2)} (N_1 - N_2) \approx 1.015 \times 10^4 \text{ cm}^{-1}$$

In principle, the peak absorption coefficient increases with the doping concentration N. However, if the concentration is too high, the screening, band bending, and many-body effects [25] will be important as they affect the energy levels and, therefore, the peak absorption wavelength. The occupation of the second level N_2 will also decrease the absorption. ∎

Example   We consider GaAs/Al$_{0.3}$Ga$_{0.7}$As superlattice structures, as shown in Fig. 9.12, with a well width W and a barrier width b. The absorption spectrum is shown in Fig. 9.13a for a structure with $W = b = 100$ Å and in Fig. 9.13b for $W = 100$ Å and $b = 20$ Å. We can see that as the barrier width b is reduced from 100 to 20 Å the interminiband width is broadened because of the strong coupling of well states. We assume a surface concentration of $N_s = 2 \times 10^{11}/\text{cm}^2$, $T = 77$ K and a linewidth $\Gamma = 15$ meV.

If we reduce the well width W to 40 Å, the intersubband transitions will occur from the bound to the continuum minibands. The resultant absorption spectrum is a broader spectrum with a longer energy tail at the high-energy side, as shown in Fig. 9.13c. Here we use the same $N_s = 2 \times 10^{11}/\text{cm}^2$ and a

(a) Bound-to-bound transition

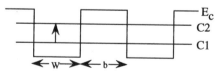

(b) Bound-to-continuum miniband transitions

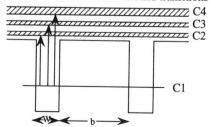

Figure 9.12. (a) Intersubband bound-to-bound transition in a superlattice. (b) Intersubband bound-to-continuum transitions in a superlattice.

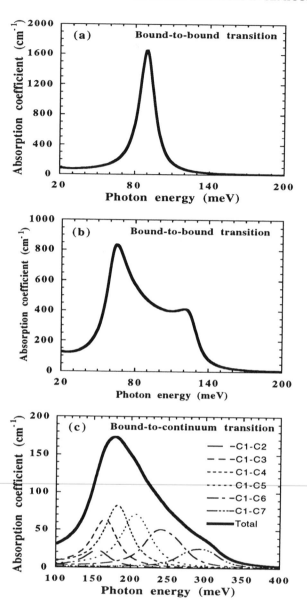

Figure 9.13. The absorption coefficients for intersubband bound-to-bound transition in a superlattice for (a) $W = b = 100$ Å (thick barriers, small miniband widths), and (b) $W = 100$ Å, $b = 20$ Å (thin barriers, large miniband widths). (c) The absorption coefficient of the intersubband bound-to-continuum miniband transitions is shown for $W = 40$ Å and $b = 300$ Å.

linewidth $\Gamma = 40$ meV. The broad spectrum is contributed by a larger Γ as well as the transitions to multiple minibands. ∎

9.6.3 Experimental Results

In Fig. 9.14, we show the experimental data for five different samples by Levine et al. [22]. The material parameters are listed as follows:

Sample	Well Width (Å)	Barrier Width (Å)	x	$N_D(10^{18}/cm^3)$
A	40	500	0.26	1
B	40	500	0.25	1.6
C	60	500	0.15	0.5
E	50	500	0.26	0.42
F	50	50	0.30	0.42
		500	0.26	

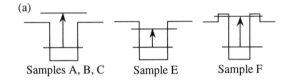

(a)

Samples A, B, C   Sample E   Sample F

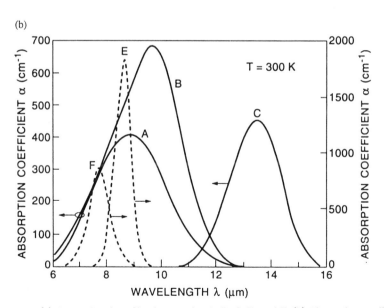

Figure 9.14. (a) Energy band profiles for samples A, B, C, E, and F. (b) Absorption coefficients for the five samples in (a). Note that the absorption coefficients of samples, A, B, and C (left vertical scale) are smaller than those of samples E and F using the right scale. (After Ref. 22.)

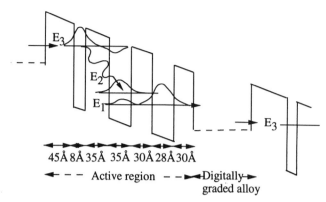

45Å 8Å 35Å 35Å 30Å 28Å 30Å

◄ ─ ─ ─ Active region ─ ─►◄Digitally►
                                   graded alloy

Figure 9.15. A period of a quantum cascade laser [27] using intersubband transition between E_3 and E_2. The barriers are $Al_{0.48}In_{0.52}As$ and the wells are $In_{0.47}Ga_{0.53}As$ materials. The calculated values are $E_3 - E_2 = 295$ meV and $E_2 - E_1 = 30$ meV.

Samples A, B, and C were designed for the bound-to-continuum transition; sample E was designed for bound-to-bound state transition and sample F for bound-to-quasibound state transition. It can be seen that the bound-to-bound transition (sample) E has a large peak absorption spectrum with a narrow linewidth. Sample F has a narrower spectrum than those of the bound-to-continuum transitions in samples A, B, and C. These samples result in different responsivities because the carrier collection by the electrodes also play an important role. For example, the responsivity of sample F is larger than that of sample E because the electrons arriving at the final (quasibound) state can easily be collected compared with those in the final bound state of sample E.

9.6.4 Intersubband Quantum Cascade Laser [27]

Intersubband lasers using a structure such as that shown in Fig. 9.15 as a period have been demonstrated [27] recently. The laser structure was grown with the $Al_{0.48}In_{0.52}As/Ga_{0.47}In_{0.53}As$ heterojunctions lattice matched to InP substrate by MBE. There are 25 periods of the active undoped regions, with graded regions consisting of an AlInAs/GaInAs superlattice with a constant period shorter than the electron thermal de Broglie wavelength. Electrons are injected through the 45-Å AlInAs barrier into the E_3 energy level of the active region, as shown in Fig. 9.15. The reduced spatial overlap of the E_3 and E_2 wave functions and the strong coupling to the E_1 level in an adjacent well ensure a population inversion between these states. The laser operates between the E_3 and E_2 levels with an energy difference $E_3 - E_2 \approx 295$ meV or an operation wavelength ≈ 4.2 μm. The applied bias field is around 100 kV/cm and the band diagram is as shown in Fig. 9.15, where the graded region is near flat-band condition. The estimated tunneling time into the first trapezoidal barrier is around 0.2 ps. Therefore, the filling of the state E_3 is very efficient, while the intersubband optical-phonon-limited relaxation time τ_{32} is estimated to be around 4.3 ps at this bias field with a reduced overlap integral. Since τ_{32} is relatively long, the population inversion can be achieved because the lower state empties with a fast relaxation time (around 0.6 ps) due to the adjacent 28-Å GaInAs well with E_1 state. The relaxation between the E_2 and E_1 states is very fast because the energy separation is equal or close to an optical phonon energy so that scattering occurs with essentially zero momentum transfer. Finally, the tunneling out of the last 30-Å barrier is extremely fast (smaller than 0.5 ps). The laser emits photons at a wavelength of 4.2 μm with a peak power above 8 mW in pulsed operation. The operation temperature was as high as 88 K.

9.7   GAIN SPECTRUM IN A QUANTUM-WELL LASER WITH VALENCE-BAND-MIXING EFFECTS

In Section 9.4, we present a simplified model for the interband gain of a quantum-well laser based on the parabolic band structures. In quantum wells, valence-band-mixing effects between the heavy-hole and light-hole subbands are important and the subband structures can be highly non-parabolic, especially for strained quantum wells, as discussed in Sections 4.8 and 4.9. In this section, we present a general theory for the gain spectrum in quantum wells, taking into account the valence-band-mixing effects [26]. The theory is applicable to both strained and unstrained quantum-well lasers.

9.7.1   General Formulation for the Gain Spectrum with Valence Band Mixing

Our starting equation follows (9.1.25) for the absorption spectrum with the delta function replaced by a Lorentzian function (9.1.31) to account for scattering broadenings,

$$\alpha(\hbar\omega) = C_0 \frac{1}{V} \sum_{\eta,\sigma} \sum_{\mathbf{k}_a} \sum_{\mathbf{k}_b} |\hat{e} \cdot \mathbf{p}_{ba}|^2 \frac{\Gamma/(2\pi)}{(E_b - E_a - \hbar\omega)^2 + (\Gamma/2)^2} (f_a - f_b)$$

$$(9.7.1a)$$

$$C_0 = \frac{\pi e^2}{n_r c \varepsilon_0 m_0^2 \omega}$$

$$(9.7.1b)$$

where the quantum numbers are $|a\rangle = |\mathbf{k}_a, \sigma\rangle$ for initial states in the valence band and $|b\rangle = |\mathbf{k}_b, \eta\rangle$ for final states in the conduction band. The summations over the spins of the initial (hole) states σ and the final (electron) states η have been written out explicitly instead of a simple factor of 2. Here $\eta = \uparrow$, or $\downarrow$ for the conduction band, and σ is more complicated. For the formulation using the block-diagonalized Hamiltonian, σ refers to U for the upper Hamiltonian with an eigenfunction in the mth subband denoted by

$$\Psi_m^U(\mathbf{k}_t, \mathbf{r}) = \frac{e^{i\mathbf{k}_t \cdot \boldsymbol{\rho}}}{\sqrt{A}} \left[g_m^{(1)}(k_t, z)|1\rangle + g_m^{(2)}(k_t, z)|2\rangle \right] \qquad (9.7.2)$$

and $\sigma = L$ for the lower Hamiltonian with an eigenfunction in the mth subband denoted by

$$\Psi_m^L(\mathbf{k}_t, \mathbf{r}) = \frac{e^{i\mathbf{k}_t \cdot \boldsymbol{\rho}}}{\sqrt{A}} \left[g_m^{(3)}(k_t, z)|3\rangle + g_m^{(4)}(k_t, z)|4\rangle \right] \qquad (9.7.3)$$

The nth conduction subband wave function is denoted by

$$\Psi_n^{cn}(\mathbf{k}_t', \mathbf{r}) = \frac{e^{i\mathbf{k}_t' \cdot \boldsymbol{\rho}}}{\sqrt{A}} \phi_n(z)|S\eta\rangle \qquad \eta = \uparrow \text{ or } \downarrow \qquad (9.7.4)$$

Here $|S\eta\rangle(= u_c(r))$ denotes the spherically symmetric wave function of the Bloch periodic part of the electron state. We have to label the quantum number $\mathbf{k}_a$ by $(\mathbf{k}_t, m)$ for the mth subband, where m refers to the quantum number for the z dependence of the wave functions. We label $\mathbf{k}_b$ as $(\mathbf{k}_t', n)$ for the nth conduction subband.

The matrix element will contain a $\mathbf{k}_t$-selection rule when we take the inner product, $\mathbf{k}_t = \mathbf{k}_t'$. Therefore, we can write

$$\alpha(\hbar\omega) = C_0 \frac{1}{V} \sum_{\sigma,\eta} \sum_{n,m} \sum_{\mathbf{k}_t} |\hat{e} \cdot \mathbf{p}_{nm}^{\eta\sigma}|^2 \frac{\Gamma/(2\pi)}{\left[E_{hm}^{en}(k_t) - \hbar\omega \right]^2 + (\Gamma/2)^2} (f_v^{\sigma m} - f_c^n)$$

$$(9.7.5)$$

where

$$E_{hm}^{en}(k_t) = E_{en}(k_t) - E_m^\sigma(k_t) + E_g \qquad (9.7.6)$$

is the energy difference between the nth electron subband and the mth hole subband at $\mathbf{k}_t$. Here

$$E_{en}(k_t) = E_{en}(0) + \frac{\hbar^2 k_t^2}{2m_e^*} \qquad (9.7.7)$$

is taken to be parabolic in the conduction subband. The nonparabolic valence subband structures are described by $E_m^\sigma(k_t)$ obtained from the upper $(\sigma = U)$ or the lower $(\sigma = L)$ Hamiltonian.

9.7.2 Evaluation of the Momentum Matrix Elements

The momentum matrix element $|\hat{e} \cdot \mathbf{p}_{nm}^{\eta\sigma}|^2$ is defined as

$$|\hat{e} \cdot \mathbf{p}_{nm}^{\eta\sigma}|^2 = |\langle \Psi_n^{c\eta} | \hat{e} \cdot \mathbf{p} | \Psi_m^{\sigma} \rangle|^2 \tag{9.7.8}$$

with the factor $\delta_{\mathbf{k}_t, \mathbf{k}_t'}$ dropped. Here we list the definitions of the old and new basis functions and their transformation relations, which have been discussed in Chapter 4.

Old bases

$$\left| \frac{3}{2}, \frac{3}{2} \right\rangle = \frac{-1}{\sqrt{2}} |(X + iY)\uparrow\rangle$$

$$\left| \frac{3}{2}, \frac{1}{2} \right\rangle = \frac{-1}{\sqrt{6}} |(X + iY)\downarrow\rangle + \frac{2}{\sqrt{6}} |Z\uparrow\rangle$$

$$\left| \frac{3}{2}, \frac{-1}{2} \right\rangle = \frac{1}{\sqrt{6}} |(X - iY)\uparrow\rangle + \frac{2}{\sqrt{6}} |Z\downarrow\rangle$$

$$\left| \frac{3}{2}, \frac{-3}{2} \right\rangle = \frac{1}{\sqrt{2}} |(X - iY)\downarrow\rangle \tag{9.7.9a}$$

New bases

$$|1\rangle = \alpha \left| \frac{3}{2}, \frac{3}{2} \right\rangle - \alpha^* \left| \frac{3}{2}, \frac{-3}{2} \right\rangle$$

$$|2\rangle = -\beta^* \left| \frac{3}{2}, \frac{1}{2} \right\rangle + \beta \left| \frac{3}{2}, \frac{-1}{2} \right\rangle$$

$$|3\rangle = \beta^* \left| \frac{3}{2}, \frac{1}{2} \right\rangle + \beta \left| \frac{3}{2}, \frac{-1}{2} \right\rangle$$

$$|4\rangle = \alpha \left| \frac{3}{2}, \frac{3}{2} \right\rangle + \alpha^* \left| \frac{3}{2}, \frac{-3}{2} \right\rangle$$

$$\alpha = \frac{1}{\sqrt{2}} e^{i(3\pi/4 - 3\phi/2)} \qquad \beta = \frac{1}{\sqrt{2}} e^{i(-\pi/4 + \phi/2)} \tag{9.7.9b}$$

The matrix element $|\hat{e} \cdot \mathbf{p}_{nm}^{\eta\sigma}|^2$ can be found as shown in the following example. For TE polarization ($\hat{e} = \hat{x}$ or $\hat{y}$) and $\eta = \uparrow, \sigma = U$, we obtain

$$|\hat{x} \cdot \mathbf{p}_{nm}^{\uparrow U}|^2 = |\langle \Psi_n^{c \uparrow} | p_x | \Psi_m^U \rangle|^2$$

$$= \delta_{\mathbf{k}_t, \mathbf{k}_t'} |\langle \phi_n | g_m^{(1)} \rangle \langle S \uparrow | p_x | 1 \rangle + \langle \phi_n | g_m^{(2)} \rangle \langle S \uparrow | p_x | 2 \rangle|^2$$

$$= \delta_{\mathbf{k}_t, \mathbf{k}_t'} |\langle S | p_x | X \rangle|^2 \left| \langle \phi_n | g_m^{(1)} \rangle \left(\frac{-\alpha}{\sqrt{2}} \right) + \langle \phi_n | g_m^{(2)} \rangle \frac{\beta}{\sqrt{6}} \right|^2$$

$$= \delta_{\mathbf{k}_t, \mathbf{k}_t'} |\langle S | p_x | X \rangle|^2 \frac{1}{4} \left\{ \langle \phi_n | g_m^{(1)} \rangle^2 + \frac{1}{3} \langle \phi_n | g_m^{(2)} \rangle^2 \right.$$

$$\left. + \frac{2}{\sqrt{3}} \cos 2\phi \langle \phi_n | g_m^{(1)} \rangle \langle \phi_n | g_m^{(2)} \rangle \right\} \quad (9.7.10)$$

where we have used the following relations:

Spin $\uparrow$

$$\langle S \uparrow | p_x | 1 \rangle = \frac{-\alpha}{\sqrt{2}} \langle S | p_x | X \rangle \qquad \langle S \uparrow | p_z | 1 \rangle = 0$$

$$\langle S \uparrow | p_x | 2 \rangle = \frac{\beta}{\sqrt{6}} \langle S | p_x | X \rangle \qquad \langle S \uparrow | p_z | 2 \rangle = -\sqrt{\frac{2}{3}} \beta^* \langle S | p_z | Z \rangle$$

$$\langle S \uparrow | p_x | 3 \rangle = \frac{\beta}{\sqrt{6}} \langle S | p_x | X \rangle \qquad \langle S \uparrow | p_z | 3 \rangle = \sqrt{\frac{2}{3}} \beta^* \langle S | p_z | Z \rangle$$

$$\langle S \uparrow | p_x | 4 \rangle = \frac{-\alpha}{\sqrt{2}} \langle S | p_x | X \rangle \qquad \langle S \uparrow | p_z | 4 \rangle = 0 \qquad (9.7.11a)$$

Spin $\downarrow$

$$\langle S \downarrow | p_x | 1 \rangle = \frac{-\alpha^*}{\sqrt{2}} \langle S | p_x | X \rangle \qquad \langle S \downarrow | p_z | 1 \rangle = 0$$

$$\langle S \downarrow | p_x | 2 \rangle = \frac{\beta^*}{\sqrt{6}} \langle S | p_x | X \rangle \qquad \langle S \downarrow | p_z | 2 \rangle = \beta \sqrt{\frac{2}{3}} \langle S | p_z | Z \rangle$$

$$\langle S \downarrow | p_x | 3 \rangle = \frac{-\beta^*}{\sqrt{6}} \langle S | p_x | X \rangle \qquad \langle S \downarrow | p_z | 3 \rangle = \beta \sqrt{\frac{2}{3}} \langle S | p_z | Z \rangle$$

$$\langle S \downarrow | p_x | 4 \rangle = \frac{\alpha^*}{\sqrt{2}} \langle S | p_x | X \rangle \qquad \langle S \downarrow | p_z | 4 \rangle = 0 \qquad (9.7.11b)$$

Using the axial approximation, the band structures are isotropic in the k_x-k_y plane; we, therefore, write

$$\frac{1}{V}\sum_{\mathbf{k}_t} = \frac{1}{L_z}\int_0^\infty \frac{k_t\,dk_t}{2\pi}\int_0^{2\pi}\frac{d\phi}{2\pi} \tag{9.7.12}$$

The term containing the $\cos 2\phi$ factor in (9.7.10) will not contribute to $\alpha(\hbar\omega)$ since its integration over ϕ vanishes. Let us pull out the factor $|\langle S|p_x|X\rangle|^2$ in the matrix element, where

$$|\langle S|p_x|X\rangle|^2 = 3M_b^2 = \frac{m_0^2}{\hbar^2}P^2 = \frac{m_0}{2}E_p \tag{9.7.13}$$

is defined in terms of the Kane's parameter P or an energy parameter E_p. The expression for $\alpha(\hbar\omega)$ can be cast in the form

$$\alpha(\hbar\omega) = C_0\frac{2}{L_z}\sum_\sigma\sum_{n,m}\int_0^\infty\frac{k_t\,dk_t}{2\pi}M_{nm}^\sigma(k_t)$$
$$\times\frac{\Gamma/(2\pi)}{\left[E_{hm}^{en}(k_t)-\hbar\omega\right]^2+(\Gamma/2)^2}(f_v^{\sigma m}-f_c^n) \tag{9.7.14}$$

Here a factor of 2 for sum over η is explicitly shown, since we find the final expression for the magnitude squared $|\hat{e}\cdot\mathbf{p}_{nm}^{\eta\sigma}|^2$ is independent of η.

TE polarization: $\hat{e} = \hat{x}$

$$M_{nm}^\sigma(k_t) = \frac{3}{4}\left[\langle\phi_n|g_m^{(1)}\rangle^2 + \frac{1}{3}\langle\phi_n|g_m^{(2)}\rangle^2\right]M_b^2 \qquad \sigma = U$$
$$= \frac{3}{4}\left[\frac{1}{3}\langle\phi_n|g_m^{(3)}\rangle^2 + \langle\phi_n|g_m^{(4)}\rangle^2\right]M_b^2 \qquad \sigma = L \tag{9.7.15}$$

TM polarization: $\hat{e} = \hat{z}$

$$M_{nm}^\sigma(k_t) = \langle\phi_n|g_m^{(2)}\rangle^2 M_b^2 \qquad \sigma = U$$
$$= \langle\phi_n|g_m^{(3)}\rangle^2 M_b^2 \qquad \sigma = L \tag{9.7.16}$$

9.7.3 Final Expression for the Gain Spectrum and Numerical Examples

The optical gain is obtained from the negative of the absorption function $g(\hbar\omega) = -\alpha(\hbar\omega)$. Therefore, we have

$$g(\omega) = C_0\frac{2}{L_z}\sum_\sigma\sum_{n,m}\int_0^\infty\frac{k_t\,dk_t}{2\pi}M_{nm}^\sigma(k_t)$$
$$\times\frac{\Gamma/(2\pi)}{\left[E_{hm}^{en}(k_t)-\hbar\omega\right]^2+(\Gamma/2)^2}(f_c^n-f_v^{\sigma m}) \tag{9.7.17}$$

Notice that the form (9.7.17) or (9.7.14) is similar to (9.4.9a) using the parabolic band model. For a symmetric quantum well without an applied electric field, we find that the two valence bands with $\sigma = U$ and $\sigma = L$ are degenerate. The matrix elements turn out to be the same for $\sigma = U$ and L. Therefore, the sum over σ can be replaced by a factor 2 and only $g_m^{(1)}$ and $g_m^{(2)}$ have to be calculated with the corresponding energy dispersion relation $E_m^U(k_t)$.

In Fig. 9.16a, we plot the matrix elements $2M(k_t)/M_b^2$ using Eqs. (9.7.15) and (9.7.16) for the TE polarization and the TM polarization for the first conduction band to the first heavy hole subband (C1-HH1) and to the first light hole subband (C1-LH1) transitions in a $In_{0.53}Ga_{0.47}As/In_{1-x}Ga_xAs_{1-y}P_y$ quantum well lattice matched to InP substrate with a well width of 60 Å. The factor of 2 here accounts for sum over σ of the hole spins in the valence band (not in the conduction band). The conduction band spin degeneracy of a factor of 2 is always included in the density of states. The barrier $In_{1-x}Ga_xAs_{1-y}P_y$ has a bandgap wavelength of 1.3 μm and its parameters can be obtained from Table K.3 in Appendix K. The valence subband structures are plotted in Fig. 9.16b, with HH1 as the top valence subband. The nonparabolic behavior is clearly seen.

The gain spectra for both TE and TM polarizations are plotted in Fig. 9.17 for two carrier concentrations n. The surface carrier concentration is $n_s = nL_z$ and the gain coefficient in a period of a total width $L_T = L_z + L_b$ is proportional to $1/L_T$ instead of $1/L_z$ in (9.7.17). Here L_b is the thickness of the barrier. We can see that for a given carrier density, the TE gain coefficient is larger than that of the TM gain coefficient because the holes will occupy the lowest subband which is heavy hole in nature, and the conduction-to-heavy hole dipole matrix is mostly TE polarized as can be seen from Fig. 9.16a. Similar behaviors for the $GaAs/Al_xGa_{1-x}As$ quantum-well lasers can be found in Ref. 26.

9.7.4 Spontaneous Emission Spectrum and Radiative Current Density

By comparing (9.2.33) with (9.2.34), the spontaneous emission rate per unit volume per unit energy interval ($s^{-1}cm^{-3}eV^{-1}$), taking into account valence-band-mixing effects in quantum wells, can be written as

$$r^{spon}(E = \hbar\omega) = \left(\frac{8\pi n_r^2 E^2}{h^3 c^2}\right) C_0 \frac{2}{L_z} \sum_\sigma \sum_{n,m} \int_0^\infty \frac{k_t\, dk_t}{2\pi} M_{nm}^\sigma(k_t)$$

$$\times \frac{\Gamma/(2\pi)}{\left[E_{hm}^{en}(k_t) - \hbar\omega\right]^2 + (\Gamma/2)^2} f_c^n(1 - f_v^{\sigma m}) \quad (9.7.18)$$

which is similar to $g(\omega)$ except for the prefactor $(8\pi n_r^2 E^2/h^3 c^2)$ and the Fermi–Dirac factors $f_c^n(1 - f_v^{\sigma m})$. The total spontaneous emission rate per

(a)

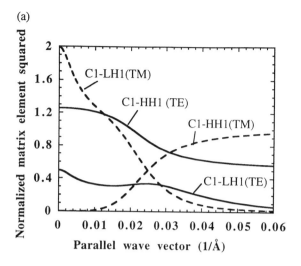

(b)

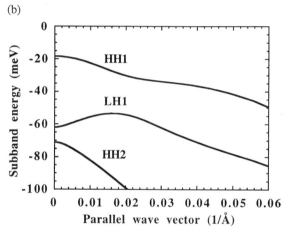

Figure 9.16. (a) The average of the square of the momentum matrix normalized by the bulk matrix element for TE and TM polarizations for the transition between the first conduction subband (C1) and the first heavy-hole subband (HH1) or the first light-hole subband (LH1). (b) The valence subband structures of the $In_{0.53}Ga_{0.47}As/In_{1-x}Ga_xAs_{1-y}P_y$ quantum well ($L_w = 60$ Å) lattice-matched to an InP substrate.

unit volume $(s^{-1}cm^{-3})$ is to sum up the average of three polarizations $(2 \times TE + TM)/3$ and then integrate over the emission spectrum to obtain

$$R_{sp} = \int_0^\infty r^{spon}(\hbar\omega)\, d(\hbar\omega) \tag{9.7.19}$$

A radiative current density (A/cm^2) for a semiconductor laser is defined as

$$J_{rad} = qdR_{sp} \tag{9.7.20}$$

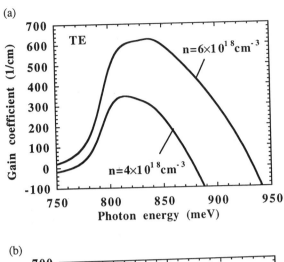

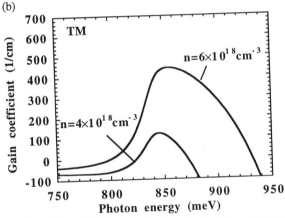

Figure 9.17. The linear gain versus the photon energy for (a) TE and (b) TM polarization at two different carrier concentrations.

where $q = 1.6 \times 10^{-19}$ C is a unit charge and d is the thickness of the active region where an injection current provides the electrons and holes for radiative recombination.

PROBLEMS

9.1 In classical physics, the equation of motion for a charge e in the presence of an electric field $\mathbf{E}$ and a magnetic flux density $\mathbf{B}$ is described by the Lorenz force equation

$$\frac{d}{dt}\mathbf{p} = \mathbf{F} = e(\mathbf{E} + \mathbf{v} \times \mathbf{B})$$

where $\mathbf{p} = m\mathbf{v}$. The magnetic flux density $\mathbf{B}$ and the electric field in classical electromagnetics are expressed in terms of the vector potential $\mathbf{A}(r, t)$ and the scalar potential $\phi(r, t)$ (see Chapter 5)

$$\mathbf{B} = \nabla \times \mathbf{A} \quad \text{and} \quad \mathbf{E} = -\nabla\phi - \frac{\partial \mathbf{A}}{\partial t}$$

Show that using the Hamiltonian in the presence of the electromagnetic field

$$H = \frac{1}{2m}(\mathbf{p} - e\mathbf{A})^2 + e\phi$$

and the equations for the Hamiltonian

$$\frac{d}{dt}r_i = \frac{\partial H}{\partial p_i} \qquad \frac{d}{dt}p_i = -\frac{\partial H}{\partial r_i}$$

leads to the classical Lorenz force equation. If we choose Coulomb gauge, we have $\nabla \cdot \mathbf{A} = 0$, and ϕ is zero since $\rho = 0$ for the optical field. The periodic potential $V(\mathbf{r})$ is the background static potential which is not coupled to the vector potential $\mathbf{A}$ for the optical field and is added to the Hamiltonian for completeness. Therefore,

$$H = \frac{1}{2m}(\mathbf{p} - e\mathbf{A})^2 + V(\mathbf{r})$$

9.2 (a) For a semiconductor laser operating at 0.8 μm with an optical output power of 10 mW coming from a 20-μm $\times$ 1-μm cross-sectional area, estimate the electric field strength E using $S = \omega A_0^2 k_{op}/2\mu = E^2 n_r/2\mu_0 c$, where $n_r = 3.4$. Calculate k_{op} and A_0.

(b) Repeat (a) for a 1.55-μm laser wavelength for the same power. Assume that the other parameters are unchanged.

9.3 Derive the relation (9.2.37) between the spontaneous emission rate per unit volume per unit energy interval $r^{spon}(E)$ and the gain coefficient $g(E)$. Check the dimensions on both sides of (9.2.37).

9.4 Derive the Bernard–Duraffourg population inversion condition.

9.5 (a) Find the quasi-Fermi levels F_c and F_v for the bulk structure at zero temperature in terms of the injection carrier densities n and p, respectively.

(b) Derive an expression for the gain spectrum $g(\hbar\omega)$ at $T = 0$ K.

9.6 Use the material parameters in Appendix K for GaAs materials, except that the band gap at zero temperature is assumed to be 1.522 eV. Neglect the light-hole band for simplicity.

(a) Find the numerical values for the quasi-Fermi levels for the electrons and holes in a bulk semiconductor in the presence of carrier injection at *zero temperature* assuming $n = p = 5 \times 10^{18}$ cm^{-3}.

(b) Calculate and plot the gain spectrum for part (a).

9.7 Repeat Problem 9.6 for a quantum-well structure with an effective well width of 120 Å. Assume an infinite well model and a constant momentum matrix (independent of k) for simplicity. Label the major transition energies in your plots.

9.8 The E-k relations of a semiconductor material with a nonparabolicity conduction band can be described as

$$\text{Conduction band:} \quad E_c(k) = E_g + \frac{\hbar^2 k_0^2}{2m_e^*}\left[\left(\frac{k}{k_0}\right)^2 + \beta\left(\frac{k}{k_0}\right)^4\right]$$

$$\text{Valence band:} \quad E_v(k) = -\frac{\hbar^2 k_0^2}{2m_h^*}\left(\frac{k}{k_0}\right)^2$$

where k_0 and β are positive material parameters describing the nonparabolicity.

(a) Derive an expression for the interband absorption spectrum $\alpha(\omega)$ for a bulk structure at zero temperature. Assume a constant momentum matrix.

(b) Plot the absorption spectrum $\alpha(\hbar\omega)$ vs. the normalized energy $(\hbar\omega - E_g)/(\hbar^2 k_0^2/2m_r)$ for (i) $\beta = 0$ and (ii) $\beta = 0.1 m_e^*/m_r$, where m_r is the reduced effective mass $(1/m_r = 1/m_e^* + 1/m_h^*)$.

9.9 Consider a GaAs quantum well with an effective well width $L = 100$ Å (assuming an infinite well model). The electron and hole energies are

$$E_{en}(k) = E_{en}(0) + \frac{\hbar^2 k_t^2}{2m_e^*}$$

$$E_{hm}(k) = E_{hm}(0) - \frac{\hbar^2 k_t^2}{2m_h^*}$$

where

$$k_t^2 = k_x^2 + k_y^2, \, m_e^* = 0.0665 m_0, \, m_h^* = 0.34 m_0.$$

The other parameters are $E_g = 1.424$ eV; $m_0 = 9.1 \times 10^{-31}$ kg, $\hbar = 1.0546 \times 10^{-34}$ J $\cdot$ s $= 0.6582 \times 10^{-15}$ eV $\cdot$ s, $q = 1.60218 \times 10^{-19}$ C, $n_r = 3.68$, and $\varepsilon_0 = 8.854 \times 10^{-12}$ F/m.

(a) Calculate the subband energies, $E_{en}(0)$ and $E_{hm}(0)$, for the first two electron subbands and hole subbands, respectively.

(b) Using the normalized wave functions and the subband energies of the first two electron and hole subbands, calculate the absorption spectrum due to the interband transitions within the infinite well model. Assume that the matrix element $|\hat{e} \cdot \mathbf{p}_{cv}|$ is a constant. (The overlap integrals between the electron wave functions $f_n(z)$ and the hole wave functions $g_m(z)$ are assumed to satisfy the relation $\int f_n(z)g_m(z)\,dz = 1$ if $n = m$, and 0 if $n \neq m$.)

(c) Assume that $|\hat{e} \cdot \mathbf{p}_{cv}| \simeq m_0 E_p/6$, and $E_p = 25.7$ eV for GaAs. Calculate the magnitudes of the absorption coefficients (1/cm) for part (b) at (i) $\hbar\omega = 1.43$ eV, (ii) $\hbar\omega = 1.55$ eV, and (iii) $\hbar\omega = 1.75$ eV.

9.10 (a) Estimate the constants $C_0 = \pi e^2/(n_r c \varepsilon_0 m_0^2 \omega)$, the momentum matrix element $|\hat{e} \cdot \mathbf{p}_{cv}|^2 = 1.5 M_b^2 = 1.5(m_0 E_p/6)$, the reduced density of states $\rho_r^{2D} = m_r/(\pi\hbar^2 L_z)$, and the absorption coefficient (TE polarization) $\alpha_0(\hbar\omega)$ for a GaAs quantum well with an effective well width $L_z = 100$ Å near the absorption edge.

(b) Repeat (a) for an $\text{In}_{0.53}\text{Ga}_{0.47}\text{As}$ quantum well with $L_z = 100$ Å near the absorption edge. Note $E_g(\text{In}_{0.53}\text{Ga}_{0.47}\text{As}) = 0.75$ eV.

9.11 (a) Plot the optical matrix elements for both TE and TM polarizations (see Table 9.1) vs. the transverse wave number k_t for a GaAs quantum well with an effective well width $L_z = 100$ Å.

(b) Compare the optical matrix elements in (a) with those of another GaAs quantum well with a smaller effective well width $L_z = 40$ Å.

9.12 If a Gaussian line-shape function is used in the expressions for optical absorptions, what is the complete expression for the Gaussian function to replace the delta function $\delta(E_c - E_v - \hbar\omega)$?

9.13 Compare the interband and the intersubband absorption coefficients in a semiconductor quantum-well structure. Discuss the optical matrix elements, the density of states and the absorption spectra.

9.14 (a) To calculate the intersubband absorption in the n-type modulation-doped $\text{GaAs}/\text{Al}_{0.3}\text{Ga}_{0.7}\text{As}$ quantum well with a well width 100 Å, we use the following procedures:

(i) The ground-state energy E_1 for the first conduction subband for a finite barrier model can be calculated numerically. We can then define an effective well width L_{eff} such that

$$E_1 = \frac{\hbar^2}{2m_e^*}\left(\frac{\pi}{L_{\text{eff}}}\right)^2$$

Find L_{eff} for this $\text{GaAs}/\text{Al}_{0.3}\text{Ga}_{0.7}\text{As}$ quantum-well structure.

(ii) Using the infinite well model with an effective well width L_{eff} from part (i), calculate the second subband energy E_2 and the intersubband dipole moment

$$M_{21} = |e| \int \phi_1(z) z \phi_2(z) \, dz$$

where the integration is over the width L_{eff}.

(iii) Calculate the absorption coefficient α for a doping density $N_D = 3 \times 10^{17}/\text{cm}^3$, assuming that only the first subband is occupied. Assume that the refractive $n_r = 3.6$, $T = 77$ K, and the linewidth is 30 meV.

(b) Repeat part (a) for an $\text{In}_{0.53}\text{Ga}_{0.47}\text{As}/\text{InP}$ quantum-well structure with a well width 100 Å and the same doping density. Use the parameters such as band gaps and the electron effective masses from the Table on the Physical Properties of GaInAs Alloys in Appendix K. Assume $\Delta E_c = 40\% \Delta E_g$. Compare the results with the $\text{GaAs}/\text{Al}_{0.3}\text{Ga}_{0.7}\text{As}$ system and comment on the differences.

REFERENCES

1. M. Planck, *Verh. deut. Phys. Gesellsch.* **2**, 202 and 237 (1900).

2. A. Einstein, *Ann. Phys.* **17**, 132 (1905); "Zur Quantentheorie der Strahlung (On the quantum theory of radiation)," *Phys. Z.* **18**, 121–128 (1917).

3. W. Heitler, *The Quantum Theory of Radiation*, Clarendon, Oxford, UK, 1954.

4. R. Loudon, *The Quantum Theory of Light*, 2d ed., Clarendon, Oxford, UK, 1986.

5. H. Haken, *Light*, Vol. 1, *Waves, Photons, Atoms*, p. 204, Chapter 7, North-Holland, Amsterdam, 1986.

6. G. H. B. Thompson, *Physics of Semiconductor Laser Devices*, Chapter 2, Wiley, New York, 1980.

7. H. C. Casey, Jr., and M. B. Panish, *Heterostructure Lasers, Part A: Fundamental Principles*, Chapter 3, Academic, Orlando, FL, 1978.

8. G. Lasher and F. Stern, "Spontaneous and stimulated recombination radiation in semiconductors," *Phys. Rev.* **133**, A553–A563 (1964).

9. S. L. Chuang, J. O'Gorman, and A. F. J. Levi, "Amplified spontaneous emission and carrier pinning in laser diodes," *IEEE J. Quantum Electron.* **29**, 1631–1639 (1993).

10. H. C. Casey, Jr., and F. Stern, "Concentration-dependent absorption and spontaneous emission of heavily doped GaAs," *J. Appl. Phys.* **47**, 631–643 (1976).

11. M. G. A. Bernard and G. Duraffourg, "Laser conditions in semiconductors," *Phys. Status Solidi* **1**, 699–703 (1961).

12. A. Yariv, *Quantum Electronics*, 3d ed., Wiley, New York, 1989.

13. E. O. Kane, "Band structure of indium antimonide," *J. Phys. Chem. Solids* **1**, 249–261 (1957).

14. M. Yamanishi and I. Suemune, "Comment on polarization dependent momentum matrix elements in quantum well lasers," *Jpn. J. Appl. Phys.* **23**, L35–L36 (1984).

15. M. Asada, A. Kameyama, and Y. Suematsu, "Gain and intervalence band absorption in quantum-well lasers," *IEEE J. Quantum Electron.* **QE-20**, 745–753 (1984).

16. R. H. Yan, S. W. Corzine, L. A. Coldren, and I. Suemune, "Corrections to the expression for gain in GaAs," *IEEE J. Quantum Electron.* **26**, 213–216 (1990).

17. H. Kobayashi, H. Iwamura, T. Saku, and K. Otsuka, "Polarization-dependent gain-current relationship in GaAs–AlGaAs MQW laser diodes," *Electron. Lett.* **19**, 166–168 (1983).

18. M. Yamada, S. Ogita, M. Yamagishi, K. Tabata, N. Nakaya, M. Asada, and Y. Suematsu, "Polarization-dependent gain in GaAs/AlGaAs multiquantum-well lasers: theory and experiment," *Appl. Phys. Lett.* **45**, 324–325 (1984).

19. E. Zielinski, F. Keppler, S. Hausser, M. H. Pilkuhn, R. Sauer, and W. T. Tsang, "Optical gain and loss processes in GaInAs/InP MQW laser structures," *IEEE J. Quantum Electron.* **25**, 1407–1416 (1989).

20. E. Zielinski, H. Schweizer, S. Hausser, R. Stuber, M. H. Pilkuhn, and G. Weimann, "Systematics of laser operation in GaAs/AlGaAs multiquantum well hetero structures," *IEEE J. Quantum Electron.* **QE-23**, 969–976 (1987).

21. L. C. West and S. J. Eglash, "First observation of an extremely large dipole infrared transition within the conduction band of a GaAs quantum well," *Appl. Phys. Lett.* **46**, 1156–1158 (1983).

22. For a review, see B. F. Levine, "Quantum-well infrared photodetectors," *J. Appl. Phys.* **74**, R1–R81 (1993).

23. M. O. Manasreh, Ed., *Semiconductor Quantum Wells and Superlattices for Long-Wavelength Infrared Detectors*, Artech House, Norwood, MA, 1993.

24. D. Ahn and S. L. Chuang, "Calculation of linear and nonlinear intersubband optical absorption in a quantum well model with an applied electric field," *IEEE J. Quantum Electron.* **QE-23**, 2196–2204 (1987).

25. S. L. Chuang, M. S. C. Luo, S. Schmitt-Rink, and A. Pinczuk, "Many-body effects on intersubband transitions in semiconductor quantum-well structures," *Phys. Rev. B* **46**, 1897–1899 (1992).

26. D. Ahn and S. L. Chuang, "Optical gain and gain suppression of quantum-well lasers with valence band mixing," *IEEE J. Quantum Electron.* **26**, 13–24 (1990).

27. J. Faist, F. Capasso, D. L. Sivco, C. Sirtori, A. L. Hutchinoson, and A. Y. Cho, "Quantum cascade laser," *Science* **264**, 553–556 (1994).

10

Semiconductor Lasers

Semiconductor lasers are important optoelectric devices for optical communication systems. For a historical account of the inventions and development of semiconductor lasers, see Refs. 1–3. The first semiconductor lasers were fabricated in 1962 using homojunctions [4–7]. These lasers had high threshold current density (e.g., $19\,000$ A$/$cm^2) and operated at cryogenic temperatures. The estimated current density [2] for room temperature operation would be $50\,000$ A$/$cm^2. The concept of heterojunction semiconductor lasers was realized in 1969–1970 with a low threshold current density (1600 A$/$cm^2) operating at room temperature [8–11]. These double-heterostructure diode lasers provide both carrier and optical confinements, which improve the efficiency for stimulated emission, and make applications to optical communication systems practical as transmitters.

In the late 1970s, the concept of quantum-well structures for semiconductor lasers was proposed and realized experimentally [12–16]. The threshold current density was reduced [16] to about 500 A$/$cm^2. The reduction of the confinement region for the electron–hole pairs to a tiny volume in the quantum-well region improves the laser performance significantly. The quantum-well structures provide a wavelength tunability by varying the well width and the barrier height. These devices are also excellent practical examples for textbook problems on basic quantum mechanics, that is, the particle in a box model.

In the late 1980s, strained quantum-well lasers were proposed to improve the laser performance [17, 18]. Superior diode lasers using strained quantum wells reduced [19] the threshold current density to 65 A$/$cm^2 at room temperature. Further reduction in threshold current density using strain effects has been possible [20]. The threshold current density of semiconductor lasers has improved by approximately one order of magnitude smaller per decade since their invention in 1962. The improvements resulted from the development of novel ideas and a realizable technology such as liquid-phase epitaxy (LPE), molecular-beam epitaxy (MBE), or metal-organic chemical-vapor deposition (MOCVD) [21].

In the 1990s, research on strained quantum-well lasers continued, with remarkable progress toward high-performance diode lasers. Quantum-wire semiconductor lasers have been realized. The potential improvement using quantum-wire and quantum-dot semiconductor lasers has been under investigation both theoretically and experimentally [22–25]. Novel structures such as surface-emitting lasers [26–28] and microdisk lasers [29, 30] have also been realized. The microdisk lasers using a small disk with a radius R of the order 10 μm and a thickness of 0.6 μm show interesting resonator physics with gallery-whispering-mode lasing characteristics. Many of these structures also make research on the cavity quantum electrodynamics practical [31].

In this chapter, we study the basic semiconductor lasers and their physical principles. Various important designs and their models are discussed. Since the research on semiconductor lasers and technology is so rapid, these physical principles and theoretical models will provide a good tool to design diode lasers with the desired characteristics, or to set up a foundation to design novel laser structures with superior performance.

10.1   DOUBLE HETEROJUNCTION SEMICONDUCTOR LASERS

10.1.1   Energy Band Diagram and Carrier Injection

Let us first consider a P–InP/p–In$_{1-x}$Ga$_x$As$_{1-y}$P$_y$/N–InP double hetero-junction [32, 33] structure. Before the formation of the heterojunctions, the energy band diagram is shown schematically in Fig. 10.1a, where the energy levels are all measured from the vacuum level. At thermal equilibrium (Fig. 10.1b), the Fermi level is a constant across the three regions. The band diagram still contains the features of the bulk properties in the regions away from the heterojunctions, assuming that the central region is wide enough. The built-in potentials are determined by the Fermi levels:

$$-qV_D^{P-p} = E_{FP} - E_{fp} = \left(E_{FP} - E_{VP} \right) - \Delta E_V - \left(E_{fp} - E_{vp} \right) \quad (10.1.1)$$

$$-qV_D^{p-N} = E_{fp} - E_{FN} = \left(E_{fp} - E_{vp} \right) + \Delta E_V - \left(E_{FN} - E_{VN} \right) \quad (10.1.2)$$

where

$$\Delta E_V = E_{vp} - E_{VP} = E_{vp} - E_{VN} \quad (10.1.3)$$

With an applied voltage V_a, the injection current is assumed to be I, and the current density is

$$J = \frac{I}{wL} \quad (10.1.4)$$

where w is the width of the active region in the lateral direction and L is the cavity length of the diode laser. The carrier concentration in the active region

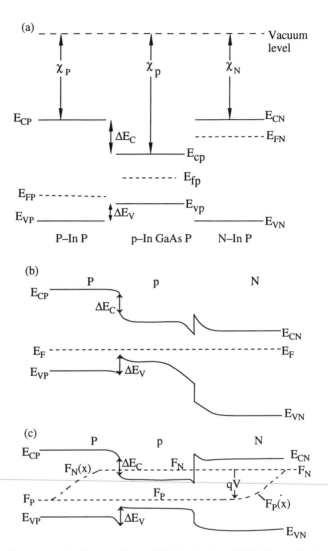

Figure 10.1. Energy band diagram for a P-InP/p-InGaAsP/N-InP double heterojunction structure (a) before contact, (b) after contact at thermal equilibrium ($V = 0$), and (c) under a forward bias ($V > 0$).

is determined by the rate equation:

$$\frac{\partial n}{\partial t} = D \nabla^2 n + \frac{J}{qd} - R(n) \qquad (10.1.5)$$

where the first term on the right-hand side accounts for carrier diffusion, the second term is due to the carrier injection into the active region with a thickness d, and the last term, $R(n)$, accounts for the carrier recombinations due to both radiative and nonradiative processes. The unit charge q is 1.6×10^{-19} C. The carrier diffusion is a consequence of intraband scattering, which is important for gain-guided structures. For an index-guided structure, we can assume for simplicity that the diffusion effect is negligible, and obtain

$$\frac{J}{qd} = R(n) \tag{10.1.6}$$

at steady state. The recombination rate is given by

$$R(n) = A_{nr}n + Bn^2 + Cn^3 + R_{st}N_{ph} \tag{10.1.7}$$

where the first term is due to nonradiative processes, Bn^2 is due to the spontaneous radiative recombination rate, Cn^3 accounts for nonradiative Auger recombinations, and the last term accounts for stimulated recombination that leads to emission of light and it is proportional to the photon density N_{ph}. The Auger term is important only for long-wavelength semiconductor lasers.

We have assumed that $n \simeq p$ in the active region, i.e., the doping concentration in the active region is small. Otherwise, we use Bnp instead of Bn^2, and Cn^2p or Cnp^2, depending on the type of Auger processes. B may depend on the carrier concentration:

$$B \simeq B_0 - B_1 n \tag{10.1.8}$$

The coefficient for the stimulated emission rate is

$$R_{st} = v_g g(n) \tag{10.1.9}$$

where v_g is the group velocity $v_g = c/n_g$, n_g is the (group) refractive index, c is the speed of light in free space, and $g(n)$ is the gain coefficient. Below or near threshold, we can ignore $R_{st}N_{ph}$ since the photon density is small, and write

$$R(n) = \frac{n}{\tau_e(n)} \tag{10.1.10}$$

where

$$\tau_e(n) = \frac{1}{A_{nr} + Bn + Cn^2} \tag{10.1.11}$$

Given the injection current density J, the carrier concentration is determined by

$$\frac{J}{qd} = A_{nr}n + Bn^2 + Cn^3 \tag{10.1.12}$$

if we know the coefficients A_{nr}, B, and C for the laser structure. Sometimes the carrier lifetime τ_e is taken as an input parameter and one determines n approximately by $n = J\tau_e/qd$. Knowing n in the active region, the quasi-Fermi level in the active region is determined by

$$n = N_c F_{1/2}\left(\frac{F_n - E_c}{k_B T}\right) \tag{10.1.13a}$$

$$N_c = 2\left(\frac{m_e^* k_B T}{2\pi\hbar^2}\right)^{3/2} \tag{10.1.13b}$$

and the hole concentration in the active region is determined by the charge neutrality condition:

$$n + N_A^- = p + N_D^+ \tag{10.1.14}$$

where the ionized acceptor and donor concentrations are

$$N_A^- = \frac{N_A}{1 + 4\exp\left[(E_A - F_p)/k_B T\right]} \tag{10.1.15}$$

$$N_D^+ = \frac{N_D}{1 + 2\exp\left[(F_n - E_D)/k_B T\right]} \tag{10.1.16}$$

E_A and E_D are the acceptor and donor energy levels, and N_A and N_D are the acceptor and donor concentrations, respectively. The quasi-Fermi level in the active region for the holes is then determined by

$$p = N_v F_{1/2}\left(\frac{E_v - F_p}{k_B T}\right) \tag{10.1.17a}$$

$$N_v = 2\left(\frac{k_B T}{2\pi\hbar^2}\right)^{3/2}\left(m_{hh}^{3/2} + m_{lh}^{3/2}\right) \tag{10.1.17b}$$

From the electron and hole concentrations, n and p, or the quasi-Fermi levels, F_n and F_p, we can calculate the gain directly using the band structure. Typically, a linear relation for the peak gain coefficient vs. the carrier density such as

$$g(n) = a(n - n_{tr}) \tag{10.1.18}$$

for a bulk semiconductor laser is obtained over a region of carrier concentration, where a is the differential gain, and n_{tr} is a transparency concentration when the gain is zero (Fig. 10.2). For quantum-well lasers, an empirical formula using a logarithmic dependence for the gain vs. carrier density or

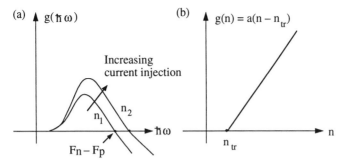

Figure 10.2.   (a) The gain spectrum for different carrier densities, $n_2 > n_1$; (b) the peak gain vs. the carrier density n.

gain vs. injection current density is usually used. This is discussed further in Sections 10.3 and 10.4.

10.1.2   Threshold Condition

The threshold condition is determined by

$$g_{th}\Gamma = \alpha_i + \alpha_m \tag{10.1.19}$$

$$\alpha_i = \alpha_0(1 - \Gamma) + \alpha_g\Gamma \tag{10.1.20}$$

$$\alpha_m = \frac{1}{2L}\ln\frac{1}{R_1 R_2} \tag{10.1.21}$$

where Γ is the optical confinement factor, α_i is the intrinsic loss due to absorptions inside the guide α_g and outside α_0, and α_m accounts for the transmissions (outputs) at the two end mirrors (facets). With an increase in the injection current density J, the gain will increase due to the increase of the carrier concentration n. When the threshold condition is reached, the carrier concentration n will be pinned at the threshold value n_{th} since the gain is pinned at the threshold gain $g_{th} = (\alpha_i + \alpha_m)/\Gamma$; therefore, $n_{th} = n_{tr} + (\alpha_i + \alpha_m)/(\Gamma a)$ and

$$J_{th} \simeq \frac{qdn_{th}}{\tau_e(n_{th})} \tag{10.1.22}$$

$$\tau_e(n_{th}) = \frac{1}{A_{nr} + Bn_{th} + Cn_{th}^2} \tag{10.1.23}$$

Below the threshold condition, the output light consists mainly of spontaneous emission and its magnitude is governed by Bn^2. A further increase in the injection current density J above threshold leads to the light emission through the stimulated emission process:

$$J = qd\left(A_{nr}n_{th} + Bn_{th}^2 + Cn_{th}^3\right) + qdR_{st}N_{ph}$$

$$= J_{th} + qdv_g g_{th}N_{ph} \qquad (10.1.24)$$

where $R_{st} = v_g g(n) = v_g g_{th}$ since n is pinned at n_{th}. Therefore,

$$N_{ph} = \frac{1}{qdv_g g_{th}}(J - J_{th}) \qquad (10.1.25)$$

The photon lifetime τ_p is defined as

$$\frac{1}{\tau_p} = v_g(\alpha_i + \alpha_m) = v_g \Gamma g_{th} \qquad (10.1.26)$$

which accounts for the loss rate of the photons in the laser cavity due to absorptions and transmissions. We obtain

$$\frac{N_{ph}}{\tau_p} = \frac{\Gamma}{qd}(J - J_{th}) \qquad (10.1.27)$$

10.1.3 Output Power and Differential Quantum Efficiency

The optical output power P_{out} vs. the injection current (L–I curve) is determined by

$$P_{out} = \left(\begin{array}{c}\text{energy of}\\\text{a photon}\end{array}\right)\left(\begin{array}{c}\text{photon}\\\text{density}\end{array}\right)\left(\begin{array}{c}\text{effective volume of the}\\\text{optical mode}\end{array}\right)\left(\begin{array}{c}\text{escape rate}\\\text{of photons}\end{array}\right)$$

$$= \hbar\omega N_{ph}(Lwd_{op})(v_g\alpha_m) \qquad (10.1.28)$$

where $v_g\alpha_m$ is the escape rate of photons. Here w is the width and d_{op} is the effective thickness of the optical mode ($d_{op} = d/\Gamma$). Therefore

$$P_{out} = \hbar\omega v_g\alpha_m \frac{\tau_p}{q}(J - J_{th})Lw$$

$$= \frac{\hbar\omega}{q}\frac{\alpha_m}{\alpha_m + \alpha_i}(I - I_{th}) \qquad (10.1.29)$$

The above relation assumes that the internal quantum efficiency η_i is one, where η_i is defined as the percentage of the injected carriers that contribute to the radiative recombinations:

$$\eta_i = \frac{Bn^2 + R_{st}N_{ph}}{An + Bn^2 + Cn^3 + R_{st}N_{ph}} \qquad (10.1.30)$$

Generally, $\eta_i < 1$, and Eq. (10.1.29) should be modified as

$$P_{out} = \frac{\hbar\omega}{q} \frac{\alpha_m}{\alpha_m + \alpha_i} \eta_i (I - I_{th}) \qquad (10.1.31)$$

The external differential quantum efficiency η_e is defined as

$$\eta_e = \frac{dP_{out}/dI}{\hbar\omega/q} = \eta_i \frac{\alpha_m}{\alpha_m + \alpha_i}$$

$$= \eta_i \frac{\ln(1/R)}{\alpha_i L + \ln(1/R)} \qquad (10.1.32a)$$

The inverse of the external differential quantum efficiency is

$$\eta_e^{-1} = \eta_i^{-1} \left[1 + \alpha_i \frac{L}{\ln(1/R)} \right] \qquad (10.1.32b)$$

A plot of η_e^{-1} vs. the cavity length L shows a linear relationship with the intercept η_i^{-1} at $L = 0$, as shown in Fig. 10.3.

In real devices, leakage current exists. The effect of leakage current I_L can be taken into account in

$$I = I_A + I_L = JwL + I_L \qquad (10.1.33)$$

where I_A is the current injected into the active region. Therefore, I_{th} is

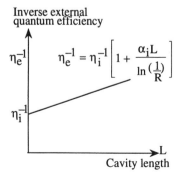

Inverse external quantum efficiency

$$\eta_e^{-1} = \eta_i^{-1} \left[1 + \frac{\alpha_i L}{\ln \left(\frac{1}{R} \right)} \right]$$

L
Cavity length

Figure 10.3. A plot of the inverse external quantum efficiency $1/\eta_e$ vs. the laser cavity length L. The intercept with the vertical axis gives the inverse intrinsic quantum efficiency $1/\eta_i$.

modified to be

$$I_{th} = J_{th}wL + I_L$$

$$= \frac{qn_{th}(wLd)}{\tau_e(n_{th})} + I_L \qquad (10.1.34)$$

Also, the leakage current I_L may also be increased with an increase in I. The output optical power P_{out} vs. the injection current I relation can be modified to be

$$P_{out} = \frac{\hbar\omega}{q} \frac{\alpha_m}{\alpha_m + \alpha_i} \eta_i (I - I_{th} - \Delta I_L) \qquad (10.1.35)$$

where ΔI_L accounts for any additional increase in the leakage current due to the increase of I. A typical output power vs. injection current relation is shown in Fig. 10.4. Below the threshold injection current, the output light intensity is negligibly small. Above threshold, the output power is linearly increasing until saturation effects occur.

There are three possible reasons for saturation [33]:

1. The leakage current increases with the injection current I, i.e., $\Delta I_L \neq 0$.
2. The threshold current I_{th} may also depend on the injection current I due to junction heating. The increase in temperature reduces the recombination lifetime τ_e. For example, the Auger recombination rate increases when the temperature is raised.
3. The internal absorption α_{int} increases with an increase in the injection current I.

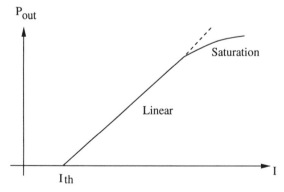

Figure 10.4. A typical diode laser output power vs. injection current density relation ($L - I$ curve).

The output power vs. the injection current becomes a nonlinear dependence when saturation occurs.

10.1.4 Leakage Current [34–36]

The leakage current can be obtained from the minority carrier concentration. For example, the hole concentration $P(x)$ in the N region satisfies

$$\frac{\partial}{\partial t}P(x) = \frac{-1}{q}\frac{\partial}{\partial x}j_P + G(x) - \frac{P - P_0}{\tau_P} \qquad (10.1.36a)$$

$$j_P(x) = -qD_P\frac{dP}{dx} \qquad (10.1.36b)$$

In the steady state, assuming no generation in the cladding regions, we find the differential equation for $P(x)$:

$$\frac{d^2P}{dx^2} - \frac{P - P_0}{D_P\tau_P} = 0 \qquad (10.1.37)$$

Let us write P as $P_N(x)$ and P_0 as P_{N0} for the hole concentrations in the N region. The solution for the hole concentration is of the form

$$P_N(x) = C_1 e^{-x/L_P} + C_2 e^{x/L_P} + P_{N0} \qquad (10.1.38)$$

where $L_P = \sqrt{D_P\tau_P}$. Assume that the hole concentration is P_N at $x = x_N = 0$. The excess carrier concentration is $\delta P_N(x) = P_N(x) - P_{N0} = P_N - P_{N0}$ at $x = 0$. The other boundary condition, $\delta P_N(x)$, vanishes at the contact $x = W_N$. The hole concentration is found to be

$$P_N(x) = (P_N - P_{N0})\frac{\sinh\left[(W_N - x)/L_P\right]}{\sinh(W_N/L_P)} + P_{N0} \qquad (10.1.39)$$

The leakage current density at $x = x_N \simeq 0^+$ is, therefore,

$$j_P(0^+) = \frac{qD_P(P_N - P_{N0})}{L_P\tanh(W_N/L_P)} \qquad (10.1.40)$$

Similarly, the leakage current density flowing into the P region is

$$j_N = \frac{qD_N(N_P - N_{P0})}{L_N\tanh(W_P/L_N)} \qquad (10.1.41)$$

where we have used the electron concentration in the P region

$$N_P(x) = (N_P - N_{P0}) \frac{\sinh[(W_P + x)/L_N]}{\sinh(W_P/L_N)} + N_{P0} \quad (10.1.42)$$

and $\delta N_P = N_P(x) - N_{P0} = 0$ at $x = -W_P$, where x is measured from the edge of the $P - p$ heterojunction and the electron concentration is $N_P(x = 0) = N_P$. The background concentrations P_{N0} and N_{P0} at thermal equilibrium are usually ignored since they are small. The total leakage current is

$$I_L = wL(j_P + j_N)$$

$$= wLq \left[\frac{D_N N_P}{L_N \tanh(W_P/L_N)} + \frac{D_P P_N}{L_P \tanh(W_N/L_P)} \right] \quad (10.1.43)$$

Note that the minority carrier concentration P_N at $x = x_N$ or N_P at $x = -x_P$ can be obtained using the band bending and (10.1.17) and (10.1.13), respectively. For P_N, we have, at $x = x_N$, the boundary of the depletion region near the $p - N$ junction

$$E_{VN} - F_P = E_{VN} - F_N + qV$$

$$= (E_{CN} - F_N) - E_{GN} + qV \quad (10.1.44)$$

as shown in Fig. 10.1c, where V is the bias voltage. Here $E_{CN} - F_N$ is determined by the doping concentration in the N region. We then use (10.1.17) to find P_N:

$$P_N = N_V F_{1/2} \left(\frac{E_{VN} - F_P}{k_B T} \right) \simeq N_V e^{(E_{VN} - F_P)/k_B T} \quad (10.1.45a)$$

$$N_V = 2 \left(\frac{k_B T}{2\pi\hbar^2} \right)^{3/2} (m_{hh}^{3/2} + m_{lh}^{3/2}) \quad (10.1.45b)$$

Similarly, for N_P at the boundary of the depletion region in the P region, we have

$$F_N - E_{CP} = qV + F_P - E_{CP}$$

$$= qV + (F_P - E_{VP}) - E_{GP} \quad (10.1.46)$$

where $F_P - E_{VP}$ is determined by the doping in the P region. The electron concentration in the P region N_P is given by (10.1.13),

$$N_P = N_C F_{1/2}\left(\frac{F_N - E_{CP}}{k_B T}\right) \simeq N_C\, e^{(F_N - E_{CP})/k_B T} \qquad (10.1.47a)$$

$$N_C = 2\left(\frac{m_e^* k_B T}{2\pi\hbar^2}\right)^{3/2} \qquad (10.1.47b)$$

The diffusion coefficients D_P and D_N in the doped wide-gap semiconductors can be taken from the mobility data based on the Einstein relation, e.g., $D_P = \mu_P kT/q$, and $D_N = \mu_N kT/q$.

10.1.5   Light-Emitting Diode vs. Laser Diode: The Role of Spontaneous Emission vs. Amplified Spontaneous Emission

In Chapter 9, we discuss the spontaneous emission $r^{\text{spon}}(\hbar\omega)$ from a forward biased p-n junction diode. The typical spontaneous emission spectrum is commonly used in light-emitting diodes (LEDs). The band gap of semiconductors is chosen such that LEDs in the visible region and the infrared region can be designed with high quantum efficiency. For a simple planar structure such as shown [37–39] in Fig. 10.5, the light emission from the top surface of the cladding layer due to a spontaneously emitted photon will be limited by a cone angle of $2\theta_c$, where θ_c is the critical angle between the p-type semiconductor and the air. Light with an angle of emission outside of this cone angle will be reflected back to the semiconductor region. Therefore, the geometric design such as a hemispherical lens or parabolic lens above the p-n junction diode has been used to optimize the light emission. Other important considerations for visible LEDs include the responsivity of human eyes. Infrared

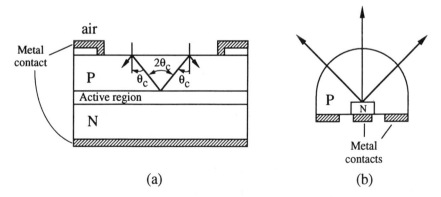

(a)   (b)

Figure 10.5.   (a) Schematic diagram of a planar LED device to show the effect of total internal refraction. (b) A semiconductor hemisphere geometry for LED operation.

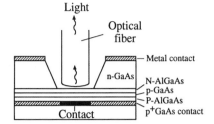

Figure 10.6. AlGaAs surface-emitting LED with an attached fiber to couple light output to an optical fiber.

LEDs for applications in optical communication are designed such that efficient coupling into optical fibers can be achieved, as shown in Fig. 10.6, with a surface emitter configuration [40]. The double heterojunction Al-GaAs/GaAs structure also provides the carrier confinement to increase the efficiency [41].

For a laser diode, the cavity design and the amplified spontaneous emission will be important for the formation of laser modes. To understand the physics of the laser diode vs. light-emitting diode operation, we consider an LED that is designed to have the same Fabry–Perot structure as an LD except that its two ends are antireflection coated. If we measure the spontaneous emission power spectrum from the top or bottom of the diode, called

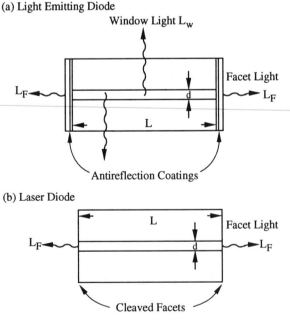

Figure 10.7. (a) The spontaneous emission coming out of the side (window light L_w) vs. the amplified spontaneous emission coming out from the facet (facet light L_F) in an LED structure. (b) A laser diode with the output power coming from both facets.

the window light L_w, it will be a broad spectrum determined by the spontaneous emission rate per unit volume $r^{\text{spon}}(\hbar\omega)$ (Fig. 10.7a) [42, 43]:

$$L_w(\hbar\omega) = \hbar\omega r^{\text{spon}}(\hbar\omega)wdL \qquad (10.1.48)$$

where w is the width, d is the thickness, and L is the cavity length of the index-guided structure. The facet light spectrum $L_F(\hbar\omega)$, taken from a facet of the LED, is the amplified spontaneous emission spectrum [42–45]:

$$L_F(\hbar\omega) = \hbar\omega \int_{z=0}^{L} r^{\text{spon}}(\hbar\omega)e^{G_n(\hbar\omega)z}\,dz\,wd$$

$$= \hbar\omega r^{\text{spon}}(\hbar\omega)\left[\frac{e^{G_n(\hbar\omega)L} - 1}{G_n(\hbar\omega)}\right]wd \qquad (10.1.49)$$

where $G_N(\hbar\omega) = \Gamma g - \alpha_i$ is the net modal gain (including all the losses due to absorptions and scatterings in the device). Note that at the transparency wavelength, $G_n(\hbar\omega) = 0$ and $L_F(\hbar\omega) = L_w(\hbar\omega)$ when no amplification by the gain action exists. With enough carrier injection such that $G_n > 0$, we see

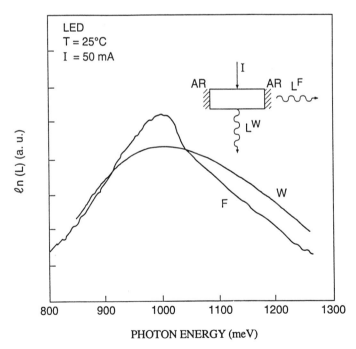

Figure 10.8. Measured spontaneous emission spectrum obtained by collecting photons through a window in the substrate L^W and the amplified spontaneous emission spectrum obtained by collecting photons from the facet L^F of an LED. (After Ref. 42).

that the photon density in the spectral region where $g(\hbar\omega)$ is positive will experience amplification, while that outside the positive gain region will experience absorption. Since the gain spectrum is narrower than that of the spontaneous emission spectrum [46] the facet light will be narrower than that of the window light. These results [42] are shown in Fig. 10.8. As a matter of fact, a comparison of these two spectra has been used to extract the gain spectrum of a laser diode structure. By measuring the two spectra $L_F(\hbar\omega)$ and $L_w(\hbar\omega)$ at the same current injection level, we can obtain the gain

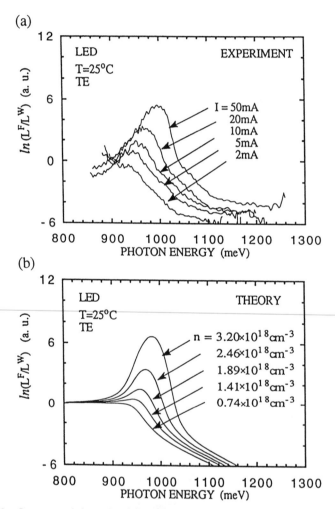

Figure 10.9. Spectrum of the ratio of facet light to window light of the LED at $T = 25°C$ at $I = 50, 20, 10, 5,$ and 2 mA for TE polarization. (a) Experimental data and (b) theoretical results are shown with arbitrary vertical scale. The corresponding carrier densities are indicated in (b). (After Ref. 43 © 1993 IEEE.)

spectrum. Furthermore, if we take the logarithmic function of the ratio of the two spectra, $\ln[L_F(\hbar\omega)/L_w(\hbar\omega)]$, it will be close to that of the gain spectrum, since it is proportional to $\ln[(e^{G_n L} - 1)/G_n] \sim (\Gamma g_n - \alpha_i)L$ if the overall gain $(G_n L \gg 1)$ is large enough (Fig. 10.9). By fitting the gain spectrum with a theoretical gain model, we can extract the carrier density n at a given injection current I, as shown in Fig. 10.10. The carrier density vs. the injection current I is a monotonically increasing function of I, as shown in Fig. 10.10a at 25°C and Fig. 10.10b at 55°C.

(a)

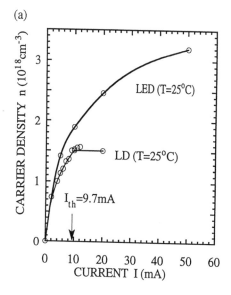

(b)

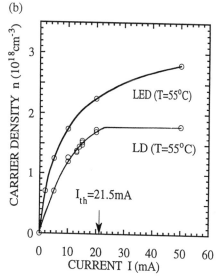

Figure 10.10. Carrier density versus injection current plotted for both LED and LD at (a) $T = 25°C$ and (b) $T = 55°C$. The solid curves are just visual guides. (After Ref. 43 © 1993 IEEE.)

10.1.6 Analysis of the Amplified Spontaneous Emission

In a laser diode, the Fabry–Perot cavity provides feedback reflections of light inside the cavity. We can understand the lasing action by looking at the amplified spontaneous emission process below threshold. The intensity of a Fabry–Perot mode inside the cavity satisfies [47, 48]

$$\frac{dI^{\pm}(z)}{dz} = \pm G_n I^{\pm}(z) \pm W(\hbar\omega) \tag{10.1.50a}$$

$$W(\hbar\omega) = \left(\tfrac{1}{2}\right)\beta r_{spon}(\hbar\omega)\hbar\omega \tag{10.1.50b}$$

where $G_n = \Gamma g - \alpha_i$ is the net modal gain, β is a spontaneous coupling factor which accounts for the fraction of the spontaneous emission photons coupled into the lasing mode, and the factor $1/2$ accounts for waves propagating in the $+z$ and $-z$ directions. The solutions for the forward and backward propagation light intensities are

$$I^{+}(z) = I_0^{+} e^{G_n z} - \frac{W(\hbar\omega)}{G_n(\hbar\omega)}$$

$$I^{-}(z) = I_0^{-} e^{-G_n z} - \frac{W(\hbar\omega)}{G_n(\hbar\omega)} \tag{10.1.51}$$

The boundary conditions at $z = 0$, $I^{+}(0) = R_1 I^{-}(0)$ and at $z = L$, $I^{-}(L) = R_2 I^{+}(L)$ determine the constants I_0^{+} and I_0^{-}. For example,

$$I_0^{+} = \frac{\left(1 + R_1 e^{G_n L}\right) - R_1\left(1 + R_2 e^{G_n L}\right)}{1 - R_1 R_2 e^{2G_n L}} \frac{W(\hbar\omega)}{G_n(\hbar\omega)} \tag{10.1.52}$$

The output intensity from facet 2 of the laser cavity is [47, 48]

$$I_2 = (1 - R_2)I^{+}(z = L)$$

$$= (1 - R_2)\frac{\left(e^{G_n L} - 1\right)\left(R_1 e^{G_n L} + 1\right)}{1 - R_1 R_2 e^{2G_n L}} \frac{W(\hbar\omega)}{G_n(\hbar\omega)} \tag{10.1.53}$$

Note that in the above analysis, only the optical intensity is considered. The phase information is ignored. We can also see that the threshold condition occurs at

$$1 - R_1 R_2 e^{2G_n L} \simeq 0 \quad \text{or} \quad G_n = \frac{1}{2L}\ln\left(\frac{1}{R_1 R_2}\right) \tag{10.1.54}$$

where the optical output approaches infinity at the lasing wavelength. In other words, the output intensity I_2 is proportional to the spontaneous emission coupling $W(\hbar\omega)$ multiplied by the transmission coefficient and is amplified by various factors in (10.1.53), including the vanishing denominator.

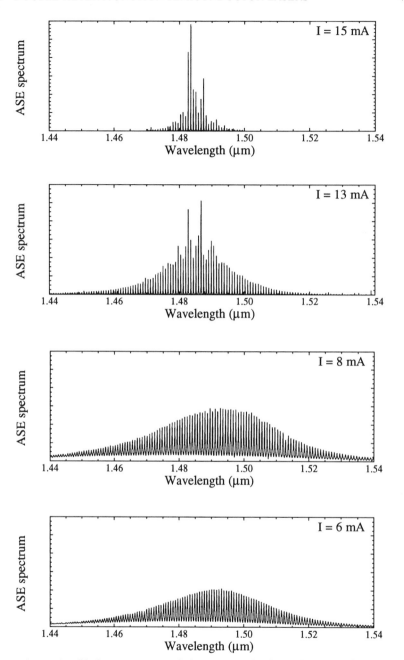

Figure 10.11. Amplified spontaneous emission spectra of a laser diode at different current injection levels, I = 6, 8, 13, and 15 mA. The vertical scales are in arbitrary units and do not have the same scales in different figures. The threshold current is 13.5 mA.

In reality, the optical output is not infinity; we may say that the net threshold gain G_n approaches (but is slightly smaller than) the mirror transmission losses $(1/2L)\ln(1/R_1R_2)$. An interesting limit occurs when both facets are antireflection coated, $R_1 = R_2 = 0$. We find

$$I_2 = \left(e^{G_nL} - 1\right)\frac{W(\hbar\omega)}{G_n(\hbar\omega)} \tag{10.1.55}$$

which is the same as the facet light power $L_F(\hbar\omega)$ per unit cross section wd as expected. We also note that the transmissivity $T_2 = 1 - R_2$ can also be written as $\ln(1/R_2)$ when $R_2 \to 1$ because $\ln[1/(1 - T_2)] \simeq \ln(1 + T_2) \simeq T_2 = (1 - R_2)$. Therefore, the mirror loss in its distributed form $\alpha_m = (1/2L)\ln(1/R_1R_2)$ is used very often.

The evolution of the facet light or amplified spontaneous emission from a laser diode as the injection current increases from below-threshold operations ($I = 6$ and 8 mA) to near-threshold ($I = 13$ mA) and above-threshold operation ($I = 15$ mA) is shown in Fig. 10.11. Note that the vertical scales are not the same for the four figures. The lasing action starts to show drastically when the injection current approaches the threshold level (13.5 mA). An expansion of the plot for the amplified spontaneous emission spectrum at 8 mA was shown in Fig. 7.17, where the Fabry–Perot mode spacing can be seen clearly to be around 7.9 Å for the laser with a cavity length $L = 370$ μm. The carrier concentrations n extracted from the gain measurement below and near the threshold can be extracted and plotted as a function of the injection current. Above threshold, the carrier concentration in the laser diode is pinned at the threshold carrier density, since most of the extra carriers injected contribute to stimulated emission and output optical power. Important issues on the mode fluctuations and temperature dependence are discussed in Refs. 42 and 43.

10.2 GAIN-GUIDED AND INDEX-GUIDED SEMICONDUCTOR LASERS

In this section, we study both gain-guided and index-guided semiconductor lasers. The difference is in the lateral confinements of the optical mode and the carriers. The index-guided semiconductor lasers have been demonstrated to have superior performance to that of the gain-guided geometry. However, the gain-guided structures are still used because of their ease of fabrication. They are investigated here because many interesting physical processes, including carrier injection and diffusion, optical gain, and spatial carrier profiles, can be extracted from these structures.

10.2.1 Stripe-Geometry Gain-Guided Semiconductor Lasers [34, 38, 49–57]

A stripe-geometry gain-guided semiconductor laser is shown in Fig. 10.12, where the metal contact is the stripe with a width S defined by processing procedure. The injected current J spreads along the lateral (i.e., y) direction. Therefore, the carrier density n distributes along the y direction determined by the current spreading and the lateral diffusion. The effective width of the carrier density distribution and the gain profile along the lateral direction are therefore wider than the stripe width S. This gain inhomogeneity along the y direction provides the guiding mechanism for the optical mode confinement in the lateral direction. An associated effect is the antiguidance due to the reduction in the refractive index variation induced by the gain profile. In this section, we study an approximate analysis method to model the gain-guided stripe-geometry semiconductor laser.

Analysis of the Optical Modal Profile and Complex Propagation Constant. We consider the TE mode with index guidance in the x direction, perpendicular to the p-n junction plane, and with gain guidance along the y direction, which is transverse to the propagation direction. The optical electric field is approximated by

$$\mathbf{E} \simeq \hat{y} E_y(x, y, z) \simeq \hat{y} F(x) G(y) e^{ik_z z} \qquad (10.2.1)$$

under scalar approximation, where $E_y(x, y, z)$ satisfies the wave equation

$$\nabla^2 E_y + \omega^2 \mu \varepsilon(x, y) E_y = 0 \qquad (10.2.2)$$

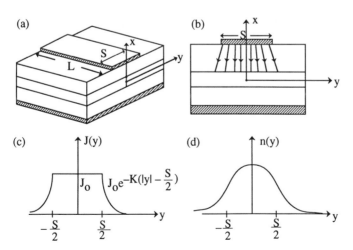

Figure 10.12. (a) A stripe-geometry gain-guided semiconductor laser, (b) cross section, (c) distribution of the current density $J(y)$, and (d) the carrier density profile $n(y)$.

where a parabolic profile in the active region is assumed because of the approximate parabolic carrier and gain profiles.

$$
\varepsilon(x, y) = \begin{cases} \varepsilon_1 & x > \dfrac{d}{2} \\[2mm] \varepsilon_2(0) - a^2 y^2 & |x| \le \dfrac{d}{2} \\[2mm] \varepsilon_1 & x < -\dfrac{d}{2} \end{cases} \qquad (10.2.3)
$$

Our goal is to find the electric field profile $F(x)G(y)$ and the complex propagation constant k_z for the inhomogeneous permittivity profile (10.2.3). We can write $\varepsilon(x, y)$ in terms of an unperturbed part, $\varepsilon^{(0)}(x)$, and a perturbed part, $\Delta\varepsilon(x, y)$,

$$
\varepsilon(x, y) = \varepsilon^{(0)}(x) + \Delta\varepsilon(x, y) \qquad (10.2.4)
$$

where

$$
\varepsilon^{(0)}(x) = \begin{cases} \varepsilon_1 & x > \dfrac{d}{2} \\[2mm] \varepsilon_2(0) & |x| \le \dfrac{d}{2} \\[2mm] \varepsilon_1 & x < -\dfrac{d}{2} \end{cases} \qquad \Delta\varepsilon(x, y) = \begin{cases} 0 & x > \dfrac{d}{2} \\[2mm] -a^2 y^2 & |x| \le \dfrac{d}{2} \\[2mm] 0 & x < -\dfrac{d}{2} \end{cases}
$$

$$
(10.2.5)
$$

The unperturbed part $\varepsilon^{(0)}(x)$ is the permittivity profile of a slab optical waveguide; the solution for the optical field was obtained in Chapter 7. Here we substitute (10.2.1) into (10.2.2) and obtain

$$
\frac{1}{F}\frac{\partial^2 F}{\partial x^2} + \omega^2\mu\varepsilon^{(0)}(x) - k_z^2 + \frac{1}{G}\frac{\partial^2 G}{\partial y^2} + \omega^2\mu\,\Delta\varepsilon = 0 \qquad (10.2.6)
$$

Assuming that the index guidance along the x direction provides a good optical confinement, we approximate $F(x)$ by the electric profile $E_y^{(0)}(x)$ of the unperturbed slab waveguide with a propagation constant β_z:

$$
F(x) \simeq E_y^{(0)}(x) \qquad (10.2.7a)
$$

where $E_y^{(0)}(x)$ satisfies

$$
\frac{\partial^2}{\partial x^2}E_y^{(0)}(x) + \left(\omega^2\mu\varepsilon^{(0)}(x) - \beta_z^2\right)E_y^{(0)}(x) = 0 \qquad (10.2.7b)
$$

Using the definition

$$v^2 = \beta_z^2 - k_z^2 \qquad (10.2.8)$$

in (10.2.6) and (10.2.7), we find

$$\frac{\partial^2 G(y)}{\partial y^2} + \left(\omega^2\mu\,\Delta\varepsilon + v^2\right)G(y) = 0 \qquad (10.2.9)$$

To solve (10.2.9), we multiply it by the optical intensity profile $I(x) = (\beta_z/2\omega\mu)|E_y^{(0)}(x)|^2$ and integrate over x from $-\infty$ to $+\infty$ to obtain

$$\frac{\partial^2 G(y)}{\partial y^2} + \left(v^2 - \Gamma\omega^2\mu a^2 y^2\right)G(y) = 0 \qquad (10.2.10)$$

noting that the wave function $F(x) = E_y^{(0)}(x)$ has been normalized ($\int_{-\infty}^{\infty}I(x)\,dx = 1$) and $\Delta\varepsilon(x,y)$ is zero outside the waveguide, which introduces the optical confinement factor $\Gamma = \int_{|x|\leq d/2}I(x)\,dx$. Equation (10.2.10) is independent of x and its solutions are the Hermite Gaussian functions (see Section 3.3):

$$\frac{d^2 u_n(t)}{dt^2} + \left(2n + 1 - t^2\right)u(t) = 0 \qquad (10.2.11a)$$

$$u_n(t) = H_n(t)e^{-t^2/2} \qquad n = 0,1,2,3,\ldots \qquad (10.2.11b)$$

where

$$\frac{d^2}{dt^2}H_n - 2t\frac{d}{dt}H_n + 2nH_n = 0 \qquad (10.2.11c)$$

If we change the variable from y to t in (10.2.10),

$$t = \xi^{1/2}y = \left(\omega^2\mu a^2\Gamma\right)^{1/4}y \qquad (10.2.12a)$$

$$\xi = \left(\omega^2\mu a^2\Gamma\right)^{1/2} = k_0 a\left(\frac{\Gamma}{\varepsilon_0}\right)^{1/2} \qquad (10.2.12b)$$

we obtain the following equation similar to (10.2.11a)

$$\frac{d^2 G}{dt^2} + \left(\frac{v^2}{\xi} - t^2\right)G = 0 \qquad (10.2.13)$$

Therefore, we find

$$G = H_n(t)e^{-t^2/2} = H_n(\xi^{1/2}y)e^{-\xi y^2/2} \tag{10.2.14a}$$

$$\nu^2 = \xi(2n + 1) = (\omega^2\mu a^2\Gamma)^{1/2}(2n + 1) \tag{10.2.14b}$$

The complex propagation constant of the gain-guided laser, k_z, is obtained from

$$k_z^2 = \beta_z^2 - \nu^2 = \beta_z^2 - (\omega^2\mu a^2\Gamma)^{1/2}(2n + 1) \tag{10.2.15}$$

for the nth lateral mode. We note that β_z depends on the index m of the TE_m mode for the slab guide geometry. Therefore, k_z is prescribed by two mode indices, m and n, which are associated with the optical field variation along the x and y directions, respectively. The optical electric field is given by

$$E_y(x, y, z) = E_y^{(0)}(x)H_n\left[(k_0a)^{1/2}\left(\frac{\Gamma}{\varepsilon_0}\right)^{1/4}y\right]e^{-k_0a(\Gamma/\varepsilon_0)^{1/2}y^2/2}\, e^{ik_z z}$$

$$\tag{10.2.16}$$

Using binomial expansion in (10.2.15), we have for $n = 0$

$$k_z \simeq \beta_z - \frac{k_0a}{2\beta_z}\left(\frac{\Gamma}{\varepsilon_0}\right)^{1/2} \tag{10.2.17}$$

Because $a = a_r + ia_i$ is complex, the constant phase front is determined by

$$k_z^r z - k_0a_i\left(\frac{\Gamma}{\varepsilon_0}\right)^{1/2}\frac{y^2}{2} = \text{constant} \tag{10.2.18}$$

which is a cylindrical parabolic surface (Fig. 10.13), where k_z^r is the real part of k_z, which is close to β_z.

The complex refractive index profile in the active region is given by

$$n_r(y) + in_i(y) = \left(\frac{\varepsilon_2(0) - a^2y^2}{\varepsilon_0}\right)^{1/2} \tag{10.2.19}$$

Writing the real and imaginary parts of $\varepsilon_2(0)$ as

$$\varepsilon_2(0) = \varepsilon_{2r}(0) + i\varepsilon_{2i}(0) \tag{10.2.20}$$

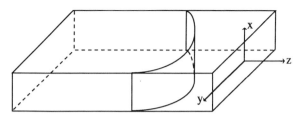

Figure 10.13. The constant phase front of a gain-guided mode is a cyclindrical parabolic surface [51].

we can expand (10.2.19) using the binomial expansion, noting that $|\varepsilon_{2i}(0)| \ll \varepsilon_{2r}(0)$, and the perturbation $|a^2 y^2| \ll \varepsilon_{2r}(0)$ in the region under or close to the stripe region. We obtain the real and imaginary parts of the index profiles

$$n_r(y) \simeq \left(\frac{\varepsilon_{2r}(0)}{\varepsilon_0}\right)^{1/2}\left[1 - \frac{a_r^2 - a_i^2}{2\varepsilon_{2r}(0)}y^2\right] \tag{10.2.21a}$$

$$n_i(y) \simeq \left(\frac{\varepsilon_{2r}(0)}{\varepsilon_0}\right)^{1/2}\left[\frac{\varepsilon_{2i}(0) - 2a_r a_i y^2}{2\varepsilon_{2r}(0)}\right] \tag{10.2.21b}$$

which depend on y with a parabolic profile.

Current Spreading and Carrier Density Profile. To account for the current spreading effect, an approximate formula for the injected current density $J(y)$ can be used:

$$J(y) = \begin{cases} J_0 & |y| \le S/2 \\ J_0 e^{-K(|y|-S/2)} & |y| \ge S/2 \end{cases} \tag{10.2.22}$$

The carrier distribution $n(y)$ along the y direction can be solved from the diffusion equation, using Eq. (2.4.18) with $\delta n(y) = n(y) - n_0 \simeq n(y)$. Since the active region is undoped, the thermal equilibrium value n_0 is very small,

$$\frac{d^2 n}{dy^2} - \frac{n}{L_n^2} = -\frac{J(y)}{qdD_n} \tag{10.2.23}$$

where $L_n = \sqrt{D_n \tau_n}$ is the diffusion length, and D_n is the electron diffusion coefficient. The solution for $n(y)$ can be put in analytical form. For example, below the stripe, $J(y) = J_0$, we write $n(y)$ as the sum of the homogeneous solutions and the particular solution:

$$n(y) = C_1 e^{-y/L_n} + C_2 e^{y/L_n} + N_0 \qquad |y| \leq \frac{S}{2} \qquad (10.2.24)$$

where $N_0 = J_0 \tau_n /(qd)$.

Since the structure is symmetric with respect to y, we have $C_1 = C_2$, that is, an even function of y. Outside the stripe region,

$$n(y) = C_3 e^{-(|y|-S/2)/L_n} + \frac{1}{1 - K^2 L_n^2} N_0 e^{-K(|y|-S/2)} \qquad (10.2.25)$$

The coefficients C_1 and C_3 are then determined by the continuities of $n(y)$ and the diffusion current density along the lateral direction $J_y(y) \propto dn/dy$ at $y = S/2$. We find

$$2C_1 = \frac{-KL_n}{1 + KL_n} e^{-S/2L_n} N_0 \qquad (10.2.26a)$$

and

$$C_3 = \frac{-KL_n}{1 + KL_n} N_0 \left(e^{-S/2L_n} \cosh \frac{S}{2L_n} + \frac{KL_n}{1 - KL_n} \right) \qquad (10.2.26b)$$

Therefore,

$$n(y) =$$

$$\begin{cases} N_0 \left[1 - \dfrac{KL_n}{1 + KL_n} e^{-S/2L_n} \cosh\left(\dfrac{y}{L_n} \right) \right] & |y| \leq \dfrac{S}{2} \\ \\ \dfrac{N_0}{1 + KL_n} \left[-KL_n \left(e^{-S/2L_n} \cosh \dfrac{S}{2L_n} + \dfrac{KL_n}{1 - KL_n} \right) e^{-(|y|-S/2)/L_n} \right. \\ \qquad\qquad \left. + \dfrac{1}{1 - KL_n} e^{-K(|y|-S/2)} \right] & |y| \geq \dfrac{S}{2} \end{cases}$$

$$(10.2.27)$$

If we ignore the current spreading outside of the stripe region, $K \to \infty$ in (10.2.22) so that $J(y) = 0$ for $|y| \geq S/2$. Equation (10.2.27) reduces to

$$n(y) = N_0 \begin{cases} \left[1 - e^{-S/2L_n} \cosh\left(\dfrac{y}{L_n} \right) \right] & |y| \leq \dfrac{S}{2} \\ \\ \sinh\left(\dfrac{S}{2L_n} \right) e^{-|y|/L_n} & |y| \geq \dfrac{S}{2} \end{cases} \qquad (10.2.28)$$

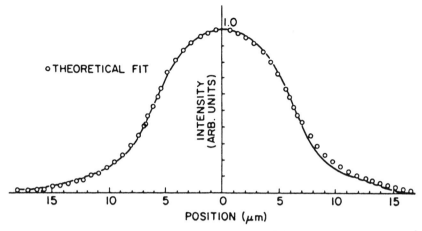

Figure 10.14. Theoretical plot of the carrier density profile $n(y)$ in a gain-guided semiconductor laser measured by the spontaneous emission intensity. The theoretical model assumes a stripe width $S = 11.7 \ \mu m$ and a diffusion length $L_n = 3.6 \ \mu m$. (After Ref. 54 © 1977 IEEE.)

where $N_0 = J_0 \tau_n / (qd) = G_0 \tau_n$. The above result (10.2.28) is the same as (2.4.21) in Section 2.4.

Previously we had an approximate relation between the gain and the carrier density for a bulk semiconductor laser:

$$g(y) = a[n(y) - n_{tr}] \qquad (10.2.29)$$

Therefore, the gain profile can be approximated using the expressions (10.2.27) or (10.2.28) for $n(y)$ in (10.2.28). Since the spontaneous emission spectrum depends on the local carrier density $n(y)$, a measurement of the emission profile as a function of y will determine the density profile. Therefore, the carrier density profile can be mapped out from the emission intensity profile, and the results agree very well [50, 54] with the theoretical curve calculated using (10.2.28), as shown in Fig. 10.14.

10.2.2 Index-Guided Semiconductor Lasers [33, 58, 59]

In the gain-guided semiconductor lasers, the carrier spreading along the lateral direction degrades the laser performance, such as a high threshold current density and a low differential quantum efficiency. To improve the laser performance, index-guided structures have been used such that a better optical confinement, and therefore, an improved stimulated emission process, can occur in the active region. One example is a ridge waveguide laser with a weakly index-guided structure, as shown in Fig. 10.15a. Another example is an etched-mesa buried heterostructure with improved optical and carrier confinements in the active region, as shown in Fig. 10.15b.

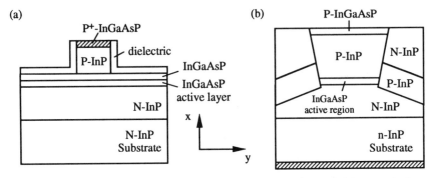

Figure 10.15. (a) A ridge waveguide laser for index guidance [33, 58]. (b) An etched-mesa buried-heterostructure index-guided laser [59].

For a ridge waveguide laser, an effective index profile $n_{eff}(y)$ along the lateral y direction can be obtained using the effective index method by solving the slab waveguide problem at a fixed cross section defined by a constant y. The guidance condition and the optical field pattern are then obtained from the waveguide theory for the index profile $n_{eff}(y)$, following the procedures described [60] in Section 7.5. The effective index method can also be applied to a rectangular dielectric waveguide structure as the active region of the etched-mesa buried-heterostructure laser. We find the optical electric field for the fundamental TE mode (or higher-order TE modes) from (7.5.3):

$$\mathbf{E} = \hat{y} E_y(x, y) \simeq \hat{y} F(x, y) G(y) \tag{10.2.30}$$

We obtain the optical confinement factor using the definition

$$\Gamma = \frac{\int\int_{\text{active region}} \mathbf{E} \times \mathbf{H}^* \cdot \hat{z} \, dx \, dy}{\int_{-\infty}^{\infty} \int_{-\infty}^{\infty} \mathbf{E} \times \mathbf{H}^* \cdot \hat{z} \, dx \, dy} \tag{10.2.31}$$

Analytical expressions for the slab waveguide geometry are derived in Chapter 7. Here we may use

$$\mathbf{E} \times \mathbf{H}^* \cdot \hat{z} \simeq -E_y H_x^* = \frac{k_z |E_y|^2}{\omega \mu} \tag{10.2.32}$$

Since $k_z / \omega \mu$ is a constant for a given mode, we can simply use

$$\Gamma = \frac{\int\int_{\text{active}} |E_y(x, y)|^2 \, dx \, dy}{\int_{-\infty}^{\infty} \int_{-\infty}^{\infty} |E_y(x, y)|^2 \, dx \, dy} \simeq \Gamma_x \Gamma_y \tag{10.2.33}$$

which is approximated by the product of the two optical confinement factors along the x direction (with a slab geometry in the effective index method) and along the y direction when the field is approximately given by

$$E_y \simeq F(x)G(y) \qquad (10.2.34)$$

It is a good approximation for a strongly index-guided structure and $F(x, y) \simeq F(x)$ if the geometry along the y direction is uniform such that the first part of the wave function $F(x, y)$ in the effective index method described in Section 7.5 can be assumed to be independent of y. Index-guided semiconductor lasers have been shown to exhibit excellent performance including the fundamental mode operation, low threshold, high quantum efficiency and low temperature sensitivity [33].

10.3 QUANTUM-WELL LASERS

Quantum-well (QW) structures [12–16, 61], as shown in Fig. 10.16, have been used as the active layer of semiconductor laser diodes with reduced threshold

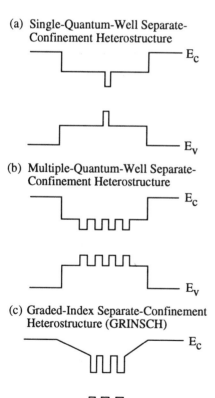

(a) Single-Quantum-Well Separate-Confinement Heterostructure

(b) Multiple-Quantum-Well Separate-Confinement Heterostructure

(c) Graded-Index Separate-Confinement Heterostructure (GRINSCH)

Figure 10.16. Band-gap profiles for (a) single-quantum-well, (b) multiple-quantum-well, and (c) graded-index separate-confinement heterostructure (GRINSCH) semiconductors lasers.

current densities as compared with those for conventional double-hetero-structure (DH) semiconductor diode lasers. Research on quantum-well physics and semiconductor lasers has been of great interest recently. For a brief history, see Ref. 3. Various designs such as single-quantum-well (SQW), multiple-quantum-well (MQW) and graded-index separate-confinement het-erostructures (GRINSCH) have been used for semiconductor lasers [21]. As we have seen in Chapter 9, quantum-well structures show quantized sub-bands and steplike densities of states. The density of states for a quasi-two-dimensional structure has been used to reduce threshold current density and improve temperature stability. Energy quantization provides another degree of freedom to tune the lasing wavelength by varying the well width and the barrier height. Scaling laws for quantum-well lasers and quantum-wire lasers show significant reduction of threshold current in reduced dimensions [24].

10.3.1 A Simplified Gain Model

The simplest model we consider is the gain spectrum based on (9.4.18) for a finite temperature, assuming a zero scattering linewidth first [62, 63]:

$$g(\hbar\omega) = \sum_{n,m} g_m [f_c^n(E_t = \hbar\omega - E_{hm}^{en}) - f_v^m(E_t = \hbar\omega - E_{hm}^{en})] H(\hbar\omega - E_{hm}^{en})$$

$$(10.3.1a)$$

where

$$g_m = C_0 |\hat{e} \cdot \mathbf{M}|^2 |I_{hm}^{en}|^2 \rho_r^{2D} \simeq C_0 |\hat{e} \cdot \mathbf{M}|^2 \rho_r^{2D} \delta_{nm} \qquad (10.3.1b)$$

$$C_0 = \frac{\pi e^2}{n_r c \varepsilon_0 m_0^2 \omega} \qquad (10.3.1c)$$

and

$$\rho_r^{2D} = \frac{m_r}{\pi \hbar^2 L_z} \qquad (10.3.1d)$$

is the reduced density of states. The overlap integral between the nth conduction subband and the mth hole subband is usually very close to unity for $n = m$, and it vanishes if $n \neq m$ because of the even–odd parity consideration. The polarization-dependent momentum matrix element is listed in Table 9.1 for conduction to heavy-hole and light-hole band transitions. The occupation factors for the electrons in the nth conduction subband and the electrons in the mth hole subband are

$$f_c^n(E_t = \hbar\omega - E_{hm}^{en}) = \frac{1}{1 + e^{[E_{en} + (m_r/m_e^*)(\hbar\omega - E_{hm}^{en}) - F_c]/k_B T}} \qquad (10.3.2a)$$

$$f_v^m(E_t = \hbar\omega - E_{hm}^{en}) = \frac{1}{1 + e^{[E_{hm} - (m_r/m_h^*)(\hbar\omega - E_{hm}^{en}) - F_v]/k_B T}} \qquad (10.3.2b)$$

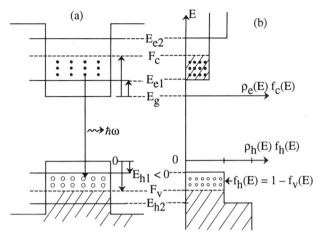

Figure 10.17. (a) Population inversion in a quantum-well structure, where $E_g + F_c - F_v > \hbar\omega > E_{e1} - E_{h1} + E_g$. (b) The product of the conduction band density of states $\rho_e(E)$ and the Fermi–Dirac occupation probability $f_c(E)$ for the calculation of the electron density n is plotted vs. the energy E in the vertical scale. Similarly, $\rho_h(E)f_n(E) = \rho_h(E)[1 - f_v(E)]$ is plotted vs. E for the energy ($E < 0$) in the valence band. Assume that the temperature T is 0 K.

Gain occurs when $f_c^n > f_v^m$, that is, the population inversion is achieved (Fig. 10.17). It also leads to $E_g + F_c - F_v > \hbar\omega$, where F_c and F_v are the quasi-Fermi levels for electrons and holes, measured from the conduction and valence band edges, respectively. Only those electrons and holes satisfying the k-selection rule contribute significantly to the gain process.

10.3.2   Determination of Electron and Hole Quasi-Fermi Levels

The quasi-Fermi levels F_c and F_v are determined by the carrier concentrations n and p which satisfy the charge neutrality condition

$$n + N_A^- = p + N_D^+ \tag{10.3.3}$$

For undoped quantum wells, $N_D^+ = 0$ and $N_A^- = 0$, we have

$$n = p$$

and

$$n = \int_0^\infty dE\rho_e(E)f_c(E) \qquad p = \int_{-\infty}^\infty dE\rho_h(E)[1 - f_v(E)]$$

$$\rho_e(E) = \frac{\rho_{es}}{L_z}\sum_{n=1}^\infty H(E - E_{cn}) \qquad \rho_{es} = \frac{m_e^*}{\pi\hbar^2}$$

$$\rho_h(E) = \frac{\rho_{hs}}{L_z}\sum_{m=1}^\infty H(E_{hm} - E) \qquad \rho_{hs} = \frac{m_h^*}{\pi\hbar^2} \tag{10.3.4}$$

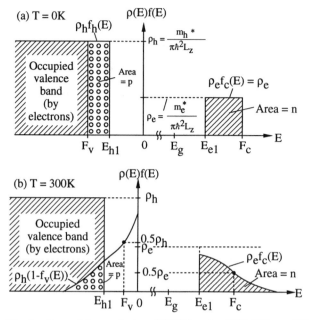

Figure 10.18. The functions of the products $\rho_e(E)f_c(E)$ and $\rho_h(E)f_h(E) = \rho_h(E)[1 - f_v(E)]$ vs. the electron energy E for the calculations of the carrier concentrations n and p, respectively, for (a) $T = 0$ K and (b) $T = 300$ K. The area under $\rho_e(E)f_c(E)$ is the electron concentration and the area under $\rho_h(E)f_h(E)$ is the hole concentration.

where $H(x)$ is the Heaviside step function; $H(x) = 1$ if $x > 0$ and $H(x) = 0$ if $x < 0$. In Fig. 10.18 we plot the products $\rho_e(E)f_c(E)$ and $\rho_h(E)[1 - f_v(E)]$ vs. the energy E for $T = 0$ and 300 K, respectively. The areas below these functions give the carrier concentrations n and p. Alternatively, it is convenient to use the surface electron and surface hole concentration, n_s and p_s, respectively:

$$n_s = \int_0^\infty dE \rho_{es}^{2D}(E) f_c(E) \qquad p_s = \int_{-\infty}^\infty dE \rho_{hs}^{2D}(E) [1 - f_v(E)]$$

$$\rho_{es}^{2D}(E) = \frac{m_e^*}{\pi \hbar^2} \sum_{n=1}^\infty H(E - E_{cn}) \qquad \rho_{hs}^{2D}(E) = \frac{m_h^*}{\pi \hbar^2} \sum_{m=1}^\infty H(E_{hm} - E)$$

$$(10.3.5)$$

where we have integrated over the well width such that the well width L_z will not appear in the surface concentrations, except for the quantized energy levels which appear as the band edges in the density of states.

10.3.3 Zero-Temperature Gain Spectrum

To understand the gain spectrum, let us consider the gain at $T = 0$ K. Remember that the model here is oversimplified, since it does not include the effects of scatterings or inhomogeneous broadenings such as those due to well-width fluctuations and the electron–hole Coulomb interactions. Our model here is to show the simplest picture for population inversion for the optical gain process. At $T = 0$ K, the Fermi–Dirac distributions are step functions. For the electrons in the conduction subband, we have

$$f_c(E) = \begin{cases} 1 & E < F_c \\ 0 & E > F_c \end{cases} \tag{10.3.6}$$

The electron concentration is simply the area under the function $\rho_e(E)f_c(E)$, which is plotted vs. E in Fig. 10.18a:

$$n = \frac{m_e^*}{\pi \hbar^2 L_z} \sum_{\substack{n = \text{occupied} \\ \text{subbands only}}} (F_c - E_{en}) \tag{10.3.7}$$

For a single (ground-state) subband case,

$$n = \frac{m_e^*}{\pi \hbar^2 L_z}(F_c - E_{e1}) \qquad n_s = \frac{m_e^*}{\pi \hbar^2}(F_c - E_{e1}) \tag{10.3.8}$$

the location of the quasi-Fermi level is linearly proportional to the carrier concentration at zero temperature. Similar results hold for holes:

$$p = \frac{m_h^*}{\pi \hbar^2 L_z}(E_{h1} - F_v) \tag{10.3.9}$$

For unstrained quantum wells, the topmost subband is a heavy-hole subband. The hole effective mass in the plane of the quantum well m_h^* is usually smaller than its value in the bulk semiconductors because of its coupling to the light-hole subbands. However, m_h^* is larger than m_e^* for most unstrained quantum wells. We find that for $n = p$,

$$m_h^*(E_{h1} - F_v) = m_e^*(F_c - E_{e1}) \tag{10.3.10}$$

which means that the areas under the function $\rho_e(E)f_c(E)$ and $\rho_h(E)f_v(E)$ should be equal. Since $m_h^* > m_e^*$, we find $E_{h1} - F_v < F_c - E_{e1}$. That is, the quasi-Fermi level F_v is closer to the hole subband edge than the separation of F_n and E_{c1}. The transparency optical energy occurs at $\hbar\omega = F_c - F_v$, where $f_c - f_v = 0$. As a matter of fact, $f_c = 1$ and $f_v = 0$ for $\hbar\omega < E_g + F_c - F_v$, and $f_c = 0$, $f_v = 1$ for $\hbar\omega > E_g + F_c - F_v$. The gain spectrum is a

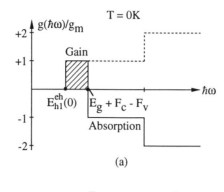

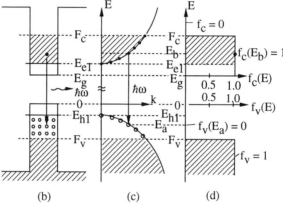

Figure 10.19. An illustration of optical gain of a quantum well at $T = 0$ K. (a) Normalized gain spectrum, (b) the energy profile in real space, (c) the energy dispersion in k space, and (d) the Fermi–Dirac function vs. the energy for the electrons in the conduction band and in the valence band.

simple rectangular window, as shown in Fig. 10.19a:

$$g(\hbar\omega) = \begin{cases} g_m & E_{h1}^{e1} < \hbar\omega < E_g + F_c - F_v \\ -g_m \sum_{n=m} H(\hbar\omega - E_{hm}^{en}) & \text{otherwise} \end{cases} \qquad (10.3.11)$$

For the optical energy between the $e1$–$h1$ transition band edge and the quasi-Fermi level separation $E_g + F_c - F_v$, the photons experience gain. Otherwise, they experience absorption with a spectrum similar to that of an unpumped quantum well with a stepwise shape. The only modification is near the band edge. The occupations of states by electrons and holes in the quantum well are shown in Fig. 10.19b in real space and in Fig. 10.19c in k

space showing the k-selection rule. The Fermi–Dirac distributions for electrons in the conduction band $f_c(E)$ and for electrons in the valence band $f_v(E)$ are shown in Fig. 10.19d, where $E_b(k) - E_a(k) = \hbar\omega$ has to be satisfied from the k-selection rule and the energy conservation condition, which lead to the population inversion factor $f_c(E_b) - f_v(E_a)$. We see that for $E_g + E_{e1} - E_{h1} < \hbar\omega < E_g + F_c - F_v$, $f_c(E_b) = 1$ and $f_v(E_a) = 0$; therefore, gain occurs. Otherwise, $f_c = 0$ and $f_v = 1$, and absorption occurs, as shown in Fig. 10.19a.

10.3.4 Finite-Temperature Gain Spectrum

At a finite temperature, the gain spectrum looks like that shown in Fig. 10.20a. The Fermi–Dirac distribution will deviate from a sharp step function and is equal to $\frac{1}{2}$ at the quasi-Fermi level. The electron concentration is the area below the function $\rho_e(E)f_c(E)$, as shown in Fig. 10.18b, which can be

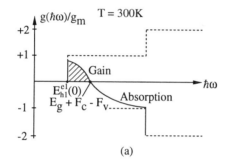

(a)

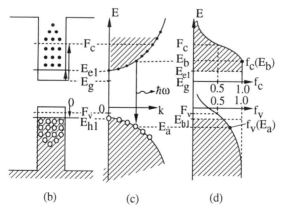

(b)   (c)   (d)

Figure 10.20. An idealized gain spectrum of a quantum well at $T = 300$ K assuming a zero linewidth: (a) normalized gain spectrum, (b) the potential energy profile in real space, (c) the parabolic band structure in k space, and (d) the Fermi–Dirac distributions (horizontal scales) vs. energy E (vertical scale).

obtained analytically:

$$n = \int_0^\infty dE \rho_e(E) f_c(E) = \sum_{n=1}^\infty n_c \ln\left(1 + e^{(F_c - E_{en})/k_B T}\right) \quad (10.3.12a)$$

where the prefactor

$$n_c = \frac{m_e^* k_B T}{\pi \hbar^2 L_z} \quad (10.3.12b)$$

is a characteristic concentration at finite temperature T. Similarly, the hole concentration is

$$p = \sum_{m=1}^\infty n_v \ln\left(1 + e^{(E_{hm} - F_v)/k_B T}\right) \quad (10.3.13a)$$

where

$$n_v = \frac{m_h^* k_B T}{\pi \hbar^2 L_z} \quad (10.3.13b)$$

which can account for heavy-hole and light-hole subbands, where m accounts for all hole subbands. The gain spectrum is given by Eq. (10.3.1a), and is plotted in Fig. 10.20a. It starts with a peak value at the transition edge $\hbar\omega = E_{h1}^{e1}(0) = E_g + E_{e1} - E_{h1}$, where $E_t = 0$, and decreases to zero at $E_g + F_c - F_v$, then becomes absorption at higher optical energies. The sharp rise in gain near the band edge compared with the soft increase in a bulk (3D) semiconductor is due to the steplike density of states in 2D vs. the slow increase of the square-root ($\sqrt{E}$) density of states in 3D. For a single subband pair, $n = e1$ and $m = h1$, we have the gain spectrum

$$g(\hbar\omega) = g_m\left\{ f_c\left[E_t = \hbar\omega - E_{h1}^{e1}(0)\right] - f_v\left[E_t = \hbar\omega - E_{h1}^{e1}(0)\right] \right\} \quad (10.3.14)$$

where g_m is given in (10.3.1b). The peak gain occurs at $E_t = 0$ or $\hbar\omega = E_{h1}^{e1}(0)$:

$$g_p = g_m\left[f_c\left(\hbar\omega = E_{h1}^{e1}\right) - f_v\left(\hbar\omega = E_{h1}^{e1}\right) \right] \quad (10.3.15a)$$

where

$$f_c\left(\hbar\omega = E_{h1}^{e1}\right) = \frac{1}{1 + e^{(E_{e1} - F_c)/k_B T}} \quad (10.3.15b)$$

$$f_v\left(\hbar\omega = E_{h1}^{e1}\right) = \frac{1}{1 + e^{(E_{h1} - F_v)/k_B T}} \quad (10.3.15c)$$

10.3.5 Peak Gain Coefficient Versus the Carrier Density

As shown in Fig. 10.21, when we increase the carrier density, the gain increases. Approximations used in the literature [62] are $f_c(E_t = 0) \simeq 1 - e^{-n/n_c}$ and $f_v(E_t = 0) \simeq e^{-p/n_v}$. These results are exact expressions if only one conduction and one valence subband are occupied. They are approximately true if we redefine [62]

$$n_c \simeq \frac{m_e^* k_B T}{\pi \hbar^2 L_z} \sum_{n=1}^{\infty} e^{(E_{e1} - E_{en})/k_B T} \qquad (10.3.16)$$

and a similar equation for holes:

$$n_v \simeq \frac{m_h^* k_B T}{\pi \hbar^2 L_z} \sum_{m=1}^{\infty} e^{(E_{hm} - E_{h1})/k_B T} \qquad (10.3.17)$$

Therefore, an approximate formula for the peak gain g_p as a function of the electron concentration n is given by

$$g_p = g_m\left(1 - e^{-n/n_c} - e^{-p/n_v}\right) \qquad (10.3.18)$$

Since $n_v/n_c = m_h^*/m_c^* \equiv R$ is the ratio of the hole and electron effective mass for a single subband occupation, we have

$$g_p = g_m\left(1 - e^{-n/n_c} - e^{-n/Rn_c}\right) \qquad (10.3.19)$$

which can be plotted as a function of n. The transparency carrier density

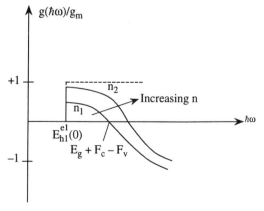

Figure 10.21. Normalized gain spectrum at two different injection levels ($n_2 > n_1$).

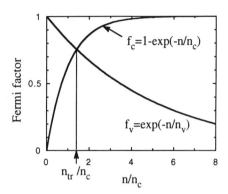

Figure 10.22. A plot of the occupation probability of electrons in the conduction band $f_c \simeq 1 - e^{-n/n_c}$ and that of electrons in the valence band $f_v \simeq e^{-n/n_v}$ (the probability of holes in the valence band $f_h = 1 - f_v$). The interception point where $f_c = f_v$ gives the transparency carrier density n_{tr}.

occurs at $g_p = 0$,

$$e^{-n_{tr}/n_c} + e^{-n_{tr}/(Rn_c)} = 1 \qquad (10.3.20)$$

If $R = 1$, we find $n_{tr} = n_c \ln 2$. We can also plot $f_c \simeq 1 - e^{-n/n_c}$ and $f_v \simeq e^{-n/n_v}$ vs. the carrier concentration n; the intersection point gives the transparency density n_{tr}, which corresponds to $g_p = 0$, as shown in Fig. 10.22. The peak gain g_p vs. the carrier density n is plotted in Fig. 10.23.

The analytical formula also leads to an expression for the differential gain [62, 64]:

$$\frac{\partial}{\partial n} g_p(n) = \frac{g_m}{n_c} \left(e^{-n/n_c} + \frac{1}{R} e^{-n/Rn_c} \right) \qquad (10.3.21)$$

Very often, the $g_p(n)$ vs. n curve is assumed to have a logarithmic shape as

$$g_p(n) = g_0 \left(1 + \ln \frac{n}{n_0} \right) \qquad (10.3.22)$$

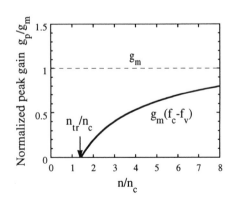

Figure 10.23. Peak gain coefficient g_p normalized by the maximum gain g_m is plotted as a function of the carrier density n.

where $g_p = g_0$ at $n = n_0$, and $g_p = 0$ at $n = n_0 e^{-1} = n_{tr}$. This is an approximate formula which can be fitted to the curve in Fig. 10.23 in principle. We can also calculate the peak gain directly vs. the current density J, taking into account the nonradiative Auger recombination, the leakage current, and the radiative recombination. We first assume a carrier concentration n, then we calculate the peak gain and

$$J_{rad} = qL_z R_{sp}(n) \tag{10.3.23a}$$

$$J_{Aug} = qL_z R_{Aug}(n) \tag{10.3.23b}$$

and the leakage current J_{leak} from the expressions in Section 10.1; then, $J = J_{rad} + J_{Aug} + J_{leak}$. Here the spontaneous emission rate per unit volume $(s^{-1} cm^{-3})$ $R_{sp}(n)$ can be calculated using the formulation with the valence-band mixings discussed in Chapter 9 or simply using the simplified model [65, 66]:

$$R_{sp}(n) \simeq Bn^2 \tag{10.3.24a}$$

Similarly,

$$R_{Aug}(n) \simeq Cn^3 \tag{10.3.24b}$$

which can also be calculated directly using the interaction Hamiltonians for Auger processes, i.e., carrier interactions via the Coulomb potential. The finite scattering linewidth can also be taken into account [67–71] by convolving the zero linewidth gain spectrum $g(\hbar\omega)$ in (10.3.14) with a Lorentzian function $L(\hbar\omega)$, i.e.,

$$g_{new}(\hbar\omega) = \int g(\hbar\omega') L(\hbar\omega - \hbar\omega') \, d\hbar\omega' \tag{10.3.25a}$$

where

$$L(x) = \frac{\gamma/\pi}{x^2 + \gamma^2} \tag{10.3.25b}$$

The procedure will lead to a peak gain vs. the current density relation:

$$g_p = g_p(J) \tag{10.3.26}$$

A common empirical formula is again a logarithmic relation [72–76]:

$$g_p(J) = g_0 \left(1 + \ln \frac{J}{J_0} \right) \tag{10.3.27}$$

We see that if $J \propto n^\beta$, where β can be between 2 and 3 for a certain range of current density, the above relation leads to

$$g_p = g_0\left(1 + \beta \ln \frac{n}{n_0}\right) = g_0\beta\left(\frac{1}{\beta} + \ln \frac{n}{n_0}\right) \qquad (10.3.28)$$

which is also a logarithmic relation, yet the coefficients have to be different from those in (10.3.22).

10.3.6 Scaling Laws for Multiple-Quantum-Well (MQW) Lasers

For a quantum-well laser lasing from only the first quantized electron and hole subbands, we use the empirical logarithmic formula for the peak gain-current density relation (10.3.27) for the rest of the discussions in this chapter:

$$g_w = g_0\left[\ln\left(\frac{J_w}{J_0}\right) + 1\right] \qquad (10.3.29)$$

where J_w and g_w are the injected current density and the peak gain coefficient of a single-quantum-well (SQW) structure. Such a relation is plotted in Fig. 10.24a. The transparency current density occurs at $J_{tr} = J_0 e^{-1}$.

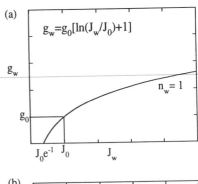

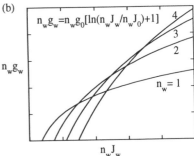

Figure 10.24. (a) Gain (g_w) vs. injection current density (J_w) relation for a single-quantum-well structure; (b) gain ($n_w g_w$) vs. injection current density ($n_w J_w$) relation for a multiple-quantum-well structure with n_w wells.

For an MQW structure with n_w quantum wells and a cavity length L, the required modal gain G_{th} at threshold condition is

$$G_{th} = n_w \Gamma_w g_w = \alpha + \frac{1}{2L} \ln \frac{1}{R_1 R_2} \qquad (10.3.30)$$

where α is the internal optical loss, Γ_w is the optical confinement factor per well, and R_1 and R_2 are the optical power reflection coefficients at both facets. We should note that g_w is a "material gain" for the quantum-well region assuming that all of the electron–photon interactions occur in the well region and $g_w \propto 1/L_z$ in the gain model. The optical confinement factor Γ_w for a single well takes into account the optical modal distribution and only that fraction of the photons inside the well provides gain. A simple approximation is, therefore [77],

$$\Gamma_w \simeq \frac{L_z}{W_{mode}} \qquad (10.3.31)$$

where W_{mode} is the "full width" at nearly half maximum of the intensity profile of the optical mode. For a separate-confinement structure as in Figs. 10.16a and b, we can calculate the optical confinement factor Γ_o for the optical waveguide part, ignoring the well region first, then use

$$\Gamma_w = \Gamma_o \frac{L_z}{W_{mode}} \qquad (10.3.32)$$

where Γ_0 takes into account the refractive index difference of the large optical waveguide for optical confinement, and L_z / W_{mode} accounts for the carrier confinement and stimulated emission. The modal gain for a single quantum well should be

$$G = \Gamma_w g_w$$
$$= \Gamma_w g_m \left[f_c \big(\hbar\omega = E_{h1}^{el}(0) \big) - f_v \big(\hbar\omega = E_{h1}^{el}(0) \big) \right] \qquad (10.3.33)$$

where

$$\Gamma_w g_m = \frac{L_z}{W_{mode}} g_m \Gamma_o$$
$$= \frac{\omega\pi}{n_r c \varepsilon_0} |\hat{e} \cdot \mathbf{M}_{ch}|^2 \frac{m_r}{\pi \hbar^2} \frac{\Gamma_o}{W_{mode}}$$
$$= \frac{2\pi m_r}{\lambda_0 n_r \varepsilon_0 \hbar^2} |\hat{e} \cdot \mathbf{M}_{ch}|^2 \frac{\Gamma_o}{W_{mode}} \qquad (10.3.34)$$

which is also the maximum achievable gain for a single quantum well. For n_w quantum wells, the modal gain is approximately $n_w \Gamma_w g_w$. The threshold current density J_{th} for the MQW structure is then

$$J_{th} = \frac{n_w J_w}{\eta} \qquad (10.3.35)$$

where η is the internal quantum efficiency of the injection current. Therefore, we can also plot the material gain $(n_w g_w)$ vs. the injection current density $(n_w J_w)$ using [72–78] (10.3.9):

$$n_w g_w = n_w g_0 \left[\ln \left(\frac{n_w J_w}{n_w J_0} \right) + 1 \right] \qquad (10.3.36)$$

assuming that the well-to-well coupling can be ignored for simplicity. This relation is plotted in Fig. 10.24b for $n_w = 1, 2, 3,$ and 4. For $n_w = 1$, as shown in Fig. 10.24a, the intersection with the horizontal axis is $J_w = J_0 e^{-1}$ and $g_w = 0$. When $J_w = J_0$, we have $g_w = g_0$. For $n_w = 2$, we find that $2g_w = 0$ occurs at $(2J_w) = 2J_0 e^{-1}$; therefore, the horizontal intersection is shifted to the right by a factor of 2. At $2J_w = 2J_0$, we obtain the vertical axis $(2g_w) = (2g_0)$, which is twice that for $n_w = 1$. Therefore, the gain–current density curve for $n_w = 2$ starts at twice the transparency current density of that for $n_w = 1$, and increases by a factor of two faster than that for $n_w = 1$. We obtain the threshold current density by [72–76] substituting (10.3.29) into (10.3.35):

$$J_{th} = \left(\frac{n_w J_0}{\eta} \right) \exp \left[\left(\frac{g_w}{g_0} \right) - 1 \right]$$

$$= \left(\frac{n_w J_0}{\eta} \right) \exp \left[\frac{1}{(n_w \Gamma_w g_0)} \left(\alpha + \frac{1}{2L} \ln \frac{1}{R_1 R_2} \right) - 1 \right] \qquad (10.3.37)$$

If we take the logarithmic function of J_{th}, we see

$$\ln J_{th} = \ln \left(\frac{n_w J_0}{\eta} \right) + \frac{\alpha}{n_w \Gamma_w g_0} + \frac{L_{opt}}{L} - 1 \qquad (10.3.38)$$

where we define an optimum cavity length parameter

$$L_{opt} \equiv \frac{1}{2} \frac{1}{(n_w \Gamma_w g_0)} \ln \left(\frac{1}{R_1 R_2} \right) \qquad (10.3.39)$$

with a dimension of length. Therefore, $\ln J_{th}$ varies linearly with $1/L$. Such a dependence has been plotted for various numbers of quantum wells as shown in Fig. 10.25a.

(a)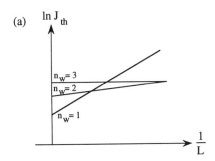

(b)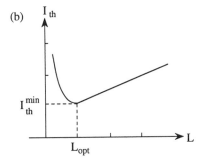

Figure 10.25. (a) Theoretical curves for the logarithm of the threshold current density, $\ln(J_{th})$ using (10.3.38), vs. the inverse cavity length for lasers with 1, 2, and 3 quantum wells. (b) A plot of the threshold current I_{th} vs. the cavity length L for $n_w = 1$ using (10.3.40).

For real device measurements, the threshold current I_{th} $(= J_{th}wL)$ is measured, where w is the width of the stripe. We have

$$I_{th} = \frac{wLn_wJ_0}{\eta} \exp\left[\frac{1}{n_w\Gamma_wg_0}\left(\alpha + \frac{1}{2L}\ln\frac{1}{R_1R_2}\right) - 1\right] \quad (10.3.40)$$

The above expression for I_{th} contains a prefactor increasing linearly with L, and an argument in the exponential function that decreases with L, resulting in a minimum of the threshold current occurring at $L = L_{opt}$ as shown in Fig. 10.25b. The minimum threshold current, I_{th}^{min}, is then given by

$$I_{th}^{min} = \frac{1}{2}\left(\frac{wJ_0}{\eta\Gamma_wg_0}\right)\ln\left(\frac{1}{R_1R_2}\right)\exp\left(\frac{\alpha}{n_w\Gamma_wg_0}\right) \quad (10.3.41)$$

If we assume that the loss coefficient α and other parameters such as Γ_w, J_0, η, and g_0 are independent of the number of wells, we find an optimum number of wells, n_{opt}, such that I_{th} in (10.3.40) is minimized ($\partial I_{th}/\partial n_w = 0$):

$$n_{opt} = \frac{1}{\Gamma_wg_0}\left[\alpha + \frac{1}{2L}\ln\left(\frac{1}{R_1R_2}\right)\right] \quad (10.3.42)$$

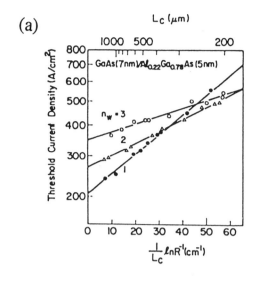

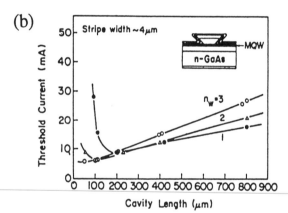

Figure 10.26. Experimental results showing (a) the linear behavior of the $\ln(J_{th})$ vs. the inverse cavity length $1/L$ following (10.3.38) for lasers with 1, 2, and 3 quantum wells, and (b) the threshold current I_{th} as a function of the cavity length L. A minimum value of the threshold current exists at an optimum cavity length L_{opt} based on (10.3.40) and (10.3.41). (After Ref. 73 © 1988 IEEE.)

10.3.7 Comparison With Experimental Data [73]

The above simple relations between the threshold current density J_{th} and the cavity length L or the number of quantum wells n_w have been demonstrated experimentally. In Fig. 10.26, we show the experimental data [73] for (a) broad area contact and (b) ridge waveguide $GaAs/Al_{0.22}Ga_{0.78}As$ separate-confinement-heterostructure (SCH) quantum-well lasers. The undoped GaAs quantum wells are 70 Å in width and the $Al_{0.22}Ga_{0.78}As$ barriers are 50 Å wide. The linear relation between $\ln(J_{th})$ and the inverse cavity length $1/L$ is clearly shown in Fig. 10.26a for three numbers of quantum wells, $n_w = 1, 2,$ and 3. The threshold current I_{th} is also shown to have a minimum at an optimum cavity length in Fig. 10.26b, in agreement with the analytical relations (10.3.40) and (10.3.41). The stripe widths are 4.0, 4.5, and 3.5 μm for $n_w = 1, 2,$ and 3, respectively in Fig. 10.26b.

10.4 STRAINED QUANTUM-WELL LASERS

In recent years, strain effects in semiconductors have been proposed and explored intensively for optoelectronic device applications [79, 80]. The tunability of strained materials makes them attractive for designing semiconductor lasers, electrooptic modulators, and photodetectors operating at a desired wavelength. The concept of band structure engineering [17, 18, 81] is further explored in strained semiconductors beyond the previous applications using quantum wells by controlling the well width and the barrier height. Using strained material systems such as $In_{1-x}Ga_xAs/InP$, $In_{1-x}Ga_xAs/In_{1-x}Ga_xAs_{1-y}P_y$, and $In_{1-x}Ga_xAs/GaAs$ quantum wells, a wide tunable range of wavelength is possible. Furthermore, the band structure including the band gap, the effective masses, and therefore the density of states can be engineered. It has been proposed that by using a compressively strained quantum-well structure, semiconductor lasers with a lower threshold current density can be achieved because of the reduced in-plane heavy-hole effective mass [17, 18, 81]. A reduced threshold carrier density also reduces the nonradiative Auger recombination, one of the limiting factors for long-wave length (1.55-μm) semiconductor lasers.

Experimentally, high-performance strained quantum-well lasers [82, 83], such as low-threshold current density [19, 20, 84–89], high-power output [90–93], and high-temperature operation [94–97], have been demonstrated with remarkable results. Theoretical models using both simplified parabolic band structures and realistic band structures of strained quantum wells have also been used to investigate the optical gains in these semiconductor lasers [98–105].

To understand the strained quantum-well systems, we will use the $In_{1-x}Ga_xAs/InP$ materials as an illustration. The ternary $In_{1-x}Ga_xAs$ alloy has a lattice constant $a(x)$, which is a linear interpolation of the lattice constants of GaAs and InAs, $a(GaAs)$, and $a(InAs)$:

$$a(x) = x\,a(GaAs) + (1 - x)a(InAs)$$

$$= 5.6533x + 6.0584(1 - x) \tag{10.4.1}$$

The lattice constant of the InP substrate a_0 is 5.8688 Å. When a thin $In_{1-x}Ga_xAs$ layer is grown on the InP substrate, at $x \simeq 0.468 \simeq 0.47$, the lattice constant of $In_{0.53}Ga_{0.47}As$ is the same as that of InP. It is the lattice-matched condition.

If $a(x) > a_0$ (i.e., $x < 0.47$), an elastic strain occurs which will be compressive in the plane of the layer resulting in a tensile strain in the direction perpendicular to the interface. We call this case biaxial compression, or just compressive strain. Similarly, if $a(x) < a_0(InP)$ (i.e., $x > 0.47$), the strain in the plane of the layer is tensile and there is a compressive strain in the perpendicular direction. We usually call this case biaxial tension, or tensile strain. Theoretical work [106, 107] shows that beyond a critical layer thickness, large densities of misfit dislocations are formed to accommodate the strain. Therefore, the strained layer dimension is always kept below the critical thickness for optoelectronic material applications.

10.4.1 The Influence of the Effective Mass on Gain and Transparency Carrier Density

A band structure calculation shows that the effective mass of the heavy-hole band along the parallel plane of a quantum well has a reduction in its effective mass. Therefore, the density of states of the heavy-hole subband is reduced. This helps to reduce the transparency carrier density required for population inversion, as shown in Fig. 10.27. The Bernard–Duraffourg population inversion condition [108] requires the separation of the quasi-Fermi levels $F_c - F_v + E_g > \hbar\omega > E_g + E_{e1} - E_{h1} (\equiv E_g^{\text{eff}})$.

For conventional semiconductors, the heavy-hole effective mass is always several times that of the electron in the conduction band. The quasi-Fermi level of the hole is usually above the valence band edge instead of being below the band edge, while the conduction band degenerates easily with electron population. Since a large separation of the quasi-Fermi levels $F_c - F_v + E_g > E_g^{\text{eff}}$ (the effective band gap of the quantum well) is required, an ideal situation will be a symmetric band structure configuration such that $m_h^* = m_e^*$, as shown in Fig. 10.27b. The population inversion condition $F_c - F_v + E_g = E_g^{\text{eff}}$ can then be satisfied easily at a smaller carrier density, as shown. If we take the simplified model for the peak gain [62, 64] from the previous section, then

$$g_p(n) = g_m(f_c - f_v)$$

$$\simeq g_m\left(1 - e^{-n/n_c} - e^{-n/Rn_c}\right) \tag{10.4.2}$$

where $R = m_h^*/m_e^*$ is the ratio of the heavy hole and the conduction band

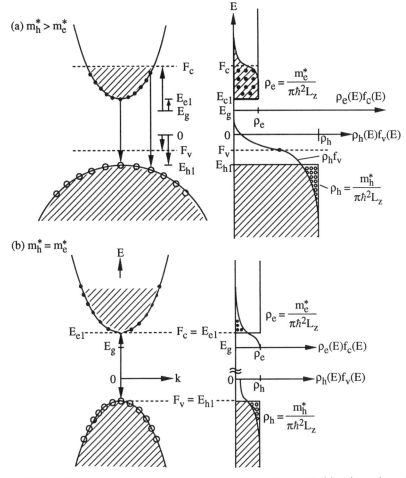

Figure 10.27. Populaion inversion in a quantum-well structure with (a) $m_h^* > m_e^*$ and (b) $m_h^* = m_e^*$.

electron effective mass. In Fig. 10.28a we plot $f_c = 1 - e^{-n/n_c}$ and $f_v = e^{-n/Rn_c}$ vs. the carrier concentration n for two values of R. We see that the intersection point $f_c = f_v$ determines the transparency density n_{tr}, at which the peak gain is zero. This transparency density n_{tr} is reduced when the mass rate is reduced from $R = 5$ to $R = 1$. The peak gain is also plotted in Fig. 10.28b for $R = 1$ and $R = 5$. We note that while the transparency density n_{tr} is smaller at $R = 1$, the maximum achievable gain also saturates at a smaller value when n increases, since g_m is proportional to the joint density of states, which is proportional to the reduced effective mass $m_r = m_e^* m_h^* / (m_e^* + m_h^*) = m_e^* R / (R + 1)$. If $m_h^* = m_e^*$, we have $m_r = m_e^*/2$. If $m_h^* = 5m_e^*$, we have $m_r = 5m_e^*/6$.

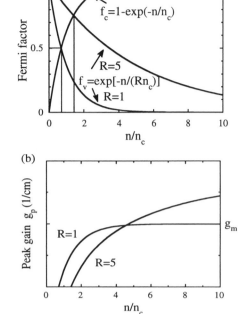

Figure 10.28. (a) Occupation probabilities $f_c = 1 - e^{-n/n_c}$ and $f_v = e^{-n/(Rn_c)}$ are plotted vs. the normalized carrier concentration n/n_c for two heavy-hole-to-electron effective mass ratios $R \ (= m_h^*/m_e^*) = 5$ and $R = 1$. (b) Peak gain coefficient vs. the normalized carrier density n/n_c for $R = 5$ and $R = 1$.

10.4.2   Strain Effects on the Band-Edge Energies

For $\mathrm{In}_{1-x}\mathrm{Ga}_x\mathrm{As}$ quantum-well layers grown on InP substrate, we have the in-plane strain [109, 110]

$$\varepsilon = \varepsilon_{xx} = \varepsilon_{yy} = \frac{a_0 - a(x)}{a_0} \tag{10.4.3}$$

where a_0 is the lattice constant of the InP. For a compressive strain, $a(x) > a_0$, therefore, the in-plane strain ε is negative. On the other hand, the in-plane strain ε is positive for a tensile strain since $a(x) < a_0$. The strain in the perpendicular direction is

$$\varepsilon_\perp = \varepsilon_{zz} = -2\frac{\sigma}{1-\sigma}\varepsilon = -2\frac{C_{12}}{C_{11}}\varepsilon \tag{10.4.4}$$

where σ is Poisson's ratio and C_{11}, C_{12} are the elastic stiffness constants. For most of the III–V compound semiconductors, $\sigma \simeq 1/3$ or $C_{12} \simeq 0.5C_{11}$.

The band gap of an unstrained $In_{1-x}Ga_xAs$ at 300 K is given by

$$E_g(x) = 0.324 + 0.7x + 0.4x^2 \text{ (eV)} \tag{10.4.5}$$

For a strained $In_{1-x}Ga_xAs$ layer, the conduction band edge is shifted by

$$\delta E_c = a_c(\varepsilon_{xx} + \varepsilon_{yy} + \varepsilon_{zz}) = 2a_c\left(1 - \frac{C_{12}}{C_{11}}\right)\varepsilon \tag{10.4.6}$$

and the valence subbands are shifted by

$$\delta E_{HH} = -P_\varepsilon - Q_\varepsilon \qquad \delta E_{LH} = -P_\varepsilon + Q_\varepsilon$$

$$P_\varepsilon = -a_v(\varepsilon_{xx} + \varepsilon_{yy} + \varepsilon_{zz}) = -2a_v\left(1 - \frac{C_{12}}{C_{11}}\right)\varepsilon$$

$$Q_\varepsilon = -\frac{b}{2}(\varepsilon_{xx} + \varepsilon_{yy} - 2\varepsilon_{zz}) = -b\left(1 + 2\frac{C_{12}}{C_{11}}\right)\varepsilon \tag{10.4.7}$$

Therefore, we have the new band edges:

$$E_c = E_g(x) + \delta E_c$$
$$E_{HH} = \delta E_{HH} = -P_\varepsilon - Q_\varepsilon$$
$$E_{LH} = \delta E_{LH} = -P_\varepsilon + Q_\varepsilon \tag{10.4.8}$$

The effective band gaps are

$$E_{C-HH} = E_g(x) + \delta E_c - \delta E_{HH} = E_g(x) + \delta E_c + P_\varepsilon + Q_\varepsilon$$

$$= E_g(x) + 2a\left(1 - \frac{C_{12}}{C_{11}}\right)\varepsilon - b\left(1 + 2\frac{C_{12}}{C_{11}}\right)\varepsilon \tag{10.4.9a}$$

$$E_{C-LH} = E_g(x) + \delta E_c - \delta E_{LH} = E_g(x) + \delta E_c + P_\varepsilon - Q_\varepsilon$$

$$= E_g(x) + 2a\left(1 - \frac{C_{12}}{C_{11}}\right)\varepsilon + b\left(1 + 2\frac{C_{12}}{C_{11}}\right)\varepsilon \tag{10.4.9b}$$

where the hydrostatic deformation potential a (eV) has been defined:

$$a = a_c - a_v$$

As can be seen from Table 10.1, we have $a_c < 0$ and $a_v > 0$, using our definitions in this section. Therefore, we have the energy shift for compressive strain (with the notation for the trace of the strain matrix $Tr(\bar{\bar{\varepsilon}}) = \varepsilon_{xx} + \varepsilon_{yy} + \varepsilon_{zz}$),

$$\delta E_c = a_c \, Tr(\bar{\bar{\varepsilon}}) > 0 \qquad a_v \, Tr(\bar{\bar{\varepsilon}}) < 0 \qquad b\left(1 + 2\frac{C_{12}}{C_{11}}\right)\varepsilon > 0$$

i.e., the conduction band edge is shifted upward by an amount δE_c, and the

Table 10.1 Physical Quantities of a Strained Semiconductor [109, 110]

Physical Quantity	Compressive	Tension
In-plane strain $\varepsilon = \varepsilon_{xx} = \varepsilon_{yy} = \dfrac{a_0 - a(x)}{a_0}$	Negative	Positive
$\varepsilon_\perp = -2\dfrac{C_{12}}{C_{11}}\varepsilon$	Positive	Negative
Conduction-band deformation potential a_c (eV)	Negative	
Valence-band deformation potential a_v (eV)	Positive	
$a = a_c - a_v$ (eV)	Negative	
Shear deformation potential b (eV)	Negative	
$\delta E_c = a_c(\varepsilon_{xx} + \varepsilon_{yy} + \varepsilon_{zz})$	Positive	Negative
$P_\varepsilon = -a_v(\varepsilon_{xx} + \varepsilon_{yy} + \varepsilon_{zz})$	Positive	Negative
$Q_\varepsilon = -\dfrac{b}{2}(\varepsilon_{xx} + \varepsilon_{yy} - 2\varepsilon_{zz})$	Negative	Positive
Heavy-hole band-edge shift (eV)	$\delta E_{HH} = -P_\varepsilon - Q_\varepsilon$	
Light-hole band-edge shift (eV)	$\delta E_{LH} = -P_\varepsilon + Q_\varepsilon$	
Effective band gaps (eV)	$E_{C-HH} = E_g(x) + \delta E_c + P_\varepsilon + Q_\varepsilon$	
	$E_{C-LH} = E_g(x) + \delta E_c + P_\varepsilon - Q_\varepsilon$	

valence band edge is shifted downward by the magnitude of $a_v \, \mathrm{Tr}(\bar{\bar{\varepsilon}})$, then split upward by $b[1 + 2(C_{12}/C_{11})]\varepsilon$ for the heavy hole and downward by $b[1 + 2(C_{12}/C_{11})]\varepsilon$ for the light hole, as shown in Fig. 10.29. For tensile strain, the in-plane strain ε is positive since $a(x)$ is smaller than the lattice constant of the substrate a_0. Therefore, the directions of the band edge shifts are opposite to those of the compressive strain.

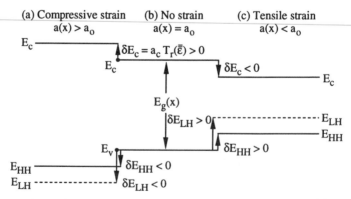

Figure 10.29. Conduction and valence band edges for semiconductors under (a) a compressive strain, (b) no strain, and (c) a tensile strain. The function $E_g(x)$ is the band gap of the unstrained $\mathrm{In}_{1-x}\mathrm{Ga}_x\mathrm{As}$ alloy. For example, $E_g(x) = 0.324 + 0.7x + 0.4x^2 e$ at 300 K. The band-edge shifts are $\delta E_c = 2a_c(1 - C_{12}/C_{11})\varepsilon$, $\delta E_{HH} = -P_\varepsilon - Q_\varepsilon$, $\delta E_{LH} = -P_\varepsilon + Q_\varepsilon$, where $P_\varepsilon = -2a_v(1 - C_{12}/C_{11})\varepsilon$, $Q_\varepsilon = -b(1 + 2C_{12}/C_{11})\varepsilon$, and the in-plane strain $\varepsilon = [a(x) - a_0]/a_0$. $a(x)$ is the lattice constant of the unstrained $\mathrm{In}_{1-x}\mathrm{Ga}_x\mathrm{As}$ alloy and a_0 is that of the substrate, which is also the in-plane lattice constant of the strained $\mathrm{In}_{1-x}\mathrm{Ga}_x\mathrm{As}$ lattice.

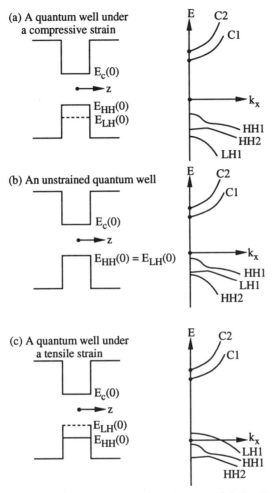

Figure 10.30. Band-edge profiles in real space along the growth (z) direction and the quantized subband dispersions in k space along the k_x direction (perpendicular to the growth direction) for a quantum well with (a) a compressive strain, (b) no strain, and (c) a tensile strain.

10.4.3 Band Structure of a Strained Quantum Well

For a quantum-well structure such as an $In_{1-x}Ga_xAs$ layer sandwiched between InP barriers, the band structures are shown in Fig. 10.30 for (a) a compressive strain ($x < 0.468$), (b) no strain ($x = 0.468$), and (c) a tensile strain ($x > 0.468$). The left-hand side shows the quantum-well band structures in real space vs. position along the growth (z) direction. The right-hand side shows the quantized subband dispersions in momentum space along the parallel (k_x) direction in the plane of the layer. These dispersion curves show the modification of the effective masses or the densities of states due to both the quantization and strain effects. More precise valence band structures can be found in Section 4.9.

Example: Gain Spectrum of a Strained Quantum Well   Using the formulation for the gain spectrum in a strained quantum well, we plot in Fig. 10.31 the gain spectrum for light with a TE ($\hat{x}$ or $\hat{y}$) or TM ($\hat{z}$) polarization for the three cases in Fig. 10.30. We consider an $In_{1-x}Ga_xAs$ quantum-well system with $In_{1-x}Ga_xAs_{1-y}P_y$ barriers (band-gap wavelength $\lambda_b = 1.3\ \mu m$) lattice-matched to InP substrate. Electronic and optical properties of similar quantum-well structures were shown in Figs. 4.22a–d, Fig. 9.16, and Fig. 9.17. As shown in Fig. 10.31a for compressive strain (quantum-well gallium mole fraction $x = 0.41$, well width = 45 Å) and Fig. 10.31b for the lattice-matched case ($x = 0.47$, $L_z = 60$ Å), the gain of the TE polarization is always larger than that of the TM polarization, because the top valence subband for these two cases is always heavy hole in nature and the optical matrix element for C1–HH1 transition prefers TE polarization. The injected holes populate mostly the ground (HH1) subband. On the other hand, Fig. 10.31c shows that the TM polarization may be dominant for the tensile strain case ($x = 0.53$, and $L_z = 15$ Å). The well widths are chosen such that all the lowest band-edge transition wavelengths are close to 1.55 μm. The gains were calculated for the same total width $L_T = L_z + L_b = 200$ Å in a unit period, where L_b is the barrier width. The same surface carrier concentration $n_s = nL_z = 3 \times 10^{12}/cm^2$ is used in all three structures.   ∎

If we recall the parabolic band model in Chapter 9, the band-edge ($k_t = 0$) momentum-matrix elements are

TE polarization

$$|\hat{x} \cdot \mathbf{M}_{c-hh}|^2 = |\hat{y} \cdot \mathbf{M}_{c-hh}|^2 = \tfrac{3}{2}M_b^2 \qquad (10.4.10a)$$

$$|\hat{x} \cdot \mathbf{M}_{c-lh}|^2 = |\hat{y} \cdot \mathbf{M}_{c-lh}|^2 = \tfrac{1}{2}M_b^2 \qquad (10.4.10b)$$

TM polarization

$$|\hat{z} \cdot \mathbf{M}_{c-hh}|^2 = 0 \qquad (10.4.10c)$$

$$|\hat{z} \cdot \mathbf{M}_{c-lh}|^2 = 2M_b^2 \qquad (10.4.10d)$$

where M_b is the optical momentum matrix element of the bulk semiconductor, which is related to the energy parameter E_p for the matrix element by $M_b^2 = m_0 E_p/6$, and E_p is usually tabulated in databooks (see Appendix K). In practice, the matrix elements depend on the transverse wave vector k_t and are very close to the above values at the band edge where $k_t = 0$, then vary

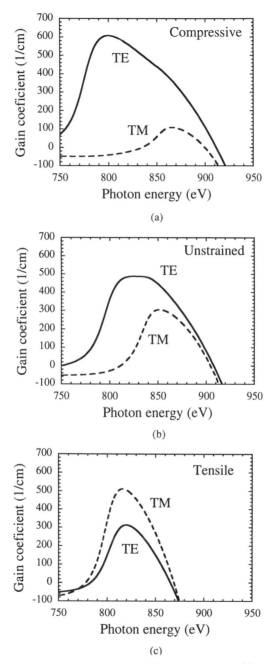

Figure 10.31. Gain spectra for both TE and TM polarizations of (a) a compressive strain ($x = 0.41$, well width = 45 Å), (b) an unstrained ($x = 0.47$, $L_z = 60$ Å), and (c) a tensile strain ($x = 0.53$, and $L_z = 115$ Å) $In_{1-x}Ga_xAs/In_{1-x}Ga_xAs_{1-y}P_y$ (barrier bandgap wavelength $\lambda_b = 1.3$ μm) quantum-well laser. These material gain coefficients were calculated for the same period $L_T = 200$ Å and surface carrier concentration $n_s = nL_z = 3 \times 10^{12}/cm^2$ in all three structures.

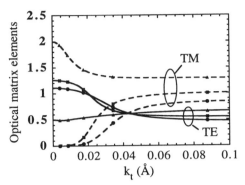

Figure 10.32. Normalized momentum matrix element $2|M_{nm}(k_t)|^2/M_b^2$ vs. the transverse k_t vector using (9.7.15) and (9.7.16) for compressive (dots), unstrained (solid squares), and tensile (solid triangles) strained quantum-well lasers. Here M_{nm} refers to $n = C1$ and m is the top valence subband (HH1 for compressive and unstrained and LH1 for tensile case).

as k_t increases, as shown in Fig. 10.32, using the valence-band mixing model in Section 9.7 and ignoring the spin–orbit split-off band. Laser actions with the above polarization characteristics, TE for compressive strain and lattice-matched quantum-well lasers and TM for tensile strain quantum-well lasers, have been reported [111]. These phenomena also agree with those in conventional externally stressed or thermally stressed diode lasers [112, 113].

In Fig. 10.33, we plot the optical modal gain vs. the surface carrier concentration $n_s = nL_z$, where L_z is the well width for three cases: (a) compressive, (b) lattice-matched, and (c) tensile strain. For long-wavelength semiconductors, the well width L_z has to be adjusted such that lasing action at 1.55 μm can be achieved. The design usually requires L_z (compressive) $< L_z$ (lattice-matched) $< L_z$ (tensile). As discussed in Section 10.3, it is the modal gain Γg that is important in determining the threshold condition, and

$$\Gamma g \propto \frac{L_z}{W_{\text{mode}}} g = \frac{1}{W_{\text{mode}}} (L_z g) \qquad (10.4.11)$$

Figure 10.33. Modal gain coefficient vs. the surface carrier concentration for compressive strain (dots), unstrained (squares), and tensile strain (triangles) $In_{1-x}Ga_xAs / In_{1-x}Ga_xAs_{1-y}P_y$ quantum-well lasers with the band-edge transition wavelengths near 1.55 μm. The solid curves are TE polarization and the dashed curve is TM polarization.

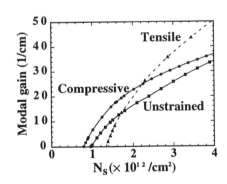

where L_z will cancel with the factor $(1/L_z)$ in g. Therefore, it is more useful to compare the modal gain Γg for three different strains if the well width L_z varies. We see that the compressive strain quantum-well laser has the smallest transparency carrier density at which $g = 0$. However, it also saturates faster than the other two cases. On the other hand, the tensile strain case (TM polarization) has a larger transparency current density, yet it increases faster, implying a larger differential gain.

We can also plot the radiative current density

$$J_{rad} = qL_z R_{sp}(n)$$

$$= qL_z \int_0^{\infty} r^{spon}(\hbar\omega) \, d(\hbar\omega) \tag{10.4.12}$$

where the well width factor L_z is canceled, since $r^{spon}(\hbar\omega) \propto 1/L_z$ from (9.7.18), and $J_{rad}(n_s)$ is a two-dimension current density with a dimension (A/cm^2). A common approximation is

$$R_{sp}(n) = Bn^2 = B'n_s^2 \tag{10.4.13}$$

where $n_s = nL_z$ and $B' = B/L_z^2$. Therefore

$$J_{rad} = qB_s n_s^2 \tag{10.4.14}$$

with $B_s = B/L_z$. A plot of J_{rad} vs. n_s for three different strain conditions is shown in Fig. 10.34 for comparison using (10.4.12) with the full valence-band mixing model. We see that the quadratic dependence (10.4.14) is a very good approximation.

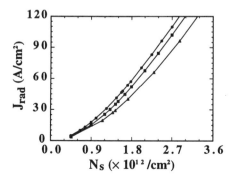

Figure 10.34. Radiative current density J_{rad} (A/cm^2) vs. the surface carrier concentration N_s $(1/cm^2)$ for compressive strain (dots), unstrained (squares), and tensile strain (triangles) quantum-well lasers.

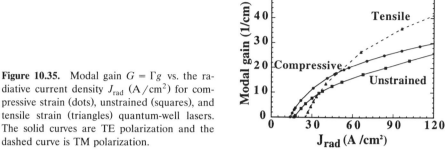

Figure 10.35. Modal gain $G = \Gamma g$ vs. the radiative current density J_{rad} (A/cm^2) for compressive strain (dots), unstrained (squares), and tensile strain (triangles) quantum-well lasers. The solid curves are TE polarization and the dashed curve is TM polarization.

10.4.4 *G-J* Relation

The peak optical gain can also be plotted vs. the radiative current density J_{rad} directly for three strains, as shown in Fig. 10.35. We see that the compressive strain case has the lowest transparency current density and its gain saturates faster as the current density is increased. In real devices, the injected current density has other losses, such as nonradiative Auger recombination and intervalence band absorptions, which have to be added to the horizontal axis to obtain the true *G-J* relation. If these loss current densities are less sensitive to strains, we would expect the threshold gain $G_{th} = \Gamma g_{th}$ to determine the threshold current density by the intersection of the *G-J* curve with a horizontal line $G = G_{th}$. We see that for a small G_{th} the compressive strain laser will have the smallest threshold current density. On the other hand, if G_{th} is large, the tensile strain laser may have the smallest threshold current density. These *G-J* curves provide good design rules for strained quantum-well lasers.

An empirical *G-J* relation using the logarithmic dependence

$$ G = n_w \Gamma_w g_w = n_w \Gamma_w g_0 \left[\ln\left(\frac{n_w J_w}{n_w J_0} \right) + 1 \right] \qquad (10.4.15) $$

as discussed in Section 10.3 has also been used to study strained quantum-well lasers with very good agreement with experimental data [75, 76, 114]. The major equations are the same as those discussed in Section 10.3 and are not repeated here. The same analysis using (10.3.35)–(10.3.42) has been applied to study strained quantum-well lasers.

Advanced issues such as Auger recombination rates [115–118], temperature sensitivity [119], and high-speed carrier transport and capture in strained quantum wells [120, 121] are under intensive investigation. Many-body effects [122, 123] due to carrier–carrier Coulomb interactions on the gain spectrum have been investigated. Many of these require more theoretical and experimental work to fully understand the physics of strained quantum-well lasers.

10.5 COUPLED LASER ARRAYS

Coherent high-power semiconductor laser arrays [124–137] have been of considerable interest recently. Shown in Fig. 10.36 are two examples of gain-guided and buried-ridge semiconductor laser arrays [125, 130]. The concept of coupled laser arrays is very similar to that of the antenna arrays. Since the amount of power of a single element semiconductor laser is limited because of its small size in the radiation aperture (facet), a coupled array is expected to deliver a higher power with a narrower radiation pattern (i.e., a

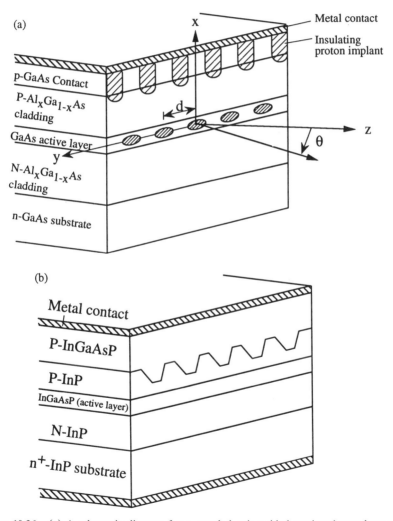

Figure 10.36. (a) A schematic diagram for a coupled gain-guided semiconductor laser array [125]. (b) Semiconductor buried-ridge laser array [130].

better directivity) if the elements are properly excited. The challenging research issues in laser arrays are how to excite the desired (fundamental) mode without wasting optical powers to different sidelobes.

In principle, for a given semiconductor laser array, the optical modes should be calculated [134] by solving Maxwell's equations for a given gain and index profile, which is determined by the injection and coupling conditions. Here we present a supermode analysis for its simplicity and ease of understanding the operation of laser arrays. The supermode analysis for a coupled laser array follows closely the coupled-mode theory [126]. The optical electric field can be written as a linear combination of the field of each element,

$$\mathbf{E}(x, y, z) = \sum_{n=1}^{N} a_n(z)\mathbf{E}^{(n)}(x, y) \tag{10.5.1}$$

where $\mathbf{E}^{(n)}(x, y)$ refers to the optical electric field in the nth element before the coupling effect is taken into account. The coupling effort is included in the coupling coefficient K assuming only nearby coupling.

10.5.1 Solution of the Coupled-Mode Equations

The coupled-mode equations for N identical equally spaced waveguides, assuming only adjacent waveguide coupling with a coupling coefficient K, can be written as

$$\frac{d}{dz}\mathbf{A}(z) = i\overline{\overline{\mathbf{M}}}\mathbf{A}(z) \tag{10.5.2}$$

where

$$\mathbf{A}(z) = \begin{bmatrix} a_1(z) \\ a_2(z) \\ \vdots \\ a_N(z) \end{bmatrix} \qquad \overline{\overline{\mathbf{M}}} = \begin{bmatrix} \beta & K & & & 0 \\ K & \beta & K & & \\ & \ddots & \ddots & \ddots & \\ & & K & \beta & K \\ 0 & & & K & \beta \end{bmatrix} \tag{10.5.3}$$

and β is the propagation constant of the individual waveguide mode. The general solutions for the normalized eigenvectors $\mathbf{A}^{(\ell)}(z)$ and their corresponding eigenvalues β_ℓ for an arbitrary N are

$$\mathbf{A}^{(\ell)}(z) = \begin{bmatrix} a_1^{(\ell)} \\ a_2^{(\ell)} \\ \vdots \\ a_N^{(\ell)} \end{bmatrix} e^{i\beta_\ell z} \qquad \ell = 1, 2, \dots, N \tag{10.5.4}$$

where

$$\beta_\ell = \beta + 2K \cos\left(\frac{\ell\,\pi}{N+1}\right) \tag{10.5.5}$$

$$a_n^{(\ell)} = \sqrt{\frac{2}{N+1}} \sin\left(\frac{n\ell\,\pi}{N+1}\right) \qquad n = 1, 2, \ldots, N \tag{10.5.6}$$

The order of β_ℓ is chosen such that $\beta_1 > \beta_2 > \beta_3 \cdots > \beta_N$. To prove this, we start with the general form (10.5.4) and substitute it into the differential Eq. (10.5.2):

$$\begin{bmatrix} \beta & K & & & 0 \\ K & \beta & K & & \\ & K & \ddots & \ddots & \\ & & & \beta & K \\ 0 & & & K & \beta \end{bmatrix} \begin{bmatrix} a_1 \\ a_2 \\ a_3 \\ \vdots \\ a_N \end{bmatrix} = \beta_\ell \begin{bmatrix} a_1 \\ a_2 \\ a_3 \\ \vdots \\ a_N \end{bmatrix} \tag{10.5.7a}$$

The above eigenequation can be put in the form

$$\begin{bmatrix} (\beta - \beta_\ell)/K & 1 & & & \\ 1 & (\beta - \beta_\ell)/K & 1 & & 0 \\ & 1 & (\beta - \beta_\ell)/K & & \\ & & \ddots & \ddots & 1 \\ 0 & & & 1 & (\beta - \beta_\ell)/K \end{bmatrix} \begin{bmatrix} a_1 \\ a_2 \\ a_3 \\ \vdots \\ a_N \end{bmatrix} = \begin{bmatrix} 0 \\ 0 \\ 0 \\ \vdots \\ 0 \end{bmatrix}$$

$$\tag{10.5.7b}$$

We use the property of the $N \times N$ determinant, $D_N(2\cos\theta)$, which is a polynomial of order N with an argument $2\cos\theta$:

$$D_N(2\cos\theta) = \det \begin{bmatrix} 2\cos\theta & 1 & & & 0 \\ 1 & 2\cos\theta & 1 & & \\ & \ddots & \ddots & \ddots & \\ & & 1 & 2\cos\theta & 1 \\ 0 & & & 1 & 2\cos\theta \end{bmatrix} = \frac{\sin[(N+1)\theta]}{\sin\theta} \tag{10.5.8}$$

which obeys the recursion relation

$$D_N(2x) = 2x D_{N-1}(2x) - D_{N-2}(2x) \tag{10.5.9}$$

where $x = \cos\theta$. This recursion relation can be proved easily using the definition of the $N \times N$ determinant and expanding it in terms of the two elements in the first row (or column). The above recursion relation (10.5.9) is satisfied by the Nth-order Chebyshev polynomial [138] $U_n(\cos\theta)$, or

$$D_N(2\cos\theta) = \frac{\sin(N+1)\theta}{\sin\theta} \qquad (10.5.10)$$

We define

$$2\cos\theta = \frac{\beta - \beta_\ell}{K} \quad \text{or} \quad \beta_\ell = \beta - 2K\cos\theta \qquad (10.5.11)$$

The eigenvalue equation can have nontrivial solutions only if the determinant is zero:

$$\frac{\sin(N+1)\theta}{\sin\theta} = 0 \qquad (10.5.12)$$

Therefore,

$$(N+1)\theta = m\pi \qquad m = 1, 2, 3, \ldots, N \qquad (10.5.13)$$

We can choose $m = N + 1 - \ell$, $\ell = 1, 2, \ldots, N$, such that

$$\theta_\ell = \frac{\ell}{N+1}\pi \qquad \theta = (\pi - \theta_\ell)$$

$$\beta_\ell = \beta + 2K\cos\theta_\ell, \qquad \ell = 1, 2, 3, \ldots, N \qquad (10.5.14)$$

and $\beta_1, \beta_2, \beta_3, \ldots, \beta_N$ are in descending order for $K > 0$.

The eigenvectors can be obtained by substituting β_ℓ into (10.5.7b):

$$\begin{cases} -2a_1\cos\theta_\ell + a_2 = 0 \\ a_{n-1} - 2a_n\cos\theta_\ell + a_{n+1} = 0 \qquad n = 2, 3, \ldots, N-1 \quad (10.5.15) \\ a_{N-1} - 2a_N\cos\theta_\ell = 0 \end{cases}$$

The above difference equation can be solved assuming the solution a_n of the form

$$a_n = r^n$$

Therefore,

$$\frac{1}{r} - 2\cos\theta_\ell + r = 0 \qquad (10.5.16)$$

or $r = \exp(\pm i\theta_\ell)$. The general solution for a_n is therefore $\exp(+in\theta_\ell)$ or

$\exp(-in\theta_\ell)$, or their linear combinations:

$$a_n = A \cos n\theta_\ell + B \sin n\theta_\ell \qquad (10.5.17)$$

Since $a_0 = 0$, we find $A = 0$. Therefore,

$$a_n = B \sin n\theta_\ell$$

The amplitude B can be found from the normalization condition:

$$\sum_{n=1}^{N} |a_n|^2 = 1 \qquad (10.5.18)$$

Since

$$\sum_{n=1}^{N} \sin^2 n\theta_\ell = \sum_{n=1}^{N} \frac{1 - \cos 2n\theta_\ell}{2}$$

$$= \frac{N}{2} - \frac{1}{2}\text{Re}\left(\sum_{n=1}^{N} e^{i2n\theta_\ell}\right)$$

$$= \frac{N}{2} - \frac{\cos[(N+1)\theta_\ell]\sin(N\theta_\ell)}{2\sin\theta_\ell}$$

$$= \frac{N+1}{2} \qquad (10.5.19)$$

where $\theta_\ell = \ell\pi/(N+1)$ has been used, we find the normalization constant $B = \sqrt{2}/\sqrt{N+1}$. We conclude that the N eigenvectors with corresponding eigenvalues β_ℓ are described by (10.5.4)–(10.5.6).

Example As an example, for $N = 5$, we have the following five eigenvectors:

$$\frac{1}{\sqrt{3}}\begin{bmatrix} 1/6 \\ \sqrt{3}/2 \\ 1/3 \\ \sqrt{3}/2 \\ 1/6 \end{bmatrix}, \frac{1}{2}\begin{bmatrix} 1 \\ 1 \\ 0 \\ -1 \\ -1 \end{bmatrix}, \frac{1}{\sqrt{3}}\begin{bmatrix} 1 \\ 0 \\ -1 \\ 0 \\ 1 \end{bmatrix}, \frac{1}{2}\begin{bmatrix} 1 \\ -1 \\ 0 \\ 1 \\ -1 \end{bmatrix}, \frac{1}{\sqrt{3}}\begin{bmatrix} 1/6 \\ -\sqrt{3}/2 \\ 1/3 \\ -\sqrt{3}/2 \\ 1/6 \end{bmatrix} \qquad (10.5.20)$$

with corresponding eigenvalues, $\beta_\ell(\ell = 1, 2, \ldots, 5)$:

$$\beta + \sqrt{3}K, \quad \beta + K, \quad \beta, \quad \beta - K, \quad \beta - \sqrt{3}K$$

Plots of the near-field patterns using (10.5.1) assuming that each element $E^{(i)}(x, y)$ is that of a single waveguide are shown in Fig. 10.37.

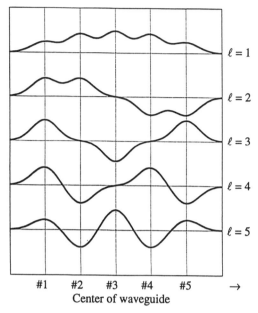

Figure 10.37. Plots of the supermodes using (10.5.1) and (10.5.20) for $N = 5$.

As we can see in Fig. 10.37, the highest-order mode $\ell = 5$ has alternating amplitudes with zero crossings between the elements. This out-of-phase mode tends to dominate the lasing behavior of the laser arrays because the overlap of the optical field with the gain profile is maximized for this mode. When the optical field goes to zero between adjacent elements, the mode experiences a minimum of optical absorption between the emitters. Therefore, the laser threshold gain is minimized for the out-of-phase mode and it will lase first with enough injection current. Lasing of this out-of-phase mode is undesirable since it gives double sidelobes in the radiation pattern. Various schemes have been used to improve the laser array performance [133, 134].

▪

10.5.2 Far-Field Radiation Pattern

To calculate the far-field radiation pattern, recall from Chapter 5 that the optical electric field under far-field approximation is given by the Fourier transform of the aperture field,

$$\mathbf{E}(\mathbf{r}) = \frac{-ik\,e^{ikr}}{4\pi r}\hat{r} \times \iint dx'\,dy'\,e^{-i\mathbf{k}\cdot\mathbf{r}'}\mathbf{M}_s(\mathbf{r}') \qquad (10.5.21)$$

where the equivalent magnetic surface current density is related to the aperture field by

$$\begin{aligned}
\mathbf{M}_s &= -2\hat{n} \times \mathbf{E}_A(\mathbf{r}') \\
&= -2\hat{z} \times \mathbf{E}_A(\mathbf{r}')
\end{aligned} \tag{10.5.22}$$

If $\mathbf{E}_A(\mathbf{r}') = \hat{y}E_A(x', y')$ for TE modes in the waveguides, we have

$$\mathbf{M}_s = 2\hat{x}E_A$$
$$\hat{r} \times \hat{x} = \sin\theta\sin\phi(-\hat{z}) + \cos\theta\hat{y} \tag{10.5.23}$$

Considering the observation point in the y-z plane, ($\phi = \pi/2$), we have

$$\mathbf{E}(\theta) \propto (-\sin\theta\hat{z} + \cos\theta\hat{y}) \int\int dx\, dy\, e^{-i\mathbf{k}\cdot\mathbf{r}}E_A(\mathbf{r}) \tag{10.5.24}$$

The aperture field is given by (10.5.1), noting that $E^{(n)}(\mathbf{r}) = E^{(1)}(\mathbf{r} - n\mathbf{d})$, where $E^{(1)}$ is the optical aperture field of the first element:

$$E_A(\mathbf{r}) = \sum_n a_n^{(\ell)}E^{(1)}(\mathbf{r} - n\mathbf{d}) \tag{10.5.25}$$

Therefore, the far-field radiation pattern is

$$|\mathbf{E}(\theta)|^2 = \underbrace{\left| \sum_{n=1}^N a_n^{(\ell)}e^{-i\mathbf{k}\cdot n\mathbf{d}} \right|^2}_{\text{Array pattern}} \underbrace{\left| \int d\mathbf{r}\, e^{-i\mathbf{k}\cdot\mathbf{r}}E^{(1)}(\mathbf{r}) \right|^2}_{\text{Unit pattern}} \tag{10.5.26}$$

Using $\mathbf{k}\cdot\mathbf{d} = kd\sin\theta$, the array pattern of the ℓth supermode, $A_\ell(\theta)$, can be obtained from

$$\begin{aligned}
A_\ell(\theta) &= \left| \sum_{n=1}^N a_n^{(\ell)}e^{-inkd\sin\theta} \right|^2 \\
&= \left| \sum_{n=1}^N \sqrt{\frac{2}{N+1}}\sin(n\theta_\ell)e^{-inkd\sin\theta} \right|^2 \\
&= \frac{1}{2(N+1)}\left| \sum_{n=1}^N e^{in(\theta_\ell - kd\sin\theta)} - \sum_{n=1}^N e^{-in(\theta_\ell + kd\sin\theta)} \right|^2 \\
&= \frac{1}{2(N+1)}\left| (-1)^\ell \frac{\sin[(N/2)(\theta_\ell - kd\sin\theta)]}{\sin\frac{1}{2}(\theta_\ell - kd\sin\theta)} \right. \\
&\qquad\qquad\qquad \left. - \frac{\sin[(N/2)(\theta_\ell + kd\sin\theta)]}{\sin\frac{1}{2}(\theta_\ell + kd\sin\theta)} \right|^2 \tag{10.5.27}
\end{aligned}$$

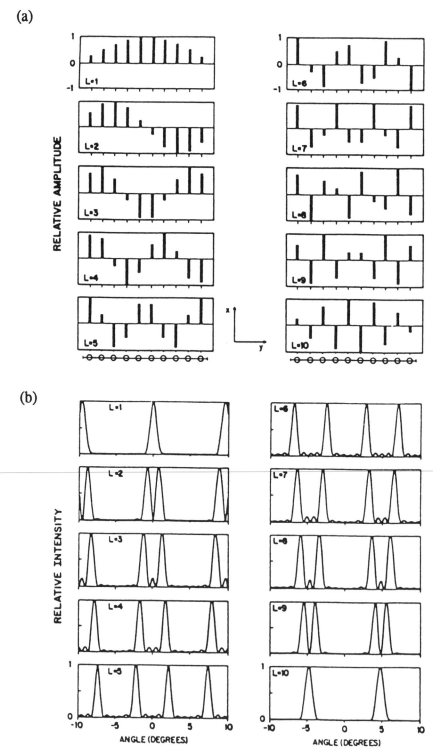

Figure 10.38. (a) The magnitudes of the 10 eigenvectors $A^{(\ell)}(z)$ for $\ell = 1, 2, 3, \ldots, 10$ of a ten-element array. (b) The corresponding far-field patterns of the ten array modes. (After Ref.

In Fig. 10.38a the magnitudes of the 10 eigenvectors $A^{(\ell)}(z)$ for $\ell = 1, 2, \ldots, 10$ of a ten-element array are shown (from Ref. 126). Each mode is labeled by ℓ. For each eigenmode ℓ, the magnitudes $a_n^{(\ell)}$, $n = 1, 2, \ldots, 10$ are represented by vertical bars. The corresponding far-field patterns are shown in Fig. 10.38b. We can see that for the fundamental ($\ell = 1$) and the highest ($\ell = 10$) order modes, the amplitudes are not uniform throughout the elements.

To improve the uniformity of the near-field profile, the two outermost waveguides are chosen [128] to have a smaller spacing to increase the coupling coefficient by a $\sqrt{2}$ factor, with all the other guides equally spaced. The coupling matrix becomes

$$
\overline{\overline{M}} =
\begin{bmatrix}
\beta & \sqrt{2}\,K & 0 & 0 & \cdot & \cdot & \cdot & 0 \\
\sqrt{2}\,K & \beta & K & 0 & \cdot & \cdot & \cdot & \cdot \\
0 & K & \beta & K & \cdot & \cdot & \cdot & \cdot \\
0 & 0 & K & \cdot & \cdot & \cdot & \cdot & \cdot \\
\cdot & \cdot & \cdot & \cdot & \cdot & \beta & K & \cdot \\
\cdot & \cdot & \cdot & \cdot & \cdot & K & \beta & \sqrt{2}\,K \\
0 & \cdot & \cdot & \cdot & \cdot & 0 & \sqrt{2}\,K & \beta
\end{bmatrix}
\tag{10.5.28}
$$

The lowest and the highest supermodes have a uniform near-field intensity envelope and the injected charges will be utilized more efficiently. These supermodes should be relatively stable with increased pumping above threshold. Another design is to use a Y-junction waveguide [129] structure such that all waveguide modes will lase in phase with equal amplitudes.

Using closely spaced antiguided waveguides [133, 134], a new class of phase-locked diode laser arrays has been developed. Fundamental array-mode operation in a diffraction-limit-beam pattern is obtained up to 200 mW. The out-of-phase mode is also shown to lase with 110 mW per uncoated facet. The threshold currents of these antiguided arrays have been shown to be in the 270- to 320-mA range with an external quantum efficiency near 30 to 35% and cw operation up to 200 mW. As pointed out before, the real optical modal profiles in the laser arrays using gain-guided [132] or antiguided [134–136] structures have to be calculated properly to compare with experimental data. The supermode analysis is presented in this section only to provide some basic understanding of the laser array characteristics.

10.6 DISTRIBUTED FEEDBACK LASERS [33, 77, 139–141]

In Section 8.6 we discussed distributed feedback (DFB) structures and studied the electric field and the reflection coefficient. The reflection coefficient depends on the coupling coefficient K and the detuning δ of the propagation constant β from the Bragg wave number $\ell\pi/\Lambda$, where ℓ refers to the order of the grating mode used in the DFB process. We can see that the periodic structures provide an optical frequency selection property. In Fabry–Perot-type semiconductor lasers, there are many longitudinal modes within the gain spectrum, which is generally broad. Although there are some side-mode suppressions in cw operation of the semiconductor lasers above threshold, the powers of the side modes increase rapidly when the laser diode is pulsed at high bit rates. In optical communication systems with common chromatic dispersion in optical fibers, these undesirable side-mode power partitions limit the band width of the information transmission rate. Therefore, a mode selectivity provided by periodic structures such as distributed feedback lasers or distributed Bragg reflectors are attractive for applications for single longitudinal-mode operation of diode lasers. A schematic diagram of a distributed feedback (DFB) semiconductor laser is illustrated in Fig. 10.39, where a grating above the active region provides the optical distributed feedback coupling. Using the coupled-mode theory in Section 8.6, we summarize the results for the optical electric field in a distributed feedback structure:

$$E_y(x) = A(z)\, e^{i\beta_0 z}U(x) + B(z)\, e^{-i\beta_0 z}U(x) \qquad (10.6.1)$$

where $U(x) = E_y^{(TE_0)}(x)$ is the mode pattern of the TE_0 mode. $\beta_0 = \ell\pi/\Lambda$ is the Bragg wave vector for the ℓth order grating. The amplitudes $A(z)$ and $B(z)$ of the forward and backward propagating waves satisfy the coupled-mode equation (8.7.18):

$$\frac{d}{dz}\begin{bmatrix} A(z) \\ B(z) \end{bmatrix} = i\begin{bmatrix} \Delta\beta & K_{ab} \\ K_{ba} & -\Delta\beta \end{bmatrix}\begin{bmatrix} A(z) \\ B(z) \end{bmatrix} \qquad (10.6.2)$$

In general, K_{ab} and K_{ba} are complex for a lossy or gain structure. Only for a lossless structure,

$$K_{ba} = -K_{ab}^* \qquad (10.6.3)$$

Define

$$q = \sqrt{(\Delta\beta)^2 + K_{ba}K_{ab}} \qquad (10.6.4)$$

Figure 10.39. A schematic diagram for a distributed feedback (DFB) semiconductor laser, where a periodic grating structure above the active region provides the optical distributed feedback process.

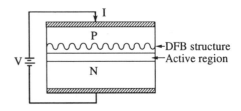

10.6.1 Solution of the Coupled-Mode Equations in DFB Structures

The general solution to the coupled-mode equations is expressed as the sum of the two eigenmodes:

$$
\begin{bmatrix} A(z) \\ B(z) \end{bmatrix} = \begin{bmatrix} A^+ \\ B^+ \end{bmatrix} e^{iqz} + \begin{bmatrix} A^- \\ B^- \end{bmatrix} e^{-iqz}
\tag{10.6.5}
$$

where

$$
\frac{B^+}{A^+} = \frac{\Delta\beta - q}{-K_{ab}} = \frac{K_{ba}}{\Delta\beta + q} \qquad \text{for the } e^{+iqz} \text{ eigenmode} \tag{10.6.6a}
$$

and

$$
\frac{B^-}{A^-} = \frac{\Delta\beta + q}{-K_{ab}} = \frac{K_{ba}}{\Delta\beta - q} \qquad \text{for the } e^{-iqz} \text{ eigenmode} \tag{10.6.6b}
$$

We can define the DFB "reflection" coefficient for the " $+$ " eigenmode

$$
r_p(q) = \frac{B^+}{A^+} \left(= \frac{K_{ba}}{\Delta\beta + q} \right) \tag{10.6.7a}
$$

and the " $-$ " eigenmode

$$
r_m(q) = \frac{A^-}{B^-} \left(= \frac{-K_{ab}}{\Delta\beta + q} \right) \tag{10.6.7b}
$$

Write the general solution (10.6.5) as

$$
A(z) = A^+ e^{iqz} + r_m(q) B^- e^{-iqz}
$$
$$
B(z) = r_p(q) A^+ e^{iqz} + B^- e^{-iqz} \tag{10.6.8}
$$

The boundary conditions at two sharp ends of the DFB structure are

1. At $z = 0$, $A(0) = r_1 B(0)$, therefore

$$
(1 - r_1 r_p) A^+ + (r_m - r_1) B^- = 0 \tag{10.6.9}
$$

2. At $z = L$, $B(L) = r_2 A(L)$, thus

$$
(r_2 - r_p) e^{i2qL} A^+ - (1 - r_2 r_m) B^- = 0 \tag{10.6.10}
$$

In general, r_1 and r_2 are complex. For nontrivial solutions of A^+ and B^-, we must have the determinantal equation satisfied:

$$(1 - r_1 r_p)(1 - r_2 r_m) + (r_2 - r_p)(r_m - r_1) e^{i2qL} = 0 \quad (10.6.11)$$

or

$$\left[\frac{r_1 - r_m(q)}{1 - r_1 r_p(q)} \right] \left[\frac{r_2 - r_p(q)}{1 - r_2 r_m(q)} \right] e^{i2qL} = 1 \quad (10.6.12)$$

Notice that the gain of the structure, in general, should be incorporated into $\Delta\beta$, K_{ab}, and K_{ba}:

$$\Delta\beta = \beta - \beta_0 \rightarrow \left(k_0 n - i\frac{G}{2} \right) - \beta_0 = \delta - i\frac{G}{2} \quad (10.6.13)$$

where $\delta = \beta_{TE_0} - \beta_0$ is the detuning parameter, $G = \Gamma g$ is the net modal gain taking into account the optical confinement factor Γ and the net material gain g.

If we further assume that the lossless (or no-gain) condition for the coupling coefficient

$$K_{ba} = -K_{ab}^*$$

is approximately valid, then

$$q \simeq \sqrt{\left(\delta - i\frac{G}{2} \right)^2 - |K|^2} \quad (10.6.14)$$

where $K \equiv K_{ab}$ has been used, and the effect of gain is introduced through the term $G/2$ in the detuning $\Delta\beta$. The eigenequation (10.6.12) should be solved numerically, in general, for the threshold gain g and the lasing mode spectrum (or detuning) $\delta \simeq k_0 n - \beta_0$, since $\beta_{TE_0} \simeq k_0 n$.

10.6.2 Laser Oscillation Conditions

If there is no distributed feedback structure, $K = 0, r_m = r_p = 0$, we recover the eigenequation for the Fabry–Perot lasers,

$$r_1 r_2 e^{i2qL} = 1 \quad (10.6.15)$$

where $q \rightarrow \delta - ig/2$. If there are no sharp boundaries at $z = 0$ and $z = L$ such that $r_1 = r_2 = 0$, we have simply

$$r_m(q) r_p(q) e^{i2qL} = 1 \quad (10.6.16)$$

Therefore, $r_p(q)$ and $r_m(q)$ simply act as reflection coefficients for the forward and backward propagating waves. The above equation simplifies to (assuming $K_{ba} = -K_{ab}^*$)

$$\frac{|K|^2}{(q + \Delta\beta)^2} e^{i2qL} = 1 \tag{10.6.17}$$

Noting that $q^2 = (\Delta\beta)^2 - |K|^2$, we have

$$q = i\,\Delta\beta \tan qL \tag{10.6.18}$$

The above result can also be obtained from the denominator of the reflection coefficient for the DFB structure (8.6.37),

$$\Delta\beta \sinh SL + iS \cosh SL = 0 \tag{10.6.19}$$

since the laser oscillation condition is satisfied without an external injected light. Noting that $q = iS$, we obtain (10.6.18) immediately.

To analyze the solutions for the oscillation condition, we write

$$\Delta\beta L = -iqL \cot(qL) \tag{10.6.20}$$

or

$$\delta L - i\frac{GL}{2} = -i\sqrt{\left(\delta L - i\frac{GL}{2}\right)^2 - (KL)^2}\,\cot\sqrt{\left(\delta L - i\frac{GL}{2}\right)^2 - (KL)^2} \tag{10.6.21}$$

The above equation has to be solved for $(\delta L, GL)$ simultaneously for each given KL.

In the high gain (weak coupling) limit, $GL \gg KL$, we have $q \simeq \Delta\beta$; therefore, (10.6.17) becomes

$$K^2 e^{i2\Delta\beta L} = (2\Delta\beta)^2 \tag{10.6.22}$$

We compare the magnitude and the phase of the above equation. We obtain [33]

$$K^2 e^{GL} = G^2 + 4\delta^2 \tag{10.6.23a}$$

and

$$\delta L = \left(m - \frac{1}{2}\right)\pi + \tan^{-1}\left(\frac{2\delta}{G}\right) \tag{10.6.23b}$$

Once the threshold modal gain G is obtained from (10.6.23a) for a given K

and δ, the threshold current density can be estimated from the model for the gain–current $(G\text{-}J)$ relation for a given semiconductor laser structure as an active double-heterostructure or a quantum-well structure. Note that the reflection coefficient for a DBR structure with a finite length L is obtained from (10.6.8)

$$\Gamma(0) = \frac{B(0)}{A(0)} = \frac{r_p(q)A^+ + B^-}{A^+ + r_m(q)B^-} \tag{10.6.24}$$

Noting that $B(L) = 0$, we have $B^- = -r_p A^+ e^{i2qL}$, or

$$\Gamma(0) = \frac{r_p\left(1 - e^{i2qL}\right)}{1 - r_m r_p\, e^{i2qL}} \tag{10.6.25a}$$

$$= \frac{iK^* \sin qL}{-i\Delta\beta \sin qL + q \cos qL} \tag{10.6.25b}$$

$$= \frac{-K^* \sinh SL}{\Delta\beta \sinh SL + iS \cosh SL} \tag{10.6.25c}$$

for $q = iS$ in a passive lossless structure. The last expression (10.6.25c) is the same as that derived in (8.6.37). We can see that, in general, $\Gamma(0)$ is a complex quantity,

$$\Gamma(0) = |\Gamma(0)|e^{i\phi} \tag{10.6.26}$$

10.6.3 Distributed Bragg Reflector (DBR) Semiconductor Laser

In Fig. 10.40 we show a distributed Bragg reflector (DBR) semiconductor laser. The frequency selectivity is provided by the reflection coefficients r_1

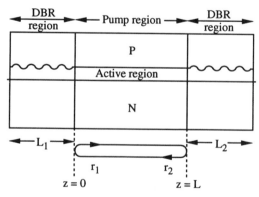

Figure 10.40. Schematic diagram of a distributed Bragg reflector (DBR) semiconductor laser.

and r_2 at both ends, and the active region is the same as a Fabry–Perot edge emitting laser. As a matter of fact, a DBR laser is just a Fabry–Perot laser with a mirror reflectivity varying with the wavelength. The lasing threshold occurs at the wavelength for which the overall reflectivity is maximum. The analysis is very similar to that of a Fabry–Perot laser except that r_1 and r_2 are complex numbers and are wavelength dependent.

The lasting condition occurs at

$$r_1 r_2 e^{i2\beta L} = 1 \qquad (10.6.27)$$

where r_1 is the reflection coefficient of the electric field looking into the left DBR reflector (with a length L_1) at $z = 0$, and r_2 is that looking into the right DBR reflector at $z = L$. The propagation constant is

$$\beta = \beta_0 - \frac{i}{2}(\Gamma g - \alpha) \qquad (10.6.28)$$

where β_0 is the propagation constant, Γ is the optical confinement factor, and α is the total loss in the guide and in the substrate:

$$\alpha = \Gamma \alpha_i + (1 - \Gamma)\alpha_s \qquad (10.6.29)$$

Note that

$$r_1 = R_1^{1/2} e^{i\phi_1} \qquad r_2 = R_2^{1/2} e^{i\phi_2} \qquad (10.6.30)$$

We obtain

$$\Gamma g_{th} = \alpha + \frac{1}{2L} \ln\left(\frac{1}{R_1 R_2}\right) \qquad (10.6.31)$$

$$2\beta_0 L + \phi_1 + \phi_2 = 2m\pi \qquad (10.6.32)$$

The reflection coefficient for a DBR region with a length L_1 has been obtained in (10.6.25c):

$$r_1 = |r_1| e^{i\phi_1} = \frac{-K^* \sinh SL_1}{\Delta\beta \sinh SL_1 + i S \cosh SL_1} \qquad (10.6.33)$$

It should be noted that for an abrupt junction between the active region and the DBR region, the reflection coefficient r_1 (and r_2) has to be multiplied by an effective coupling coefficient C_0 (with $|C_0| \le 1$) to account for the imperfect power transmission across the junction, that is, $r_1 \to C_0 r_1$, and the threshold condition (10.6.31) becomes [139]

$$\Gamma g_{th} = \alpha + \frac{1}{2L} \ln \frac{1}{R_1 R_2 C_0^4} \qquad (10.6.34)$$

From the threshold gain (10.6.31) or (10.6.34), the threshold current density J_{th} is then obtainable from the gain–current model for a given diode laser structure.

Stable single-mode operation of semiconductor lasers using DFB and DBR structures has been achieved. The temperature stability of these structures has also been demonstrated. In a conventional Fabry–Perot semiconductor laser, the lasing wavelength as a function of temperature is dominated by the temperature coefficient of the peak gain wavelength $d\lambda_p/dT$, which is approximately 0.5 nm/degree for long-wavelength InGaAsP/InP double-heterostructure lasers [139]. On the other hand, for DBR and DFB lasers, the lasing wavelength is dominated by the temperature coefficient of the refractive index, dn/dT, which is approximately 0.1 nm/degree. Therefore, a stable fixed-mode operation of a DBR or DFB laser over a wide range of temperature (> 100 degrees) has been achieved [139].

10.7 SURFACE-EMITTING LASERS

Recently, vertical-cavity surface-emitting lasers (VCSEL) [26–28, 142–147] have been developed with performances comparable to those of conventional Fabry–Perot edge-emitting diode lasers. A few schematic diagrams are shown in Fig. 10.41. The light output is orthogonal to the plane of the wafers. This change in the direction of laser emission drastically changes the design, fabrication scalability, and array configurability of semiconductor lasers.

The concept of surface emission from a semiconductor laser dates back to 1965 when Melngailis reported [148] on a vertical cavity structure. Later different groups reported on the grating surface emission [149, 150] and 45°-corner turning mirrors etched into the device [151]. Major work on VCSELs was done [147, 152] after 1977. More recently many improvements with significant progress on the performances of the VCSELs have been demonstrated. Because of their short cavity lengths, the surface-emitting lasers have inherent single-frequency operation. They have other advantages such as wafer scale probing and ease of fiber coupling. Therefore, VCSELs have potential applications for optical interconnects, optical communication, and optical processing.

10.7.1 Threshold Condition

The threshold analysis of a VCSEL can be modified from that of a DBR semiconductor laser with a complex propagation constant $\beta = \beta_r + i(\alpha - \Gamma g)/2$:

$$r_1 r_2 \, e^{i2\beta L} = 1 \qquad (10.7.1)$$

where r_1 and r_2 are the complex reflection coefficients of the electric field

(a)

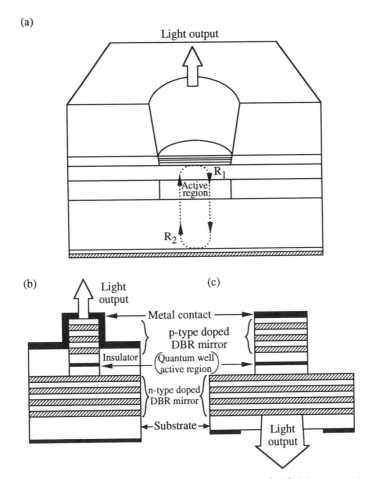

Figure 10.41. (a) Schematic diagram of a surface-emitting laser [147]. (b) Front and (c) back surface-emitting lasers using etching through the active region [142].

for the DBR mirrors at the top and bottom of the vertical cavity. The threshold condition is

$$\Gamma g = \alpha + \frac{1}{2L}\ln\left(\frac{1}{R_1 R_2}\right) \qquad (10.7.2)$$

using $R_1 = |r_1|^2$ and $R_2 = |r_2|^2$. Here the optical confinement factor Γ accounts for both the longitudinal and transverse confinements

$$\Gamma = \left(\gamma \frac{d}{L}\right)\xi \qquad (10.7.3)$$

where d is the active layer thickness, L is the cavity length, $\gamma = 2$ if the thin active layer is placed at the maximum of the standing wave, and $\gamma = 1$ for a thick active layer. The factor d/L accounts for the longitudinal optical confinement, and ξ is the transverse optical confinement factor. The symbol α accounts for the optical absorptions inside and outside the active region, plus any diffraction loss due to the mode mismatches when the laser mode propagates to the reflecting mirrors. This diffraction loss depends on the size of the diameter of the active region and the locations of the mirrors.

10.7.2 Carrier Injection and Optical Profile

Depending on the fabrication processes and the structures of the surface-emitting lasers, the current distribution, the carrier density profile, and the optical mode pattern vary. For example, as shown in Fig. 10.42a, the ion-implanted regions act as insulators and the injected current into the active layer will have a distribution as shown in Fig. 10.42b. The guidance of the surface-emitting laser in the transverse direction is the same as that for the gain-guided stripe geometry diode laser, except that here the injection profile has a circular cross section instead of a long stripe geometry, and the light propagation is along the vertical direction instead of the parallel

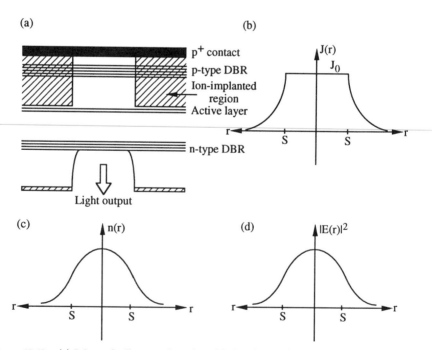

Figure 10.42. (a) Schematic diagram of a gain-guided surface-emitting laser with ion-plantation regions, (b) current density, (c) carrier density, and (d) fundamental optical mode profile [153].

direction to the active layer. The carrier density profile (Fig. 10.42c) can be solved from the diffusion equation given the current density profile. The fundamental optical mode pattern is shown in Fig. 10.42d once the gain profile is obtained

$$g(r) \simeq g'[n(r) - n_{\mathrm{tr}}]$$ (10.7.4a)

for bulk lasers, where g' is the differential gain and n_{tr} is the transparency carrier density. For quantum-well lasers, one may use

$$g(r) \simeq g_0 \left\{ \ln\left[\frac{n(r)}{n_0} \right] + 1 \right\}$$ (10.7.4b)

where g_0 and n_0 are constant parameters. Approximate analytic formulas can be found in Ref. 153.

For the surface-emitting laser in Fig. 10.41b, the device has been etched all the way through the active region to provide lateral carrier confinement. The top structure has a cylindrical semiconductor region with air as the outside region. The interface scattering and diffraction losses also provide a certain degree of mode selectivity and the fundamental (EH_{11}, HE_{11} or TEM_{00}) mode of a rectangular or cylindrical waveguide is considered to be very important.

In the following, we show the experimental results from Ref. 144. In Fig. 10.43, we show the near-field patterns of a 15-μm-square VCSEL with a structure (using a shallow implant range of about 1 μm with a proton dosage of $3 \times 10^{15}/\mathrm{cm}^2$ at 100 keV) at various current injection levels above threshold. At a small injection current above threshold condition, the fundamental TEM_{00} mode (called the EH_{00} mode in Chapter 7) lases with a full-width–half-maximum of 7.8 μm, as shown in Fig. 1043a, and its lasing wavelength is 9500.85 Å, as shown in the emission spectrum in Fig. 10.44a. As the current increases, the lasing mode changes to a TEM_{01} mode, shown in Fig. 10.43b, with a lasing wavelength 9505.90 Å, shown in Fig. 10.44b, and then shifts to a TEM_{10} mode lasing at 9517.40 Å, shown in Fig. 10.43c and in Fig. 10.44c over a small range of current. At an even higher current level, both the TEM_{00} and TEM_{11} modes lase with the total near-field pattern shown in Fig. 10.43d, and their emission wavelengths are 9516.92 and 9523.80 Å, respectively, as shown in Fig. 10.44d. The mode pattern in Fig. 10.43d can be spectrally resolved due to the wavelength difference of the two modes and is shown in Fig. 10.43e, where the TEM_{00} and TEM_{11} modes are clearly separated in the near-field image.

The red shift of the lasing wavelength is caused by the junction heating as the injection current increases. The narrow linewidths of the lasing spectra also indicate that the laser emits a single transverse mode, since the wavelength separations between two adjacent transverse modes are around 1 to 2 Å for the gain-guided VCSELs.

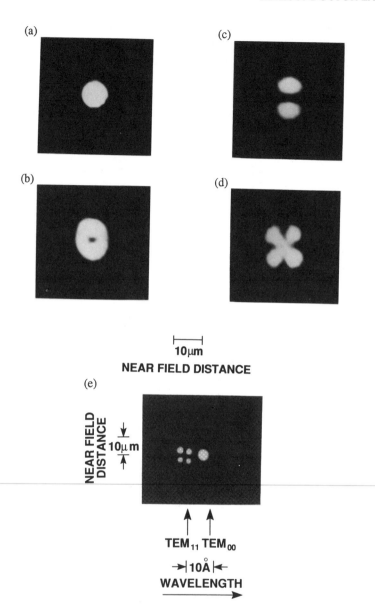

Figure 10.43. The cw near-field patterns of a 15-μm square VCSEL emitting (a) a TEM_{00} (fundamental) mode with a full-width–half-maximum of 7.8 μm at a small current above threshold, (b) a TEM_{01} and (c) a TEM_{10} mode at higher currents, and (d) both TEM_{00} and TEM_{11} modes at even higher currents. The mode pattern in (d) is spectrally resolved to show the near-field image in (e), where the TEM_{00} and TEM_{11} modes are clearly separated due to their wavelength differences, shown in Fig. 10.44. (After Ref. 144 © 1991 IEEE.)

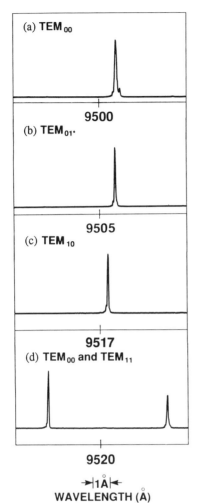

Figure 10.44. The lasing spectra at different current injection levels corresponding to those in Figs. 10.43a–d. The lasing wavelengths are (a) 9500.85 Å, (b) 9505.90 Å, and (c) 9517.40 Å. In (d), the two lasing wavelengths for the TEM_{00} and TEM_{11} modes are 9516.92 and 9523.80 Å, respectively. (After Ref. 144 © 1991 IEEE.)

10.7.3 Junction Heating

It should be noted that the junction heating is important in surface-emitting lasers. A simplified model to account for the temperature rise due to the injected current from a circular disk above an infinite substrate is [145]

$$\Delta T_{jct} = \frac{P_{IV} - P_{opt}}{4\sigma S} \qquad (10.7.5)$$

where P_{IV} is the total input electric power (the current–voltage product), P_{opt} is the optical light output, σ is the thermal conductivity of the semiconductor ($= 0.45$ W/cm°C for GaAs), and S is the disk radius.

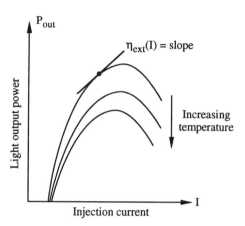

Figure 10.45. A plot of the light output of a surface-emitting laser vs. the injection current for different temperatures. The slope at a particular current level I is the differential quantum efficiency, which can be negative at a high current level or a high temperature.

10.7.4 Optical Output and Differential Quantum Efficiency

The optical output power is

$$P_{\text{out}} = \frac{\hbar\omega}{q}\eta\big[I - I_{\text{th}}(I, T_{\text{jct}})\big] \qquad (10.7.6)$$

where η depends on two factors: (1) the injection efficiency accounting for the fraction of injected carriers contributing to the emission process (some of the carriers can recombine in the undoped confinement regions where the carriers do not interact with the optical field), and (2) the optical efficiency accounting for the fraction of generated photons that are transmitted out of the cavity. Note that the threshold current depends on the injection current as well as on the junction temperature T_{jct}. Models for the laser output taking into account the temperature heating have also been developed in Refs. 154–156.

The differential quantum efficiency is then current-dependent:

$$\eta_{\text{ext}}(I) = \frac{q}{\hbar\omega}\frac{dP_{\text{out}}}{dI} = \eta\left(1 - \frac{dI_{\text{th}}}{dI}\right) \qquad (10.7.7)$$

We see that $\eta_{\text{ext}}(I)$ can be negative if $dI_{\text{th}}/dI > 1$. The light output P_{out} vs. the injection current I will have a negative slope in this case, as shown in Fig. 10.45. Experimental data showing these negative external quantum efficiency can be found in Refs. 145, 146, and 154.

PROBLEMS

10.1 Plot the energy band diagram for a P-$Al_{0.3}Ga_{0.7}As$/i-GaAs/N-$Al_{0.3}Ga_{0.7}As$ double heterojunction under zero bias with the active GaAs layer thickness $d = 1 \ \mu m$. Assume that $N_A = 1 \times 10^{18} \ cm^{-3}$ in the P region and $N_D = 1 \times 10^{18} \ cm^{-3}$ in the N region. Use the depletion approximation.

10.2 Plot the energy band diagram for a P-InP/i-InGaAsP/N-InP double heterostructure under zero bias with the active layer $d = 1 \ \mu m$ using the depletion approximation. Assume that $N_A = 1 \times 10^{18} \ cm^{-3}$ in the P region and $N_D = 1 \times 10^{18} \ cm^{-3}$ in the N region. The active InGaAsP layer has a band gap energy 1.0 eV and is lattice-matched to InP. Assume that the band edge discontinuity ratio $\Delta E_c / \Delta E_v$ is $60/40$.

10.3 A semiconductor laser operating at 1.55 μm has an internal quantum efficiency $\eta_i = 0.75$, intrinsic absorption coefficient $\alpha_i = 8 \ cm^{-1}$, and the mirror reflectivity $R = 0.3$ on both facets.

  (a) Plot the inverse external quantum efficiency $1/\eta_e$ vs. the cavity length L.

  (b) Assuming that the threshold current is 3.5 mA at $L = 500 \ \mu m$, plot the optical output power vs. the injection current I. Label the slope of the curve.

  (c) Repeat (b) for $I_{th} = 4$ mA and $L = 1000 \ \mu m$.

10.4 If the mirror reflectivities are R_1 and R_2 at the two ends of the semiconductor laser facets, the threshold condition is given by

$$\Gamma g_{th} = \alpha_i + \frac{1}{2L} \ln \frac{1}{R_1 R_2}$$

  (a) Find the expressions for the optical output powers, P_1 and P_2, from facets 1 and 2.

  (b) Plot P_1 vs. R_1 for $0.1 \le R_1 \le 1.0$ and P_2 vs. R_2 for $0.1 \le R_2 \le 1.0$. You may assume that $\alpha_i = 10 \ cm^{-1}$ and $L = 500 \ \mu m$.

  (c) Compare P_1 with P_2 if $R_1 = 0.3$ and $R_2 = 0.5$.

10.5 **(a)** Derive the carrier density profile $n(y)$ in (10.2.27) and show that if we ignore the current spreading outside of the stripe region, $n(y)$ is given by (10.2.28).

  (b) Plot $n(y)$ using (10.2.28) and show that it agrees with the experimental data in Fig. 10.14.

10.6 Describe the mode patterns $|E_y(x, y)|^2$ using (10.2.16) for a gain-guided semiconductor laser.

10.7 **(a)** Explain the physical reasons why $\ln J_{th}$ of a quantum-well laser increases linearly with the inverse cavity length $1/L$ as shown in Fig. 10.25a. What determines the slope of the line?

(b) How do you find the cavity length L such that the threshold current I_{th} is minimized? What are the factors determining this minimum threshold current?

(c) Find typical numerical values of some quantum-well lasers and replot Figs. 10.25a and b.

10.8 Plot the gain spectrum using (10.3.14) for a 100-Å GaAs/Al$_{0.3}$Ga$_{0.7}$As quantum-well laser assuming a parabolic band model and an infinite barrier approximation to calculate the subband energies and wave functions. What is the maximum gain coefficient g_m?

10.9 **(a)** Calculate the parameters n_c and n_v for a 100-Å GaAs/Al$_{0.3}$Ga$_{0.7}$As quantum-well laser at $T = 300$ K. Assume single subband occupation.

(b) Find the transparency carrier density n_{tr} for part (a).

10.10 Discuss how the number of quantum wells can be optimized to minimize the threshold current density at a given threshold gain value using Fig. 10.24.

10.11 Calculate the conduction, heavy-hole and light-hole band-edge energies for In$_{1-x}$Ga$_x$As/In$_{0.52}$Al$_{0.48}$As quantum-well structures where In$_{0.52}$Al$_{0.48}$As is lattice-matched to the InP substrate, assuming the gallium mole fraction (a) $x = 0.37$, (b) $x = 0.47$, and (c) $x = 0.57$ in the well region. Use the physical parameters in Appendix K.

10.12 Discuss the advantages of compressively strained and tensile strained quantum-well lasers based on the effective mass, the density of states, and the optical momentum matrix elements.

10.13 Compare the normalized optical momentum matrix element in Fig. 10.32 with the analytical results in Table 9.1 in Section 9.5 using a parabolic band model.

10.14 Using the coupled-mode equations in Section 10.5 assuming only nearby-element coupling, write the explicit eigenvalues and eigenvectors for (a) three coupled waveguides and (b) four coupled waveguides. Plot the modal profiles of the supermodes similar to those in Fig. 10.37.

10.15 Derive the array pattern for the ℓ th supermode, $A_\ell(\theta)$, in (10.5.27).

10.16 Derive the oscillation condition (10.6.21) for a DFB laser. Discuss the solution for the phase δL and the gain GL.

10.17 Derive the lasing conditions for the DBR semiconductor laser in Fig. 10.40.

10.18 Compare the threshold condition for a surface-emitting laser with that for a DBR semiconductor laser. What are their differences and similarities?

10.19 Explain why the optical output power vs. injection current ($L-I$) curve of a surface-emitting laser has a negative differential quantum efficiency at a high injection level.

10.20 Compare the current density, carrier concentration, and optical modal profiles of a gain-guided surface-emitting laser (Fig. 10.42) with those of the gain-guided stripe-geometry semiconductor laser in Section 10.2 (Fig. 10.12).

REFERENCES

1. *IEEE J. Quantum Electron.*, special issue on *Semiconductor Lasers*, guest editor, P. D. Dapkus, **QE-23**, June 1987. See the Historical Section, pp. 651–692, which includes the reminiscent accounts on the invention of semiconductor lasers by R. N. Hall, M. I. Nathan, N. Holonyak, Jr., and R. H. Rediker. It also contains the previously unpublished notes of J. von Neumann, written in 1953, which outlined remarkably his ideas on "the photon-disequilibrium-amplification scheme."

2. H. C. Casey, Jr. and M. B. Panish, *Heterostructure Lasers, Part A: Fundamental Principles*, Academic, New York, 1978.

3. C. H. Henry, "The origin of quantum wells and the quantum well lasers," Chapter 1 in *Quantum Well Lasers*, P. S. Zory, Jr., Ed. Academic, New York, 1993.

4. R. N. Hall, G. E. Fenner, J. D. Kingsley, T. J. Soltys, and R. O. Carlson, "Coherent light emission from GaAs junctions," *Phys. Rev. Lett.* **9**, 366–368 (1962).

5. M. I. Nathan, W. P. Dumke, G. Burns, F. H. Dill, Jr., and G. Lasher, "Stimulated emission of radiation from GaAs p-n junctions," *Appl. Phys. Lett.* **1**, 62–64 (1962).

6. N. Holonyak, Jr. and S. F. Bevacqua, "Coherent (visible) light emission from $Ga(As_{1-x}P_x)$ junctions," *Appl. Phys. Lett.* **1**, 82–83 (1962).

7. T. M. Quist, R. H. Rediker, R. J. Keyes, W. E. Krag, B. Lax, A. L. McWhorter, and H. J. Zeiger, "Semiconductor maser of GaAs," *Appl. Phys. Lett.* **1**, 91–92 (1962).

8. H. Kroemer, "A proposed class of heterojunction injection lasers," *Proc. IEEE* **51**, 1782–1783 (1963).

9. H. Kressel and H. Nelson, "Close confinement gallium arsenide p-n junction lasers with reduced optical loss at room temperature," *RCA Rev.* **30**, 106–113 (1969).

10. Zh. I. Alferov, V. M. Andreev, E. L. Portnoi, and M. K. Trukan, "AlAs–GaAs heterojunction injection lasers with a low room-temperature threshold," *Sov. Phys. Semicond.* **3**, 1107–1110 (1970).

11. I. Hayashi, M. B. Panish, P. W. Foy, and S. Sumski, "Junction lasers which operate continuously at room temperature," *Appl. Phys. Lett.* **17**, 109–111 (1970).

12. R. Dingle, W. Wiegmann, and C. H. Henry, "Quantum states of confined carriers in very thin $Al_xGa_{1-x}As$–$GaAs$–$Al_xGa_{1-x}As$ heterostructures," *Phys. Rev. Lett.* **33**, 827–830 (1974).

13. R. D. Dupuis, P. D. Dapkus, N. Holonyak, Jr., E. A. Rezek, and R. Chin, "Room-temperature laser operation of quantum-well $Ga_{1-x}Al_xAs$–$GaAs$ laser diodes grown by metal organic chemical vapor deposition," *Appl. Phys. Lett.* **32**, 295–297 (1978).

14. R. D. Dupuis, P. D. Dapkus, N. Holonyak, Jr., and R. M. Kolbas, "Continuous room-temperature multiple-quantum-well $Al_xGa_{1-x}As$–$GaAs$ injection lasers grown by metalorganic chemical vapor deposition," *Appl. Phys. Lett.* **35**, 221–223 (1979).

15. N. Holonyak, Jr., R. M. Kolbas, R. D. Dupuis, and P. D. Dapkus, "Quantum-well heterostructure lasers," *IEEE J. Quantum Electron.* **QE-16**, 170–186 (1980).

16. W. T. Tsang, "A graded-index waveguide separate-confinement laser with very low threshold and a narrow Gaussian beam," *Appl. Phys. Lett.* **39**, 134–137 (1981).

17. E. Yablonovitch and E. O. Kane, "Reduction of lasing threshold current density by the lowering of valence band effective mass," *J. Lightwave Technol.* **LT-4**, 504–506 (1986); "Erratum" **LT-4**, 961 (1986).

18. A. R. Adams, "Band-structure engineering for low-threshold high-efficiency semiconductor lasers," *Electron. Lett.* **22**, 249–250 (1986).

19. H. K. Choi and C. A. Wang, "InGaAs/AlGaAs strained quantum well diode lasers with extremely low threshold current density and high efficiency," *Appl. Phys. Lett.* **57**, 321–323 (1990).

20. R. L. Williams, M. Dion, F. Chatenoud, and K. Dzurko, "Extremely low threshold current strained InGaAs/AlGaAs lasers by molecular beam epitaxy," *Appl. Phys. Lett.* **58**, 1816–1818 (1991).

21. W. T. Tsang, Vol. Ed., *Lightwave Communications Technology*, in R. K. Willardson and A. C. Beer, Eds., *Semiconductors and Semimetals*, Vol. 22, Part A-E, Academic, New York, 1985.

22. Y. Arakawa and A. Yariv, "Quantum well lasers: gain, spectra, dynamics," *IEEE J. Quantum Electron.* **QE-22**, 1887–1899 (1986).

23. E. Kapon, "Quantum wire lasers," *Proc. IEEE* **80**, 398–410 (1992).

24. A. Yariv, "Scaling laws and minimum threshold currents for quantum-confined semiconductor lasers," *Appl. Phys. Lett.* **53**, 1033–1035 (1988).

25. W. Wegscheider, L. N. Pfeiffer, M. M. Dignam, A. Pinczuk, K. W. West, S. L. McCall, and R. Hull, "Lasing from excitons in quantum wires," *Phys. Rev. Lett.* **71**, 4071–4074 (1993).

26. K. Iga, S. Ishkawa, S. Ohkouchi, and T. Nishimura, "Room temperature pulsed oscillation of GaAlAs/GaAs surface emitting junction laser," *IEEE J. Quantum Electron.* **QE-21**, 315–320 (1985).

27. S. W. Corzine, R. S. Geels, R. H. Yan, J. W. Scott, L. A. Coldren, and P. L. Gourley, "Efficient, narrow-linewidth distributed-Bragg-reflector surface-emitting laser with periodic gain," *IEEE Photon. Technol. Lett.* **1**, 321–323 (1989).

28. G. A. Evans and J. M. Hammer, Eds., *Surface Emitting Semiconductor Lasers and Arrays*, Academic, San Diego, 1993.

29. S. L. McCall, A. F. J. Levi, R. E. Slusher, S. J. Pearton, and R. A. Logan, "Whispering-gallery mode microdisk lasers," *Appl. Phys. Lett.* **60**, 289–291 (1992).

30. A. F. J. Levi, R. E. Slusher, S. L. McCall, T. Tanbun-Ek, D. L. Coblentz, and S. J. Pearton, "Room temperature operation of microdisc lasers with submilliamp threshold current," *Electron. Lett.* **28**, 1010–1011 (1992).

31. Y. Yamamoto and R. E. Slusher, "Optical processes in microcavities," *Physics Today* **46**, 66–73 (1993).

32. M. Yano, H. Imai, and M. Takusagawa, "Analysis of electrical, threshold, and temperature characteristics of InGaAsP/InP double-heterojunction lasers," *IEEE J. Quantum Electron.* **QE-17**, 1954–1963 (1981).

33. G. P. Agrawal and N. K. Dutta, *Long-Wavelength Semiconductor Lasers*, Van Nostrand Reinhold, New York, 1986.

34. H. C. Casey, Jr. and M. B. Panish, *Heterostructure Lasers, Part B: Materials and Operating Characteristics*, Academic, Orlando, FL, 1978.

35. A. R. Goodwin, J. R. Peters, M. Pion, G. H. B. Thomson, and J. E. A. Whiteway, "Threshold temperature characteristics of double heterostructure $Ga_{1-x}Al_xAs$ lasers," *J. Appl. Phys.* **46**, 3126–3131 (1975).

36. P. J. Anthony, J. R. Pawlik, V. Swaminathan, and W. T. Tsang, "Reduced threshold current temperature dependence in double heterostructure lasers due to separate *p-n* and heterojunctions," *IEEE J. Quantum Electron.* **QE-19**, 1030–1034 (1983).

37. C. J. Nuese and J. I. Pankove, "Light-emitting diodes—LEDs," Chapter 2 in J. I. Pankov, Ed., *Display Devices*, Springer, Berlin, 1980.

38. S. Sze, *Physics of Semiconductor Devices*, Wiley, New York, 1981.

39. J. Wilson and J. F. B. Hawkes, *Optoelectronics: An Introduction*, Prentice-Hall, Englewood Cliffs, NJ, 1983.

40. C. A. Burrus and B. I. Miller, "Small-area, double heterostructure aluminum–gallium arsenide electroluminescent diode sources for optical-fiber transmission lines," *Opt. Commun.* **4**, 307–309 (1971).

41. M. Ettenberg, H. Kressel, and J. P. Wittke, "Very high radiance edge-emitting LED," *IEEE J. Quantum Electron.* **QE-12**, 360–364 (1976).

42. J. O'Gorman, S. L. Chuang, and A. F. J. Levi, "Carrier pinning by mode fluctuations in laser diodes," *Appl. Phys. Lett.* **62**, 1454–1456 (1993).

43. S. L. Chuang, J. O'Gorman, and A. F. J. Levi, "Amplified spontaneous emission and carrier pinning in laser diodes," *IEEE J. Quantum Electron.*, special issue on *Semiconductor Lasers* **29**, 1631–1639 (1993).

44. B. Zee, "Broadening mechanism in semiconductor (GaAs) lasers: limitations to single mode power emission," *IEEE J. Quantum Electron.* **QE-14**, 727–736 (1978).

45. B. W. Hakki and T. L. Paoli, "Gain spectra in GaAs double-heterostructure injection lasers," *J. Appl. Phys.* **46**, 1299–1306 (1975).

46. G. Lasher and F. Stern, "Spontaneous and stimulated recombination radiation in semiconductors," *Phys. Rev.* **133**, A553–A563 (1964).

47. G. H. F. Thompson, *Physics of Semiconductor Laser Devices*, Wiley, New York, 1980.

48. M. J. Adams, J. V. Collins, and I. D. Henning, "Analysis of semiconductor laser optical amplifiers," *IEE Proc.* **132**, pt. J, 58–63 (1985).

49. H. Yonezu, I. Sakuma, K. Kobayashi, T. Kamejima, M. Ueno, and Y. Nannichi, "A GaAs–Al$_x$Ga$_{1-x}$As double heterostructure planar stripe laser," *Jpn. J. Appl. Phys.* **12**, 1585–1592, (1973).

50. B. W. Hakki, "Carrier and gain spatial profiles in GaAs stripe geometry lasers," *J. Appl. Phys.* **44**, 5021–5029 (1973).

51. D. D. Cook and F. R. Nash, "Gain-induced guiding and astigmatic output beam of GaAs lasers, *J. Appl. Phys.* **46**, 1660–1672 (1975).

52. N. Chinone, "Nonlinearity in power-output-current characteristics of stripe-geometry injection lasers," *J. Appl. Phys.* **48**, 3237–3243 (1977).

53. W. T. Tsang, "The effects of lateral current spreading, carrier out-diffusion, and optical mode losses on the threshold current density of GaAs–Al$_x$Ga$_{1-x}$As stripe-geometry DH lasers," *J. Appl. Phys.* **49**, 1031–1044 (1978).

54. T. L. Paoli, "Waveguiding in a stripe-geometry junction laser," *IEEE J. Quantum Electron.* **QE-13**, 662–668 (1977).

55. W. Streifer, D. R. Scifres, and R. D. Burnham, "Analysis of gain-induced waveguiding in stripe-geometry diode lasers," *IEEE J. Quantum Electron.* **QE-14**, 418–427 (1978).

56. W. B. Joyce, "Current-crowded carrier confinement in double-heterostructure lasers," *J. Appl. Phys.* **51**, 2394–2401 (1980).

57. W. Streifer, R. D. Burnham, and D. R. Scifres, "An analytic study of (GaAl)As gain guided lasers at threshold," *IEEE J. Quantum Electron.* **QE-18**, 856–864 (1982).

58. I. P. Kaminow, L. W. Stulz, J. S. Ko, A. G. Dentai, R. E. Nahory, J. C. DeWinter, and R. L. Hartman, "Low-threshold InGaAsP ridge waveguide lasers at 1.3 μm," *IEEE J. Quantum Electron.* **QE-19**, 1312–1319 (1983).

59. H. Hirao, M. S. Tsuji, K. Mizuishi, A. Doi, and M. Nakamura, "Long wavelength InGaAsP/InP lasers for optical fiber communication systems," *J. Opt. Commun.* **1**, 10–14 (1980).

60. H. Kawaguchi and T. Kawakami, "Transverse-mode control in an injection laser by a strip-loaded waveguide," *IEEE J. Quantum Electron.* **QE-13**, 556–560 (1977).

61. M. Asada, A. Kameyama, and Y. Suematsu, "Gain and intervalence band absorption in quantum well lasers," *IEEE J. Quantum Electron.* **20**, 745–753 (1984).

62. K. J. Vahala and C. E. Zah, "Effect of doping on the optical gain and the spontaneous noise enhancement factor in quantum well amplifiers and lasers studied by simple analytical expressions," *Appl. Phys. Lett.* **52**, 1945–1947 (1988).

63. K. Y. Lau, "Ultralow threshold quantum well lasers," Chapter 4, and "Dynamics of quantum well lasers," Chapter 5, in *Quantum Well Lasers*, P. S. Zory, Jr., Ed., Academic, San Diego (1993).

64. A. Ghiti, M. Silver, and E. P. O'Reilly, "Low threshold current and high differential gain in ideal tensile- and compressive-strained quantum-well lasers," *J. Appl. Phys.* **71**, 4626–4628 (1992).

65. M. Asada and Y. Suematsu, "The effects of loss and nonradiative recombination on the temperature dependence of threshold current in 1.5–1.6 μm GaInAsP/InP lasers," *IEEE J. Quantum Electron.* **QE-19**, 917–923 (1983).

66. B. Sermage, D. S. Chemla, D. Sivco, and A. Y. Cho, "Comparison of Auger recombination in GaInAs–AlInAs multiple quantum well structure and in bulk GaInAs," *IEEE J. Quantum Electron.* **QE-22**, 774–780 (1986).

67. M. Asada, "Intraband relaxation time in quantum-well lasers," *IEEE J. Quantum Electron.* **25**, 2019–2026 (1989).

68. E. Zielinski, H. Schweizer, S. Hausser, R. Stuber, M. H. Pilkuhn, and G. Weimann, "Systematics of laser operation in GaAs/AlGaAs multiquantum well heterostructures," *IEEE J. Quantum Electron.* **QE-23**, 966–976 (1987).

69. E. Zielinski, F. Keppler, K. Streubel, F. Scholz, R. Sauer, and W. T. Tsang, "Optical gain in strain-free and strained layer $Ga_x In_{1-x} As / InP$ superlattices," *Superlattices and Microstructures* **5**, 555–559 (1989).

70. P. Blood, S. Colak, and A. I. Kucharska, "Influence of broadening and high-injection effects on GaAs–AlGaAs quantum well lasers," *IEEE J. Quantum Electron.* **24**, 1593–1604 (1988).

71. A. I. Kucharska and D. J. Robbins, "Lifetime broadening in GaAs–AlGaAs quantum well lasers," *IEEE J. Quantum Electron.* **26**, 443–448, 1990.

72. P. W. A. McIlroy, A. Kurobe, and Y. Uematsu, "Analysis and application of theoretical gain curves to the design of multi-quantum-well lasers," *IEEE J. Quantum Electron.* **QE-21**, 1958–1963 (1985).

73. A. Kurobe, H. Furuyama, S. Naritsuka, N. Sugiyama, Y. Kokubun, and M. Nakamura, "Effects of well number, cavity length, and facet reflectivity on the reduction of threshold current of GaAs/AlGaAs multiquantum well lasers," *IEEE J. Quantum Electron.* **QE-24**, 635–640 (1988).

74. M. Rosenzweig, M.Möhrle, H. Düser, and H. Venghaus, "Threshold-current analysis of InGaAs–InGaAsP multiquantum-well separate-confinement lasers," *IEEE J. Quantum Electron.* **27**, 1804–1811 (1991).

75. J. S. Osinski, K. M. Dzurko, S. G. Hummel, and P. D. Dapkus, "Optimization of stripe width for low-threshold operation of quantum well laser diodes," *Appl. Phys. Lett.* **56**, 2487–2489 (1990).

76. J. S. Osinski, Y. Zou, P. Grodzinski, A. Mathur, and P. D. Dapkus, "Low-threshold-current-density 1.5 μm laser using compressively strained InGaAsP quantum wells," *IEEE Photon. Technol. Lett.* **4**, 10–13 (1992).

77. A. Yariv, *Quantum Electronics*, 3d ed., Wiley, New York, 1989.

78. M. Yamada, K. Tabata, S. Ogita, and M. Yamagishi, "Calculation of lasing gain and threshold current in GaAs–AlGaAs multi-quantum-well lasers," *Trans. IECE Jpn.* **E68**, 102–108 (1985).

79. G. C. Osbourn, "$In_xGa_{1-x}As-In_yGa_{1-y}As$ strained-layer superlattices: A proposal for useful, new electronic materials," *Phys. Rev. B* **27**, 5126–5128 (1983).

80. T. P. Pearsall, Vol. Ed., *Strained Layer Superlattices: Physics*, Vol. 32, 1990; and *Strained Layer Superlattices: Materials Science and Technology*, Vol. 33, 1991, in R. K. Willardson and A. C. Beer, Eds., *Semiconductor and Semimetals*, Academic, New York.

81. E. Yablonovitch and E. O. Kane, "Band structure engineering of semiconductor lasers for optical communications," *J. Lightwave Technol.* **6**, 1292–1299 (1988).

82. P. J. A. Thijs, L. F. Tiemeijer, P. I. Kuindersma, J. J. M. Binsma, and T. Van Dongen, "High-performance 1.5 μm wavelength InGaAs–InGaAsP strained quantum well lasers and amplifiers," *IEEE J. Quantum Electron.* **27**, 1426–1439 (1991).

83. C. E. Zah, R. Bhat, F. J. Favire, Jr., S. G. Menocal, N. C. Andreadakis, K. W. Cheung, D. M. D. Hwang, M. A. Koza, and T. P. Lee, "Low-threshold 1.5 μm compressive-strained multiple- and single-quantum-well lasers," *IEEE J. Quantum Electron.* **27**, 1440–1450 (1991).

84. K. J. Beernink, P. K. York, and J. J. Coleman, "Dependence of threshold current density on quantum well composition for strained-layer InGaAs–GaAs lasers by metalorganic chemical vapor deposition," *Appl. Phys. Lett.* **55**, 2585–2587 (1989).

85. T. Tanbun-Ek, R. A. Logan, H. Temkin, S. N. G. Chu, N. A. Olsson, A. M. Sergent, and K. W. Wecht, "Growth and characterization of continuously graded index separate confinement heterostructure (GRIN-SCH) InGaAs–InP long wavelength strained layer quantum-well lasers by metalorganic vapor phase epitaxy," *IEEE J. Quantum Electron.* **26**, 1323–1327 (1990).

86. C. E. Zah, R. Bhat, K. W. Cheung, N. C. Andreadakis, F. J. Favire, S. G. Menocal, E. Yablonovitch, D. M. Hwang, M. Koza, T. J. Gmitter, and T. P. Lee, "Low-threshold (≤ 92 A/cm^2) 1.6 μm strained-layer single-quantum-well laser diodes optically pumped by a 0.8-μm laser diode," *Appl. Phys. Lett.* **57**, 1608–1609 (1990).

87. H. Temkin, N. K. Dutta, T. Tanbun-Ek, R. A. Logan, and A. M. Sergent, "InGaAs/InP quantum well lasers with sub-mA threshold current," *Appl. Phys. Lett.* **57**, 1610–1612 (1990).

88. W. T. Tsang, M. C. Wu, T. Tanbun-Ek, R. A. Logan, S. N. G. Chu, and A. M. Sergent, "Low-threshold and high-power output 1.5-μm InGaAs/InGaAsP separate confinement multiple quantum well laser grown by chemical beam epitaxy," *Appl. Phys. Lett.* **57**, 2065–2067 (1990).

89. A. Kasukawa, R. Bhat, C. E. Zah, M. A. Koza, and T. P. Lee, "Very low threshold current density 1.5-μm GaInAs/AlGaInAs graded-index separate-confinement-heterostructure strained quantum well laser diodes growth by organometallic chemical vapor deposition," *Appl. Phys. Lett.* **59**, 2486–2488 (1991).

90. T. Tanbun-Ek, R. A. Logan, N. A. Olsson, H. Temkin, A. M. Sergent, and K. W. Wecht, "High power output 1.48–1.51 μm continuously graded index separate confinement strained quantum well lasers," *Appl. Phys. Lett.* **57**, 224–226 (1990).

91. M. Joma, H. Horikawa, Y. Matsui, and T. Kamijoh, "High-power 1.48-μm multiple quantum-well lasers with strained quaternary wells entirely grown by metalorganic vapor phase epitaxy," *Appl. Phys. Lett.* **58**, 2220–2222 (1991).

92. A. Moser, A. Oosenbrug, E. E. Latta, Th. Forster, and M. Gasser, "High-power operation of strained InGaAs/AlGaAs single quantum well lasers," *Appl. Phys. Lett.* **59**, 2642–2644 (1991).

93. H. K. Choi, C. A. Wang, D. F. Kolesar, R. L. Aggarwal, and J. N. Walpole, "Higher-power, high-temperature operation of AlInGaAs–AlGaAs strained single-quantum-well diode lasers," *IEEE Photon. Technol. Lett.* **3**, 857–859 (1991).

94. H. Temkin, T. Tanbun-Ek, R. A. Logan, D. A. Cebula, and A. M. Sergent, "High temperature operation of lattice matched and strained InGaAs–InP quantum well lasers," *IEEE Photon. Technol. Lett.* **3**, 100–102 (1991).

95. P. L. Derry, R. J. Fu, C. S. Hong, E. Y. Chan, and L. Figueroa, "Analysis of the high temperature characteristics of InGaAs–AlGaAs strained quantum-well lasers," *IEEE J. Quantum Electron.* **28**, 2698–2705 (1992).

96. H. Temkin, D. Coblentz, R. A. Logan, J. P. van der Ziel, T. Tanbun-Ek, R. D. Yadvish, and A. M. Sergent, "High temperature characteristics of InGaAsP/InP laser structures," *Appl. Phys. Lett.* **62**, 2402–2404 (1993).

97. H. Nobuhara, K. Tanaka, T. Yamamoto, T. Machida, T. Fujii, and K. Wakao, "High-temperature operation of InGaAs/InGaAsP compressive-strained QW lasers with low threshold currents," *IEEE Photon. Technol. Lett.* **5**, 961–962 (1993).

98. D. Ahn and S. L. Chuang, "Optical gain in a strained-layer quantum-well laser," *IEEE J. Quantum Electron.* **24**, 2400–2406 (1988).

99. T. C. Chong and C. G. Fonstad, "Theoretical gain of strained-layer semiconductor lasers in the large strain regime," *IEEE J. Quantum Electron.* **25**, 171–178 (1989).

100. E. P. O'Reilly and A. R. Adams, "Band-structure engineering in strained semiconductor lasers," *IEEE J. Quantum Electron.* **30**, 366–379 (1994).

101. S. W. Corzine, R. H. Yan, and L. A. Coldren, "Theoretical gain in strained InGaAs/AlGaAs quantum wells including valence-band mixing effects," *Appl. Phys. Lett.*, **57**, 2835–2837 (1990).

102. N. K. Dutta, J. Lopata, D. L. Sivco, and A. Y. Cho, "Temperature dependence of threshold of strained quantum well lasers," *Appl. Phys. Lett.* **58**, 1125–1127 (1991).

103. I. Suemune, "Theoretical study of differential gain in strained quantum well structures," *IEEE J. Quantum Electron.* **27**, 1149–1159 (1991).

104. E. P. O'Reilly, G. Jones, A. Giti, and A. R. Adams, "Improved performance due to suppression of spontaneous emission in tensile-strain semiconductor lasers," *Electron. Lett.* **27**, 1417–1419 (1991).

105. L. F. Lester, S. D. Offsey, B. K. Ridley, W. J. Schaff, B. A. Foreman, and L. F. Eastman, "Comparison of the theoretical and experimental differential gain in strained layer InGaAs/GaAs quantum well lasers," *Appl. Phys. Lett.* **59**, 1162–1164 (1991).

106. J. W. Matthews and A. E. Blakeslee, "Defects in epitaxial multilayers," *J. Crystal Growth* **27**, 118 (1974).

107. R. People and J. C. Bean, "Calculation of critical layer thickness versus lattice mismatch for Ge_xSi_{1-x}/Si strained-layer heterostructures," *Appl. Phys. Lett.* **47**, 322–324 (1985); "Erratum," *Appl. Phys. Lett.* **49**, 229 (1986).

108. M. G. A. Bernard and G. Duraffourg, "Laser conditions in semiconductors," *Phys. Status Solidi* **1**, 699–703 (1961).

109. S. L. Chuang, "Efficient band-structure calculations of strained quantum wells," *Phys. Rev. B* **43**, 9649–9661 (1991).

110. C. Y. P. Chao and S. L. Chuang, "Spin–orbit-coupling effects on the valence-band structure of strained semiconductor quantum wells," *Phys. Rev. B* **46**, 4110–4122 (1992).

111. T. Tanbun-Ek, N. A. Olsson, R. A. Logan, K. W. Wecht, and A. M. Sergent, "Measurements of the polarization dependent of the gain of strained multiple quantum well InGaAs–InP lasers," *IEEE Photon Technol. Lett.* **3**, 103–105 (1991).

112. N. B. Patel, J. E. Ripper, and P. Brosson, "Behavior of threshold current and polarization of stimulated emission of GaAs injection lasers under uniaxial stress," *IEEE J. Quantum Electron.* **QE-9**, 338–341 (1973).

113. J. M. Liu and Y. C. Chen, "Digital optical signal processing with polarization-bistable semiconductor lasers," *IEEE J. Quantum Electron.* **QE-21**, 298–306 (1985).

114. J. S. Osinski, P. Grodzinski, Y. Zou, and P. D. Dapkus, "Threshold current analysis of compressive strain (0–1.8%) in low-threshold, long-wavelength quantum well lasers," *IEEE J. Quantum Electron.* **29**, 1576–1585 (1993).

115. Y. Zou, J. S. Osinski, P. Grodzinski, P. D. Dapkus, W. Rideout, W. F. Sharfin, and F. D. Crawford, "Effect of Auger recombination and differential gain on the temperature sensitivity of 1.5-μm quantum well lasers," *Appl. Phys. Lett.* **62**, 175–177 (1993).

116. G. Fuchs, C. Schiedel, A. Hangleiter, V. Harle, and F. Scholz, "Auger recombination in strained and unstrained InGaAs/InGaAsP multiple quantum-well lasers," *Appl. Phys. Lett.* **62**, 396–398 (1993).

117. M. C. Wang, K. Kash, C. E. Zah, R. Bhat, and S. L. Chuang, "Measurement of nonradiative Auger and radiative recombination rates in strained-layer quantum-well systems," *Appl. Phys. Lett.* **62**, 166–168 (1993).

118. Y. Zou, J. S. Osinski, P. Grodzinski, P. D. Dapkus, W. C. Rideout, W. F. Sharfin, and F. D. Crawford, "Experimental study of Auger recombination, gain and temperature sensitivity of 1.5 μm compressively strained semiconductor lasers," *IEEE J. Quantum Electron.* **29**, 1565–1575 (1993).

119. J. O'Gorman, A. F. J. Levi, T. Tanbun-Ek, D. L. Coblentz, and R. A. Logan, "Temperature dependence of long wavelength semiconductor lasers," *Appl. Phys. Lett.* **60**, 1058–1060 (1992).

120. D. Morris, B. Deveaud, A. Regreny, and P. Auvray, "Electron and hole capture in multiple-quantum-well structures," *Phys. Rev. B* **47**, 6819–6822 (1993).

121. R. Nagarajan, T. Fukushima, M.Ishikawa, J. E. Bowers, R. S. Geels, and L. A. Coldren, "Transport limits in high-speed quantum-well lasers: experiment and theory," *IEEE Photon. Technol. Lett.* **4**, 121–123 (1992).

122. S. Schmitt-Rink, C. Ell, and H. Haug, "Many-body effects in the absorption, gain, and luminescence spectra of semiconductor quantum-well structures," *Phys. Rev. B* **33**, 1183–1189 (1986).

123. M. F. Pereira, Jr., S. W. Koch, and W. W. Chow, "Many-body effects in the gain spectra of strained quantum wells," *Appl. Phys. Lett.* **59**, 2941–2943 (1991).

124. D. R. Scifres, W. Striefer, and R. D. Burnham, "High-power coupled-multiple-stripe phase-locked injection laser," *Appl. Phys. Lett.* **34**, 259–261 (1979).

125. D. R. Scifres, R. D. Burnham, C. Lindstrom, W. Streifer, and T. L. Paoli, "Phase-locked (GaAl)As laser-emitting 1.5-W cw per mirror," *Appl. Phys. Lett.* **42**, 645–647 (1983).

126. J. K. Butler, D. E. Ackley, and D. Botez, "Coupled-mode analysis of phase-locked injection laser arrays," *Appl. Phys. Lett.* **44**, 293–295, 1984; and "Erratum," 44, p. 935 (1984).

127. S. Wang, J. Z. Wilcox, M. Jansen, and J. J. Yang, "In-phase locking in diffraction-coupled phased-array diode lasers," *Appl. Phys. Lett.* **48**, 1770–1772 (1986).

128. W. Streifer, M. Osiński, D. R. Scifres, D. F. Welch, and P. S. Cross, "Phased-array lasers with a uniform, stable supermode," *Appl. Phys. Lett.* **49**, 1496–1498 (1986).

129. W. Streifer, P. S. Cross, D. F. Welch, and D. R. Scifres, "Analysis of a Y-junction semiconductor laser array," *Appl. Phys. Lett.* **49**, 58–60 (1986).

130. E. Kapon, Z. Rav-Noy, S. Margalit, and A. Yariv, "Phase-locked arrays of buried-ridge InP/InGaAsP diode lasers," *J. Lightwave Technol.* **LT-4**, 919–925 (1986).

131. G. R. Hadley, J. P. Hohimer, and A. Owyoung, "High-order ($\nu > 10$) eigen-modes in ten-stripe gain-guided diode laser arrays," *Appl. Phys. Lett.* **49**, 684–686 (1986).

132. W. K. Marshall and J. Katz, "Direct analysis of gain-guided phase-locked semiconductor laser arrays," *IEEE J. Quantum Electron.* **QE-22**, 827–832 (1986).

133. I. J. Mawst, D. Botez, T. J. Roth, G. Peterson, and J. J. Yang, "Diffraction-coupled, phase-locked arrays of antiguided, quantum-well lasers grown by metalorganic chemical vapour deposition," *Electron. Lett.* **24**, 958–959 (1988).

134. D. Botez, L. Mawst, P. Hayashida, G. Peterson, and T. J. Roth, "High-power, diffraction-limited-beam operation from phase-locked diode-laser arrays of closely spaced "leaky" waveguides (antiguides)," *Appl. Phys. Lett.* **53**, 464–466 (1988).

135. G. R. Hadley, "Index-guided arrays with a large index step," *Opt. Lett.* **14**, 308–310 (1989).

136. D. Botez, L. J. Mawst, G. L. Peterson, and T. J. Roth, "Phase-locked arrays of antiguides: modal content and discrimination," *IEEE J. Quantum Electron.* **26**, 482–495 (1990).

137. M. Sakamoto, J. G. Endriz, and D. R. Scifres, "120W CW output power from a monolithic AlGaAs(800nm) laser diode array mounted on a diamond heat sink," *Electron. Lett.* **28**, 197–199 (1992).

138. M. Abramowitz and I. A. Stegun, Eds., *Handbook of Mathematical Functions*, Dover, New York, 1972.

139. Y. Suematsu, K. Kishino, S. Arai, and F. Koyama, "Dynamic single-mode semiconductor lasers with a distributed reflector," Chapter 4 in R. K. Willardson and A. C. Beer, Eds., *Semiconductors and Semimetals*, Vol. 22, Academic, San Diego, 1985.

140. H. Kogelnik and C. V. Shank, "Stimulated emission in a periodic structure," *Appl. Phys. Lett.* **18**, 152–154 (1971).

141. S. Wang, "Design consideration of the DBR injection laser and the waveguiding structure for integrated optics," *IEEE J. Quantum Electron.* **QE-13**, 176–186 (1977).

142. J. L. Jewell, J. P. Harbison, A. Scherer, Y. H. Lee, and L. T. Florez, "Vertical-cavity surface-emitting lasers: design, growth, fabrication, characterization," *IEEE J. Quantum Electron.* **27**, 1332–1346 (1991).

143. R. S. Geels, S. W. Corzine, and L. A. Coldren, "InGaAs vertical-cavity surface-emitting lasers," *IEEE J. Quantum Electron.* **27**, 1359–1367 (1991).

144. C. J. Chang-Hasnain, J. P. Harbison, G. Hasnain, A. C. Von Lehmen, L. T. Florez, and N. G. Stoffel, "Dynamic, polarization, and transverse mode characteristics of vertical cavity surface emitting lasers," *IEEE J. Quantum Electron.* **27**, 1402–1409 (1991).

145. J. W. Scott, R. S. Geels, S. W. Corzine, and L. A. Coldren, "Modeling temperature effects and spatial hole burning to optimize vertical-cavity surface-emitting laser performance," *IEEE J. Quantum Electron.* **29**, 1295–1308 (1993).

146. K. Tai, K. F. Huang, C. C. Wu, and J. D. Wynn, "Continuous wave visible InGaP/InGaAlP quantum well surface emitting laser diodes," *Electron. Lett.* **29**, 1314–1316 (1993).

147. H. Soda, K. Iga, C. Kitahara, and Y. Suematsu, "GaInAsP/InP surface emitting injection lasers," *Jpn. J. Appl. Phys.* **18**, 2329–2330 (1979).

148. I. Melngailis, "Longitudinal injection-plasma laser of InSb," *Appl. Phys. Lett.* **6**, 59–60 (1965).

149. R. D. Burnham, D. R. Scifres, and W. Striefer, "Single-heterostructure distributed-feedback GaAs-diode lasers," *IEEE J. Quantum Electron.* **QE-11**, 439–449 (1975).

150. Zh. I. Alferov, V. M. Andreyev, S. A. Gurevich, R. F. Kazarinov, V. R. Larionov, M. N. Mizerov, and E. L. Portnoy, "Semiconductor lasers with the light output through the diffraction grating on the surface of the waveguide layer," *IEEE J. Quantum Electron.* **QE-11**, 449–451 (1975).

151. A. J. Springthorpe, "A novel double-heterostructure *p-n* junction laser," *Appl. Phys. Lett.* **31**, 524–525 (1977).

152. K. Iga and F. Koyama, "Vertical-cavity surface emitting lasers and arrays," Chapter 3 in G. A. Evans and J. M. Hammer, Eds., *Surface Emitting Semiconductor Lasers and Arrays*, Academic, San Diego, 1993.

153. N. K. Dutta, "Analysis of current spreading, carrier diffusion, and transverse mode guiding in surface emitting lasers," *J. Appl. Phys.* **68**, 1961–1963 (1990).

154. G. Hasnain, K. Tai, L. Yang, Y. H. Wang, R. J. Fischer, J. D. Wynn, B. Weir, N. K. Dutta, and A. Y. Cho, "Performance of gain-guided surface emitting

lasers with semiconductor distributed Bragg reflectors," *IEEE J. Quantum Electron.* **27**, 1377–1385 (1991).

155. R. Michalzik and K. J. Ebeling, "Modeling and design of proton-implanted ultralow-threshold vertical-cavity laser diodes," *IEEE J. Quantum Electron.* **29**, 1963–1974 (1993).

156. W. Nakwaski and M. Osiński, "Thermal analysis of GaAs–AlGaAs etched-well surface-emitting double-heterostructure lasers with dielectric mirrors," *IEEE J. Quantum Electron.* **29**, 1981–1995 (1993).

PART IV

Modulation of Light

11

Direct Modulation
of Semiconductor Lasers

For a semiconductor laser, the output power of light intensity P increases linearly with the injection current above threshold, as discussed in Section 10.1, Eq. (10.1.31):

$$P = \frac{\hbar\omega}{q}\eta_i\frac{\ln(1/R)}{\alpha L + \ln 1/R}(I - I_{th})$$

Therefore, for an injection current with a dc component and a small signal ac modulation,

$$I = I_0 + i(t)$$

we expect the optical output power to have corresponding dc and ac components (Fig. 11.1):

$$P(t) = P_0 + p(t)$$

In this chapter, we study the direct current modulation of diode lasers and some intrinsic effects, such as relaxation oscillations, modulation speed, and the laser linewidth theory. Our goal is to understand how the laser light output characteristics vary as we increase the modulation frequency of the injection current.

11.1 RATE EQUATIONS AND LINEAR GAIN ANALYSIS

Assume that only one mode is lasing; then the rate equations for the carrier density N ($1/\text{cm}^3$) and the photon density S ($1/\text{cm}^3$) can be written as [1–4]

$$\frac{dN}{dt} = \frac{J}{qd} - \frac{N}{\tau} - vg(N)S \tag{11.1.1}$$

$$\frac{dS}{dt} = \Gamma vg(N)S - \frac{S}{\tau_p} + \beta R_{sp} \tag{11.1.2}$$

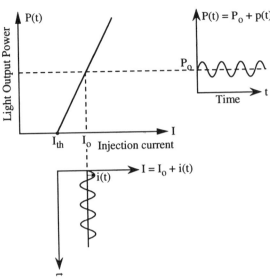

Figure 11.1. For an injection current $I = I_0 + i(t)$, the optical output power is $P(t) = P_0 + p(t)$, where (I_0, P_0) is the bias point for the direct current modulation of the semiconductor laser.

where J = the injection current density (A/cm^2)

q = a unit charge $(1.6 \times 10^{-19}$ Coulomb)

d = the thickness of the active region (cm)

τ = carrier lifetime (s)

$v = c/n_r$ is the group velocity of light (cm/s)

$g(N)$ = the gain coefficient (1/cm)

Γ = optical confinement factor

τ_p = photon lifetime (s)

β = the spontaneous emission factor

R_{sp} = the spontaneous emission rate per unit volume $(cm^{-3} s^{-1})$

In the first rate equation (11.1.1) for the carrier density N, the first term J/qd is the injected number of carriers per unit volume per second. The current density J is the current I divided by the cross-section area of carrier injection. The recombination rate

$$R(N) = \frac{N}{\tau} \qquad (11.1.3)$$

accounts for the carrier loss due to radiative and nonradiative recombinations. The time constant τ generally depends on N, but is assumed to be a constant for simplicity. The term $vg(N)S$ is the carrier loss due to stimulated emissions.

In the second rate equation (11.1.2) for photon density S, the first term $\Gamma vg(N)S$, is the increasing rate of the number of photons per unit volume due to stimulated emissions. The second term, $-S/\tau_p$, is the decreasing rate of the photon density due to absorptions and transmission through the mirrors of the laser cavity,

$$\frac{1}{\tau_p} = v\left(\alpha_i + \frac{1}{L}\ln\frac{1}{R}\right) \tag{11.1.4}$$

where τ_p has the physical meaning of the average photon lifetime in the cavity. Photons disappear from the cavity via absorption processes or transmission out of the end facets. The last term, βR_{sp}, is the fraction of spontaneous emission entering the lasing mode, which is generally very small. The spontaneous emission factor β can be calculated with a simple analytical expression for index-guided laser modes with a plane phase front [5]. For a gain-guided laser with a cylindrical constant phase front, the spontaneous emission is shown to be enhanced by another factor [6, 7].

It is noted [3] that the optical confinement factor Γ appears only in (11.1.2). This is because the optical mode extends beyond the active region d by the factor Γ. The two rate equations ensure that the total carrier and photon numbers are balanced taking into account the difference in the thicknesses (d versus d/Γ).

11.1.1 Linear Gain Theory

If we assume that the photon density S is not high so that the nonlinear gain saturation effect can be ignored, we have a simplified gain model:

$$g(N) = g_0 + g'(N - N_0) \tag{11.1.5}$$

where $g_0 = g(N_0)$, and $g' = \partial g/\partial N$ is the differential gain at $N = N_0$. Assume that

$$J(t) = J_0 + j(t)$$
$$N(t) = N_0 + n(t)$$
$$S(t) = S_0 + s(t) \tag{11.1.6}$$

where $i(t)$, $n(t)$, and $s(t)$ are small signals compared with their corresponding dc values, I_0, N_0, and S_0, respectively.

dc Solutions. The photon density S_0 and the carrier density N_0 at steady state are

$$S_0 = \frac{\beta R_{\text{sp}}}{(1/\tau_p) - \Gamma v g_0} \tag{11.1.7}$$

$$N_0 = \tau \left(\frac{J_0}{qd} - v g_0 S_0 \right) \tag{11.1.8}$$

Note that for negligible βR_{sp}, the inverse photon lifetime is approximately

$$\frac{1}{\tau_p} \simeq \Gamma v g_0 \tag{11.1.9}$$

ac Analysis. The small-signal equations are

$$\frac{d}{dt} n(t) = \frac{j(t)}{qd} - \frac{n(t)}{\tau} - v \left[g' S_0 n(t) + g_0 s(t) \right] \tag{11.1.10}$$

$$\frac{d}{dt} s(t) = \Gamma v \left[g' S_0 n(t) + g_0 s(t) \right] - \frac{s(t)}{\tau_p} \tag{11.1.11}$$

For a current injection with a microwave modulation angular frequency ω

$$j(t) = \text{Re}\left[j(\omega) e^{-i\omega t} \right] \tag{11.1.12}$$

where $j(\omega)$ is the (complex) phasor of $j(t)$, we find the solution by substituting

$$n(t) = \text{Re}\left[n(\omega) e^{-i\omega t} \right]$$
$$s(t) = \text{Re}\left[s(\omega) e^{-i\omega t} \right] \tag{11.1.13}$$

into (11.1.10) and (11.1.11). We find the complex amplitude $n(\omega)$ for the carrier density

$$n(\omega) = \frac{1}{\Gamma v g' S_0} \left(-i\omega - \Gamma v g_0 + \frac{1}{\tau_p} \right) s(\omega)$$

$$\simeq \left(\frac{-i\omega}{\Gamma v g' S_0} \right) s(\omega) \tag{11.1.14}$$

where (11.1.9) has been used, and the complex magnitude $s(\omega)$ for the photon density

$$s(\omega) = \frac{\Gamma vg'S_0[j(\omega)/qd]}{D(\omega)}$$

(11.1.15)

where the frequency dependent denominator $D(\omega)$ is

$$D(\omega) = -\omega^2 - i\omega\left(\frac{1}{\tau} + vg'S_0\right) + vg'\frac{S_0}{\tau_p}$$

(11.1.16)

Its magnitude square is

$$|D(\omega)|^2 = \left(\omega^2 - vg'\frac{S_0}{\tau_p}\right)^2 + \omega^2\left(\frac{1}{\tau} + vg'S_0\right)^2$$

(11.1.17)

Since the above function depends on ω^2, if we let $y = \omega^2$, the minimum occurs at $(\partial/\partial y)|D(\omega)|^2 = 0$, we find

$$\omega_r^2 = vg'\frac{S_0}{\tau_p} - \frac{1}{2}\left(\frac{1}{\tau} + vg'S_0\right)^2$$

(11.1.18)

where the last term is usually negligible compared with the first term and

$$\omega_r \simeq \sqrt{vg'\frac{S_0}{\tau_p}}$$

(11.1.19)

is used. The relaxation frequency $f_r = \omega_r/2\pi$ is

$$f_r = \frac{1}{2\pi}\sqrt{vg'\frac{S_0}{\tau_p}}$$

(11.1.20)

which is proportional to $\sqrt{S_0}$ or the square root of the optical output power since it is linearly proportional to the photon density S_0.

The frequency response function is

$$\left|\frac{s(\omega)}{j(\omega)}\right| = \frac{\Gamma vg'S_0/qd}{|D(\omega)|}$$

$$\simeq \frac{(\Gamma\tau_p/qd)\omega_r^2}{\left[\left(\omega^2 - \omega_r^2\right)^2 + \omega^2\left(1/\tau + \tau_p\omega_r^2\right)^2\right]^{1/2}}$$

(11.1.21)

which has a flat response at low frequencies (low pass) and peaks at $\omega = \omega_r$, then rolls off as the frequency increases further. The bandwidth or 3-dB frequency f_{3dB} ($= \omega_{3dB}/2\pi$) occurs at $|D(\omega_{3dB})| = \sqrt{2}\,\omega_r^2$, or

$$\left(\omega_{3dB}^2 - \omega_r^2\right)^2 + \omega_{3dB}^2\left(\frac{1}{\tau} + \tau_p\omega_r^2\right)^2 = 2\omega_r^4 \qquad (11.1.22)$$

since $|D(\omega = 0)| = \omega_r^2$.

Example The frequency response for a distributed feedback semiconductor laser [8] is shown in Fig. 11.2 for different optical output powers at 5, 10, 15, and 20 mW. The peak response occurs at the relaxation oscillation frequency f_r and decreases to -3 dB at f_{3dB} compared with the normalized response at low frequencies (0 dB). The bandwidth increases as the optical output power P_0 increases with a $\sqrt{P_0}$ dependence, since P_0 is proportional to the photon density S_0, as shown in Fig. 11.3. The 3-dB frequency f_{3dB} vs. the square root of the output power $\sqrt{P_0}$ are shown with the relaxation frequency f_r. Since the output power P_0 is proportional to $I - I_{th}$, the plot, f_r vs. $\sqrt{I - I_{th}}$ also shows a linear relationship. Therefore, f_r can also be plotted vs. $\sqrt{I - I_{th}}$ for semiconductor lasers and a linear relationship is shown [9]. Improvements of the modulation bandwidth using quantum wells and quantum wires or strained quantum wells have been discussed [4, 10, 11]. ■

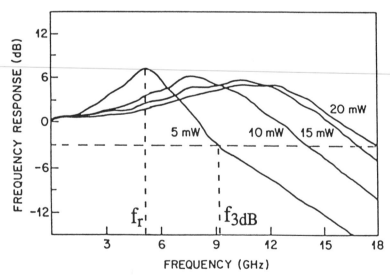

Figure 11.2. Small signal frequency response of a distributed feedback laser at different output powers. The submount temperature was 20°C. The peak response determines the relaxation frequency f_r and the -3 dB response gives the 3 dB frequency f_{3dB}. (After Ref. 8.)

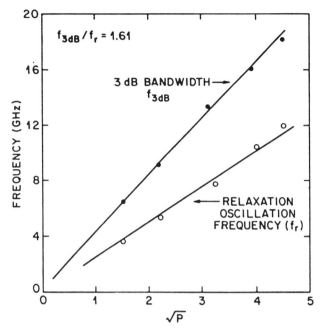

Figure 11.3. The relaxation oscillation frequency and the 3-dB bandwidth of the distributed feedback laser in Fig. 11.2 are plotted as a function of square root of output power P (or P_0 in the text). (After Ref. 8.)

11.2 HIGH-SPEED MODULATION RESPONSE WITH NONLINEAR GAIN SATURATION [12, 13]

11.2.1 Nonlinear Gain Saturation

The gain model $g(N)$ can be taken from

$$g(N) = \frac{g_0 + g'(N - N_0)}{1 + \varepsilon S} \tag{11.2.1}$$

where $g_0 = g(N_0)$, $g' = (\partial g / \partial N)_{N = N_0}$ is the differential gain at N_0. The factor $1 + \varepsilon S$ accounts for nonlinear gain saturation, which is important when the photon density is high. The factor ε is called the gain suppression coefficient. The steady-state solution at $I = I_0$ is obtained from $d/dt = 0$ in the rate equations:

$$\frac{J_0}{qd} = \frac{N_0}{\tau} + \frac{v g_0 S_0}{1 + \varepsilon S_0} \tag{11.2.2}$$

$$\Gamma v \frac{g_0 S_0}{1 + \varepsilon S_0} = \frac{S_0}{\tau_p} - \beta R_{\text{sp}} \tag{11.2.3}$$

If $\varepsilon \neq 0$, the general solution of (11.2.3) is

$$
S_0 = \frac{\tau_p}{2\varepsilon} \left[-\left(\frac{1}{\tau_p} - \Gamma v g_0 - \varepsilon \beta R_{sp} \right) \right.
$$

$$
\left. + \sqrt{\left(\frac{1}{\tau_p} - \Gamma v g_0 - \varepsilon \beta R_{sp} \right)^2 + 4\left(\frac{\varepsilon}{\tau_p} \right) \beta R_{sp}} \right] \quad (11.2.4)
$$

and N_0 is obtained from (11.2.2) using the above S_0. Using the linearized expression by substituting (11.1.6) into (11.2.1), we obtain

$$
g(N) = \frac{g_0}{1 + \varepsilon S_0} + \frac{g'n(t)}{1 + \varepsilon S_0} - \frac{g_0}{(1 + \varepsilon S_0)^2} \varepsilon s(t) \quad (11.2.5)
$$

The ac responses $n(t)$ and $s(t)$ satisfy the following equations:

$$
\frac{d}{dt} n(t) = \frac{j(t)}{qd} - \frac{n(t)}{\tau} - \left(\frac{v g'S_0}{1 + \varepsilon S_0} \right) n(t) - \frac{v g_0}{(1 + \varepsilon S_0)^2} s(t) \quad (11.2.6)
$$

and

$$
\frac{d}{dt} s(t) = \left(\frac{\Gamma v g'S_0}{1 + \varepsilon S_0} \right) n(t) + \frac{\Gamma v g_0}{(1 + \varepsilon S_0)^2} s(t) - \frac{s(t)}{\tau_p} \quad (11.2.7)
$$

which can be derived by keeping the first-order terms of $n(t)$ and $s(t)$ in (11.1.1) and (11.1.2) and using (11.2.5).

11.2.2 Sinusoidal Steady-State Solution of the Small-Signal Equations

Using $d/dt \rightarrow -i\omega$ in (11.2.7) and (11.2.3), we relate s to n

$$
\left[-i\omega(1 + \varepsilon S_0) + \frac{\varepsilon S_0}{\tau_p} + \frac{\beta R_{sp}}{S_0} \right] s(\omega) = (\Gamma v g'S_0) n(\omega) \quad (11.2.8)
$$

We then use (11.2.6) and find

$$
\left(-i\omega + \frac{1}{\tau} + \frac{v g'S_0}{1 + \varepsilon S_0} \right) n(\omega) = \frac{j(\omega)}{qd} - \frac{v g_0}{(1 + \varepsilon S_0)^2} s(\omega) \quad (11.2.9)
$$

Therefore, the small signal photon density function is

$$s(\omega) = \Gamma vg'S_0(j(\omega)/qd)/$$

$$\left\{ \left[-i\omega + \frac{1}{\tau} + \frac{vg'S_0}{(1 + \varepsilon S_0)} \right] \left[-i\omega(1 + \varepsilon S_0) + \frac{\varepsilon S_0}{\tau_p} + \frac{\beta R_{sp}}{S_0} \right] \right.$$

$$\left. + \frac{vg'}{(1 + \varepsilon S_0)} \left(\frac{S_0}{\tau_p} - \beta R_{sp} \right) \right\} \quad (11.2.10)$$

Define

$$\omega_r^2 = vg'\left(\frac{S_0}{\tau_p} - \beta R_{sp} \right) \simeq vg'\frac{S_0}{\tau_p} \quad (11.2.11)$$

which is the same as (11.1.19). We obtain the frequency response function for semiconductor lasers

$$\left| \frac{s(\omega)}{j(\omega)} \right|^2 \simeq \left| \frac{\omega_r^2 \Gamma \tau_p/qd}{-\omega^2 + \omega_r^2 - i\omega\gamma} \right|^2$$

$$= \left(\frac{\Gamma \tau_p}{qd} \right)^2 \frac{\omega_r^4}{\left(\omega^2 - \omega_r^2 \right)^2 + \omega^2 \gamma^2} \quad (11.2.12)$$

where a damping factor γ is defined:

$$\gamma \simeq \frac{1}{\tau} + vg'S_0 + \frac{\varepsilon S_0}{\tau_p} + \beta \frac{R_{sp}}{S_0}$$

$$\simeq vg'S_0 \left(1 + \frac{\varepsilon}{vg'\tau_p} \right) + \frac{1}{\tau}$$

$$= Kf_r^2 + \frac{1}{\tau} \quad (11.2.13)$$

Here a K factor (ns)

$$K = 4\pi^2 \left(\tau_p + \frac{\varepsilon}{vg'} \right) \quad (11.2.14)$$

and $\omega_r = 2\pi f_r$ have been used.

The 3-dB cutoff frequency occurs at

$$\left(\omega_{3dB}^2 - \omega_r^2 \right)^2 + \omega_{3dB}^2 \gamma^2 = 2\omega_r^4 \quad (11.2.15)$$

The maximum possible bandwidth occurs when the following condition is satisfied:

$$2\omega_r^2 = \gamma^2 = K^2 f_r^4 \qquad (11.2.16)$$

such that the frequency response function is a monotonic decreasing function, (11.2.12) $\propto \omega_r^4/(\omega^4 + \omega_r^4)$. We then have the maximum relaxation frequency by solving (11.2.16) for f_r:

$$f_{r,\max} = \frac{2\pi\sqrt{2}}{K} \qquad (11.2.17)$$

By fitting the frequency response function (11.2.12) to the experimental data, the damping factor γ can be found together with the relaxation frequency f_r. Equation (11.2.13), $\gamma = K f_r^2 + 1/\tau$, shows that a linear relation with a slope K holds if the damping factor γ is plotted vs. f_r^2 and the intercept with the vertical axis gives the inverse carrier lifetime $1/\tau$.

Example Figure 11.4 shows the experimental results [14] for two strained quantum-well lasers with (1) a slope $K = 0.22$ ns for a tensile strain laser with four quantum wells, and a maximum bandwidth $f_{r,\max} = 2\pi\sqrt{2}/K = 40$ GHz, and (2) a slope with $K = 0.58$ ns for a compressive strain laser with four quantum wells and $f_{r,\max} = 15$ GHz. Both lines intercept the vertical axis at $1/\tau \simeq 5$ GHz, and the carrier lifetime τ at threshold is 0.2 ns for both

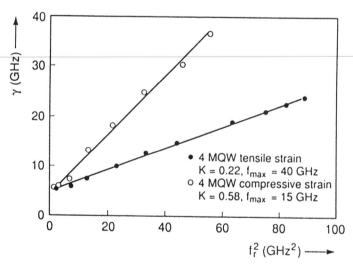

Figure 11.4. Experimental results showing the linear relation between the damping factor γ and f_r^2 for two quantum-well lasers, $\gamma = K f_r^2 + 1/\tau$. The slope gives the K factor (in ns) and the intercept with the γ axis gives the inverse carrier lifetime $1/\tau$ at threshold. (After Ref. 14.)

samples. The difference of the slopes of the two laser structures can be explained from the difference in the differential gain of over a factor of 2, since $f_r^2/P_0 = 3.6$ GHz^2/mW for the compressive strain laser and $f_r^2/P_0 = 7.7$ GHz^2/mW for the tensile strain sample. ∎

Since the damping factor depends on the K factor, relaxation frequency, and the inverse carrier lifetime, the experimental data provide very good guidance for the design of high-speed semiconductor lasers. The K factor, $K = 4\pi^2(\tau_p + (\varepsilon/vg'))$, can also be used to determine the nonlinear gain suppression coefficient ε. Theoretical models and experimental data on strained and unstrained quantum-well lasers have been presented with interesting results [15–24]. The measured nonlinear gain suppression coefficient ε ranges from 2 to 13 × 10^{-17} cm^3 for InGaAs or InGaAsP materials in the quantum wells [15, 17, 18]. Various physical mechanisms such as the well-barrier hole burning effects [20], carrier heating and spectral hole burning [21], carrier transport [22, 23], and carrier capture by and escape from quantum wells [25–27] have been investigated and shown to affect by varying degrees the high-speed modulation of semiconductor quantum-well lasers. More work is in progress to understand the ultimate limit on the high-speed modulation of semiconductor lasers.

11.3 SEMICONDUCTOR LASER SPECTRAL LINEWIDTH AND THE LINEWIDTH ENHANCEMENT FACTOR

The spectral properties of semiconductor lasers have been investigated since the early 1980s. Experiments by Fleming and Mooradian [28] showed that the laser spectral linewidth has a Lorentzian shape and the linewidth is inversely proportional to the optical output power. However, the magnitude of the linewidth was much larger than they had expected from conventional theories. A model proposed by Henry [29, 30] explained the phenomena by noting that the semiconductor laser is similar to a detuned oscillator, and there is a spectrum linewidth enhancement due to the coupling between the amplitude and phase fluctuations of the optical field. A linewidth enhancement factor α_e is introduced

$$\alpha_e = \frac{\partial n'/\partial N}{\partial n''/\partial N} \tag{11.3.1}$$

and the laser linewidth has a broadening enhancement by an amount $1 + \alpha_e^2$. Here n' and n'' are the real and imaginary parts of the refractive index due to the carrier injection into the active region and N is the carrier density. A more formal derivation has been given by Vahala and Yariv [31]. In this section, we present the model of Henry.

11.3.1 Basic Equations for the Optical Intensity and Phase in the Presence of Spontaneous Emission

Consider an optical field $E(z, t)$ given by

$$E(z, t) = E(t) \, e^{i(kz - \omega t)} \qquad (11.3.2)$$

$$E(t) = \sqrt{I(t)} \, e^{i\phi(t)} \qquad (11.3.3)$$

where $I(t)$ represents the intensity and $\phi(t)$ the phase of laser field. We assume $I(t) = E(t)E^*(t)$ has been normalized such that it represents the average number of photons in the cavity. The time-dependent magnitude $E(t)$ is a complex phasor, assuming that its time variation is much slower than the optical frequency ω. The phasor $E(t)$ is plotted in the complex plane as a vector with a magnitude $\sqrt{I(t)}$ and a phase $\phi(t)$, as shown in Fig. 11.5. The basic assumption is that a random spontaneous emission alters $E(t)$ by ΔE, which adds a unit magnitude (one photon) and a phase θ, which is random:

$$\Delta E = e^{i(\phi + \theta)} \qquad (11.3.4)$$

There are two contributions to the phase change $\Delta \phi$:

$$\Delta \phi = \Delta \phi' + \Delta \phi'' \qquad (11.3.5)$$

where $\Delta \phi'$ is due to the out-of-phase component of ΔE, and $\Delta \phi''$ is due to the intensity change which is coupled to the phase change. To obtain the first contribution, $\Delta \phi'$, we note from Fig. 11.5, that $\sqrt{I} \, \Delta \phi' \simeq \sin \theta$, or

$$\Delta \phi' = \frac{\sin \theta}{\sqrt{I}} \qquad (11.3.6)$$

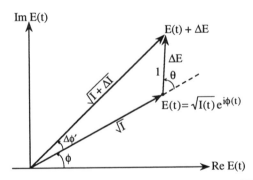

Figure 11.5. A plot on the complex optical electric field [29] $E(t) = \sqrt{I(t)} \exp[i\phi(t)]$ domain showing that its magnitude $\sqrt{I}$ and phase ϕ can be changed by the spontaneous emission of a photon (magnitude is one since the intensity I has been normalized to represent the photon number in the cavity) with a phase change $\Delta \phi'$.

To obtain the second contribution, $\Delta\phi''$, due to the intensity change, we start from the wave equation

$$\frac{\partial^2}{\partial z^2}E(z,t) = \frac{1}{c^2}\frac{\partial^2}{\partial t^2}\varepsilon_r E(z,t) \tag{11.3.7}$$

where c is speed of light in free space and ε_r is the relative permittivity $\varepsilon_r = \varepsilon/\varepsilon_0$ of the semiconductor. We obtain

$$\frac{2i\omega}{c^2}\varepsilon_r\frac{\partial E(t)}{\partial t} = -\left(\frac{\omega^2}{c^2}\varepsilon_r - k^2\right)E(t) \tag{11.3.8}$$

neglecting the term $\partial^2 E(t)/\partial t^2$, since $E(t)$ is a slowly varying function. We can also express ε_r in terms of the complex refractive index

$$\sqrt{\varepsilon_r} = n' + in'' \tag{11.3.9}$$

and

$$\frac{\omega}{c}\sqrt{\varepsilon_r} = \frac{\omega}{c}n' + i\frac{\omega}{c}n''$$

$$k = \frac{\omega}{c}n' \qquad \frac{1}{2}(g-\alpha) = -\frac{\omega}{c}n'' \tag{11.3.10}$$

where g is the gain coefficient, and α is the absorption coefficient of the optical intensity. At threshold, the gain is balanced by the absorption, $(g=\alpha)$, $n'' = 0$, and ε_r is real. Changes in carrier density N will cause n' and n'' to deviate from the threshold values:

$$\varepsilon_r = (n' + \Delta n' + i\Delta n'')^2$$

$$\simeq n'^2 + 2in'\,\Delta n''(1 - i\alpha_e) \tag{11.3.11}$$

where we have defined a linewidth enhancement factor α_e as the ratio of the change in the real part of the refractive index to the change in the imaginary part:

$$\alpha_e = \frac{\Delta n'}{\Delta n''} \tag{11.3.12}$$

Therefore

$$\frac{\partial E}{\partial t} = -\frac{\omega\Delta n''}{n'}(1 - i\alpha_e)E(t)$$

$$= \left(\frac{g-\alpha}{2}\right)v(1 - i\alpha_e)E(t) \tag{11.3.13}$$

where the group velocity $v = c/n'$ has been used, and we ignore the dispersion effect for simplicity. If we include the effect of the material dispersion, (11.3.8) has to be modified [29] and the result for (11.3.13) still holds with the group velocity given by $v = c/(n' + \omega\, \partial n'/\partial\omega)$, which is derived using $k = \omega n'/c$ and $v = (\partial k/\partial\omega)^{-1}$. Substituting $E(t) = \sqrt{I(t)}\, e^{i\phi(t)}$ into (11.3.13) and separating the real and imaginary parts, we find

$$\frac{1}{2}\frac{dI}{dt} = \left(\frac{g - \alpha}{2}\right)vI \qquad (11.3.14a)$$

$$\frac{d\phi}{dt} = -\left(\frac{g - \alpha}{2}\right)v\alpha_e \qquad (11.3.14b)$$

Therefore, we obtain

$$\frac{d\phi}{dt} = \frac{-\alpha_e}{2I}\frac{dI}{dt} \qquad (11.3.15)$$

Initially at $t = 0$, $I(0) = I + \Delta I$, and at $t = \infty$, the relaxation oscillations die out, $I(\infty) = I$, and we obtain

$$\Delta\phi'' = \frac{\alpha_e}{2I}\Delta I$$

$$= \frac{\alpha_e}{2I}\left(1 + 2\sqrt{I}\cos\theta\right) \qquad (11.3.16)$$

which can be derived from the relation among the three sides of the triangle in Fig. 11.5. The total phase change is then

$$\Delta\phi = \Delta\phi' + \Delta\phi''$$

$$= \frac{\alpha_e}{2I} + \frac{1}{\sqrt{I}}(\sin\theta + \alpha_e\cos\theta) \qquad (11.3.17)$$

The ensemble average of the spontaneous emission events at a time duration t is contributed from the constant term $(\alpha_e/2I)$ multiplied by the total number of the events, $R_{sp}t$:

$$\langle\Delta\phi\rangle = \frac{\alpha_e}{2I}R_{sp}t \qquad (11.3.18)$$

since $\langle\sin\theta\rangle = \langle\cos\theta\rangle = 0$. Here R_{sp} is the spontaneous emission rate (1/s). Equation (11.3.18) gives an angular frequency shift:

$$\Delta\omega = \frac{d}{dt}\langle\Delta\phi\rangle = \frac{\alpha_e}{2I}R_{sp} \qquad (11.3.19)$$

The total phase fluctuation for $R_{sp}t$ spontaneous events gives the variance:

$$
\langle \Delta\phi^2 \rangle = \left\langle \frac{1}{I}(\sin\theta + \alpha_e \cos\theta)^2 \right\rangle
$$

$$
= \frac{1 + \alpha_e^2}{2I} R_{sp}|t| \tag{11.3.20}
$$

We use an absolute value for $|t|$ since $\langle \Delta\phi^2 \rangle$ is a positive quantity. Therefore, we found the mean $\langle \Delta\phi \rangle$ and its variance $\langle \Delta\phi^2 \rangle$ as above.

11.3.2 Power Spectrum and Semiconductor Laser Spectral Linewidth

The power spectrum of the laser is the Fourier transform of the correlation function:

$$
W(\omega) = \int_{-\infty}^{\infty} dt \, e^{i\omega t} \langle E^*(t)E(0) \rangle
$$

$$
= \int_{-\infty}^{\infty} dt \, e^{i\omega t} \langle [I(t)I(0)]^{1/2} e^{-i\Delta\phi(t)} \rangle
$$

$$
\simeq I(0) \int_{-\infty}^{\infty} dt \, e^{i\omega t} \langle e^{-i\Delta\phi(t)} \rangle \tag{11.3.21}
$$

where the small intensity fluctuation is neglected and the amplitude function for the field $E(t) = [I(t)]^{1/2} e^{i\phi(t)}$ has been used, which does not include the central frequency of the laser, that is, ω in (11.3.2), and

$$
\Delta\phi(t) = \phi(t) - \phi(0)
$$

Since the spontaneous emission events are random, the phase ϕ should have a Gaussian probability distribution function, $P(\Delta\phi) =$ a Gaussian function. The ensemble average for a Gaussian distribution is [30, 32]

$$
\langle e^{-i\Delta\phi(t)} \rangle = \int_{-\infty}^{\infty} d(\Delta\phi) P(\Delta\phi) e^{-i\Delta\phi}
$$

$$
= e^{-\langle \Delta\phi^2 \rangle/2} \tag{11.3.22}
$$

Using the result for the variance $\langle \Delta\phi^2 \rangle$ in (11.3.20), we can define a coherence time as

$$
\frac{1}{t_c} \equiv \frac{1 + \alpha_e^2}{4I} R_{sp} \tag{11.3.23}
$$

such that $\langle \Delta\phi^2 \rangle/2 = |t|/t_c$. The power spectrum (11.3.21) gives a Lorentzian function

$$W(\omega) \simeq I(0) \frac{2t_c}{1 + \omega^2 t_c^2} \tag{11.3.24}$$

with a full width at half-maximum (FWHM) of

$$\Delta\omega = \frac{2}{t_c} = \frac{\left(1 + \alpha_e^2\right)}{2I} R_{\text{sp}} \tag{11.3.25}$$

or

$$\Delta f = \frac{\left(1 + \alpha_e^2\right)}{4\pi I} R_{\text{sp}} \tag{11.3.26}$$

The number of photons I in the laser cavity (the photon density multiplied by the volume) is related to the optical output power by (10.1.28)

$$P_0 = \tfrac{1}{2}(I\hbar\omega)(v\alpha_m) \tag{11.3.27}$$

where $I\hbar\omega$ is the total photon energy and $v\alpha_m$ is the escaping rate of the photon out of the cavity with a length L, where

$$\alpha_m = \frac{1}{L}\ln\frac{1}{R} \tag{11.3.28}$$

and R is the mirror reflectivity at both ends. We find

$$\Delta f = \frac{\hbar\omega v \alpha_m R_{\text{sp}}}{8\pi P_0}\left(1 + \alpha_e^2\right)$$

$$= \frac{\hbar\omega v R_{\text{sp}} \ln(1/R)}{8\pi P_0 L}\left(1 + \alpha_e^2\right) \tag{11.3.29}$$

which are commonly used to explain the spectral linewidth of semiconductor lasers. Alternatively, the spontaneous emission rate $(1/s)$ is related to the gain coefficient $(1/cm)$ by a dimensionless spontaneous emission factor n_{sp}:

$$R_{\text{sp}} = vgn_{\text{sp}} \tag{11.3.30}$$

where

$$n_{\text{sp}} = \frac{1}{1 - e^{(\hbar\omega - \Delta F)/k_B T}} \tag{11.3.31}$$

and $\Delta F = F_c - F_v$ is the separation of the quasi-Fermi levels between the electron and the hole. Note that the photon number I and the spontaneous emission rate R_{sp} $(1/s)$ are defined for the whole volume of the active region.

11.3.3 Linewidth Enhancement Factor in Semiconductor Lasers

Experimental data [33] for the laser linewidth Δf depending on the optical output power with an inverse law are shown in Fig. 11.6. The surprisingly large linewidth measured in this set of data was explained by Henry [29] using the correction factor $(1 + \alpha_e^2)$, where $\alpha_e \simeq 5$. More measurements have been done for various semiconductor lasers including index-guided double-hetero-structure, unstrained and strained quantum-well lasers, with reduced linewidth enhancement factor [34, 35] α_e.

In a semiconductor laser, the injected carrier-induced refractive index change is associated with the change in the gain. The linewidth enhancement factor α_e can be directly expressed in terms of the differential change of the refractive index per injected carrier vs. the differential gain:

$$\alpha_e = \frac{dn'}{dn''}$$

$$= -\frac{4\pi}{\lambda} \frac{dn/dN}{dg/dN} \qquad (11.3.32)$$

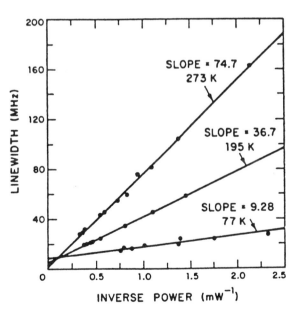

Figure 11.6. Semiconductor laser linewidth versus inverse power at three temperatures exhibiting the linear behavior. (After Ref. 33.) The magnitude of the large linewidth was explained [29] using the correction factor $1 + \alpha_e^2$ with $\alpha_e \simeq 5$ at room temperature.

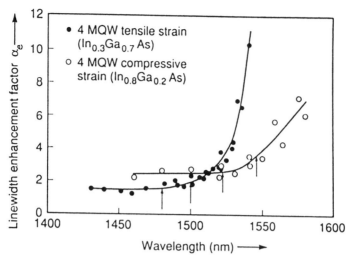

Figure 11.7. The linewidth enhancement factor α_e vs. wavelength of two types of strained MQW lasers. The lasing wavelength is indicated by the arrows. (After Ref. 14.)

where we have used $dn''/dN = (-1/2)(dg/dN)/(2\pi/\lambda)$ in (11.3.10) with $\omega/c = 2\pi/\lambda$, and λ as the wavelength in free space. The linewidth enhancement factor varies from about 1.5 to 10 and is dependent on the lasing wavelength. Experimental data of α_e for two strained quantum-well lasers are shown in Fig. 11.7. The high-speed modulation characteristics of these lasers are shown in Fig. 11.4.

The dependence of α_e on the loss or the Fermi levels of semiconductor lasers has also been shown to be important [36]. For strained quantum wells, theoretical analysis [37] shows that α_e can be reduced to about 1.1 using tensile strains with TM polarization of semiconductor quantum-well lasers. Similar behavior to the spectral broadening in semiconductor lasers also occurs in an intensity modulator using the electroabsorption effects [38–40]. This is because a phase modulation due to the change in the refractive index is associated with the change in the absorption coefficient.

PROBLEMS

11.1 Estimate the photon lifetime for a semiconductor laser assuming the following parameters: $\alpha_i = 10$ cm^{-1}, refractive index $= 3.2$, and cavity length $L = 200$ μm. Discuss the effects on photon lifetime if we increase the cavity length or reflectivity by a factor of two.

11.2 Derive (a) the dc solutions for S_0 and N_0 and (b) the ac solutions for $s(t)$ and $n(t)$ in the linear gain theory.

11.3 Discuss how one may improve the relaxation frequency f_r of a semiconductor laser.

11.4 Derive an expression for the 3-dB angular frequency ω_{3dB} for the small signal frequency response of a semiconductor laser using the linear gain theory.

11.5 Derive the dc solutions for S_0 and N_0 in the nonlinear gain theory for the frequency response of a semiconductor laser.

11.6 Derive the ac solutions for $s(t)$ and $n(t)$ in the nonlinear gain theory.

11.7 Discuss the effect of decreasing the damping factor γ.

11.8 Plot the frequency response curve when the condition $2\omega_r^2 = \gamma^2$ is satisfied.

11.9 Derive (11.2.13) and discuss the approximations used.

11.10 Derive (11.3.13) and (11.3.14) taking into account the material dispersion.

11.11 Derive (11.3.22) for a Gaussian distribution function $P(\Delta\phi)$.

11.12 Describe the factors determining the spectral linewidth of a semiconductor laser. How may one decrease the semiconductor laser linewidth?

REFERENCES

1. T. Ikegami and Y. Suematsu, "Resonance-like characteristics of the direct modulation of a junction laser," *Proc. IEEE* **55**, 122–123 (1967).

2. A. Yariv, *Quantum Electronics*, 3d ed., Wiley, New York, 1989.

3. K. Lau and A. Yariv, "High-frequency current modulation of semiconductor lasers," Chapter 2, in W. T. Tsang, Vol. Ed., *Lightwave Communications Technology*, Vol. 22, Part B, in R. K. Willardson and A. C. Beer, Eds., *Semiconductors and Semimetals*, Academic, New York, 1985.

4. K. Y. Lau, "Ultralow threshold quantum well lasers," Chapter 4, and "Dynamics of quantum well lasers," Chapter 5, in P. S. Zory, Jr., Ed., *Quantum Well Lasers*, Academic, San Diego, 1993.

5. T. P. Lee, C. A. Burrus, J. A. Copeland, A. G. Dentai, and D. Marcuse, "Short-cavity InGaAsP injection lasers: Dependence of mode spectra and single-longitudinal-mode power on cavity length," *IEEE J. Quantum Electron.* **QE-18**, 1101–1112 (1982).

6. K. Petermann, "Calculated spontaneous emission factor for double-heterostructure injection lasers with gain-induced waveguiding," *IEEE J. Quantum Electron.* **QE-15**, 566–570 (1979).

7. W. Streifer, D. R. Scifres, and R. D. Burnham, "Spontaneous emission factor of narrow-strip gain-guided diode lasers," *Electron. Lett.* **17**, 933 (1981).

8. N. K. Dutta, S. J. Wang, A. B. Piccirilli, R. F. Karlicek, Jr., R. L. Brown, M. Washington, U. K. Chakrabarti, and A. Gnauck, "Wide-bandwidth and

high-power InGaAsP distributed feedback lasers," *J. Appl. Phys.* **66**, 4640–4644 (1989).

9. K. Uomi, M. Aoki, T. Tsuchiya, and A. Takai, "Dependence of high-speed properties on the number of quantum wells in 1.55 μm InGaAs–InGaAsP MQW λ/4-shifted DFB lasers," *IEEE J. Quantum Electron.* **29**, 355–360 (1993).

10. Y. Arakawa, K. Vahala, A. Yariv, and K. Lau, "Enhanced modulation bandwidth of GaAlAs double heterostructure lasers in high magnetic fields: Dynamic response with quantum wire effects," *Appl. Phys. Lett.* **47**, 1142–1144 (1985).

11. Y. Arakawa and A. Yariv, "Quantum well lasers: gain, spectra, dynamics," *IEEE J. Quantum Electron.* **QE-22**, 1887–1899 (1986).

12. R. Olshansky, P. Hill, V. Lanzisera, and W. Powazinik, "Frequency response of 1.3 μm InGaAsP high speed semiconductor lasers," *IEEE J. Quantum Electron.* **QE-23**, 1410–1418 (1987).

13. R. Olshansky, P. Hill, V. Lanzisera, and W. Powazinik, "Universal relationship between resonant frequency and damping rate of 1.3 μm InGaAsP semiconductor lasers," *Appl. Phys. Lett.* **50**, 653–655 (1987).

14. L. F. Tiemeijer, P. J. A. Thijs, P. J. de Waard, J. J. M. Binsma, and T. V. Dongen, "Dependence of polarization, gain, linewidth enhancement factor, and K factor on the sign of the strain of InGaAs/InP strained-layer multiquantum well lasers," *Appl. Phys. Lett.* **58**, 2738–2740 (1991).

15. T. Fukushima, J. E. Bowers, R. A. Logan, T. Tanbun-Ek, and H. Temkin, "Effect of strain on the resonant frequency and damping factor in InGaAs/InP multiple quantum well lasers," *Appl. Phys. Lett.* **58**, 1244–1246 (1991).

16. S. D. Offsey, W. J. Schaff, L. F. Lester, L. F. Eastman, and S. K. McKernan, "Strained-layer InGaAs–GaAs–AlGaAs lasers grown by molecular beam epitaxy for high-speed modulation," *IEEE J. Quantum Electron.* **27**, 1455–1462 (1991).

17. H. Yasaka, K. Takahata, N. Yamamoto, and M. Naganuma, "Gain saturation coefficients of strained-layer multiple quantum-well distributed feedback lasers," *IEEE Photon. Technol. Lett.* **3**, 879–882 (1991).

18. J. Zhou, N. Park, J. W. Dawson, K. Vahala, M. A. Newkirk, U. Koren, and B. I. Miller, "Highly nondegenerate four-wave mixing and gain nonlinearity in a strained multiple-quantum-well optical amplifier," *Appl. Phys. Lett.* **62**, 2301–2303 (1993).

19. S. R. Chinn, "Measurement of nonlinear gain suppression and four-wave mixing in quantum well lasers," *Appl. Phys. Lett.* **59**, 1673–1675 (1991).

20. W. Rideout, W. F. Sharfin, E. S. Koteles, M. O. Vassell, and B. Elman, "Well-barrier hole burning in quantum well lasers," *IEEE Photon. Technol. Lett.* **3**, 784–786 (1991).

21. A. Uskov, J. Mork, and J. Mark, "Theory of short-pulse gain saturation in semiconductor laser amplifiers," *IEEE Photon. Technol. Lett.* **4**, 443–446 (1992).

22. R. Nagarajan, T. Fukushima, S. W. Corzine, and J. E. Bowers, "Effects of carrier transport on high-speed quantum well lasers," *Appl. Phys. Lett.* **59**, 1835–1837 (1991).

23. A. P. Wright, B. Garrett, G. H. B. Thompson, and J. E. A. Whiteaway, "Influence of carrier transport on wavelength chirp of InGaAs/InGaAsP MQW lasers," *Electron. Lett.* **28**, 1911–1913 (1992).

24. E. Meland, R. Holmstrom, J. Schlafer, R. B. Lauer, and W. Powazinik, "Extremely high-frequency (24 GHz) InGaAsP diode lasers with excellent modulation efficiency," *Electron. Lett.* **26**, 1827–1829 (1990).

25. P. W. M. Blom, J. E. M. Haverkort, P. J. van Hall, and J. H. Wolter, "Carrier–carrier scattering induced capture in quantum-well lasers," *Appl. Phys. Lett.* **62**, 1490–1492 (1993).

26. D. Morris, B. Deveaud, A. Regreny, and P. Auvray, "Electron and hole capture in multiple-quantum-well structures," *Phys. Rev. B* **47**, 6819–6822 (1993).

27. S. C. Kan, D. Vassilovski, T. C. Wu, and K. Y. Lau, "Quantum capture limited modulation bandwidth of quantum well, wire, and dot lasers," *Appl. Phys. Lett.* **62**, 2307–2309 (1993).

28. M. W. Fleming and A. Mooradian, "Fundamental line broadening of single-mode (GaAl)As diode lasers," *Appl. Phys. Lett.* **38**, 511 (1981).

29. C. H. Henry, "Theory of the linewidth of semiconductor lasers," *IEEE J. Quantum Electron.* **QE-18**, 259–264 (1982).

30. C. H. Henry, "Spectral properties of semiconductor lasers," Chapter 3, in W. T. Tsang, Vol. Ed., *Lightwave Communications Technology*, Vol. 22, Part B, in R. K. Willardson and A. C. Beers, Eds., *Semiconductors and Semimetals*, Academic, New York, 1985.

31. K. Vahala and A. Yariv, "Semiclassical theory of semiconductor lasers, Part I," *IEEE J. Quantum Electron.* **QE-19**, 1096–1101 (1983).

32. M. Lax, "Classical noise, V, Noise in self-sustained oscillators," *Phys. Rev.* **160**, 290–307 (1967).

33. D. Welford and A. Mooradian, "Output power and temperature dependence of the linewidth of single-frequency cw (GaAl)As diode lasers," *Appl. Phys. Lett.* **40**, 865–867 (1982).

34. Y. Asai, J. Ohya, and M. Ogura, "Spectral linewidth and resonant frequency characteristics of InGaAsP/InP multiquantum well lasers," *IEEE J. Quantum Electron.* **25**, 662–667 (1989).

35. N. K. Dutta, H. Temkin, T. Tanbun-Ek, and R. Logan, "Linewidth enhancement factor for InGaAs/InP strained quantum well lasers," *Appl. Phys. Lett.* **57**, 1390–1391 (1990).

36. Y. Arakawa and A. Yariv, "Fermi energy dependence of linewidth enhancement factor of GaAlAs buried heterostructure lasers," *Appl. Phys. Lett.* **47**, 905–907 (1985).

37. Y. Huang, S. Arai, and K. Komori, "Theoretical linewidth enhancement factor of $Ga_{1-x}In_xAs/GaInAsP/InP$ strained-quantum-well structures," *IEEE Photon. Technol. Lett.* **5**, 142–145 (1993).

38. F. Koyama and K. Iga, "Frequency chirping of external modulation and its reduction," *Electron. Lett.* **21**, 1065–1066 (1985).

39. Y. Noda, M. Suzuki, Y. Kushiro, and S. Akiba, "High-speed electroabsorption modulator with strip-loaded GaInAsP planar waveguide," *J. Lightwave Technol.* **LT-4**, 1445–1453 (1986).

40. T. H. Wood, "Multiple quantum well (MQW) waveguide modulators," *J. Lightwave Technol.* **6**, 743–757 (1988).

12

Electrooptic and Acoustooptic Modulators

In this chapter, we discuss electrooptic effects and modulators. The bulk electrooptic effects are discussed first, and their applications as amplitude and phase modulators are presented. These devices using waveguide structures are then shown. The basic idea is that the optical refractive index of electrooptic materials, such as $LiNbO_3$, KH_2PO_4, or GaAs, and ZnS semiconductors, can be changed by an applied electric field. Therefore, an incident optical field propagating through the crystal with a proper polarization experiences efficient electrooptic effects. The transmitted field changes in either phase or polarization, which can be used in the designs of phase modulators as well as amplitude modulators. We then discuss scattering of light by sound and present a coupled-mode analysis for acoustooptic modulators.

12.1 ELECTROOPTIC EFFECTS AND AMPLITUDE MODULATORS [1, 2]

To understand the electrooptic effects, we consider a crystal described by the constitutive relation associating the displacement vector **D** to the electric field **E** by a permittivity tensor $\bar{\bar{\varepsilon}}$,

$$\mathbf{D} = \bar{\bar{\varepsilon}} \cdot \mathbf{E}$$

$$\bar{\bar{\varepsilon}} = \begin{bmatrix} \varepsilon_{xx} & \varepsilon_{xy} & \varepsilon_{xz} \\ \varepsilon_{yx} & \varepsilon_{yy} & \varepsilon_{yz} \\ \varepsilon_{zx} & \varepsilon_{zy} & \varepsilon_{zz} \end{bmatrix} \tag{12.1.1}$$

and the permeability is μ_0. Let us define the inverse of the permittivity tensor $\bar{\bar{\varepsilon}}$ as $\bar{\bar{K}}$:

$$\bar{\bar{K}} = \bar{\bar{\varepsilon}}^{-1} \tag{12.1.2}$$

The index ellipsoid of the crystal is described by

$$\varepsilon_0 \sum_{ij} K_{ij} x_i x_j = 1 \tag{12.1.3}$$

where $x_1 = x$, $x_2 = y$, and $x_3 = z$ for convenience. For most crystals, $\bar{\bar{\varepsilon}}$ is symmetric due to the symmetry property of the structure. Therefore, $\bar{\bar{\varepsilon}}$ can be diagonalized to be

$$\bar{\bar{\varepsilon}} = \begin{bmatrix} \varepsilon_x & 0 & 0 \\ 0 & \varepsilon_y & 0 \\ 0 & 0 & \varepsilon_z \end{bmatrix} \qquad (12.1.4)$$

The coordinate system in which $\bar{\bar{\varepsilon}}$ is diagonalized is called the principal system. In this system, $\varepsilon_0 \bar{\bar{K}}$ is a diagonal matrix with the diagonal elements equal to the reciprocals of the square of the refractive indices of the three characteristic polarizations along the direction of the principal axes:

$$\varepsilon_0 \bar{\bar{K}} = \varepsilon_0 \begin{bmatrix} 1/\varepsilon_x & 0 & 0 \\ 0 & 1/\varepsilon_y & 0 \\ 0 & 0 & 1/\varepsilon_z \end{bmatrix} = \begin{bmatrix} 1/n_x^2 & 0 & 0 \\ 0 & 1/n_y^2 & 0 \\ 0 & 0 & 1/n_z^2 \end{bmatrix} \quad (12.1.5)$$

and the index ellipsoid (12.1.3) is

$$\varepsilon_0 \left(\frac{x_1^2}{\varepsilon_x} + \frac{x_2^2}{\varepsilon_y} + \frac{x_3^2}{\varepsilon_z} \right) = \frac{x^2}{n_x^2} + \frac{y^2}{n_y^2} + \frac{z^2}{n_z^2} = 1 \qquad (12.1.6)$$

where $n_i = \sqrt{\varepsilon_i/\varepsilon_0}$, $i = x$, y, and z.

12.1.1 Electrooptic Effects

In a linear electrooptic material, the index ellipsoid is changed in the presence of an applied electric field $\mathbf{F}$, and K_{ij} becomes $K_{ij} + \Delta K_{ij}$, where the change ΔK_{ij} is linearly proportional to the electric field:

$$\varepsilon_0 \Delta K_{ij} = \sum_{k=1}^{3} r_{ijk} F_k \qquad (12.1.7)$$

The linear electrooptic effect is also called the Pockels effect, after Friedrich Pockels [3] (1865–1913), who described it in 1893. These r_{ijk} coefficients are also called Pockels coefficients. Equation (12.1.7) can also be generalized to include the quadratic electrooptic effects, which are usually smaller than the linear effects,

$$\varepsilon_0 \Delta K_{ij} = \sum_{k=1}^{3} r_{ijk} F_k + \sum_{k,\ell=1}^{3} S_{ijk\ell} F_k F_\ell \qquad (12.1.8)$$

However, for materials with centrosymmetry, the index ellipsoid function must be an even function of the applied electric field, since it must remain invariant upon the sign reversal of the electric field. Therefore, r_{ijk} vanishes and the quadratic electrooptic effects dominate. The quadratic electrooptic effect is also called the Kerr effect, after John Kerr [3] (1824–1907), who discovered the effect in 1875. From the symmetry property of the crystal, the following matrix correspondence is defined:

$$\begin{bmatrix} 11 & 12 & 13 \\ & 22 & 23 \\ & & 33 \end{bmatrix} \leftrightarrow \begin{bmatrix} 1 & 6 & 5 \\ & 2 & 4 \\ & & 3 \end{bmatrix} \qquad (12.1.9)$$

that is, we have $(ij) \leftrightarrow I = 1, 2, \ldots, 6$, and $r_{ijk} = r_{jik} \equiv r_{Ik}$. Note that r_{Ik} is a 6×3 matrix, and Eq. (12.1.7) can be rewritten as

$$\varepsilon_0 \begin{bmatrix} (\Delta K)_1 \\ (\Delta K)_2 \\ (\Delta K)_3 \\ (\Delta K)_4 \\ (\Delta K)_5 \\ (\Delta K)_6 \end{bmatrix} = \begin{bmatrix} r_{11} & r_{12} & r_{13} \\ r_{21} & r_{22} & r_{23} \\ r_{31} & r_{32} & r_{33} \\ r_{41} & r_{42} & r_{43} \\ r_{51} & r_{52} & r_{53} \\ r_{61} & r_{62} & r_{63} \end{bmatrix} \begin{bmatrix} F_1 \\ F_2 \\ F_3 \end{bmatrix} \qquad (12.1.10)$$

Depending on the symmetry of the crystal, many of the matrix elements r_{Ik} may vanish. We usually refer to some databook [4] or reference tables [5–8] for the crystal symmetry and nonvanishing r_{Ik} values. A few important electrooptic materials are shown in Table 12.1, for illustration purposes.

Example   The potassium dihydrogen phosphate (KDP or KH_2PO_4) crystal is uniaxial in the absence of an applied field:

$$\varepsilon_0 \sum_{ij} K_{ij} x_i x_j = \frac{x^2}{n_o^2} + \frac{y^2}{n_o^2} + \frac{z^2}{n_3^2} = 1 \qquad (12.1.11)$$

where $n_x = n_y = n_o$, and $n_z = n_e$. With an applied field, we have $r_{Ik} = 0$ except for r_{63} and $r_{41} = r_{52}$. Therefore, Eq. (12.1.10) gives $\varepsilon_0(\Delta K)_4 = r_{41}F_1$, $\varepsilon_0(\Delta K)_5 = r_{52}F_2$, and $\varepsilon_0(\Delta K)_6 = r_{63}F_3$. Note that $x_1 = x, x_2 = y, x_3 = z$, and use the mapping table in (12.1.9):

$$\varepsilon_0 \sum_{ij} \Delta K_{ij} x_i x_j = 2r_{41}F_x yz + 2r_{52}F_y xz + 2r_{63}F_z yx \qquad (12.1.12)$$

where a factor 2 accounts for the symmetric property of the matrix ΔK_{ij}.

Therefore, the new index ellipsoid $\varepsilon_0 \sum_{ij}(K_{ij} + \Delta K_{ij})x_i x_j = 1$ becomes

$$\frac{x^2}{n_o^2} + \frac{y^2}{n_o^2} + \frac{z^2}{n_e^2} + 2r_{41}F_x yz + 2r_{52}F_y xz + 2r_{63}F_z xy = 1 \quad (12.1.13)$$

In general, the above index ellipsoid may not have the principal axes along the x, y, or z directions any more, as will be shown in the following examples. ∎

12.1.2 Longitudinal Amplitude Modulator

Consider the setup as in Fig. 12.1 with the applied electric field along the propagation direction (z) of light $\mathbf{F} = \hat{z}F_z$. The index ellipsoid is

$$\frac{x^2}{n_o^2} + \frac{y^2}{n_o^2} + 2r_{63}F_z xy + \frac{z^2}{n_e^2} = 1 \quad (12.1.14)$$

The above equation shows that the principal axes along the x and y directions are rotated because of the cross term, $2r_{63}F_z xy$. We have to find

Table 12.1 A Few Electrooptic Materials With Their Parameters [1, 4, 6, 9]

Point-Group Symmetry	Material	Refractive Index n_o	Refractive Index n_e	Wavelength $\lambda_0(\mu m)$	Nonzero Electrooptic Coefficients (10^{-12} m / V)
$3m$	LiNbO$_3$	2.297	2.208	0.633	$r_{13} = r_{23} = 8.6, r_{33} = 30.8$
					$r_{42} = r_{51} = 28, r_{22} = 3.4$
					$r_{12} = r_{61} = -r_{22}$
32	Quartz (SiO$_2$)	1.544	1.553	0.589	$r_{41} = -r_{52} = 0.2$
					$r_{62} = r_{21} = -r_{11} = 0.93$
$\bar{4}2m$	KH$_2$PO$_4$ (KDP)	1.5115	1.4698	0.546	$r_{41} = r_{52} = 8.77, r_{63} = 10.3$
		1.5074	1.4669	0.633	$r_{41} = r_{52} = 8, r_{63} = 11$
$\bar{4}2m$	NH$_4$H$_2$PO$_4$ (ADP)	1.5266	1.4808	0.546	$r_{41} = r_{52} = 23.76, r_{63} = 8.56$
		1.5220	1.4773	0.633	$r_{41} = r_{52} = 23.41, r_{63} = 7.828$
$\bar{4}2m$	KD$_2$PO$_4$ (KD*P)	1.5079	1.4683	0.546	$r_{41} = r_{52} = 8.8, r_{63} = 26.8$
$\bar{4}3m$	GaAs	3.60	$= n_o$	0.9	$r_{41} = r_{52} = r_{63} = 1.1$
		3.42	$= n_o$	1.0	$r_{41} = r_{52} = r_{63} = 1.5$
		3.34	$= n_o$	10.6	$r_{41} = r_{52} = r_{63} = 1.6$
$\bar{4}3m$	InP	3.29	$= n_o$	1.06	$r_{41} = r_{52} = r_{63} = 1.45$
		3.20	$= n_o$	1.35	$r_{41} = r_{52} = r_{63} = 1.3$
$\bar{4}3m$	ZnSe	2.60	$= n_o$	0.633	$r_{41} = r_{52} = r_{63} = 2.0$
$\bar{4}3m$	β-ZnS	2.36	$= n_o$	0.6	$r_{41} = r_{52} = r_{63} = 2.1$

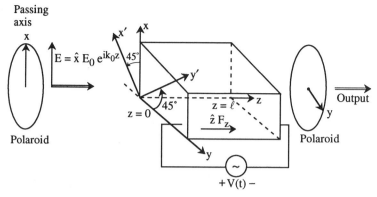

Figure 12.1. A longitudinal amplitude modulator in which an applied electric field is biased along the direction of optical wave propagation.

the new principal optical axes such that the index ellipsoid can be described by a diagonal $(K + \Delta K)_{ij}$ matrix.

A coordinate rotation of 45° on the x-y plane gives

$$x = \frac{1}{\sqrt{2}}(x' + y')$$

$$y = \frac{1}{\sqrt{2}}(-x' + y') \qquad (12.1.15)$$

Substituting (12.1.15) into (12.1.14), we find

$$x'^2\left(\frac{1}{n_o^2} - r_{63}F_z\right) + y'^2\left(\frac{1}{n_o^2} + r_{63}F_z\right) + \frac{z^2}{n_e^2} = 1 \qquad (12.1.16)$$

We may rewrite the index ellipsoid as

$$\frac{x'^2}{n_x'^2} + \frac{y'^2}{n_y'^2} + \frac{z^2}{n_e^2} = 1 \qquad (12.1.17)$$

where

$$n_x' = n_o\frac{1}{\left(1 - r_{63}n_o^2F_z\right)^{1/2}} \simeq n_o + \frac{r_{63}}{2}n_o^3F_z \qquad (12.1.18a)$$

and, similarly,

$$n_y' \simeq n_o - \frac{r_{63}}{2}n_o^3F_z \qquad (12.1.18b)$$

We note that in the new coordinate system $(x'-y'-z)$, the matrices $(K + \Delta K)_{ij}$ and ε_{ij} are diagonalized:

$$\varepsilon_0\left(\overline{\overline{\mathbf{K}}} + \Delta\overline{\overline{\mathbf{K}}}\right) = \begin{bmatrix} 1/n_x'^2 & 0 & 0 \\ 0 & 1/n_y'^2 & 0 \\ 0 & 0 & 1/n_e^2 \end{bmatrix} \qquad \overline{\overline{\varepsilon}} = \begin{bmatrix} \varepsilon_x' & 0 & 0 \\ 0 & \varepsilon_y' & 0 \\ 0 & 0 & \varepsilon_z \end{bmatrix} \quad (12.1.19)$$

where $\varepsilon_x' = n_x'^2\varepsilon_0$, $\varepsilon_y' = n_y'^2\varepsilon_0$ and $\varepsilon_z = n_e^2\varepsilon_0$. For a plane wave propagating in the $+z$ direction and polarized in the $\hat{x}'$ direction in a crystal described by diagonal permittivity tensor (12.1.19),

$$\mathbf{E} = \hat{x}'E_0\,e^{i\beta z} \qquad (12.1.20)$$

It is easy to show (Problem 12.2) from Maxwell's equations that the propagation constant β is

$$\beta = \omega\sqrt{\mu_0\varepsilon_x'} = kn_x' \qquad (12.1.21)$$

where $k = \omega\sqrt{\mu_0\varepsilon_0} = 2\pi/\lambda_0$ and λ_0 is the wavelength in free space. Similarly, for

$$\mathbf{E} = \hat{y}'E_0\,e^{i\beta z} \qquad (12.1.22)$$

we find the propagation constant is $\beta = kn_y'$.

The incident optical field, after passing through the polarizer, can be expressed as

$$\begin{aligned} \mathbf{E} &= \hat{x}E_0\,e^{ikz} \\ &= \hat{x}'\frac{E_0}{\sqrt{2}}e^{ikz} + \hat{y}'\frac{E_0}{\sqrt{2}}e^{ikz} \end{aligned} \qquad (12.1.23)$$

in free space. Upon hitting the surface at $z = 0$, the wave is decomposed into two orthonormal polarizations along the $\hat{x}'$ and $\hat{y}'$ directions; each satisfies all of the Maxwell's equations independently since both are characteristic polarizations of the crystal. The propagation constants of the $\hat{x}'$ and $\hat{y}'$ components are kn_x' and kn_y', respectively. Neglecting the reflections at the surfaces $z = 0$ and $z = \ell$, the optical field at $z = \ell$ can be written as

$$\mathbf{E} = \hat{x}'\frac{E_0}{\sqrt{2}}e^{ikn_x'\ell} + \hat{y}'\frac{E_0}{\sqrt{2}}e^{ikn_y'\ell} \qquad (12.1.24)$$

The transmitted field passing through the second polaroid is the y component of $\mathbf{E}$ in (12.1.24), or $\hat{y} \cdot \mathbf{E}$, using $\hat{x}' = (\hat{x} - \hat{y})/\sqrt{2}$ and $\hat{y}' = (\hat{x} + \hat{y})/\sqrt{2}$:

$$
\begin{aligned}
E_y &= \frac{E_0}{2}\left(-e^{ikn'_x\ell} + e^{ikn'_y\ell}\right) \\
&= \frac{E_0}{2}e^{ikn_o\ell}\left(-e^{ikn_o^3r_{63}F_z\ell/2} + e^{-ikn_o^3r_{63}F_z\ell/2}\right)
\end{aligned}
\tag{12.1.25}
$$

The transmitted power intensity divided by the incident power intensity is proportional to

$$
\frac{P_t}{P_i} = \frac{|E_y|^2}{|E_0|^2} = \sin^2\left(\frac{kn_o^3}{2}r_{63}F_z\ell\right)
\tag{12.1.26}
$$

Noting that $F_z(t)\ell = V(t)$ is the applied voltage, we define V_π, the voltage yielding a phase difference of π between the two characteristic polarizations, that is, $k(n'_x - n'_y)\ell = \pi$,

$$
V_\pi = \frac{\pi}{kn_o^3r_{63}} = \frac{\lambda_0}{2n_o^3r_{63}}
\tag{12.1.27}
$$

and obtain

$$
\frac{P_t}{P_i} = \sin^2\left(\frac{\pi V}{2V_\pi}\right)
\tag{12.1.28}
$$

By varying $V(t) = F_z(t)\ell$, the output light intensity is modulated. To obtain a linear response, $V(t)$ has to be biased near $V_\pi/2$ where the transmission factor is 50%, as shown in Fig. 12.2, i.e.

$$
V(t) = \frac{V_\pi}{2} + V_0 \sin \omega_m t
\tag{12.1.29}
$$

and the transmission factor is

$$
\begin{aligned}
\frac{P_t}{P_i} &= \sin^2\left(\frac{\pi}{4} + \frac{\pi V_0}{2V_\pi}\sin \omega_m t\right) \\
&= \frac{1}{2}\left[1 + \sin\left(\frac{\pi V_0}{V_\pi}\sin \omega_m t\right)\right] \\
&\simeq \frac{1}{2}\left(1 + \frac{\pi V_0}{V_\pi}\sin \omega_m t\right) \qquad \text{if } \pi V_0 \ll V_\pi
\end{aligned}
\tag{12.1.30}
$$

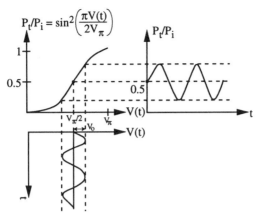

Figure 12.2.   Transmission of a light intensity in a longitudinal amplitude electrooptical modulator. The applied voltage $V(t)$ is biased at a dc value $V_\pi/2$.

We see that for a small input signal $\pi V_0 \ll V_\pi$, and a bias at the 50% point, $V_\pi/2$, a linear response can be achieved. However, since V_π is typically very large, a better way is to add a quarter-wave plate between the electrooptic crystal and the output polaroid, with the two principal axes of the plate along the x' and y' directions such that an extra phase difference of $\pi/2$ is introduced between the x' and y' components of (12.1.24):

$$\mathbf{E} = \frac{E_0}{\sqrt{2}}\left(\hat{x}'\, e^{i\pi/2}\, e^{ikn'_x\ell} + \hat{y}'\, e^{ikn'_y\ell} \right) \tag{12.1.31}$$

$$\frac{P_t}{P_i} = \frac{|E_y|^2}{|E_0|^2} = \sin^2\left(\frac{\pi}{4} + \frac{kn_o^3}{2} r_{63} F_z(t)\, \ell \right)$$

$$= \sin^2\left(\frac{\pi}{4} + \frac{\pi}{2}\frac{V(t)}{V_\pi} \right) \tag{12.1.32}$$

In this case, the modulation voltage is $V(t) = V_0 \sin \omega_m t$ instead of (12.1.29). No dc bias is necessary.

12.1.3   Transverse Amplitude Modulator

A transverse amplitude modulator is shown in Fig. 12.3 in which the applied field is biased in a direction perpendicular to the propagation direction of light. The incident optical electric field after passing through the polaroid is

$$\mathbf{E}_1 = \frac{\hat{x}' + \hat{z}}{\sqrt{2}} E_0\, e^{iky'} \tag{12.1.33}$$

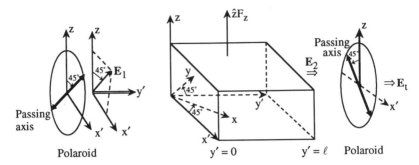

Figure 12.3. A transverse amplitude modulator. The bias field $\hat{z}F_z$ is perpendicular to the direction of optical wave propagation.

propagating in free space. Note that the direction of propagation has been chosen to be along the y' direction of the electrooptic crystal, which makes an angle 45° with the principal x and y axes of the unbiased crystal. After passing through the crystal, the x' and z components gain different phases at $y' = \ell$:

$$\mathbf{E}_2 = \hat{x}'\frac{E_0}{\sqrt{2}}e^{ikn'_x\ell} + \hat{z}\frac{E_0}{\sqrt{2}}e^{ikn_e\ell} \qquad (12.1.34)$$

Here, we still use KDP crystal in our analysis. The transmitted field through the second polaroid with the passing axis given by $(-\hat{x}' + \hat{z})/\sqrt{2}$ is

$$E_t = \mathbf{E}_2 \cdot \left(\frac{-\hat{x}' + \hat{z}}{\sqrt{2}}\right) = \frac{E_0}{2}\left(-e^{ikn'_x\ell} + e^{ikn_e\ell}\right)$$

$$= \frac{E_0}{2}e^{ik\ell(n'_x+n_e)/2}\left(-e^{ik\ell(n'_x-n_e)/2} + e^{-ik\ell(n'_x-n_e)/2}\right) \quad (12.1.35)$$

Therefore, we find the transmission factor

$$\frac{P_t}{P_i} = \frac{|E_t|^2}{|E_0|^2} = \sin^2\phi \qquad (12.1.36)$$

where

$$\phi = k\left(\frac{n'_x - n_e}{2}\right)\ell$$

$$= k\left(\frac{n_o - n_e}{2}\right)\ell + k\frac{n_o^3 r_{63}}{4}F_z(t)\ell \qquad (12.1.37)$$

Here the time-dependent field $F_z(t)$ modulates the phase $\phi(t)$, which determines the output light intensity as a function of time. Since $F_z(t) = V(t)/d$, where d is the thickness of the crystal, we find that the phase difference between the two characteristic polarizations ($\hat{x}'$ and $\hat{z}$ components) is, from (12.1.37),

$$k(n'_x - n_e)\ell = 2\phi = k(n_o - n_e)\ell + k\frac{n_o^3 r_{63}}{2}\frac{V(t)}{d}\ell \quad (12.1.38)$$

Again we can define a half-wave voltage V_π as the voltage required to introduce an extra phase shift to π:

$$V_\pi = \frac{2\pi}{kn_o^3 r_{63}}\left(\frac{d}{\ell}\right) = \frac{\lambda_0}{n_o^3 r_{63}}\left(\frac{d}{\ell}\right) \quad (12.1.39)$$

We can see that the factor d/ℓ can be chosen to be small; therefore, the half-wave voltage V_π is reduced compared with that for a longitudinal amplitude modulator. We write

$$\phi = \frac{k\ell}{2}(n_o - n_e) + \frac{\pi}{2}\frac{V(t)}{V_\pi} \quad (12.1.40)$$

The transmission factor is then

$$\frac{P_t}{P_i} = \sin^2\left[\frac{k\ell}{2}(n_o - n_e) + \frac{\pi}{2}\frac{V(t)}{V_\pi}\right]$$

$$= \frac{1}{2}\left\{1 - \cos\left[k\ell(n_o - n_e) + \frac{\pi V(t)}{V_\pi}\right]\right\} \quad (12.1.41)$$

A linear response is obtainable if we choose

$$\frac{k\ell}{2}(n_o - n_e) = \frac{\pi}{4} \quad (12.1.42)$$

and the input signal $V(t) = V_0 \sin \omega_{mt}$ is small, $V_0 \ll V_\pi$. The transmission factor is the same as (12.1.32) and is similar to Fig. 12.2.

12.2 PHASE MODULATORS

12.2.1 Optical Phase Modulation

Consider an incident optical field propagating along the z direction and passing through the polaroid as shown in Fig. 12.4:

$$\mathbf{E} = \hat{x}'E_0\, e^{ikz} \quad (12.2.1)$$

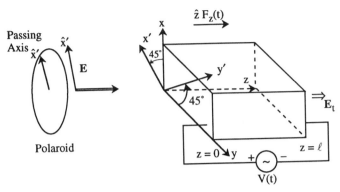

Figure 12.4. A longitudinal phase modulator with an applied electric field $\hat{z}F_z(t)$ along the propagation direction of optical field **E**.

Since the polaroid has been aligned such that the passing axis is along one of the characteristic polarizations of the electrooptic crystal, $\hat{x}'$, the transmitted field at $z = \ell$ is simply

$$\mathbf{E}_t = \hat{x}'E_0\, e^{ikn'_x\ell} \tag{12.2.2}$$

$$= \hat{x}'E_0\, e^{ikn_o\ell}\, e^{ikn_o^3 r_{63}F_z(t)\ell/2} \tag{12.2.3}$$

Therefore, if we modulate the applied electric field as

$$F_z(t) = F_0 \sin \omega_m t \tag{12.2.4}$$

we find

$$\mathbf{E}_t = \hat{x}'E_0\, e^{ikn_o\ell + i\delta \sin \omega_m t} \tag{12.2.5}$$

where

$$\delta = k\frac{r_{63}}{2}n_o^3 F_0 \ell \tag{12.2.6}$$

The electric field in the time domain is

$$\mathbf{E}_t(\ell, t) = \mathrm{Re}\big(\mathbf{E}_t\, e^{-i\omega t}\big)$$

$$= \hat{x}'E_0 \cos\big(kn_o\ell + \delta \sin \omega_m t - \omega t\big) \tag{12.2.7}$$

The phase of the output optical field is modulated by the factor $\delta \sin \omega_m t$. If we use the mathematical identity [10]

$$e^{i\delta \cos \phi} = \sum_{m=-\infty}^{\infty} J_m(\delta)\, e^{im(\phi + \pi/2)} \tag{12.2.8}$$

or, equivalently,

$$e^{i\delta \sin \phi} = \sum_{m=-\infty}^{\infty} J_m(\delta) e^{im\phi} \qquad (12.2.9)$$

Equation (12.2.7) can also be written as

$$E_t(\ell, t) = \text{Re}(E_t e^{-i\omega t}) = \text{Re}(\hat{x}'E_0 e^{i(kn_o\ell - \omega t)} e^{i\delta \sin \omega_m t})$$

$$= \text{Re}\left[\hat{x}'E_0 e^{i(kn_o\ell - \omega t)} \sum_{m=-\infty}^{\infty} J_m(\delta) e^{im\omega_m t}\right]$$

$$= \hat{x}'E_0 \sum_{m=-\infty}^{\infty} J_m(\delta) \cos[kn_o\ell - (\omega - m\omega_m)t] \qquad (12.2.10)$$

The output optical field contains, in addition to the fundamental frequency ω with an amplitude $J_0(\delta)E_0$, various sidebands with frequencies, $\omega \pm \omega_m$, $\omega \pm 2\omega_m, \ldots$, etc., with corresponding amplitudes $\mp J_1(\delta)E_0, J_2(\delta)E_0, \ldots$. Note that $J_{-m}(\delta) = (-1)^m J_m(\delta)$. If the magnitude $\delta \simeq 2.4048$, the root of the zeroth-order Bessel function ($J_0(\delta) = 0$), all of the power in the fundamental frequency ω is transferred to the nonzero-order harmonics.

Example LiNbO$_3$ has a 3m point-group symmetry and from Table 12.1 its r_{Ik} matrix has the form [1, 11]

$$\bar{\bar{r}} = \begin{bmatrix} 0 & -r_{22} & r_{13} \\ 0 & r_{22} & r_{13} \\ 0 & 0 & r_{33} \\ 0 & r_{51} & 0 \\ r_{51} & 0 & 0 \\ -r_{22} & 0 & 0 \end{bmatrix} \quad \text{where} \quad \begin{matrix} r_{12} = -r_{22} & r_{23} = r_{13} \\ r_{42} = r_{51} & r_{61} = -r_{22} \end{matrix} \qquad (12.2.11)$$

The r values at a wavelength $\lambda_0 = 0.633~\mu$m are

$$r_{13} = 8.6 \times 10^{-12}~\text{m/V} \qquad r_{33} = 30.8 \times 10^{-12}~\text{m/V}$$

$$r_{22} = 3.4 \times 10^{-12}~\text{m/V} \qquad r_{51} = 28.0 \times 10^{-12}~\text{m/V}$$

The refractive indices of LiNbO$_3$ have a uniaxial form:

$$n_x = n_y = n_o = 2.297 \quad \text{and} \quad n_z = n_e = 2.208$$

For an applied electric field $\mathbf{F} = \hat{x}F_1 + \hat{y}F_2 + \hat{z}F_3$, the index ellipsoid is described by

$$\varepsilon_0 \sum_{i,j} (K_{ij} + \Delta K_{ij}) x_i x_j = 1$$

or using the symmetry in the $\bar{\bar{r}}$ matrix,

$$x^2\left(\frac{1}{n_o^2} - r_{22}F_2 + r_{13}F_3\right) + y^2\left(\frac{1}{n_o^2} + r_{22}F_2 + r_{13}F_3\right)$$

$$+ z^2\left(\frac{1}{n_e^2} + r_{33}F_3\right) + 2yzr_{51}F_2 + 2zxr_{51}F_1 - 2xyr_{22}F_1 = 1 \quad (12.2.12)$$

Since r_{33} is the largest coefficient, an applied electric field along the z direction will be most efficient for the electrooptic control. Therefore, for $F_1 = F_2 = 0$, the index ellipsoid is given by

$$x^2\left(\frac{1}{n_o^2} + r_{13}F_3\right) + y^2\left(\frac{1}{n_o^2} + r_{13}F_3\right) + z^2\left(\frac{1}{n_e^2} + r_{33}F_3\right) = 1 \quad (12.2.13)$$

or

$$\frac{x^2}{n_x^2} + \frac{y^2}{n_y^2} + \frac{z^2}{n_z^2} = 1 \qquad (12.2.14a)$$

where the new refractive indices are

$$n_x = n_y \simeq n_o - \tfrac{1}{2}n_o^3 r_{13}F_3 \qquad (12.2.14b)$$

$$n_z \simeq n_e - \tfrac{1}{2}n_e^3 r_{33}F_3 \quad \blacksquare \qquad (12.2.14c)$$

12.2.2 X-cut LiNbO₃ Crystal

For an X-cut LiNbO$_3$ crystal, as shown in Fig. 12.5a, two electrodes are placed symmetrically on both sides of the waveguides such that the bias field $\mathbf{F} = \hat{z}F_3$ is along the z direction and the index ellipsoid is described by (12.2.14). An incident optical electric field with TE polarization will transmit as

$$\mathbf{E} = \hat{z}E_0\, e^{ikn_z y} \underset{(y=\ell)}{=} \hat{z}E_0\, e^{ikn_e\ell - i(1/2)n_e^3 r_{33}F_3\ell} \qquad (12.2.15a)$$

Similarly, for TM polarization,

$$\mathbf{E} = \hat{x}E_0\, e^{ikn_x y} \underset{(y=\ell)}{=} \hat{x}E_0\, e^{ikn_o\ell - i(1/2)n_o^3 r_{13}F_3\ell} \qquad (12.2.15b)$$

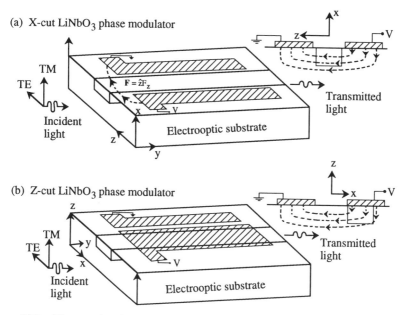

Figure 12.5. Electrooptic phase modulator using (a) X-cut LiNbO$_3$ substrate, where the electrodes are placed symmetrically on both sides of the waveguide, such that the bias field is along the z direction; and (b) Z-cut LiNbO$_3$ substrate, where one electrode is placed directly above the waveguide such that the bias field in the waveguide is along the positive (or negative) z direction for the most efficient phase modulator since r_{33} is the largest electrooptic coefficient.

Therefore, TE polarization should be used for most efficient phase modulation since $r_{33} > r_{13}$.

12.2.3 Z-cut LiNbO$_3$ Crystal

For a Z-cut LiNbO$_3$ crystal as shown in Fig. 12.5b, the electrodes are placed such that the waveguide is below one of the two electrodes where the field is perpendicular to the Z-cut surface and $\mathbf{F} = \hat{z}F_3$. The optical transmission field will be

$$\mathbf{E} = \hat{x}E_0\, e^{ikn_o\ell - i(1/2)n_o^3 r_{13} F_3 \ell} \qquad \text{(TE polarization)} \qquad (12.2.16a)$$

and

$$\mathbf{E} = \hat{z}E_0\, e^{ikn_e\ell - i(1/2)n_e^3 r_{33} F_3 \ell} \qquad \text{(TM polarization)} \qquad (12.2.16b)$$

In this case, TM polarization is preferred for most efficient phase modulation. For more discussions on the polarization and electrode designs, including the TE and TM polarization conversions, see Refs. 11–16.

12.3   ELECTROOPTIC EFFECTS IN WAVEGUIDE DEVICES

In Chapter 8, we discussed optical directional couplers using parallel wave-guides. Here we discuss briefly the applications of electrooptic effects in waveguide structures. We have discussed the use of KDP and $LiNbO_3$ in electrooptic amplitude and phase modulators. Some of these materials, such as $LiNbO_3$, as electrooptic crystals have been used in many commercial devices. Here, we discuss the use of GaAs in electrooptic waveguide devices. For integrated optoelectronics, semiconductor materials, especially III–V compounds, are attractive because many active and passive components, such as semiconductor lasers, photodetectors, and field effect transistors, are made of these compound semiconductors. However, considerable research work is still necessary for integrating passive and active devices with desired opera-tion characteristics.

Example   GaAs at 10.6 μm wavelength $n_o = 3.34, r_{41} = r_{52} = r_{63} = 1.6 \times 10^{-12}$ m/V. All other r components are zero.

$$\bar{\bar{r}} = \begin{bmatrix} 0 & 0 & 0 \\ 0 & 0 & 0 \\ 0 & 0 & 0 \\ r_{41} & 0 & 0 \\ 0 & r_{52} & 0 \\ 0 & 0 & r_{63} \end{bmatrix} \qquad (12.3.1)$$

For a biased field $\mathbf{F} = \hat{x}F_1 + \hat{y}F_2 + \hat{z}F_3$, the new index ellipsoid is

$$\frac{x^2}{n_o^2} + \frac{y^2}{n_o^2} + \frac{z^2}{n_o^2} + 2r_{63}F_1 yz + 2r_{63}F_2 xz + 2r_{63}F_3 xy = 1 \quad (12.3.2)$$

If we choose $F_1 = F_2 = 0, \mathbf{F} = \hat{z}F_3$, we then have

$$\frac{x^2}{n_o^2} + \frac{y^2}{n_o^2} + 2r_{63}F_3 xy + \frac{z^2}{n_o^2} = 1 \qquad (12.3.3)$$

This is similar to that of the KDP materials, except that the crystal in the absence of field is isotropic, or $n_e = n_o$. Again a rotation of 45° in the x-y plane gives $x = (x' + y')/\sqrt{2}, y = (-x' + y')/\sqrt{2}$,

$$\frac{x'^2}{n_{x'}^2} + \frac{y'^2}{n_{y'}^2} + \frac{z'^2}{n_o^2} = 1 \qquad (12.3.4a)$$

where

$$n_{x'} = n_o + \tfrac{1}{2}n_o^3 r_{63} F_3 \qquad (12.3.4b)$$

$$n_{y'} = n_o - \tfrac{1}{2}n_o^3 r_{63} F_3 \qquad (12.3.4c)$$

The previous analysis for longitudinal amplitude modulators for LiNbO$_3$ is applicable to GaAs materials. ■

In the following examples, we consider (a) a Mach–Zehnder interferometric waveguide modulator, (b) a directional coupler modulator, and (c) a $\Delta\beta$-phase-reversal directional coupler, as shown in Fig. 12.6. The input

(a) A Mach-Zehnder interferometric waveguide modulator

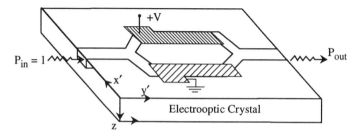

(b) A directional coupler modulator

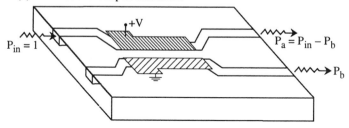

(c) A $\Delta\beta$-phase-reversal directional coupler

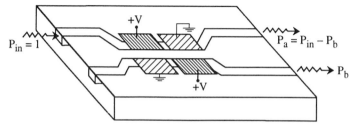

Figure 12.6. Waveguide electrooptic devices with electrode designs. (a) A Mach–Zehnder interferometric modulator, (b) a directional coupler modulator, and (c) a $\Delta\beta$-phase-reversal directional coupler.

power P_{in} is taken as 1 in each case. The electrodes are designed such that the applied electric field is along the z direction, $\mathbf{F} = \pm \hat{z} F_3$, and the bias fields in two waveguides are opposite in signs. Therefore, the difference between the two propagation factors is approximately

$$
\begin{aligned}
\Delta \beta &\simeq k \left(n_o + \frac{1}{2} n_o^3 r_{63} F_3 \right) - k \left(n_o - \frac{1}{2} n_o^3 r_{63} F_3 \right) \\
&= k n_o^3 r_{63} F_3 \\
&\simeq k n_o^3 r_{63} \frac{V(t)}{d}
\end{aligned}
\tag{12.3.5}
$$

where an effective width d is defined for the electric field ($F_3 \simeq V/d$). The important point is that $\Delta \beta \propto V(t)$. We will study the transmission characteristics of these devices as a function of the detuning factor $\Delta \beta \ell$.

12.3.1   Mach–Zehnder Interferometric Waveguide Modulator

As shown in Fig. 12.6a, a single waveguide is branched into two arms for a distance ℓ and combined again into one arm as the output waveguide. The waveguide dimension can be chosen to guide the fundamental mode only.

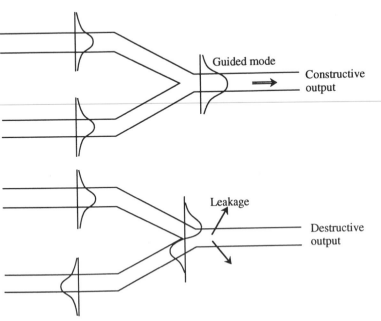

Figure 12.7. An illustration of the Y-junction constructive and destructive outputs in a Mach–Zehnder interferometric waveguide modulator.

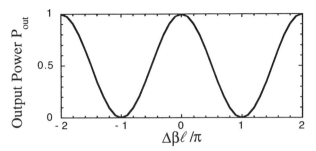

Figure 12.8. The output power from a Mach–Zehnder interferometric waveguide modulator as a function of the mismatch factor $\Delta\beta L$.

With a proper choice of the polarization of the incident wave and the electrode design, the transmitted intensity is

$$P_{out} = \frac{1}{4}|e^{i\beta_a \ell} + e^{i\beta_b \ell}|^2$$

$$= \cos^2\left(\frac{\Delta\beta}{2}\ell\right) \qquad (12.3.6)$$

where the output intensity is normalized such that the peak transmission factor is 1 for perfect power transmission. One way to understand this transmission behavior is that if the guided modes are in phase at the exit of the Y junction (Fig. 12.7), they add up constructively and transmit with the maximum power. If they are out of phase by 180°, they will cancel each other. Anther way to look at this is that if they add up to a first-order mode, it will leak out over a very short distance, since the waveguide is designed to guide the fundamental mode only, resulting in a destructive output. A plot of the output power P_{out} vs. the mismatch $\Delta\beta L$ is shown in Fig. 12.8. The interferometric behavior is clearly seen.

12.3.2 Directional Coupler Modulator

For an incident optical beam into waveguide a in a directional coupler modulator, the output power is

$$P_b = |b(\ell)|^2 = \frac{|K|^2}{\psi^2}\sin^2(\psi\ell) \qquad (12.3.7a)$$

$$\psi = \sqrt{\left(\frac{\Delta\beta}{2}\right)^2 + K^2} \qquad (12.3.7b)$$

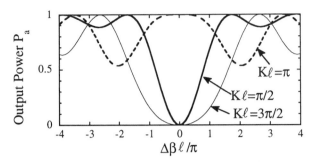

Figure 12.9. The output power from waveguide a as a function of $\Delta\beta L$ for $K\ell = \pi/2$, π, and $3\pi/2$ for a directional coupler modulator.

and

$$P_a = P_{\text{in}} - P_b = 1 - |b(\ell)|^2 \tag{12.3.8}$$

where the input power is assumed to be 1. Since $\Delta\beta = \beta_a - \beta_b = kn_o^3 r_{63} F_3$ $\simeq kn_o^3 r_{63} V/d$, we plot the output power P_a vs. $\Delta\beta\ell$. Suppose we design the modulator with a length ℓ such that $P_a = 0$, and $P_b = 1$, at $\Delta\beta = 0$, i.e., $K\ell = \pi/2$. To switch to $P_a = 1$, and $P_b = 0$, we require at least $\Delta\beta\ell = \sqrt{3}\,\pi$, assuming the field-induced change in the refractive index affects the coupling coefficient negligibly. (Otherwise, we can calculate the field-dependent K and still use the expressions for P_a and P_b in (12.3.7) and (12.3.8) to find the output powers.) To switch from a cross state to a parallel state, the applied voltage has to be large enough such that $\Delta\beta\ell = \sqrt{3}\,\pi$ is satisfied. A plot of P_a vs. $\Delta\beta\ell$ for $K\ell = \pi/2$ is shown as the thick solid curve in Fig. 12.9. We also plot P_a vs. $\Delta\beta\ell$ for $K\ell = \pi$, and $K\ell = 3\pi/2$. We see that complete switching from the $\otimes$ state to the $\ominus$ state is possible (for $K\ell = \pi/2$ or $3\pi/2$). For $K\ell = \pi$, where we start with the $\ominus$ state at $\Delta\beta\ell = 0$, it is impossible to switch to the $\otimes$ state simply by changing $\Delta\beta\ell$ alone. This fact can also be checked with the switching diagram in Fig. 8.19.

12.3.3 Δβ-Phase-Reversal Directional Coupler [11–17]

The $\Delta\beta$-phase-reversal directional coupler is shown in Fig. 12.6c, and its analysis can be found in Section 8.6. The switching diagram is shown in Fig. 8.21. Suppose we start with the parallel state at $K\ell = \pi$, $\Delta\ell = 0$, where $\Delta = (\beta_a - \beta_b)/2$. The output power is

$$P_a = |a(\ell)|^2$$

$$= \left| \cos^2\left(\frac{\psi\ell}{2}\right) + \frac{\Delta^2 - K^2}{\psi^2} \sin^2\left(\frac{\psi\ell}{2}\right) \right|^2 \tag{12.3.9}$$

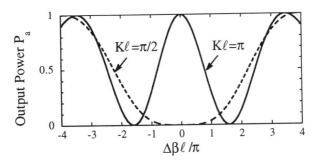

Figure 12.10.   The output power P_a of a $\Delta\beta$-phase-reversal directional coupler as a function of $\Delta\beta\ell$ for $K\ell = \pi/2$ and π.

We plot P_a vs. $\Delta\beta\ell$ as the solid curve in Fig. 12.10 for $K\ell = \pi$ and also for $K\ell = \pi/2$ (dashed curve) for comparison.

12.4   SCATTERING OF LIGHT BY SOUND: RAMAN–NATH AND BRAGG DIFFRACTIONS

The refractive index of a medium can be caused by a mechanical strain produced by an acoustic wave; this is called the acoustooptic effect. A sound wave creates a sinusoidal perturbation of the density, or strain or pressure of the material. The induced change in refractive index can be described as

$$\Delta n(\mathbf{r}, t) = \Delta n \cos(\mathbf{k}_s \cdot \mathbf{r} - \omega_s t) \qquad (12.4.1)$$

with ω_s = the angular frequency, $\mathbf{k}_s$ = the wave vector, $k_s = 2\pi/\lambda_s$, λ_s = wavelength, and $v_s = \omega_s/k_s$ is the velocity of sound in the medium.

12.4.1   Raman–Nath Diffraction [9]

Here the length of the interaction between the light and the acoustic wave ℓ is small:

$$\ell \ll \frac{kn}{k_s^2} \qquad (12.4.2)$$

where $k = 2\pi/\lambda_0$, n = the refractive index of the medium, and λ_0 = the optical wavelength in free space. This is called the Raman–Nath regime of diffraction (Fig. 12.11). In this case, the thin region in which the acoustic wave propagates acts like a phase grating, and the diffracted lights can go to many different directions determined by the generalized Snell's law for a

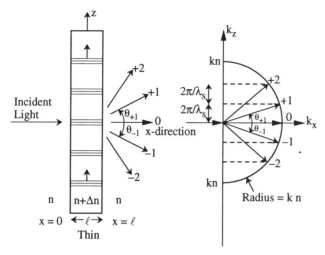

Figure 12.11. Raman–Nath diffraction. The interaction length ℓ is very short and the thin region acts like an optical phase grating with the period equal to the acoustic wavelength λ_s.

grating:

$$kn \sin \theta_m = m \frac{2\pi}{\lambda_s} \qquad m = \text{an integer}$$

$$\theta_m = \sin^{-1}\left(m \frac{\lambda_0}{n\lambda_s}\right) \qquad (12.4.3)$$

A simple analysis of the diffraction efficiency for this case is to consider $\mathbf{E} = \hat{y}E_y(x, z, t)$ at $x = \ell$

$$E_y(x = \ell, z, t) = E_0\, e^{i[kn(z,t)\ell - \omega t]} = E_0\, e^{i(kn\ell - \omega t)} e^{ik\Delta n\ell \cos(k_s z - \omega_s t)} \qquad (12.4.4)$$

where

$$n(z, t) = n + \Delta n \cos(k_s z - \omega_s t) \qquad (12.4.5)$$

has been used. We then write the field at $x = \ell$ using the mathematical identity [10]

$$e^{iu \cos \phi} = \sum_{m=-\infty}^{\infty} i^m J_m(u)\, e^{im\phi} \qquad (12.4.6)$$

and set

$$u = k\Delta n\ell \qquad (12.4.7)$$

$$\phi = k_s z - \omega_s t \qquad (12.4.8)$$

The electric field at $x = \ell$ then becomes

$$E_y\left(x = \ell, z, t\right) = E_0\, e^{i(kn\ell - \omega t)} \sum_{m=-\infty}^{\infty} i^m J_m\left(\frac{2\pi}{\lambda_0}\Delta n\,\ell\right) e^{im(k_s z - \omega_s t)}$$

$$= E_0\, e^{ikn\ell} \sum_{m=-\infty}^{\infty} i^m J_m\left(\frac{2\pi}{\lambda_0}\Delta n\,\ell\right) e^{imk_s z}\, e^{-i(\omega + m\omega_s)t} \quad (12.4.9)$$

Since for $x \geq \ell$, the electric field has to satisfy the wave equation in the medium described by the refractive index n, we should have the solution of the form

$$E_y\left(x \geq \ell, z, t\right) = \sum_{m=-\infty}^{\infty} E_m\, e^{ik_{xm}(x-\ell) + ik_{zm}z - i\omega_m t} \quad (12.4.10)$$

where

$$\omega_m = \omega + m\omega_s \qquad k_{zm} = mk_s$$

$$k_{xm} = \sqrt{\left(\frac{\omega_m}{c}n\right)^2 - k_{zm}^2} \quad (12.4.11)$$

and

$$E_m = E_0\, e^{ikn\ell}i^m J_m\left(\frac{2\pi}{\lambda_0}\Delta n\,\ell\right) \quad (12.4.12)$$

We note that $\omega_s \ll \omega$; therefore, $\omega_m \simeq \omega$, and

$$k_{xm} = \sqrt{\left(\frac{\omega}{c}n\right)^2 - (mk_s)^2} \quad (12.4.13)$$

The diffraction angle of the mth order is therefore

$$kn \sin\theta_m \simeq mk_s \quad (12.4.14)$$

or

$$\theta_m = \sin^{-1}\left(\frac{mk_s}{kn}\right) = \sin^{-1}\left(m\frac{\lambda_0}{n\lambda_s}\right) \quad (12.4.15)$$

12.4.2 Bragg Diffraction

When the interaction length ℓ between the optical and acoustic waves is long compared with kn/k_s^2, we have the Bragg diffraction. In this case, the incident $\mathbf{k}_i$ vector has to come from a particular direction satisfying the

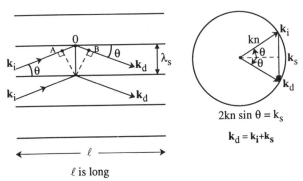

$$2kn \sin \theta = k_s$$

$$k_d = k_i + k_s$$

ℓ is long

Figure 12.12. Bragg diffraction. When the interaction length ℓ is long, a particular angle of incidence with one diffracted beam satisfying the Bragg condition $2kn \sin \theta = k_s$ will be observed. ($\overline{AO} + \overline{OB} = \lambda / n$ for constructive interference. Therefore, $2\lambda_s \sin \theta = \lambda / n$.)

Bragg condition:

$$2kn \sin \theta = k_s \qquad (12.4.16)$$

where θ is the angle of incidence, which is also the angle of diffraction. There is only one diffracted beam determined by the above Bragg condition (see Fig. 12.12). Our analysis of the Bragg diffraction is presented in Section 12.5.

12.5 COUPLED-MODE ANALYSIS FOR BRAGG ACOUSTOOPTIC WAVE COUPLER [1, 2]

The analysis for Bragg diffraction can be based on the coupled-mode theory. We start with the Maxwell equations:

$$\nabla \times \mathbf{E} = -\frac{\partial}{\partial t} \mu \mathbf{H}$$

$$\nabla \times \mathbf{H} = \frac{\partial}{\partial t} \mathbf{D} \qquad (12.5.1)$$

where the displacement vector is

$$\mathbf{D} = \varepsilon_0 n^2 (\mathbf{r}, t) \mathbf{E} \qquad (12.5.2)$$

and the refractive index variation is

$$n(\mathbf{r}, t) = n + \Delta n(\mathbf{r}, t)$$

$$\Delta n(\mathbf{r}, t) = \Delta n \cos(\mathbf{k}_s \cdot \mathbf{r} - \omega_s t) \qquad (12.5.3)$$

Here the background refractive index n and the amplitude of variation Δn are independent of the position and t.

Consider a TE polarized wave $\mathbf{E} = \hat{y}E_y$ and assume that both the acoustic wave and the optical wave propagate in the x-z plane ($\mathbf{r} = x\hat{x} + z\hat{z}$). This solution $\mathbf{E} = \hat{y}E_y(x, z, t)$ satisfies Gauss's law because

$$\nabla \cdot \mathbf{D} = \nabla \cdot \left[\varepsilon_0 n^2(\mathbf{r}, t)\mathbf{E}\right] = \frac{\partial}{\partial y}\left[\varepsilon_0 n^2(x, z, t)E_y(x, z, t)\right] = 0 \quad (12.5.4)$$

and $\nabla \cdot \mathbf{E} = \hat{y} \cdot \nabla E_y(x, z, t) = 0$ too. The wave equation is derived from (12.5.1) and (12.5.2):

$$\nabla \times (\nabla \times \mathbf{E}) = -\mu_0 \varepsilon_0 \frac{\partial^2}{\partial t^2}\left[n^2 + 2n\,\Delta n(\mathbf{r}, t)\right]\mathbf{E} \quad (12.5.5)$$

$$\nabla^2 \mathbf{E} - \frac{n^2}{c^2}\frac{\partial^2 \mathbf{E}}{\partial t^2} = \frac{2n}{c^2}\frac{\partial^2}{\partial t^2}\left[\Delta n(\mathbf{r}, t)\mathbf{E}\right] \quad (12.5.6)$$

We assume the incident electric field to be

$$\mathbf{E}_i(\mathbf{r}, t) = \hat{y}\tfrac{1}{2}E_i(\mathbf{r})\,e^{i(\mathbf{k}_i \cdot \mathbf{r} - \omega_i t)} + \text{c.c.} \quad (12.5.7)$$

and the diffracted electric field to be

$$\mathbf{E}_d(\mathbf{r}, t) = \hat{y}\tfrac{1}{2}E_d(\mathbf{r})\,e^{i(\mathbf{k}_d \cdot \mathbf{r} - \omega_d t)} + \text{c.c.} \quad (12.5.8)$$

The variation of the refractive index can be put in the form

$$\Delta n(\mathbf{r}, t) = \Delta n \cos(\mathbf{k}_s \cdot \mathbf{r} - \omega_s t)$$
$$= \frac{\Delta n}{2}e^{i(\mathbf{k}_s \cdot \mathbf{r} - \omega_s t)} + \frac{\Delta n}{2}e^{-i(\mathbf{k}_s \cdot \mathbf{r} - \omega_s t)} \quad (12.5.9)$$

Then

$$\nabla^2 \mathbf{E}_i = \hat{y}\frac{1}{2}\left[-k_i^2 E_i + 2i\mathbf{k}_i \cdot \nabla E_i + \nabla^2 E_i\right]e^{i(\mathbf{k}_i \cdot \mathbf{r} - \omega_i t)} + \text{c.c.}$$
$$\approx \hat{y}\frac{1}{2}\left(-k_i^2 E_i + 2ik_i\frac{\partial E_i}{\partial r_i}\right)e^{i(\mathbf{k}_i \cdot \mathbf{r} - \omega_i t)} + \text{c.c.} \quad (12.5.10)$$

where the second derivative of E_i has been ignored, since we assume that the amplitude $E_i(r)$ is slowly varying compared with the $\exp(i\mathbf{k}_i \cdot \mathbf{r})$ dependence and r_i is now along the direction of $\mathbf{k}_i$. A similar expression holds for $\nabla^2 E_d$.

The term containing the product of $\Delta n(\mathbf{r}, t)\mathbf{E}$ will give rise to four terms:

$$\Delta n(\mathbf{r}, t)\mathbf{E}_i = \frac{\Delta n}{4} E_i(r) \left(e^{i(\mathbf{k}_i + \mathbf{k}_s)\cdot\mathbf{r} - i(\omega_i + \omega_s)t} + e^{i(\mathbf{k}_i - \mathbf{k}_s)\cdot\mathbf{r} - i(\omega_i - \omega_s)t} \right) + \text{c.c.}$$

$$(12.5.11)$$

and a similar expression holds for $\Delta n(\mathbf{r}, t)\mathbf{E}_d$. Noting that the total electric field $\mathbf{E} = \mathbf{E}_i + \mathbf{E}_d$, we compare the terms of the same spatial and time variations and find

$$\mathbf{k}_d = \mathbf{k}_i + \mathbf{k}_s \qquad \omega_d = \omega_i + \omega_s \qquad (12.5.12)$$

or

$$\mathbf{k}_d = \mathbf{k}_i - \mathbf{k}_s \qquad \omega_d = \omega_i - \omega_s \qquad (12.5.13)$$

These results are illustrated in Fig. 12.13. Equation (12.5.12) shows the conservations of momentum and energy for a photon with initial wave vector $\mathbf{k}_i$ absorbing a phonon with a wave vector $\mathbf{k}_s$ resulting in a final photon state with momentum $\hbar\mathbf{k}_d = \hbar\mathbf{k}_i + \hbar\mathbf{k}_s$ and energy $\hbar\omega_d = \hbar\omega_i + \hbar\omega_s$. Similarly, (12.5.13) corresponds to the emission of a phonon from the incident photon. Here $\hbar$ is the Planck constant. Also noting that $k_i = (\omega_i/c)n$, $k_d = (\omega_d/c)n$, we find from (12.5.6) and (12.5.10)

$$i\mathbf{k}_i \cdot \nabla E_i = ik_i\frac{\partial E_i}{\partial r_i} = -\frac{\omega_i^2 n}{2c^2}\Delta n E_d(r)$$

Since $\mathbf{r}_i$ is along the direction of $\mathbf{k}_i$, and $\mathbf{r}_d$ is along the direction of $\mathbf{k}_d$, we take $\mathbf{r} = x\hat{x}$, and

$$r_i \cos\theta = x \qquad r_d \cos\theta = x \qquad (12.5.14)$$

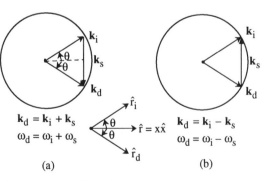

(a)   (b)

Figure 12.13. The diagrams for the diffraction of light by sound: (a) $\mathbf{k}_d = \mathbf{k}_i + \mathbf{k}_s$, $\omega_d = \omega_i + \omega_s$ and (b) $\mathbf{k}_d = \mathbf{k}_i - \mathbf{k}_s$, $\omega_d = \omega_i - \omega_s$.

We obtain

$$\frac{dE_i}{dx} = iK_i E_d \qquad K_i = \frac{\omega_i \Delta n}{2c \cos \theta} \qquad (12.5.15a)$$

$$\frac{dE_d}{dx} = iK_d E_i \qquad K_d = \frac{\omega_d \Delta n}{2c \cos \theta} \qquad (12.5.15b)$$

Since $\omega_s \ll \omega_i, \omega_d$ we have $\omega_d \simeq \omega_i \equiv \omega$ and $K_i \simeq K_d \equiv K$:

$$K = \frac{\omega \Delta n}{2c \cos \theta} \qquad (12.5.16)$$

The solutions for the coupled-mode equation given the initial conditions $E_i(0)$ and $E_d(0)$ are

$$E_i(x) = E_i(0)\cos Kx + iE_d(0)\sin Kx$$
$$E_d(x) = E_d(0)\cos Kx + iE_i(0)\sin Kx \qquad (12.5.17)$$

If initially, $E_d(0) = 0$, the field amplitudes are

$$E_i(x) = E_i(0)\cos Kx$$
$$E_d(x) = iE_i(0)\sin Kx \qquad (12.5.18)$$

The energy $|E_i(0)|^2$ is coupled to $|E_d(x)|^2$ and backward during the interaction as the optical waves propagate along the x direction, as shown in Fig. 12.14. We can write that the diffraction efficiency at a length ℓ is

$$\eta = \frac{|E_d(\ell)|^2}{|E_i(0)|^2} = \sin^2 K\ell \qquad (12.5.19)$$

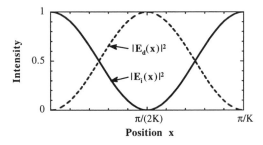

Figure 12.14. The coupling of energies between the incident and diffracted optical waves in an acoustooptic medium in which a sound wave propagates.

PROBLEMS

12.1 Calculate the voltage parameter $V_\pi = \lambda_0/(2n_o^3 r_{63})$ for the materials and wavelengths with the nonzero r_{63} coefficients in Table 12.1.

12.2 Show from Maxwell's equations that for a permittivity tensor in the principal axis system,

$$\bar{\bar{\varepsilon}} = \begin{bmatrix} \varepsilon_x & 0 & 0 \\ 0 & \varepsilon_y & 0 \\ 0 & 0 & \varepsilon_z \end{bmatrix}$$

(a) a plane wave polarized along the principal axis $\hat{x}$ and propagating along the z direction,

$$\mathbf{E} = \hat{x} E_0\, e^{i\beta z}$$

the propagation constant is $\beta = \omega\sqrt{\mu\varepsilon_x}$;

(b) a plane wave of the form $\mathbf{E} = \hat{y} E_0\, e^{i\beta z}$ will have a propagation constant $\beta = \omega\sqrt{\mu\varepsilon_y}$.

12.3 For a longitudinal amplitude modulator as shown in Fig. 12.1,

(a) if the bias voltage is $V(t) = (0.5 + 0.1\sin \omega_m t)V_\pi$, plot the output light intensity as a function of time.

(b) Repeat part (a) if $V(t) = 0.5 V_\pi \sin \omega_m t$.

12.4 Modify the design in Fig. 12.1 by adding a quarter-wave plate such that the transfer function (12.1.32) can be realized with a linear response and the bias voltage $V(t)$ will not require a dc bias voltage.

12.5 For the transverse modulator shown in Fig. 12.3, plot the transmission factor P_t/P_i vs. time, assuming that

$$\frac{k\ell}{2}(n_o - n_e) = \frac{\pi}{4} \quad \text{and} \quad V(t) = 0.1 V_\pi \sin \omega_m t$$

12.6 A quarter-wave plate is added immediately after the first polaroid in the transverse amplitude modulator in Fig. 12.3, and the electrooptical material is GaAs ($n_e = n_o = 3.42$), assuming that the wavelength λ_0 is 1.0 μm. The electric field $\mathbf{E}_1$ is circularly polarized

$$\mathbf{E}_1 = \left(\hat{x}' + \hat{z}\, e^{i\pi/2}\right)\frac{E_0}{\sqrt{2}} e^{iky'}$$

before impinging on the GaAs crystal.

(a) Find the electric field $\mathbf{E}_2$ at $y' = \ell$ in Fig. 12.3.

(b) Find the transmitted field $\mathbf{E}_1$ after passing the exit polaroid.

(c) Obtain the transmission factor P_t/P_i and plot it vs. time for $V(t) = (V_\pi/4)\sin \omega_m t$.

12.7 Consider a transverse electrooptic modular, as shown in Fig. 12.15. The incident electric field is randomly polarized and only half of its power passes through the polaroid. The crystal is a KDP with an ac electric field applied in the z direction, and the refractive index ellipsoid is described by n_o on the x-y plane and n_e along the z axis before the ac field is applied.

(a) Find the expressions for the optical electric fields $\mathbf{E}_1$ and $\mathbf{E}_2$.

(b) Find the expressions for the electric fields $\mathbf{E}_3$ and $\mathbf{E}_4$ after they have been reflected from the perfect mirror.

(c) Assume that the applied ac electric field across the modulator in this problem is $F_z(t) = F_{z0}\cos \omega t$. Find the ratio of the output optical intensity to the incident optical intensity as a function of time. Use a graphical approach to illustrate your solution assuming that

$$\frac{k\ell}{2}(n_e - n_o) = \frac{\pi}{8} \qquad \frac{kn_o^3}{2}r_{63}F_{z0}\ell = \frac{\pi}{8}$$

(d) If we have a dc applied field, $F_z = E_0$, find the value E_0 such that the incident light P_{in} is completely absorbed by the system.

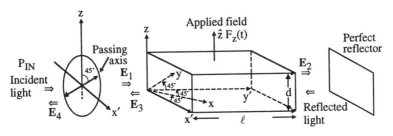

Figure 12.15.

12.8 (a) For A GaAs transverse modulator, derive the index ellipsoid for

$$\mathbf{F} = F_1\hat{x} + F_2\hat{y} + F_3\hat{z}$$

(b) If $\mathbf{F}$ is along the (111) direction, i.e.,

$$\mathbf{F} = \frac{(\hat{x} + \hat{y} + \hat{z})F_0}{\sqrt{3}}$$

design a transverse modulator and calculate the voltage parameter V_π.

12.9 Discuss the design of a phase modulator using GaAs compared with that for LiNbO$_3$ used in the text.

12.10 For a GaAs phase modulator, compare the longitudinal configuration in Fig. 12.4 vs. a possible transverse configuration such that the direction of the applied electric field **F** is perpendicular to the direction of the optical wave propagation.

12.11 Derive (12.3.5) and (12.3.6).

12.12 (a) Check the output power P_a in Fig. 12.9 using (12.3.8) for $K\ell = \pi/2$.

(b) Plot P_a vs. $(\Delta\beta\ell/\pi)$ for $K\ell = 2\pi$.

12.13 Plot the output power P_a vs. $\Delta\beta\ell/\pi$ for a $\Delta\beta$-phase-reversal coupler using (12.3.9) for $K\ell = 3\pi/2$.

12.14 Derive (12.4.9)–(12.4.12).

12.15 Derive the coupled-mode equations in (12.5.15a) and (12.5.15b).

REFERENCES

1. A. Yariv, *Optical Electronics*, 3d ed. Holt-Rinehart & Winston, New York, 1985.

2. H. A. Haus, *Waves and Fields in Optoelectronics*, Prentice-Hall, Englewood Cliffs, NJ, 1984.

3. B. E. A. Saleh and M. C. Teich, *Fundamentals of Photonics*, Wiley, New York, 1991.

4. S. Adachi, *Physical Properties of III–V Semiconductor Compounds*, Wiley, New York, 1992.

5. K. H. Hellwege, Ed., *Landolt-Börnstein Numerical Data and Functional Relationships in Science and Technology*, New Series, Group III **17a**, Springer, Berlin, 1982; Groups III–V **22a**, Springer, Berlin, 1986.

6. K. Tada and N. Suzuki, "Linear electrooptical properties of InP," *Jpn. J. Appl. Phys.* **19**, 2295–2296 (1980); and N. Suzuki and K. Tada, "Electrooptic properties and Raman scattering in InP," *Jpn. J. Appl. Phys.* **23**, 291–295 (1984).

7. S. Adachi and K. Oe, "Linear electro-optic effects in zincblende-type semiconductors: Key properties of InGaAsP relevant to device design," *J. Appl. Phys.* **56**, 74–80 (1984); and "Quadratic electrooptic (Kerr) effects in zincblende-type semiconductors: Key properties of InGaAsP relevant to device design," *J. Appl. Phys.* **56**, 1499–1504 (1984).

8. S. Adachi, *Properties of Indium Phosphide*, INSPEC, The Institute of Electrical Engineers, London, 1991.

9. A. K. Ghatak and K. Thyagarajan, *Optical Electronics*, Cambridge University Press, Cambridge, UK, 1989.

10. M. Abramowitz and I. A. Stegun, Eds., *Handbook of Mathematical Functions with Formulas, Graphs, and Mathematical Tables*, Chapter 9, Dover, New York, 1972.

11. S. Thaniyavarn, "Optical modulation: Electrooptical Devices," Chapter 4 in K. Chang, Ed., *Handbook of Microwave and Optical Components*, Vol. 4 of *Fiber and Electro-Optical Components*, Wiley, New York, 1991.

12. H. Nishihara, M. Haruna, and T. Suhara, *Optical Integrated Circuits*, McGraw-Hill, New York, 1989.

13. T. Tamir, Ed., *Guided-Wave Optoelectronics*, 2d ed., Springer, Berlin, 1990.

14. R. C. Alferness, "Guided-wave devices for optical communication," *IEEE J. Quantum Electron.* **QE-17**, 946–959 (1981).

15. O. G. Ramer, "Integrated optic electrooptic modulator electrode analysis," *IEEE J. Quantum Electron.* **QE-18**, 386–392 (1982).

16. D. Marcuse, "Optimal electrode design for integrated optics modulators," *IEEE J. Quantum Electron.* **QE-18**, 393–398 (1982).

17. H. Kogelnik and R. V. Schmidt, "Switched directional couplers with alternating $\Delta\beta$," *IEEE J. Quantum Electron.* **QE-12**, 396–401 (1976).

13

Electroabsorption Modulators

Electroabsorption effects near the semiconductor band edges have been an interesting research subject for many years. These include the interband photon-assisted tunneling or Franz–Keldysh effects [1–3] and the exciton absorption effects [4–9]. With the recent development of research in semiconductor quantum-well structures, optical absorptions in quantum wells have been shown to exhibit a drastic change by an applied electric field [10–13]. While previous excitonic electroabsorptions in bulk semiconductors were mostly observed at low temperatures, sharp excitonic absorption spectra in quantum wells have been observed at room temperature. These so-called quantum-confined Stark effects (QCSE) [11, 12] show a significant amount of change of the absorption coefficient with an applied voltage bias because of the enhanced exciton binding energy in a quasi-two-dimensional structure using quantum wells. The quantum-well barriers confine both the electrons and holes within the wells; therefore, the exciton binding energy is increased and the exciton is more difficult to ionize. The analytic solutions for pure two-dimensional and three-dimensional hydrogen models in Chapter 3 show that the exciton binding energy of the 1s ground state is four times larger in the 2D case than in the 3D case [14]. The sharp excitonic absorption spectrum with a small scattering linewidth shows the possibility of a big change of the absorption coefficient by an applied voltage bias. The change in the absorption coefficient can be as large as 10^4 cm^{-1} in GaAs$/$Al$_x$Ga$_{1-x}$As quantum wells [10–13]. Interesting quantum-well electroabsorption modulators at room temperature have been the subject of intensive research recently.

In this chapter we discuss the theory for electroabsorptions with and without excitonic effects. In Section 13.1 we present the effective mass theory for a two-particle system: an electron–hole pair. The general formulation for the optical absorption due to an electron–hole pair is discussed. We show that a change of variables from the electron and hole position coordinates r_e and r_h to their difference coordinates $r = r_e - r_h$ and their center-of-mass coordinates R lead to possible analytical solutions [4, 5, 8, 9] when the interaction potential is due to (1) an electric field only, which leads to electroabsorption effects in which a light is incident, or (2) the Coulomb interaction between the electron and the hole, which gives the excitonic absorption when a light is incident, or (3) both an electric field bias and the exciton effects.

Case (1), the Franz–Keldysh effect, is discussed in Section 13.2. The exciton effects, case (2), are presented in Section 13.3. Both have analytical solutions for direct band-gap semiconductors near the absorption edge. Case (3) in a quantum well, presented in Section 13.4, is called the quantum-confined Stark effect. The general solutions are obtained using two methods: One is based on a numerical solution of the Schrödinger equation in the momentum space for an electron–hole pair confined in a quantum well with an applied electric field [15]. The other method is based on a variational method [11, 12], which is commonly used in the literature because of its relative simplicity and accuracy especially for the bound state energy of the 1s excitons. Device applications including quantum-well electroabsorption modulators [16, 17] and self-electrooptic effect devices (SEEDs) [18–20] are discussed in Sections 13.5 and 13.6, respectively.

13.1 GENERAL FORMULATION FOR OPTICAL ABSORPTION DUE TO AN ELECTRON–HOLE PAIR

In Chapter 9, we derive the general formula for absorption coefficient in SI units:

$$\alpha(\hbar\omega) = C_0 \frac{2}{V} \sum_{i,f} |\langle f|e^{i\mathbf{k}_{op}\cdot\mathbf{r}}\,\hat{e}\cdot\mathbf{p}|i\rangle|^2 \,\delta(E_f - E_i - \hbar\omega)\big[f(E_i) - f(E_f)\big]$$

$$\text{(13.1.1a)}$$

$$C_0 = \frac{\pi e^2}{n_r c \varepsilon_0 m_0^2 \omega}$$

$$\text{(13.1.1b)}$$

The absorption coefficient depends on the initial state $|i\rangle$ with corresponding energy E_i and the final state $|f\rangle$ with corresponding energy E_f. The summation over the initial and final states taking into account the Fermi occupation factor $f(E)$ of these states gives the overall absorption spectrum. We also note that the delta function accounts for the energy conservation and the matrix element in (13.1.1a) takes into account the momentum conservation automatically, as has been discussed in Chapter 9, where no interaction between the electrons and holes is considered.

13.1.1 Two-Particle Wave Function and the Effective Mass Equation

To describe an electron–hole pair state, the two-particle wave function $\psi(\mathbf{r}_e, \mathbf{r}_h)$ for an electron at position $\mathbf{r}_e$ and a hole at position $\mathbf{r}_h$ can be expressed as a linear combination of the direct product of the single (uncorrelated) electron and hole Bloch functions, $\psi_{c\mathbf{k}_e}(\mathbf{r}_e)$ and $\psi_{v-\mathbf{k}_h}(\mathbf{r}_h)$, respectively:

$$\psi(\mathbf{r}_e,\mathbf{r}_h) = \sum_{\mathbf{k}_e} \sum_{\mathbf{k}_h} A(\mathbf{k}_e,\mathbf{k}_h)\psi_{c\mathbf{k}_e}(\mathbf{r}_e)\psi_{v-\mathbf{k}_h}(\mathbf{r}_h) \qquad (13.1.2)$$

where $A(\mathbf{k}_e,\mathbf{k}_h)$ represents the amplitude function. Note that the Bloch functions $\psi_{c\mathbf{k}_e}(\mathbf{r}_e)$ and $\psi_{v-\mathbf{k}_h}(\mathbf{r}_h)$ contain both the slowly varying plane-wave-like envelope and fast-varying Bloch periodic functions. In the effective mass approximation for electron and hole pairs, an envelope function $\Phi(\mathbf{r}_e,\mathbf{r}_h)$ is defined as the inverse Fourier transform of the amplitude function $A(\mathbf{k}_e,\mathbf{k}_h)$:

$$\Phi(\mathbf{r}_e,\mathbf{r}_h) = \sum_{\mathbf{k}_e} \sum_{\mathbf{k}_h} A(\mathbf{k}_e,\mathbf{k}_h)\frac{e^{i\mathbf{k}_e\cdot\mathbf{r}_e}}{\sqrt{V}}\frac{e^{i\mathbf{k}_h\cdot\mathbf{r}_h}}{\sqrt{V}} \qquad (13.1.3)$$

which is the plane-wave expansion of the two-particle wave function. The Fourier transform of the wave function $\Phi(\mathbf{r}_e,\mathbf{r}_h)$ is

$$A(\mathbf{k}_e,\mathbf{k}_h) = \int d^3\mathbf{r}_e\int d^3\mathbf{r}_h\Phi(\mathbf{r}_e,\mathbf{r}_h)\frac{e^{-i\mathbf{k}_e\cdot\mathbf{r}_e}}{\sqrt{V}}\frac{e^{-i\mathbf{k}_h\cdot\mathbf{r}_h}}{\sqrt{V}} \qquad (13.1.4)$$

The major difference between ψ and Φ in (13.1.2) and (13.1.3) is the basis functions used in their expansions. In the envelope function Φ for the electron–hole pair states, the fast-varying Bloch periodic parts $u_c(\mathbf{r})$ and $u_v(\mathbf{r})$, which behave like $|iS\rangle$ and $|\frac{3}{2},\pm\frac{3}{2}\rangle$ or $|\frac{3}{2},\pm\frac{1}{2}\rangle$, have been dropped from the basis functions and only the plane-wave parts are kept. The envelope wave function $\Phi(\mathbf{r}_e,\mathbf{r}_h)$ satisfies the effective mass equation

$$\left[E_g + E_c(-i\nabla_e) - E_v(-i\nabla_h) + V(\mathbf{r}_e,\mathbf{r}_h)\right]\Phi(\mathbf{r}_e,\mathbf{r}_h) = E\Phi(\mathbf{r}_e,\mathbf{r}_h) \qquad (13.1.5)$$

where we have replaced $\mathbf{k}_e$ in the dispersion relation $E_c \equiv E_c(\mathbf{k}_e)$ by the differential operator $-i\nabla_e$ for the $\mathbf{r}_e$ variables, and $\mathbf{k}_h$ in $E_v \equiv E_v(\mathbf{k}_h)$ by $-i\nabla_h$ for the $\mathbf{r}_h$ variables. Using the parabolic model, we have $E_c(\mathbf{k}_e) = \hbar^2 k_e^2/2m_e^*$ and $E_v(\mathbf{k}_h) = -\hbar^2 k_h^2/2m_h^*$.

The interaction potential $V(\mathbf{r}_e,\mathbf{r}_h)$ may be of the form

$$(1) \qquad\qquad V(\mathbf{r}_e,\mathbf{r}_h) = e\mathbf{F}\cdot(\mathbf{r}_e - \mathbf{r}_h) \qquad (13.1.6)$$

It is the potential energy of a free electron and a free hole in the presence of a uniform electric field $\mathbf{F}$. This will lead to the Franz–Keldysh effect [1–3] for the optical absorption, as discussed in Section 13.2. The interaction potential can also be of the form

(2)
$$V(\mathbf{r}_e, \mathbf{r}_h) = -\frac{e^2}{4\pi\varepsilon_s|\mathbf{r}_e - \mathbf{r}_h|} \qquad (13.1.7)$$

It is the Coloumb interaction between an electron at $\mathbf{r}_e$ and a hole at $\mathbf{r}_h$. Here ε_s is the permittivity of the semiconductor. This potential leads to the exciton effect [4–9] in the optical absorption, which is discussed in Section 13.3.

13.1.2 Solution of the Two-Particle Effective-Mass Equation

In general, for $V(\mathbf{r}_e, \mathbf{r}_h) = V(\mathbf{r}_e - \mathbf{r}_h)$, which depends only on the difference between the electron and hole position vectors, we may change the variables into the difference coordinate and the center-of-mass coordinate system, $\mathbf{r}$ and $\mathbf{R}$, respectively, as shown in Fig. 13.1a.

$$\mathbf{r} = \mathbf{r}_e - \mathbf{r}_h \qquad \mathbf{R} = \frac{m_e^*\mathbf{r}_e + m_h^*\mathbf{r}_h}{M} \qquad (13.1.8)$$

where $M = m_e^* + m_h^*$. The corresponding Fourier transform variables of $\mathbf{r}$

(a) Real Space

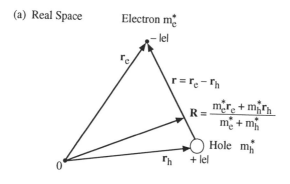

(b) Momentum Space

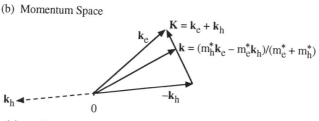

Figure 13.1. (a) An illustration of the electron position vector $\mathbf{r}_e$, the hole position vector $\mathbf{r}_h$, the difference coordinate vector $\mathbf{r} = \mathbf{r}_e - \mathbf{r}_h$, and the center-of-mass coordinate vector $\mathbf{R} = (m_e^*\mathbf{r}_e + m_h^*\mathbf{r}_h)/(m_e^* + m_h^*)$. (b) The relations between the wave vectors, $\mathbf{k}_e$, $\mathbf{k}_h$, $\mathbf{k}$, and $\mathbf{K}$ in the Fourier transform space.

and $\mathbf{R}$ in the momentum space are (Fig. 13.1b)

$$\mathbf{k} = \frac{m_h^* \mathbf{k}_e - m_e^* \mathbf{k}_h}{M} \quad \text{and} \quad \mathbf{K} = \mathbf{k}_e + \mathbf{k}_h \qquad (13.1.9)$$

respectively, which can also be checked using

$$\exp(i\mathbf{k} \cdot \mathbf{r} + i\mathbf{K} \cdot \mathbf{R}) = \exp(i\mathbf{k}_e \cdot \mathbf{r}_e + i\mathbf{k}_h \cdot \mathbf{r}_h)$$

We can also express the above relations as

$$\mathbf{r}_e = \mathbf{R} + \frac{m_h^*}{M}\mathbf{r} \qquad \mathbf{r}_h = \mathbf{R} - \frac{m_e^*}{M}\mathbf{r} \qquad (13.1.10)$$

and

$$\mathbf{k}_e = \frac{m_e^*}{M}\mathbf{K} + \mathbf{k} \qquad \mathbf{k}_h = \frac{m_h^*}{M}\mathbf{K} - \mathbf{k} \qquad (13.1.11)$$

From the corresponding differential operators

$$\mathbf{k}_e = -i\nabla_e \qquad \mathbf{k}_h = -i\nabla_h$$
$$\mathbf{k} = -i\nabla_r \qquad \mathbf{K} = -i\nabla_R \qquad (13.1.12)$$

we obtain

$$\frac{\hbar^2 k_e^2}{2m_e^*} + \frac{\hbar^2 k_h^2}{2m_h^*} = \frac{\hbar^2 K^2}{2M} + \frac{\hbar^2 k^2}{2m_r} \qquad (13.1.13)$$

and

$$-\frac{\hbar^2}{2m_e^*}\nabla_e^2 - \frac{\hbar^2}{2m_h^*}\nabla_h^2 = -\frac{\hbar^2}{2M}\nabla_R^2 - \frac{\hbar^2}{2m_r}\nabla_r^2 \qquad (13.1.14)$$

Here m_r is the reduced effective mass, defined by $1/m_r = (1/m_e^*) + (1/m_h^*)$. Therefore, the effective mass equation (13.1.5) becomes

$$\left[-\frac{\hbar^2}{2M}\nabla_R^2 - \frac{\hbar^2}{2m_r}\nabla_r^2 + V(\mathbf{r}) - (E - E_g) \right]\Phi(\mathbf{R},\mathbf{r}) = 0 \quad (13.1.15)$$

The solution to the above equation can then be obtained using the method of the separation of variables, noting that the $\mathbf{R}$ dependence is a simple free particle wave function,

$$\Phi(\mathbf{R},\mathbf{r}) = \frac{e^{i\mathbf{K}\cdot\mathbf{R}}}{\sqrt{V}}\phi(\mathbf{r}) \qquad (13.1.16)$$

where $\phi(\mathbf{r})$ satisfies

$$\left[-\frac{\hbar^2}{2m_r}\nabla^2 + V(\mathbf{r}) - \mathcal{E}\right]\phi(r) = 0 \tag{13.1.17}$$

and the energy $\mathcal{E}$ is

$$\mathcal{E} = E - E_g - \frac{\hbar^2 K^2}{2M} \tag{13.1.18}$$

Define the Fourier transform pair as

$$\phi(\mathbf{r}) = \sum_{\mathbf{k}} a(\mathbf{k})\frac{e^{i\mathbf{k}\cdot\mathbf{r}}}{\sqrt{V}} \tag{13.1.19a}$$

$$a(\mathbf{k}) = \int d^3r\, \phi(\mathbf{r})\frac{e^{-i\mathbf{k}\cdot\mathbf{r}}}{\sqrt{V}} \tag{13.1.19b}$$

We find

$$\Phi(\mathbf{r}_e,\mathbf{r}_h) = \Phi(\mathbf{R},\mathbf{r}) = \frac{e^{i\mathbf{K}\cdot\mathbf{R}}}{\sqrt{V}}\phi(\mathbf{r})$$

$$= \frac{e^{i\mathbf{K}\cdot\mathbf{R}}}{\sqrt{V}}\sum_{\mathbf{k}} a(\mathbf{k})\frac{e^{i\mathbf{k}\cdot(\mathbf{r}_e-\mathbf{r}_h)}}{\sqrt{V}} \tag{13.1.20}$$

13.1.3   Optical Matrix Element of the Two-Particle Transition Picture

The optical matrix element between the ground state (all electrons are in the valence band) and the final state (the electron–hole pair state) is described by [4, 5, 13]

$$\langle f|e^{i\mathbf{k}_{op}\cdot\mathbf{r}}\,\hat{e}\cdot\mathbf{p}|i\rangle$$

$$= \sum_{\mathbf{k}_e}\sum_{\mathbf{k}_h} A^*(\mathbf{k}_e,\mathbf{k}_h)\langle c,\mathbf{k}_e|e^{i\mathbf{k}_{op}\cdot\mathbf{r}}\,\hat{e}\cdot\mathbf{p}|v,-\mathbf{k}_h\rangle$$

$$\cong \sum_{\mathbf{k}_e}\sum_{\mathbf{k}_h} A^*(\mathbf{k}_e,\mathbf{k}_h)\,\hat{e}\cdot\mathbf{p}_{cv}(\mathbf{k}_e)\,\delta_{\mathbf{k}_e+\mathbf{k}_h,\mathbf{k}_{op}}$$

$$\simeq \sum_{\mathbf{k}} A^*(\mathbf{k},-\mathbf{k})\,\hat{e}\cdot\mathbf{p}_{cv}(\mathbf{k}) \tag{13.1.21}$$

where the long wavelength (or dipole) approximation $\mathbf{k}_{op} \simeq 0$ has been used.

Therefore, the **k** selection rule

$$\mathbf{k}_e + \mathbf{k}_h = \mathbf{k}_{op} \simeq 0 \qquad (13.1.22)$$

has been adopted.

Comparing (13.1.20) with the definition in (13.1.3), and using $\mathbf{K} = \mathbf{k}_e + \mathbf{k}_h \simeq 0$, we find that the matrix element is

$$\langle f|e^{i\mathbf{k}_{op}\cdot\mathbf{r}}\hat{e}\cdot\mathbf{p}|i\rangle \simeq \hat{e}\cdot\mathbf{p}_{cv}\sum_{\mathbf{k}}A^*(\mathbf{k},-\mathbf{k})$$

$$= \hat{e}\cdot\mathbf{p}_{cv}\sum_{\mathbf{k}}a^*(\mathbf{k})$$

$$= \hat{e}\cdot\mathbf{p}_{cv}\sqrt{V}\phi^*(0) \qquad (13.1.23)$$

where we have assumed that $\hat{e}\cdot\mathbf{p}_{cv}$ is independent of **k**.

13.1.4 Absorption Formula

The absorption coefficient is given by substituting the matrix element (13.1.23) into (13.1.1)

$$\alpha(\omega) = C_0|\hat{e}\cdot\mathbf{p}_{cv}|^2 2\sum_n|\phi_n(0)|^2\,\delta(\mathscr{E}_n + E_g - \hbar\omega) \qquad (13.1.24)$$

where n corresponds to the discrete and continuum states of $\phi(r)$ satisfying the effective mass equation in the difference coordinate system, (13.1.17). The equation for a Coulomb potential is the Schrödinger equation for a hydrogen atom, and its solutions for both bound and continuum states have been presented in Chapter 3.

13.1.5 Physical Interpretation of $2|\phi_n(0)|^2$ and Relation to Density of States

Consider the case of a free electron and a free hole without Coulomb interaction, that is, $V(\mathbf{r}) = 0$ in (13.1.17). With $\mathbf{K} = 0$, we have $\mathscr{E} = E - E_g$. Therefore, we use a new energy E measured from the band gap E_g for convenience:

$$-\frac{\hbar^2}{2m_r}\nabla^2\phi(\mathbf{r}) = E\phi(\mathbf{r}) \qquad (13.1.25)$$

In the discrete picture, we have the wave function and the energy spectrum

$$\phi_n(\mathbf{r}) = \frac{e^{ik_x x + ik_y y + ik_z z}}{\sqrt{V}}$$

$$E_n = \frac{\hbar^2 k^2}{2m_r} \tag{13.1.26}$$

where $n \equiv (n_x, n_y, n_z)$, $k_x = n_x 2\pi/L$, $k_y = n_y 2\pi/L$, $k_z = n_z 2\pi/L$ and the normalization rule

$$\int \phi_n^*(\mathbf{r}) \phi_{n'}(\mathbf{r}) \, d^3 r = \delta_{nn'} = \begin{cases} 1 & \text{if } n = n' \\ 0 & \text{if } n \neq n' \end{cases} \tag{13.1.27}$$

have been used. We check $\phi_n(0) = 1/\sqrt{V}$ and

$$2 \sum_n |\phi_n(0)|^2 \delta(E_n - E) = 2 \left(V \int \frac{d^3 k}{(2\pi)^3} \right) \frac{1}{V} \delta\left(\frac{\hbar^2 k^2}{2m_r} - E \right)$$

$$= \frac{1}{2\pi^2} \left(\frac{2m_r}{\hbar^2} \right)^{3/2} \sqrt{E} = \rho_r^{3D}(E) \tag{13.1.28}$$

which is the three-dimensional reduced density of states, where $\sum_n = [V/2\pi)^3] \int d^3 k$ in the discrete picture has been used. (See Problem 13.1 for an alternative approach.)

13.1.6 Optical Absorption Spectrum for Interband Free Electron–Hole Transitions

The optical absorption is given by integrating (13.1.24)

$$\alpha(\hbar\omega) = A_0 \rho_r^{3D}(\hbar\omega - E_g) \tag{13.1.29a}$$

where

$$A_0 = C_0 |\hat{e} \cdot \mathbf{p}_{cv}|^2 = \frac{\pi e^2 |\hat{e} \cdot \mathbf{p}_{cv}|^2}{n_r c \varepsilon_0 m_0^2 \omega} \tag{13.1.29b}$$

$$\rho_r^{3D}(\hbar\omega - E_g) = \frac{1}{2\pi^2} \left(\frac{2m_r}{\hbar^2} \right)^{3/2} \sqrt{\hbar\omega - E_g} \tag{13.1.29c}$$

which gives the absorption coefficient due to a free electron and hole. The momentum-matrix element of a bulk semiconductor is

$$|\hat{e} \cdot \mathbf{p}_{cv}|^2 = M_b^2 = \frac{m_0}{6} E_p \qquad (13.1.30)$$

where the energy parameter E_p (in electron volt) for the matrix element is tabulated in Table K.2 in Appendix K.

13.2 FRANZ–KELDYSH EFFECT [1–3, 21–26]

Let us consider the case of a uniform applied electric field, $V(\mathbf{r}) = e\mathbf{F} \cdot \mathbf{r}$. The Schrödinger equation for the wave function $\phi(\mathbf{r})$ in the difference coordinate system (13.1.17) (with $K = 0$) is

$$\left(\frac{-\hbar^2}{2m_r} \nabla^2 + e\mathbf{F} \cdot \mathbf{r} \right) \phi(\mathbf{r}) = E\phi(\mathbf{r}) \qquad (13.2.1)$$

Assume that the applied field is in the z direction, $\mathbf{F} = \hat{z}F$ (Fig. 13.2a). The solution can be written as

$$\phi(\mathbf{r}) = \frac{e^{ik_x x + ik_y y}}{\sqrt{A}} \phi(z) \qquad (13.2.2)$$

where the z-dependent wave function $\phi(z)$ satisfies

$$\left(\frac{-\hbar^2}{2m_r} \frac{d^2}{dz^2} + eFz \right) \phi(z) = E_z \phi(z) \qquad (13.2.3)$$

and the total energy E is related to E_g for the z-dependent wave function

$$E = \frac{\hbar^2}{2m_r} \left(k_x^2 + k_y^2 \right) + E_z \qquad (13.2.4)$$

13.2.1 Solution of the Schrödinger Equation for a Uniform Electric Field

The solution of the Schrödinger equation (13.2.3) with a uniform field can be obtained by a change of variable:

$$Z = \left(\frac{2m_r eF}{\hbar^2} \right)^{1/3} \left(z - \frac{E_z}{eF} \right) \qquad (13.2.5)$$

Therefore,

$$\frac{d^2\phi(Z)}{dZ^2} - Z\phi(Z) = 0 \tag{13.2.6}$$

The Airy functions [27] $Ai(Z)$ or $Bi(Z)$ are the solutions. Since the wave function has to decay as z approaches $+\infty$ (because of the potential $+eFz$), the Airy function $Ai(Z)$ has to be chosen. The energy spectrum is continuous since the potential is not bounded as $z \to -\infty$. Therefore, the (real) wave function satisfying the normalization condition

$$\int_{-\infty}^{\infty} dz \, \phi_{E_{z1}}(z)\phi_{E_{z2}}(z) = \delta(E_{z1} - E_{z2}) \tag{13.2.7}$$

for a continuum spectrum is

$$\phi_{E_z}(z) = \left(\frac{2m_r}{\hbar^2}\right)^{1/3} \frac{1}{(eF)^{1/6}} Ai\left[\left(\frac{2m_r eF}{\hbar^2}\right)^{1/3}\left(z - \frac{E_z}{eF}\right)\right] \tag{13.2.8}$$

To prove that $\phi_{E_z}(z)$ satisfies the normalization condition, we use the integral representation [27] of the Airy function:

$$Ai(t) = \int_{-\infty}^{\infty} \frac{dk}{2\pi} e^{i(kt + k^3/3)} \tag{13.2.9}$$

Therefore,

$$\int_{-\infty}^{\infty} Ai(t - \alpha_1) Ai(t - \alpha_2)\, dt$$

$$= \int_{-\infty}^{\infty} \frac{dk}{2\pi} \int_{-\infty}^{\infty} \frac{dk'}{2\pi} \int_{-\infty}^{\infty} dt\, e^{i(t-\alpha_1)k + i(k^3/3)} e^{i(t-\alpha_2)k' + i(k'^3/3)}$$

$$= \int_{-\infty}^{\infty} \frac{dk}{2\pi} e^{-i(\alpha_1 - \alpha_2)k}$$

$$= \delta(\alpha_1 - \alpha_2) \tag{13.2.10}$$

where the identity

$$\int_{-\infty}^{\infty} dt\, e^{i(k+k')t} = 2\pi\,\delta(k + k') \tag{13.2.11}$$

has been used. Using

$$t = \left(\frac{2m_r eF}{\hbar^2}\right)^{1/3} z \qquad \alpha = \left(\frac{2m_r eF}{\hbar^2}\right)^{1/3} \frac{E_z}{eF}$$

and

$$\delta(\alpha_1 - \alpha_2) = \left(\frac{2m_r}{\hbar^2 e^2 F^2}\right)^{-1/3} \delta(E_{z1} - E_{z2})$$

in (13.2.10), we obtain the normalization condition (13.2.7).

13.2.2 Summation of the Density of States and Absorption Spectrum

Since the quantum number is determined by (k_x, k_y, E_z) as described above in the wave function (13.2.2) and the corresponding energy spectrum (13.2.4), the sum over all the states n for the absorption spectrum in (13.1.24) has to be replaced by the sum over all the quantum numbers:

$$\sum_n \rightarrow \sum_{k_x} \sum_{k_y} \int dE_z$$

where the sum over the energy E_z is an integral since E_z is a continuous spectrum and a delta function normalized rule (13.2.7) has been adopted. Therefore,

$$\alpha(\hbar\omega) = A_0 2 \sum_{k_x k_y} \int dE_z |\phi(\mathbf{r} = 0)|^2 \delta\left[\frac{\hbar^2}{2m_r}(k_x^2 + k_y^2) + E_z + E_g - \hbar\omega\right]$$

$$(13.2.12)$$

where A_0 is given by (13.1.29b). Since

$$\frac{2}{A} \sum_{k_x k_y} = 2\int \frac{d^2 \mathbf{k}_t}{(2\pi)^2} = \frac{m_r}{\pi\hbar^2} \int dE_t \qquad (13.2.13)$$

where $E_t = \hbar^2(k_x^2 + k_y^2)/2m_r = \hbar^2 k_t^2/2m_r$, we carry out the integration over E_t with the delta function and obtain the expression for the absorption coefficient:

$$\alpha(\hbar\omega) = A_0 \frac{m_r}{\pi\hbar^2} \int_{E_z = -\infty}^{\hbar\omega - E_g} dE_z |\phi_{E_z}(z = 0)|^2$$

$$= A_0 \frac{m_r}{\pi\hbar^2} \int_{-\infty}^{\hbar\omega - E_g} dE_z \left(\frac{2m_r}{\hbar^2}\right)^{2/3} \frac{1}{(eF)^{1/3}} Ai^2\left[\left(\frac{2m_r}{\hbar^2 e^2 F^2}\right)^{1/3}(-E_z)\right]$$

$$(13.2.14)$$

Let

$$\hbar\theta_F = \left(\frac{\hbar^2 e^2 F^2}{2m_r}\right)^{1/3} \qquad \tau = -\frac{E_z}{\hbar\theta_F} \qquad \eta = \frac{E_g - \hbar\omega}{\hbar\theta_F} \qquad (13.2.15)$$

We find the absorption coefficient

$$\begin{aligned}\alpha(\hbar\omega) &= \frac{A_0}{2\pi}\left(\frac{2m_r}{\hbar^2}\right)^{3/2}\sqrt{\hbar\theta_F}\int_\eta^\infty d\tau Ai^2(\tau) \\ &= \frac{A_0}{2\pi}\left(\frac{2m_r}{\hbar^2}\right)^{3/2}\sqrt{\hbar\theta_F}\left[-\eta Ai^2(\eta) + Ai'^2(\eta)\right] \quad (13.2.16)\end{aligned}$$

where $Ai'(\eta)$ is the derivative of $Ai(\eta)$ with respect to η.

It is interesting to show that in the limit when $F \to 0$, we have

$$\lim_{F\to 0}\left[\pi\sqrt{\hbar\theta_F}\int_\eta^\infty Ai^2(\tau)\,d\tau\right] = \sqrt{\hbar\omega - E_g}$$

for $\hbar\omega > E_g$ as expected, since

$$\alpha(\hbar\omega) \underset{F\to 0}{\to} \alpha_0(\hbar\omega) = A_0\frac{1}{2\pi^2}\left(\frac{2m_r}{\hbar^2}\right)^{3/2}\sqrt{\hbar\omega - E_g} \quad (13.2.17)$$

Notice that the prefactor A_0 depends on the bulk momentum-matrix element $|\hat{e}\cdot \mathbf{p}_{cv}|^2 = M_b^2$, which can be determined experimentally [24] by fitting the measured absorption spectrum with the above theoretical results with $F = 0$ and $F \neq 0$. The Franz–Keldysh absorption spectrum (13.2.16) is plotted schematically in Fig. 13.2b (solid curve) and compared with the zero-field

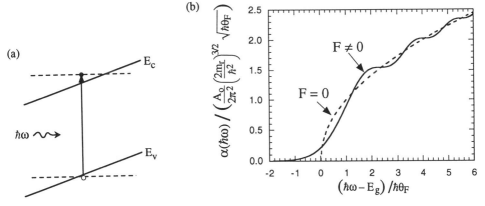

Figure 13.2. (a) Franz–Keldysh effect or photo-assisted absorption in a bulk semiconductor with a uniform electric field bias. (b) Absorption spectrum for a finite field $F \neq 0$ (solid curve). The dashed line is the free electron and hole absorption spectrum without an applied electric field.

spectrum (dashed curve) using (13.2.17). It shows the Franz–Keldysh oscillation phenomena in the absorption spectrum above the band gap and the
exponentially decaying behavior below the band gap.

13.3   EXCITON EFFECT [4–9, 10–15]

When we consider the Coulomb interaction between the electron and the
hole

$$V(\mathbf{r}) = \frac{-e^2}{4\pi\varepsilon_s r} \tag{13.3.1}$$

the wave function $\phi(\mathbf{r})$ satisfies the Schrödinger equation for the hydrogen
atom

$$\left[-\frac{\hbar^2}{2m_r}\nabla^2 + V(\mathbf{r})\right]\phi(\mathbf{r}) = E\phi(\mathbf{r}) \tag{13.3.2}$$

The solutions $\phi(\mathbf{r})$ for both the three-dimensional (3D) and the two-dimensional (2D) cases have been studied in Section 3.4 and Appendix A and the
results are tabulated in Table 3.1.

13.3.1   3D Exciton [4, 5]

We use the most general formula (13.1.24) for the absorption coefficient:

$$\alpha(\hbar\omega) = A_0 \sum_n 2|\phi_n(0)|^2 \delta(\hbar\omega - E_g - E_n) \tag{13.3.3a}$$

$$A_0 = \frac{\pi e^2|\hat{e} \cdot \mathbf{p}_{cv}|^2}{n_r c\varepsilon_0 \omega m_0^2} \tag{13.3.3b}$$

where the summation over n includes both the bound and continuum states.
The wave functions $\phi_n(\mathbf{r})$ have been normalized properly for both bound and
continuum states as discussed in Sections 13.1, 13.2, and 3.4.

For bound state contributions, we have the oscillator strength

$$|\phi_n(0)|^2 = \frac{1}{\pi a_o^3 n^3} \qquad a_o = \frac{4\pi\varepsilon_s\hbar^2}{m_r e^2} = \text{the exciton Bohr radius}$$

and the exciton binding energy

$$E_n = -\frac{R_y}{n^2} \qquad R_y = \frac{m_r e^4}{2\hbar^2 (4\pi\varepsilon_s)^2} = \text{the exciton Rydberg energy}$$

Therefore,

$$\alpha_B(\hbar\omega) = A_0 \sum_{n=1}^{\infty} \left(\frac{2}{\pi a_o^3 n^3}\right) \delta\left(\hbar\omega - E_g + \frac{R_y}{n^2}\right)$$

$$= A_0 \sum_{n=1}^{\infty} \frac{2}{R_y \pi a_o^3 n^3} \delta\left(\varepsilon + \frac{1}{n^2}\right) \tag{13.3.4}$$

where

$$\varepsilon = \frac{\hbar\omega - E_g}{R_y} \tag{13.3.5}$$

is a normalized energy measured from the band gap E_g.
    For the continuum-state contributions, we obtain

$$\alpha_C(\hbar\omega) = A_0 \int_0^{\infty} dE \left[\frac{e^{\pi/\sqrt{E/R_y}}}{2\pi R_y a_o^3 \sinh\left(\pi/\sqrt{E/R_y}\right)}\right] \delta(\hbar\omega - E_g - E)$$

$$= \left(\frac{A_0}{2\pi^2 R_y a_o^3}\sqrt{\varepsilon}\right)\left[\frac{(\pi/\sqrt{\varepsilon})\, e^{\pi/\sqrt{\varepsilon}}}{\sinh\left(\pi/\sqrt{\varepsilon}\right)}\right] \tag{13.3.6}$$

where the first bracket is $A_0 \rho_r^{3D}(E = \hbar\omega - E_g)$, and the second bracket is called the Sommerfeld enhancement factor for the 3D case [4–9, 28]:

$$S_{3D}(\varepsilon) = \frac{(\pi/\sqrt{\varepsilon})\, e^{\pi/\sqrt{\varepsilon}}}{\sinh\left(\pi/\sqrt{\varepsilon}\right)}$$

$$= \frac{2\pi/\sqrt{\varepsilon}}{1 - e^{-2\pi/\sqrt{\varepsilon}}} \tag{13.3.7}$$

As $\varepsilon \to \infty$, we find

$$\alpha_C(\hbar\omega) \to \frac{A_0}{2\pi^2 R_y a_o^3}\sqrt{\varepsilon}\, e^{\pi/\sqrt{\varepsilon}} \to \frac{A_0}{2\pi^2 R_y a_o^3}\left(\sqrt{\varepsilon} + \pi\right) \tag{13.3.8}$$

which approaches the 3D joint density-of-states in an interband transition without the exciton effects, $\sqrt{(\hbar\omega - E_g)/R_y}$, plus a constant π. As $\varepsilon \to 0$,

$$\alpha_C(\hbar\omega) \to \frac{A_0}{2\pi^2 R_y a_o^3}(2\pi) \tag{13.3.9}$$

which gives a finite value in contrast to the vanishing result of the interband absorption at $\hbar\omega = E_g$.

The total absorption due to both the bound and continuum states is given by

$$\alpha(\hbar\omega) = \frac{A_0}{2\pi^2 R_y a_o^3}\left[4\pi \sum_{n=1}^{\infty} \frac{1}{n^3}\delta\left(\frac{\hbar\omega - E_g}{R_y} + \frac{1}{n^2}\right)\right.$$
$$\left. + S_{3D}\left(\frac{\hbar\omega - E_g}{R_y}\right)\sqrt{\frac{\hbar\omega - E_g}{R_y}}\right] \tag{13.3.10}$$

If we include the finite linewidth due to scatterings by replacing the delta function by a Lorentzian function, $\delta(x) = (\gamma/\pi)/(x^2 + \gamma^2)$, where γ is the half-linewidth normalized by Rydberg if x is a normalized energy, we find

$$\alpha(\hbar\omega) = \frac{A_0}{2\pi^2 R_y a_o^3}\left[4\sum_{n=1}^{\infty}\frac{\gamma/n^3}{\left(\varepsilon + \frac{1}{n^2}\right)^2 + \gamma^2} + \int_0^{\infty}\frac{d\varepsilon'}{\pi}\frac{\gamma S_{3D}(\varepsilon')\sqrt{\varepsilon'}}{(\varepsilon - \varepsilon')^2 + \gamma^2}\right]$$
$$\tag{13.3.11}$$

13.3.2 2D Exciton [14, 29]

The absorption spectrum for a two-dimensional structure with exciton effects can also be obtained using (13.1.24):

$$\alpha(\hbar\omega) = A_0 \sum_n 2|\phi_n(0)|^2 \delta(\hbar\omega - E_g - E_n) \tag{13.3.12}$$

For bound state contributions, we use

$$|\phi_n(0)|^2 = \frac{1}{\pi a_o^2\left(n - \frac{1}{2}\right)^3} \qquad E_n = -\frac{R_y}{\left(n - \frac{1}{2}\right)^2} \tag{13.3.13}$$

and obtain

$$\alpha_B(\hbar\omega) = A_0 \sum_{n=1}^{\infty} \frac{2}{R_y a_o^2 \pi (n - \frac{1}{2})^3} \delta\left[\varepsilon + \frac{1}{(n - \frac{1}{2})^2}\right] \quad (13.3.14)$$

where $\varepsilon = (\hbar\omega - E_g)/R_y$ again. The continuum-state contributions give

$$\alpha_C(\hbar\omega) = A_0 \int_0^{\infty} dE \left[\frac{S_{2D}(E)}{2\pi R_y a_o^2}\right] \delta(E + E_g - \hbar\omega)$$

$$= A_0 \frac{S_{2D}(\varepsilon)}{2\pi R_y a_o^2}$$

$$= A_0 \frac{m_r}{\pi \hbar^2} S_{2D}(\varepsilon) \quad (13.3.15)$$

where the two-dimensional Sommerfeld enhancement factor is

$$S_{2D}(\varepsilon) = \frac{2}{1 + e^{-2\pi/\sqrt{\varepsilon}}} \quad (13.3.16)$$

The total absorption is the sum of $\alpha_B(\hbar\omega)$ and $\alpha_C(\hbar\omega)$:

$$\alpha(\hbar\omega) = \frac{A_0}{2\pi R_y a_o^2} \left\{ 4 \sum_{n=1}^{\infty} \frac{1}{(n - \frac{1}{2})^3} \delta\left[\varepsilon + \frac{1}{(n - \frac{1}{2})^2}\right] + S(\varepsilon) \right\} \quad (13.3.17)$$

Notice that without the Sommerfeld enhancement factor, $S_{2D}(\varepsilon)$ is set to 1 and $\alpha_f(\hbar\omega) = A_0/(2\pi R_y a_o^2)$. If we include the finite linewidth effect, we have

$$\alpha(\hbar\omega) = \frac{A_0}{2\pi R_y a_o^2} \left\{ 4 \sum_{n=1}^{\infty} \frac{1}{\pi(n - \frac{1}{2})^3} \frac{\gamma}{\left[\varepsilon + \frac{1}{(n - \frac{1}{2})^2}\right]^2 + \gamma^2} \right.$$

$$\left. + \int_0^{\infty} \frac{d\varepsilon'}{\pi} \frac{\gamma S_{2D}(\varepsilon')}{(\varepsilon' - \varepsilon)^2 + \gamma^2} \right\} \quad (13.3.18)$$

Table 13.1   Absorption Coefficients due to Exciton Bound and Continuum States

$$\varepsilon = \frac{\hbar\omega - E_g}{R_y} \qquad A_0 = \frac{\pi e^2 |\hat{e} \cdot \mathbf{p}_{cv}|^2}{n_r c \varepsilon_0 \omega m_0^2} \qquad a_o = \frac{\hbar^2}{m_r}\left(\frac{4\pi\varepsilon_s}{e^2}\right) \qquad R_y = \frac{m_r e^4}{2\hbar^2(4\pi\varepsilon_s)^2}$$

Bound States	Continuum States

Two-Dimensional Exciton

ZERO LINEWIDTH

$$A_0 \sum_{n=1}^{\infty} \left[\frac{2}{\pi a_o^2 \left(n - \frac{1}{2}\right)^3}\right] \frac{1}{R_y} \delta\left(\varepsilon + \frac{1}{\left(n - \frac{1}{2}\right)^2}\right) \qquad \frac{A_0}{2\pi R_y a_o^2} S_{2D}(\varepsilon)$$

$$S_{2D}(\varepsilon) = \frac{2}{1 + \exp\left(-2\pi/\sqrt{\varepsilon}\right)}$$

FINITE LINEWIDTH

$$A_0 \sum_{n=1}^{\infty} \left[\frac{2}{\pi a_o^2 \left(n - \frac{1}{2}\right)^3}\right] \frac{1}{R_y} \frac{\gamma/\pi}{\left[\varepsilon + \frac{1}{\left(n - \frac{1}{2}\right)^2}\right]^2 + \gamma^2} \qquad \frac{A_0}{2\pi R_y a_o^2} \int_0^{\infty} \frac{d\varepsilon'}{\pi} \frac{\gamma S_{2D}(\varepsilon')}{(\varepsilon' - \varepsilon)^2 + \gamma^2}$$

Three-Dimensional Exciton

ZERO LINEWIDTH

$$A_0 \sum_{n=1}^{\infty} \left(\frac{2}{\pi a_o^3 n^3}\right) \frac{1}{R_y} \delta\left(\varepsilon + \frac{1}{n^2}\right) \qquad \frac{A_0}{2\pi^2 R_y a_o^3} \sqrt{\varepsilon}\, S_{3D}(\varepsilon)$$

$$S_{3D}(\varepsilon) = \frac{2\pi/\sqrt{\varepsilon}}{1 - e^{-2\pi/\sqrt{\varepsilon}}}$$

FINITE LINEWIDTH

$$A_0 \sum_{n=1}^{\infty} \left(\frac{2}{\pi a_o^3 n^3}\right) \frac{1}{R_y} \frac{\gamma/\pi}{\left(\varepsilon + \frac{1}{n^2}\right)^2 + \gamma^2} \qquad \frac{A_0}{2\pi^2 R_y a_o^3} \int_0^{\infty} \frac{d\varepsilon'}{\pi} \frac{\gamma\sqrt{\varepsilon'}\, S_{3D}(\varepsilon')}{(\varepsilon' - \varepsilon)^2 + \gamma^2}$$

The above results for both 2D and 3D excitons are summarized in Table 13.1. The absorption spectra for a finite linewidth and zero linewidth are plotted in Fig. 13.3 for comparison.

13.3.3   Experimental Results for 3D and Quasi-2D Excitons

Experimentally, the linewidth γ is always finite and increases with temperature. For example, in Fig. 13.4a, we show the absorption spectra [30] of a bulk (3D) GaAs at four different temperatures, $T = 21, 90, 186$, and 294 K. We can see that the exciton linewidth is broadened with an increasing temperature. The absorption edge has a red shift since the GaAs bandgap

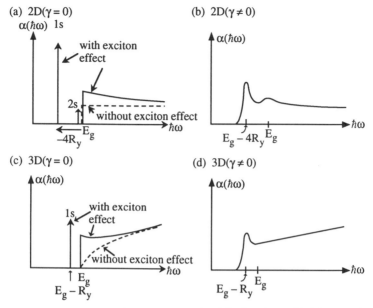

Figure 13.3. Absorption spectra for a two-dimensional (2D) exciton with (a) a zero linewidth and (b) a finite linewidth, and a three-dimensional (3D) exciton with (c) a zero linewidth and (d) a finite linewidth.

decreases with increasing temperature [31]:

$$E_g(T) = E_g(0) - \frac{aT^2}{T + b} \quad \text{(eV)} \tag{13.3.19}$$

where $E_g(0) = 1.519$ eV, $a = 5.405 \times 10^{-4}$ eV/K, and $b = 204$ K. For comparison, we show the exciton absorption spectra [32] of an $In_{0.53}Ga_{0.47}As/In_{0.52}Al_{0.48}As$ (lattice-matched to InP substrate) quantum-well structure at different temperatures in Fig. 13.4b. The quantum-well structure has a quasi-two-dimensional character since the electron and hole wave functions are confined in the z direction with a finite well width instead of being restricted to the x-y plane as in the "pure" 2D case. It is expected that the binding energy of the 1s exciton in the quasi-2D structure will be between the 3D value $(= R_y)$ and 2D value $(= 4R_y)$. Furthermore, we also see the splitting of the heavy-hole (HH) exciton and light-hole (LH) exciton in a quasi-2D structure, while we do not observe the HH and LH exciton splittings in a bulk GaAs sample because of the degeneracy of the HH and LH bands at the zone center of the valence-band structure. The energies and absorption spectra of the HH and LH excitons are further investigated in Section 13.4.

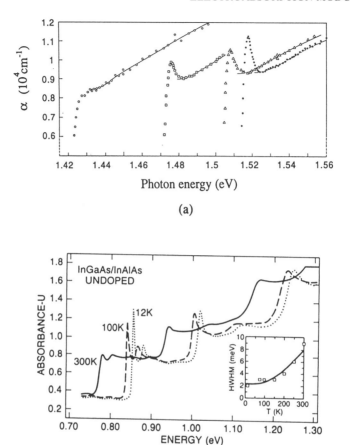

Figure 13.4. (a) Band-edge absorption spectra of a bulk GaAs sample at $T = 294$ K (circles), 186 K (squares), 90 K (triangles), and 21 K (dots). (After Ref. 30.) (b) Absorption spectra of an $In_{0.53}Ga_{0.47}As / In_{0.52}Al_{0.48}As$ quantum-well sample at $T = 300$ K (solid curve), 100 K (dashed curve), and 12 K (dotted curve). The insert shows the half-width at half-maximum (HWHM) of the first absorption peak as a function of temperature. Squares are measured data; curve is calculated. (After Ref. 32 © 1988 IEEE.)

The insert in Fig. 13.4b shows the measured (squares) half-width at half-maximum (HWHM) of the first heavy-hole absorption line as a function of temperature. The solid line is a fit to the expression

$$\frac{\Gamma}{2}(= \text{HWHM}) = \Gamma_0 + \frac{\Gamma_{ph}}{\exp\left(\dfrac{\hbar\omega_{LO}}{k_BT}\right) - 1} \qquad (13.3.20)$$

where $\Gamma_0 = 2.3$ meV accounts for the inhomogeneous broadenings such as scatterings by interface roughness and alloy fluctuations, and the second term represents the homogeneous broadening due to InGaAs longitudinal optical (LO) phonon scatterings with $\hbar\omega_{LO} = 35$ meV and $\Gamma_{ph} = 15.3$ meV.

13.4   QUANTUM CONFINED STARK EFFECT (QCSE) [11–13, 15]

In this section, we consider the exciton absorption in a quantum-well structure in the presence of a uniform applied electric field. The effective mass equation, similar to (13.1.5), can be written as [12]

$$\left(H_e - H_h + E_g - \frac{e^2}{4\pi\varepsilon_s|\mathbf{r}_e - \mathbf{r}_h|} \right)\Phi(\mathbf{r}_e,\mathbf{r}_h) = E\Phi(\mathbf{r}_e,\mathbf{r}_h) \quad (13.4.1)$$

where

$$H_e = -\frac{\hbar^2}{2m_e^*}\nabla_e^2 + V_e(\mathbf{r}_e) \qquad (13.4.2a)$$

$$-H_h = -\frac{\hbar^2}{2m_h^*}\nabla_h^2 + V_h(\mathbf{r}_h) \qquad (13.4.2b)$$

The electron potential $V_e(\mathbf{r}_e)$ or the hole potential $V_h(\mathbf{r}_h)$ may also include the effect of the electric field $|e|\mathbf{F}\cdot\mathbf{r}_e$ or $-|e|\mathbf{F}\cdot\mathbf{r}_h$. The interaction between the electron and the hole is due to the Coulomb potential.

Let us assume that the quantum well is grown along the z axis and the uniform electric field is also applied along the z direction, the electron and hole potential can be written as (Fig. 13.5)

$$V_e(\mathbf{r}_e) = V_e(z_e) \qquad (13.4.3a)$$

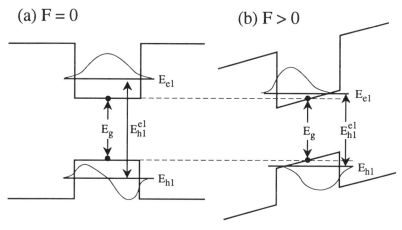

Figure 13.5. Quantum-well energy subbands and wave functions (a) in the absence of an applied electric field, and (b) in the presence of an applied electric field.

and

$$V_h(\mathbf{r}_h) = V_h(z_h) \tag{13.4.3b}$$

13.4.1 Solution of the Electron–Hole Effective-Mass Equation With Excitonic Effects

Using the transformations for the difference coordinate and the center-of-mass coordinate systems for the x and y components,

$$\boldsymbol{\rho} = \boldsymbol{\rho}_e - \boldsymbol{\rho}_h \qquad \mathbf{R}_t = \frac{m_e^* \boldsymbol{\rho}_e + m_h^* \boldsymbol{\rho}_h}{M} \tag{13.4.4}$$

where $M = m_e^* + m_h^*$, and $\boldsymbol{\rho} = x\hat{x} + y\hat{y}$, we obtain after following similar procedures as in (13.1.8) to (13.1.15)

$$\left[-\frac{\hbar^2}{2M}\nabla_{\mathbf{R}_t}^2 - \frac{\hbar^2}{2m_r}\nabla_\rho^2 - \frac{\hbar^2}{2m_e^*}\frac{d^2}{dz_e^2} - \frac{\hbar^2}{2m_h^*}\frac{d^2}{dz_h^2} \right.$$
$$\left. + V_e(z_e) + V_h(z_h) - \frac{e^2}{4\pi\varepsilon_s|\mathbf{r}_e - \mathbf{r}_h|} - (E - E_g) \right] \Phi(\mathbf{r}_e, \mathbf{r}_h) = 0 \tag{13.4.5}$$

Since the dependence on $\mathbf{R}_t$ comes from only the leading kinetic energy term, the solution can be written as [33]

$$\Phi(\mathbf{r}_e, \mathbf{r}_h) = \frac{e^{i\mathbf{K}_t \cdot \mathbf{R}_t}}{\sqrt{A}} F(\boldsymbol{\rho}, z_e, z_h) \tag{13.4.6}$$

where the exciton envelope function $F(\boldsymbol{\rho}, z_e, z_h)$ satisfies

$$\left[-\frac{\hbar^2}{2m_r}\nabla_\rho^2 + H(z_e) - H(z_h) - \frac{e^2}{4\pi\varepsilon_s|\mathbf{r}_e - \mathbf{r}_h|} \right.$$
$$\left. - \left(E - E_g - \frac{\hbar^2 K_t^2}{2M} \right) \right] F(\boldsymbol{\rho}, z_e, z_h) = 0 \tag{13.4.7}$$

The center of mass of the electron–hole pair is thus moving freely with a kinetic energy $\hbar^2 K_t^2/(2M)$. Here, identical relations to (13.1.8) to (13.1.11), for the parallel components of the real-space and momentum-space vectors exist, e.g.,

$$\mathbf{K}_t = \mathbf{k}_{et} + \mathbf{k}_{ht} \qquad \mathbf{k}_t = \frac{m_h^*\mathbf{k}_{et} - m_e^*\mathbf{k}_{ht}}{M} \tag{13.4.8}$$

Note that

$$H(z_e) = -\frac{\hbar^2}{2m_e^*}\frac{d^2}{dz_e^2} + V_e(z_e) \qquad -H(z_h) = -\frac{\hbar^2}{2m_h^*}\frac{d^2}{dz_h^2} + V_h(z_h)$$

$$(13.4.9)$$

which are simply one-dimensional Schrödinger equations for a particle-in-a-box model. Let us consider

$$H(z_e)f_n(z_e) = E_{en}f_n(z_e) \qquad H(z_h)g_m(z_h) = E_{hm}g_m(z_h) \quad (13.4.10)$$

where $f_n(z_e)$ and $g_m(z_h)$ are the free-electron and the free-hole wave functions in the absence of any interactions. With the Coulomb interaction term, the solution $F(\rho, z_e, z_h)$ is more complicated. However, using the completeness properties of the solutions $\{f_n(z_e)\}$ and $\{g_m(z_h)\}$, we can expand the exciton envelope function as [12, 15, 19]

$$F(\rho, z_e, z_h) = \sum_n \sum_m \phi_{nm}(\rho)f_n(z_e)g_m(z_h) \qquad (13.4.11)$$

The Fourier transform pairs for the ρ dependence can be written as

$$\phi_{nm}(\rho) = \sum_{k_t} G_{nm}(k_t)\frac{e^{ik_t\cdot\rho}}{\sqrt{A}}$$

$$G_{nm}(k_t) = \int d^2\rho\,\phi_{nm}(\rho)\frac{e^{-ik_t\cdot\rho}}{\sqrt{A}} \qquad (13.4.12)$$

The envelope function $F(\rho, z_e, z_h)$ becomes

$$F(\rho, z_e, z_h) = \sum_{nm}\sum_{k_t} G_{nm}(k_t)\frac{e^{ik_t\cdot\rho}}{\sqrt{A}}f_n(z_e)g_m(z_h) \qquad (13.4.13)$$

The complete envelope function is obtained from (13.4.6), noting that all possible K_t can be included:

$$\Phi(r_e, r_h) = \sum_{K_t}\sum_{k_t}\sum_{nm} G_{nm}(k_t)\frac{e^{iK_t\cdot R_t}}{\sqrt{A}}\frac{e^{ik_t\cdot\rho}}{\sqrt{A}}f_n(z_e)g_m(z_h)$$

$$= \sum_{k_{et}}\sum_{k_{ht}}\sum_{nm} G_{nm}(k_t)\frac{e^{ik_{et}\cdot\rho_e}}{\sqrt{A}}\frac{e^{ik_{ht}\cdot\rho_h}}{\sqrt{A}}f_n(z_e)g_m(z_h) \qquad (13.4.14)$$

Comparing (13.4.13) and (13.4.14) with (13.1.2) and (13.1.3), we find that the original electron–hole pair state can be written as [8, 9, 15, 34]

$$\Psi(\mathbf{r}_e, \mathbf{r}_h) = \sum_{\mathbf{k}_{et} n} \sum_{\mathbf{k}_{ht} m} G_{nm}(\mathbf{k}_t) \psi_{n\mathbf{k}_{et}}(\mathbf{r}_e) \psi_{m-\mathbf{k}_{ht}}(\mathbf{r}_h) \quad (13.4.15a)$$

$$\psi_{n\mathbf{k}_{et}}(\mathbf{r}_e) = \frac{e^{i\mathbf{k}_{et}\cdot\boldsymbol{\rho}_e}}{\sqrt{A}} f_n(z_e) u_c(\mathbf{r}_e) \quad (13.4.15b)$$

$$\psi_{m-\mathbf{k}_{ht}}(\mathbf{r}_h) = \frac{e^{-i\mathbf{k}_{ht}\cdot\boldsymbol{\rho}_h}}{\sqrt{A}} g_m(z_h) u_v(\mathbf{r}_h) \quad (13.4.15c)$$

13.4.2 Optical Matrix Element for Excitonic Transitions in a Quantum Well

The optical matrix element is obtained using $|i\rangle = |$ground state$\rangle$, $|f\rangle = |\Psi(\mathbf{r}_e, \mathbf{r}_h)\rangle$, and

$$\langle f|\hat{e}\cdot\mathbf{p}|i\rangle = \sum_{nm}\sum_{\mathbf{k}_{et}}\sum_{\mathbf{k}_{ht}} G_{nm}^*(\mathbf{k}_t)\langle\psi_{n\mathbf{k}_{et}}(\mathbf{r})|\hat{e}\cdot\mathbf{p}|\psi_{m-\mathbf{k}_{ht}}(\mathbf{r})\rangle$$

$$= \sum_{nm}\sum_{\mathbf{K}_t}\sum_{\mathbf{k}_t} G_{nm}^*(\mathbf{k}_t)\hat{e}\cdot\mathbf{p}_{cv}I_{nm}\delta_{\mathbf{k}_{et}+\mathbf{k}_{ht},0}$$

$$= \sum_{nm}\sum_{\mathbf{k}_t} G_{nm}^*(\mathbf{k}_t)\hat{e}\cdot\mathbf{p}_{cv}I_{nm}$$

$$= \sum_{nm}\left[\sqrt{A}\,\phi_{nm}^*(\boldsymbol{\rho}=0)\right]\hat{e}\cdot\mathbf{p}_{cv}I_{nm} \quad (13.4.16)$$

where $I_{nm} = \int_{-\infty}^{\infty} f_n^*(z)g_m(z)dz$, and $\mathbf{p}_{cv} = \langle u_c(\mathbf{r})|\mathbf{p}|u_v(\mathbf{r})\rangle$. The above matrix element can be expressed alternatively in terms of $F(\boldsymbol{\rho}, z_e, z_h)$:

$$\langle f|\hat{e}\cdot\mathbf{p}|i\rangle = \int_{-\infty}^{\infty} dz F^*(\boldsymbol{\rho}=0, z, z)\sqrt{A}\,\hat{e}\cdot\mathbf{p}_{cv} \quad (13.4.17)$$

Noting that $\mathbf{K}_t = \mathbf{k}_{et} + \mathbf{k}_{ht} = 0$ from (13.4.16), we substitute the expression (13.4.11) for $F(\boldsymbol{\rho}, z_e, z_h)$ into (13.4.7), and obtain

$$\sum_{n'm'}\left[-\frac{\hbar^2}{2m_r}\nabla_\rho^2 + E_{en'} - E_{hm'} - \frac{e^2}{4\pi\varepsilon_s|\mathbf{r}_e-\mathbf{r}_h|} - (E - E_g)\right]$$

$$f_{n'}(z_e)g_{m'}(z_h)\phi_{n'm'}(\boldsymbol{\rho}) = 0 \quad (13.4.18)$$

Multiplying the above equation by $f_n^*(z_e)$ and $g_m^*(z_h)$ and integrating over z_e and z_h, we obtain

$$\left[-\frac{\hbar^2}{2m_r}\nabla_\rho^2 - V_{nm}(\rho)\right]\phi_{nm}(\boldsymbol{\rho}) = E_{ex}\phi_{nm}(\boldsymbol{\rho}) \quad (13.4.19)$$

where

$$V_{nm}(\rho) \equiv \sum_{n'm'} \int dz_e f_n^*(z_e) f_{n'}(z_e) \int dz_h g_m^*(z_h) g_{m'}(z_h) \frac{e^2}{4\pi\varepsilon_s \left[\rho^2 + |z_e - z_h|^2\right]^{1/2}}$$

$$\simeq \int dz_e |f_n(z_e)|^2 \int dz_h |g_m(z_h)|^2 \frac{e^2}{4\pi\varepsilon_s \left[\rho^2 + |z_e - z_h|^2\right]^{1/2}} \qquad (13.4.20)$$

$$E_{ex} = E - \left(E_g + E_{en} - E_{hm}\right) \qquad (13.4.21)$$

We have ignored coupling among different subbands so that only $n' = n$ and $m' = m$ are taken into account in (13.4.20). The wave function $\phi_{nm}(\rho)$ satisfies the normalization condition in a two-dimensional space:

$$\int d^2\rho |\phi_{nm}(\rho)|^2 = \int_0^\infty 2\pi\rho \, d\rho |\phi_{nm}(\rho)|^2 = 1 \qquad (13.4.22)$$

This was derived using

$$\int d^3r_e \int d^3r_h |\Phi(r_e, r_h)|^2 = 1 \qquad (13.4.23)$$

and $d^3r_e \, d^3r_h = d^2R_t \, d^2\rho \, dz_e \, dz_h$; therefore,

$$\int d^2\rho \int dz_e \int dz_h |F(\rho, z_e, z_h)|^2 = 1 \qquad (13.4.24)$$

13.4.3 Variational Method for Exciton Problem

Two different methods are commonly used to find the solution for the exciton equation (13.4.19). The most common approach is a variational method, which is very useful to find the bound state solution. Noting that the 1s state solution of $\phi(\rho)$ behaves like $e^{-\rho/a_0}$ or $e^{-\rho/(2a_0)}$, the following variational form is assumed in the variational approach [12]:

$$E_{ex}(\lambda) = \frac{\langle \phi | - \left(\hbar^2/2m_r\right)\nabla_\rho^2 - V(\rho)|\phi\rangle}{\langle \phi | \phi\rangle} \qquad (13.4.25)$$

where

$$\phi(\rho) = \sqrt{\frac{2}{\pi}} \frac{1}{\lambda} e^{-\rho/\lambda} \qquad (13.4.26)$$

which satisfies the normalization condition $\langle \phi | \phi \rangle = 1$ in (13.4.22). We find

$$E_{ex}(\lambda) = \frac{\hbar^2}{2m_r\lambda^2}$$

$$-\frac{e^2}{4\pi\varepsilon_s}\frac{4}{\lambda^2}\int dz_e |f_n(z_e)|^2 \int dz_h |g_m(z_h)|^2 \int_0^\infty \rho\,d\rho\,\frac{e^{-2\rho/\lambda}}{\left[\rho^2 + (z_e - z_h)^2\right]^{1/2}}$$

(13.4.27)

It is convenient to write in terms of the normalized parameters

$$\beta = \frac{a_o}{\lambda} \qquad a_o = \frac{4\pi\varepsilon_s\hbar^2}{m_r e^2}$$

(13.4.28)

and the normalized exciton binding energy

$$\varepsilon_{ex} = \frac{E_{ex}}{R_y} \qquad R_y = \frac{m_r e^4}{2\hbar^2(4\pi\varepsilon_s)^2}$$

(13.4.29)

$$\varepsilon_{ex} = \beta^2 - 4\beta \int dz_e |f_n(z_e)|^2 \int dz_h |g_m(z_h)|^2 G\left(\frac{2\beta|z_e - z_h|}{a_0}\right)$$

(13.4.30)

where the function $G(x)$ is defined as an integral

$$G(x) = \int_0^\infty dt\,\frac{te^{-t}}{\left(t^2 + x^2\right)^{1/2}}$$

(13.4.31)

which is a smooth monotonic function and can be approximated [55] analytically. It has the properties that $G(0) = 1$ and $G(\infty) = 0$. Typically, for a quantum-well problem, $|z_e - z_h|$ is finite and the argument in $G(x)$ is over a finite range. The pure two-dimensional limit can be obtained by ignoring the z dependence and using $G(0) = 1$,

$$\varepsilon_{ex} = \beta^2 - 4\beta$$

(13.4.32)

Thus we find the minimum at $\beta = 2$ and $\varepsilon_{ex} = -4$ as expected for a pure 2D exciton binding energy.

13.4.4 Momentum-Space Solution for the Exciton Problem

Alternatively, the real-space exciton differential equation (13.4.19) can also be written in terms of the momentum space integral equation [15, 35]:

$$\left[+ \frac{\hbar^2 k_t^2}{2m_r} G_{nm}(\mathbf{k}_t) - \sum_{n'm'} \sum_{\mathbf{k}'_t} V_{nn',mm'}(|\mathbf{k}_t - \mathbf{k}'_t|) G_{nm}(\mathbf{k}'_t) \right] = E_{ex} G_{nm}(\mathbf{k}_t)$$

$$(13.4.33)$$

where

$$V_{nn',mm'}(q) = \frac{e^2}{2\varepsilon_s qA} \int dz_e \int dz_h f_n^*(z_e) f_{n'}(z_e) g_m^*(z_h) g_{m'}(z_h) e^{-q|z_e - z_h|}$$

$$(13.4.34)$$

A direct numerical solution of the above equation is discussed in Ref. 15.

13.4.5 Optical Absorption Spectrum with Exciton Effects in Quantum Wells

The solution to the eigenvalue equation (13.4.19) is a set of exciton binding energies and corresponding eigenfunctions for 1s, 2s, 3s, and continuum states. We denote the quantum number for the exciton state as x. The absorption coefficient for a quantum-well structure can be obtained by substituting the matrix element $\langle f | \hat{e} \cdot \mathbf{p} | i \rangle$ into (13.1.1)

$$\alpha(\hbar\omega) = C_0 \frac{2}{V} \sum_x \left| \sqrt{A} \sum_{nm} \phi_{nm}^x(\rho = 0) \hat{e} \cdot \mathbf{p}_{cv} I_{nm} \right|^2 \delta(E_x - \hbar\omega)$$

$$C_0 \equiv \frac{\pi e^2}{n_r c \varepsilon_0 \omega m_0^2} \qquad\qquad (13.4.35)$$

where the exciton transition energy is

$$E_x = E_{hm}^{en} + E_{ex} \qquad\qquad (13.4.36a)$$

and the band edge transition energy is

$$E_{hm}^{en} = E_g + E_{en} - E_{hm} \qquad\qquad (13.4.36b)$$

The exciton binding energy E_{ex} is a discretized set of 1s, 2s, 3s, ... states and continuum-state energies. Assuming that there is no mixing between different subbands, consider only the pair $n = C1$ and $m = HH1$, for example. We may drop the summation over nm and treat each pair of n and m independently. This assumption is valid only if the subband energy difference is much larger than the exciton binding energy. The absorption coefficient becomes

$$\alpha(\hbar\omega) = C_0 \frac{2}{L} \sum_x |\phi^x_{nm}(\rho = 0)|^2 |\hat{e} \cdot \mathbf{p}_{cv}|^2 |I_{nm}|^2 \, \delta(E_x - \hbar\omega) \quad (13.4.37)$$

Exciton Discrete (1s)-State Contribution. For the 1s state, we find

$$\alpha(\hbar\omega) = C_0 \frac{2}{L} \left(\frac{2}{\pi}\frac{1}{\lambda^2}\right) |\hat{e} \cdot \mathbf{p}_{cv}|^2 |I_{nm}|^2 \frac{\gamma/\pi}{(E_x - \hbar\omega)^2 + \gamma^2} \quad (13.4.38)$$

where a finite linewidth 2γ has been assumed and the delta function is replaced by a Lorentzian function and $x = 1s$ state. The matrix element $|\hat{e} \cdot \mathbf{p}_{cv}|^2$ is obtained [15, 36] from (9.5.20)–(9.5.23) in Section 9.5.

For TE polarization ($\hat{e} = \hat{x}$ or $\hat{y}$)

$$|\hat{e} \cdot \mathbf{p}_{cv}|^2 = M_b^2 \begin{cases} \frac{3}{2} & \text{Heavy-hole exciton} \\ \frac{1}{2} & \text{Light-hole exciton} \end{cases} \quad (13.4.39)$$

For TM polarization ($\hat{e} = \hat{z}$)

$$|\hat{e} \cdot \mathbf{p}_{cv}|^2 = M_b^2 \begin{cases} 0 & \text{Heavy-hole exciton} \\ 2 & \text{Light-hole exciton} \end{cases} \quad (13.4.40)$$

where M_b^2 is the bulk matrix element discussed in Chapter 9:

$$M_b^2 \left(\cong \frac{m_0^2 E_g(E_g + \Delta)}{6m_e^*(E_g + 2\Delta/3)} \right) = \frac{m_0}{6} E_p \quad (13.4.41)$$

Exciton Continuum-State Contributions. For continuum-state contributions, the wave function $|\phi^x(\rho = 0)|^2$ gives a Sommerfeld enhancement factor [13–15]. Note that the sum over the states x becomes the sum over a continuum distribution of states k_x and k_y,

$$2\sum_x = 2\sum_{k_x}\sum_{k_y} = A\frac{m_r}{\pi\hbar^2}\int dE_t \quad (13.4.42)$$

and $E_{ex} \to \hbar^2 k_t^2/(2m_r) = E_t$. We obtain the absorption coefficient due to the continuum states

$$\alpha_C(\hbar\omega) = C_0 \frac{m_r}{\pi\hbar^2 L} M_b^2 |I_{nm}|^2 \int_0^\infty dE_t\, M(E_t)|\phi^x(0)|^2 \frac{\gamma/\pi}{(E_x - \hbar\omega)^2 + \gamma^2}$$

(13.4.43)

where $E_x = E_{hm}^{en} + E_t$ and $|\phi^x(0)|^2$ is usually approximated by the Sommerfeld enhancement factor for a 2D exciton:

$$|\phi^x(0)|^2 \simeq \frac{s_0}{1 + e^{-2\pi/(ka_o)}}$$

(13.4.44)

where $1 \le s_0 \le 2$. For a pure 2D exciton, $s_0 = 2$ and

$$ka_o = \sqrt{E_t/R_y}$$

(13.4.45)

The matrix-element $M(E_t)$ is defined [15, 37, 38] as $|\hat{e} \cdot \mathbf{p}_{cv}|^2 = M(E_t)M_b^2$, which has been derived in Section 9.5 and is tabulated in Table 9.1.

TE polarization

$$M(E_t) = \begin{cases} \frac{3}{4}(1 + \cos^2\theta_{nm}) & \text{Heavy-hole exciton} \\ \frac{1}{4}(5 - 3\cos^2\theta_{nm}) & \text{Light-hole exciton} \end{cases}$$

(13.4.46)

TM polarization

$$M(E_t) = \begin{cases} 0 & \text{Heavy-hole exciton} \\ \frac{1}{2}(1 + 3\cos^2\theta_{nm}) & \text{Light-hole exciton} \end{cases}$$

(13.4.47)

where $\cos^2\theta_{nm} \simeq (E_{en} + |E_{hm}|)/(E_{en} + |E_{hm}| + E_t)$. The heavy-hole exciton contribution to the TM polarization is taken to be zero instead of $(\frac{3}{2})\sin^2\theta_{nm}$ because a rigorous valence-band mixing model shows that the heave-hole exciton has a negligible contribution to the TM case [39–43]. Note that $k_o = 1/a_o$ and

$$\frac{m_r}{\pi\hbar^2} = \frac{k_o^2}{2\pi R_y}$$

(13.4.48)

Total Absorption Spectrum. The complete absorption spectrum can be written as the sum of the bound-state and continuum-state contributions [15]:

$$\alpha(\omega) = C_0 \frac{2}{L} M_b^2 |I_{nm}|^2 \frac{k_o^2}{2\pi R_y} \left[4 \sum_{\substack{x = \text{discrete} \\ \text{states}}} M(0) |a_o \phi^x(0)|^2 \frac{R_y \gamma}{(E_x - \hbar\omega)^2 + \gamma^2} \right.$$

$$\left. + \int_0^\infty \frac{dE_t}{\pi} M(E_t) |\phi^x(0)|^2 \frac{\gamma}{(E_{hm}^{en} - E_t - \hbar\omega)^2 + \gamma^2} \right] \quad (13.4.49)$$

Let us look at the band-edge transition energy E_{hm}^{en}:

$$E_{hm}^{en} = E_g + E_{en} - E_{hm} \quad (13.4.50)$$

for the electron subband n and the hole subband m in the presence of an applied electric field. Note that the subband energies are measured from the band edges at the center of the quantum well (positive for electrons and negative for holes).

13.4.6 Perturbation Method

A simple second-order perturbation theory shows [44] that (see the example in Section 3.5 and its references)

$$E_n = E_n^{(0)} + \frac{C_n m^* e^2 F^2 L_{\text{eff}}^4}{\hbar^2}$$

$$C_n = \frac{32}{\pi^6} \sum_{m \neq n} \frac{|1 - (-1)^{n-m}|^2 m^2 n^2}{(n^2 - m^2)^5}$$

$$= \frac{n^2 \pi^2 - 15}{24 n^4 \pi^4} \quad (13.4.51)$$

in an infinite quantum-well model assuming an effective well width of L_{eff}. Therefore, the band-edge transition energy for the first conduction and the first heavy-hole subband is determined by

$$E_{h1}^{el}(F) = E_{h1}^{el}(F = 0) + C_1(m_e^* + m_h^*)e^2 F^2 L_{\text{eff}}^4 / \hbar^2 \quad (13.4.52)$$

where $C_1 = -2.19 \times 10^{-3}$. As an example, for a GaAs/Al$_x$Ga$_{1-x}$As quantum well with an effective well width of 100 Å, the above change in the transition energy is $E_{h1}^{el}(F) - E_{h1}^{el}(F = 0) = -11.8$ meV at $F = 100$ kV/cm assuming that $m_e^* = 0.0665 m_0$ and $m_{hh}^* = 0.34 m_0$. The exciton binding energy E_{ex} also depends on the electric field strength. In general, the band-edge

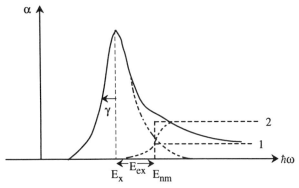

Figure 13.6. Exciton absorption spectrum of a quantum-well structure in the absence of an applied electric field. The contributions due to the discrete and continuum states are shown separately as dashed lines.

shift due to the electric field is quite clear and appears stronger than the binding energy change with the electric field. Variational methods instead of the above perturbation method for the band-edge energies have also been used [44–47].

13.4.7 Exciton Absorption Spectrum and Comparison With Experimental Data

Theoretical absorption spectrum for a single pair of transitions from the first conduction subband to the first heavy-hole subband is shown in Fig. 13.6, where the discrete 1s bound state $E_x = E_{1s}$ contributes as a Lorentzian spectrum, and the steplike density of states enhanced by the Sommerfeld factor contributes as the high-energy tail.

With an applied electric field, the quantum confined Stark effects can be measured from the shift of the peak absorption coefficient as a function of the applied electric field. Figure 13.7 shows the polarization-dependent optical absorption spectra for (a) TE and (b) TM polarizations [17, 48] with estimated electric fields [13]. Theoretical results using the parabolic band model [15] presented in this section and using a valence-band mixing model [43] have been used to successfully match these experimental data, as shown in Fig. 13.8. To understand these data, we make the following observations:

1. The exciton absorption peak energy depends on the band-edge transition energy $E_{h1}^{e1}(F)$, which shifts quadratically as a function of the field, minus the amount of the 1s state exciton binding energy E_{1s}, which is about 8 meV for the heavy hole exciton. The transition energies of the heavy-hole and light-hole exciton peaks vs. the applied electric field are shown in Fig. 13.9. The binding energy for a bulk (3D) GaAs R_y is

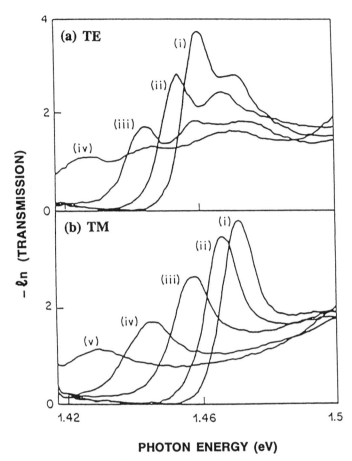

Figure 13.7. Experimental absorption spectra of a $GaAs/Al_{0.3}Ga_{0.7}As$ quantum-well waveguide modulator as a function of field: (a) TE polarization for (i) 0 kV/cm, (ii) 60 kV/cm, (iii) 100 kV/cm, (iv) 150 kV/cm. (b) TM polarization for (i) 0 kV/cm, (ii) 60 kV/cm, (iii) 110 kV/cm, (iv) 150 kV/cm, and (v) 200 kV/cm. (After Ref. 48.) [13, 17].

about 4.2 meV and is $4R_y \simeq 16.8$ meV for a pure 2D exciton. A quasi-two-dimensional quantum-well structure gives a binding energy corresponding to an "effective dimension" between 2D and 3D. Therefore, the binding energy of the quantum well is somewhere between 2D and 3D.

As a matter of fact, since analytical solutions for the hydrogen atom equation in an α-dimensional space have analytical solutions for an integer α such as 2 and 3, the idea is to extend the general result for a given α using the concept of analytical continuation. Using this approach, the oscillator strength for the bound and continuum states can

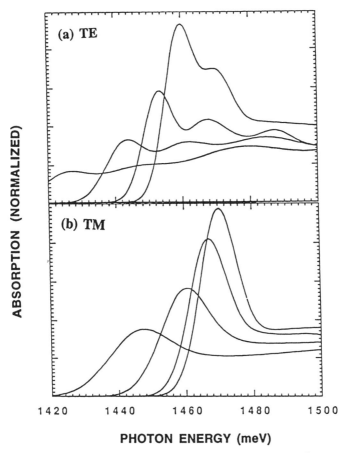

Figure 13.8. Theoretically calculated absorption spectra for the quantum-well waveguide modulator in Fig. 13.7: (a) TE polarization, (b) TM polarization. (After Ref. 43.)

be obtained using the analytical formulas once the "effective dimension" is determined. The effective dimension can be extracted by comparing the binding energy calculated variationally, as discussed in this section, with the analytical formula for binding energy [49].

2. The TE oscillator strength for the heavy-hole exciton is approximately three times that of the light-hole exciton. However, since the light-hole exciton transition energy is already in the continuum states of the heavy-hole transition, the spectrum would not show a 3:1 ratio.

3. The TM polarization spectra show that it is the light-hole exciton transition which is dominant for this polarization as a result of the optical momentum matrix-selection rule.

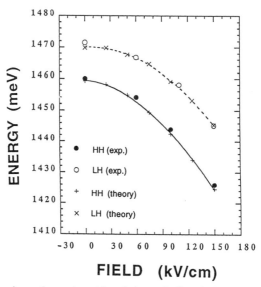

Figure 13.9. Comparison of experimental and theoretically calculated exciton energies vs. the electric field. (After Ref. 43.)

4. At a fixed optical energy $\hbar\omega$ of the incident light, the absorption coefficient can change drastically especially when $\hbar\omega$ is near an exciton peak absorption energy. This enhanced change of the absorption by an applied voltage is discussed further in the next section.

13.5 INTERBAND ELECTROABSORPTION MODULATOR

The electroabsorption modulators can be designed [10, 50–56] using a waveguide configuration (a) or transverse transmission (or reflection) configuration (b). Consider the modulator structures shown in Figs. 13.10a and b. Suppose the operating optical energy $\hbar\omega_0$ is chosen near the exciton peak when a voltage V is applied (Fig. 13.10c). The transmission coefficient is proportional to

$$T(V) = e^{-\alpha(V)L} \tag{13.5.1}$$

For the waveguide modulator, $\alpha(V)$ is the absorption coefficient of the waveguide region multiplied by the optical confinement factor Γ, and L is the total length of the guide for the waveguide modulator. For the transverse transmission modulator, $\alpha(V)$ is the average absorption coefficient of the multiple-quantum-well region, and L is the total thickness of the MQW region. The couplings or reflections at the facets are ignored here for

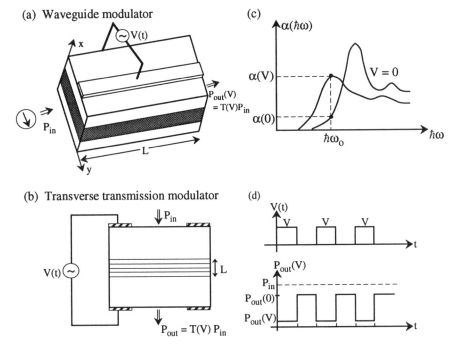

Figure 13.10. (a) A waveguide electroabsorption modulator. (b) A transverse transmission electroabsorption modulator. (c) The absorption coefficient $\alpha(\hbar\omega)$ of a quantum-well modulator at two different bias voltages, V and 0, for example. (d) With a bias voltage $V(t)$ across the modulator, the transmitted optical power (solid square wave) is modulated. Dashed line is the input power P_{in}.

convenience. The on/off ratio (or contrast ratio) $R_{on/off}$ is defined as (Fig. 13.10d)

$$R_{on/off} = \frac{P_{out}(V_{on} = 0)}{P_{out}(V_{off} = V)} = \frac{T(0)}{T(V)} \qquad (13.5.2a)$$

or in decibels

$$R_{on/off}(dB) = 10 \log \frac{T(V_{on} = 0)}{T(V_{off} = V)}$$
$$= 4.343[\alpha(V) - \alpha(0)]L \qquad (13.5.2b)$$

Therefore, in principle, the magnitude of the extinction ratio or the on/off ratio can be made as large as possible by increasing the cavity length L. However, we note that the insertion loss L_{in} is defined as

$$L_{in} = \frac{P_{in} - P_{out}(0)}{P_{in}} = 1 - T(0) = 1 - e^{-\alpha(0)L} \qquad (13.5.3)$$

at the transmission (on) state. Since $\alpha(0)$ is always finite, a large cavity length L will decrease the transmissivity $T(0)$ exponentially. We may achieve an infinite on/off ratio using an infinitely long cavity and obtain no light transmission even for the on-state. If $T(0) \rightarrow 0$, the insertion loss approaches 100%. Naturally, this is not desirable and an optimum design, which maximizes the extinction ratio and minimizes the insertion loss, is necessary. Note that V_{on} is set to zero here only for illustration purposes. For high-speed switching, it is usually set a nonzero reverse bias voltage for fast carrier sweep-out of the absorption region.

Another useful figure of merit in the design of the electroabsorption modulator is the change in absorption coefficient per unit applied voltage.

$$\frac{\Delta \alpha}{\Delta V} = \frac{\alpha(V_{off}) - \alpha(V_{on})}{\Delta V} \tag{13.5.4}$$

where the change in voltage $\Delta V = |V_{on} - V_{off}|$. This occurs because the on/off ratio per unit applied voltage is

$$\frac{R_{on/off}}{\Delta V} = 4.343 \frac{[\alpha(V) - \alpha(0)]L}{\Delta V}$$

$$= 4.343 \frac{\Delta \alpha}{\Delta F} \tag{13.5.5}$$

where the electric field across the multiple-quantum-well region is approximately given by $F = V/L$, if the built-in voltage V_{bi} is ignored. (Otherwise $\Delta F = (V - V_{bi})/L$ has to be used.) For more discussions on the figure of merits, such as $\Delta \alpha / (\Delta F)^2, C(\Delta V)^2$, and the optimization of the contrast ratio, drive voltage, bandwidth, and total insertion loss, see Ref. 56.

13.6 SELF-ELECTROOPTIC EFFECT DEVICES (SEEDs) [18–20, 50–56]

Interesting optical switch devices such as the self-electrooptic effect devices have been demonstrated using the quantum confined Stark effects. These devices show optical bistability and consist of a multiquantum-well p-i-n diode structure with different possible loads, such as a resistor (R-SEED), a constant current source, or another p-i-n multiple-quantum-well diode (Symmetric or S-SEED). In this section, we discuss basic physical principles of an R-SEED. As shown in Fig. 13.11a, the circuit equation is given by Kirchoff's voltage law:

$$V_0 = V + RI_R \tag{13.6.1}$$

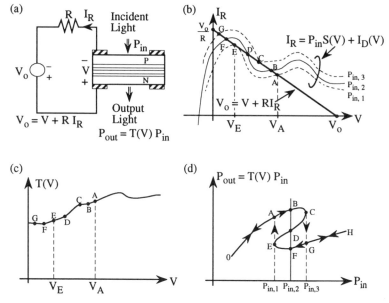

Figure 13.11. (a) A circuit diagram for a self-electrooptic effect device with a resistor load (R-SEED) in the presence of an incident laser light at the input P_{in}. (b) A graphical solution for the photocurrent response $I_R = S(V)P_{in} + I_D(V)$ and the load line $V_0 = V + RI_R$. (c) The transmission $T(V)$ of the laser light passing through the MQW *p-i-n* diode is plotted as a function of the voltage drop V across the diode. (d) The switch diagram for the optical output power vs. the optical input power with the arrows showing the path of switching.

where V is the voltage drop across the diode. The reverse bias voltage V and the reversed current I_R are defined as shown in Fig. 13.11a. For an incident laser light with a power P_{in}, the current I_R is the sum of the photocurrent $S(V)P_{in}$ and the dark current $I_D(V)$:

$$I_R = P_{in}S(V) + I_D(V) \tag{13.6.2}$$

where $S(V)$ is the responsivity of the diode and is defined as

$$S(V) = q\frac{\eta_{int}A}{\hbar\omega} \tag{13.6.3}$$

where

$$A \simeq (1 - R_p)(1 - e^{-\alpha L} + R_n e^{-\alpha L}) \tag{13.6.4}$$

is the absorbance, that is, the fraction of absorbed power for a unit incident power, assuming a single path absorption. The absorbance can be derived by noting that the reflectivity at the front surface R_p, plus the transmission after passing through the substrate $(1 - R_p)(1 - R_n)e^{-\alpha L}$, plus the absorbance A, must equal 1. At steady state, the voltage drop across the multiple-quantum-well diode V and the current I_R is determined by the simultaneous solutions of the above two equations. A graphical illustration of these two equations is plotted in Fig. 13.11b, where the intersection points stand for the possible solutions for I_R and V.

As can be seen from Fig. 13.11b, when the optical input power P_{in} is increased, the photocurrent is increased, in general, and the number of intersection points varies. For an optical power equal to $P_{in,1}$, we have two intersection points A and E with corresponding voltages V_A and V_E. Therefore, there are two possible output states: $P_{out} = T(V_A)P_{in,1}$, which is higher in output power, and $P_{out} = T(V_E)P_{in,1}$, which is lower in output power because of the transmissivity $T(V_A) > T(V_E)$ (Fig. 13.11c). Once we increase the optical input power to $P_{in,2}$, we find that there are three intersection points, B, D, and F, with corresponding voltages V_B, V_D, and V_F. Therefore, the optical output powers through the diode are given by three possible values: $P_{out} = T(V_B)P_{in,2} > T(V_D)P_{in,2} > T(V_F)P_{in,2}$. Increasing the optical input power to $P_{in,3}$, we have only two intersection points, C and G, with corresponding voltages, V_C and V_G. Therefore, we obtain the output powers $P_{out} = T(V_C)P_{in,3} > T(V_G)P_{in,3}$.

The switching curve and the directions are shown in Fig. 13.11d. At a small input power, $P_{in} < P_{in,1}$, the optical transmission power $P_{out} = T(V)P_{in}$ is monotonically increasing with the input power P_{in}, since the photocurrent is small for a small incident optical power. Therefore, most of the reverse bias voltage is across the diode and V is close to V_0 ($V_A < V < V_0$). In this range of voltage, the transmission coefficient is rather flat or monotonic. As the input optical power exceeds $P_{in,1}$, $P_{in,2}$, and $P_{in,3}$, photocurrent will appear in the circuit, and the voltage drop across the diode will drop from V_C to V_G, where the transmission will drop from a high value $T(V_C)$ to a low value $T(V_G)$. After $P_{in} > P_{in,3}$, the output power starts to increase again, since $P_{out} = T(V)P_{in}$ increases as P_{in} increases. This switch sequence is therefore $A \rightarrow B \rightarrow C \rightarrow G \rightarrow H$.

In the reverse direction, if we switch down the optical input power from a large value of P_{in} at H, it will go through $H \rightarrow G \rightarrow F \rightarrow E \rightarrow A \rightarrow 0$, since at point E the voltage has to switch from V_E to V_A as we decrease the input power $P_{in} \leq P_{in,1}$. Therefore, the optical output power will switch from a low state E to a high state A due to the large transmission coefficient $T(V_A) > T(V_E)$.

Symmetric self-electrooptic effect devices [53] (S-SEEDs) using another p-i-n multiple-quantum-well diode as the load, and field-effect transistor self-electrooptic effect devices (F-SEEDs) [20] have also been demonstrated to show interesting physics and applications. For example, the S-SEED can act as a differential logic gate capable of NOR, OR, NAND, and AND functions. These devices made by maximizing the ratio of the absorption coefficients in the high and low states while minimizing the change in electric field can give nearly optimum performance [53]. From the *I-V* curve of a SEED (Figs. 13.11a and b), we see that negative differential conductivity exists. A SEED oscillator [19, 52] can also be designed by a series connection of a SEED and an *L-C* resonator circuit and the SEED is optically pumped to produce a negative electric conductance in the photocurrent response. For example, oscillators with oscillation frequencies from 8.5 to 110 MHz have been demonstrated [52]. For the 8.5-MHz oscillator, frequency tuning by changing the bias voltage of the SEED has a tuning rate 16.7 kHz/V. This frequency tuning is caused by the change in the capacitance in the depletion layer of the SEED with the voltage change. The capacitance can also be changed optically by changing the optical power coupled to the SEED.

PROBLEMS

13.1 In this problem, we use an alternative approach to show the physical interpretation of $2|\phi_n(0)|^2$ as the density of states. We start with (13.1.25).

$$-\frac{\hbar^2}{2m_r}\nabla^2\phi(\mathbf{r}) = E\phi(\mathbf{r})$$

For convenience, we use the normalized distance $\underline{r} = r/a_o$, the normalized wave number $\kappa = k/k_o = ka_o$, and normalized energy $\underline{E} = E/R_y = \kappa^2$, where $a_o = 4\pi\varepsilon_s\hbar^2/m_r e^2$ is the exciton Bohr radius, and $R_y = m_r e^4/2\hbar^2(4\pi\varepsilon_s)^2$ is the exciton Rydberg energy.

(a) Show that the solution is of the form

$$\phi(\mathbf{r}) = R(\underline{r})Y_{\ell m}(\theta, \varphi)$$

where $Y_{\ell m}(\theta, \varphi)$ are the spherical harmonics as shown in the hydrogen atom case. $R(\underline{r})$ satisfies

$$\left[\frac{d^2}{d\underline{r}^2} - \frac{\ell(\ell+1)}{\underline{r}^2} + \kappa^2\right]\underline{r}R(\underline{r}) = 0$$

The solution is the spherical Bessel function

$$R_{\kappa\ell}(\underline{r}) = \sqrt{\frac{2}{\pi}}\,\kappa j_\ell(\kappa\underline{r})$$

(b) Show that $R_{\kappa\ell}(\underline{r})$ above satisfies the $\delta(\kappa - \kappa')$ normalization rule

$$\int_0^\infty R_{\kappa\ell}^*(r)R_{\kappa'\ell}(\underline{r})\underline{r}^2\,d\underline{r} = \delta(\kappa - \kappa')$$

(c) Show that in physical units, the wave function can be written as

$$R_{E\ell}(r) = \frac{1}{(R_y a_o^3)^{1/2}} \sqrt{\frac{ka_0}{\pi}} \, j_\ell(kr)$$

satisfying the $\delta(E - E')$ normalization rule

$$\int_0^\infty R_{E\ell}^*(r) R_{E'\ell}(r) r^2 \, dr = \delta(E - E')$$

where $E = \kappa^2 R_y = \hbar^2 k^2/(2m_r)$ has been used.

(d) The wave function at $r = 0$ does not vanish only if $\ell = 0$. The complete wave function is

$$\phi_{E\ell m}(\mathbf{r}) = R_{E\ell}(r) Y_{\ell m}(\theta, \varphi)$$

Show that the wave function at the origin is

$$2|\phi_{E00}(0)|^2 = \frac{2}{4\pi^2} \frac{k}{R_y a_o^2}$$

$$= \frac{1}{2\pi^2} \left(\frac{2m_r}{\hbar^2}\right)^{3/2} \sqrt{E}$$

which is exactly the 3D reduced density of states, denoted by $\rho_r^{3D}(E)$.

13.2 Derive Eq. (13.1.13).

13.3 Plot the Franz–Keldysh absorption spectrum for GaAs with an applied field $F = 100$ kV/cm at $T = 300$ K.

13.4 Compare the oscillation strengths and the exciton binding energies of the bound states of the 2D and 3D excitons.

13.5 Compare the absorption spectra of the continuum-state contributions of the 2D and 3D excitons.

13.6 **(a)** Show that for interband absorption in a pure two-dimensional structure without exciton effects, the absorption spectrum taking into account the finite linewidth broadening is given by

$$\alpha(\hbar\omega) = \frac{A_0}{2\pi R_y a_o^2} \int_0^\infty \frac{d\varepsilon'}{\pi} \frac{\gamma}{(\varepsilon' - \varepsilon)^2 + \gamma^2}$$

where $\varepsilon = (\hbar\omega - E_G)/R_y$.

(b) Carry out the integration analytically and plot the absorption spectrum $\alpha(\hbar\omega)$.

13.7 Derive (13.4.17).

13.8 Show that $G(x)$ defined in (13.4.31) gives $G(0) = 1$ and $G(\infty) = 0$.

13.9 Use the perturbation result (13.4.52) for the band-edge transition energy $E_{h1}^{el}(F)$ to compare with the data shown in Fig. 13.7. Estimate the exciton binding energies for the heavy-hole and light-hole excitons separately. Discuss the accuracy of this simple method.

13.10 Compare the advantages and disadvantages of the waveguide modulator vs. the transverse transmission modulator.

13.11 Discuss the physics and operation principles of a SEED.

REFERENCES

1. W. Franz, *Z. Naturforsch.* **13a**, 484 (1958).

2. L. V. Keldysh, "The effect of a strong electric field on the optical properties of insulating crystals," *Soviet Phys. JETP* **34**, 788–790 (1958).

3. K. Tharmalingam, "Optical absorption in the presence of a uniform field," *Phys. Rev.* **130**, 2204–2206 (1963).

4. R. J. Elliot, "Intensity of optical absorption by excitons," *Phys. Rev.* **108**, 1384–1389 (1957).

5. R. J. Elliot, "Theory of Excitons: I," pp. 269–293 in C. G. Kuper and G. D. Whitfield, Eds., *Polarons and Excitons, Scottish Universities' Summer School*, Plenum, New York, 1962.

6. R. S. Knox, "Theory of Excitons," in *Solid State Physics*, Suppl. 5, Academic, New York, 1963.

7. J. D. Dow and D. Redfield, "Electroabsorption in semiconductors: The excitonic absorption edge," *Phys. Rev. B* **1**, 3358–3371 (1970).

8. E. J. Johnson, "Absorption near the fundamental edge," Chapter 6, pp. 153–258 in R. K. Williardson and A. C. Beer, Eds., *Semiconductors and Semimetals*, Vol. 3, Academic, New York, 1967.

9. J. O. Dimmock, "Introduction to the theory of exciton states in semiconductors," Chapter 7, pp. 259–319 in R. K. Williardson and A. C. Beer, Eds., *Semiconductors and Semimetals*, Vol. 3, Academic, New York, 1967.

10. T. H. Wood, C. A. Burrus, D. A. B. Miller, D. S. Chemla, T. C. Damen, A. C. Gossard, and W. Wiegmann, "High-speed optical modulation with GaAs/GaAlAs quantum wells in a *p-i-n* diode structure," *Appl. Phys. Lett.* **44**, 16–18 (1984).

11. D. A. B. Miller, D. S. Chemla, T. C. Damen, A. C. Gossard, W. Wiegmann, T. H. Wood, and C. A. Burrus, "Band-edge electroabsorption in quantum well structures: The quantum-confined Stark effect," *Phys. Rev. Lett.* **53**, 2173–2176 (1984).

12. D. A. B. Miller, D. S. Chemla, T. C. Damen, A. C. Gossard, W. Wiegmann, T. H. Wood, and C. A. Burrus, "Electric field dependence of optical absorption near the band gap of quantum well structures," *Phys. Rev. B* **32**, 1043–1060 (1985).

13. S. Schmitt-Rink, D. S. Chemla, and D. A. B. Miller, "Linear and nonlinear optical properties of semiconductor quantum wells," *Adv. Phys.* **38**, 89–188 (1989).

14. M. Shinada and S. Sugano, "Interband optical transitions in extremely anisotropic semiconductors, I: Bound and unbound exciton absorption," *J. Phys. Soc. Jpn.* **21**, 1936–1946 (1966).

15. S. L. Chuang, S. Schmitt-Rink, D. A. B. Miller, and D. S. Chemla, "Exciton Green's-function approach to optical absorption in a quantum well with an applied electric field," *Phys. Rev. B* **43**, 1500–1509 (1991).

16. T. H. Wood, "Multiple quantum well (MQW) waveguide modulators," *J. Lightwave Technol.* **6**, 743–757 (1988).

17. D. A. B. Miller, J. S. Weiner, and D. S. Chemla, "Electric-field dependence of linear optical properties in quantum well structures: Waveguide electroabsorption and sum rules," *IEEE J. Quantum Electron.* **QE-22**, 1816–1830 (1986).

18. D. A. B. Miller, D. S. Chemla, T. C. Damen, A. C. Gossard, W. Wiegmann, T. H. Wood, and C. A. Burrus, "Novel hybrid optically bistable switch: The quantum well self-electro-optic effect device," *Appl. Phys. Lett.* **45**, 13–15 (1984).

19. D. A. B. Miller, D. S. Chemla, T. C. Damen, T. H. Wood, C. A. Burrus, A. C. Gossard, and W. Wiegmann, "The quantum well self-electrooptic effect device: Optoelectronic bistability and oscillation, and self-linearized modulation," *IEEE J. Quantum Electron.* **QE-21**, 1462–1476 (1985).

20. D. A. B. Miller, M. D. Feuer, T. Y. Chang, S. C. Shunk, J. E. Henry, D. J. Burrows, and D. S. Chemla, "Field-effect transistor self-electrooptic effect device: Integrated photodiode, quantum well modulator and transistor," *IEEE Photon. Technol. Lett.* **1**, 62–64 (1989).

21. D. E. Aspnes, "Electric-field effects on the dielectric constant of solids," *Phys. Rev.* **153**, 972–982 (1967).

22. D. E. Aspnes and N. Bottka, "Electric-field effects on the dielectric function of semiconductors and insulators," pp. 459–543 in R. K. Williardson and A. C. Beer, Eds., *Semiconductors and Semimetals*, Vol. 9, *Modulation Techniques*, Academic, New York, 1972.

23. M. Cardona, *Modulation Spectroscopy*, pp. 165–275 in *Solid State Physics*, Suppl. 11, Academic, New York, 1969.

24. B. R. Bennett and R. A. Soref, "Electrorefraction and electroabsorption in InP, GaAs, GaSb, InAs, and InSb," *IEEE J. Quantum Electron.* **QE-23**, 2159–2166 (1987).

25. D. A. B. Miller, D. S. Chemla, and S. Schmitt-Rink, "Relation between electroabsorption in bulk semiconductors and in quantum wells: The quantum-confined Franz–Keldysh effect," *Phys. Rev. B* **33**, 6976–6982 (1986).

26. H. Shen and F. H. Pollak, "Generalized Franz–Keldysh theory of electromodulation," *Phys. Rev. B* **42**, 7097–7102 (1990).

27. M. Abramowitz and I. A. Stegun, *Handbook of Mathematical Functions*, Chapter 10, Dover, New York, 1972.

28. F. Bassani and G. Pastori Parravicini, *Electronic States and Optical Transitions in Solids*, Pergamon, Oxford, UK, 1975.

29. C. Y. P. Chao and S. L. Chuang, "Analytical and numerical solutions for a two-dimensional exciton in momentum space," *Phys. Rev. B* **43**, 6530–6543 (1991).

30. M. D. Sturge, "Optical absorption of gallium arsenide between 0.6 and 2.75 eV," *Phys. Rev.* **127**, 768–773 (1962).

31. J. S. Blakemore, "Semiconducting and other major properties of gallium arsenide," *J. Appl. Phys.* **53**, R123–R181 (1982).

32. G. Livescu, D. A. B. Miller, D. S. Chemla, M. Ramaswamy, T. Y. Chang, N. Sauer, A. C. Gossard, and J. H. English, "Free carrier and many-body effects in absorption spectra of modulation-doped quantum wells," *IEEE J. Quantum Electron.* **24**, 1677–1689 (1988).

33. G. Bastard and J. A. Brum, "Electronic states in semiconductor heterostructures," *IEEE J. Quantum Electron.* **QE-22**, 1625–1644 (1986).

34. G. D. Sanders and Y. C. Chang, "Theory of photoabsorption in modulation-doped semiconductor quantum wells," *Phys. Rev. B* **35**, 1300–1315 (1987).

35. R. Zimmermann, "On the dynamic Stark effect of excitons, the low field limit," *Phys. Status Solidi B* **146**, 545–554 (1988).

36. Y. Kan, H. Nagai, M. Yamanishi, and I. Suemune, "Field effects on the refractive index and absorption coefficient in AlGaAs quantum well structures and their feasibility for electrooptic device applications," *IEEE J. Quantum Electron.* **QE-23**, 2167–2180 (1987).

37. M. Yamanishi and I. Suemune, "Comment on polarization dependent momentum matrix elements in quantum well lasers," *Jpn. J. Appl. Phys.* **23**, L35–L36 (1984).

38. M. Asada, A. Kameyama, and Y. Suematsu, "Gain and intervalence band absorption in quantum-well lasers," *IEEE J. Quantum Electron.* **QE-20**, 745–753 (1984).

39. B. Zhu and K. Huang, "Effect of valence-band hybridization on the exciton spectra in GaAs–Ga$_{1-x}$Al$_x$As quantum wells," *Phys. Rev. B* **36**, 8102–8108 (1987).

40. G. E. W. Bauer and T. Ando, "Exciton mixing in quantum wells," *Phys. Rev. B* **38**, 6015–6030 (1988).

41. H. Chu and Y. C. Chang, "Theory of line shapes of exciton resonances in semiconductor superlattices," *Phys. Rev. B* **39**, 10861–10871 (1989).

42. L. C. Andreani and A. Pasquarello, "Accurate theory of excitons in GaAs–Ga$_{1-x}$Al$_x$As quantum wells," *Phys. Rev. B* **42**, 8928–8938 (1990).

43. C. Y. P. Chao and S. L. Chuang, "Momentum-space solution of exciton excited states and heavy-hole–light-hole mixing in quantum wells," *Phys. Rev. B* **48**, 8210–8221 (1993).

44. M. Matsuura and T. Kamizato, "Subbands and excitons in a quantum well in an electric field," *Phys. Rev. B* **33**, 8385–8389 (1986).

45. G. Bastard, E. E. Mendez, L. L. Chang, and L. Esaki, "Variational calculations on a quantum well in an electric field," *Phys. Rev. B* **28**, 3241–3245 (1983).

46. S. Nojima, "Electric field dependence of the exciton binding energy in GaAs/Al$_x$Ga$_{1-x}$As quantum wells," *Phys. Rev. B* **37**, 9087–9088 (1988).

47. D. Ahn and S. L. Chuang, "Variational calculations of subbands in a quantum well with uniform electric field: Gram–Schmidt orthogonalization approach," *Appl. Phys. Lett.* **49**, 1450–1452 (1986).

48. J. S. Weiner, D. A. B. Miller, D. S. Chemla, T. C. Damen, C. A. Burrus, T. H. Wood, A. C. Gossard, and W. Wiegmann, "Strong polarization-sensitive electroabsorption in GaAs/AlGaAs quantum well waveguides," *Appl. Phys. Lett.* **47**, 1148–1150 (1985).

49. P. Lefebvre, P. Christol, and H. Mathieu, "Unified formulation of excitonic absorption spectra of semiconductor quantum wells, superlattices and quantum wires," *Phys. Rev. B* **48**, 17308–17315 (1993).

50. H. S. Cho and P. R. Prucnal, "Effect of parameter variations on the performance of GaAs/AlGaAs multiple-quantum-well electroabsorption modulators," *IEEE J. Quantum Electron.* **25**, 1682–1690 (1989).

51. G. Lengyel, K. W. Jelley, and R. W. H. Engelmann, "A semi-empirical model for electroabsorption in GaAs/AlGaAs multiple quantum well modulator structures," *IEEE J. Quantum Electron.* **26**, 296–304 (1990).

52. C. R. Giles, T. H. Wood, and C. A. Burrus, "Quantum-well SEED optical oscillators," *IEEE J. Quantum Electron.* **26**, 512–518 (1990).

53. A. L. Lentine, D. A. B. Miller, L. M. F. Chirovsky, and L. A. D'Asaro, "Optimization of absorption in symmetric self-electrooptic effect devices: A system perspective," *IEEE J. Quantum Electron.* **27**, 2431–2439 (1991).

54. P. J. Mares and S. L. Chuang, "Comparison between theory and experiment for InGaAs/InP self-electrooptic effect devices," *Appl. Phys. Lett.* **61**, 1924–1926 (1992).

55. P. J. Mares and S. L. Chuang, "Modeling of self-electrooptic-effect devices," *J. Appl. Phys.* **74**, 1388–1397 (1993).

56. M. K. Chin and W. S. C. Chang, "Theoretical design optimization of multiple-quantum-well electroabsorption waveguide modulators," *IEEE J. Quantum Electron.* **29**, 2476–2488, 1993.

PART V

Detection of Light

14

Photodetectors

Photodetectors play important roles in optical communication systems. The major physical mechanism of photodetectors is the absorption of photons, which changes the electric properties of the electronic system, such as the generation of a photocurrent in a photoconductor or a photovoltage in a photovoltaic detector. The performance of a photodetector depends on the optical absorption process, the carrier transport, and the interaction with the circuit system.

For intrinsic optical absorptions, such as interband processes in a direct semiconductor, the general theory for the absorption spectrum has been presented in Chapter 9. The interband absorption creates electron–hole pairs. The carrier transport of these electrons and holes after generation depends on the design of the photodetectors. In this chapter, we study photoconductors, photodiodes using p-n junctions and p-i-n structures, avalanche photodiodes, and intersubband quantum-well photodetectors. Our focus is on the understanding of the physical processes of the carrier generation and transport. We derive the relation between the optical power and the photocurrents. Some important noises in these photodetectors are also discussed. More extensive treatment of photodetectors can be found in Refs. 1–5.

14.1 PHOTOCONDUCTORS

14.1.1 Photoconductivity

Consider a uniform p-type semiconductor with a uniform optical illumination. The total electron and hole carrier concentrations, n and p, deviate from their thermal equilibrium values, n_0 and p_0, by the excess carrier concentrations, δn and δp, respectively, due to the optical excitation:

$$n = n_0 + \delta n \qquad p = p_0 + \delta p \qquad (14.1.1)$$

Since $p_0 \gg n_0$ in an extrinsic p-type semiconductor, the net thermal recombination rate (2.3.14) can be reduced to

$$R_n^{\text{net}} = \frac{\delta n}{\tau_n} \qquad (14.1.2)$$

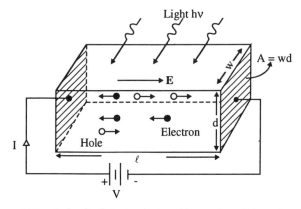

Figure 14.1. A simple photo conductor with an external bias voltage V.

where the low-level injection condition, $\delta n, \delta p \ll p_0$, has been assumed. Thus the electron concentration will satisfy the rate equation

$$\frac{\partial}{\partial t} n = G_0 - \frac{\delta n}{\tau_n} \qquad (14.1.3)$$

where G_0 is the net optical generation rate. For a uniform semiconductor with a dc voltage bias V, as shown in Fig. 14.1, the electron and hole current densities are given by only the drift components, since there is no diffusion current due to the lack of spatial dependence ($\partial/\partial x \equiv 0$):

$$J_n = q\mu_n nE \qquad J_p = q\mu_p pE \qquad (14.1.4)$$

The total current density is

$$J = J_n + J_p = q(\mu_n n + \mu_p p)E \equiv \sigma E \qquad (14.1.5)$$

where the conductivity is

$$\sigma = q(\mu_n n + \mu_p p) \qquad (14.1.6)$$

The photoconductivity $\Delta\sigma$ is defined as the difference between the conductivity when there is an optical injection and the dark conductivity σ_0:

$$\Delta\sigma = \sigma - \sigma_0 = q(\mu_n \delta n + \mu_p \delta p) \qquad (14.1.7)$$

$$\sigma_0 = q(\mu_n n_0 + \mu_p p_0) \qquad (14.1.8)$$

The total current I is the current density J multiplied by the cross-sectional area $A = wd$.

$$I = JA = \sigma EA = \frac{\sigma AV}{\ell} \qquad (14.1.9)$$

where the electric field $E = V/\ell$, with ℓ the length of the sample, has been used. The photocurrent ΔI is defined as the difference between the total current in the presence of optical excitation and the dark current $I_0 = \sigma_0 AV/\ell$.

$$\Delta I = I - I_0 = \Delta\sigma\frac{A}{\ell}V = q(\mu_n\,\delta n + \mu_p\,\delta p)\frac{A}{\ell}V \qquad (14.1.10)$$

14.1.2 Photocurrent Responses in the Time Domain

Case 1: A Constant Light Intensity. If G_0 is independent of time, $\partial/\partial t = 0$ at steady state, we obtain

$$\delta n = G_0\tau_n \qquad (14.1.11)$$

and $\delta p = \delta n$, since each broken bond creates one electron–hole pair. Thus, the photocurrent is given by

$$\Delta I = q(\mu_n + \mu_p)G_0\tau_n\frac{AV}{\ell} \simeq q\mu_n G_0\tau_n\frac{AV}{\ell} \qquad (14.1.12)$$

since usually $\mu_n \gg \mu_p$. The above expression can also be expressed in terms of the transit time of the electrons:

$$\tau_t = \frac{\ell}{v_n} = \frac{\ell}{\mu_n E} = \frac{\ell^2}{\mu_n V} \qquad (14.1.13)$$

such that

$$\Delta I = q(G_0\ell A)\frac{\tau_n}{\tau_t} \qquad (14.1.14)$$

The above expression has a good physical interpretation. The term $G_0\ell A$ is the total number of electron–hole pairs created per second in the sample with a volume ℓA. The ratio τ_n/τ_t gives the photoconductive gain, which is determined by how fast the electrons can transit across the electrodes and contribute to the photocurrent in the circuit before they can recombine with holes.

As shown in Fig. 14.1, when an electron–hole pair is created, the photocurrent will be small if the electron and the hole immediately recombine before they can be collected by the electrodes (i.e., the recombination lifetime τ_n is much shorter than the transit time). On the other hand, when the transit time τ_t is short, a significant number of photogenerated electrons will be able to reach the left electrode before they recombine with holes in the semiconductor. To preserve charge neutrality in the semiconductor, the left electrode will provide the same number of holes at the same rate as the electrons reaching that electrode per second, resulting in the photoconductor current measured by the external circuit. For example, $\tau_n/\tau_t = 10$ is equivalent to having ten round-trips taken by the electron before it disappears by recombination in the photoconductor.

The optical generation rate is equal to the number of injected photons per second, or photon flux ($P_{opt}/h\nu$) per unit volume (ℓwd) multiplied by the quantum efficiency η:

$$G_0 = \eta \frac{P_{opt}/h\nu}{\ell wd} \tag{14.1.15}$$

where w is the width, d is the depth of the sample ($wd = A$), P_{opt} is the optical power (W) of the injected light, and η is the quantum efficiency or the fraction of photons creating electron–hole pairs. If the surface reflections and the finite thickness of the detector are considered, the quantum efficiency η is just the intrinsic quantum efficiency η_i multiplied by the absorbance derived in (5.3.51):

$$\eta = \eta_i(1 - R)(1 - e^{-\alpha d}) \tag{14.1.16}$$

where R is the optical reflectivity between the air and semiconductor, and α is the absorption coefficient of the optical intensity. The injected primary photocurrent I_{ph} is defined as

$$I_{ph} = q\eta \frac{P_{opt}}{h\nu} \tag{14.1.17}$$

The photocurrent is

$$\Delta I = q\left(\frac{\eta P_{opt}}{h\nu}\right)\frac{\tau_n}{\tau_t} = I_{ph}\frac{\tau_n}{\tau_t} \tag{14.1.18}$$

Again the photoconductive gain is given by

$$\frac{\Delta I}{I_{ph}} = \frac{\tau_n}{\tau_t} \tag{14.1.19}$$

If there are more electrons traveling across the electrodes before recombining with the holes, there will be more photocurrent appearing in the external circuit. The current responsivity R_λ (A/W) is the photocurrent response per unit optical incident power

$$R_\lambda = \frac{\Delta I}{P_{opt}} = q \frac{\eta}{h\nu} \frac{\tau_n}{\tau_t} \tag{14.1.20}$$

and depends on the operation wavelength λ.

Case 2: Transient Response. If the constant light illumination is switched off at $t = 0$, that is, $G_0(t) = G_0$, $t \leq 0$ and 0 for $t > 0$, the response for $t \geq 0$ will be

$$\frac{\partial}{\partial t} \delta n(t) = -\frac{\delta n(t)}{\tau_n} \tag{14.1.21}$$

or

$$\begin{aligned}
\delta n(t) &= \delta n(0) e^{-t/\tau_n} \\
&= G_0 \tau_n\, e^{-t/\tau_n}
\end{aligned} \tag{14.1.22a}$$

and the photocurrent response obtained from (14.1.10) or (14.1.12) is

$$\Delta I(t) = q\mu_n G_0 \tau_n \left(\frac{AV}{\ell} \right) e^{-t/\tau_n} \tag{14.1.22b}$$

The result is shown in Fig. 14.2a. The decay time constant is the minority carrier recombination lifetime τ_n. If the light injection is an impulse function,

$$G_0(t) = g_0 \delta(t) \tag{14.1.23}$$

Then,

$$\frac{\partial}{\partial t} \delta n(t) = g_0 \delta(t) - \frac{\delta n(t)}{\tau_n} \tag{14.1.24}$$

Integrating the above equation from $t = 0_-$ to 0_+ will give

$$\delta n(0_+) - \delta n(0_-) = g_0$$

Assuming that initially $\delta n(0_-) = 0$, that is, the semiconductor is at thermal equilibrium before $t = 0$, and the excess carrier concentration is zero, we obtain

$$\delta n(t) = g_0\, e^{-t/\tau_n} \qquad \text{for } t \geq 0 \tag{14.1.25}$$

This result is plotted in Fig. 14.2b.

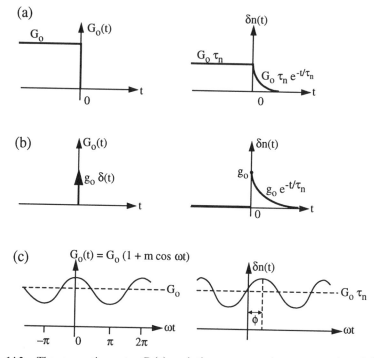

Figure 14.2. The generation rate $G_0(t)$ and the excess carrier concentration $\delta n(t)$ for (a) switch-off, (b) impulse response, and (c) sinusoidal steady-state response of a photoconductor.

Case 3: Sinusoidal Steady-State Response. If the optical intensity is modulated by a sinusoidal signal such that

$$G_0(t) = G_0 \cos \omega t \qquad (14.1.26)$$

the sinusoidal steady-state response can be found by letting

$$\delta n(t) = \mathrm{Re}\big(\delta n\, e^{-i\omega t}\big) \qquad (14.1.27)$$

where Re means the real part of the following quantity. We have

$$\frac{\partial}{\partial t}\delta n(t) = G_0 \cos \omega t - \frac{\delta n(t)}{\tau_n} \qquad (14.1.28)$$

Using $G_0 \cos \omega t = \mathrm{Re}(G_0\, e^{-i\omega t})$, we obtain

$$\delta n = \frac{G_0 \tau_n}{1 - i\omega\tau_n} \qquad (14.1.29)$$

Thus

$$\delta n(t) = \text{Re}\left(\frac{G_0 \tau_n}{1 - i\omega\tau_n} e^{-i\omega t}\right)$$

$$= \frac{G_0 \tau_n}{\sqrt{1 + \omega^2 \tau_n^2}} \cos(\omega t - \phi) \qquad (14.1.30)$$

where $\phi = \tan^{-1}(\omega\tau_n)$ is the phase delay in the ac response.

On the other hand, if the optical intensity or the optical generation rate consists of a dc and an ac component,

$$G_0(t) = G_0(1 + m \cos \omega t) \qquad (14.1.31)$$

where m is the modulation index, the response of the excess carrier concentration will be

$$\delta n(t) = G_0 \tau_n \left[1 + \frac{m}{\sqrt{1 + \omega^2 \tau_n^2}} \cos(\omega t - \phi)\right] \qquad (14.1.32)$$

where ϕ is the same as in (14.1.30). The above results are shown in Fig. 14.2c.

Using (14.1.15)–(14.1.17), we find that for an optical input power

$$P(t) = P_{\text{opt}}(1 + m \cos \omega t) \qquad (14.1.33)$$

the photocurrent response is

$$I(t) = I_p \left[1 + \frac{m}{\sqrt{1 + \omega^2 \tau_n^2}} \cos(\omega t - \phi)\right] \qquad (14.1.34a)$$

$$I_p = q\eta \frac{P_{\text{opt}}}{h\nu} \frac{\tau_n}{\tau_t} \qquad (14.1.34b)$$

In other words, for a root-mean-square (rms) optical power of which we replace the cosine function by $1/\sqrt{2}$,

$$P_{\text{rms}} = \frac{mP_{\text{opt}}}{\sqrt{2}} \qquad (14.1.35)$$

the rms photocurrent signal is

$$i_p = q\eta \frac{P_{\text{rms}}}{h\nu}\left(\frac{\tau_n}{\tau_t}\right)\frac{1}{\sqrt{1 + \omega^2 \tau_n^2}} \qquad (14.1.36)$$

14.1.3   Noises in Photoconductors [6]

In this subsection, we present some fundamental concepts including the spectral density function for noises in photodetectors. Important noises such as generation–recombination or shot and thermal noise are discussed.

Spectral Density Function $S(f)$. The signal such as the photocurrent in the time domain $i(t)$ of the photoconductor is related to its Fourier transform $i(f)$ in the frequency domain by

$$i(t) = \int_{-\infty}^{\infty} i(f)e^{-i2\pi ft}\, df \tag{14.1.37a}$$

$$i(f) = \int_{-\infty}^{\infty} i(t)e^{i2\pi ft}\, dt \tag{14.1.37b}$$

The average power P over a time duration T is

$$P = \frac{1}{T}\int_{-T/2}^{T/2} i^2(t)\, dt = \frac{1}{T}\int_{-\infty}^{\infty} |i(f)|^2\, df$$

$$= \frac{2}{T}\int_{0}^{\infty} |i(f)|^2\, df = \int_{0}^{\infty} S(f)\, df \tag{14.1.38}$$

where we have used

$$\lim_{T\to\infty} \int_{-T/2}^{T/2} e^{i2\pi(f-f')t}\, dt = \delta(f-f') \tag{14.1.39}$$

and defined the spectral density function $S(f)$ as

$$S(f) = \frac{2}{T}|i(f)|^2 \tag{14.1.40}$$

Shot Noise. For a sequence of events such as collection of charges within a time interval T, say

$$i(t) = \sum_{i=1}^{N_t} h(t-t_i) \qquad 0 \le t \le T \tag{14.1.41}$$

where N_t = total number of events within time T, we obtain the Fourier transform $i(f)$:

$$i(f) = \sum_{i=1}^{N_t} h(f)\exp(i2\pi ft_i) \tag{14.1.42}$$

The ensemble average of $|i(f)|^2$ is

$$\langle |i(f)|^2 \rangle = \left\langle |h(f)|^2 \left\{ N_t + \sum_{i \neq j}^{N_t} \sum_{j}^{N_t} \exp\left[i2\pi f(t_j - t_i) \right] \right\} \right\rangle$$

$$= |h(f)|^2 \langle N_t \rangle = \langle N \rangle T |h(f)|^2 \qquad (14.1.43)$$

where $\langle N \rangle = \langle N_t \rangle / T$ is the average number of events occurring per second. The random distributions of t_i and t_j give the cancellation in the second term in (14.1.43) when $i \neq j$. The spectral density function is

$$S(f) = \frac{2}{T} \langle |i(f)|^2 \rangle = 2 \langle N \rangle |h(f)|^2 \qquad (14.1.44)$$

Now we consider an injected electron between two capacitor plates with a separation ℓ. The current is

$$h(t) = \frac{q}{\ell} v(t) \qquad 0 \leq t \leq \tau_t \qquad (14.1.45)$$

where $v(t)$ is the instantaneous velocity of the electron and τ_t is the transit time. The average current $\langle I \rangle$ is related to the average number of electron injections per second $\langle N \rangle$ by

$$\langle I \rangle = q \langle N \rangle \qquad (14.1.46)$$

The Fourier transform of $h(t)$ reduces to

$$h(f) = \int_0^{\tau_t} \frac{q}{\ell} v(t) e^{i2\pi f t} \, dt \simeq \int \frac{q}{\ell} \frac{dx(t)}{dt} \, dt = q \qquad (14.1.47)$$

assuming the frequency is low enough such that $f\tau_t \ll 1$. The spectral density function is

$$S(f) = 2 \langle N \rangle |h(f)|^2 = 2 \langle N \rangle q^2 = 2q \langle I \rangle \qquad (14.1.48)$$

The power of this shot noise within a frequency interval between f and $f + \Delta f$ associated with the current is denoted by

$$\langle i_S^2(f) \rangle \equiv S(f) \Delta f = 2q \langle I \rangle \Delta f \qquad (14.1.49)$$

Generation–Recombination Noise. In a photoconductor, the photogenerated carriers have a finite lifetime τ, which is a random variable with a mean value $\langle \tau \rangle$ denoted as τ_n for the electrons. The photocurrent due to *one* injected electron is

$$h(t) = \begin{cases} qv/\ell & 0 \le t \le \tau \\ 0 & \text{otherwise} \end{cases} \qquad (14.1.50)$$

Here we use v as the mean drift velocity and $\tau_t = \ell/v$ as the mean transit time. The Fourier transform of $h(t)$ is

$$h(f) = \int_0^\tau e^{i2\pi ft} \frac{q}{\tau_t} \, dt = \frac{q}{\tau_t} \left(\frac{e^{i2\pi f\tau} - 1}{i2\pi f} \right)$$

$$|h(f)|^2 = \left(\frac{q}{\tau_t} \right)^2 \frac{1}{(2\pi f)^2} (2 - e^{-i2\pi f\tau} - e^{i2\pi f\tau}) \qquad (14.1.51)$$

If we assume that the probability function of the random variable τ obeys Poisson statistics

$$p(\tau) = \frac{1}{\tau_n} e^{-\tau/\tau_n} \qquad (14.1.52)$$

with an average $\langle \tau \rangle = \int_0^\infty \tau p(\tau) \, d\tau = \tau_n$, we find

$$\left\langle |h(f)|^2 \right\rangle = \int_0^\infty |h(f)|^2 p(\tau) \, d\tau = \frac{2q^2(\tau_n/\tau_t)^2}{1 + (2\pi f\tau_n)^2} \qquad (14.1.53)$$

Since the average photocurrent with the photoconductive gain τ_n/τ_t due to $\langle N \rangle$ injected (primary) electrons per second is

$$I_p \equiv \langle I \rangle = \langle N \rangle q \frac{\tau_n}{\tau_t} \qquad (14.1.54)$$

the spectral density function due to $\langle N \rangle$ injected electrons per second is obtained from (14.1.44), (14.1.53), and (14.1.54)

$$S(f) = 2\langle N \rangle \left\langle |h(f)|^2 \right\rangle = \frac{4qI_p(\tau_n/\tau_t)}{1 + (2\pi f\tau_n)^2} \qquad (14.1.55)$$

The generation–recombination noise current is, therefore,

$$\langle i_{GR}^2 \rangle \equiv S(f)\Delta f = \frac{4qI_p(\tau_n/\tau_t)\Delta f}{1 + (2\pi f\tau_n)^2} \qquad (14.1.56)$$

Thermal Noise (also **Johnson Noise** or **Nyquist Noise**) [6–8]. In a photode-tector, the random thermal motion of charge carriers contribute to a thermal noise current. In other words, the thermal noise of a resistor R results in a random electric current $i(t)$, which is characterized by a power spectral density at temperature T:

$$S(f) = \frac{4}{R} \frac{hf}{e^{(hf/k_BT)} - 1} \tag{14.1.57}$$

This thermal noise current adds to the photocurrent signal $i_p(t)$ and affects the clarity of the detected signals. At low frequency $hf \ll k_BT$, we have

$$S(f) \simeq \frac{4k_BT}{R} \tag{14.1.58}$$

The thermal noise is described by

$$\langle i_T^2 \rangle = \int_0^{\Delta f} S(f)\, df \simeq \frac{4k_BT\Delta f}{R} \tag{14.1.59}$$

where the bandwidth Δf of the circuit is assumed to be much smaller than k_BT/h.

Signal-to-Noise Ratio. The power signal-to-noise ratio is therefore [8]

$$\left(\frac{S}{N}\right)_{\text{power}} = \frac{i_p^2}{\langle i_T^2 \rangle + \langle i_{GR}^2 \rangle} = \frac{\eta m^2 (P_{\text{opt}}/h\nu)}{8\Delta f\left[1 + (k_BT/qRI_p)(\tau_t/\tau_n)(1 + \omega^2\tau_n^2)\right]} \tag{14.1.60}$$

Noise-Equivalent Power (NEP). The noise-equivalent power is defined as the power that corresponds to the incident rms optical power ($p_{\text{rms}} = m\, P_{\text{opt}}/\sqrt{2}$) required such that the signal-to-noise ratio is one in a bandwidth of 1 Hz.

Detectivity (D^*). The detectivity is defined as

$$D^* = \frac{\sqrt{A\Delta f}}{\text{NEP}} \text{ cm (Hz)}^{1/2}/\text{W} \tag{14.1.61}$$

where A is the detector cross-sectional area in cm^2. More discussions on photoconductive detectors such as Hg$_x$Cd$_{1-x}$Te can be found in Ref. 9.

14.1.4 *n-i-p-i* Superlattice Photoconductor [10, 11]

Recently, modulation doping in semiconductors has been introduced for novel device applications. An interesting example is a GaAs doping superlattice used as a photoconductor. For an extensive review of the compositional and doping superlattices, see Refs. 10 and 11. Here we consider a GaAs doped periodically *n*-type and *p*-type separated by intrinsic regions as shown in Fig. 14.3a. The electric field profile $E(z)$ can be obtained by noting that $dE(z)/dz = \rho(z)/\varepsilon$, where $\rho(z) = +qN_D$ in the *n*-doped regions, $-qN_A$ in the *p*-doped regions, and zero in the intrinsic regions. Therefore, $E(z)$ is either a linear profile with a positive slope qN_D/ε in *n* regions, a negative slope $-qN_A/\varepsilon$ in *p* regions, or a zero slope in the intrinsic regions (Fig. 14.3b). The potential profile for the conduction band edge $E_C(z) = -q\varphi(z) = +q\int_{-\infty}^{z} E(z')\,dz'$ is the integral of the electric field profile and is shown in Fig. 14.3c.

For an incident light with energy above the band gap, the photogenerated carriers will fall to the band edge and will drift or diffuse to the valleys of

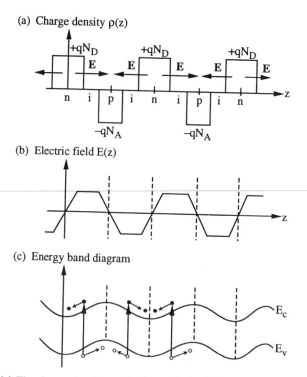

Figure 14.3. (a) The charge density profile due to ionized donors and acceptors in an *n-i-p-i* doping superlattice using the depletion approximation. (b) The electric field profile due to the charge distribution in (a). Notice that the electric field $\mathbf{E} = \hat{z}E(z)$ is alternating between positive and negative directions. (c) The energy band diagram of the *n-i-p-i* superlattice. The photogenerated electron–hole pairs are separated in real space because of the band profiles.

each band, as shown in Fig. 14.3c. The electrons are separated from the holes in real space, resulting in a very long recombination lifetime τ_n. This enhancement of a long lifetime has been found to be many orders of magnitude larger than the bulk carrier lifetime. Since the photocurrent response is proportional to $G_0\tau_n$, an extremely large responsivity using the *n-i-p-i* superlattice as a photoconductor can be designed.

14.2   *p-n* JUNCTION PHOTODIODES [12–15]

Consider a *p-n* junction photodiode as shown in Fig. 14.4a. The charge distribution $\rho(x)$, the electric field $E(x)$ and the potential energy profile under the depletion approximation have been discussed in Chapter 2. Here we investigate the photocurrent response if the diode is illuminated by a uniform light intensity, described by a generation rate $G(x, t)$, which is the number of electron–hole pairs created per unit time per unit volume. Let us focus on the *n* side of the diode.

The charge continuity equation is given by (2.4.2)

$$\frac{\partial p_n}{\partial t} = G(x, t) - \frac{\delta p_n}{\tau_p} - \frac{1}{q}\frac{\partial}{\partial x}J_p(x) \qquad (14.2.1)$$

(a) A p-n junction photodiode

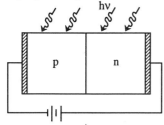

(b) Charge density

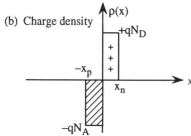

(c) Electric field

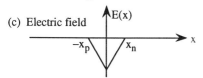

Figure 14.4. (a) A *p-n* junction diode under the illumination of a uniform light. (b) The charge distribution $\rho(x)$ under depletion approximation. (c) The electric field $E(x)$ obtained from Gauss's law.

where $p_n = p_{n0} + \delta p_n$ is the total hole concentration in the n region, p_{n0} is the hole concentration in the absence of any electric or optical injection, and δp_n is the excess hole concentration due to the external injections. The minority (hole) current density in the quasi-neutral region ($x \geq x_n$) is dominated by the diffusion component [12–14], as discussed in Chapter 2:

$$J_p(x) \simeq -qD_p \frac{\partial p_n}{\partial x} \tag{14.2.2}$$

Since p_{n0} is independent of x and t, we have at steady state, if $G(x,t) = G_0$ is independent of x and t,

$$D_p \frac{\partial^2}{\partial x^2} \delta p_n - \frac{\delta p_n}{\tau_p} = -G_0 \tag{14.2.3}$$

The above equation can be solved by summing the homogeneous and particular solutions:

$$\delta p_n(x) = \underbrace{c_1 e^{-(x-x_n)/L_p} + c_2 e^{(x-x_n)/L_p}}_{\text{homogeneous solution}} + \underbrace{G_0 \tau_p}_{\text{particular solution}} \tag{14.2.4}$$

where $L_p = \sqrt{D_p \tau_p}$ is the diffusion length for holes. The particular solution is due to the optical generation. If the n region is very long, we can set $c_2 = 0$; otherwise, $\delta p_n(x \to \infty) \to +\infty$, which is unphysical. We expect as $x \to +\infty$ that $\delta p_n(x)$ will approach $G_0 \tau_p$ = total photogenerated holes. At $x = x_n$, the hole concentration is pinned by the voltage bias V with the exponential dependence

$$p_n(x = x_n) = p_{n0} e^{qV/k_BT} \tag{14.2.5}$$

if the Boltzmann statistics are assumed. Therefore, we obtain

$$\delta p_n(x_n) = p_{n0}(e^{qV/k_BT} - 1) \tag{14.2.6}$$

and

$$\delta p_n(x) = \left[p_{n0}(e^{qV/k_BT} - 1) - G_0 \tau_p \right] e^{-(x-x_n)/L_p} + G_0 \tau_p \tag{14.2.7}$$

The current density $J_p(x)$ is

$$J_p(x) \simeq -qD_p \frac{\partial}{\partial x} \delta p_n(x)$$

$$= q \frac{D_p}{L_p} \left[p_{n0}(e^{qV/k_BT} - 1) - G_0 \tau_p \right] e^{-(x-x_n)/L_p} \tag{14.2.8}$$

We obtain $J_p(x)$ at the boundary of the depletion region x_n as

$$J_p(x_n) = q\frac{D_p}{L_p}p_{n0}\left(e^{qV/k_BT} - 1\right) - qG_0L_p \qquad (14.2.9)$$

where the first term is due only to the voltage bias, and the last term is due to optical generation. We see that only that portion of the photogenerated holes within a diffusion length L_p away from the depletion boundary can diffuse (and survive) to the depletion region and be swept across the depletion region by the electric field and collected as the photocurrent by the external circuits. This means that the majority of the carriers on the p side have to supply this current immediately. A parallel (or dual) approach for the electron current density at $x = -x_p$ gives

$$J_n(-x_p) = q\frac{D_n}{L_n}n_{p0}\left(e^{qV/k_BT} - 1\right) - qG_0L_n \qquad (14.2.10)$$

The total current I is the sum of $J_p(x_n)$ and $J_n(-x_p)$ multiplied by the cross-sectional area of the diode A:

$$I = A\left[J_p(x_n) + J_n(-x_p)\right]$$
$$= I_0\left(e^{qV/k_BT} - 1\right) - qAG_0(L_p + L_n) \qquad (14.2.11)$$

where

$$I_0 = qA\left(\frac{D_p}{L_p}p_{n0} + \frac{D_n}{L_n}n_{p0}\right) \qquad (14.2.12)$$

is the diode reverse current. The last term, $-qAG_0(L_p + L_n)$, is the photocurrent of the diode, which is proportional to the generation rate of the electron–hole pairs, G. The *I-V* curves of a photodiode with and without the illumination of light are plotted in Fig. 14.5. When $G_0 = 0$, the diode reverse

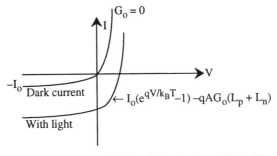

Figure 14.5.   The *I-V* curves of a photodiode with and without illumination.

current $-I_0$ is the dark current, which is usually very small compared with the photocurrent $I_{ph} \simeq -qAG_0(L_p + L_n)$ under a reverse bias condition. Therefore, the photocurrent is proportional to the generation rate G_0, which is proportional to the incident optical power P_{opt}.

The problems with the p-n junction photodiodes are as follows:

1. Optical absorption within the diffusion lengths L_p and L_n is very small, that is, over narrow regions of L_p and L_n near the depletion region. Since L_p and L_n are very small, the contributions of the photocurrents are not effective.

2. The diffusion process is slow, which results in a slow photoresponse if the optical intensity varies with time.

3. The junction capacitance C_j is simply $\varepsilon A / x_w$, where x_w is the total depletion width derived in (2.5.24) simplified for a homojunction diode with $\varepsilon_p = \varepsilon_N = \varepsilon$:

$$C_j = A \left[\frac{q\varepsilon}{2(V_0 - V)} \frac{N_D N_A}{N_D + N_A} \right]^{1/2} \tag{14.2.13}$$

which can slow down the response by the RC_j time delay. For example, if $A = (1 \text{ mm})^2$, $\varepsilon = 11.7\varepsilon_0$, $N_D = 10^{15}$ cm$^{-3} \ll N_A$ for a p^+-n photodiode, and $-V = 10V \gg V_0$, we obtain $C_j \simeq 30$ pF and the 3-dB cutoff frequency $f_{3dB} = 1/(2\pi RC_j) = 100$ MHz for $R = 50 \ \Omega$.

$R_0 A$ Product [15]. A useful figure of merit for the p-n junction photodiodes is the $R_0 A$ product. Since the photodiode is operated at zero-bias voltage in many direct detection applications, the differential resistance at zero-bias voltage R_0 multiplied by the junction area A is commonly used:

$$(R_0 A)^{-1} = \frac{1}{A} \frac{dI}{dV} \bigg|_{V=0} = \frac{dJ}{dV} \bigg|_{V=0} \tag{14.2.14}$$

So far, we have derived the dark current I_0 contributed by the diffusion processes in this section. Using (14.2.11) and (14.2.12), we obtain

$$\begin{aligned}
(R_0 A)^{-1} &= \frac{q^2}{k_B T} \left(\frac{D_p}{L_p} p_{n0} + \frac{D_n}{L_n} n_{p0} \right) \\
&= \frac{q^2}{k_B T n_i^2} \left(\frac{D_p N_D}{L_p} + \frac{D_n N_A}{L_n} \right) \\
&= \frac{q}{n_i^2} \sqrt{\frac{q}{k_B T}} \left(N_D \sqrt{\frac{\mu_p}{\tau_p}} + N_A \sqrt{\frac{\mu_n}{\tau_n}} \right) \tag{14.2.15}
\end{aligned}$$

where we have used the relations $p_{n0} = N_D/n_i^2$, $n_{p0} = N_A/n_i^2$, and the Einstein relations $D_p/\mu_p = D_n/\mu_n = k_B T/q$. The first term in (14.2.15) is the contribution to $1/(R_0 A)$ from the diffusion current on the n side of the photodiode, and the second term is from the diffusion current on the p side. There can also be other contributions to $R_0 A$ products, such as the generation–recombination current in the space-charge region, the surface leakage current, and the interband tunneling current, which depend on the material properties, device geometry, and surface conditions.

For a root-mean-square photon flux density Φ (number of photons per second per unit area) of monochromatic radiation at a wavelength λ, we can write the rms photocurrent

$$i_p = q\eta\Phi A = q\eta \frac{P_\lambda}{h\nu} \tag{14.2.16}$$

where A is the photodetector illumination area, $\Phi = P_\lambda/(h\nu A)$, P_λ is the root-mean-square input optical power at λ ($P_\lambda = P_{\text{rms}} = mP_{\text{opt}}/\sqrt{2}$), and η is the quantum efficiency of the photodiode including the effects of the internal quantum efficiency, the reflection, and the absorption depth. The current responsivity is therefore

$$R_\lambda = \frac{i_p}{P_\lambda} = \eta \frac{q}{h\nu} \tag{14.2.17}$$

The signal-to-noise ratio S/N (current) is

$$\frac{S}{N} = \frac{i_p}{\sqrt{\langle i_n^2 \rangle}} = \frac{R_\lambda P_\lambda}{\sqrt{\langle i_n^2 \rangle}} \tag{14.2.18}$$

The detectivity for the above S/N is defined as

$$D_\lambda^* = \frac{R_\lambda \sqrt{A\,\Delta f}}{\sqrt{\langle i_n^2 \rangle}} \quad \text{cm (Hz)}^{1/2}/\text{W} \tag{14.2.19}$$

For a photodiode at thermal equilibrium (i.e., no externally applied voltage and no illumination of light), the thermal noise depends on the zero bias resistance R_0 using (14.1.59):

$$\langle i_n^2 \rangle = \frac{4k_B T\,\Delta f}{R_0} \tag{14.2.20}$$

When not in thermal equilibrium, the *I-V* curve is

$$I(V) = I_0\left(e^{(qV/k_BT)} - 1\right) - I_{ph} \qquad (14.2.21)$$

$$I_{ph} = q\eta\Phi_B A \qquad (14.2.22)$$

where Φ_B is the photon flux density due to the background radiation. The mean-squared shot noise current has contributions from three additive terms [15]: (1) a forward current, which depends on voltage, $I_0\exp(qV_B/k_BT)$, (2) a reverse diode saturation current, and (3) the background radiation induced photocurrent. Since these shot noise currents fluctuate independently, the total mean-squared shot noise current is

$$\langle i_n^2 \rangle = 2q\left(I_0 e^{qV/k_BT} \Delta f + I_0\,\Delta f + I_{ph}\,\Delta f\right) \qquad (14.2.23)$$

At an operation voltage $V = 0$, $R_0^{-1} = (dI/dV)_{V=0} = qI_0/k_BT$ and (14.2.23) can be written as

$$\langle i_n^2 \rangle = \left(\frac{4k_BT}{R_0} + 2q^2\eta\Phi_B A\right)\Delta f \qquad (14.2.24)$$

The detectivity at zero bias voltage is then obtained from (14.2.19)

$$D_\lambda^* = \left(\frac{q\eta}{h\nu}\right)\frac{1}{\left[(4k_BT/R_0 A) + 2q^2\eta\Phi_B\right]^{1/2}} \qquad (14.2.25)$$

For a thermally limited case, i.e., when the thermal noise is dominant over the background radiation induced signal and other noises, we have

$$(D_\lambda^*)_T = \frac{q\eta}{h\nu}\sqrt{\frac{R_0 A}{4k_BT}} \qquad (14.2.26)$$

which relates the $R_0 A$ product to the thermally limited detectivity. If the photodiode is background radiation limited, which means that the background radiation-induced photocurrent is dominant, we have

$$(D_\lambda^*)_{BLIP} = \frac{1}{h\nu}\sqrt{\frac{\eta}{2\Phi_B}} \qquad (14.2.27)$$

which is the detectivity of the background limited infrared photodetector (BLIP).

14.3   *p-i-n* PHOTODIODES [8]

To enhance the responsivity of the photodiode, an intrinsic region used as the major absorption layer is added (Fig. 14.6a). For a light injected from the p^+ side with an optical power intensity I_{opt} (W/cm^2), the generation rate is

$$G(x) = (1 - R)\eta_i \left(\frac{I_{opt}}{h\nu} \right) \alpha e^{-\alpha x} \qquad (14.3.1)$$

The optical power intensity I_{opt} is the incident optical power P_{opt} divided by the area A ($I_{opt} = P_{opt}/A$). Note that the total injected number of electrons

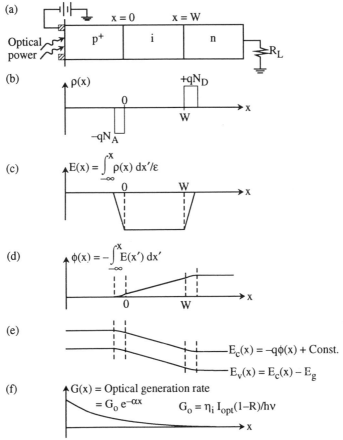

Figure 14.6. (a) A *p-i-n* photodiode under optical illumination from the p^+ side, (b) the charge density $\rho(x)$ under depletion approximation, (c) the static electric field profile $E(x)$, (d) the electrostatic potential $\varphi(x)$, (e) the conduction and valence band edge profiles, and (f) the optical generation rate $G(x)$.

per unit area per second is

$$S_0 = \int_0^\infty G(x)\,dx = (1-R)\eta_i \frac{I_{opt}}{h\nu} \tag{14.3.2}$$

where $I_{opt}/h\nu$ is the number of photons injected per unit area per second, and η_i is the internal quantum efficiency for the probability of creating an electron–hole pair for each incident photon. The energy band profiles for the p-i-n diode can be obtained graphically based on the well-known depletion approximation, as shown in Figs. 14.6b–e for the charge density $\rho(x)$, the electric field $E(x)$, the potential $\phi(x)$, and the band diagram.

At steady state, the total photocurrent consists of both a drift and a diffusion current:

$$J = J_{dr} + J_{diff} \tag{14.3.3}$$

Considering the p^+ region to be of negligible thickness, we look at the contribution in the intrinsic region $0 < x < W$:

$$J_{dr} = -q\int_0^W G(x)\,dx = -qS_0\left(1 - e^{-\alpha W}\right)$$
$$= -q\frac{I_{opt}}{h\nu}\eta_i(1-R)\left(1 - e^{-\alpha W}\right) \tag{14.3.4}$$

where the minus sign accounts for the fact that the draft current flows in the $-x$ direction. The above expression also shows that an increase of $W \gg 1/\alpha$ enhances the photocurrent because of the increasing amount of absorption.

For $x > W$, the analysis is similar to that of the p-n junction diode, where the hole (minority) current density on the n side is due only to diffusion:

$$J_{diff} \simeq -qD_p\frac{\partial}{\partial x}P_n(x) \tag{14.3.5}$$

$$0 = \frac{\partial P_n}{\partial t} = G(x) - \frac{\delta P_n}{\tau_p} - \frac{1}{q}\frac{\partial}{\partial x}J_p(x) \tag{14.3.6}$$

Therefore, we solve

$$\frac{\partial^2}{\partial x^2}\delta P_n - \frac{1}{L_p^2}\delta P_n = -\frac{1}{D_P}G(x) \tag{14.3.7}$$

The solution for $\delta P_n(x)$ consists of the homogeneous solution and the particular solution

$$\delta P_n(x) = \underbrace{Ae^{-(x-W)/L_p}}_{\substack{\text{homogeneous}\\\text{solution}}} + \underbrace{Ce^{-\alpha x}}_{\substack{\text{particular}\\\text{solution}}} \tag{14.3.8}$$

and we have discarded the term $e^{+(x-W)/L_p}$ in the homogeneous part since $\delta P_n(x \to \infty)$ should be finite. C is contributed by $G(x)$ and is obtained by substituting $Ce^{-\alpha x}$ into (14.3.7):

$$C = \frac{S_0}{D_p} \frac{\alpha L_p^2}{1 - \alpha^2 L_p^2} \tag{14.3.9}$$

The coefficient A is then determined by the boundary condition

$$\delta P_n(W) = P_{n0}\left(e^{qV/k_B T} - 1\right) \simeq -P_{n0} \tag{14.3.10}$$

which is pinned by the reverse bias voltage. Therefore,

$$A = -P_{n0} - Ce^{-\alpha W} \tag{14.3.11}$$

and

$$\delta P_n(x) = -\left(P_{n0} + Ce^{-\alpha W}\right)e^{-(x-W)/L_p} + Ce^{-\alpha x} \tag{14.3.12}$$

The hole current density on the n side is

$$J_{\text{diff}} = -qD_p \frac{d}{dx}P_n(x)\bigg|_{x=W}$$

$$= qD_p\alpha\left(1 - \frac{1}{\alpha L_p}\right)C_1 e^{-\alpha W} - q\frac{D_p}{L_p}P_{n0}$$

$$= -q\left(\frac{S_0 \alpha L_p}{1 + \alpha L_p}e^{-\alpha W} + P_{n0}\frac{D_p}{L_p}\right) \tag{14.3.13}$$

The total current density is

$$J = J_{\text{dr}} + J_{\text{diff}} \qquad (\text{at } x = W)$$

$$= -qS_0\left(1 - \frac{e^{-\alpha W}}{1 + \alpha L_p}\right) - qP_{n0}\frac{D_p}{L_p} \tag{14.3.14}$$

The quantum efficiency is

$$\eta = \frac{J/q}{I_{\text{opt}}/h\nu} = \eta_i(1 - R)\left(1 - \frac{e^{-\alpha W}}{1 + \alpha L_p}\right) \tag{14.3.15}$$

neglecting the contribution of the diffusion term in (14.3.14). Note that if $W \to \infty$, the current density is dominated by

$$J = -qS_0 \tag{14.3.16}$$

and $\eta = \eta_i(1 - R)$ as expected. Long-wavelength *p-i-n* photodiodes for high-speed receiver applications are of great interest for optical communication systems. For high-sensitivity optical receivers, photodiodes with a small junction area around a few tens of microns in diameter and a low doping depletion region of a few microns are desired [16]. At low doping, the center absorption region can be depleted at a small bias voltage and reduce the tunneling leakage current. Most of the long-wavelength photodetectors use $In_{0.47}Ga_{0.53}As$ grown on an InP substrate as the absorption region. The leakage current in InGaAs *p-i-n* diodes is dominated by the interband tunneling at high reverse-bias voltages and generation–recombination processes at low voltages [17, 18]. Hybrid *p-i-n*/FET receivers have been assembled for high-speed photoreceivers, which offer better sensitivity than other *p-i-n* photodiode receivers [16, 19]. Ultrawideband *p-i-n* photodetectors have also been shown to have a great potential for high-speed applications [20] with an impulse response in the picosecond scale.

14.4 AVALANCHE PHOTODIODES

To enhance the photocurrent response, some built-in multiplication processes may be utilized such that more photocurrents can be extracted in the external circuits at a given optical illumination. Ideally, we would have a single carrier-type photomultiplier; the carrier concentration will grow if impact ionization occurs in a region with a large electric field. In semiconductors, both electrons and holes can impact ionize more electron–hole pairs, as shown in Fig. 14.7a. A schematic diagram for an avalanche diode is shown in Fig. 14.7b. A feedback process occurs since electrons and holes travel in

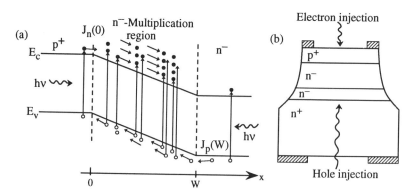

Figure 14.7. (a) The energy band diagram for an avalanche photodiode with the electron and hole ionization coefficients α_n and β_p. The electron and hole injections are given by $J_n(0)$ and $J_p(W)$. (b) A schematic diagram for an avalanche photodiode.

opposite directions. Let us define

α_n = the electron ionization coefficient $(1/\text{cm})$
= the number of electron–hole pairs generated by one incident electron per unit distance

β_p = the hole ionization coefficient $(1/\text{cm})$
= the number of electron–hole pairs generated by one incident hole per unit distance

In general, $\alpha_n \neq \beta_p$ for most semiconductors. They are functions of the applied electric field E with an exponential dependence:

$$\alpha_n(E) = \alpha_0 e^{-C_n/E} \quad \text{and} \quad \beta_p(E) = \beta_0 e^{-C_p/E}$$

where the constants α_0, β_0, C_n, and C_p depend on the materials. For a general overview of the fundamentals of avalanche photodiodes, see Refs. 21–23.

14.4.1 Ideal Avalanche Photodiode—Single Carrier Type Capable of Ionizing Collisions [22]

We now consider the special case in which only electrons can impact ionize. Suppose we have an incident current density $J_n(x)$ at a plane located at position x (Fig. 14.8a). Over an incremental distance Δx, the total generated electron–hole pairs is $\alpha_n \Delta x$ multiplied by $J_n(x)$, since $\alpha_n \Delta x$ is the number of ionized electrons per incident electron in a distance Δx. Therefore, the current density $J_n(x + \Delta x)$ is the sum of the incident (or primary) current density and the ionized (or secondary) current density:

$$J_n(x + \Delta x) = J_n(x) + \alpha_n \Delta x J_n(x) \tag{14.4.1}$$

(a) Electron impact ionizations only   (b) Hole impact ionizations only

$J_n(x)$ $J_n(x + \Delta x)$ $J_p(x - \Delta x)$ $J_p(x)$

$\leftarrow \Delta x \rightarrow$

x $x + \Delta x$ $x - \Delta x$ x

Figure 14.8. Schematic diagrams for only (a) electron and (b) hole impact ionizations.

or

$$\frac{\mathrm{d}}{\mathrm{d}x} J_n(x) = \alpha_n J_n(x) \qquad (14.4.2)$$

Its solution is

$$J_n(x) = J_n(0) e^{\int_0^x \alpha_n(x')\,\mathrm{d}x'} \qquad (14.4.3)$$

if $\alpha_n \equiv \alpha_n(x)$ is not uniform, since the electric field $E(x)$ may not be uniform. The multiplication factor M_n for the electrons is defined as

$$M_n = \frac{J_n(W)}{J_n(0)} = e^{\int_0^W \alpha_n(x')\,\mathrm{d}x'} \qquad (14.4.4)$$

For a uniform α_n, we have

$$J_n(x) = J_n(0) e^{\alpha_n x} \qquad (14.4.5)$$

and

$$M_n = e^{\alpha_n W} \qquad (14.4.6)$$

Here the multiplication factor is finite since W is a finite width.

A similar procedure for holes propagating in the $-x$ direction, as shown in Fig. 14.8b, leads to

$$-\frac{\mathrm{d}}{\mathrm{d}x} J_p(x) = \beta_p J_p(x) \qquad (14.4.7)$$

and its solution is

$$J_p(x) = J_p(W) e^{\int_x^W \beta_p(x')\,\mathrm{d}x'} \qquad (14.4.8)$$

The hole multiplication ratio is defined as

$$M_p = \frac{J_p(0)}{J_p(W)} = e^{\int_0^W \beta_p(x')\,\mathrm{d}x'} \qquad (14.4.9)$$

For the special case in which β_p is independent of the position x, we have

$$J_p(x) = J_p(W) e^{\beta_p(W-x)} \qquad (14.4.10)$$

and

$$M_p = e^{\beta_p W} \qquad (14.4.11)$$

14.4.2 Both Electron and Hole Capable of Impact Ionization

Let us write the complete coupled equations when both electrons and holes cause impact ionizations in the presence of optical generation as well:

$$\frac{d}{dx}J_n(x) = \alpha_n(x)J_n(x) + \beta_p(x)J_p(x) + qG(x) \quad (14.4.12a)$$

$$-\frac{d}{dx}J_p(x) = \alpha_n(x)J_n(x) + \beta_p(x)J_p(x) + qG(x) \quad (14.4.12b)$$

Note that each impact ionization process creates an electron–hole pair, as does the optical generation rate $G(x)$ per unit volume. We see that the above two equations lead to

$$\frac{d}{dx}\left[J_n(x) + J_p(x)\right] = 0 \quad (14.4.13)$$

or the total current density

$$J = J_n(x) + J_p(x) = \text{constant} \quad (14.4.14)$$

which is independent of the position, as it should be since this is a one-dimensional problem. Any current passing through a surface at x has to be the same at steady state. The two coupled first-order differential equations can be solved using one of the two variables, for example, $J_n(x)$:

$$\frac{d}{dx}J_n(x) - \left[\alpha_n(x) - \beta_p(x)\right]J_n(x) = \beta_p J + qG(x) \quad (14.4.15)$$

where J is independent of x and is determined later by the boundary conditions. Noting that the first-order differential equation

$$\frac{d}{dx}y(x) + p(x)y(x) = Q(x) \quad (14.4.16)$$

has a solution of the form

$$y(x) = \frac{\int_{x_0}^{x} dx' Q(x') e^{\int_{x_0}^{x'} p(x'') dx''} + y(x_0)}{\exp \int_{x_0}^{x} p(x') dx'} \quad (14.4.17)$$

we obtain by setting all initial conditions at $x_0 = 0$:

$$J_n(x) = \frac{J\int_0^x dx' \beta_p(x') e^{-\varphi(x')} + q\int_0^x G(x') e^{-\varphi(x')} dx' + J_n(0)}{e^{-\varphi(x)}} \quad (14.4.18)$$

where

$$\varphi(x) = \int_0^x \left[\alpha_n(x') - \beta_p(x')\right] dx' \qquad (14.4.19)$$

We then match the boundary conditions at $x = W$. Suppose $J_n(0)$ and $J_p(W)$ are given; we then find

$$J_n(W) = \frac{J\int_0^W \beta_p(x')e^{-\varphi(x')} dx' + q\int_0^W G(x')e^{-\varphi(x')} dx' + J_n(0)}{e^{-\varphi(W)}} \qquad (14.4.20)$$

Using

$$J_n(W) = J - J_p(W) \quad \text{and} \quad \varphi(W) - \varphi(x') = \int_{x'}^W \left[\alpha_n(x'') - \beta_p(x'')\right] dx'' \qquad (14.4.21)$$

we find J immediately:

$$J = \frac{J_p(W) + J_n(0)e^{\varphi(W)} + q\int_0^W dx'\, G(x')e^{\int_{x'}^W [\alpha_n(x'') - \beta_p(x'')] dx''}}{1 - \int_0^W dx'\, \beta_p(x')e^{\int_{x'}^W [\alpha_n(x'') - \beta_p(x'')] dx''}} \qquad (14.4.22)$$

Therefore, we have

$$J_n(x) = J e^{\varphi(x)} \int_0^x dx'\, \beta_p(x')e^{-\varphi(x')} + q e^{\varphi(x)} \int_0^x G(x')e^{-\varphi(x')} dx' + J_n(0)e^{\varphi(x)} \qquad (14.4.23)$$

where J is given by (14.4.22). The hole current density is then

$$J_p(x) = J - J_n(x) \qquad (14.4.24)$$

Alternatively, we can start with (14.4.12b) using J_p as the variable and find

$$J_p(x) = \frac{\int_x^W dx'\, [\alpha_n(x')J + qG(x')]e^{\int_x^W [\alpha_n(x'') - \beta_p(x'')] dx''} + J_p(W)}{e^{\int_x^W [\alpha_n(x') - \beta_p(x')] dx'}} \qquad (14.4.25)$$

and obtain another expression for J using the boundary condition at $x = 0$, $J_p(0) = J - J_n(0)$:

$$J = \frac{J_n(0) + J_p(W)e^{-\varphi(W)} + q\int_0^W dx'\, G(x')e^{-\int_0^{x'} [\alpha_n(x'') - \beta_p(x'')] dx''}}{1 - \int_0^W dx'\, \alpha_n(x')e^{-\int_0^{x'} [\alpha_n(x'') - \beta_p(x'')] dx''}} \qquad (14.4.26)$$

Let us consider three special cases with only electron injection at $x = 0$, or hole injection at $x = W$, or optical injection $G(x)$ at a position x.

Case 1: Only Electron Injection at $x = 0$ ($J_p(W) = 0$ and $G(x) = 0$). The electron multiplication ratio is obtained from (14.4.22):

$$M_n = \frac{J}{J_n(0)} = \frac{e^{\varphi(W)}}{1 - \int_0^W dx' \beta_p(x') e^{\int_{x'}^W [\alpha_n(x'') - \beta_p(x'')] dx''}} \quad (14.4.27a)$$

An alternative form, if we start with $J_p(x)$ from the beginning and find J following the above procedures, is

$$M_n = \frac{J}{J_n(0)} = \frac{1}{1 - \int_0^W dx' \alpha_n(x') e^{-\int_0^{x'} [\alpha_n(x'') - \beta_p(x'')] dx''}} \quad (14.4.27b)$$

Case 2: Only Hole Injection at $x = W$ ($J_n(0) = 0$ and $G(x) = 0$). The hole multiplication ratio is, using (14.4.26),

$$M_p = \frac{J}{J_p(W)} = \frac{e^{-\varphi(W)}}{1 - \int_0^W dx' \, \alpha_n(x') e^{-\int_0^{x'} [\alpha_n(x'') - \beta_p(x'')] dx''}} \quad (14.4.28a)$$

which is also a dual form of (14.4.27a). A dual form of (14.4.27b) is

$$M_p = \frac{J}{J_p(W)} = \frac{1}{1 - \int_0^W dx' \beta_p(x') e^{\int_{x'}^W [\alpha_n(x'') - \beta_p(x'')] dx''}} \quad (14.4.28b)$$

For example, if we consider $\alpha_n = \beta_p = $ constant, and $G(x) = 0$, we find

$$J_n(x) = J\beta_p x + J_n(0) \quad (14.4.29)$$

Using $J_n(W) = J - J_p(W)$, we find

$$J = \frac{J_n(0) + J_p(W)}{1 - \beta_p W} \quad (14.4.30)$$

The results of $J_n(x)$ and $J_p(x)$ are plotted in Fig. 14.9.

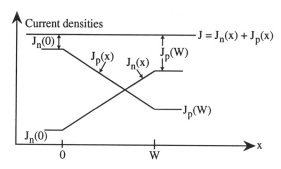

Figure 14.9. The electron and hole current densities, $J_n(x)$ and $J_p(x)$, and the total current density J as a function of position x for a special case $\alpha_n = \beta_p = $ constant and the only injections are determined by the values $J_n(0)$ and $J_p(W)$.

Case 3: Only Optical Injection $G(x') = G_0\delta(x' - x)$ $(J_n(0) = 0, J_p(W) = 0)$. We find that the total current density is

$$J = \frac{qG_0 e^{\int_x^W [\alpha_n(x'') - \beta_p(x'')] dx''}}{1 - \int_0^W dx' \beta_p(x') e^{\int_{x'}^W [\alpha_n(x'') - \beta_p(x'')] dx''}} \qquad (14.4.31)$$

and the multiplication rate depends on the position of the optical injection x:

$$M(x) = \frac{J}{qG_0} = \frac{e^{\int_x^W [\alpha_n(x') - \beta_p(x')] dx'}}{1 - \int_0^W dx' \beta_p(x') e^{\int_{x'}^W [\alpha_n(x'') - \beta_p(x'')] dx''}} \qquad (14.4.32)$$

We can also check two special optical injection positions $x = 0$ and $x = W$. We find

$$M(x = 0) = M_n \qquad (14.4.33)$$

and

$$M(x = W) = M_p \qquad (14.4.34)$$

These are the same expressions as in (14.4.27a) and (14.4.28b), as expected.

By controlling the electron injection in the p region or the hole injection in the n region, the multiplication factors $M_n(V)$ and $M_p(V)$ as a function of the reverse biased voltage V can be determined [24–31]. They usually increase very slowly at a low reverse bias and show an exponentially increasing behavior above a certain large bias voltage, as shown in Fig. 14.10. These reverse bias voltages can be of the order of 30 or 40 V. Once M_n and M_p are measured, α_n and β_p are supposed to be found from (14.4.27) and (14.4.28). If both α_n and β_p are independent of the position x, the integrations can be carried out analytically. We can express α_n and β_p in terms of M_n and M_p:

$$\alpha_n = \left(\frac{M_n - 1}{M_n - M_p}\right) \frac{1}{W} \ln \frac{M_n}{M_p} \qquad (14.4.35)$$

$$\beta_p = \left(\frac{M_p - 1}{M_n - M_p}\right) \frac{1}{W} \ln \frac{M_n}{M_p} \qquad (14.4.36)$$

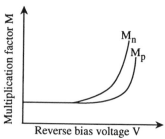

Figure 4.10. Multiplication factors for the electron and hole injections, M_n and M_p, respectively, are plotted as a function of the reverse bias voltage.

From the above discussions, we see that in order to determine the impact ionization coefficients from the multiplication of photocurrent, proper experimental conditions [24–26] have to be met: (1) $M_n(V)$ and $M_p(V)$ for pure electron injection and pure hole injection, respectively, have to be measured in the same diode, but not in complementary p^+n and n^+p devices, (2) the primary injected photocurrent (without multiplication) must be determined accurately as a function of bias voltage, and (3) the electric field should be slowly varying in space and uniform in the active region. The dependence of the electric field on the position and bias voltage must be accurately known. It is not easy to meet all of these conditions without approximations. Experimental data for Si [27], GaAs [24, 25, 28], InP [26, 29, 30], InGaAs, and InGaAsP have been reported, for example,

GaAs [25]
$$\alpha_n = 1.899 \times 10^5 e^{-(5.75 \times 10^5/E)^{1.82}} \ (1/\text{cm})$$
$$\beta_p = 2.215 \times 10^5 e^{-(6.57 \times 10^5/E)^{1.75}} \ (1/\text{cm})$$

InP [26]
(1) $240 \, \text{kV/cm} < E < 380 \, \text{kV/cm}, \ N = 1.2 \times 10^{15} \, \text{cm}^{-3}$
$$\alpha_n = 1.12 \times 10^7 e^{-3.11 \times 10^6/E} \ (1/\text{cm})$$
$$\beta_p = 4.76 \times 10^6 e^{-2.55 \times 10^6/E} \ (1/\text{cm})$$

(2) $360 \, \text{kV/cm} < E < 560 \, \text{kV/cm}, \ N = 3.0 \times 10^{16} \, \text{cm}^{-3}$
$$\alpha_n = 2.93 \times 10^6 e^{-2.64 \times 10^6/E} \ (1/\text{cm})$$
$$\beta_p = 1.62 \times 10^6 e^{-2.11 \times 10^6/E} \ (1/\text{cm})$$

(3) $530 \, \text{kV/cm} < E < 770 \, \text{kV/cm}, \ N = 1.2 \times 10^{17} \, \text{cm}^{-3}$
$$\alpha_n = 2.32 \times 10^5 e^{-7.16 \times 10^{11}/E^2} \ (1/\text{cm})$$
$$\beta_p = 2.48 \times 10^5 e^{-6.23 \times 10^{11}/E^2} \ (1/\text{cm})$$

InP [30]
$$\alpha_n = 5.55 \times 10^6 e^{-3.10 \times 10^6/E} \ (1/\text{cm})$$
$$\beta_p = 1.98 \times 10^6 e^{-2.29 \times 10^6/E} \ (1/\text{cm})$$

InGaAsP ($E_g = 0.92$ eV) [30]
$$\alpha_n = 3.37 \times 10^6 e^{-2.29 \times 10^6/E} \ (1/\text{cm})$$
$$\beta_p = 2.94 \times 10^6 e^{-2.40 \times 10^6/E} \ (1/\text{cm})$$

$In_{0.53}Ga_{0.47}As$ [30]
$$\alpha_n = 2.27 \times 10^6 e^{-1.13 \times 10^6/E} \ (1/\text{cm})$$
$$\beta_p = 3.95 \times 10^6 e^{-1.45 \times 10^6/E} \ (1/\text{cm})$$

where the electric field E is in V/cm in the above expressions. For strained $In_{0.2}Ga_{0.8}As$ and $In_{0.15}Ga_{0.63}Al_{0.22}As$ channels embedded in $Al_{0.3}Ga_{0.7}As$ material, the α_n and β_p values have been found to be higher in $In_{0.2}Ga_{0.8}As$ channels and lower in the $In_{0.15}Ga_{0.63}Al_{0.22}As$ channels compared with the unstrained GaAs channels using hole injection in a lateral p-i-n diode configuration [31].

For an rms optical power (14.1.35)

$$P_{\text{rms}} = \frac{mP_{\text{opt}}}{\sqrt{2}} \tag{14.4.37}$$

where m is the microwave modulation depth in (14.1.34), and P_{opt} is the optical input power. The multiplied rms photocurrent response is

$$i_P = q\eta \frac{P_{rms}}{h\nu} M \qquad (14.4.38)$$

where $M = M_n$ for the electron multiplication and $M = M_p$ for the hole multiplication.

Excess Noise. Since the multiplication processes are random, an excess-noise factor can be defined for the multiplication factor M, which is treated as a random variable,

$$F(M) = \frac{\langle M^2 \rangle}{\langle M \rangle^2} \qquad (14.4.39)$$

where $\langle \ \rangle$ means ensemble average. It has been shown that these excess-noise factors [32, 33] can be written in terms of the ratio of the impact ionization coefficients α_n / β_p.

$$F_n = \frac{\beta_p}{\alpha_n}\langle M_n \rangle + \left(1 - \frac{\beta_p}{\alpha_n}\right)\left(2 - \frac{1}{\langle M_n \rangle}\right) \qquad \text{electron injection} \quad (14.4.40a)$$

$$F_p = \frac{\alpha_n}{\beta_p}\langle M_p \rangle + \left(1 - \frac{\alpha_n}{\beta_p}\right)\left(2 - \frac{1}{\langle M_p \rangle}\right) \qquad \text{hole injection} \quad (14.4.40b)$$

Note that if no avalanche multiplications exist, $\langle M_n \rangle$, $\langle M_p \rangle$, F_n, and F_p are all equal to unity. In Fig. 14.11, we plot F_n vs. $\langle M_n \rangle$ for various $\beta_n / \alpha_n = k$ ratios. We see the increase of the excess-noise factor F_n with increasing β_p / α_n for electron injection. The physical reason for this is that, in the case

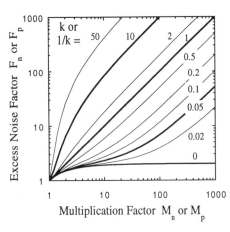

Figure 14.11. The excess-noise factor F vs. the multiplication factor $\langle M \rangle$ for different values of the ratio of the electron and hole ionization coefficients. For electron injection, $k = \beta_p / \alpha_n$ and for hole injection, $\alpha_n / \beta_p = 1/k$ should be used in the ratio.

of electron injection from the p^+ region in Fig. 14.7, the secondary electron–hole pairs also cause impact ionizations. The holes propagate in the opposite direction to that of the electrons. Therefore, if β_p/α_n is increased, the backpropagating holes will create more impact ionization currents. The measured amplified current at the end electrodes will have more fluctuating signals since these secondary or higher-order impact ionization processes will contain more random characteristics. For hole injection, $1/k$ should be used in the same diagram. The multiplication or gain noise is given by

$$\langle i_M^2 \rangle = 2q\langle M \rangle I_p F(M) I_p \tag{14.4.41}$$

The mean-squared shot-noise current after multiplication is generalized from the shot noise in (14.1.49) by adding the factor $\langle M^2 \rangle$

$$\langle i_S^2 \rangle = 2q(I_P + I_B + I_D)\langle M^2 \rangle B$$
$$= 2q(I_P + I_B + I_D)\langle M \rangle^2 FB \tag{14.4.42}$$

where I_P is the average steady-state photocurrent (14.1.34b), I_B is the background current, I_D is the dark current [34–36], and $B = \Delta f$ is the bandwidth. The thermal noise is

$$\langle i_T^2 \rangle = \frac{4k_B TB}{R_{\text{eq}}} \tag{14.4.43}$$

where $1/R_{\text{eq}} = 1/R_j + 1/R_L + 1/R_i$, accounting for the junction resistance R_j, the external load resistance R_L and the input resistance R_i of the following amplifier of the photodiode.

For the modulation depth $m = 1$, the signal-to-noise ratio (power) for the avalanche diode is [8, 21]

$$\frac{S}{N} = \frac{i_P^2}{\langle i_S^2 \rangle + \langle i_T^2 \rangle} = \frac{\frac{1}{2}(q\eta P_{\text{opt}}/h\nu)^2\langle M \rangle^2}{2q(I_P + I_B + I_D)\langle M \rangle^2 FB + 4k_B TB/R_{\text{eq}}} \tag{14.4.44}$$

Since $F(M) \geq 1$ and is a monotonically increasing function of the average multiplication ratio $\langle M \rangle$, the above signal-to-noise ratio can be optimized at a particular value of $\langle M \rangle$.

High-speed detections using avalanche photodiodes and their time dependence or frequency response have been investigated [37–40]. For $M_0 > \alpha_n/\beta_p$, the frequency-dependent multiplication factor is

$$M(\omega) = \frac{M_0}{\left[1 + (M_0\omega\tau_t)^2\right]^{1/2}} \tag{14.4.45}$$

where τ_t is an effective transit time through the avalanche region. The effects of the avalanche buildup time have also been reported [39].

Separate Absorption and Multiplication (SAM) APD [29, 30, 41, 42]. Various contributions to the dark current of an APD have been investigated, which include the generation–recombination via midgap traps in the depletion region, tunneling of the carriers across the band gap and a surface leakage current across the *p-n* junction in InGaAs APD, for example. When the reverse bias is above a certain value before the breakdown voltage V_B, it has been found that the tunneling current is dominant, unless the doping density N_D in the absorption region can be reduced below a certain value in which the tunneling current can be reduced to be smaller than the generation–recombination current. A separate absorption and multiplication structure has also been proposed [41] to reduce the tunneling current. The geometry is shown in Fig. 14.12a, where a low-field InGaAs region is used as

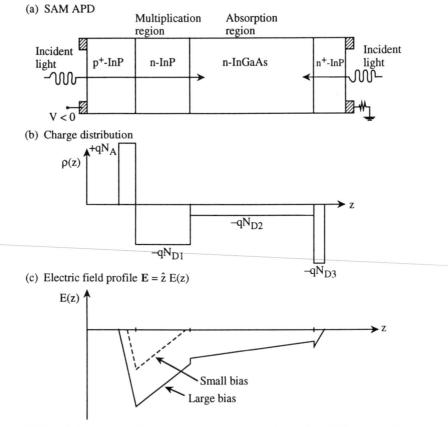

Figure 14.12. (a) A schematic diagram for a separate absorption and multiplication avalanche photodiode (SAM APD), where the absorption occurs at the narrow bandgap InGaAs region and the photogenerated carriers are swept into the InP multiplication region where the electric field is larger. (b) Charge density profile $\rho(z)$ under a large reverse bias. (c) The electric field profile (solid lines) under a large reverse bias. Dashed lines show the electric field profile for a small bias voltage.

the absorption region and the photogenerated carriers are swept into the high-field InP binary region where avalanche multiplications occur [16]. Since the InP layer has a larger band gap than that of the InGaAs absorption region, the tunneling current can be reduced. The electric field profile $E(z) = \hat{z}E(z)$ can be obtained using the charge density profile $\rho(z)$ based on the depletion approximation at a large reverse bias voltage since $\partial \varepsilon E(z)/\partial z = \rho(z)$. Since ε is slightly different in InGaAs and InP layers, a slight discontinuity in $E(z)$ occurs at the InGaAs/InP interface.

Multiple-Quantum-Well APD. Heterostructure avalanche photodiodes have been fabricated for high-speed low dark current operations [41–43] since the late 1970s. Research on quantum-well photodiodes was started in the early 1980s. A plot of the excess factor F shows that the excess noise factor is minimized if $\beta_p/\alpha_n \ll 1$ using electron injection or $\alpha_n/\beta_p \ll 1$ for hole injection. In those limits, an ideal single-carrier-type multiplication process will dominate, and the excess noise caused by the feedback process of the impact ionization caused by the secondary electrons or holes in the opposite direction can be minimized. Ideas using multiple-quantum-well (MQW) structures for APD applications have been proposed and explored both theoretically and experimentally [44–52]. For example, consider GaAs/Al$_x$Ga$_{1-x}$As MQW structures as the impact ionization region for electron injection (Fig. 14.13). Since $\Delta E_c/\Delta E_v \simeq 2{:}1$ ($\Delta E_c \simeq 0.67\,\Delta E_g$, $\Delta E_v \simeq 0.33\,\Delta E_g$), the electrons coming from the left barrier region will gain a larger kinetic energy ΔE_c when entering the barrier region than that of the holes traveling in the opposite direction. Therefore, the electron impact ionization coefficient α_n will be enhanced compared with β_p. We expect $\alpha_n/\beta_p \gg 1$. The excess noise factor is expected to be minimized. Measurements of the effective ionization coefficients α_n and β_p show an enhancement of α_n/β_p from 2 in a bulk GaAs to about 8 in a GaAs/Al$_x$Ga$_{1-x}$As

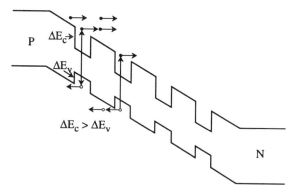

Figure 14.13. A multiple-quantum-well avalanche photodiode using GaAs/Al$_x$Ga$_{1-x}$As with the property that the ratio of the impact ionization coefficients α_n/β_p is much larger than one since ΔE_c is larger than ΔE_v.

multiple-quantum-well structure [45, 46] and $M_n = 10$ at an electric field $E = 250$ kV/cm. It has also been reported [47] that this ratio α_n / β_p varies (not monotonically) with the aluminum mole fraction x. At higher values of x (≥ 0.45) above the onset of indirect electron transitions, the noise is increased.

Intersubband Avalanche Photomultiplier. Avalanche photomultipliers using an intersubband type (bound-to-continuum state transition) in an n-type doped or p-type doped multiple-quantum-well structure have also been proposed and investigated both theoretically [50, 51] and experimentally [52–54]. The idea is to introduce electrons in the quantum-well regions by doping the wells n type. Incident photogenerated carriers will "impact ionize" those carriers confined in the wells and kick them out of the wells via Coulomb interaction contributing to the avalanche multiplication current. Since this is a single-carrier type of photomultiplication, the excess noise is expected to be minimized. Experimental results on this intersubband avalanche multiplication have been reported [52–54].

14.5 INTERSUBBAND QUANTUM-WELL PHOTODETECTORS

In Section 9.6, we discussed intersubband absorption in a quantum-well structure. To provide carriers for the intersubband transitions, donors for n-type electronic transitions have to be introduced in the quantum wells (or barriers) to provide free electrons which will be confined in the well regions at steady state without a bias voltage. When an incident infrared radiation illuminates the QW detector, electrons may absorb the photon energy and jump to a higher energy subband and be collected by the electrodes with an applied voltage. Theory and experiments on intersubband absorption and quantum-well intersubband photodetectors (QWIP) [55–87] have been investigated for long wavelength applications, which may be competitive with HgCdTe detectors. The advantages include the mature GaAs growth and processing technologies for high uniformity and reproducibility. For an extensive review of the subject, see Refs. 86 and 87 and the references therein.

For n-type multiple quantum-well photodetectors, the optical matrix selection rule shows that the optical polarization must have a component along the growth (z) axis, i.e., it must be TM polarized, as discussed in Section 9.6. For TE polarized light, the absorption is expected to be very small. However, for p-type doped quantum-well photodiodes, the valence-band mixing effects due to the heavy-hole and light-hole states show that the x- and y-polarized light can have as large an absorption coefficient as the z-polarized light [87–90]. Therefore a normal incidence geometry is possible for p-type QWIP.

In this section, we discuss mostly n-type QWIPs because of their potential for 3 to 5-μm and 8 to 12-μm photodetector applications. In Fig. 14.14, we

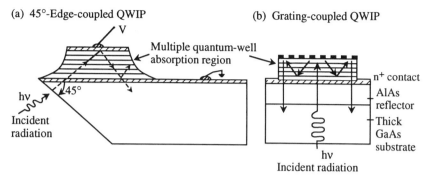

Figure 14.14. Schematic diagrams of (a) a 45°-edge-coupled quantum-well infrared photodetector (QWIP) and (b) a two-dimensional grating-coupled QWIP.

show two examples using (a) a 45°-coupled QWIP and (b) a two-dimensional grating-coupled QWIP. These designs provide the necessary polarization selection rule such that the infrared radiation will have a component along the growth direction of the multiple quantum-well absorption region. Our theory in Section 9.6 shows that the absorption spectrum is a Lorentzian function:

$$\alpha(\hbar\omega) = \alpha_0 \frac{\Gamma/(2\pi)}{(E_{21} - \hbar\omega)^2 + (\Gamma/2)^2} \qquad (14.5.1)$$

where the intersubband energy $E_{21} = E_2 - E_1$ is the subband spacing in a simple single-particle model as presented in Section 9.6. If the Coulomb interactions and screening effects are included, E_{21} will have a slight shift due to the many-body effects [91–94]. The measured absorption spectrum [87] for a bound-to-bound transition is shown in Fig. 14.15 as a Lorentzian shape. Note that the absorption is dependent on $|\langle\phi_2|z|\phi_1\rangle|^2$; therefore, a rotation of the polarization as a function of the polarization angle ϕ measured from the TM polarization wave incident at a fixed angle of incidence set at the Brewster angle $\theta_B = 73°$ shows the polarization dependence [95, 96] in Fig. 14.16 $\alpha \cos^2\phi$, as expected from the theory. At $\phi = 90°$, the incident light is polarized along the TE direction (i.e., the polarization of the incident radiation is perpendicular to the plane of incidence; therefore, it is parallel to the quantum-well plane of the QWIP), and no absorption occurs for this polarization.

Photoconductive Gain in Quantum Wells. For the bound-to-bound state transition shown in Fig. 14.17a, the photoexcited electrons in the E_2 level have to get out of the well either by tunneling or by thermionic emission (or by other phonon or impurity scattering processes). It is therefore desirable to design the E_2 level to be close to the barrier energy such that the absorption

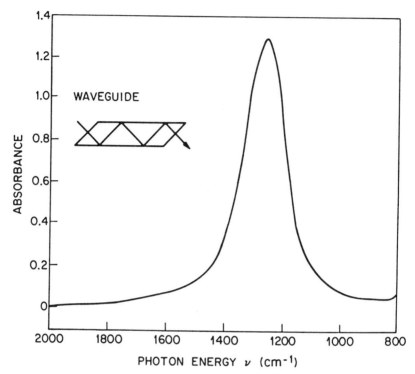

Figure 14.15. Measured QWIP absorption spectrum for a multipass waveguide geometry. (After Ref. 87.)

oscillator strength (i.e., the intersubband dipole moment) and the escape probability can be optimized.

For a bound-to-continuum state transition (Fig. 14.17b), the electrons have a greater probability to transport into the barrier region and be collected by the electrode and contribute to the photocurrent, although the intersubband dipole matrix between the ground state E_1 wave function and the highly oscillatory continuum-state wave function may be smaller. For a simplified analysis [87, 97, 98], we look at Fig. 14.17b, where p_c is treated as an effective capture probability for an incident current I_p for both the case (a) bound-to-bound and the case (b) bound-to-continuum transitions. We obtain $p_c I_p$ as the fraction of incident current captured by the well and $(1 - p_c) I_p$ as the remaining current transmitted to the next period. The incident infrared radiation creates a photocurrent i_p:

$$i_p = q\eta_w \Phi A \tag{14.5.2}$$

where Φ is the photon flux density (number of photons per second per unit area), and A is the area of the photon illumination area and η_w is the net quantum efficiency of a single well including the effect of the escape probability from the well.

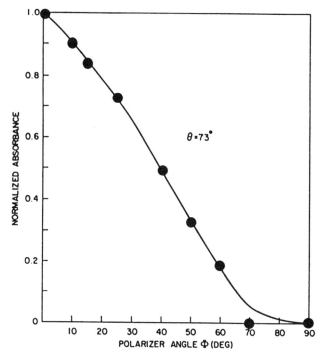

Figure 14.16. Experimental results for the polarization selection rule showing the peak absorption vs. the polarization angle ϕ where $\phi = 0°$ is TM polarization and $\phi = 90°$ is TE polarization, all at an angle of incidence $\theta_B = 73°$, the Brewster angle. (After Ref. 95.)

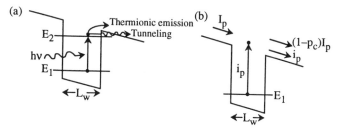

Figure 14.17. (a) Bound-to-bound state transition and (b) bound-to-continuum state transition in a biased quantum-well infrared photodetector. The well width L_w in (b) is designed to be small enough such that only one bound state exists in the quantum well and the second level E_2 is pushed into the continuum.

From current continuity, we have

$$I_p = (1 - p_c)I_p + i_p \tag{14.5.3}$$

Therefore,

$$i_p = p_c I_p \tag{14.5.4}$$

The total net photocurrent is

$$I_p = q\eta\Phi Ag \tag{14.5.5}$$

where g is defined to be the overall photoconductive gain, and η is the overall quantum efficiency of the MQW photodetector consisting of N_w quantum wells. We have $\eta \simeq N_w\eta_w$ if $\eta \ll 1$ since η is proportional to the absorbance of the structure.

From (14.5.2) to (14.5.5), we find

$$g = \frac{1}{p_c}\frac{\eta_w}{\eta} \simeq \frac{1}{p_c N_w} \tag{14.5.6}$$

which gives the value for gain. The capture probability p_c is found to decrease almost exponentially with the applied voltage [87].

Dark Current [87, 99, 100]. A simple model for the bias-dependent dark current I_D is to take the "effective" number of electrons $n^*(V)$, which tunnel out of the well or are thermally excited out of the well into the continuum states, multiplied by the average transport velocity $v(V)$, the cross-sectional area of the detector A and the electron charge q:

$$I_D(V) = qn^*(V)v(V)A \tag{14.5.7}$$

where

$$n^*(V) = \frac{m_e^*}{\pi\hbar^2 L_p}\int_{E_1}^{\infty}f(E)T(E,V)\,dE \tag{14.5.8}$$

$$f(E) = \frac{1}{1 + e^{(E - E_F)/k_B T}} \tag{14.5.9}$$

where E_F is the Fermi level measured from the conduction band edge (same as the first subband level E_1), and L_p is the length of a period. $T(E,V)$ is the tunneling probability through the triangular barrier with a bias voltage V. The velocity is

$$v = \frac{\mu F}{\sqrt{1 + (\mu F/v_s)^2}} \tag{14.5.10}$$

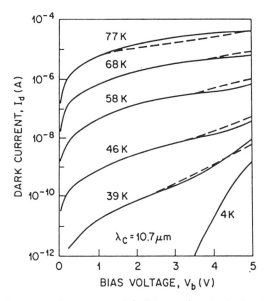

Figure 14.18. Dark currents from measured (solid curves) and calculated (dashed) data as a function of the bias voltage at various temperatures for a QWIP with a cutoff wavelength 10.7 μm. (After Ref. 99.)

where μ is the electron mobility, F is the average field determined by the bias voltage V and the overall MQW width, and v_s is the saturation drift velocity.

The above simplified model has been used to explain the dark current with a very good agreement, as shown in Fig. 14.18 for a 10.7 μm QWIP [99]. An even more simplified model [100, 101] assumes that

$$T(E) = \begin{cases} 1 & E > E_B \\ 0 & E < E_B \end{cases} \tag{14.5.11}$$

where E_B is the barrier height on the right-hand side of the quantum well. We obtain

$$n^* = \frac{m_e^* k_B T}{\pi \hbar^2 L_p} e^{-(E_C - E_F)/k_B T} \tag{14.5.12}$$

where $E_C = E_B - E_1$ is the spectral cutoff energy. The dark current becomes

$$I_D = q v A \left(\frac{m_e^* k_B T}{\pi \hbar^2 L_p} \right) e^{-(E_C - E_F)/k_B T} \tag{14.5.13}$$

and the Fermi level E_F is determined from

$$N_D = \left(\frac{m_e^* k_B T}{\pi \hbar^2 L_w} \right) \ln\left(1 + e^{(E_F - E_1)/k_B T}\right) \qquad (14.5.14)$$

Since $I_D/T \propto e^{-(E_C - E_F)k_B T}$, plotting $\ln(I_D/T)$ vs. $(E_C - E_F)$ should give a slope of $-1/k_B T$, and the result of $E_C - E_F$ can also be compared with the optically measured spectral cutoff energy E_C. This simple model has been reported to agree with experimental observations [87, 99, 102].

PROBLEMS

14.1 Consider a photoconductor (Fig. 14.1) that is an extrinsic semiconductor bar with a thickness $d = 0.1$ mm, a width $w = 1$ mm, a length $\ell = 4$ mm, and an acceptor doping concentration 10^{15} cm^{-3}. Assume that the electron mobility $\mu_n = 3000$ cm^2 V^{-1} s^{-1} ($\gg \mu_p$) and the applied voltage $V = 4$ V.

   (a) If the photoconductor is illuminated by a uniform steady light such that the optical generation rate of electrons is G_n, find an expression for the photocurrent $I_{ph} = I - I_0$, where I_0 is the dark current, and I is the current when there is illumination of light. Find a numerical value for I_{ph} if $G_n = 10^{16}$ cm^{-3} s^{-1} and $\tau_n = 10^{-3}$ s.

   (b) If the photoconductor is illuminated by a uniform light with a sinusoidal time variation, that is, $G_n(t) = g \cos \omega t$, show that the photocurrent is given by the form

$$i(t) = \frac{I_p}{\sqrt{1 + \omega^2 \tau_n^2}} \cos(\omega t - \phi)$$

   What are I_p and ϕ in terms of g and τ_n, etc.? What determines the 3-dB cutoff frequency in the frequency response of the photocurrent?

   (c) If the light has a dc (steady) and an ac component as may be used in optical communication, $G_n(t) = G_0(1 + m \cos \omega t)$, where the constant m is usually called the modulation index, find an expression for the photocurrent using the results from parts (a) and (b).

14.2 Explain why the photoconductive gain τ_n/τ_t can be much larger than 1.

14.3 Derive (14.1.34a) and (14.1.34b).

14.4 Derive the junction capacitance C_j for a heterojunction using the depletion approximation in Section 2.5 for a p-N junction.

14.5 Derive the $R_0 A$ product in (14.2.15).

14.6 Replot Figs. 14.6a–e for a p^+-n^--n^+ photodiode, where a plus superscript means heavy doping concentration and a minus superscript means light doping concentration.

14.7 Derive (14.4.26) and (14.4.27a).

14.8 An avalanche photodiode with the electron and hole ionization coefficients α_n and β_p is assumed to have a uniform field in the impact ionization region such that α_n and β_p are *independent of the position* x. The electron and hole injections are given by $J_n(0)$ and $J_p(W)$, and the generation rate due to optical injection is $G(x)$.

(a) Write the two equations for the electron and hole current densities and solve for *the hole current density* as a function of x in terms of the injection conditions $J_n(0)$ and $J_p(W)$. Find the total current density J.

(b) We assume that the electric field in the avalanche region is uniform such that α_n and β_p are independent of the position x.
   (i) If $G(x) = 0$ for all x, and $J_n(0) = 0$, find the multiplication factor for holes, M_p, defined by

$$M_p = \frac{J}{J_p(W)}$$

   (ii) On the other hand, if $G(x) = 0$ for all x, $J_p(W) = 0$, find the multiplication factor for electrons

$$M_n = \frac{J}{J_n(0)}$$

(c) Using the results in (b) for M_n and M_p show that α_n and β_p can be determined from (14.4.35) and (14.4.36) once M_n and M_p are measured:

$$\alpha_n = \left(\frac{M_n - 1}{M_n - M_p}\right)\frac{1}{W}\ln\frac{M_n}{M_p}$$

$$\beta_p = \left(\frac{M_p - 1}{M_n - M_p}\right)\frac{1}{W}\ln\frac{M_n}{M_p}$$

14.9 Discuss the physics for the excess noise factor $F(M) = \langle M^2\rangle/\langle M\rangle^2$. How can this excess noise be minimized?

14.10 Derive (14.5.12) and (14.5.13).

14.11 Discuss the polarization selection rule for an *n*-type doped quantum-well infrared detector using intersubband transitions. Why are the configurations such as a 45°-edge-coupled structure or a grating-coupled structure used in the designs of these intersubband photodetectors?

REFERENCES

1. R. K. Willardson and A. C. Beer, Eds., *Semiconductors and Semimetals*, Vol. 5, *Infrared Detectors*, Academic, New York, 1970.

2. R. K. Willardson and A. C. Beer, Eds., *Semiconductors and Semimetals*, Vol. 12, *Infrared Detectors II*, Academic, New York, 1977.

3. A. Rogalski, Ed., *Selected Papers on Semiconductor Infrared Detectors*, SPIE Milestone Series, Vol. MS66, SPIE Optical Engineering Press, Bellingham, WA, 1992.

4. W. T. Tsang, Vol. Ed., *Lightwave Communications Technology*, Vol. 22, Part D, in R. K. Willardson and A. C. Beer, Eds., *Semiconductors and Semimetals*, Academic, New York, 1985.

5. J. D. Vincent, *Fundamentals of Infrared Detector Operation and Testing*, Wiley, New York, 1990.

6. A. Yariv, *Optical Electronics*, 3d ed., Holt, Rinehart & Winston, New York, 1985.

7. B. E. A. Saleh and M. C. Teich, *Fundamentals of Photonics*, Wiley, New York, 1991.

8. S. M. Sze, *Physics of Semiconductor Devices*, 2d ed., Wiley, New York, 1981.

9. R. K. Willardson and A. C. Beer, Eds., *Semiconductors and Semimetals*, Vol. 18, *Mercury Cadmium Telluride*, Academic, New York, 1981.

10. G. H. Döhler, "Doping superlattices (*n-i-p-i* crystals)," *IEEE J. Quantum Electron.* **QE-22**, 1682–1695 (1986).

11. R. A. Street, G. H. Döhler, J. N. Miller, and P. P. Ruden, "Luminescence of *n-i-p-i* heterostructures," *Phys. Rev. B* **33**, 7043–7046 (1986).

12. P. E. Gray, D. DeWitt, A. R. Boothroyd, and J. F. Gibbons, *Physical Electronics and Circuit Models of Transistors*, Vol. 2 in *Semiconductor Electronics Education Committee Books*, Wiley, New York, 1964.

13. G. W. Neudeck, *The PN Junction Diode*, Vol. 2 in G. W. Neudeck and R. F. Pierret, Eds., *Modular Series on Solid State Devices*, Addison-Wesley, Reading, MA, 1983.

14. B. G. Streetman, *Solid State Electronic Devices*, Prentice-Hall, Englewood Cliffs, NJ, 1980.

15. M. B. Reine, A. K. Sood and T. J. Tredwell, "Photovoltaic infrared detectors," *Mercury Cadmium Telluride*, Vol. 18, in R. K. Willardson and A. C. Beer, Eds., *Semiconductors and Semimetals*, Academic, New York, 1981.

16. M. Brain and T. P. Lee, "Optical receivers for lightwave communication systems," *J. Lightwave Technol.* **LT-3**, 1281–1300 (1985).

17. S. R. Forrest, R. F. Leheny, R. E. Nahory, and M. A. Pollack, "$In_{0.53}Ga_{0.47}As$ photodiodes with dark current limited by generation–recombination and tunneling," *Appl. Phys. Lett.* **37**, 322–325 (1980).

18. S. R. Forrest, "Performance of $In_x Ga_{1-x} As_y P_{1-y}$ photodiodes with dark current limited by diffusion, generation-recombination and tunneling," *IEEE J. Quantum Electron.* **QE-17**, 217–226 (1981).

19. M. C. Brain, "Comparison of available detectors for digital optical fiber systems for the 1.2–1.55 μm wavelength range," *IEEE J. Quantum Electron.* **QE-18**, 219–224 (1982).

20. J. E. Bowers and C. A. Burrus, Jr., "Ultrawide-band long-wavelength *p-i-n* photodetectors," *J. Lightwave Technol.* **LT-5**, 1339–1350 (1987).

21. G. E. Stillman and C. M. Wolfe, "Avalanche photodiodes," pp. 291–393, *Infrared Detectors II*, Vol. 12, in R. K. Willardson and A. C. Beer, Eds., *Semiconductors and Semimetals*, Academic, New York, 1977.

22. P. P. Webb, R. J. McIntyre, and J. Conradi, "Properties of avalanche photodiodes," *RCA Rev.* **35**, 234–278 (1974).

23. T. P. Lee and T. Li, "Photodetectors," pp. 593–626 in S. E. Miller and A. G. Chynoweth, Eds., *Optical Fiber Telecommunications*, Academic, New York, 1979.

24. G. E. Stillman, "Impactionization coefficients in InGaAs," pp. 76–83 in P. Bhattacharya, Ed., *Properties of Lattice-Matched and Strained Indium Gallium Arsenide*, INSPEC, Institute of Electrical Engineers, London, UK, 1993.

25. G. E. Bulman, V. M. Robbins, K. F. Brennan, K. Hess, and G. E. Stillman, "Experimental determination of impact ionization coefficients in (100) GaAs," *IEEE Electron. Device Lett.* **4**, 181–185 (1983).

26. L. W. Cook, G. E. Bulman, and G. E. Stillman, "Electron and hole impact ionization coefficients in InP determined by photomultiplication measurements," *Appl. Phys. Lett.* **40**, 589–591 (1982).

27. M. H. Woods, W. C. Johnson, and M. A. Lampert, "Use of a Schottky barrier to measure impact ionization coefficients in semiconductors," *Solid State Electron.* **16**, 381–394 (1973).

28. H. Ando and H. Kanbe, "Ionization coefficient measurement in GaAs by using multiplication noise characteristics," *Solid State Electron.* **24**, 629–634 (1981).

29. F. Osaka, T. Mikawa, and T. Kaneda, "Impact ionization coefficients of electrons and holes in (100)-oriented $Ga_{1-x} In_x As_y P_{1-y}$," *IEEE J. Quantum Electron.* **QE-21**, 1326–1338 (1985).

30. F. Osaka and T. Mikawa, "Excess noise design of InP/GaInAsP/GaInAs avalanche photodiodes," *IEEE J. Quantum Electron.* **QE-22**, 471–478 (1986).

31. Y. C. Chen and P. K. Bhattacharya, "Impact ionization coefficients for electrons and holes in strained $In_{0.2}Ga_{0.8}As$ and $In_{0.15}Ga_{0.63}Al_{0.22}As$ channels embedded in $Al_{0.3}Ga_{0.7}As$," *J. Appl. Phys.* **73**, 465–467 (1993).

32. R. J. McIntyre, "Multiplication noise in uniform avalanche diodes," *IEEE Trans. Electron. Devices* **ED-13**, 164–168 (1966).

33. R. J. McIntyre, "The distribution of gains in uniformly multiplying avalanche photodiodes," *IEEE Trans. Electron. Devices* **ED-19**, 703–712 (1972).

34. H. Ando, H. Kanbe, M. Ito, and T. Kaneda, "Tunneling current in InGaAs and optimum design for InGaAs/InP avalanche photodiode," *Jpn. J. Appl. Phys.* **19**, L277–L280 (1980).

35. S. R. Forrest, O. K. Kim, and R. G. Smith, "Analysis of the dark current and photoresponse of $In_{0.53}Ga_{0.47}As$/InP avalanche photodiodes," *Solid State Electron.* **26**, 951–968 (1983).

36. F. Capasso, A. Y. Cho, and P. W. Foy, "Low-dark-current low-voltage 1.3–1.6 μm avalanche photodiode with high-low electric field profile and separate absorption and multiplication regions by molecular beam epitaxy," *Electron. Lett.* **20**, 635–637 (1984).

37. C. A. Lee, R. L. Batdorf, W. Wiegmann, and G. Kaminsky, "Time dependence of avalanche processes in silicon," *J. Appl. Phys.* **38**, 2787–2796 (1967).

38. R. B. Emmons, "Avalanche-photodiode frequency response," *J. Appl. Phys.* **38**, 3705–3714 (1967).

39. H. Ando and H. Kanbe, "Effect of avalanche build-up time on avalanche photodiode sensitivity," *IEEE J. Quantum Electron.* **QE-21**, 251–255 (1985).

40. G. Kahraman, B. E. A. Saleh, W. L. Sargeant, M. C. Teich, "Time and frequency response of avalanche photodiodes with arbitrary structure," *IEEE Trans. Electron. Devices* **39**, 553–560 (1992).

41. K. Nishida, K. Taguchi, and Y. Matsumoto, "In GaAsP heterostructure avalanche photodiodes with high avalanche gain," *Appl. Phys. Lett.* **35**, 251–252 (1979).

42. N. Susa, H. Nakagome, O. Mikami, H. Ando, and H. Kanbe, "New InGaAs/InP avalanche photodiode structure for the 1–1.6 μm wavelength region," *IEEE J. Quantum Electron.* **QE-16**, 864–870 (1980).

43. G. E. Stillman, V. M. Robbins, and N. Tabatabaie, "III–V compound semiconductor devices: optical detectors," *IEEE Trans. Electron. Devices* **ED-31**, 1643–1655 (1984).

44. R. Chin, N. Holonyak, Jr., G. E. Stillman, J. Y. Tang, and K. Hess, "Impact ionization in multilayered heterojunction structures," *Electron. Lett.* **16**, 467–469 (1980).

45. F. Capasso, W. T. Tsang, A. L. Hutchinson, and G. F. Williams, "Enhancement of electron impact ionization in a superlattice: A new avalanche photodiode with a large ionization rate ratio," *Appl. Phys. Lett.* **40**, 38–40 (1982).

46. F. Capasso, W. T. Tsang, and G. F. Williams, "Staircase solid-state photomultipliers and avalanche photodiodes with enhanced ionization rates ratio," *IEEE Trans. Electron. Devices* **ED-30**, 381–390 (1983).

47. T. Kagawa, H. Iwamura, and O. Mikami, "Dependence of the GaAs/AlGaAs superlattice ionization rate on Al content," *Appl. Phys. Lett.* **54**, 33–35 (1989).

48. T. Tanoue and H. Sakaki, "A new method to control impact ionization rate ratio by spatial separation of avalanching carriers in multilayered heterostructures," *Appl. Phys. Lett.* **41**, 67–70 (1982).

49. K. Brennan, "Theory of electron and hole impact ionization in quantum well and staircase superlattice avalanche photodiode structures," *IEEE Trans. Elecron. Devices* **ED-32**, 2197–2205 (1985).

50. S. L. Chuang and K. Hess, "Impact ionization across the conduction band edge discontinuity of quantum-well heterostructures," *J. Appl. Phys.* **59**, 2885–2894 (1986).

51. S. L. Chuang and K. Hess, "Tunneling-assisted impact ionization for a superlattice," *J. Appl. Phys.* **61**, 1510–1515 (1987).

52. F. Capasso, J. Allam, A. Y. Cho, K. Mohammed, R. J. Malik, A. L. Hutchinson, and D. Sivco, "New avalanche multiplication phenomenon in quantum well superlattices: Evidence of impact ionization across the band-edge discontinuity," *Appl. Phys. Lett.* **48**, 1294–1296 (1986).

53. M. Toivonen, M. Jalonen, A. Salokatve, and M. Pessa, "Unipolar avalanche multiplication phenomenon in multiquantum well structures," *Appl. Phys. Lett.* **62**, 1664–1666 (1993).

54. B. F. Levine, K. K. Choi, C. G. Bethea, J. Walker, and R. J. Malik, "Quantum well avalanche multiplication initiated by 10 μm intersubband absorption and photoexcited tunneling," *Appl. Phys. Lett.* **51**, 934–936 (1987).

55. A. Kamgar, P. Kneschaurek, G. Dorda, and J. F. Koch, "Resonance spectroscopy of electronic levels in a surface accumulation layer," *Phys. Rev. Lett.* **32**, 1251–1254 (1974).

56. S. J. Allen, Jr., D. C. Tsui, and B. Vinter, "On the absorption of infrared radiation by electrons in semiconductor inversion layers," *Solid State Commun.* **20**, 425–428 (1976).

57. L. L. Chang, L. Esaki, and G. A. Sai-Halasz, "Infrared optical devices of layered structure," *IBM Tech. Disclosure Bull.* **20**, 2019–2020 (1977).

58. L. Esaki and H. Sakaki, "New photoconductor," *IBM Tech. Disc. Bull.* **20**, 2456 (1977).

59. T. Ando, A. B. Fowler, and F. Stern, "Electronic properties of two-dimensional systems," *Rev. Mod. Phys.* **54**, 437–672 (1982).

60. L. C. Chiu, J. S. Smith, S. Margalit, A. Yariv, and A. Y. Cho, "Application of internal photoemission from quantum-well and heterojunction superlattices to infrared photodetectors," *Infrared Phys.* **23**, 93–97 (1983).

61. D. D. Coon and R. P. G. Karunasiri, "New mode of IR detection using quantum wells," *Appl. Phys. Lett.* **45**, 649–651 (1984).

62. L. C. West and S. J. Eglash, "First observation of an extremely large-dipole infrared transition within the conduction band of a GaAs quantum well," *Appl. Phys. Lett.* **46**, 1156–1158 (1985).

63. A. Harwit and J. S. Harris, Jr., "Observation of Stark shifts in quantum well intersubband transitions," *Appl. Phys. Lett.* **50**, 685–687 (1987).

64. D. Ahn and S. L. Chuang, "Calculation of linear and nonlinear intersubband optical absorption in a quantum well model with an applied electric field," *IEEE J. Quantum Electron.* **QE-23**, 2196–2204 (1987).

65. E. J. Roan and S. L. Chuang, "Linear and nonlinear intersubband electroabsorptions in a modulation-doped quantum well." *J. Appl. Phys.* **69**, 3249–3260 (1991).

66. K. Bajema, R. Merlin, F. Y. Juang, S. C. Hong, J. Singh, and P. K. Bhattacharya, "Stark effect in GaAs-Al$_x$Ga$_{1-x}$As quantum wells: Light scattering by intersubband transitions," *Phys. Rev. B* **36**, 1300–1302 (1987).

67. K. K. Choi, B. F. Levine, C. G. Bethea, J. Walker, and R. J. Malik, "Multiple quantum well 10μm GaAs/Al$_x$Ga$_{1-x}$As infrared detector with improved responsivity," *Appl. Phys. Lett.* **50**, 1814–1816 (1987).

68. M. Nakayama, H. Kuwahara, H. Kato, and K. Kubota, "Intersubband transitions in GaAs-Al$_x$Ga$_{1-x}$As modulation-doped superlattices," *Appl. Phys. Lett.* **51**, 1741–1743 (1987).

69. A. D. Wieck, J. C. Maan, U. Merkt, J. P. Kotthaus, K. Ploog, and G. Weimann, "Intersubband energies in GaAs-Ga$_{1-x}$Al$_x$As heterojunctions," *Phys. Rev. B* **35**, 4145–4148 (1987).

70. M. Zachau, "Electronic eigenstates and intersubband transitions in doped heterostructure superlattices," *Semicond. Sci. Technol.*, 879–885 (1988).

71. Z. Ikonic, V. Milanovic, and D. Tjapkin, "Bound-free intraband absorption in GaAs–Al$_x$Ga$_{1-x}$As semiconductor quantum wells," *Appl. Phys. Lett.* **54**, 247–249 (1989).

72. X. Zhou, P. K. Bhattacharya, G. Hugo, S. C. Hong, and E. Gulari, "Intersubband absorption in strained InGaAs/AlGaAs ($0 < x < 0.15$) multiquantum wells," *Appl. Phys. Lett.* **54**, 855–856 (1989).

73. J. D. Ralston, M. Ramsteiner, B. Dischler, M. Maier, G. Brandt, P. Koidl, and D. J. As, "Intersubband transitions in partially interdiffused GaAs/AlGaAs multiple quantum-well structures," *Appl. Phys. Lett.* **70**, 2195–2199 (1991).

74. F. H. Julien, P. Vagos, J. M. Lourtioz, and D. D. Yang, "Novel all-optical 10 μm waveguide modulator based on intersubband absorption in GaAs/AlGaAs quantum wells," *Appl. Phys. Lett.* **59**, 2645–2647 (1991).

75. E. Rosencher, E. Martinet, F. Luc, Ph. Bois, and E. Bockenhoff, "Discrepancies between photocurrent and absorption spectroscopies in intersubband photoionization from GaAs/AlGaAs multiquantum wells," *Appl. Phys. Lett.* **59**, 3255–3257 (1991).

76. K. K. Choi, L. Fotiadis, M. Taysing-Lara, and W. Chang, "Infrared absorption and photoconductive gain of quantum well infrared photodetectors," *Appl. Phys. Lett.* **60**, 592–594 (1992).

77. L. S. Yu, Y. H. Wang, and S. L. Li, "Low dark current step-bound-to-miniband transition InGaAs/GaAs/AlGaAs multiquantum-well infrared detector," *Appl. Phys. Lett.* **60**, 992–994 (1992).

78. W. Chen and T. G. Andersson, "Intersubband transitions for differently shaped quantum wells under an applied electric field," *Appl. Phys. Lett.* **60**, 1591–1593 (1992).

79. H. C. Liu, "Noise gain and operating temperature of quantum well infrared photodetectors," *Appl. Phys. Lett.* **61**, 2703–2705 (1992).

80. H. C. Liu, "Dependence of absorption spectrum and responsivity on the upper state position in quantum well intersubband photodetectors," *J. Appl. Phys.* **73**, 3062–3067 (1993).

81. J. P. Peng, Y. M. Mu, and X. C. Shen, "Performance of the multiquantum well infrared photodetector," *J. Appl. Phys.* **74**, 1421–1425 (1993).

82. M. Helm, W. Hilber, T. Fromherz, F. M. Peeters, K. Alavi, and R. N. Pathak, "Infrared absorption in superlattices: a probe of the miniband dispersion and the structure of the impurity band," *Phys. Rev. B* **48**, 1601–1606 (1993).

83. R. P. G. Karunasiri, J. S. Park, and K. L. Wang, "$Si_{1-x}Ge_x$/Si multiple quantum well infrared detector," *Appl. Phys. Lett.* **59**, 2588–2590 (1991).

84. E. Rosencher, B. Vinter and B. Levine, Eds., *Intersubband Transitions in Quantum Wells*, Sept. 9–14, Cargese, France, Plenum, New York, 1992.

85. H. C. Liu, B. F. Levine and J. Y. Andersson, Eds., *Quantum Well Intersubband Transition Physics and Devices*, Plenum, New York, 1994.

86. M. O. Manasreh, Ed., *Semiconductor Quantum Wells and Superlattices for Long-Wavelength Infrared Detectors*, Artech House, Norwood, MA 1993.

87. B. F. Levine, "Quantum-well infrared photodetectors," *J. Appl. Phys.* **74**, R1–R81 (1993).

88. Y. C. Chang and R. B. James, "Saturation of intersubband transitions in p-type semiconductor quantum wells," *Phys. Rev. B* **39**, 12672–12681 (1989).

89. B. F. Levine, S. D. Gunapala, J. M. Kuo, S. S. Pei, and S. Hui, "Normal incidence hole intersubband absorption long wavelength $GaAs/Al_xGa_{1-x}As$ quantum well infrared photodetectors," *Appl. Phys. Lett.* **59**, 1864–1866 (1991).

90. P. Man and D. S. Pan, "Analysis of normal-incident absorption in p-type quantum-well infrared photodetectors," *Appl. Phys. Lett.* **61**, 2799–2801 (1992).

91. K. M. S. V. Bandara, D. D. Coon, Byungsung O, Y. F. Lin, and M. H. Francombe, "Exchange interactions in quantum well subbands," *Appl. Phys. Lett.* **53**, 1931–1933 (1988).

92. M. O. Manasreh, F. Szmulowicz, T. Vaughan, K. R. Evans, C. E. Stutz, and D. W. Fischer, "Origin of the blueshift in the intersubband infrared absorption in $GaAs/Al_{0.3}Ga_{0.7}As$ multiple quantum wells," *Phys. Rev. B* **43**, 9996–9999 (1991).

93. J. W. Choe, Byungsung O, K. M. S. V. Bandara, and D. D. Coon, "Exchange interaction effects in quantum well infrared detectors and absorbers," *Appl. Phys. Lett.* **56**, 1679–1681 (1990).

94. S. L. Chuang, M. S. C. Luo, S. Schmitt-Rink, and A. Pinczuk, "Many-body effects on intersubband transitions in semiconductor quantum-well structures," *Phys. Rev. B* **46**, 1897–1900 (1992).

95. B. F. Levine, R. J. Malik, J. Walker, K. K. Choi, C. G. Bethea, D. A. Kleinman, and J. M. Vandenberg, "Strong 8.2 μm infrared intersubband absorption in doped GaAs/AlAs quantum well waveguides," *Appl. Phys. Lett.* **50**, 273–275 (1987).

96. J. Y. Andersson and G. Landgren, "Intersubband transitions in single AlGaAs/GaAs quantum wells studied by Fourier transform infrared spectroscopy," *J. Appl. Phys.* **64**, 4123–4127 (1988).

97. H. C. Liu, "Photoconductive gain mechanism of quantum-well intersubband infrared detectors," *Appl. Phys. Lett.* **60**, 1507–1509 (1992).

98. B. F. Levine, A. Zussman, S. D. Gunapala, M. T. Asom, J. M. Kuo, and W. S. Hobson, "Photoexcited escape probability, optical gain, and noise in quantum well infrared photodetectors," *J. Appl. Phys.* **72**, 4429–4443 (1992).

99. B. F. Levine, C. G. Bethea, G. Hasnain, V. O. Shen, E. Pelve, R. R. Abbott, and S. J. Hsieh, "High sensitivity low dark current 10 μm GaAs quantum well infrared photodetectors," *Appl. Phys. Lett.* **56**, 851–853 (1990).

100. M. A. Kinch and A. Yariv, "Performance limitations of GaAs/AlGaAs infrared superlattices," *Appl. Phys. Lett.* **55**, 2093–2095 (1989).

101. S. D. Gunapala, B. F. Levine, L. Pfeiffer, and K. West, "Dependence of the performance of GaAs/AlGaAs quantum well infrared photodetectors on doping and bias," *Appl. Phys. Lett.* **69**, 6517–6520 (1991).

102. B. F. Levine, A. Zussman, J. M. Kuo, and J. de Jong, "19 μm cutoff long-wavelength GaAs/Al$_x$Ga$_{1-x}$As quantum-well infrared photodetectors," *Appl. Phys. Lett.* **71**, 5130–5135 (1992).

Appendix A

The Hydrogen Atom (3D and 2D Exciton Bound and Continuum States) [1–5]

A.1 THREE-DIMENSIONAL (3D) CASE

The hydrogen atom is a two-particle system described by

$$H = -\frac{\hbar^2}{2m_1}\nabla_1^2 - \frac{\hbar^2}{2m_2}\nabla_2^2 + V(\mathbf{r}_1 - \mathbf{r}_2) \tag{A.1}$$

A general method to solve this problem is to change the variables to the center-of-mass coordinates $\mathbf{R}$ and the difference coordinates $\mathbf{r}$:

$$\mathbf{R} = \frac{m_1\mathbf{r}_1 + m_2\mathbf{r}_2}{M} \tag{A.2}$$

$$\mathbf{r} = \mathbf{r}_1 - \mathbf{r}_2 \tag{A.3}$$

We define the total mass M

$$M = m_1 + m_2 \tag{A.4}$$

and the reduced mass m_r

$$\frac{1}{m_r} = \frac{1}{m_1} + \frac{1}{m_2} \tag{A.5}$$

The first two operators in H can be rewritten as

$$-\frac{\hbar^2}{2m_1}\nabla_1^2 - \frac{\hbar^2}{2m_2}\nabla_2^2 = -\frac{\hbar^2}{2M}\nabla_{\mathbf{R}}^2 - \frac{\hbar^2}{2m_r}\nabla_{\mathbf{r}}^2 \tag{A.6}$$

where $\nabla_{\mathbf{R}}$ and $\nabla_{\mathbf{r}}$ are the gradients with respect to $\mathbf{R}$ and $\mathbf{r}$. Equation (A.6) can be derived using the fact that

$$
\frac{\partial}{\partial x_1} = \frac{\partial X}{\partial x_1}\frac{\partial}{\partial X} + \frac{\partial x}{\partial x_1}\frac{\partial}{\partial x} = \frac{m_1}{M}\frac{\partial}{\partial X} + \frac{\partial}{\partial x}
$$

$$
\frac{\partial}{\partial x_2} = \frac{\partial X}{\partial x_2}\frac{\partial}{\partial X} + \frac{\partial x}{\partial x_2}\frac{\partial}{\partial x} = \frac{m_2}{M}\frac{\partial}{\partial X} - \frac{\partial}{\partial x} \quad \text{etc.}
$$

We can show the relation in (A.6) for all terms involving $\partial^2/\partial x_1^2$, $\partial^2/\partial x_2^2$, $\partial^2/\partial X^2$ and $\partial^2/\partial x^2$, then use an analogy for other components.

The original Schrödinger equation

$$
H\psi(\mathbf{r}_1,\mathbf{r}_2) = E_T\psi(\mathbf{r}_1,\mathbf{r}_2) \tag{A.7}
$$

can be separated into two parts if we let

$$
\psi(\mathbf{r}_1,\mathbf{r}_2) = f(\mathbf{R})\psi(\mathbf{r})
$$

where

$$
-\frac{\hbar^2}{2M}\nabla_\mathbf{R}^2 f(\mathbf{R}) = E_K f(\mathbf{R}) \tag{A.8}
$$

$$
\left[-\frac{\hbar^2}{2m_r}\nabla_r^2 + V(\mathbf{r})\right]\psi(\mathbf{r}) = E\psi(\mathbf{r}) \tag{A.9}
$$

and the total energy E_T is

$$
E_T = E_K + E \tag{A.10}
$$

The eigenenergy and eigenfunction of the first equation (A.8) in the center-of-mass coordinates are simply those of a free "particle" with a mass M:

$$
E_K = \frac{\hbar^2 \mathbf{K}^2}{2M} \tag{A.11}
$$

and

$$
f(\mathbf{R}) = \frac{e^{i\mathbf{K}\cdot\mathbf{R}}}{\sqrt{V}} \tag{A.12}
$$

Once $\psi(\mathbf{r})$ is solved from (A.9), the complete solution is given by

$$
\psi(\mathbf{r}_1,\mathbf{r}_2) = \frac{e^{i\mathbf{K}\cdot\mathbf{R}}}{\sqrt{V}}\psi(\mathbf{r}) \tag{A.13}
$$

Solutions of $\psi(\mathbf{r})$ for the Coulomb Potential

Using the definition of $\nabla_{\mathbf{r}}^2$

$$\nabla_{\mathbf{r}}^2 = \frac{1}{r}\frac{\partial^2}{\partial r^2}r + \frac{1}{r^2 \sin\theta}\frac{\partial}{\partial\theta}\left(\sin\theta\frac{\partial}{\partial\theta}\right) + \frac{1}{r^2 \sin^2\theta}\frac{\partial^2}{\partial\varphi^2} \quad \text{(A.14)}$$

the solution to Equation (A.9) with the Coulomb potential

$$V(r) = -\frac{e^2}{4\pi\varepsilon r} \quad \text{(A.15)}$$

can be obtained by the separation of variables:

$$\psi(\mathbf{r}) = R(r)\Theta(\theta)\Phi(\varphi) = R(r)Y(\theta,\varphi) \quad \text{(A.16)}$$

and

$$\frac{d^2}{d\varphi^2}\Phi + m^2\Phi = 0 \quad \text{(A.17)}$$

$$\frac{1}{\sin\theta}\frac{d}{d\theta}\left(\sin\theta\frac{d\Theta}{d\theta}\right) - \frac{m^2}{\sin^2\theta}\Theta + \lambda\Theta = 0 \quad \text{(A.18)}$$

$$-\frac{\hbar^2}{2m_r}\frac{1}{r}\frac{d^2}{dr^2}(rR) + \frac{\hbar^2}{2m_r}\frac{\lambda}{r^2}R + [V(r) - E]R(r) = 0 \quad \text{(A.19)}$$

Spherical Harmonics Solutions $Y_{\ell m}(\theta,\varphi)$

It is easy to see that the solution for Φ is simply

$$\Phi(\varphi) = \frac{1}{\sqrt{2\pi}}e^{im\varphi} \quad \text{(A.20)}$$

where m is an integer, since $\Phi(\varphi)$ should be periodic in φ with a period 2π. A change of variable from θ to ξ,

$$\xi = \cos\theta, \qquad P(\xi) \equiv \Theta(\theta)$$

leads to

$$\frac{d}{d\xi}\left[(1 - \xi^2)\frac{dP}{d\xi}\right] - \frac{m^2}{1 - \xi^2}P + \lambda P = 0 \quad \text{(A.21)}$$

One notes that the above equation is even in ξ, that is, if $\xi \to -\xi$, or equivalently, $\theta \to \pi - \theta$, the same equation is obtained. Thus the solution is symmetric or antisymmetric with respect to the x-y plane.

If $m = 0$, the above equation becomes

$$\frac{d}{d\xi}\left[(1 - \xi^2)\frac{dP}{d\xi}\right] + \lambda P = 0 \qquad (A.22)$$

which can be easily solved using the series expansion:

$$P(\xi) = \sum_{k=0}^{\infty} a_k \xi^k \qquad (A.23)$$

A recursive formula is then obtained by substituting (A.23) into (A.22):

$$a_{k+2} = \frac{k(k + 1) - \lambda}{(k + 1)(k + 2)} a_k \qquad (A.24)$$

If the series does not terminate at some finite value of k, $a_{k+2}/a_k \to k/(k + 2)$, and the series will not converge at $P(\xi = 1)$. Thus λ must be an integer $\lambda = \ell(\ell + 1)$ for some finite value ℓ, that is, $k = \ell$, where ℓ is an integer. The solution is

$$P_\ell(\xi) = \frac{1}{2^\ell} \sum_{k=0}^{[\ell/2]} \frac{(-1)^k (2\ell - 2k)!}{(\ell - k)! k! (\ell - 2k)!} \xi^{\ell - 2k} \qquad (A.25)$$

where $[\ell/2]$ means the largest integer $\leq \ell/2$.

In general, the solution to (A.21) is given by the associated Legendre functions satisfying

$$\frac{d}{d\xi}\left[(1 - \xi^2)\frac{dP_\ell^m}{d\xi}\right] - \frac{m^2}{1 - \xi^2} P_\ell^m(\xi) + \ell(\ell + 1)P_\ell^m(\xi) = 0 \quad (A.26)$$

where

$$P_\ell^m(\xi) = (1 - \xi^2)^{m/2} \frac{d^m P_\ell(\xi)}{d\xi^m} = \frac{1}{2^\ell \ell!}(1 - \xi^2)^{m/2} \frac{d^{\ell+m}}{d\xi^{\ell+m}}(\xi^2 - 1)^\ell$$

$$(A.27)$$

for positive $m \leq \ell$. For negative m with $|m| \leq \ell$, $|m|$ should be used in (A.27) since the differential equation (A.26) should give the same results for $\pm m$. The first few polynomials are

$$\ell = 0, \quad P_0^0(\xi) = 1$$

$$\ell = 1, \quad P_1^0(\xi) = \xi$$

$$P_1^1(\xi) = \sqrt{1 - \xi^2}$$

$$\ell = 2, \quad P_2^0(\xi) = \tfrac{1}{2}(3\xi^2 - 1)$$

$$P_2^1(\xi) = 3\xi\sqrt{1 - \xi^2}$$

$$P_2^2(\xi) = 3(1 - \xi^2) \tag{A.28}$$

Noting that

$$\int_{-1}^{1} [P_\ell^m(\xi)]^2 \, d\xi = \frac{2}{2\ell + 1} \frac{(\ell + m)!}{(\ell - m)!} \tag{A.29}$$

with $0 \le m \le \ell$, the normalized solutions for the angular dependence can be obtained:

$$Y_{\ell m}(\theta, \varphi) = \sqrt{\frac{2\ell + 1}{4\pi} \frac{(\ell - |m|)!}{(\ell + |m|)!}} \, (-1)^{(m + |m|)/2} P_\ell^{|m|}(\cos \theta) \, e^{im\varphi} \tag{A.30}$$

and

$$\int_{\phi=0}^{2\pi} \int_{\theta=0}^{\pi} |Y_\ell^m(\theta, \varphi)|^2 \sin \theta \, d\theta \, d\varphi = 1 \tag{A.31}$$

The first few spherical harmonics are

$\ell = 0$ (s orbit)

$$Y_{00} = \frac{1}{\sqrt{4\pi}} \tag{A.32a}$$

$\ell = 1$ (p orbits)

$$Y_{10}(\theta, \varphi) = \sqrt{\frac{3}{4\pi}} \cos\theta = \sqrt{\frac{3}{4\pi}} \frac{z}{r} \equiv |Z\rangle$$

$$Y_{1\pm 1}(\theta, \varphi) = \mp\sqrt{\frac{3}{8\pi}} \sin\theta\, e^{\pm i\varphi} = \mp\sqrt{\frac{3}{8\pi}} \frac{x \pm iy}{r}$$

$$\equiv \mp\frac{1}{\sqrt{2}} |X \pm iY\rangle \qquad (A.32b)$$

$\ell = 2$ (d orbits)

$$Y_{20}(\theta, \varphi) = \sqrt{\frac{5}{16\pi}} (3\cos^2\theta - 1) = \sqrt{\frac{5}{16\pi}} \left(\frac{3z^2}{r^2} - 1\right)$$

$$Y_{2\pm 1}(\theta, \varphi) = \mp\sqrt{\frac{15}{8\pi}} \sin\theta \cos\theta\, e^{\pm i\varphi} = \mp\sqrt{\frac{15}{8\pi}} \frac{(x \pm iy)z}{r^2}$$

$$Y_{2\pm 2}(\theta, \varphi) = \sqrt{\frac{15}{32\pi}} \sin^2\theta\, e^{\pm 2i\varphi} = \sqrt{\frac{15}{32\pi}} \frac{(x \pm iy)^2}{r^2} \qquad (A.32c)$$

Radial Functions for the Bound and Continuum States

The radial function $R(r)$ satisfies

$$-\frac{\hbar^2}{2m_r} \frac{1}{r} \frac{d^2}{dr^2}(rR) + \frac{\hbar^2}{2m_r} \frac{\ell(\ell+1)}{r^2} R + [V(r) - E]R = 0 \quad (A.33)$$

which may be rewritten in the form

$$\left\{\frac{d^2}{dr^2} - \frac{\ell(\ell+1)}{r^2} + \frac{2m_r}{\hbar^2}[E - V(r)]\right\}u(r) = 0 \qquad (A.34)$$

where

$$u(r) = rR(r) \qquad (A.35)$$

Let us define the Bohr radius

$$a_o = \frac{4\pi\varepsilon\hbar^2}{e^2 m_r} \qquad (A.36)$$

and

$$k_o = \frac{1}{a_o}$$

The Rydberg energy R_y is defined as

$$R_y = \frac{\hbar^2 k_o^2}{2m_r} = \frac{m_r e^4}{2(4\pi\varepsilon)^2 \hbar^2} \tag{A.37}$$

In the following, we consider the bound and continuum states.

Case 1:   3D Bound State Solutions $(E < 0)$. For bound states, we define the variables

$$\gamma^2 = -\frac{R_y}{E} \tag{A.38}$$

$$\rho = \frac{2r}{\gamma a_o} \tag{A.39}$$

Equation (A.34) reduces to

$$\left[\frac{d^2}{d\rho^2} - \frac{\ell(\ell+1)}{\rho^2} - \frac{1}{4} + \frac{\gamma}{\rho} \right] u(\rho) = 0 \tag{A.40}$$

If we look at the asymptotic behavior of the function $u(\rho)$ at $\rho \to \infty$ from (A.40), we find $(d^2/d\rho^2 - \frac{1}{4})u(\rho) = 0$ and the solution should behave as $\exp(\rho/2)$ or $\exp(-\rho/2)$. The latter term $\exp(-\rho/2)$ should be chosen, since the former blows up as ρ approaches infinity. If we look at $\rho \to 0$, the differential equation (A.40) is

$$\left[\frac{d^2}{d\rho^2} - \frac{\ell(\ell+1)}{\rho^2} \right] u(\rho) = 0 \tag{A.41}$$

for $\ell \neq 0$, and its solution is either $\rho^{\ell+1}$ or $\rho^{-\ell}$. The former should be chosen since $u(\rho)$ should be a regular function at the origin. Thus, in general, we may assume $u(\rho)$ of the form

$$u(\rho) = e^{-\rho/2} \rho^{\ell+1} f(\rho) \tag{A.42}$$

Equation (A.40) reduces to

$$\rho\frac{d^2f(\rho)}{d\rho^2} + (2\ell + 2 - \rho)\frac{df(\rho)}{d\rho} + (\gamma - \ell - 1)f(\rho) = 0 \quad (A.43)$$

By checking a series solution to the above equation, it is found that $(\gamma - \ell - 1)$ must be an integer N, $N = 0, 1, 2, \ldots$. Or

$$\gamma = N + \ell + 1 \equiv n \quad (A.44)$$

Using n as the principal quantum number, we have

$$\ell = 0, 1, 2, \ldots, n - 1 \quad (A.45)$$

The solutions of the differential equation of the form

$$\rho\frac{d^2}{d\rho^2}f + (\beta - \rho)\frac{df}{d\rho} - \alpha f = 0 \quad (A.46)$$

are the confluent hypergeometric functions,

$$F(\alpha, \beta; \rho) = 1 + \frac{\alpha}{\beta}\frac{\rho}{1!} + \frac{\alpha(\alpha + 1)}{\beta(\beta + 1)}\frac{\rho^2}{2!} + \frac{\alpha(\alpha + 1)(\alpha + 2)}{\beta(\beta + 1)(\beta + 2)}\frac{\rho^3}{3!} + \cdots$$
$$(A.47)$$

Note that $F(\alpha, \beta; 0) = 1$.

We can identify that $\alpha = -n + \ell + 1$ and $\beta = 2\ell + 2$ and obtain the solution for $u(\rho)$:

$$u(\rho) = e^{-\rho/2}\rho^{\ell+1}F(-n + \ell + 1, 2\ell + 2; \rho) \quad (A.48)$$

We can check from the definition in (A.47) that the polynomial F in (A.48) terminates after a finite number of terms since $\ell \leq n - 1$. Alternatively, the associated Laguerre polynomials $L_n^m(\rho)$ are used. It is related to F by

$$L_n^m(\rho) = (-1)^m\frac{(n!)^2}{m!(n - m)!}F(-(n - m), m + 1; z)$$

$$= \frac{n!}{(n - m)!}e^z\frac{d^n}{dz^n}(e^{-z}z^{n-m})$$

$$= (-1)^m\frac{n!}{(n - m)!}e^z z^{-m}\frac{d^{n-m}}{dz^{n-m}}(e^{-z}z^n) \quad (A.49)$$

Note that $L_n^m(0) = (-1)^m(n!)^2/[m!(n - m)!]$. Therefore, $u(r)$ can also be written in the form

$$u(\rho) = C e^{-\rho/2} \rho^{\ell+1} L_{n+\ell}^{2\ell+1}(\rho) \tag{A.50}$$

where C is a constant to be determined by the normalization condition. Using the integral identity

$$\int_0^\infty e^{-\rho} \rho^{2\ell+2} \left[L_{n+\ell}^{2\ell+1}(\rho) \right]^2 d\rho = \frac{2n[(n+\ell)!]^3}{(n-\ell-1)!} \tag{A.51}$$

or

$$\int_0^\infty e^{-\rho} \rho^{2\ell+2} \left[F(-n+\ell+1, 2\ell+2; \rho) \right]^2 d\rho = \frac{2n[(2\ell+1)!]^2 (n-\ell-1)!}{(n+\ell)!} \tag{A.52}$$

we find that the radial wave function satisfying the normalization condition

$$\int_0^\infty R^2(r) r^2 \, dr = 1 \tag{A.53}$$

is

$$
\begin{aligned}
R_{n\ell}(r) &= -\left(\frac{2}{na_o}\right)^{3/2} \left\{ \frac{(n-\ell-1)!}{2n[(n+\ell)!]^3} \right\}^{1/2} e^{-r/(na_o)} \left(\frac{2r}{na_o}\right)^\ell L_{n+\ell}^{2\ell+1}\left(\frac{2r}{na_o}\right) \\
&= \frac{(2/na_o)^{3/2}}{(2\ell+1)!} \left[\frac{(n+\ell)!}{2n(n-\ell-1)!} \right]^{1/2} e^{-r/na_o} \left(\frac{2r}{na_o}\right)^\ell \\
&\quad \times F\left(-n+\ell+1, 2\ell+2; \frac{2r}{na_o}\right) \tag{A.54}
\end{aligned}
$$

As $r \to 0$, $R_{n\ell}(r) \to (2r/na_o)^\ell$, which is nonzero only when $\ell = 0$. The complete wave functions are

$$\psi_{n\ell m}(r, \theta, \varphi) = R_{n\ell}(r) Y_{\ell m}(\theta, \varphi) \tag{A.55}$$

The first few wave functions are

$$\psi_{100} = \frac{1}{\sqrt{\pi}} \left(\frac{1}{a_o}\right)^{3/2} e^{-r/a_o} \qquad \text{(1s state)}$$

$$\psi_{200} = \frac{1}{2\sqrt{2\pi}} \left(\frac{1}{a_o}\right)^{3/2} \left(1 - \frac{r}{2a_o}\right) e^{-r/2a_o} \qquad \text{(2s state)}$$

$$\psi_{210} = \frac{1}{2\sqrt{2\pi}} \left(\frac{1}{a_o}\right)^{3/2} \left(\frac{r}{2a_o}\right) e^{-r/2a_o} \cos\theta \qquad \text{(2p state)}$$

$$\psi_{21\pm1} = \frac{1}{4\sqrt{\pi}} \left(\frac{1}{a_o}\right)^{3/2} \left(\frac{r}{2a_o}\right) e^{-r/2a_o} \sin\theta \, e^{\pm i\varphi} \qquad \text{(2p states)} \quad \text{(A.56)}$$

Note that at $r = 0$, $\psi \neq 0$ only if $\ell = m = 0$ (s states). We have

$$R_{n0}(0) = \frac{2}{(na_o)^{3/2}} \qquad Y_{00} = \frac{1}{\sqrt{4\pi}}$$

and

$$|\psi_{n00}(r = 0)|^2 = \frac{1}{\pi a_o^3 n^3} \qquad (A.57)$$

The wave function at $r = 0$ will be useful when we study the exciton effects on optical absorption in Chapter 13.

Case 2: 3D Continuum-State Solutions $(E > 0)$. When the energy is positive, the solutions will not be quantized. Instead they become continuous. The solution to (A.34) for the radial wave function is obtained following the change of variables as in (A.38) and (A.39),

$$\gamma^2 = -\frac{R_y}{E} < 0 \qquad (A.58)$$

$$\rho = \frac{2r}{\gamma a_o} = 2i\kappa \underset{\sim}{r} \qquad (A.59)$$

We choose $1/\gamma = i\kappa$, where $\kappa = \sqrt{E/R_y} = k/k_o$ is the normalized wave number, since

$$E = \frac{\hbar^2 k^2}{2m_r} \qquad (A.60)$$

and the Rydberg is

$$R_y = \frac{\hbar^2 k_o^2}{2m_r} = \frac{\hbar^2}{2m_r a_o^2} \tag{A.61}$$

and $\underset{\sim}{r} = r/a_o$ is the normalized radial distance. The solution $u(\rho)$ has the same form as in (A.48),

$$u(\rho) = \text{const.}\, e^{-\rho/2}\rho^{\ell+1} F\left(\ell + 1 + \frac{i}{\kappa}, 2\ell + 2; \rho\right) \tag{A.62}$$

We therefore look for the normalized radial wave function of the form

$$R_{\kappa\ell}(\underset{\sim}{r}) = N_{\kappa\ell}(2\kappa\underset{\sim}{r})^\ell \exp(-i\kappa\underset{\sim}{r}) F\left(\ell + 1 + \frac{i}{\kappa}, 2\ell + 2; 2i\kappa\underset{\sim}{r}\right) \tag{A.63}$$

such that

$$\int_0^\infty R_{\kappa\ell}^*(\underset{\sim}{r}) R_{\kappa'\ell}(\underset{\sim}{r})\underset{\sim}{r}^2\, d\underset{\sim}{r} = \delta(\kappa - \kappa') \tag{A.64}$$

in the $\underset{\sim}{r}$ space. Later on, we change the normalized variable $\underset{\sim}{r}$ to the real space distance r, and the $\delta(\kappa - \kappa')$ normalization rule to the $\delta(E - E')$ normalization rule.

We note the asymptotic behavior of F,

$$F(\alpha, \gamma; z) = \frac{\Gamma(\gamma)}{\Gamma(\alpha)} e^z z^{\alpha-\gamma} + \frac{\Gamma(\gamma)}{\Gamma(\gamma - \alpha)}(-z)^{-\alpha} \tag{A.65}$$

as $|z| \to \infty$. We thus have

$$R_{\kappa\ell}(\underset{\sim}{r}) = N_{\kappa\ell}(2\kappa\underset{\sim}{r})^\ell \exp(-i\kappa\underset{\sim}{r})\left[\frac{\Gamma(2\ell + 2)}{\Gamma(\ell + 1 + i/\kappa)}\exp(2i\kappa\underset{\sim}{r})(2i\kappa\underset{\sim}{r})^{-\ell-1+i/\kappa}\right.$$

$$\left. + \frac{\Gamma(2\ell + 2)}{\Gamma(\ell + 1 - i/\kappa)}(-2i\kappa\underset{\sim}{r})^{-\ell-1-i/\kappa}\right]$$

$$= N_{\kappa\ell}(2\kappa\underset{\sim}{r})^\ell (2\ell + 1)!\left[\frac{\exp(i\kappa\underset{\sim}{r})(2i\kappa\underset{\sim}{r})^{-\ell-1+i/\kappa}}{\Gamma(\ell + 1 + i/\kappa)} + \text{c.c.}\right] \tag{A.66}$$

where c.c. means complex conjugate of the first term in the large bracket. We use $\Gamma(z^*) = [\Gamma(z)]^*$, $2i\kappa\underline{r} = \exp[i\,\pi/2 + \ln(2\kappa\underline{r})]$, and

$$(2i\kappa\underline{r})^{-\ell-1+i/\kappa} = (2i\kappa\underline{r})^{-\ell-1}(2i\kappa\underline{r})^{i/\kappa}$$

$$= (2\kappa\underline{r})^{-\ell-1}\,e^{-i(\ell+1)\pi/2}\exp\left[-\frac{\pi}{2\kappa} + \frac{i}{\kappa}\ln(2\kappa\underline{r})\right] \quad (A.67)$$

The radial function can be written as

$$R_{\kappa\ell}(\underline{r}) = \frac{b_{\kappa\ell}}{r}\cos\left[\kappa\underline{r} + \frac{1}{\kappa}\ln(2\kappa\underline{r}) - \delta_{\kappa\ell}\right] \qquad \text{for } \underline{r} \to \infty \quad (A.68a)$$

$$b_{\kappa\ell} = N_{\kappa\ell}\frac{(2\ell + 1)!}{\kappa}\frac{e^{-\pi/(2\kappa)}}{|\Gamma(\ell + 1 + i/\kappa)|} \quad (A.68b)$$

and the phase factor is

$$\delta_{\kappa\ell} = (\ell + 1)\frac{\pi}{2} + \arg[\Gamma(\ell + 1 + i/\kappa)] \quad (A.68c)$$

The normalization condition is determined by the asymptotic behavior of $R_{\kappa\ell}$ as $\underline{r} \to \infty$. Since $\kappa\underline{r} \gg (1/\kappa)\ln(2\kappa\underline{r})$ and $\delta_{\kappa\ell}$, we find

$$\int_0^\infty R^*_{\kappa\ell}(\underline{r}) R_{\kappa'\ell}(\underline{r})\underline{r}^2\,d\underline{r} = b^*_{\kappa\ell}b_{\kappa'\ell}\int_0^\infty \cos(\kappa\underline{r})\cos(\kappa'\underline{r})\,d\underline{r}$$

$$= |b_{\kappa\ell}|^2\frac{\pi}{2}\delta(\kappa - \kappa') \quad (A.69)$$

Therefore, we choose $b_{\kappa\ell} = \sqrt{2/\pi}$ to satisfy the $\delta(\kappa - \kappa')$ normalization rule (A.64). In deriving (A.69), we have expressed $\cos(\kappa r)$ as the sum of two exponentials and use

$$\int_0^\infty \exp[i(\kappa - \kappa')\underline{r}]\,d\underline{r} = \pi\delta(\kappa - \kappa') \quad (A.70)$$

noting that κ, κ' are both positive. From $b_{\kappa\ell}$, we find

$$N_{\kappa\ell} = \sqrt{\frac{2}{\pi}}\frac{\kappa\,e^{\pi/(2\kappa)}|\Gamma(\ell + 1 + i/\kappa)|}{(2\ell + 1)!} \quad (A.71)$$

and $R_{\kappa\ell}(\underline{r})$ is given by (A.66). The quantity $|\Gamma(\ell + 1 + i/\kappa)|$ can be evaluated noting that for $\ell = 0$,

$$
\left|\Gamma\left(1 + \frac{i}{\kappa}\right)\right|^2 = \Gamma\left(1 + \frac{i}{\kappa}\right)\Gamma\left(1 - \frac{i}{\kappa}\right) = \frac{i}{\kappa}\Gamma\left(\frac{i}{\kappa}\right)\Gamma\left(\frac{1 - i}{\kappa}\right)
$$

$$
= \frac{\pi/\kappa}{\sinh(\pi/\kappa)} \tag{A.72}
$$

For $\ell > 0$, $\Gamma(\ell + 1 + i/\kappa) = (\ell + i/\kappa)(\ell - 1 + i/\kappa)\cdots(1 + i/\kappa)\Gamma(1 + i/\kappa)$, and

$$
\left|\Gamma\left(\ell + 1 + \frac{i}{\kappa}\right)\right|^2 = \Gamma\left(\ell + 1 + \frac{i}{\kappa}\right)\Gamma\left(\ell + 1 - \frac{i}{\kappa}\right)
$$

$$
= \left[\prod_{s=1}^{\ell}\left(s^2 + \frac{1}{\kappa^2}\right)\right]\frac{\pi/\kappa}{\sinh(\pi/\kappa)} \tag{A.73}
$$

Therefore, we obtain

$$
N_{\kappa\ell} = \sqrt{\frac{2}{\pi}}\,\frac{\kappa\,e^{\pi/(2\kappa)}}{(2\ell + 1)!}\left[\prod_{s=1}^{\ell}\left(s^2 + \frac{1}{\kappa^2}\right)\right]^{1/2}\left[\frac{\pi/\kappa}{\sinh(\pi/\kappa)}\right]^{1/2} \tag{A.74}
$$

If we change the normalization rule to the physical quantities in terms of $R_{E\ell}(r)$ in the real space r and use the energy normalization rule,

$$
\int_0^{\infty} R_{E\ell}(r)R_{E'\ell}(r)r^2\,dr = \delta(E - E') \tag{A.75}
$$

we obtain

$$
R_{E\ell}(r) = \frac{C_{E\ell}}{\left(R_y a_o^3\right)^{1/2}(2\ell + 1)!}(2kr)^{\ell}\,e^{-ikr}F\left(\ell + 1 + \frac{i}{ka_o}, 2\ell + 2; 2ikr\right)
$$

$$
C_{E\ell} = \left[\prod_{s=1}^{\ell}\left(s^2 + \frac{1}{\kappa^2}\right)\frac{e^{\pi/\kappa}}{\sinh(\pi/\kappa)}\right]^{1/2} \tag{A.76}
$$

where we note that $kr = \kappa\underline{r}$, $E = \hbar^2 k^2/(2m_r)$, $E/R_y = \kappa^2$, and

$$
\delta(\kappa - \kappa') = 2\kappa\delta\left(\frac{E}{R_y} - \frac{E'}{R_y}\right) = 2\kappa R_y\,\delta(E - E') \tag{A.77}
$$

Again, as $r \to 0$, $R_{E\ell}(r) \to (2kr)^{\ell}$, which vanishes except for $\ell = 0$. The complete wave function is given by

$$\psi_{E\ell m}(\mathbf{r}) = R_{E\ell}(r)Y_{\ell m}(\theta, \phi) \tag{A.78}$$

where the spherical harmonics $Y_{\ell m}(\theta, \phi)$ are given in (A.30). At $\mathbf{r} = 0$, we find $\psi_{E\ell m}(\mathbf{r}) \neq 0$ only if $\ell = m = 0$. Therefore, using $C_{E0} = [e^{\pi/\kappa}/\sinh(\pi/\kappa)]^{1/2}$ and $Y_{00} = 1/\sqrt{4\pi}$, we obtain

$$|\psi_{E00}(\mathbf{r} = 0)|^2 = \frac{1}{R_y a_o^3}\left[\frac{e^{\pi/\kappa}}{\sinh(\pi/\kappa)}\right]\frac{1}{4\pi} \tag{A.79}$$

A.2 TWO-DIMENSIONAL (2D) CASE

In this case, the position vectors of the two particles $\mathbf{r}_1, \mathbf{r}_2$, the center-of-mass coordinates $\mathbf{R}$ and the difference vector $\mathbf{r}$ are all in the x-y plane. All the expressions still follow the equations (A.1)–(A.11), except that only the x-y dependence exists, i.e.,

$$V(\mathbf{r}) = \frac{-e^2}{4\pi\varepsilon r} \tag{A.80}$$

where $\mathbf{r} = x\hat{x} + y\hat{y} = (x_1 - x_2)\hat{x} + (y_1 - y_2)\hat{y}$. The solution is

$$\psi(\mathbf{r}_1, \mathbf{r}_2) = \frac{e^{i\mathbf{K}\cdot\mathbf{R}}}{\sqrt{A}}\psi(\mathbf{r}) \tag{A.81}$$

where A is the cross-section area, and $\psi(\mathbf{r})$ satisfies

$$\left[\frac{-\hbar^2}{2m_r}\left(\frac{1}{r}\frac{d}{dr}r\frac{d}{dr} + \frac{1}{r^2}\frac{\partial^2}{\partial\phi^2}\right) - \frac{e^2}{4\pi\varepsilon r}\right]\psi(\mathbf{r}) = E\psi(\mathbf{r}) \tag{A.82}$$

The solution is of the form

$$\psi(\mathbf{r}) = R(r)\frac{e^{im\varphi}}{\sqrt{2\pi}} \tag{A.83}$$

and the radial function satisfies

$$\left[\frac{1}{r}\frac{d}{dr}r\frac{d}{dr} - \frac{m^2}{r^2} + \frac{2m_r}{\hbar^2}\left(E + \frac{e^2}{4\pi\varepsilon r}\right)\right]R(r) = 0 \tag{A.84}$$

Case 1:   2D Bound State Solutions $(E < 0)$. Using a change of variables, the same as those in the three-dimensional case (A.38) and (A.39),

$$\gamma^2 = -\frac{R_y}{E} \tag{A.85}$$

$$\rho = \frac{2r}{\gamma a_o} \tag{A.86}$$

we obtain

$$\left(\frac{d^2}{d\rho^2} + \frac{1}{\rho}\frac{d}{d\rho} - \frac{m^2}{\rho^2} + \frac{\gamma}{\rho} - \frac{1}{4} \right) R(\rho) = 0 \tag{A.87}$$

As $\rho \to \infty$, we find the dominant terms above are $(d^2/d\rho^2 - \frac{1}{4})R(\rho) = 0$. Therefore, we set $R(\rho) = e^{-\rho/2}h(\rho)$, where $h(\rho)$ satisfies

$$\left[\frac{d^2}{d\rho^2} + \frac{(1 - \rho)}{\rho}\frac{d}{d\rho} + \frac{1}{\rho}\left(\gamma - \frac{1}{2} - \frac{m^2}{\rho} \right) \right] h(\rho) = 0 \tag{A.88}$$

As $\rho \to 0$, we find that the dominant terms are

$$\left(\frac{d^2}{d\rho^2} + \frac{1}{\rho}\frac{d}{d\rho} - \frac{m^2}{\rho^2} \right) h(\rho) = 0 \tag{A.89}$$

Therefore, $R(\rho)$ behaves like $\rho^{|m|}$. We then assume that $h(\rho) = \rho^{|m|}f(\rho)$, and find that $f(\rho)$ satisfies

$$\left[\rho\frac{d^2}{d\rho^2} + (2|m| + 1 - \rho)\frac{d}{d\rho} + \left(\gamma - \frac{1}{2} - |m| \right) \right] f(\rho) = 0 \tag{A.90}$$

If we compare the above with the definition of the confluence hypergeometric function $F(\alpha, \beta; \rho)$ in (A.46) and (A.47)

$$\left[\rho\frac{d^2}{d\rho^2} + (\beta - \rho)\frac{d}{d\rho} - \alpha \right] F(\alpha, \beta; \rho) = 0 \tag{A.91}$$

we obtain

$$f(\rho) = F\left(|m| + \tfrac{1}{2} - \gamma, 2|m| + 1; \rho\right) \tag{A.92}$$

The above polynomial should not approach infinity faster than any finite power of polynomials, and the acceptable solution is only when $|m| + \frac{1}{2} - \gamma$

is a negative integer, i.e., $\gamma \geq |m| + \frac{1}{2}$, or use $\gamma = n - \frac{1}{2}$, $n = 1, 2, 3, \ldots$ and $n \geq |m| + 1$. The radial function is given by

$$R(\rho) = e^{-\rho/2}\rho^{|m|}F(|m| + 1 - n, 2|m| + 1; \rho) \qquad (A.93)$$

Since $\gamma = n - \frac{1}{2}$, we find the energy levels

$$E_n = -\frac{R_y}{\gamma^2} = -\frac{R_y}{\left(n - \frac{1}{2}\right)^2} \qquad n = 1, 2, 3, \ldots \qquad (A.94)$$

Using the relations

$$L_n^m(z) = (-1)^m \frac{(n!)^2}{m!(n-m)!}F[-(n-m), m+1; z] \qquad (A.95)$$

and

$$\int_0^\infty e^{-\rho}\rho^{2m}\left[L_{n+m}^{2m}(\rho)\right]^2 \rho \, d\rho = (2n+1)\frac{(n+m)!}{(n-m)!} \qquad (A.96)$$

we obtain the normalized radial wave function $R_{nm}(r)$ satisfying

$$\int_0^\infty R_{nm}(r)R_{nm}(r) \, dr = 1 \qquad (A.97)$$

$$
\begin{aligned}
R_{nm}(r) &= \frac{4}{a_o}\sqrt{\frac{(n-|m|-1)!}{(2n-1)^3(n+|m|-1)!}} \\
&\quad \times e^{-\rho/2}\rho^{|m|}L_{n+|m|-1}^{2|m|}(\rho) \\
&= \frac{4}{a_o}\sqrt{\frac{(n+|m|-1)!}{(2n-1)^3(n-|m|-1)!}} \\
&\quad \times \frac{e^{-\rho/2}}{(2|m|)!}\rho^{|m|}F(-n+|m|+1, 2|m|+1; \rho) \qquad (A.98)
\end{aligned}
$$

$$\rho = \frac{2r}{\left(n - \frac{1}{2}\right)a_o} \qquad n = 1, 2, 3, \ldots \quad \text{and} \quad |m| \leq n - 1$$

$$(A.99)$$

The complete wave function is

$$\psi_{nm}(\mathbf{r}) = R_{nm}(r)\frac{e^{im\phi}}{\sqrt{2\pi}} \tag{A.100}$$

As $r \to 0$, $R_{nm}(r) \to \rho^{|m|}$, which does not vanish only if $m = 0$. We obtain

$$|\psi_{n0}(\mathbf{r} = 0)|^2 = |R_{n0}(0)|^2 \frac{1}{2\pi}$$

$$= \frac{1}{\pi a_o^2 \left(n - \frac{1}{2}\right)^3} \tag{A.101}$$

The above result is $1/(n - \frac{1}{2})^3$ per area of the circle determined by the Bohr radius a_0.

Case 2: 2D Continuum State Solutions $(E > 0)$. The procedure is very similar to that in the 3D case:

$$\gamma^2 = -\frac{R_y}{E} < 0 \tag{A.102}$$

$$\rho = \frac{2r}{\gamma a_o} = 2i\kappa\underline{r} \tag{A.103}$$

We choose

$$\frac{1}{\gamma} = i\kappa$$

and define $\underline{r} = r/a_0$ again. The radial function $R(\rho)$ behaves like

$$R(\rho) = \text{Const.}\, e^{-\rho/2}\rho^{|m|}F\left(|m| + \frac{1}{2} + \frac{i}{\kappa}, 2|m| + 1; \rho\right) \tag{A.104}$$

If we assume that

$$R_{\kappa m}(\underline{r}) = N_{\kappa m}\exp(-i\kappa\underline{r})(2\kappa\underline{r})^{|m|}F\left(|m| + \frac{1}{2} + \frac{i}{\kappa}, 2|m| + 1; \rho\right) \tag{A.105}$$

and follow the same steps as that in the 3D case, we obtain

$$N_{\kappa m} = \sqrt{\frac{\kappa}{\pi}} \frac{e^{\pi/(2\kappa)}}{(2|m|)!}\left|\Gamma\left(|m| + \frac{1}{2} + \frac{i}{\kappa}\right)\right| \tag{A.106}$$

and $R_{\kappa m}(\underset{\sim}{r})$ satisfies the $\delta(\kappa - \kappa')$ normalization rule

$$\int_0^\infty R_{\kappa m}^*(\underset{\sim}{r}) R_{\kappa' m}(\underset{\sim}{r}) \underset{\sim}{r} \, d\underset{\sim}{r} = \delta(\kappa - \kappa') \tag{A.107}$$

The only minor difference between the 2D and 3D radial wave functions is that in 2D, as $\underset{\sim}{r} \to \infty$, the prefactor in front of the cosine function is given by $1/\sqrt{\underset{\sim}{r}}$.

$$R_{\kappa m}(\underset{\sim}{r}) = \frac{b_{\kappa m}}{\sqrt{\underset{\sim}{r}}} \cos\left[\kappa \underset{\sim}{r} + \frac{1}{\kappa}\ln(2\kappa\underset{\sim}{r}) - \delta_{\kappa m}\right] \tag{A.108a}$$

where

$$b_{\kappa m} = N_{\kappa m}(2|m|)! \sqrt{\frac{2}{\kappa}} \frac{e^{-\pi/(2\kappa)}}{\left|\Gamma(|m| + \frac{1}{2} + i/\kappa)\right|} \tag{A.108b}$$

$$\delta_{\kappa m} = \left(|m| + \frac{1}{2}\right)\frac{\pi}{2} + \arg\left[\Gamma\left(|m| + \frac{1}{2} + \frac{i}{\kappa}\right)\right] \tag{A.108c}$$

and $b_{\kappa m} = \sqrt{2/\pi}$ is determined by the $\delta(\kappa - \kappa')$ normalization rule. If we change back to the physical quantities in the real space r instead of the normalized distance $\underset{\sim}{r} = r/a_0$, and use the $\delta(E - E')$ normalization rule, noting that

$$\frac{E}{R_y} = \kappa^2, \quad \delta(\kappa - \kappa') = 2\kappa R_y \, \delta(E - E') \tag{A.109}$$

we find for

$$\int_0^\infty R_{Em}^*(r) R_{E'm}(r) r \, dr = \delta(E - E') \tag{A.110}$$

$$R_{Em}(r) = \frac{R_{\kappa m}(\underset{\sim}{r})}{\sqrt{2\kappa R_y} \, a_o} \tag{A.111}$$

Using

$$\left|\Gamma\left(\frac{1}{2} + \frac{i}{\kappa}\right)\right|^2 = \Gamma\left(\frac{1}{2} + \frac{i}{\kappa}\right)\Gamma\left(\frac{1}{2} - \frac{i}{\kappa}\right) = \frac{\pi}{\cosh(\pi/\kappa)} \quad \text{(A.112)}$$

and

$$\left|\Gamma\left(|m| + \frac{1}{2} + \frac{i}{\kappa}\right)\right|^2 = \prod_{s=1}^{|m|}\left[\left(s - \frac{1}{2}\right)^2 + \frac{1}{\kappa^2}\right]\frac{\pi}{\cosh(\pi/\kappa)} \quad \text{(A.113)}$$

we obtain

$$R_{Em}(r) = \frac{C_{Em}}{\sqrt{2R_y}\,a_o(2|m|)!}(2kr)^{|m|}\,e^{-ikr}F\left(|m| + \frac{1}{2} + \frac{i}{\kappa}, 2|m| + 1; 2ikr\right)$$

$$\text{(A.114a)}$$

$$C_{Em} = \left\{\sum_{s=1}^{|m|}\left[\left(s - \frac{1}{2}\right)^2 + \frac{1}{\kappa^2}\right]\frac{e^{\pi/\kappa}}{\cosh(\pi/\kappa)}\right\}^{1/2} \quad \text{(A.114b)}$$

Note that $E = \hbar^2 k^2/2m_r$, and $\kappa = k/k_o = ka_o$. The complete wave function is given by

$$\psi_{Em}(\mathbf{r}) = R_{Em}(r)\frac{e^{im\varphi}}{\sqrt{2\pi}} \quad \text{(A.115)}$$

The wave function approaches $(2kr)^{|m|}$ as $r \to 0$, and it vanishes at $r = 0$ except for $m = 0$:

$$|\psi_{E0}(\mathbf{r} = 0)|^2 = \left|\frac{1}{\sqrt{2\pi}}R_{E0}(0)\right|^2$$

$$= \frac{1}{R_y a_o^2}\left[\frac{e^{\pi/\kappa}}{\cosh(\pi/\kappa)}\right]\frac{1}{4\pi}$$

$$= \frac{1}{R_y a_o^2 2\pi[1 + \exp(-2\pi/\kappa)]} \quad \text{(A.116)}$$

Important results of this appendix are summarized in Table 3.1.

REFERENCES

1. For both bound and continuum state solutions in the three-dimensional case see L. D. Landau and E. M. Lifshitz, *Quantum Mechanics*, 3d ed., Pergamon, Oxford, UK 1977, p. 117, and H. A. Bethe and E. E. Salpeter, *Quantum Mechanics of One- and Two-Electron Atoms*, Springer, Berlin, 1957.

2. For an n-dimensional space, $n \geq 2$, see M. Bander and C. Itzykson, "Group theory and the hydrogen atom (I) and (II)," *Rev. Mod. Phys.* **38**, 330–345 (1966), and **38**, 346–358 (1966).

3. C. Y. P. Chao and S. L. Chuang, "Analytical and numerical solutions for a two-dimensional exciton in momentum space," *Phys. Rev. B* **43**, 6530–6543 (1991).

4. E. Menzbacher, *Quantum Mechanics*, 2d ed., Wiley, New York, 1970.

5. H. Haug and S. W. Koch, *Quantum Theory of the Optical and Electronic Properties of Semiconductors*, World Scientific, Singapore, 1990.

Appendix B

Proof of the Effective Mass Theory

B.1 SINGLE BAND

To prove the effective mass theory (4.4.5) in Section 4.4, let us write the Bloch function

$$\psi_{n\mathbf{k}}(\mathbf{r}) \equiv |n\,\mathbf{k}\rangle \qquad (B.1)$$

satisfying (4.4.3). Since $\{\psi_{n\mathbf{k}}(\mathbf{r})\}$ is a complete set of basis functions, we may expand the solution $\psi(\mathbf{r})$ in terms of these functions:

$$\psi(\mathbf{r}) = \sum_n \int_{\text{B.Z.}} \frac{d^3\mathbf{k}}{(2\pi)^3} a_n(\mathbf{k}) \psi_{n\mathbf{k}}(\mathbf{r})$$

$$= \sum_n \int_{\text{B.Z.}} \frac{d^3\mathbf{k}}{(2\pi)^3} a_n(\mathbf{k}) |n\,\mathbf{k}\rangle \qquad (B.2)$$

where the integration is over the first Brillouin zone (BZ) in $\mathbf{k}$ space. Multiplying (4.4.4) by $\psi_{n\mathbf{k}}^*(\mathbf{r})$ and integrating over the volume, we find

$$(E_n(\mathbf{k}) - E)a_n(\mathbf{k}) + \sum_{n'} \int_{\text{B.Z.}} \frac{d^3\mathbf{k}'}{(2\pi)^3} \langle n\mathbf{k}|U|n'\mathbf{k}'\rangle a_{n'}(\mathbf{k}') = 0 \quad (B.3)$$

where

$$\langle n\mathbf{k}|U|n'\mathbf{k}'\rangle = \int d^3\mathbf{r}\, \psi_{n\mathbf{k}}^*(\mathbf{r}) U(\mathbf{r}) \psi_{n'\mathbf{k}'}(\mathbf{r}) \qquad (B.4)$$

and the orthonormal relation $\langle n\mathbf{k}|n'\mathbf{k}'\rangle = \delta_{n,n'}\delta(\mathbf{k} - \mathbf{k}')$ has been used. Let us consider the Fourier expansion of $U(\mathbf{r})$

$$U(\mathbf{r}) = \int \tilde{U}_{\mathbf{k}} e^{i\mathbf{k}\cdot\mathbf{r}} \frac{d^3\mathbf{k}}{(2\pi)^3} \qquad (B.5)$$

We find

$$\langle n\mathbf{k}|U|n'\mathbf{k}'\rangle = \int d^3r\, e^{-i(\mathbf{k}-\mathbf{k}')\cdot\mathbf{r}}u_{n\mathbf{k}}^*(\mathbf{r})u_{n'\mathbf{k}'}(\mathbf{r})U(\mathbf{r}) \tag{B.6}$$

Since the product $u_{n\mathbf{k}}^*(\mathbf{r})u_{n'\mathbf{k}'}(\mathbf{r})$ is periodic in $\mathbf{r}$, we may expand this product in terms of a Fourier series

$$u_{n\mathbf{k}}^*(\mathbf{r})u_{n'\mathbf{k}'}(\mathbf{r}) = \sum_{\mathbf{G}} C(n\mathbf{k}, n'\mathbf{k}', \mathbf{G})e^{i\mathbf{G}\cdot\mathbf{r}} \tag{B.7}$$

where $\mathbf{G}$ sums over all reciprocal lattice vectors. Therefore,

$$\langle n\mathbf{k}|U|n'\mathbf{k}'\rangle = \sum_{\mathbf{G}} C(n\mathbf{k}, n'\mathbf{k}', \mathbf{G})\tilde{U}_{\mathbf{k}-\mathbf{k}'-\mathbf{G}} \tag{B.8}$$

Here $\mathbf{k}$ and $\mathbf{k}'$ are in the first Brillouin zone.

We assume the following:

1. The perturbation $|U(\mathbf{r})|$ is small enough that there is no mixing between the bands (single-band case, $n = n'$).
2. $U(\mathbf{r})$ is slowly varying in $\mathbf{r}$. Thus $\tilde{U}_{\mathbf{k}}$ is falling off rapidly for large $\mathbf{k}$,

$$\left|\tilde{U}_{\mathbf{k}-\mathbf{k}'-\mathbf{G}}\right|_{\mathbf{G}\neq 0} \ll |\tilde{U}_{\mathbf{k}-\mathbf{k}'}| \tag{B.9}$$

   or equivalently, $u_{n\mathbf{k}}^* u_{n\mathbf{k}'} \simeq 1$ is assumed.
3. The integration over $\mathbf{k}$ is mainly contributed from $\mathbf{k} = \mathbf{k}_0$ (an extremum or zone center) since we are interested in the behavior of energy near $\mathbf{k}_0$. For convenience, we take $\mathbf{k}_0 = \mathbf{0}$.

The resultant equation for $a_n(\mathbf{k})$ becomes

$$\left(E_n(\mathbf{k}) - E\right)a_n(\mathbf{k}) + \int \frac{d^3k'}{(2\pi)^3}\tilde{U}_{\mathbf{k}-\mathbf{k}'}a_n(\mathbf{k}') = 0 \tag{B.10}$$

Define an envelope function $F(\mathbf{r})$

$$F(\mathbf{r}) = \int a_n(\mathbf{k}')e^{i\mathbf{k}'\cdot\mathbf{r}}\frac{d^3k'}{(2\pi)^3} \tag{B.11}$$

We obtain

$$\left[E_n(-i\nabla) + U(\mathbf{r})\right]F(\mathbf{r}) = EF(\mathbf{r}) \tag{B.12}$$

from the inverse Fourier transform of (B.10). The above equation is the effective mass equation in real space.

The total wave function is

$$\psi(\mathbf{r}) = \int a_n(\mathbf{k}')\psi_{n\mathbf{k}'}(\mathbf{r})\frac{d^3k'}{(2\pi)^3}$$

$$\simeq \int a_n(\mathbf{k}')e^{i\mathbf{k}'\cdot\mathbf{r}}u_{n\mathbf{k}_0}(\mathbf{r})\frac{d^3k'}{(2\pi)^3}$$

$$= F(\mathbf{r})u_{n\mathbf{k}_0}(\mathbf{r}) \tag{B.13}$$

using the leading order approximation for $u_{n\mathbf{k}}(\mathbf{r})$.

Using the $\mathbf{k}\cdot\mathbf{p}$ theory in Section 4.1, we have

$$E_n(\mathbf{k}) = E_n(0) + \sum_{\alpha,\beta} D^{\alpha\beta}k_\alpha k_\beta \tag{B.14}$$

Thus, the effective mass equation can also be written as

$$\left[\sum_{\alpha,\beta} D^{\alpha\beta}\left(-i\frac{\partial}{\partial x_\alpha}\right)\left(-i\frac{\partial}{\partial x_\beta}\right) + U(\mathbf{r})\right]F(\mathbf{r}) = [E - E_n(0)]F(\mathbf{r}) \tag{B.15}$$

where $D^{\alpha\beta}$ is given by (4.1.13). Alternatively, $D^{\alpha\beta} = (\hbar^2/2)(1/m^*)_{\alpha\beta}$, and we can write (B.15) as the effective mass equation (4.4.5).

B.2 DEGENERATE BANDS

To prove the effective mass theory (4.4.11) for degenerate bands [1], we choose a complete set of basis functions:

$$|n\,\mathbf{k}\rangle \equiv \chi_{n\mathbf{k}}(\mathbf{r}) \equiv e^{i\mathbf{k}\cdot\mathbf{r}}u_{n0}(\mathbf{r}) \tag{B.16}$$

Expand the wave function $\psi(\mathbf{r})$ for (4.4.9) in terms of these basis functions:

$$\psi(\mathbf{r}) = \sum_n \int \frac{d^3k}{(2\pi)^3}a_n(\mathbf{k})\chi_{n\mathbf{k}}(\mathbf{r}) \tag{B.17}$$

We proceed as before and obtain

$$\left[E_j(0) + \frac{\hbar^2k^2}{2m_0}\right]a_j(\mathbf{k}) + \sum_i \frac{\hbar}{m_0}\mathbf{k}\cdot\mathbf{p}_{ji}a_i(\mathbf{k}) + \int \frac{d^3k'}{(2\pi)^3}\tilde{U}_{\mathbf{k}-\mathbf{k}'}a_j(\mathbf{k}') = Ea_j(\mathbf{k}) \tag{B.18}$$

For the choice of the basis functions as in Section 4.3 for the Luttinger–Kohn Hamiltonian, $\mathbf{p}_{ji} = 0$ for all $i, j = 1, \ldots, 6$. That is why we have to go to a second-order perturbation theory.

Define

$$F_j(\mathbf{r}) = \int \frac{d^3 k}{(2\pi)^3} e^{i\mathbf{k}\cdot\mathbf{r}} a_j(\mathbf{k}) \tag{B.19}$$

We obtain in $\mathbf{k}$ space

$$\sum_{j'=1}^{6} \sum_{\alpha,\beta} \left(D_{jj'}^{\alpha\beta} k_\alpha k_\beta \right) a_{j'}(\mathbf{k}) + \int \frac{d^3 k'}{(2\pi)^3} \tilde{U}_{\mathbf{k}-\mathbf{k}'} a_j(\mathbf{k}') = \left[E - E_j(0) \right] a_j(\mathbf{k}) \tag{B.20}$$

Therefore, we obtain the effective mass equation in real space:

$$\sum_{j'=1}^{6} \left[\sum_{\alpha,\beta} D_{jj'}^{\alpha\beta} \left(-i \frac{\partial}{\partial x_\alpha} \right) \left(-i \frac{\partial}{\partial x_\beta} \right) + U(\mathbf{r}) \delta_{jj'} \right] F_{j'}(\mathbf{r}) = \left[E - E_j(0) \right] F_j(\mathbf{r}) \tag{B.21}$$

where

$$D_{jj'}^{\alpha\beta} = \frac{\hbar^2}{2m_0} \left[\delta_{jj'} \delta_{\alpha\beta} + \sum_{\gamma} \frac{p_{j\gamma}^\alpha p_{\gamma j'}^\beta + p_{j\gamma}^\beta p_{\gamma j'}^\alpha}{m_0 (E_0 - E_\gamma)} \right] \tag{B.22}$$

and the wave function in the effective-mass theory is given by

$$\psi(\mathbf{r}) = \sum_{j=1}^{6} F_j(\mathbf{r}) u_{j0}(\mathbf{r}) \tag{B.23}$$

for the leading-order approximation for $u_{j\mathbf{k}}(\mathbf{r}) \simeq u_{j0}(\mathbf{r})$. Notice that if $U(\mathbf{r}) \equiv 0$, the solution to (B.21) is indeed the plane wave function, that is, $F_j(\mathbf{r}) = a_j(\mathbf{k}) e^{i\mathbf{k}\cdot\mathbf{r}}$, $j = 1, 2, \ldots, 6$, where $\{a_j(\mathbf{k}), j = 1, \ldots, 6\}$ is an eigenvector of (4.3.16).

REFERENCE

1. J. M. Luttinger and W. Kohn, "Motion of electrons and holes in perturbed periodic fields," *Phys. Rev.* **97**, 869–883 (1955).

Appendix C

Derivations of the Pikus–Bir Hamiltonian for a Strained Semiconductor

In this appendix, we derive the Pikus–Bir Hamiltonian [1, 2] discussed in Section 4.5. Using the coordinate transformation between the uniformly deformed coordinates and the undeformed coordinates (4.5.2) and (4.5.4), we obtain

$$r'_i = r_i + \sum_{j=1}^{3} \varepsilon_{ij} r_j \tag{C.1a}$$

noting that both (x', y', z') and (x, y, z) are components using the same basis vectors $\hat{x}$, $\hat{y}$, and $\hat{z}$, where $r_1 = x$, $r_2 = y$, $r_3 = z$, and $r'_1 = x'$, $r'_2 = y'$, $r'_3 = z'$. For example, $x' = x + \varepsilon_{xx} x + \varepsilon_{xy} y + \varepsilon_{xz} z$. In a one-dimensional case with strain along the x direction only, we have $\varepsilon_{xx} = (x' - x)/x$ and $\varepsilon_{xy} = \varepsilon_{xz} = 0$. If a shear-type strain is introduced, in the x-y plane, we have $\varepsilon_{xy} = (x' - x)/y$, when $\varepsilon_{xx} = \varepsilon_{xz} = 0$. In vector form, we have

$$\mathbf{r}' = \mathbf{r} + \bar{\bar{\varepsilon}} \cdot \mathbf{r} = \left(1 + \bar{\bar{\varepsilon}}\right) \cdot \mathbf{r} \tag{C.1b}$$

The inverses of (C.1a) and (C.1b) are

$$r_i = r'_i - \sum_j \varepsilon_{ij} r'_j \tag{C.2a}$$

$$\mathbf{r} = \left(1 - \bar{\bar{\varepsilon}}\right) \cdot \mathbf{r}' \tag{C.2b}$$

In the uniformly deformed crystal, the potential is still periodic except that the function $V(\mathbf{r}')$ is a different potential from $V_0(\mathbf{r})$. We write the Schrödinger equation for the Bloch function in the deformed crystal coordinate system $\mathbf{r}'$:

$$\left[\frac{\mathbf{p}'^2}{2m_0} + V(\mathbf{r}')\right]\psi_{n\mathbf{k}'}(\mathbf{r}') = E_n(\mathbf{k}')\psi_{n\mathbf{k}'}(\mathbf{r}') \tag{C.3}$$

Using the chain rule and (C.2b), we have

$$\frac{\partial}{\partial r_i'} = \sum_j \frac{\partial r_j}{\partial r_i'} \frac{\partial}{\partial r_j} = \frac{\partial}{\partial r_i} - \sum_j \varepsilon_{ji} \frac{\partial}{\partial r_j} \tag{C.4a}$$

or

$$\mathbf{p}' = \mathbf{p} \cdot \left(1 - \bar{\bar{\varepsilon}}\right) \tag{C.4b}$$

Similarly,

$$p'^2 = p^2 - 2 \sum_{i,j} p_i \varepsilon_{ij} p_j \tag{C.5}$$

The periodic potential can be expanded as

$$V\left[\left(1 + \bar{\bar{\varepsilon}}\right) \cdot \mathbf{r}\right] = V_0(\mathbf{r}) + \sum_{i,j} V_{ij} \varepsilon_{ij} \tag{C.6}$$

where

$$V_{ij} = \left. \frac{\partial V}{\partial \varepsilon_{ij}} \right|_{\varepsilon_{ij} \to 0}$$

Therefore, Eq. (C.3) becomes

$$\left[H_0 + H_\varepsilon\right] \psi_{n\mathbf{k}'}\left[\left(1 + \bar{\bar{\varepsilon}}\right) \cdot \mathbf{r}\right] = E_n(\mathbf{k}') \psi_{n\mathbf{k}'}\left[\left(1 + \bar{\bar{\varepsilon}}\right) \cdot \mathbf{r}\right] \tag{C.7}$$

where

$$H_0 = \frac{p^2}{2m_0} + V_0(\mathbf{r}) \tag{C.8}$$

$$H_\varepsilon \equiv \sum_{\alpha, \beta} \hat{D}^{\alpha\beta} \varepsilon_{\alpha\beta} = \sum_{\alpha, \beta} \left(-\frac{1}{m_0} p_\alpha p_\beta + V_{\alpha\beta}\right) \varepsilon_{\alpha\beta} \tag{C.9}$$

Noting that we intend to treat H_ε as a perturbation due to strain, and H_0 contains $V_0(\mathbf{r})$ with a period equal to that of the undeformed crystal, we write the Bloch function as

$$\begin{aligned} \psi_{n\mathbf{k}'}\left[\left(1 + \bar{\bar{\varepsilon}}\right) \cdot \mathbf{r}\right] &= e^{i\mathbf{k}' \cdot \mathbf{r}'} u_{n\mathbf{k}'}(\mathbf{r}') \\ &= e^{i\mathbf{k}' \cdot (1 + \bar{\bar{\varepsilon}}) \cdot \mathbf{r}} u_{n\mathbf{k}'}\left[\left(1 + \bar{\bar{\varepsilon}}\right) \cdot \mathbf{r}\right] \\ &= e^{i\mathbf{k} \cdot \mathbf{r}} u_{n\mathbf{k}}^s(\mathbf{r}) \end{aligned} \tag{C.10}$$

where $\mathbf{k} = (1 + \bar{\bar{\varepsilon}}) \cdot \mathbf{k}'$ has been used, and we rename $u_{n\mathbf{k}'}[(1 + \bar{\bar{\varepsilon}}) \cdot \mathbf{r}]$ as $u_{n\mathbf{k}}^s(\mathbf{r})$, for the strained Bloch periodic part, which is to be determined.

Substituting (C.10) into (C.7), we obtain

$$H_0\left[e^{i\mathbf{k}\cdot\mathbf{r}}u_{n\mathbf{k}}^s(\mathbf{r})\right] = e^{i\mathbf{k}\cdot\mathbf{r}}\left(H_0 + \frac{\hbar}{m_0}\mathbf{k}\cdot\mathbf{p} + \frac{\hbar^2 k^2}{2m_0}\right)u_{n\mathbf{k}}^s(\mathbf{r}) \quad \text{(C.11)}$$

$$H_\varepsilon\left[e^{i\mathbf{k}\cdot\mathbf{r}}u_{n\mathbf{k}}^s(\mathbf{r})\right] = e^{i\mathbf{k}\cdot\mathbf{r}}\left(H_\varepsilon - \frac{2\hbar}{m_0}\sum_{\alpha,\beta}k_\alpha\varepsilon_{\alpha\beta}P_\beta\right)u_{n\mathbf{k}}^s(\mathbf{r}) \quad \text{(C.12)}$$

Therefore, we have the equation for the wave function $u_{n\mathbf{k}}^s(\mathbf{r})$:

$$[H_0 + H']u_{n\mathbf{k}}^s(\mathbf{r}) = \left(E - \frac{\hbar^2 k^2}{2m_0}\right)u_{n\mathbf{k}}^s(\mathbf{r}) \quad \text{(C.13a)}$$

$$H' = H_k + H_\varepsilon + H_{\varepsilon k} \quad \text{(C.13b)}$$

$$H_k = \frac{\hbar}{m_0}\mathbf{k}\cdot\mathbf{p} \quad \text{(C.13c)}$$

$$H_\varepsilon = \sum_{\alpha,\beta}\hat{D}^{\alpha\beta}\varepsilon_{\alpha\beta} \quad \text{(C.13d)}$$

$$H_{\varepsilon k} \equiv -2\frac{\hbar}{m_0}\sum_{\alpha,\beta}k_\alpha\varepsilon_{\alpha\beta}P_\beta \quad \text{(C.13e)}$$

The above result is similar to that in the original paper of Pikus and Bir, if we note that $\mathbf{k}' = (1 - \bar{\bar{\varepsilon}})\cdot\mathbf{k}$. Our conventions are that primes $\mathbf{k}'$ and $\mathbf{r}'$ are used for the deformed crystal and $\mathbf{k}$ and $\mathbf{r}$ are used for the undeformed crystal. Since the perturbation term H_k vanishes in the first order $(H_k)_{nm} = 0$ because of parity consideration, we seek perturbation terms second order in k and first order in $\varepsilon_{\alpha\beta}$.

C.1   SINGLE-BAND CASE

For example, a single conduction band is probably the simplest case. The perturbation theory for the single subband leads to

$$u_{n\mathbf{k}}^s(\mathbf{r}) \simeq u_{n0}(\mathbf{r})$$
$$H_0 u_{n0}(\mathbf{r}) = E_n(0)u_{n0}(\mathbf{r})$$

$$E = E_n(0) + \frac{\hbar^2 k^2}{2m_0} + (H_\varepsilon)_{nn} + \frac{\hbar^2}{m_0^2}\sum_{n'\neq n}\frac{(\mathbf{k}\cdot\mathbf{p})_{nn'}(\mathbf{k}\cdot\mathbf{p})_{n'n}}{E_n - E_{n'}}$$

$$+ \frac{\hbar}{m_0}\sum_{n'\neq n}\frac{(\mathbf{k}\cdot\mathbf{p})_{nn'}(H_\varepsilon)_{n'n} + (H_\varepsilon)_{nn'}(\mathbf{k}\cdot\mathbf{p})_{n'n}}{E_n - E_{n'}} \quad \text{(C.14a)}$$

$$= E_n(0) + (H_\varepsilon)_{nn} + \sum_{\alpha,\beta}D^{\alpha\beta}k_\alpha k_\beta \quad \text{(C.14b)}$$

where $D^{\alpha\beta}$ is the same as (4.1.13). We have used the fact that the term in the perturbation theory

$$\sum_{n' \neq n} \frac{H'_{nn'} H'_{n'n}}{E_n - E_{n'}}$$

where $H' = (\hbar/m_0)\mathbf{k} \cdot \mathbf{p} + H_\varepsilon + H_{\varepsilon k}$, has nonzero contribution only from the first term $(\hbar/m_0)\mathbf{k} \cdot \mathbf{p}$ because of parity consideration, and we keep terms up to the second order in k and the first order in ε. The diagonal term $\langle n|H_\varepsilon|n \rangle$ leads to

$$\langle n|H_\varepsilon|n \rangle = a_c(\varepsilon_{xx} + \varepsilon_{yy} + \varepsilon_{zz}) \tag{C.15}$$

because of the isotropic nature of the conduction band edge. Therefore, the single-band (conduction band) strained dispersion relation is given by

$$E = E_n(0) + \frac{\hbar^2}{2} \sum_{\alpha,\beta} \left(\frac{1}{m^*}\right)_{\alpha\beta} k_\alpha k_\beta + a_c(\varepsilon_{xx} + \varepsilon_{yy} + \varepsilon_{zz}) \tag{C.16}$$

and

$$u^s_{nk}(\mathbf{r}) = u_{n0}(\mathbf{r}) \tag{C.17}$$

C.2   DEGENERATE BANDS

We assume in general that

$$u^s_{nk}(\mathbf{r}) = \sum_j^A a_j(\mathbf{k}) u_{j0}(\mathbf{r}) + \sum_\gamma^B a_\gamma(\mathbf{k}) u_{\gamma 0}(\mathbf{r}) \tag{C.18}$$

as in Section 4.4, where class A consists of those bands of interest, such as the degenerate valence bands: two heavy-hole, two light-hole and two spin-split-off bands. Class B consists of those bands outside of class A. We follow the same procedures in Löwdin's perturbation method presented in Section 3.6 and obtain

$$U^A_{jj'} = H_{jj'} + \sum_{\gamma \neq jj'}^B \frac{H'_{j\gamma} H'_{\gamma j'}}{E_0 - E_\gamma} \tag{C.19a}$$

$$\sum_{j'}^A \left(U^A_{jj'} - E\delta_{jj'}\right) a_{j'}(\mathbf{k}) = 0 \tag{C.19b}$$

from (3.6.12) and (3.6.5), respectively. The result turns out to be almost the same as that in the Luttinger–Kohn Hamiltonian:

$$U_{jj'}^A = \delta_{jj'} E_j(0) + \sum_{\alpha, \beta} D_{jj'}^{\alpha\beta} k_\alpha k_\beta$$

$$+ \sum_{\alpha, \beta} \hat{D}_{jj'}^{\alpha\beta} \varepsilon_{\alpha\beta} + \frac{\hbar}{m_0} \sum_\gamma^B \frac{\mathbf{k} \cdot \mathbf{p}_{j\gamma}(H_\varepsilon)_{\gamma j'} + (H_\varepsilon)_{j\gamma} \mathbf{k} \cdot \mathbf{p}_{\gamma j'}}{E_0 - E_\gamma} \quad \text{(C.20)}$$

or

$$H = H^{LK} + H_\varepsilon \quad \text{(C.21)}$$

The first two terms in (C.20), $E_j(0)\delta_{jj'} + \sum_{\alpha, \beta} D_{jj'}^{\alpha\beta} k_\alpha k_\beta \equiv H_{jj'}^{LK}$, are exactly the elements of the Luttinger–Kohn Hamiltonian, and have been expressed in matrix form in (4.3.14). The third term in (C.20)

$$(H_\varepsilon)_{jj'} = \sum_{\alpha, \beta} \hat{D}_{jj'}^{\alpha\beta} \varepsilon_{\alpha\beta} \quad \text{(C.22)}$$

is the linear strain Hamiltonian. The last term in (C.20) vanishes for a lattice of diamond type, in which there is a center of inversion, and either $p_{j\gamma}$ or $(H_\varepsilon)_{\gamma j}$ vanishes. For crystals that do not have a center of inversion, this term may not be zero. It means that an extremum in k space can be shifted due to deformation. However, this term is considered negligible for most III–V compounds. We can also express the third term due to strain by identifying

$$k_\alpha k_\beta \leftrightarrow \varepsilon_{\alpha\beta} \quad \text{(C.23a)}$$

Therefore [3],

$$\frac{\hbar^2 \gamma_1}{2m_0} \leftrightarrow D_v^d = -a_v \quad \text{(C.23b)}$$

$$\frac{\hbar^2 \gamma_2}{2m_0} \leftrightarrow \frac{D_u}{3} = -\frac{b}{2} \quad \text{(C.23c)}$$

$$\frac{\hbar^2 \gamma_3}{2m_0} \leftrightarrow \frac{D_u'}{3} = \frac{-d}{2\sqrt{3}} \quad \text{(C.23d)}$$

where the constants D_v^d, D_u, and D_u' come from different components of $\hat{D}_{jj'}^{\alpha\beta}$. In matrix form, we have [3, 4]

$$
H = - \begin{bmatrix}
P+Q & -S & R & 0 & -\dfrac{1}{\sqrt{2}}S & \sqrt{2}R \\[6pt]
-S^+ & P-Q & 0 & R & -\sqrt{2}Q & \sqrt{\tfrac{3}{2}}S \\[6pt]
R^+ & 0 & P-Q & S & \sqrt{\tfrac{3}{2}}S^+ & \sqrt{2}Q \\[6pt]
0 & R^+ & S^+ & P+Q & -\sqrt{2}R^+ & -\dfrac{1}{\sqrt{2}}S^+ \\[6pt]
-\dfrac{1}{\sqrt{2}}S^+ & -\sqrt{2}Q & \sqrt{\tfrac{3}{2}}S & -\sqrt{2}R & P+\Delta & 0 \\[6pt]
\sqrt{2}R^+ & \sqrt{\tfrac{3}{2}}S^+ & \sqrt{2}Q & -\dfrac{1}{\sqrt{2}}S & 0 & P+\Delta
\end{bmatrix}
\begin{matrix}
|\tfrac{3}{2},\tfrac{3}{2}\rangle \\[6pt]
|\tfrac{3}{2},\tfrac{1}{2}\rangle \\[6pt]
|\tfrac{3}{2},-\tfrac{1}{2}\rangle \\[6pt]
|\tfrac{3}{2},-\tfrac{3}{2}\rangle \\[6pt]
|\tfrac{1}{2},\tfrac{1}{2}\rangle \\[6pt]
|\tfrac{1}{2},-\tfrac{1}{2}\rangle
\end{matrix}
$$

$$\text{(C.24)}$$

where

$$
\begin{aligned}
P = P_k + P_\varepsilon \qquad & Q = Q_k + Q_\varepsilon \\
R = R_k + R_\varepsilon \qquad & S = S_k + S_\varepsilon
\end{aligned}
\qquad \text{(C.25a)}
$$

$$
P_k = \left(\frac{\hbar^2}{2m_0}\right)\gamma_1\left(k_x^2 + k_y^2 + k_z^2\right)
$$

$$
Q_k = \left(\frac{\hbar^2}{2m_0}\right)\gamma_2\left(k_x^2 + k_y^2 - 2k_z^2\right)
$$

$$
R_k = \left(\frac{\hbar^2}{2m_0}\right)\sqrt{3}\left[-\gamma_2\left(k_x^2 - k_y^2\right) + 2i\gamma_3 k_x k_y\right]
$$

$$
S_k = \left(\frac{\hbar^2}{2m_0}\right)2\sqrt{3}\,\gamma_3(k_x - ik_y)k_z
\qquad \text{(C.25b)}
$$

$$
P_\varepsilon = -a_v(\varepsilon_{xx} + \varepsilon_{yy} + \varepsilon_{zz}) \qquad Q_\varepsilon = -\frac{b}{2}(\varepsilon_{xx} + \varepsilon_{yy} - 2\varepsilon_{zz})
$$

$$
R_\varepsilon = \frac{\sqrt{3}}{2}b(\varepsilon_{xx} - \varepsilon_{yy}) - id\varepsilon_{xy} \qquad S_\varepsilon = -d(\varepsilon_{xz} - i\varepsilon_{yz})
\qquad \text{(C.25c)}
$$

where the wave vector $\mathbf{k}$ is interpreted as a differential operator $-i\nabla$; ε_{ij} is the symmetric strain tensor; γ_1, γ_2, and γ_3 are the Luttinger inverse mass parameters; a_v, b, and d are the Pikus–Bir deformation potentials; and Δ is the spin–orbit split-off energy. The basis function $|j, m\rangle$ denotes the Bloch wave function at the zone center and is listed in Eq. (4.3.3). Here the energy zero is taken to be at the top of the unstrained valence band. The Hamiltonian (C.24) has been used extensively to study the strain effects on the band structures of semiconductors.

REFERENCES

1. G. E. Pikus and G. L. Bir, "Effects of deformation on the hole energy spectrum of germanium and silicon," *Sov. Phys.–Solid State* **1**, 1502–1517 (1960).

2. G. L. Bir and G. E. Pikus, *Symmetry and Strain-Induced Effects in Semiconductors*, Wiley, New York, 1974.

3. S. L. Chuang, "Efficient band-structure calculations of strained quantum wells using a two-by-two Hamiltonian," *Phys. Rev. B* **43**, 9649–9661 (1991). (Note that a_v in this paper should be taken as $-a_v$ to compare with data in the literature.)

4. C. Y. P. Chao and S. L. Chuang, "Spin–orbit-coupling effects on the valence-band structure of strained semiconductor quantum wells," *Phys. Rev. B.* **46**, 4110–4122 (1992).

Appendix D

Semiconductor Heterojunction Band Lineups in the Model-Solid Theory [1]

Semiconductor heterojunctions and superlattices have been under intensive investigation both theoretically and experimentally for the past three decades. The potential device applications using heterojunctions are tremendous. In this appendix, we discuss a simplified model to determine the energy band lineups of semiconductor heterojunctions based on the model-solid theory [1–4]. The goal is to develop a reliable model to predict band offsets for a wide variety of heterojunctions without the need for difficult calculations such as in the local-density-functional theory or ab initio pseudopotential method. The relation of the model-solid theory to the fully self-consistent first-principles calculations can be found in Refs. 2 and 3.

The major idea is to set up an absolute reference energy level. All calculated energies can then be put on an absolute energy scale, allowing us to derive band lineups. In the model-solid theory, an average energy over the three uppermost valence bands (the heavy-hole, the light-hole, and the spin–orbit split-off bands) $E_{v,av}$ is obtained from theory and is referred to as the absolute energy level. The values of $E_{v,av}$ for different semiconductors are usually tabulated [1] (Table K.2 in Appendix K) so that no calculations for these values are necessary. These results should be compared with those of the first-principle calculations whenever possible to justify the model. An estimate of the maximum possible error is about 0.1 eV. Band offsets should be checked with experimental data such as those in Refs. 5–19. The model-solid theory provides a simple guideline for estimating the band offsets for materials, especially ternary compounds with varying compositions for which experimental data may not be always available.

D.1 UNSTRAINED SEMICONDUCTORS

If materials A and B have the same lattice constants, we may have an ideal heterojunction and there is no strain in the semiconductors. For this case, the heavy-hole and light-hole band edges (E_{HH} and E_{LH}) are degenerate at the zone center, and their energy position is denoted as E_v:

$$E_v = E_{v,av} + \frac{\Delta}{3} \tag{D.1}$$

where Δ is the spin–orbit splitting energy, and the spin–orbit split-off band-edge energy E_{SO} is

$$E_{SO} = E_v - \Delta = E_{v,av} - \frac{2\Delta}{3} \qquad (D.2)$$

The conduction band edge is obtained by adding the band-gap energy E_g to E_v:

$$E_c = E_v + E_g \qquad (D.3)$$

Note that in the model-solid theory, the spin–orbit splitting energy Δ and the band-gap energy E_g are taken from experimental results. The only input provided by the model-solid theory is the tabulated $E_{v,av}$ value. This $E_{v,av}$ value is essentially the same as the p-state energy E_p in Fig. 4.3a.

With the above results, the band lineups between materials A and B are shown in Fig. D.1. We have

$$\Delta E_g = E_g^A - E_g^B$$

and the band-edge discontinuities are

$$\Delta E_c = E_c^A - E_c^B \qquad \Delta E_v = E_y^B - E_v^A \qquad (D.4)$$

$$\Delta E_c + \Delta E_v = \Delta E_g \qquad (D.5)$$

The partition ratios of the band-edge discontinuities, $Q_c = \Delta E_c / \Delta E_g$ and $Q_v = \Delta E_v / \Delta E_g$, are obtained from this theory and can also be compared with experimental data.

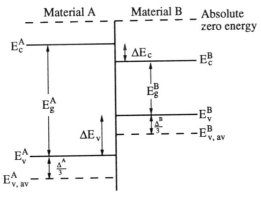

Figure D.1. Band lineups in the model-solid theory. $E_{v,av}$ in each material region is obtained from the model-solid theory and is tabulated in Appendix K. The bandgap energy E_g and the spin–orbit splitting Δ of each material are taken from experimental results.

D.2   STRAINED SEMICONDUCTORS

If a material A with a lattice constant a is grown on a substrate with a lattice constant a_0 along the z direction, we have

$$\varepsilon_{xx} = \varepsilon_{yy} = \frac{a_0 - a}{a} \tag{D.6a}$$

and

$$\varepsilon_{zz} = -2\frac{C_{12}}{C_{11}}\varepsilon_{xx} \tag{D.6b}$$

The band-edge shifts are

$$\Delta E_{v,\,\mathrm{av}} = a_v(\varepsilon_{xx} + \varepsilon_{yy} + \varepsilon_{zz}) \equiv -P_\varepsilon \tag{D.7a}$$
$$\Delta E_c = a_c(\varepsilon_{xx} + \varepsilon_{yy} + \varepsilon_{zz}) \equiv P_c \tag{D.7b}$$

The position of the average energy of the valence bands $E_{v,\,\mathrm{av}}$ under strain is shifted from its unstrained position $E^0_{v,\,\mathrm{av}}$ in (D.1) by $-P_\varepsilon$:

$$E_{v,\,\mathrm{av}} = E^0_{v,\,\mathrm{av}} - P_\varepsilon \tag{D.8}$$

We thus have the center of the valence-band-edge energy

$$E_v = E_{v,\,\mathrm{av}} + \frac{\Delta}{3} = E^0_v - P_\varepsilon \tag{D.9}$$

The heavy-hole, light-hole, and spin–orbit split-off band edges are

$$E_{\mathrm{HH}} = E^0_v - P_\varepsilon - Q_\varepsilon \tag{D.10}$$
$$E_{\mathrm{LH}} = E^0_v - P_\varepsilon - \frac{\Delta}{2} + \frac{Q_\varepsilon}{2} + \frac{1}{2}[\Delta^2 + 2\Delta Q_\varepsilon + 9Q_\varepsilon^2]^{1/2} \tag{D.11}$$
$$E_{\mathrm{SO}} = E^0_v - P_\varepsilon - \frac{\Delta}{2} + \frac{Q_\varepsilon}{2} - \frac{1}{2}[\Delta^2 + 2\Delta Q_\varepsilon + 9Q_\varepsilon^2]^{1/2} \tag{D.12}$$

The conduction band edge is shifted by P_c given by (D.7b):

$$E_c = E^0_v + E_g(x) + P_c \tag{D.13}$$

Note that in the limit of a large spin–orbit split-off energy $\Delta \gg |Q_\varepsilon|$, we can ignore the coupling of the spin–orbit split-off band and

$$E_{\mathrm{LH}} \simeq E^0_v - P_\varepsilon + Q_\varepsilon \tag{D.14a}$$
$$E_{\mathrm{SO}} \simeq E^0_v - P_\varepsilon - \Delta \tag{D.14b}$$

For a ternary alloy such as $A_x B_{1-x} C$ with a lattice constant $a(x)$,

$$a(x) = xa(AC) + (1 - x)a(BC) \qquad (D.15)$$

which is a linear interpolation of the lattice constants $a(AC)$ and $a(BC)$ of the binary compound semiconductors, we use the following formula to calculate an energy level $E \; (= E^0_{v,\,av}$ for example):

$$E(A_x B_{1-x} C) = xE(AC) + (1 - x)E(BC)$$
$$+ 3x(1 - x)[-a_v(AC) + a_v(BC)]\frac{\Delta a}{a_0} \qquad (D.16)$$

where the last term accounts for a *strain* contribution to the ternary alloy, and $\Delta a = a(AC) - a(BC)$ is the different between the lattice constants of two compounds AC and BC. Once $E^0_{v,\,av}$ is determined the band-edge energies for the strained ternary compound can be calculated following (D.6)–(D.13).

Many theoretical parameters for the electronic and optical properties such as those listed in Table K.2 in Appendix K can be found in the data books compiled by various groups (such as Refs. 20–23), review papers (such as Refs. 24–27), and research papers (Refs. 28–37).

Example Ga As / Al As Heterojunction   GaAs and AlAs have almost the same lattice constants. Therefore, the heterojunction has a negligible strain. We see from Table K.2 in Appendix K that

$$E_{v,\,av}(\text{GaAs}) = -6.92 \text{ eV}, \Delta(\text{GaAs}) = 0.34 \text{ eV}, E_g(\text{GaAs}) = 1.52 \text{ eV}$$

$$E_{v,\,av}(\text{AlAs}) = -7.49 \text{ eV}, \Delta(\text{AlAs}) = 0.28 \text{ eV}, E_g^{\text{dir}}(\text{AlAs}) = 3.13 \text{ eV}$$

Therefore,

$$E_v(\text{GaAs}) = -6.92 + \frac{0.34}{3} = -6.81 \text{ eV}$$

$$E_v(\text{AlAs}) = -7.49 + \frac{0.28}{3} = -7.40 \text{ eV}$$

$$\Delta E_v = -6.81 + 7.40 = 0.59 \text{ eV}$$

Also, the band-gap discontinuity is $\Delta E_g = 1.61$ eV, and the valence-band discontinuity ratio is $\Delta E_v / \Delta E_g = 0.37$.   ∎

Example   *In$_{0.53}$Ga$_{0.47}$As/InP Heterojunction*

$$a(\text{GaAs}) = 5.6533 \text{ Å}, \, a(\text{InAs}) = 6.0584 \text{ Å}, \, a(\text{InP}) = 5.8688 \text{ Å}$$

$$E_{v,\text{av}}(\text{In}_{1-x}\text{Ga}_x\text{As}) = xE_{v,\text{av}}(\text{GaAs}) + (1-x)E_{v,\text{av}}(\text{InAs})$$

$$+ 3x(1-x)\left[-a_v(\text{GaAs}) + a_v(\text{InAs})\right]\frac{\Delta a}{a}$$

$$\Delta a = 5.6533 - 6.0584 = -0.4051 \text{ Å}$$

$$a(\text{In}_{1-x}\text{Ga}_x\text{As}) = xa(\text{GaAs}) + (1-x)a(\text{InAs})$$

$$\Delta(\text{In}_{1-x}\text{Ga}_x\text{As}) \simeq x\Delta(\text{GaAs}) + (1-x)\Delta(\text{InAs})$$

For $x = 0.47$, $\text{In}_{0.53}\text{Ga}_{0.47}\text{As}$ is lattice matched to InP. Therefore, we do not have the strain terms ($P_\varepsilon = 0, Q_\varepsilon = 0$). We obtain

$$E_{v,\text{av}}(\text{In}_{0.53}\text{Ga}_{0.47}\text{As}) = -6.779 \text{ eV}, \, \Delta = 0.361 \text{ eV}$$

Using $E_{v,\text{av}}(\text{InP}) = -7.04$ eV and $E_v(\text{InP}) = -7.003$ eV, we find $\Delta E_v = 0.344$ eV. From room temperature data for the band gap, $E_g(\text{In}_{0.53}\text{Ga}_{0.47}\text{As}) = 0.73$ eV, $E_g(\text{InP}) = 1.35$ eV, and $\Delta E_g = 0.62$ eV, we obtain the ratio $\Delta E_v / \Delta E_g = 0.55 = 55\%$.   ∎

D.3   SOME EXPERIMENTAL REPORTS ON BAND-EDGE DISCONTINUITIES

There has been a considerable amount of experimental data on band offsets, mostly on unstrained systems. For strained semiconductors, the band offsets are complicated by the deformation potentials, which also shift the conduction and valence-band edges. Therefore, fewer data are available for strained heterojunctions.

1. GaAs/Al$_x$Ga$_{1-x}$As system

$$E_g(\text{GaAs}) = 1.424 \text{ eV}   (300 \text{ K})$$

$$E_g(\text{Al}_x\text{Ga}_{1-x}\text{As}) = 1.424 + 1.247x \text{ eV}   (300 \text{ K})$$

$$\Delta E_g(x) = 1.247x \text{ eV}$$

$$\Delta E_c = 0.67\Delta E_g \qquad \Delta E_v = 0.33 \, \Delta E_g$$

$$(\Delta E_c = 0.69 \, \Delta E_g, \Delta E_v \cong 0.31 \, \Delta E_g, \text{Ref. 11})$$

2. $In_{0.53}Ga_{0.47}As / InP$ ($\sim$ 0 K) [18]

$$E_g(InP) = 1.423 \text{ eV}$$

$$E_g(In_{0.53}Ga_{0.47}As) = 0.811 \text{ eV}$$

$$\Delta E_g = 0.612 \text{ eV}$$

$$\Delta E_c = 0.252 \text{ eV} = 0.41 \, \Delta E_g,$$

$$\Delta E_v = 0.360 \text{ eV} = 0.59 \, \Delta E_g$$

3. $In_{0.52}Al_{0.48}As / InP$ ($\sim$ 0 K) [18]

$$E_g(InP) = 1.423 \text{ eV}$$

$$E_g(In_{0.52}Al_{0.48}As) = 1.511 \text{ eV}$$

$$\Delta E_g = 0.088 \text{ eV}$$

$$\Delta E_c = 0.252 \text{ eV} = 2.86 \, \Delta E_g,$$

$$\Delta E_v = -0.164 \text{ eV} \quad (\text{Type II})$$

The above results for $In_{0.53}Ga_{0.47}As / InP / In_{0.52}Al_{0.48}As$ band offsets and their transitivity relations are illustrated [18] in Fig. D.2. The transitivity relations give $\Delta E_c = 0.504$ eV $= 0.72 \, \Delta E_g$, and $\Delta E_v = 0.196$ eV $= 0.28 \, \Delta E_g$ for an $In_{0.53}Ga_{0.47}As / In_{0.52}Al_{0.48}As$ heterojunction.

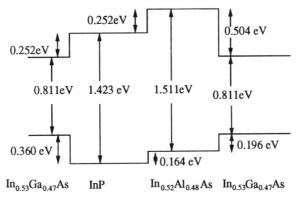

Figure D.2. Band offsets and transitivity of $In_{0.53}GA_{0.47}As / InP$ ($\Delta E_c = 0.41 \, \Delta E_g$, $\Delta E_v = 0.59 \, \Delta E_g$) and $In_{0.52}Al_{0.48}As / InP$ ($\Delta E_c = 0.252$ eV $= 2.86 \, \Delta E_g > \Delta E_g = 0.088$ eV) at low temperatures (0 K) [18].

4. $In_{1-x}Ga_xAs_yP_{1-y}/InP$ lattice-matched system [22]

For $In_{1-x}GA_xAs_yP_{1-y}$ quaternary semiconductor lattice matched to InP substrate,

$$x = \frac{0.1896y}{0.4176 - 0.0125y}$$

$$E_g = \left(In_{1-x}Ga_xAs_yP_{1-y}\right) = 1.35 - 0.775y + 0.149y^2 \text{ eV}$$

$$\Delta E_g(y) = 0.775y - 0.149y^2 \text{ eV}$$

$$\Delta E_v(y) = 0.502y - 0.152y^2 \text{ eV}$$

$$\Delta E_c(y) = \Delta E_g(y) - \Delta E_v(y) = 0.273y + 0.003y^2 \text{ eV}$$

where $\Delta E_v(y)$ was determined experimentally. ∎

Some reports on strained $In_xGa_{1-x}As/InP$, $In_xGa_{1-x}As/In_{0.52}Al_{0.48}As$, InGaAs/InGaAsP, and $In_xGa_{1-x}As/GaAs$ can be found in Refs. 7–9, 14, 15, and 19.

REFERENCES

1. C. G. Van de Walle, "Band lineups and deformation potentials in the model-solid theory," *Phys. Rev. B* **39**, 1871–1883 (1989).

2. C. G. Van de Walle and R. M. Martin, "Theoretical calculations of semiconductor heterojunction discontinuities," *J. Vac. Sci. Technol. B* **4**, 1055–1059 (1986).

3. C. G. Van de Walle and R. M. Martin, "Theoretical study of band offsets at semiconductor interfaces," *Phys. Rev. B* **35**, 8154–8165 (1987).

4. C. G. Van de Walle, K. Shahzad, and D. J. Olego, "Strained-layer interfaces between II–VI compound semiconductors," *J. Vac. Sci. Technol. B* **6**, 1350–1353 (1988).

5. R. People, K. W. Wecht, K. Alavi, and A. Y. Cho, "Measurement of the conduction-band discontinuity of molecular beam epitaxial grown $In_{0.52}Al_{0.48}As/In_{0.53}Ga_{0.47}As$, *N-n* heterojunction by *C-V* profiling," *Appl. Phys. Lett.* **43**, 118–120 (1983).

6. R. C. Miller, A. C. Gossard, D. A. Kleinman, and O. Munteanu, "Parabolic quantum wells with the GaAs-$Al_xGa_{1-x}As$ system," *Phys. Rev. B* **29**, 3740–3743 (1984).

7. R. People, "Effects of coherency strain on the band gap of pseudomorphic $In_xGa_{1-x}As$ on (001)InP," *Appl. Phys. Lett.* **50**, 1604–1606 (1987).

8. R. People, "Band alignments for pseudomorphic InP/$In_xGa_{1-x}As$ heterostructures for growth on (001)InP," *J. Appl. Phys.* **62**, 2551–2553 (1987).

9. G. Ji, D. Huang, U. K. Reddy, T. S. Henderson, R. Houdré, and H. Morkoç, "Optical investigation of highly strained InGaAs–GaAs multiple quantum wells," *J. Appl. Phys.* **62**, 3366–3373 (1987).

10. C. D. Lee and S. R. Forrest, "Effects of lattice mismatch on $In_xGa_{1-x}As/InP$ heterojunctions," *Appl. Phys. Lett.* **57**, 469–471 (1990).

11. L. Hrivnák, "Determination of Γ electron and light hole effective masses in $Al_xGa_{1-x}As$ on the basis of energy gaps, band-gap offsets, and energy levels in $Al_xGa_{1-x}As/GaAs$ quantum wells," *Appl. Phys. Lett.* **56**, 2425–2427 (1990).

12. B. R. Nag and S. Mukhopadhyay, "Band offset in $InP/Ga_{0.47}In_{0.53}As$ heterostructures," *Appl. Phys. Lett.* **58**, 1056–1058 (1991).

13. M. S. Hybertsen, "Band offset transitivity at the $InGaAs/InAlAs/InP(001)$ heterointerfaces," *Appl. Phys. Lett.* **58**, 1759–1761 (1991).

14. B. Jogai, "Valence-band offset in strained $GaAs–In_xGa_{1-x}As$ superlattices," *Appl. Phys. Lett.* **59**, 1329–1331 (1991).

15. J. H. Huang, T. Y. Chang, and B. Lalevic, "Measurement of the conduction-band discontinuity in pseudomorphic $In_xGa_{1-x}As/In_{0.52}Al_{0.48}As$ heterostructures," *Appl. Phys. Lett.* **60**, 733–735 (1992).

16. S. Tiwari and D. J. Frank, "Empirical fit to band discontinuities and barrier heights in III–V alloy systems," *Appl. Phys. Lett.* **60**, 630–632 (1992).

17. R. F. Kopf, M. H. Herman, M. L. Schnoes, A. P. Perley, G. Livescu, and M. Ohring, "Band offset determination in analog graded parabolic and triangular quantum wells of GaAs/AlGaAs and GaInAs/AlInAs," *J. Appl. Phys.* **71**, 5004–5011 (1992).

18. J. Bohrer, A. Krost, T. Wolf, and D. Bimberg, "Band offsets and transitivity of $In_xGa_{1-x}As/In_{1-y}Al_yAs/InP$ heterostructures," *Phys. Rev. B* **47**, 6439–6443 (1993).

19. M. Nido, K. Naniwae, T. Terakado, and A. Suzuki, "Band-gap discontinuity control for InGaAs/InGaAsP multiquantum-well structures by tensile-strained barriers," *Appl. Phys. Lett.* **62**, 2716–2718 (1993).

20. K. H. Hellwege, Ed., *Landolt-Börnstein Numerical Data and Functional Relationships in Science and Technology*, New Series, Group III **17a**, Springer, Berlin, 1982; Groups III–V **22a**, Springer, Berlin, 1986.

21. For a brief version of the data book in Ref. 20, see: O. Madelung, Ed., *Semiconductors, Group IV Elements and III–V Compounds*, in R. Poerschke, Ed., *Data in Science and Technology*, Springer, Berlin, 1991.'

22. S. Adachi, *Physical Properties of III–V Semiconductor Compounds*, Wiley, New York, 1992.

23. S. Adachi, *Properties of Indium Phosphide*, INSPEC, The Institute of Electrical Engineers, London, 1991.

24. J. S. Blakemore, "Semiconducting and other major properties of gallium arsenide," *J. Appl. Phys.* **53**, R123–R181 (1982).

25. S. Adachi, "Material parameters of $In_xGa_{1-x}As_yP_{1-y}$ and related binaries," *J. Appl. Phys.* **53**, 8775–8792 (1982).

26. S. Adachi and K. Oe, "Internal strain and photoelastic effects in $Ga_{1-x}Al_xAs/GaAs$ and $In_xGa_{1-x}As_yP_{1-y}/InP$ crystals," *J. Appl. Phys.* **54**, 6620–6627 (1983).

27. S. Adachi, "GaAs, AlAs, and $Ga_{1-x}Al_xAs$: Material parameters for use in research and device applications," *J. Appl. Phys.* **58**, R1–R28 (1985).

28. P. Lawaetz, "Valence-band parameters in cubic semiconductors," *Phys. Rev. B* **4**, 3460–3467 (1971).

29. R. E. Nahory, M. A. Pollack, and W. D. Johnston, Jr., "Band gap versus composition and demonstration of Vegard's law for $In_{1-x}Ga_xAs_yP_{1-y}$ lattice matched to InP," *Appl. Phys. Lett.* **33**, 659–661 (1978).

30. K. Alavi and R. L. Aggarwal, "Interband magnetoabsorption of $In_{0.53}Ga_{0.47}As$," *Phys. Rev. B* **21**, 1311–1315 (1980).

31. A. Raymond, J. L. Robert, and C. Bernard, "The electron effective mass in heavily doped GaAs," *J. Phys. C: Solid State Phys.* **12**, 2289–2293 (1979).

32. L. G. Shantharama, A. R. Adams, C. N. Ahmad, and R. J. Nicholas, "The $\mathbf{k} \cdot \mathbf{p}$ interaction in InP and GaAs from the band-gap dependence of the effective mass," *J. Phys. C: Solid State Phys.* **17**, 4429–4442 (1984).

33. W. Stolz, J. C. Maan, M. Altarelli, L. Tapfer, and K. Ploog, "Absorption spectroscopy on $Ga_{0.47}In_{0.53}As/Al_{0.48}In_{0.52}As$ multi-quantum-well hetero-structures. I. Excitonic transitions," *Phys. Rev. B* **36**, 4301–4309 (1987).

34. W. Stolz, J. C. Maan, M. Altarelli, L. Tapfer, and K. Ploog, "Absorption spectroscopy on $Ga_{0.47}In_{0.53}As/Al_{0.48}In_{0.52}As$ multi-quantum-well hetero-structures; II: Subband structure," *Phys. Rev. B* **36**, 4310–4315 (1987).

35. L. W. Molenkamp, R. Eppenga, G. W. 't Hooft, P. Dawson, C. T. Foxon, and K. J. Moore, "Determination of valence-band effective-mass anisotropy in GaAs quantum wells by optical spectroscopy," *Phys. Rev. B* **38**, 4314–4317 (1988).

36. D. Gershoni and H. Temkin, "Optical properties of III–V strained-layer quantum wells," *J. Luminescence* **44**, 381–398 (1989).

37. R. Sauer, S. Nilsson, P. Roentgen, W. Heuberger, V. Graf, A. Hangleiter, and R. Spycher, "Optical study of extended-molecular flat islands in lattice-matched $In_{0.53}Ga_{0.47}As/InP$ and $In_{0.53}Ga_{0.47}As/In_{1-x}Ga_xAs_yP_{1-y}$ quantum wells grown by low-pressure metal-organic vapor-phase epitaxy with different interruption cycles," *Phys. Rev. B* **46**, 9525–9537 (1992).

Appendix E

Kramers–Kronig Relations

The induced electric polarization **P** in a material is due to the response of the medium to an electric field **E** (ignoring the spatial dependence on **r**):

$$\mathbf{P}(t) = \varepsilon_0 \int_{-\infty}^{t} \chi(t - \tau)\mathbf{E}(\tau)\, d\tau$$

$$= \varepsilon_0 \int_{0}^{\infty} \chi(\tau)\mathbf{E}(t - \tau)\, d\tau \qquad (E.1)$$

The integration over τ is from $-\infty$ to t in the first expression, since the response **P**(t) at time t comes from the excitation field before t. In other words, the function $\chi(\tau)$ has the property that

$$\chi(\tau) = 0 \qquad \text{for } \tau < 0 \qquad (E.2)$$

i.e., the system is causal. Since

$$\mathbf{D}(t) = \varepsilon_0 \mathbf{E}(t) + \mathbf{P}(t) \qquad (E.3)$$

we find in the frequency domain

$$\mathbf{D}(\omega) = \varepsilon(\omega)\mathbf{E}(\omega) \qquad (E.4)$$

by taking the Fourier transform of (E.3), where

$$\varepsilon(\omega) = \varepsilon_0 \left[1 + \int_{0}^{\infty} \chi(\tau)e^{i\omega\tau}\, d\tau \right] \qquad (E.5)$$

Since $\chi(\tau)$ is a real function (it is in the time domain), we see that

$$\varepsilon(-\omega) = \varepsilon^*(\omega) \qquad (E.6)$$

If we write $\varepsilon(\omega)$ in terms of its real and imaginary parts,

$$\varepsilon(\omega) = \varepsilon'(\omega) + i\varepsilon''(\omega) \qquad (E.7)$$

we find that

$$\varepsilon'(-\omega) = \varepsilon'(\omega) \tag{E.8}$$

$$\varepsilon''(-\omega) = -\varepsilon''(\omega) \tag{E.9}$$

that is, the real part of $\varepsilon(\omega)$ is an even function, and the imaginary part is an odd function of ω. Another physical property of $\varepsilon(\omega)$ is that, in the high-frequency limit, $\varepsilon(\omega)$ tends to ε_0, because the polarization processes $P(t)$, which are responsible for χ, cannot occur when the field changes sufficiently rapidly [1]:

$$\varepsilon(\infty) \rightarrow \varepsilon_0 \tag{E.10}$$

Define an integral I in the complex ω' plane,

$$I = \frac{1}{2\pi i} \oint_C \frac{\varepsilon(\omega') - \varepsilon_0}{\omega' - \omega} d\omega' \tag{E.11}$$

where the closed contour C is shown in Fig. E.1. It is in the upper ω' plane. The function $\varepsilon(\omega)$ is analytic in the upper-half plane, since when $\omega' = \omega_R + i\omega_I$, the integrand in (E.5) includes an exponentially decreasing factor $e^{-\omega_I \tau}$ when $\omega_I > 0$. Since the function $\chi(\tau)$ is finite (for a physical process) throughout $0 < \tau < \infty$, the integral in (E.5) converges.

Since the function $\varepsilon(\omega') - \varepsilon_0$ is analytic in the upper ω' plane, the closed contour does not have any poles; therefore, the integral I vanishes using Cauchy's theorem. By breaking the contour into three parts—(i) on the real axis, $-R < \omega' < \omega - \delta$, $\omega + \delta < \omega' < R$, (ii) along the infinitesimally small semicircle near ω with a radius δ, and (iii) along the big semicircle with a large radius R—we find that the contribution due to part (iii) is zero as $R \rightarrow \infty$. Therefore, as $\delta \rightarrow 0$, the integration along part (i) is the principle

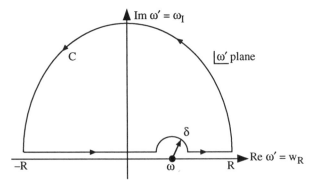

Figure E.1. The integration contour on the upper-half ω' plane for the derivation of the Kramers–Kronig relation.

value (denoted by P) of the integration along the real axis from $-\infty$ to $+\infty$, and we obtain

$$\frac{1}{2\pi i}P\int_{-\infty}^{\infty}\frac{\varepsilon(\omega')-\varepsilon_0}{\omega'-\omega}\,d\omega' + \frac{1}{2\pi i}\int_{\delta}\frac{\varepsilon(\omega')-\varepsilon_0}{\omega'-\omega}\,d\omega' = 0 \qquad (E.12)$$

The second integral equals $-\frac{1}{2}[\varepsilon(\omega)-\varepsilon_0]$, which can be evaluated simply by a change of variable from ω' to θ, $\omega'-\omega = \delta e^{i\theta}$ at a constant radius δ. We find

$$\varepsilon(\omega) - \varepsilon_0 = \frac{1}{\pi i}P\int_{-\infty}^{\infty}\frac{\varepsilon(\omega')-\varepsilon_0}{\omega'-\omega}\,d\omega' \qquad (E.13)$$

We separate (E.13) into real and imaginary parts and find

$$\varepsilon'(\omega) - \varepsilon_0 = \frac{1}{\pi}P\int_{-\infty}^{\infty}\frac{\varepsilon''(\omega')}{\omega'-\omega}\,d\omega' \qquad (E.14)$$

$$\varepsilon''(\omega) = -\frac{1}{\pi}P\int_{-\infty}^{\infty}\frac{\varepsilon'(\omega')-\varepsilon_0}{\omega'-\omega}\,d\omega' \qquad (E.15)$$

The above results are the Kramers–Kronig relations, which relate the real and the imaginary parts of $\varepsilon(\omega)$ to each other. If we make use of the even property of the real part $\varepsilon'(\omega)$ and the odd property of the imaginary part $\varepsilon''(\omega)$, we obtain alternatively

$$\varepsilon'(\omega) - \varepsilon_0 = \frac{2}{\pi}P\int_{0}^{\infty}\frac{\omega'\varepsilon''(\omega')}{\omega'^2-\omega^2}\,d\omega' \qquad (E.16)$$

$$\varepsilon''(\omega) = -\frac{2\omega}{\pi}P\int_{0}^{\infty}\frac{\varepsilon'(\omega')-\varepsilon_0}{\omega'^2-\omega^2}\,d\omega' \qquad (E.17)$$

In a homogeneous, isotropic medium with a complex permittivity function $\varepsilon(\omega) = \varepsilon'(\omega) + i\varepsilon''(\omega)$ and a permeability μ_0, the propagation constant of an electromagnetic wave at an angular frequency ω is

$$k(\omega) = \omega\sqrt{\mu_0\varepsilon(\omega)}$$

$$= \omega\sqrt{\mu_0\varepsilon_0}\left[\varepsilon'_r(\omega) + i\varepsilon''_r(\omega)\right]^{1/2}$$

$$= \frac{\omega}{c}\bar{n} \qquad (E.18)$$

where we have used the real and the imaginary parts of the relative permittivity

$$\varepsilon_r(\omega) = \frac{\varepsilon(\omega)}{\varepsilon_0}$$

$$\varepsilon_r'(\omega) = \frac{\varepsilon'(\omega)}{\varepsilon_0} \qquad \varepsilon_r''(\omega) = \frac{\varepsilon''(\omega)}{\varepsilon_0} \qquad (E.19)$$

and the complex refractive index

$$\bar{n} = n + i\kappa \qquad (E.20)$$

We obtain the relations

$$n^2 - \kappa^2 = \varepsilon_r'(\omega) \qquad (E.21a)$$

$$2n\kappa = \varepsilon_r''(\omega) \qquad (E.21b)$$

Here the real and the imaginary parts of the refractive index $n(\omega)$ and $\kappa(\omega)$ can be expressed as

$$n^2 = \tfrac{1}{2}\left[\varepsilon_r'(\omega) + \sqrt{\varepsilon_r'^2(\omega) + \varepsilon_r''^2(\omega)}\right] \qquad (E.22a)$$

$$\kappa^2 = \tfrac{1}{2}\left[-\varepsilon_r'(\omega) + \sqrt{\varepsilon_r'^2(\omega) + \varepsilon_r''^2(\omega)}\right] \qquad (E.22b)$$

Experimental data for n and κ of GaAs and InP semiconductors as a function of optical energy and wavelength are tabulated and plotted in Appendix J.

REFERENCE

1. L. D. Landau, E. M. Lifshitz, and L. P. Pitaevskii, *Electrodynamics of Continuous Media*, 2d ed., Pergamon, New York, 1984.

Appendix F

Poynting's Theorem and Reciprocity Theorem

F.1 POYNTING'S THEOREM

The power conservation is a very useful law. From Maxwell's equations in the time domain

$$\nabla \times \mathbf{E} = -\frac{\partial}{\partial t}\mathbf{B} \tag{F.1}$$

$$\nabla \times \mathbf{H} = \mathbf{J} + \frac{\partial}{\partial t}\mathbf{D} \tag{F.2}$$

Dot-multiplying (F.1) by $\mathbf{H}$ and (F.2) by $\mathbf{E}$, and taking the difference, we obtain

$$\nabla \cdot (\mathbf{E} \times \mathbf{H}) = -\mathbf{H} \cdot \frac{\partial}{\partial t}\mathbf{B} - \mathbf{E} \cdot \frac{\partial \mathbf{D}}{\partial t} - \mathbf{E} \cdot \mathbf{J} \tag{F.3}$$

where $\nabla \cdot (\mathbf{E} \times \mathbf{H}) = \mathbf{H} \cdot \nabla \times \mathbf{E} - \mathbf{E} \cdot \nabla \times \mathbf{H}$ has been used. Define the Poynting vector as

$$\mathbf{S} = \mathbf{E} \times \mathbf{H} \tag{F.4}$$

which gives the instantaneous energy flux density (W/m^2). For an isotropic medium, $\mathbf{D} = \varepsilon\mathbf{E}$ and $\mathbf{B} = \mu\mathbf{H}$, the electric and magnetic energy densities are

$$w_e = \frac{\varepsilon}{2}\mathbf{E} \cdot \mathbf{E}$$

$$w_m = \frac{\mu}{2}\mathbf{H} \cdot \mathbf{H} \tag{F.5}$$

Therefore, Poynting's theorem in the time domain is simply

$$\nabla \cdot \mathbf{S} = -\frac{\partial}{\partial t}(w_e + w_m) - \mathbf{E} \cdot \mathbf{J} \tag{F.6}$$

If integrating over a volume V enclosed by a surface S, we obtain Poynting's theorem in the form

$$\iint_S \mathbf{E} \times \mathbf{H} \cdot d\mathbf{S} = -\frac{\partial}{\partial t} \iiint_V (w_e + w_m)\, dV - \iiint_V \mathbf{E} \cdot \mathbf{J}\, dV \quad \text{(F.7)}$$

i.e., the power flow out of the surface S equals the decreasing rate of the stored electric and magnetic energies plus the power supplied by the source, $-\iiint \mathbf{E} \cdot \mathbf{J}\, dV$.

A complex Poynting's theorem can also be derived from Maxwell's equations in frequency domain:

$$\nabla \cdot \left(\tfrac{1}{2}\mathbf{E} \times \mathbf{H}^*\right) = -i\omega\left(\tfrac{1}{2}\mathbf{E} \cdot \mathbf{D}^* - \tfrac{1}{2}\mathbf{B} \cdot \mathbf{H}^*\right) - \tfrac{1}{2}\mathbf{E} \cdot \mathbf{J}^* \quad \text{(F.8)}$$

If $\mathbf{J} = \mathbf{J}_d + \mathbf{J}_f$, where $\mathbf{J}_d$ accounts for dissipation, e.g., $\mathbf{J}_d = \sigma\mathbf{E}$ is the conduction current density in a conductor, we then have

$$\nabla \cdot \left(\frac{1}{2}\mathbf{E} \times \mathbf{H}^*\right) + i\frac{\omega}{2}(\mathbf{E} \cdot \mathbf{D}^* - \mathbf{B} \cdot \mathbf{H}^*) + \frac{1}{2}\mathbf{E} \cdot \mathbf{J}_d^* = \frac{-1}{2}\mathbf{E} \cdot \mathbf{J}_f^* \quad \text{(F.9)}$$

where the right-hand side is the time-averaged power supplied by the source J_f, and the terms on the left-hand side are the time-averaged power flux, the difference in the electric and magnetic stored energy density, and the time-average dissipated power, respectively.

F.2   RECIPROCITY THEOREM [1, 2]

Consider two sources $\mathbf{J}^{(1)}$ and $\mathbf{J}^{(2)}$ producing two sets of fields in the same medium described by $\mathbf{D} = \varepsilon\mathbf{E}$ and $\mathbf{B} = \mu\mathbf{H}$:

$$\nabla \times \mathbf{E}^{(1)} = -\frac{\partial}{\partial t}\mathbf{B}^{(1)} \quad \text{(F.10a)}$$

$$\nabla \times \mathbf{H}^{(1)} = \mathbf{J}^{(1)} + \frac{\partial}{\partial t}\mathbf{D}^{(1)} \quad \text{(F.10b)}$$

$$\nabla \times \mathbf{E}^{(2)} = -\frac{\partial}{\partial t}\mathbf{B}^{(2)} \quad \text{(F.11a)}$$

$$\nabla \times \mathbf{H}^{(2)} = \mathbf{J}^{(2)} + \frac{\partial}{\partial t}\mathbf{D}^{(2)} \quad \text{(F.11b)}$$

If we take $\mathbf{H}^{(2)} \cdot$ (F.10a) $- \mathbf{E}^{(1)} \cdot$ (F.11b) and subtract by $\mathbf{H}^{(1)} \cdot$ (F.11a) $- \mathbf{E}^{(2)} \cdot$ (F.10b), we find

$$\nabla \cdot \left(\mathbf{E}^{(1)} \times \mathbf{H}^{(2)} - \mathbf{E}^{(2)} \times \mathbf{H}^{(1)}\right) = -\mathbf{E}^{(1)} \cdot \mathbf{J}^{(2)} + \mathbf{E}^{(2)} \cdot \mathbf{J}^{(1)} \quad \text{(F.12)}$$

Integrating over an infinite volume and using the divergence theorem, we obtain

$$\oint_S \left(\mathbf{E}^{(1)} \times \mathbf{H}^{(2)} - \mathbf{E}^{(2)} \times \mathbf{H}^{(1)} \right) \cdot d\mathbf{S} = \int_V \mathbf{E}^{(2)} \cdot \mathbf{J}^{(1)} \, dv - \int_V \mathbf{E}^{(1)} \cdot \mathbf{H}^{(2)} \, dv$$

$$(F.13)$$

Using the property in which the surface integral on the left-hand side goes to zero as the radius of the surface goes to infinity, we obtain

$$\int_{V_1} \mathbf{E}^{(2)} \cdot \mathbf{J}^{(1)} \, dv = \int_{V_2} \mathbf{E}^{(1)} \cdot \mathbf{J}^{(2)} \, dv \qquad (F.14)$$

Since the two sources $\mathbf{J}_1$ and $\mathbf{J}_2$ are distributed over finite regions, such as those of two dipoles, the volume integrals are only over the regions of the two sources V_1 and V_2, respectively. $\mathbf{E}^{(1)}$ and $\mathbf{E}^{(2)}$ would then be the electric fields due to $\mathbf{J}^{(1)}$ and $\mathbf{J}^{(2)}$, evaluated at the positions of the other sources occupying volumes V_2 and V_1, respectively. This is the reciprocity relation.

We may generalize the above relation to field solutions in two different media, described by $\varepsilon^{(1)}(\mathbf{r})$ and $\varepsilon^{(2)}(\mathbf{r})$, and the same permeability μ. We have, using the frequency domain representation,

$$\nabla \cdot \left(\mathbf{E}^{(1)} \times \mathbf{H}^{(2)} - \mathbf{E}^{(2)} \times \mathbf{H}^{(1)} \right) = i\omega \left[\varepsilon^{(2)}(\mathbf{r}) - \varepsilon^{(1)}(\mathbf{r}) \right] \mathbf{E}^{(1)} \cdot \mathbf{E}^{(2)}$$

$$+ \mathbf{E}^{(2)} \cdot \mathbf{J}^{(1)} - \mathbf{E}^{(1)} \cdot \mathbf{J}^{(2)} \qquad (F.15)$$

If we integrate the above equation over an infinite volume, we obtain similar relations to (F.14) except for the extra term due to the difference between two dielectric functions. However, if we apply the above relation to dielectric waveguide structures which are translationally invariant in the z direction, and integrate over only a volume of a disk shape with a thickness $\Delta z \to 0$ and a radius $R \to \infty$ (see Fig. F.1), we find

$$\iiint_V \nabla \cdot \mathbf{A} \, dx \, dy \, dz = \left(\iint_{S_1} + \iint_{S_2} + \iint_{S_\infty} \right) \mathbf{A} \cdot d\mathbf{S}$$

$$= \iint_{\text{at } z + \Delta z} \mathbf{A} \cdot \hat{z} \, dx \, dy + \iint_{\text{at } z} \mathbf{A} \cdot (-\hat{z}) \, dx \, dy$$

$$\simeq \left(\frac{\partial}{\partial z} \iint \mathbf{A} \cdot \hat{z} \, dx \, dy \right) \cdot \Delta z \qquad (F.16)$$

where $\mathbf{A} = \mathbf{E}^{(1)} \times \mathbf{H}^{(2)} - \mathbf{E}^{(2)} \times \mathbf{H}^{(1)}$ and the fields at S_∞ vanish. Since $\mathbf{J}^{(1)}$

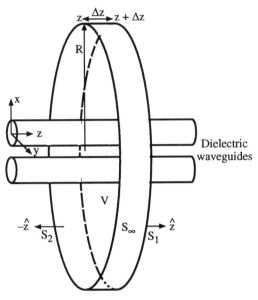

Figure F.1. The volume V for the space between z and $z + \Delta z$ enclosed by the surface $S = S_1 + S_2 + S_\infty$ for the derivation of the reciprocity relation for dielectric waveguide structures.

and $\mathbf{J}^{(2)}$ are zero in the dielectric waveguides, we obtain a reciprocity relation for the waveguide:

$$\frac{\partial}{\partial z} \iint \left(\mathbf{E}^{(1)} \times \mathbf{H}^{(2)} - \mathbf{E}^{(2)} \times \mathbf{H}^{(1)} \right) \cdot \hat{z} \, dx \, dy$$

$$= i\omega \iint \left[\varepsilon^{(2)}(x, y) - \varepsilon^{(1)}(x, y) \right] \mathbf{E}^{(1)} \cdot \mathbf{E}^{(2)} \, dx \, dy \qquad \text{(F.17)}$$

REFERENCES

1. L. D. Landau, E. M. Lifshitz, and L. P. Pitaevskii, *Electrodynamics of Continuous Media*, 2d ed., Pergamon, New York, 1984.
2. S. L. Chuang, "A coupled-mode formulation by reciprocity and a variational principle," *J. Lightwave Technol.* **LT-5**, 5–15 (1987), and "A coupled-mode theory for multiwave guide systems satisfying the reciprocity theorem and power conservation," *J. Lightwave Technol.* **LT-5**, 174–183 (1987).

Appendix G

Light Propagation in Gyrotropic Media—Magnetooptic Effects [1–3]

In this appendix, we present a general formulation to find the electromagnetic fields with the characteristic polarizations of gyrotropic media. The magnetooptic effects are then investigated.

G.1 MAXWELL'S EQUATIONS AND CHARACTERISTIC EQUATION

The constitutive relations for gyrotropic media are given by [1]

$$\mathbf{D} = \bar{\bar{\varepsilon}} \cdot \mathbf{E} \tag{G.1a}$$

$$\bar{\bar{\varepsilon}} = \begin{bmatrix} \varepsilon & i\varepsilon_g & 0 \\ -i\varepsilon_g & \varepsilon & 0 \\ 0 & 0 & \varepsilon_z \end{bmatrix} \tag{G.1b}$$

$$\mathbf{B} = \mu \mathbf{H} \tag{G.2}$$

One example is a plasma with an externally applied dc magnetic field in the $\hat{z}$ direction:

$$\varepsilon = \varepsilon_0 \left(1 - \frac{\omega_p^2}{\omega^2 - \omega_c^2} \right) \tag{G.3a}$$

$$\varepsilon_g = \varepsilon_0 \left[\frac{-\omega_p^2 \omega_c}{\omega(\omega^2 - \omega_c^2)} \right] \tag{G.3b}$$

$$\varepsilon_z = \varepsilon_0 \left(1 - \frac{\omega_p^2}{\omega^2} \right) \tag{G.3c}$$

where

$$\omega_c = \frac{qB_0}{m} \tag{G.4}$$

is the cyclotron frequency, and

$$\omega_p = \sqrt{\frac{ne^2}{m\varepsilon_0}} \tag{G.5}$$

is the plasma frequency for a carrier density n. Here m is the electron mass. For the free carrier effects in semiconductors, the effective mass of the carriers should be used.

Note that the geometry of the medium considered is rotation-invariant around the z axis. Thus, we may consider the propagation wave vector $\mathbf{k}$ to be in the x-z plane without loss of generality:

$$\mathbf{k} = \hat{x}k_x + \hat{z}k_z \tag{G.6}$$

We repeat Maxwell's equations for plane-wave solutions here:

$$\mathbf{k} \times \mathbf{E} = \omega\mathbf{B} \tag{G.7a}$$

$$\mathbf{k} \times \mathbf{H} = -\omega\mathbf{D} \tag{G.7b}$$

$$\mathbf{k} \cdot \mathbf{B} = 0 \tag{G.7c}$$

$$\mathbf{k} \cdot \mathbf{D} = 0 \tag{G.7d}$$

Using (G.7a), (G.7b), and (G.2), we obtain

$$\mathbf{k} \times (\mathbf{k} \times \mathbf{E}) = \omega\mu\mathbf{k} \times \mathbf{H} = -\omega^2\mu\bar{\bar{\varepsilon}} \cdot \mathbf{E}$$

Or, equivalently, the above vector equation can be written in a matrix representation following Table 6.1 in Section 6.3 of the text:

$$\begin{bmatrix} 0 & -k_z & 0 \\ +k_z & 0 & -k_x \\ 0 & k_x & 0 \end{bmatrix}^2 \begin{bmatrix} E_x \\ E_y \\ E_z \end{bmatrix} = -\omega^2\mu \begin{bmatrix} \varepsilon & +i\varepsilon_g & 0 \\ -i\varepsilon_g & \varepsilon & 0 \\ 0 & 0 & \varepsilon_z \end{bmatrix} \begin{bmatrix} E_x \\ E_y \\ E_z \end{bmatrix} \tag{G.8}$$

We carry out the square of the matrix on the left-hand side, and move to the right to obtain

$$\begin{bmatrix} k_z^2 - \omega^2\mu\varepsilon & -i\omega^2\mu\varepsilon_g & -k_xk_z \\ i\omega^2\mu\varepsilon_g & k_x^2 + k_z^2 - \omega^2\mu\varepsilon & 0 \\ -k_xk_z & 0 & k_x^2 - \omega^2\mu\varepsilon_z \end{bmatrix} \begin{bmatrix} E_x \\ E_y \\ E_z \end{bmatrix} = 0 \tag{G.9}$$

The dispersion relation is obtained by setting the determinant of the matrix to zero for nontrivial solutions for the electric field. We obtain the characteristic equation after some algebra

$$(k_x^2 + k_z^2 - \omega^2\mu\varepsilon)(-\omega^2\mu\varepsilon k_x^2 - \omega^2\mu\varepsilon_z k_z^2 + \omega^4\mu^2\varepsilon\varepsilon_z)$$

$$- \omega^4\mu^2\varepsilon_g^2(k_x^2 - \omega^2\mu\varepsilon_z) = 0 \qquad \text{(G.10)}$$

Rewrite the wave vector $\mathbf{k}$ in terms of the angle θ with respect to the z axis,

$$\mathbf{k} = \hat{x}k_x + \hat{z}k_z = \hat{x}k \sin\theta + \hat{z}k \cos\theta$$

and define the constants

$$K = \omega\sqrt{\mu\varepsilon} \qquad \text{(G.11a)}$$

$$K_z = \omega\sqrt{\mu\varepsilon_z} \qquad \text{(G.11b)}$$

$$K_g = \omega\sqrt{\mu\varepsilon_g} \qquad \text{(G.11c)}$$

Equation (G.10) reduces to

$$k^4(K^2 \sin^2\theta + K_z^2 \cos^2\theta) - k^2\big[(K^4 - K_g^4)\sin^2\theta + K^2K_z^2 \cos^2\theta + K^2K_z^2\big]$$

$$+ (K^4 - K_g^4)K_z^2 = 0 \qquad \text{(G.12)}$$

which has the form

$$k^4 A - k^2 B + C = 0$$

The solution k^2 is easily obtained from

$$k^2 = \frac{B \pm \sqrt{B^2 - 4AC}}{2A} \qquad \text{(G.13)}$$

where

$$A = K^2 \sin^2\theta + K_z^2 \cos^2\theta$$

$$B = (K^4 - K_g^4)\sin^2\theta + K^2K_z^2(1 + \cos^2\theta)$$

$$C = (K^4 - K_g^4)K_z^2$$

Using (G.9) again, we have

$$
\begin{bmatrix}
k^2 \cos^2 \theta - K^2 & -iK_g^2 & -k^2 \sin \theta \cos \theta \\
iK_g^2 & k^2 - K^2 & 0 \\
-k^2 \sin \theta \cos \theta & 0 & k^2 \sin^2 \theta - K_z^2
\end{bmatrix}
\begin{bmatrix}
E_x \\
E_y \\
E_z
\end{bmatrix}
= 0 \quad \text{(G.14)}
$$

where k^2 is given by the two possible roots in (G.13). Since the above determinant of the matrix is zero, Eq. (G.14) contains three algebraic equations, which are linearly dependent. Using the second and the third equations in (G.14), we find

$$
\frac{E_x}{E_y} = \frac{k^2 - K^2}{-iK_g^2} \quad \text{(G.15a)}
$$

$$
\frac{E_x}{E_z} = \frac{k^2 \sin^2 \theta - K_z^2}{k^2 \sin \theta \cos \theta} \quad \text{(G.15b)}
$$

where two possible values of k^2 are given by the roots in (G.13).

G.2   SPECIAL CASES

Case 1: $\theta = 0$ (the wave is propagating parallel to the magnetic field). Equation (G.14) becomes

$$
\begin{bmatrix}
k^2 - K^2 & -iK_g^2 & 0 \\
iK_g^2 & k^2 - K^2 & 0 \\
0 & 0 & -K_z^2
\end{bmatrix}
\begin{bmatrix}
E_x \\
E_y \\
E_z
\end{bmatrix}
= 0 \quad \text{(G.16)}
$$

We have

$$
E_z = 0, \text{ since } K_z^2 \neq 0 \quad \text{(G.17)}
$$

and

$$
\left(k^2 - K^2\right)^2 - K_g^4 = 0 \quad \text{(G.18)}
$$

We obtain two roots for k^2,

$$
k_\pm^2 = K^2 \pm K_g^2 = \omega^2 \mu \left(\varepsilon \pm \varepsilon_g\right) \quad \text{(G.19)}
$$

Substituting the above roots back into (G.16), we find

$$\frac{E_y}{E_x} = \frac{k^2 - K^2}{iK_g^2} = \frac{\pm K_g^2}{iK_g^2} = \mp i \qquad \text{for } k^2 = \omega^2 \mu(\varepsilon \pm \varepsilon_g) \quad \text{(G.20)}$$

Combining (G.17) and (G.20), we see that if the wave is propagating parallel to the magnetic field, the wave will be circularly polarized. Let us assume that the wave is propagating in the $+\hat{z}$ direction; we have either

$$\text{(a)} \quad \frac{E_y}{E_x} = -i \qquad k = \omega\sqrt{\mu(\varepsilon + \varepsilon_g)} \qquad \text{(G.21)}$$

the wave is left-hand circularly polarized (LHCP), or

$$\text{(b)} \quad \frac{E_y}{E_x} = +i \qquad k = \omega\sqrt{\mu(\varepsilon - \varepsilon_g)} \qquad \text{(G.22)}$$

the wave is right-hand circularly polarized (RHCP).

Case 2: $\theta = \pi/2$ (the wave is propagating perpendicularly to the magnetic field). Equation (G.14) becomes

$$\begin{bmatrix} -K^2 & -iK_g^2 & 0 \\ iK_g^2 & k^2 - K^2 & 0 \\ 0 & 0 & k^2 - K_z^2 \end{bmatrix} \begin{bmatrix} E_x \\ E_y \\ E_z \end{bmatrix} = 0 \qquad \text{(G.23)}$$

Possible nontrivial solutions for the electric field will be

(a) $$\qquad\qquad\qquad\qquad k^2 = K_z^2 \qquad E_z \neq 0 \qquad\qquad\qquad\qquad \text{(G.24)}$$

If $k^2 = K_z^2$ is true, we find that $E_x = 0$, $E_y = 0$ from the first two equations in (G.23). Therefore, we obtain that the characteristic polarization is linear polarization in the z direction $\mathbf{E} = \hat{z}E_z$ and the propagation constant is $k = K_z = \omega\sqrt{\mu\varepsilon_z}$.
(b) $-K^2(k^2 - K^2) - K_g^4 = 0$, which leads to

$$k^2 = K^2 - \frac{K_g^4}{K^2} \qquad \text{(G.25)}$$

If (G.25) is true, then $k^2 \neq K_z^2$, $E_z = 0$, and

$$E_x = -i\frac{K_g^2}{K^2}E_y \qquad \text{(G.26)}$$

The wave is generally elliptically polarized in the x-y plane with a propagation constant $k = \sqrt{(K^4 - K_g^4)/K^2}$. The wave vector is always on the x-y plane since $\theta = \pi/2$. Furthermore, if $\varepsilon_g = 0$, the medium becomes uniaxial. Equations (G.24) to (G.26) show that either the electric field is polarized in the z direction and $k^2 = K_z^2$, or polarized in the y direction (since $E_x = 0$, $E_z = 0$ from (G.26)) and $k^2 = K^2$. Both waves are linearly polarized propagating with different velocities. This birefringence is called the Cotton–Mouton effect.

G.3   FARADAY ROTATION

Let us consider a slab of gyrotropic medium with a dc magnetic field applied in the $+\hat{z}$ direction and the wave propagated parallel to the dc magnetic field. This is the special case 1, $\theta = 0$, discussed before, and the two characteristic polarizations are left- and right-hand circularly polarized with corresponding propagation constants given by (G.21) and (G.22).

Consider an incident plane wave as shown in Fig. G.1 with

$$\mathbf{E} = \hat{x} E_0 e^{ikz} \qquad (G.27)$$

Upon striking the interface at $z = 0$, the wave will break up into two circularly polarized waves:

$$\mathbf{E} = (\hat{x} - i\hat{y})\frac{E_0}{2}e^{ik_+ z} + (\hat{x} + i\hat{y})\frac{E_0}{2}e^{ik_- z} \qquad (G.28)$$

These two circularly polarized waves propagate with two different wave numbers, k_+ and k_-. As discussed before, $k_\pm = \omega\sqrt{\mu(\varepsilon \pm \varepsilon_g)}$. At $z = d$, we have

$$
\begin{aligned}
\mathbf{E} &= (\hat{x} - i\hat{y})\frac{E_0}{2}e^{ik_+ d} + (\hat{x} + i\hat{y})\frac{E_0}{2}e^{ik_- d} \\
&= \hat{x} E_0\left(\frac{e^{ik_+ d} + e^{ik_- d}}{2}\right) + \hat{y} E_0\left(\frac{-ie^{ik_+ d} + ie^{ik_- d}}{2}\right) \qquad (G.29)
\end{aligned}
$$

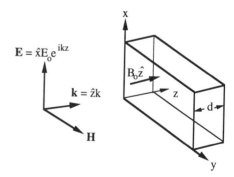

$$\mathbf{E} = \hat{x} E_0 e^{ikz}$$

$$\mathbf{k} = \hat{z} k$$

Figure G.1. A linearly polarized plane wave **E** incident on a gyrotropic medium experiences Faraday rotation after passing through the medium at $z = d$.

Thus, we have the ratio of the $\hat{y}$ component to the $\hat{x}$ component of the electric field

$$\frac{E_y}{E_x} = -i\frac{e^{ik_+d} - e^{ik_-d}}{e^{ik_+d} + e^{ik_-d}}$$

$$= \tan\left[\frac{(k_+ - k_-)d}{2}\right] \tag{G.30}$$

which is a real number. Thus the electric field at $z = d$ is again linearly polarized making an angle

$$\theta_F = \frac{(k_+ - k_-)}{2}d \tag{G.31}$$

with the $\hat{x}$ axis. We conclude that the incident linearly polarized wave (in the $\hat{x}$ direction) is rotated by an angle θ_F at $x = d$, which is called the Faraday rotation.

Since the Faraday angle θ_F depends on the difference between $k_+ = \omega\sqrt{\mu(\varepsilon + \varepsilon_g)}$ and $k_- = \omega\sqrt{\mu(\varepsilon - \varepsilon_g)}$, the carrier density n and the effective mass of the electrons can be measured from the magnetooptic effects using the Faraday rotation as discussed above. Another setup is called the Voigt configuration for which the propagation direction is perpendicular to the direction of the applied dc magnetic field $\hat{z}B_0$. The incident wave is chosen to be linearly polarized at an angle of 45° with respect to the static magnetic field. The transmitted wave does not experience a rotation; it becomes elliptically polarized, however. The phase angle or the amount of ellipticity is determined by the difference of the propagation constants of the two characteristic polarizations, which is related to the plasma frequency ω_p and the cyclotron frequency ω_c. For more discussions on the magnetooptic effects and their measurements in semiconductors, see Ref. 3.

REFERENCES

1. J. A. Kong, *Electromagnetic Wave Theory*, Wiley, New York, 1990.
2. K. C. Yeh and C. H. Liu, *Theory of Ionospheric Waves*, Academic, New York, 1972.
3. K. Seeger, *Semiconductor Physics*, Springer, Berlin, 1982.

Appendix H

Formulation of the Improved Coupled-Mode Theory

In Appendix F we derived a general reciprocity relation for waveguide systems. The relation is

$$\frac{\partial}{\partial z} \iint \left(\mathbf{E}^{(1)} \times \mathbf{H}^{(2)} - \mathbf{E}^{(2)} \times \mathbf{H}^{(1)} \right) \cdot \hat{z} \, dx \, dy$$

$$= i\omega \iint \left[\varepsilon^{(2)}(x, y) - \varepsilon^{(1)}(x, y) \right] \mathbf{E}^{(1)} \cdot \mathbf{E}^{(2)} \, dx \, dy \qquad (\text{H.1})$$

where $\mathbf{E}^{(1)}(x, y, z)$ and $\mathbf{H}^{(1)}(x, y, z)$ are a set of solutions in the medium described by $\varepsilon^{(1)}(x, y)$ *everywhere* in space. Similarly, $\mathbf{E}^{(2)}(x, y, z)$ and $\mathbf{H}^{(2)}(x, y, z)$ are another set of solutions in another medium described by $\varepsilon^{(2)}(x, y)$ everywhere in space.

Let us consider three different media.

Medium A: Waveguide *a* Only

Suppose only waveguide a exists in the whole space (Fig. H1a). The medium is described by

$$\varepsilon^{(a)}(x, y)$$

The solution is

$$\left[\mathbf{E}_t^{(a)}(x, y) + \mathbf{E}_z^{(a)}(x, y) \right] e^{i\beta_a z}$$
$$\left[\mathbf{H}_t^{(a)}(x, y) + \mathbf{H}_z^{(a)}(x, y) \right] e^{i\beta_a z} \qquad (\text{H.2})$$

for the forward propagation mode and

$$\left[\mathbf{E}_t^{(a)}(x, y) - \mathbf{E}_z^{(a)}(x, y) \right] e^{-i\beta_a z}$$
$$\left[-\mathbf{H}_t^{(a)}(x, y) + \mathbf{H}_z^{(a)}(x, y) \right] e^{-i\beta_a z} \qquad (\text{H.3})$$

for the same mode as (H.2) propagating in the $-z$ direction. Note the

686

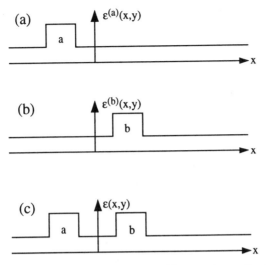

Figure H.1. Three different permittivity functions of interest for applications in deriving the improved coupled-mode theory using the reciprocity theorem. (a) only waveguide a exists in the whole space, (b) only waveguide b exists in the whole space, and (c) both waveguides a and b exist.

relations between the field components of the forward propagation modes and those of the backward propagation modes. These relations can be obtained from Maxwell's equations and can also be checked from the z component of the Poynting vector for the power flow. We see that there is a sign change before $\mathbf{H}_t^{(a)}(x, y)$ in (H.3). For example, the TE modes in a slab waveguide have a solution $\mathbf{E} = \hat{y}E_y(x)e^{i\beta z}$ and $\mathbf{H} = \nabla \times \mathbf{E}/i\omega\mu = (-\hat{x}i\beta E_y + \hat{z}(\partial/\partial x)E_y)/i\omega\mu$ for the forward propagation modes. The field expressions for the backward propagation modes are $\mathbf{E} = \hat{y}E_y(x)e^{-i\beta z}$ and $\mathbf{H} = (\hat{x}i\beta E_y + \hat{z}(\partial/\partial x)E_y)/i\omega\mu$, with a sign change in the x component.

Medium B: Waveguide b Only

Suppose only waveguide b exists in the whole space (Fig. H.1b). The permittivity function is described by

$$\varepsilon^{(b)}(x, y)$$

The field solution is

$$\begin{aligned}
\left[\mathbf{E}_t^{(b)}(x, y) + \mathbf{E}_z^{(b)}(x, y)\right] e^{i\beta_b z} \\
\left[\mathbf{H}_t^{(b)}(x, y) + \mathbf{H}_z^{(b)}(x, y)\right] e^{i\beta_b z}
\end{aligned} \tag{H.4}$$

for a mode guided in the z direction and

$$\left[\mathbf{E}_t^{(b)}(x,y) - \mathbf{E}_z^{(b)}(x,y)\right] e^{-i\beta_b z}$$
$$\left[-\mathbf{H}_t^{(b)}(x,y) + \mathbf{H}_z^{(b)}(x,y)\right] e^{-i\beta_b z} \tag{H.5}$$

for the same mode as (H.4) propagating in the $-z$ direction.

Medium C: Both Waveguides a and b Exist—Coupled Waveguides

In this case, the medium is described by

$$\varepsilon(x,y)$$

which is shown in Fig. H.1c. The field solutions can be written as the superposition of two individual waveguide modes:

$$\mathbf{E}(x,y,z) = a(z)\mathbf{E}^{(a)}(x,y) + b(z)\mathbf{E}^{(b)}(x,y)$$

$$\mathbf{H}(x,y,z) = a(z)\mathbf{H}^{(a)}(x,y) + b(z)\mathbf{H}^{(b)}(x,y) \tag{H.6}$$

Let us consider three applications of the general reciprocity relation given by Eq. (H.1), which requires two sets of field expressions in two media, $\varepsilon^{(1)}(x,y)$ and $\varepsilon^{(2)}(x,y)$.

Application 1

Suppose we choose the first medium to be the coupled waveguide system described as medium C above,

$$\varepsilon^{(1)} = \varepsilon(x,y)$$

and the field solutions $\mathbf{E}^{(1)}$ and $\mathbf{H}^{(1)}$, as in (H.6), the coupled-mode solutions.
    Choose the second medium to be only waveguide a, $\varepsilon^{(2)}(x,y) = \varepsilon^{(a)}(x,y)$ and choose the solutions to be the corresponding guided mode propagating in the $-z$ direction (H.3).

$$\mathbf{E}^{(2)} = \left[\mathbf{E}_t^{(a)}(x,y) - \mathbf{E}_z^{(a)}(x,y)\right] e^{-i\beta_a z}$$

$$\mathbf{H}^{(2)} = \left[-\mathbf{H}_t^{(a)}(x,y) + \mathbf{H}_z^{(a)}(x,y)\right] e^{-i\beta_a z} \tag{H.7}$$

We obtain from the reciprocity relation (H.1)

$$\frac{d}{dz}\left[\iint a(z)e^{-i\beta_a z}\left(-\mathbf{E}_t^{(a)}\times\mathbf{H}_t^{(a)}-\mathbf{E}_t^{(a)}\times\mathbf{H}_t^{(a)}\right)\cdot\hat{z}\,dx\,dy\right.$$

$$\left.+\iint b(z)e^{-i\beta_a z}\left(-\mathbf{E}_t^{(b)}\times\mathbf{H}_t^{(a)}-\mathbf{E}_t^{(a)}\times\mathbf{H}_t^{(b)}\right)\cdot\hat{z}\,dx\,dy\right]$$

$$=-i\omega\iint\Delta\varepsilon^{(a)}(x,y)\left[\mathbf{E}_t^{(a)}\cdot\mathbf{E}_t^{(a)}-\mathbf{E}_z^{(a)}\cdot\mathbf{E}_z^{(a)}\right]dx\,dy\,a(z)e^{-i\beta_a z}$$

$$-i\omega\iint\Delta\varepsilon^{(a)}(x,y)\left[\mathbf{E}_t^{(a)}\cdot\mathbf{E}_t^{(b)}-\mathbf{E}_z^{(a)}\cdot\mathbf{E}_z^{(b)}\right]dx\,dy\,b(z)e^{-i\beta_a z}$$

$$\text{(H.8)}$$

Or

$$\frac{d}{dz}a(z)+C\frac{db(z)}{dz}=i(\beta_a+K_{aa})a(z)+i(\beta_a C+K_{ba})b(z)\quad\text{(H.9)}$$

where

$$C=\frac{C_{ab}+C_{ba}}{2}\tag{H.10}$$

$$C_{pq}=\frac{1}{2}\iint\mathbf{E}_t^{(q)}\times\mathbf{H}_t^{(p)}\cdot\hat{z}\,dx\,dy\tag{H.11}$$

$$K_{pq}=\frac{\omega}{4}\iint\Delta\varepsilon^{(q)}(x,y)\left[\mathbf{E}_t^{(p)}\cdot\mathbf{E}_t^{(q)}-\mathbf{E}_z^{(p)}\cdot\mathbf{E}_z^{(q)}\right]dx\,dy\tag{H.12}$$

$$\Delta\varepsilon^{(q)}(x,y)=\varepsilon(x,y)-\varepsilon^{(q)}(x,y)\tag{H.13}$$

Note that the fields have been normalized such that C_{aa} and $C_{bb}=1$.

Application 2

Choose $\varepsilon^{(1)}(x,y)$, $\mathbf{E}^{(1)}$ and $\mathbf{H}^{(1)}$ to be the coupled waveguide system and the field solutions as in Application 1. Choose the second medium to be waveguide b only in the whole space

$$\varepsilon^{(2)}(x,y)=\varepsilon^{(b)}(x,y)$$

with solutions of the guided mode propagating in the $-z$ direction:

$$\mathbf{E}^{(2)} = \left[\mathbf{E}_t^{(b)}(x, y) - \mathbf{E}_z^{(b)}(x, y)\right]e^{-i\beta_b z}$$

$$\mathbf{H}^{(2)} = \left[-\mathbf{H}_t^{(b)}(x, y) + \mathbf{H}_z^{(b)}(x, y)\right]e^{-i\beta_b z} \tag{H.14}$$

We obtain

$$C\frac{d}{dz}a(z) + \frac{d}{dz}b(z) = i(\beta_b C + K_{ab})a(z) + i(\beta_b + K_{bb})b(z) \tag{H.15}$$

Therefore, we find

$$C\frac{d}{dz}\begin{bmatrix} a(z) \\ b(z) \end{bmatrix} = i\mathbf{Q}\begin{bmatrix} a(z) \\ b(z) \end{bmatrix} \tag{H.16}$$

where

$$\mathbf{C} = \begin{bmatrix} 1 & C \\ C & 1 \end{bmatrix} \tag{H.17}$$

$$\mathbf{Q} = \begin{bmatrix} K_{aa} & K_{ba} \\ K_{ab} & K_{bb} \end{bmatrix} + \begin{bmatrix} \beta_a & \beta_a C \\ \beta_b C & \beta_b \end{bmatrix} \tag{H.18}$$

Application 3

Choose

$$\varepsilon^{(1)}(x, y) = \varepsilon^{(a)}(x, y)$$

$$\mathbf{E}^{(1)} = \left[\mathbf{E}_t^{(a)}(x, y) + \mathbf{E}_z^{(a)}(x, y)\right]e^{i\beta_a z}$$

$$\mathbf{H}^{(1)} = \left[\mathbf{H}_t^{(a)}(x, y) + \mathbf{H}_z^{(a)}(x, y)\right]e^{i\beta_a z} \tag{H.19}$$

and

$$\varepsilon^{(2)}(x, y) = \varepsilon^{(b)}(x, y)$$

$$\mathbf{E}^{(2)} = \left[\mathbf{E}_t^{(b)}(x, y) - \mathbf{E}_z^{(b)}(x, y)\right]e^{-i\beta_b z}$$

$$\mathbf{H}^{(2)} = \left[-\mathbf{H}_t^{(b)}(x, y) + \mathbf{H}_z^{(b)}(x, y)\right]e^{-i\beta_b z} \tag{H.20}$$

We obtain

$$i(\beta_a - \beta_b)\left[-\iint \mathbf{E}_t^{(a)} \times \mathbf{H}_t^{(b)} \cdot \hat{z} \, dx \, dy - \iint \mathbf{E}_t^{(b)} \times \mathbf{H}_t^{(a)} \cdot \hat{z} \, dx \, dy\right]$$

$$= i\omega \iint \left(\varepsilon^{(b)} - \varepsilon^{(a)}\right) \cdot \left[\mathbf{E}_t^{(a)} \cdot \mathbf{E}_t^{(b)} - \mathbf{E}_z^{(a)} \cdot \mathbf{E}_z^{(b)}\right] dx \, dy \qquad \text{(H.21)}$$

Since $\varepsilon^{(b)} - \varepsilon^{(a)} = (\varepsilon - \varepsilon^{(a)}) - (\varepsilon - \varepsilon^{(b)}) = \Delta\varepsilon^{(a)} - \Delta\varepsilon^{(b)}$, we find

$$(\beta_b - \beta_a)C = K_{ba} - K_{ab} \qquad \text{(H.22)}$$

which is an *exact* relation. This relation also shows that for an asymmetric coupling system ($\beta_b \neq \beta_a$), the coupling coefficients are not equal ($K_{ba} \neq K_{ab}$). Only if the overlap integral C is small in the weak coupling case can K_{ba} be approximately equal to K_{ab}. Since

$$Q_{11} = \beta_a + K_{aa} \qquad Q_{22} = \beta_b + K_{bb}$$

$$Q_{12} = K_{ba} + \beta_a C = K_{ab} + \beta_b C$$

$$Q_{21} = K_{ab} + \beta_b C = K_{ba} + \beta_a C \qquad \text{(H.23)}$$

Q can also be written as

$$\mathbf{Q} = \begin{bmatrix} K_{aa} & K_{ab} \\ K_{ba} & K_{bb} \end{bmatrix} + \begin{bmatrix} \beta_a & \beta_b C \\ \beta_a C & \beta_b \end{bmatrix} \qquad \text{(H.24)}$$

Therefore,

$$\mathbf{C}\frac{d}{dz}\begin{bmatrix} a(z) \\ b(z) \end{bmatrix} = i[\mathbf{K} + \mathbf{CB}]\begin{bmatrix} a(z) \\ b(z) \end{bmatrix} \qquad \text{(H.25)}$$

where

$$\mathbf{B} = \begin{bmatrix} \beta_a & 0 \\ 0 & \beta_b \end{bmatrix} \qquad \text{(H.26)}$$

The coupled-mode equation can be written as

$$\frac{d}{dz}\begin{bmatrix} a(z) \\ b(z) \end{bmatrix} = i\mathbf{M}\begin{bmatrix} a(z) \\ b(z) \end{bmatrix} \qquad \text{(H.27)}$$

$$\mathbf{M} = \mathbf{B} + \mathbf{C}^{-1}\mathbf{K} \qquad \text{(H.28)}$$

If

$$\mathbf{M} = \begin{bmatrix} \gamma_a & k_{ab} \\ k_{ba} & \gamma_b \end{bmatrix} \tag{H.29}$$

then

$$\gamma_a = \beta_a + \frac{K_{aa} - CK_{ba}}{1 - C^2}$$

$$\gamma_b = \beta_b + \frac{K_{bb} - CK_{ab}}{1 - C^2}$$

$$k_{ab} = \frac{K_{ab} - CK_{bb}}{1 - C^2}$$

$$k_{ba} = \frac{K_{ba} - CK_{aa}}{1 - C^2} \tag{H.30}$$

It is straightforward to show that

$$k_{ab} - k_{ba} = (\gamma_a - \gamma_b)C \tag{H.31}$$

which is an *exact* relation similar to (H.22).

Appendix I

Density-Matrix Formulation
of Optical Susceptibility

The density-matrix theory plays an important role in applications to linear and nonlinear optical properties of materials in quantum electronics. The basic idea is that the density-matrix formulation provides a most convenient method to predict the expectation values of physical quantities when the exact wave function is unknown.

I.1 DENSITY-MATRIX THEORY [1, 2]

Assume that $\psi(\mathbf{r}, t)$ is the wave function of the material system under a perturbation Hamiltonian, which can be due to an electromagnetic field or other excitations. The density-matrix operator ρ is defined as the ensemble average of the form

$$\rho \equiv \overline{|\psi\rangle\langle\psi|} \tag{I.1}$$

where an overbar means the ensemble average. Explicitly, we can expand the wave function $\psi(r, t)$ using a complete set of wave functions $\phi_n(\mathbf{r})$:

$$\psi(\mathbf{r}, t) = \sum_n c_n(t)\phi_n(\mathbf{r}) \tag{I.2}$$

Though not required, $\{\phi_n(\mathbf{r})\}$ are usually chosen to be the solutions of the unperturbed Hamiltonian H_0, which describes the electronic states of the material system in the absence of any perturbation. The ensemble average of a physical quantity P is given by

$$\begin{aligned}
\langle P \rangle &= \overline{\langle \psi(\mathbf{r}, t)|\mathbf{P}|\psi(\mathbf{r}, t)\rangle} \\
&= \sum_{m,n} \overline{c_m^*(t)c_m(t)}[\phi_m(\mathbf{r}), \mathbf{P}\phi_n(\mathbf{r})] \\
&= \sum_{m,n} \rho_{nm}\mathbf{P}_{mn} \\
&= \mathrm{Tr}(\rho\mathbf{P})
\end{aligned} \tag{I.3}$$

which is the trace of the matrix product ρ_{nm} and P_{mn}, where

$$P_{mn} = \langle \phi_m(\mathbf{r}) | \mathbf{P} | \phi_n(\mathbf{r}) \rangle = \int \phi_m^*(\mathbf{r}) \mathbf{P} \phi_n(\mathbf{r}) \, d^3\mathbf{r}$$

$$\begin{aligned}
\rho_{nm} &= \langle \phi_n | \rho | \phi_m \rangle \\
&= \overline{\langle \phi_n | \psi \rangle \langle \psi | \phi_m \rangle} \\
&= \overline{c_m^*(t) c_n(t)}
\end{aligned} \tag{I.4}$$

Note that $\rho_{nm} = \rho_{mn}^*$ by definition, and $\rho_{nn} = \overline{|c_n(t)|^2}$ has the physical meaning of the probability of finding the particle in the level n.

The time-evolution of the density operator can be derived as follows. Use

$$H\psi(\mathbf{r}, t) = i\hbar \frac{\partial}{\partial t} \psi(\mathbf{r}, t) \tag{I.5}$$

or

$$\sum_n c_n(t) H\phi_n(\mathbf{r}) = i\hbar \sum_n \frac{\partial c_n(t)}{\partial t} \phi_n(\mathbf{r}) \tag{I.6}$$

Multiplied by $\phi_m^*(\mathbf{r})$ and integrated over the whole space, the above equation becomes

$$i\hbar \frac{\partial c_m(t)}{\partial t} = \sum_n c_n(t) H_{mn} \tag{I.7}$$

Thus

$$\begin{aligned}
i\hbar \frac{\partial}{\partial t} \rho_{nm} &= i\hbar \overline{\frac{\partial c_n(t)}{\partial t} c_m^*(t)} + i\hbar \overline{c_n(t) \frac{\partial c_m^*(t)}{\partial t}} \\
&= \sum_k H_{nk} \overline{c_k(t) c_m^*(t)} - \sum_k \overline{c_n(t) c_k^*(t)} H_{mk}^* \\
&= \sum_k (H_{nk} \rho_{km} - \rho_{nk} H_{km}) \\
&= [H, \rho]_{nm}
\end{aligned} \tag{I.8}$$

where $H_{mk}^* = H_{km}$ has been used. We conclude that

$$i\hbar \frac{\partial}{\partial t} \rho = [H, \rho] = H\rho - \rho H \tag{I.9}$$

Here $H = H_0 + H'$ is the total Hamiltonian, which consists of an unperturbed part H_0 and a perturbation potential H', accounting for any external perturbation.

I.2 DENSITY-MATRIX APPROACH TO THE OPTICAL PROCESSES

An advantage of the density-matrix approach is that it gives general expressions for the linear and nonlinear optical susceptibilities [2]. This approach has been shown to be very useful to study quantum electronic processes in different material systems. The density operator ρ satisfies the equation of motion as derived in (I.9):

$$\frac{\partial}{\partial t}\rho = \frac{-i}{\hbar}[H,\rho] \tag{I.10}$$

where the Hamiltonian operator consists of three parts:

$$H = H_0 + H' + H_{\text{random}} \tag{I.11}$$

H_0 is the unperturbed Hamiltonian, H' accounts for the interaction such as the electron–photon interaction, for which

$$H' = -\mathbf{M} \cdot \mathbf{E}(t) = -\sum_{j=1}^{3} M^j E_j(t) \tag{I.12}$$

where $\mathbf{M}$ is the dipole operator,

$$\mathbf{M} = e\mathbf{r} \tag{I.13}$$

and $e = -|e|$ for electrons and $+|e|$ for holes. $\mathbf{E}$ is the electric field. H_{random} includes the relaxation effects due to incoherent scattering processes:

$$\frac{\partial}{\partial t}\rho = \frac{-i}{\hbar}[H_0 + H',\rho] + \left(\frac{\partial\rho}{\partial t}\right)_{\text{relax}} \tag{I.14}$$

where

$$\left(\frac{\partial\rho}{\partial t}\right)_{\text{relax}} = \frac{-i}{\hbar}[H_{\text{random}},\rho - \rho^{(0)}] \tag{I.15}$$

which can be written in terms of the T_1 and T_2 time constants:

$$\frac{\partial}{\partial t}\left(\rho_{uu} - \rho_{uu}^{(0)}\right)_{\text{relax}} = -\frac{\rho_{uu} - \rho_{uu}^{(0)}}{(T_1)_{uu}} \tag{I.16}$$

$$\left(\frac{\partial}{\partial t}\rho_{uv}\right)_{\text{relax}} = -\frac{\rho_{uv}}{(T_2)_{uv}} \qquad \text{for } u \neq v \tag{I.17}$$

for the initial distributions $\rho_{uv}^{(0)} = \rho_{uu}^{(0)}\delta_{uv}$, which are diagonal for each state. It is convenient to use

$$\gamma_{uu} = \frac{1}{(T_1)_{uu}} \tag{I.18a}$$

$$\gamma_{uv} = \gamma_{uv} = \frac{1}{(T_2)_{uv}} \qquad \text{for } u \neq v \tag{I.18b}$$

Taking the uv component of the equation of motion (I.14) and using

$$\langle u|H_0\rho - \rho H_0|v\rangle = (E_u - E_v)\rho_{uv} \tag{I.19}$$

we obtain the density-matrix equation in the presence of an optical excitation:

$$\frac{\partial}{\partial t}\rho_{uv} = \frac{-i}{\hbar}(E_u - E_v)\rho_{uv} + \frac{i}{\hbar}\sum_{u'}\sum_j (M^j_{uu'}\rho_{u'v} - \rho_{uu'}M^j_{u'v})E_j(t)$$

$$- \gamma_{uv}(\rho_{uv} - \rho^{(0)}_{uv}) \tag{I.20}$$

In general, we may define

$$\omega_{uv} = \frac{E_u - E_v}{\hbar} \tag{I.21}$$

and consider the interaction term $H' = -\mathbf{M} \cdot \mathbf{E}(t)$ as a small perturbation. The perturbation series gives

$$\rho = \rho^{(0)} + \rho^{(1)}(\mathbf{E}) + \rho^{(2)}(\mathbf{E}^2) + \cdots \tag{I.22}$$

One obtains

$$\frac{\partial}{\partial t}\rho^{(n+1)}_{uv} = (-i\omega_{uv} - \gamma_{uv})\rho^{(n+1)}_{uv} + \frac{i}{\hbar}\sum_{u'}\sum_j (M^j_{uu'}\rho^{(n)}_{u'v} - \rho^{(n)}_{uu'}M^j_{u'v})E_j(t)$$

$$\tag{I.23}$$

for $n \geq 0$. Consider an optical field given by

$$\mathbf{E}(t) = \sum_{\alpha=1}^N \mathbf{E}(\omega_\alpha)e^{-i\omega_\alpha t} \tag{I.24}$$

For example,

$$E(t) = E(\omega)\cos \omega t = \frac{E(\omega)}{2}e^{-i\omega t} + \frac{E(\omega)}{2}e^{+i\omega t} \qquad (I.25)$$

Thus, $N = 2$, $\omega_1 = \omega$, $\omega_2 = -\omega$, and $E(\omega) = E(\omega)/2$, $E(-\omega) = E(\omega)/2$. Note that the zeroth-order solution is simply the initial condition

$$\rho_{uv}^{(0)} = \rho_{uu}^{(0)}\delta_{uv} \qquad (I.26)$$

or

$$\rho_{aa}^{(0)} = f(E_a) \qquad (I.27a)$$

$$\rho_{bb}^{(0)} = f(E_b) \qquad (I.27b)$$

as can be seen from (I.16). The general density-matrix equation (I.23) starts with the first-order $\rho_{uv}^{(1)}(t)$ as the unknowns. The first-order solution is obtained by substituting the zeroth-order solutions into (I.23). Let

$$\rho_{uv}^{(1)}(t) = \sum_{\alpha=1}^{N} \rho_{uv}^{(1)}(\omega_\alpha)e^{-i\omega_\alpha t} \qquad (I.28)$$

If only a single frequency ω is considered,

$$\rho_{uv}^{(1)}(t) = \rho_{uv}^{(1)}(\omega)e^{-i\omega t} + \rho_{uv}^{(1)}(-\omega)e^{+i\omega t} \qquad (I.29)$$

By matching the $e^{-i\omega_\alpha t}$ dependence in (I.23), one obtains

$$\rho_{uv}^{(1)}(\omega_\alpha) = \frac{\rho_{uu}^{(0)} - \rho_{vv}^{(0)}}{\hbar(\omega_\alpha - \omega_{uv} + i\gamma_{uv})} \sum_{j} M_{uv}^{j}E_j(\omega_\alpha) \qquad (I.30)$$

for $u = a, b$ and $v = a, b$, using the two-level system shown in Fig. 9.1 with $E_b > E_a$. In terms of the components, we have

$$\rho_{aa}^{(1)} = \rho_{bb}^{(1)} = 0 \qquad (I.31)$$

and

$$\rho_{ba}^{(1)}(t) = \rho_{ba}^{(1)}(\omega)e^{-i\omega t} + \rho_{ba}^{(1)}(-\omega)e^{+i\omega t} \qquad (I.32)$$

$$\rho_{ba}^{(1)}(\omega) = \frac{\rho_{aa}^{(0)} - \rho_{bb}^{(0)}}{\hbar(\omega_{ba} - \omega - i\gamma_{ba})} M_{ba} \cdot E(\omega) \qquad (I.33)$$

and

$$\rho_{ba}^{(1)}(-\omega) = \frac{\rho_{aa}^{(0)} - \rho_{bb}^{(0)}}{\hbar(\omega_{ba} + \omega - i\gamma_{ba})} \mathbf{M}_{ba} \cdot \underset{\sim}{\mathbf{E}}(-\omega) \tag{I.34}$$

Similarly

$$\rho_{ab}^{(1)}(t) = \rho_{ab}^{(1)}(\omega)e^{-i\omega t} + \rho_{ab}^{(1)}(-\omega)e^{+i\omega t} \tag{I.35}$$

where $\rho_{ab}^{(1)}(\omega)$ is obtained by exchanging a with b from $\rho_{ba}^{(1)}(\omega)$ in (I.33), and $\rho_{ab}^{(1)}(-\tilde{\omega})$ is obtained from $\rho_{ba}^{(1)}(-\omega)$. The polarization per unit volume is calculated from the trace of the matrix product of the dipole moment matrix $\mathbf{M}$ and the density matrix:

$$\begin{aligned}
\mathbf{P}(t) &= \frac{1}{V} \mathrm{Tr}[\rho(t)\mathbf{M}] \\
&= \frac{1}{V} \sum_{uv} \rho_{uv}(t)\mathbf{M}_{vu} \\
&= \frac{1}{V}(\rho_{aa}\mathbf{M}_{aa} + \rho_{ab}\mathbf{M}_{ba} + \rho_{ba}\mathbf{M}_{ab} + \rho_{bb}\mathbf{M}_{bb}) \tag{I.36}
\end{aligned}$$

The ith component of the polarization density, to the first order in the optical electric field, is given by

$$\begin{aligned}
P_i(t) &= \frac{1}{V}\left(M_{ba}^i \rho_{ab}^{(1)}(\omega) + M_{ab}^i \rho_{ba}^{(1)}(\omega)\right)e^{-i\omega t} \\
&\quad + \frac{1}{V}\left[M_{ba}^i \rho_{ab}^{(1)}(-\omega) + M_{ab}^i \rho_{ba}^{(1)}(-\omega)\right]e^{+i\omega t} \tag{I.37}
\end{aligned}$$

The definition of the electric susceptibility X_{ij} is obtained using

$$\mathbf{P}(t) = \varepsilon_0 \overline{\overline{\mathbf{X}}}(\omega) \cdot \underset{\sim}{\mathbf{E}}(\omega)e^{-i\omega t} + \varepsilon_0 \overline{\overline{\mathbf{X}}}(-\omega) \cdot \underset{\sim}{\mathbf{E}}(-\omega)e^{+i\omega t} \tag{I.38}$$

or in component form

$$P_i(t) = \sum_j \varepsilon_0 X_{ij}(\omega)E_j(\omega)e^{-i\omega t} + \sum_j \varepsilon_0 X_{ij}(-\omega)E_j(-\omega)e^{+i\omega t} \tag{I.39}$$

Substituting (I.32)–(I.35) into (I.37) and comparing with (I.39), we obtain

$$\varepsilon_0 X_{ij}(\omega) = \frac{1}{V} \left[\frac{M_{ba}^i M_{ab}^j}{\hbar(\omega_{ba} + \omega + i\gamma_{ab})} + \frac{M_{ab}^i M_{ba}^j}{\hbar(\omega_{ba} - \omega - i\gamma_{ba})} \right] \left(\rho_{aa}^{(0)} - \rho_{bb}^{(0)} \right)$$

(I.40)

$$\varepsilon_0 X_{ij}(-\omega) = \frac{1}{V} \left[\frac{M_{ba}^i M_{ab}^j}{\hbar(\omega_{ba} + \omega + i\gamma_{ab})} + \frac{M_{ab}^i M_{ba}^j}{\hbar(\omega_{ba} - \omega - i\gamma_{ba})} \right] \left(\rho_{aa}^{(0)} - \rho_{bb}^{(0)} \right)$$

(I.41)

We find that the relation

$$X_{ij}(-\omega) = X_{ji}^*(\omega)$$

(I.42)

is observed. This can also be checked with the symmetry property of the complex permittivity function $\varepsilon'(\omega) + i\varepsilon''(\omega)$, discussed in Eqs. (E.8) and (E.9) in Appendix E, in which the real part $\varepsilon'(\omega)$ is an even function and the imaginary part $\varepsilon''(\omega)$ is an odd function of ω.

When a distribution of states is considered, for example, the energy bands

$$E_a = E(\mathbf{k}_a)$$

(I.43a)

and

$$E_b = E(\mathbf{k}_b)$$

(I.43b)

for two sets of wave vectors $\mathbf{k}_a$ and $\mathbf{k}_b$. We have to sum over the density of states in the k space. Assume that $E_b > E_a$ and $\hbar\omega$ is close to $E_b - E_a$. The second term in $\varepsilon_0 X_{ij}(\omega)$ will be the resonant contribution:

$$\varepsilon_0 X_{ij}(\omega) = \frac{1}{V} \sum_{s_a, k_a} \sum_{s_b, k_b} \frac{M_{ab}^i M_{ba}^j}{(E_b - E_a - \hbar\omega) - i\hbar\gamma_{ba}} \left(\rho_{aa}^{(0)} - \rho_{bb}^{(0)} \right)$$ (I.44)

The spins s_a and s_b are considered. If the states $|a\rangle$ and $|b\rangle$ do not include the mixing of spins, one finds that the spin summation gives simply a factor of 2 since the dipole operator $\mathbf{M}$ is independent of spin and

$$M_{ab}^i = \langle s_a, \phi_a | M^i | \phi_b, s_b \rangle = \langle \phi_a | M^i | \phi_b \rangle \delta_{s_a s_b}$$

(I.45)

that is, the spins s_a and s_b must be the same. Thus, the spin selection rule gives a factor of 2 when summing over spins.

$$\varepsilon_0 X_{ij}(\omega) = \frac{2}{V} \sum_{k_a} \sum_{k_b} \frac{M_{ab}^i M_{ba}^j}{(E_b - E_a - \hbar\omega) - i\hbar\gamma_{ba}}(f_a - f_b) \qquad (\text{I.46})$$

For a plane wave propagating in an anisotropic medium with

$$\varepsilon_{ij}(\omega) = (\varepsilon_b)_{ij} + \varepsilon_0 X_{ij}(\omega) \qquad (\text{I.47})$$

where $(\varepsilon_b)_{ij}$ is the background dielectric tensor, the solutions are in the form of ordinary and extraordinary waves with corresponding propagation constants.

Let us consider the simple case when $(\varepsilon_b)_{ij}$ and X_{ij} are diagonal:

$$\varepsilon = \varepsilon_b + \varepsilon_0 X(\omega) \qquad (\text{I.48})$$

That is, the induced polarization has the same direction as the applied electric field direction, $i = j$:

$$M_{ba}^i = \hat{e} \cdot \mathbf{M}_{ba} \qquad (\text{I.49})$$

$$\varepsilon_0 X(\omega) = \frac{2}{V} \sum_{k_a} \sum_{k_b} \frac{|\hat{e} \cdot \mathbf{M}_{ba}|^2}{(E_b - E_a - \hbar\omega) - i\hbar\gamma_{ba}}(f_a - f_b) \qquad (\text{I.50})$$

Therefore

$$\varepsilon_1 = \varepsilon_b + \text{Re}(\varepsilon_0 X)$$

$$= \varepsilon_b + \frac{2}{V} \sum_{k_a} \sum_{k_b} \frac{|\hat{e} \cdot \mathbf{M}_{ba}|^2 (E_b - E_a - \hbar\omega)}{(E_b - E_a - \hbar\omega)^2 + (\Gamma/2)^2}(f_a - f_b) \qquad (\text{I.51})$$

$$\varepsilon_2 = \text{Im}(\varepsilon_0 X)$$

$$= \frac{2}{V} \sum_{k_a} \sum_{k_b} \frac{|\hat{e} \cdot \mathbf{M}_{ba}|^2 (\Gamma/2)}{(E_b - E_a - \hbar\omega)^2 + (\Gamma/2)^2}(f_a - f_b) \qquad (\text{I.52})$$

where $\hbar\gamma_{ba} = \Gamma/2$ is the half linewidth. We see that the expression for ε_2 in (I.52) is the same as (9.1.38) if the conversion (9.1.31) is used. The expression for ε_1 in (I.51) when $\Gamma = 0$ reduces to (9.1.37) if we take the resonance approximation $E_b - E_a \simeq \hbar\omega$ in (9.1.37), $(E_b - E_a)^2 - (\hbar\omega)^2 \simeq (E_b - E_a - \hbar\omega)2\hbar\omega$. Note that the expression (9.1.40) for $\varepsilon_1(\omega)$ generally gives a better convergent result than (I.51) for semiconductors using a two-band model, for which the sum over the energy can be slowly convergent since the integrand in (I.51) is only inversely proportional to energy, and we are integrating over the energy. The propagation constant is

$$k = \omega\sqrt{\mu_0\varepsilon} = \omega\sqrt{\mu[\varepsilon_b + \varepsilon_0 X(\omega)]}$$

$$\simeq \omega\sqrt{\mu\varepsilon_b}\left\{1 + \frac{\varepsilon_0}{2\varepsilon_b}[X'(\omega) + iX''(\omega)]\right\}$$

$$\equiv k_0 n_r + k_0\,\Delta n + i\frac{\alpha}{2} \tag{I.53}$$

where

$$n_r = \sqrt{\frac{\varepsilon_b}{\varepsilon_0}} = \text{the background dielectric constant} \tag{I.54}$$

$$\Delta n = \frac{1}{2n_r}X' = \frac{1}{2n_r\varepsilon_0}\text{Re}(\varepsilon_0 X) = \text{the change in refractive index} \tag{I.55}$$

and the "intensity" absorption coefficient $\alpha = 2\,\text{Im}(k)$ is

$$\alpha = \frac{\omega}{n_r c\varepsilon_0}\,\text{Im}(\varepsilon_0 X) \tag{I.56}$$

REFERENCES

1. A. Yariv, *Quantum Electronics*, 3d ed., Wiley, New York, 1989.
2. Y. R. Shen, *The Principles of Nonlinear Optics*, Wiley, New York, 1984.

Appendix J

Optical Constants of GaAs and InP

Table J.1   Optical Constants of GaAs

Optical Energy $\hbar\omega$ (eV)	Wavelength λ (μm)	Refractive Index n	Extinction Coefficient κ	Absorption Coefficient $\alpha = 4\pi\kappa/\lambda$ (10^4 cm^{-1})	References*
0.5	2.4797	3.3240			a, b
0.6	2.0664	3.3378			
0.7	1.7712	3.3543			
0.8	1.5498	3.3737			
0.9	1.3776	3.3965			
1.0	1.2399	3.4232			
1.1	1.1271	3.4546			
1.2	1.0332	3.4920			
1.3	0.9537	3.5388			
1.35	0.9184	3.5690			
1.4	0.8856	3.6140	0.0017	0.0240	a, c
1.42	0.8731		0.0271	0.3900	
1.425	0.8701		0.0554	0.8001	
1.43	0.8670		0.0572	0.8290	
1.435	0.8640		0.0557	0.8101	
1.44	0.8610		0.0568	0.8290	
1.45	0.8551		0.0612	0.8994	
1.47	0.8434		0.0664	0.9893	
1.5	0.8266	3.666	0.080	1.216	a, d
1.6	0.7749	3.700	0.091	1.476	
1.7	0.7293	3.742	0.112	1.930	
1.8	0.6888	3.785	0.151	2.755	
1.9	0.6526	3.826	0.179	3.447	
2.0	0.6199	3.878	0.211	4.277	
2.1	0.5904	3.940	0.240	5.108	
2.2	0.5636	4.013	0.276	6.154	
2.3	0.5391	4.100	0.320	7.460	
2.4	0.5166	4.205	0.371	9.025	
2.5	0.4959	4.333	0.441	11.174	
2.6	0.4769	4.492	0.539	14.204	

*References are indicated on the first row for all items below.

[a] E. D. Palik, "Gallium arsenide (GaAs)," pp. 429–443 in E. D. Palik, Ed., *Handbook of Optical Constants of Solids*, Academic, New York, 1985.

[b] A. N. Pikhtin and A. D. Yas'kov, "Dispersion of the refractive index of semiconductors with diamond and zinc-blende structures," *Sov. Phys. Semicond.* **12**, 622 (1978).

[c] H. C. Casey, D. D. Sell, and K. W. Wecht, "Concentration dependence of the absorption coefficient for n- and p-type GaAs between 1.3 and 1.6 eV," *J. Appl. Phys.* **46**, 250 (1975).

[d] D. E. Aspnes and A. A. Studna, "Dielectric functions and optical parameters of Si, Ge, GaP, GaAs, GaSb, InP, InAs, and InSb from 1.5 to 6.0 eV," *Phys. Rev. B* **27**, 985 (1983).

The refractive index n, the extinction coefficient κ, and the absorption coefficient α of GaAs as a function of optical energy are plotted in Fig. J.1.

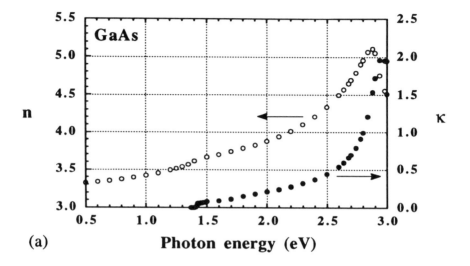

(a)

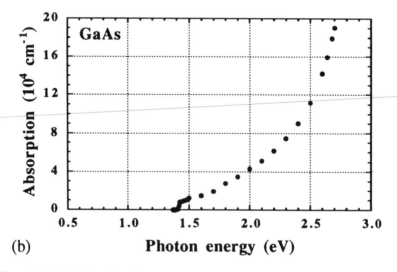

(b)

Figure J.1. (a) Real and imaginary parts, n and κ, of the complex refractive index and (b) the absorption coefficient α of GaAs vs. photon energy.

Table J.2   Optical Constants of InP

Optical Energy $\hbar\omega$ (eV)	Wavelength λ (μm)	Refractive Index n	Extinction Coefficient κ	Absorption Coefficient $\alpha = 4\pi\kappa/\lambda$ (10^4 cm^{-1})	References
0.6	2.0664	3.129			a, b
0.7	1.7712	3.146			
0.8	1.5498	3.167			
0.9	1.3776	3.191			
1.0	1.2399	3.22			
1.1	1.1271	3.254			
1.2	1.0332	3.297			
1.25	0.9919	3.324			
1.272	0.975	3.346	0.0000113	0.0001456	a, c, d
1.301	0.953	3.362	0.000281	0.003705	a, b, d
1.326	0.935	3.385	0.0059	0.07930	a, c, d
1.333	0.930	3.390	0.0109	0.1473	
1.340	0.925	3.396	0.0355	0.4822	
1.345	0.9218	3.399	0.0571	0.7791	
1.5	0.8266	3.456	0.203	3.086	a, e
1.6	0.7749	3.467	0.218	3.535	
1.7	0.7293	3.476	0.242	4.170	
1.8	0.6888	3.492	0.270	4.926	
1.9	0.6526	3.517	0.293	5.739	
2.0	0.6199	3.549	0.317	6.426	
2.1	0.5904	3.585	0.347	7.386	
2.2	0.5636	3.629	0.380	8.473	
2.3	0.5391	3.682	0.416	9.697	
2.4	0.5166	3.745	0.457	11.117	
2.5	0.4959	3.818	0.511	12.948	

[a]O. J. Glembocki and H. Piller, "Indium Phosphide (InP)," pp. 429–443 in E. D. Palik, Ed., *Handbook of Optical Constants of Solids*, Academic, New York, 1985.

[b]A. N. Pikhtin and A. D. Yas'kov, "Dispersion of the refractive index of semiconductors with diamond and zinc-blende structures," *Sov. Phys. Semicond.* **12**, 622 (1978).

[c]G. D. Pettit and W. J. Turner, "Optical Constant," *J. Appl. Phys.* **36**, 2081 (1965).

[d]B. O. Seraphin and H. E. Bennett, "Refractive index of InP," pp. 499–543 in R. K. Willardson and A. C. Beer, Eds., *Semiconductors and Semimetals*, Vol. 3, Academic, New York, 1967.

[e]D. E. Aspnes and A. A. Studna, "Dielectric functions and optical parameters of Si, Ge, GaP, GaAs, GaSb, InP, InAs, and InSb from 1.5 to 6.0 eV," *Phys. Rev. B* **27**, 985 (1983).

The refractive index n, the extinction coefficient κ, and the absorption coefficient α of InP as a function of optical energy are shown in Fig. J.2.

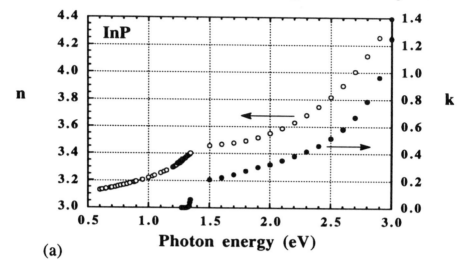

(a)

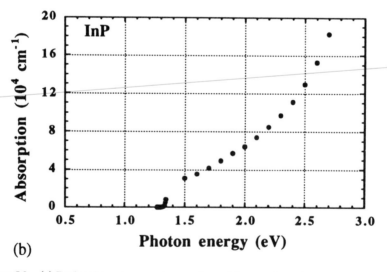

(b)

Figure J.2. (a) Real and imaginary parts, n and κ, of the complex refractive index and (b) the absorption coefficient α of InP vs. photon energy.

Appendix K

Electronic Properties of Si, Ge, and a Few Binary, Ternary, and Quaternary Compounds

Table K.1 Important Physical Properties[a,b]

Physical Properties	Si	Ge	GaAs	InAs	InP	GaP
Band gap E_g (eV)	(Indirect)	(Indirect)	(Direct)	(Direct)	(Direct)	(Indirect)
0 K	1.170	0.744	1.519	0.42	1.424	2.350
300 K	1.124	0.664	1.424	0.354	1.344	2.272
Dielectric constant ε_s (F/m)	$11.9\varepsilon_0$	$16.2\varepsilon_0$	$13.1\varepsilon_0$	$15.15\varepsilon_0$	$12.56\varepsilon_0$	$11.11\varepsilon_0$
Effective mass (m^*/m_0)						
Electrons						
Longitudinal m_l^*/m_0	0.9163	1.59	0.0665	0.023	0.077	0.254
Transverse m_t^*/m_0	0.1905	0.0823				4.8
Holes						
Heavy, m_{hh}^*/m_0	0.537	0.284	0.50	0.40	0.60	0.67
Light, m_{lh}^*/m_0	0.153	0.043	0.087	0.026	0.12	0.17
Intrinsic carrier concentration						
n_i (cm^{-3}) at 300 K	1.02×10^{10}	2.33×10^{13}	2.1×10^{6}	1.3×10^{15}	1.2×10^{8}	
Lattice constant (Å)	5.43102	5.6579	5.6533	6.0584	5.8688	5.4505
Minority carrier lifetime(s)						
at 300 K	2.5×10^{-3}	10^{-3}	$\sim 10^{-8}$			
Mobility (cm^2/Vs) at 300 K						
Electron	1450	3900	9200	$2 \sim 3.3 \times 10^4$	5370	160
Hole	370	1800	400	$100 \sim 450$	150	135

[a]O. Madelung, Ed., *Semiconductors, Group IV Elements and III–V Compounds*, in R. Poerschke, Ed., *Data in Science and Technology*, Springer, Berlin, 1991.
[b]S. Sze, *Physics of Semiconductor Devices*, Wiley, New York, 1982.

Table K.2 Important Band Structure Parameters[a-f] for GaAs, AlAs, InAs, InP, and GaP

	Materials				
	GaAs	AlAs	InAs	InP	GaP
Parameters					
a_0 (Å)	5.6533	5.6600	6.0584	5.8688	5.4505
E_g (eV)					
0 K	1.519	3.13	0.42	1.424	2.90
		2.229*			2.35*
300 K	1.424	3.03	0.354	1.344	2.78
		2.168*			2.27*
Δ (eV)	0.34	0.28	0.38	0.11	0.08
$E_{v,\text{av}}$ (eV)	−6.92	−7.49	−6.67	−7.04	−7.40
Optical matrix parameter E_p (eV)	25.7 (25.0)[f]	21.1	22.2	20.7 (16.7)[f]	22.2
Deformation potentials (eV)					
a_c (eV)	−7.17	−5.64	−5.08	−5.04	−7.14
a_v (eV)	1.16	2.47	1.00	1.27	1.70
$a = a_c - a_v$ (eV)	−8.33	−8.11	−6.08	−6.31	−8.83
b (eV)	−1.7	−1.5	−1.8	−1.7	−1.8
d (eV)	−4.55	−3.4	−3.6	−5.6	−4.5
C_{11} (10^{11} dyne/cm^2)	11.879	12.5	8.329	10.11	14.05
C_{12} (10^{11} dyne/cm^2)	5.376	5.34	4.526	5.61	6.203
C_{44} (10^{11} dyne/cm^2)	5.94	5.42	3.96	4.56	7.033
Effective masses					
m_e^*/m_0	0.067	0.15	0.023	0.077	0.25
m_{hh}^*/m_0	0.50	0.79	0.40	0.60	0.67
m_{lh}^*/m_0	0.087	0.15	0.026	0.12	0.17
$m_{hh,z}/m_0 = \dfrac{1}{\gamma_1 - 2\gamma_2}$	0.333	0.478	0.263	0.606	0.326
$m_{lh,z}/m_0 = \dfrac{1}{\gamma_1 + 2\gamma_2}$	0.094	0.208	0.027	0.121	0.199
γ_1	6.8 (6.85)	3.45	20.4	4.95	4.05
γ_2	1.9 (2.1)	0.68	8.3	1.65	0.49
γ_3	2.73 (2.9)	1.29	9.1	2.35	1.25

*Indirect band gap, $E_g(X)$ value.

[a] C. G. Van de Walle, "Band lineups and deformation potentials in the model-solid theory," *Phys. Rev. B* **39**, 1871–1883 (1989).

[b] P. Lawaetz, "Valence-band parameters in cubic semiconductors," *Phys. Rev. B.* **4**, 3460–3467 (1971).

[c] S. Adachi, "GaAs, AlAs, and Ga$_{1-x}$Al$_x$As: Material parameters for use in research and device applications," *J. Appl. Phys.* **58**, R1–R28 (1985).

[d] O. Madelung, Ed., *Semiconductors, Group IV Elements and III–V Compounds*, in R. Poerschke, Ed., *Data in Science and Technology*, Springer, Berlin, 1991.

[e] K. H. Hellwege, Ed., *Landolt–Börnstein Numerical Data and Functional Relationships in Science and Technology*, New Series, Group III **17a**, springer, Berlin, 1982; Groups III–V **22a**, Springer, Berlin, 1986.

[f] L. G. Shantharama, A. R. Adams, C. N. Ahmad and R. J. Nicholas, "The k · p interaction in InP and GaAs from the band-gap dependence of the effective mass," *J. Phys. C: Solid State Phys.* **17**, 4429–4442 (1984).

Table K.3   Important Band Structure Parameters
for $Al_xGa_{1-x}As$, $In_{1-x}Ga_xAs$, $Al_xIn_{1-x}As$, and $Ga_xIn_{1-x}As_yP_{1-y}$ Compounds[a-e]

General Interpolation Formula for Ternary Compound Parameters P:	Ref.
$$P(A_xB_{1-x}C) = xP(AC) + (1-x)P(BC)$$	

$Al_xGa_{1-x}As$

$E_g(\Gamma) = 1.424 + 1.247x$ (eV)	at 300 K	for $x < 0.4$	a
$\quad\quad\quad 1.519 + 1.447x - 0.15x^2$ (eV)	at 0 K	for $x < 0.4$	b
$m_e^*/m_0 = 0.067 + 0.083x$			a
$m_{hh}^*/m_0 = 0.50 + 0.29x$ (Density of states mass)			interpol.
$m_{lh}^*/m_0 = 0.087 + 0.063x$			f
$m_{so}^*/m_0 = 0.15 + 0.09x$			a
$\gamma_i(x) = x\gamma_i(AlAs) + (1-x)\gamma_i(GaAs)$	(for calculating transport masses)		interpol.

$In_{1-x}Ga_xAs$

$E_g(\Gamma) = 0.36 + 0.505x + 0.555x^2$ (eV)	at 300 K	c
$\quad\quad\quad 0.324 + 0.7x + 0.4x^2$ (eV)	at 300 K	a
$\quad\quad\quad 0.422 + 0.7x + 0.4x^2$ (eV)	at 2 K	
$m_e^*/m_0 = 0.025(1-x) + 0.071x - 0.0163x(1-x)$		

or

$$1/m_e^*(x) = x/m_e^*(GaAs) + (1-x)/m_e^*(InAs)$$

$In_{0.53}Ga_{0.47}As$

$E_g(\Gamma) = 0.813$ (eV)	at 2 K	a
$\quad\quad\quad 0.75$ (eV)	at 300 K	
$m_e^*/m_0 = 0.041$		
$m_{hh}^*/m_0 = 0.465$	//[001]	
$\quad\quad\quad 0.56$	//[110]	
$m_{lh}^*/m_0 = 0.0503$		

$Al_xIn_{1-x}As$

$E_g(\Gamma) = 0.36 + 2.35x + 0.24x^2$ (eV)	at 300 K		a
$\quad\quad\quad 0.357 + 2.29x$ (eV)	at 300 K	for $0.44 < x < 0.54$	
$\quad\quad\quad 0.447 + 2.22x$ (eV)	at 4 K	for $0.44 < x < 0.54$	

$Al_{0.48}In_{0.52}As$

$E_g(\Gamma) = 1.508$ (eV)	at 4K	a
$\quad\quad\quad 1.450$ (eV)	at 300 K	
$m_e^*/m_0 = 0.075$		d
$m_{hh}^*/m_0 = 0.41$		
$m_{lh}^*/m_0 = 0.096$		

Table K.3 (*Continued*)

General Interpolation Formula for Quaternary Compound Parameters P:

$$P(A_x B_{1-x} C_y D_{1-y}) = xy\,P(AC) + (1-x)(1-y)P(BD) + (1-x)y\,P(BC) + x(1-y)P(AD)$$

$Ga_x In_{1-x} As_y P_{1-y}$

$$E_g(x,y) = 1.35 + 0.668x - 1.068y + 0.758x^2 + 0.078y^2$$
$$- 0.069xy - 0.322x^2y + 0.03xy^2 \ (eV) \quad \text{at } 300\text{ K} \tag{a}$$

$$m_e^*(x,y)/m_0 = 0.08 - 0.116x + 0.026y - 0.059xy + (0.064 - 0.02x)y^2$$
$$+ (0.06 + 0.032y)x^2$$

$$a(x,y) = 5.8688 - 0.4176x + 0.1896y + 0.0125xy \quad (\text{Å})$$

Lattice-Matched to InP:

$$x = \frac{0.1894y}{(0.4184 - 0.013y)} \tag{a}$$

$$E_g(y) = 13.5 - 0.775y + 0.149y^2 \ (eV) \quad \text{at } 298\text{ K}$$
$$1.425 - 0.7668y + 0.149y^2 \ (eV) \quad \text{at } 4.2\text{ K}$$

$$m_e^*/m_0 = 0.080 - 0.039y \tag{e}$$
$$m_{hh}^*/m_0 = 0.46$$
$$m_{lh}^*/m_0 = 0.12 - 0.099y + 0.030y^2$$
$$m_{so}^*/m_0 = 0.21 - 0.01y - 0.05y^2$$

$In_{1-x-y} Al_x Ga_y As$

$$E_g(x,y) = 0.36 + 2.093x + 0.629y + 0.577x^2 + 0.436y^2$$
$$+ 1.013xy - 2.0xy(1 - x - y) \ (eV) \quad \text{at } 300\text{ K}$$

Lattice-Matched to InP:

$$(In_{0.52}Al_{0.48})_z(In_{0.53}Ga_{0.47})_{1-z}As$$
$$x = 0.48z \qquad 0.983x + y = 0.468$$
$$E_g(z) = 0.76 + 0.49z + 0.20z^2 \ (eV) \quad \text{at } 300\text{ K}$$
$$m_e^*/m_0 = 0.0427 + 0.0328z$$

[a] K. H. Hellwege, Ed., *Landolt–Börnstein Numerical Data and Functional Relationships in Science and Technology*, New Series, Group III, **17a**, Springer, Berlin, 1982; Groups III–V **22a**, Springer, Berlin, 1986.

[b] M. El Allai, C. B. Sorensen, E. Veje, and P. Tidemand-Petersson, "Experimental determination of the GaAs and $Ga_{1-x}Al_xAs$ band-gap energy dependence on temperature and aluminum mole fraction in the direct band-gap region," *Phys. Rev. B* **48**, 4398–4404 (1993).

[c] S. Adachi, "Material parameters in $In_xGa_{1-x}As_yP_{1-y}$ and related binaries," *J. Appl. Phys.* **53**, 8775–8792 (1982).

[d] P. Bhattacharya, Ed., *Properties of Lattice-Matched and Strained Indium Gallium Arsenide*, INSPEC, Institute of Electrical Engineers, London, U.K., 1993.

[e] S. Adachi, *Physical Properties of III–V Semiconductor Compounds*, Wiley, New York, 1992.

[f] H. C. Casey, Jr., and M. B. Panish, *Heterostructure Lasers Part A: Fundamental Principles*, Academic Press, Orlando, FL, 1978.

Index

Physical Constants	Symbol	Numerical Values
Speed of light in free space	c	$= 2.9979 \times 10^8$ m/s
Permittivity in free space	ε_0	$= 8.8542 \times 10^{-12}$ F/m
		$\left(\approx \dfrac{1}{36\pi} \times 10^{-9} \text{ F/m} \right)$
Permeability in free space	μ_0	$= 4\pi \times 10^{-7}$ H/m
Boltzmann constant	k_B	$= 1.3807 \times 10^{-23}$ J/K
Elementary charge	q or e	$= 1.60219 \times 10^{-19}$ C
Free electron mass	m_0	$= 9.1095 \times 10^{-31}$ kg
Planck constant	h	$= 6.6262 \times 10^{-34}$ Js
Reduced Planck constant	$\hbar = \dfrac{h}{2\pi}$	$= 1.05459 \times 10^{-34}$ Js
		$= 6.5822 \times 10^{-16}$ eVs
Angstrom unit	1 Å	$= 10^{-10}$ m $= 10^{-8}$ cm $= 10^{-4}$ μm
Bohr radius	$a_0 = \dfrac{4\pi\varepsilon_0 \hbar^2}{e^2 m_0}$	$= 0.529177$ Å
Ryberg energy	R_y	$= \dfrac{m_0 e^4}{2(4\pi\varepsilon_0)^2 \hbar^2} = \dfrac{\hbar^2}{2m_0}\left(\dfrac{1}{a_0}\right)^2 = 13.6058$ eV
Energy unit (electron-volt)	1 eV	$= 1.60219 \times 10^{-19}$ J
Thermal energy at 300 K	$k_B T$	$= 25.853$ meV

Useful Formulas and Physical Quantities

Photon energy	$\hbar\omega = h\,\dfrac{c}{\lambda} = \dfrac{1.2398}{\lambda}$ eV, where λ is in (μm)
Rydberg of an exciton	$R_y = \dfrac{m_r e^4}{2(4\pi\varepsilon_s)^2 \hbar^2} = \dfrac{(m_r/m_0)}{(\varepsilon_s/\varepsilon_0)^2} \times 13.6058$ eV
Bohr radius of an exciton	$a_0 = \dfrac{4\pi\varepsilon_s \hbar^2}{e^2 m_r} = \dfrac{\varepsilon_s/\varepsilon_0}{m_r/m_0} \times 0.529177$ Å
Quantized subband energy in a quantum well (infinite barrier model)	$E = \dfrac{\hbar^2}{2m^*}\left(\dfrac{n\pi}{L}\right)^2$
	$= \dfrac{n^2}{(m^*/m_0)}\dfrac{1}{L^2} \times 37.6033$ eV (L is in Å)
Conduction-band density parameter	$N_c = 2\left(\dfrac{m_e^* k_B T}{2\pi\hbar^2}\right)^{3/2} = 2.51 \times 10^{19}\left(\dfrac{m_e^*}{m_0}\dfrac{T}{300}\right)^{3/2}$ cm^{-3}